T5-CUL-927

SINGLET OXYGEN

This is Volume 40 of
ORGANIC CHEMISTRY
A series of monographs
Editor: HARRY H. WASSERMAN

A complete list of the books in this series appears at the end of the volume.

SINGLET OXYGEN

HARRY H. WASSERMAN

Department of Chemistry
Yale University
New Haven, Connecticut

ROBERT W. MURRAY

Department of Chemistry
University of Missouri–St. Louis
St. Louis, Missouri

ACADEMIC PRESS
New York San Francisco London 1979
A Subsidiary of Harcourt Brace Jovanovich, Publishers

ACADEMIC PRESS, INC.
111 Fifth Avenue, New York, New York 10003

United Kingdom Edition published by
ACADEMIC PRESS, INC. (LONDON) LTD.
24/28 Oval Road, London NW1 7DX

Library of Congress Cataloging in Publication Data

Main entry under title:

Singlet oxygen.

Includes bibliographies.
1. Oxidation. 2. Oxygen. I. Wasserman, Harry H.
II. Murray, Robert Wallace, Date
QD281.09S528 547'.23 77–25737
ISBN 0–12–736650–4

PRINTED IN THE UNITED STATES OF AMERICA

79 80 81 82 9 8 7 6 5 4 3 2 1

Contents

List of Contributors ix
Preface xi
Introductory Remarks xiii

1 Singlet Oxygen Electronic Structure and Photosensitization

Michael Kasha and Dale E. Brabham

I. Introduction 1
II. Primitive Representations of Oxygen Molecule States 2
III. Real and Complex Orbitals for O_2 5
IV. The Molecular State Functions for O_2 12
V. Sensitization Mechanisms for Singlet Oxygen 24
VI. Summary 31
References 32

2 Gaseous Singlet Oxygen

E. A. Ogryzlo

I. Introduction 35
II. Singlet Oxygen from Electrical Discharge 36
III. The Quenching of Singlet Oxygen 43
IV. Reactions of Singlet Oxygen 48
V. Energy-Pooling Reactions of Singlet Oxygen 55
References 56

3 Chemical Sources of Singlet Oxygen

Robert W. Murray

4 Solvent and Solvent Isotope Effects on the Lifetime of Singlet Oxygen

David R. Kearns

5 Quenching of Singlet Oxygen

Christopher S. Foote

6 1,2-Cycloaddition Reactions of Singlet Oxygen

A. Paul Schaap and K. A. Zaklika

VI. Reactions of Singlet Oxygen with Olefins in the Gas Phase 217
VII. Photooxygenation of Various Unsaturated Functional Groups 218
VIII. Rearrangement of Endoperoxides to 1,2-Dioxetanes 221
IX. Theoretical and Mechanistic Considerations 224
References 238

7 The 1,2-Dioxetanes

Paul D. Bartlett and Michael E. Landis

I. Physical Properties and Synthetic Methods 244
II. Decomposition of 1,2-Dioxetanes 255
III. The Mechanism of Dioxetane Decomposition 264
IV. Rearrangements and Alternative Cleavages 266
V. Catalysts for the Decomposition of 1,2-Dioxetanes 269
VI. Nonluminescent Reactions of 1,2-Dioxetanes 274
References 283

8 Ene-Reactions with Singlet Oxygen

K. Gollnick and H. J. Kuhn

I. General Aspects 287
II. Liquid Phase Ene-Reactions 288
III. Tabular Survey of the Ene-Reaction of Various Olefins with Singlet Oxygen 342
References 419

9 Reactions of Singlet Oxygen with Heterocyclic Systems

Harry W. Wasserman and Bruce H. Lipshutz

I. Introduction 430
II. Furans 431
III. Pyrroles 447
IV. Indoles 467
V. Imidazoles 481
VI. Purines 490
VII. Oxazoles 498
VIII. Thiazoles 502
IX. Thiopenes 503
References 506

10 The Oxidations of Electron-Rich Aromatic Compounds

Isao Saito and Teruo Matsuura

I.	Oxidation of Polycyclic Aromatic Hydrocarbons	511
II.	Oxidations of Monocyclic Aromatic Compounds	531
III.	Oxidations of Phenolic and Enolic Systems	551
	References	569

11 Role of Singlet Oxygen in the Degradation of Polymers

Martin L. Kaplan and Anthony M. Trozzolo

I.	Introduction	575
II.	Degradation of Polymers	577
III.	Biopolymers and Related Species	589
	References	592

12 Biological Roles of Singlet Oxygen

Norman I. Krinsky

I.	Introduction	597
II.	Identification of $^1O_2$ in Biological Systems	599
III.	Biological Systems	606
IV.	Summary	635
	References	636

Author Index 643
Subject Index 667

List of Contributors

Numbers in parentheses indicate the pages on which the authors' contributions begin.

PAUL D. BARTLETT (243), Department of Chemistry, Texas Christian University, Fort Worth, Texas 76129

DALE E. BRABHAM (1), Departamento de Biologia CCEN, Universidade Federal da Paraiba, Joao Pessoa PB, Brazil

CHRISTOPHER S. FOOTE (139), Department of Chemistry, University of California, Los Angeles, California 90024

K. GOLLNICK (287), Institut für Organische Chemie der Universität München, Karlstrasse 23, D-8000 Munchen 2, West Germany

MARTIN L. KAPLAN (575), Bell Laboratories, Murray Hill, New Jersey 07974

MICHAEL KASHA (1), Department of Chemistry and Institute of Molecular Biophysics, Florida State University, Tallahassee, Florida 32306

DAVID R. KEARNS (115), Department of Chemistry, Revelle College, University of California-San Diego, La Jolla, California 92093

NORMAN I. KRINSKY (597), Department of Biochemistry and Pharmacology, Tufts University School of Medicine, Boston, Massachusetts 02111

H. J. KUHN (287), Institut für Strahlenchemie im Max-Planke-Institut für Kohlenforschung, Mulheim-Ruhr. D-4330 Federal Republic of Germany

MICHAEL E. LANDIS* (243), Department of Chemistry, Texas Christian University, Fort Worth, Texas 76129

BRUCE H. LIPSHUTZ (429), Department of Chemistry, Yale University, New Haven, Connecticut 06520

TERUO MATSUURA (511), Department of Synthetic Chemistry, Faculty of Engineering, Kyoto University, Kyoto 606, Japan

ROBERT W. MURRAY (59), Department of Chemistry, University of Missouri–St. Louis, St. Louis, Missouri 63121

E. A. OGRYZLO (35), Department of Chemistry, University of British Columbia, Vancouver, British Columbia, Canada

* Present address: Southern Illinois University, Edwardsville, Illinois 62026.

ISAO SAITO (511), Department of Synthetic Chemistry, Faculty of Engineering, Kyoto University, Kyoto 606, Japan

A. PAUL SCHAAP (173), Department of Chemistry, Wayne State University, Detroit, Michigan 48202

ANTHONY M. TROZZOLO (575), Department of Chemistry, University of Notre Dame, Notre Dame, Indiana 46556

HARRY H. WASSERMAN (429), Department of Chemistry, Yale University, New Haven, Connecticut 06520

K. A. ZAKLIKA (173), Department of Chemistry, Wayne State University, Detroit, Michigan 48202

Preface

The field of singlet oxygen chemistry has developed dramatically during the past 15 years. As is outlined in the Introductory Remarks by M. Kasha, this period has witnessed an enormous increase in interest in the role of singlet oxygen in physical, chemical, and biological systems. These recent research papers have included investigations on the mechanisms of photooxygenation, spectroscopic studies on singlet oxygen, development of chemical methods for its generation, exploration of singlet oxygen reactivity with new types of substrates, physical studies on quenching and solvent effects, applications in organic synthesis, and exploration of the participation of singlet oxygen in environmental processes.

More recently biochemists and biologists have become more interested in singlet oxygen chemistry in connection with the possible role that this species may play in biological systems. Among these roles are the reactions involved in the destruction of cells and tissues in the presence of light, oxygen, and a dye (photodynamic action), phagocytosis, red blood cell damage, and certain cancer-inducing processes.

This appears to be an opportune time to gather together in a single volume a series of essays on the varied aspects of singlet oxygen chemistry, written by workers who are active in the field. We believe that this collection includes most of the important areas of current research interest. Some topics have not been covered, for example, the role of singlet oxygen in environmental chemistry. However, in such cases the subject has been recently and adequately treated in more specialized reviews.

We have attempted to organize and present the material so that it proceeds generally from discussions of physical and physical organic aspects, methods of generating singlet oxygen and the various reactions of singlet oxygen to descriptions of the involvement or possible involvement of singlet oxygen in polymer chemistry and biochemistry.

We hope that this volume will prove valuable to the active researchers in the field or those contemplating entering the field including organic, physical organic, physical, and biochemists. The coverage included should also make it of interest to graduate students and those teaching specialized topics to graduate students.

HARRY H. WASSERMAN
ROBERT W. MURRAY

Introductory Remarks: The Renascence of Research on Singlet Molecular Oxygen*

Oxygen as a chemical species has undergone two centuries of investigation. One might have imagined that it would have held no great new secrets to reveal, but research on singlet molecular oxygen is effecting a revolution in the understanding of the role of oxygen in physical, chemical, and biological systems, with special features in photon- and radiation-induced reactions.

Singlet molecular oxygen has been recognized to exist only since 1924. In the period until 1963 the research on singlet oxygen was sparse, with one or two papers appearing per year. During this "astrophysical period" singlet oxygen was regarded as a "rare" excited species largely of importance in atmospheric physics. Then the terrestrial significance of its chemical role (without the agency of radiation) became recognized in the "chemical period" since 1963.

The explosive development of research on singlet molecular oxygen is reflected in Fig. 1, which plots the number of papers on singlet molecular oxygen cited in the introductory sections of Schaap's critical reprint volume (1976). This volume reviews early papers and background papers and then reviews papers largely in the field of organic chemical reactions up to 1973. The appendix lists 26 reviews on singlet molecular oxygen and three published International Symposia, including the general one held by the New York Academy of Sciences (Annals, 1970).

In order to ascertain the growth of this subject in the period 1971 to the present, use was made of our National Science Foundation chemical literature research project. Using the CHEMCON DATA BASE (ACS) and CHEM 7071 DATA BASE, with *singlet*, *delta*, or *sigma oxygen* as key words, the number of titles found were 117 in 1971, 104 in 1972, 104 in 1973, 98 in 1975, and incomplete in 1976. If *photooxidation* were included, the number would soar to above 200 papers per year. The titles include topics

* The preparation of this paper was supported by the Division of Biomedical and Environmental Research, U. S. Energy Research and Development Administration, under Contract No. EY-76-S-05-2690.

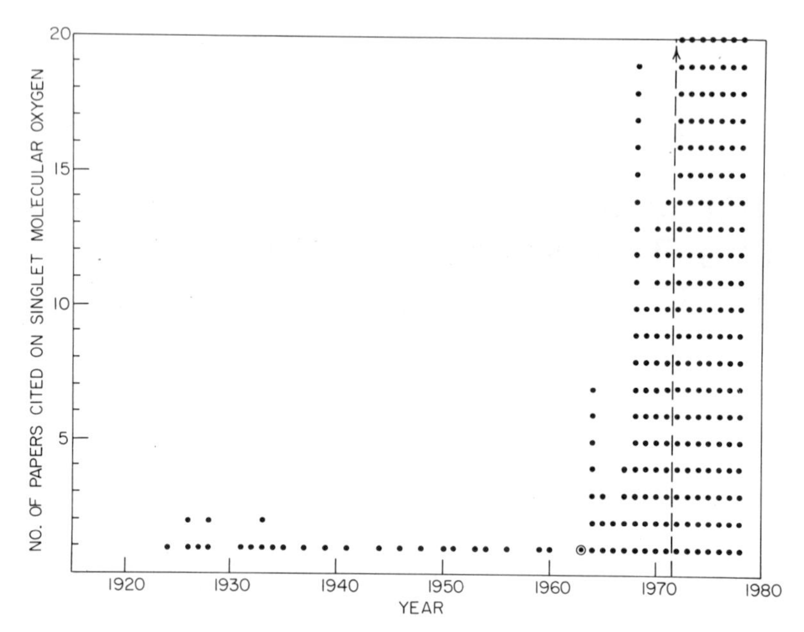

Fig. 1. Growth of literature on singlet oxygen research. Each point represents a publication as cited in the Schaap (1976) volume notes. After 1971 the chemistry computer literature search indicates more than 100 papers annually.

in atmospheric physics, basic spectroscopic physics, laser studies, inorganic chemical reactions, organic chemical reactions, biochemical reactions, industrial polymer reactions, and, recently, very strong development of biological and radiation biological reactions and applications. This unusual eruption of research on singlet molecular oxygen in its spectroscopic, physical, chemical, and biological aspects is a sign of the decades of oversight and an indication of the relevance of this subject in the full understanding of the participation of singlet molecular oxygen in what was once thought to be the exclusive domain of ordinary or triplet molecular oxygen.

The two hundredth anniversary of Joseph Priestley's discovery of oxygen was celebrated August 1 and 2, 1974, by the American Chemical Society in an Oxygen Bicentennial Symposium at Pennsylvania State University, not far from the place where the English refugee Priestley had settled in Colonial America. In 1977 it would be appropriate to celebrate the two hundredth anniversary of Carl Wilhelm Scheele's publication in Sweden of his earlier isolation and study of oxygen gas.

One especially interesting aspect of Priestley's research (Ihde, 1961) is the experimental detector he used in determining the presence of oxygen and the focus it gave to his work: *a live mouse.* When he placed a mouse on a platform inside a bell jar in which the gas had been used as an atmosphere for combustion, he discovered that the mouse could not live. But if in the same deoxygenated air bean or mint seedlings were grown, the detector-mouse survived happily. Thus, Priestley was the first to define the ecological cycle of the animal respiration versus plant photosynthesis in the atmosphere.

It is interesting to survey the titles of the papers presented at the Priestley Symposium (1974). Sir Derek Barton gave the evening Priestley Centennial Award Address on forbidden reactions of triplet oxygen, proposing a mechanism for the riddle of reactions of normal oxygen with unsaturated aliphatic hydrocarbons, which G. N. Lewis had noted much earlier. The symposium proper reflected the further trends in current research on oxygen: Singlet Molecular Oxygen Sensitized Luminescence (M. Kasha), Singlet Oxygen and Its Quenchers in Organic and Biological Chemistry (C. S. Foote), Reactions of Oxygen with Hemes and Heme Proteins (T. G. Traylor, C. K. Chang, J. Geibel, and D. Epstein), and Superoxide, Superoxide Dismutase, and Oxygen Toxicity (J. M. McCord).

Thus, the Oxygen Bicentennial Symposium focused mainly on distinctions between triplet and singlet molecular oxygen, superoxide ion, and especially the biological consequences of reaction of these species in the context of life processes. The anachronistic feature of the symposium lay in the reflection of the intense and belated chemical and biological explorations of singlet molecular oxygen since 1963.

Why this topic was so neglected and the specific steps of its renascence are worth surveying briefly as an unusual example of the evolution of a scientific research field. Accidental aspects of singlet molecular oxygen rediscovery figure prominently in this story, and although these human aspects are seldom recorded, they offer research planners and research funders a prototype to ponder on regarding "sponsorship of organized research." Apparently, the whole range of key discoveries could not have been more unplanned or more disorganized (or unproposed), and yet evidently few fields have proved so fertile in such a short period.

The chemical period in singlet molecular oxygen research starts with a chance discovery made by Howard Seliger at Johns Hopkins University. Seliger was developing a photoelectric method for spectral determination of weak bio- and chemiluminescences. It was Seliger's practice to have on the bench before him a beaker containing the biological extract to be tested. two smaller beakers containing alternative oxidants, hydrogen peroxide and sodium hypochlorite solution, and a hypodermic syringe in his hand (proving him a biophysicist). The lights were turned off after the hypodermic syringe was charged with one of the oxidants, a wait ensued while the experimenter's eyes got dark-adapted, then he took aim and discharged the syringe. A blue-green glow was usually observed from the reaction with the biological extract, and the broad chemiluminescence spectrum was then determined. One day Seliger repeated this procedure, to be greeted by a volcanic orange-red flash instead of the usual dull blue-green glow. Upon turning on the room lights, he discovered that he had missed his aim in the dark and had dischaged the syringe of hydrogen peroxide into the hypochlorite! By the most fortuitous of circumstances I was at Johns Hopkins that week at the now famous Light and Life Conference (1960) of the McCollum–Pratt Institute. Seliger called me downstairs to see the phenomenon and to call on my spectroscopist guildship for help in interpretation. He had determined the emission wavelength of the one observed peak to be 6348 Å, with a bandwidth of 170 Å by his apparatus. We discussed various principal and subsidiary species that could arise from the reaction of hydrogen peroxide with hypochlorite ion, but the spectral properties of likely possibilities were not relevant. For example, molecular oxygen in itself has no electronic transitions at 6348 Å, whereas atomic oxygen (O I) exhibits red auroral lines at 6300 and 6363 Å.

It did not seem at all possible that atomic oxygen could be the cause of the red glow in the solution chemiluminescence, and there was no way of correlating definitely the near infrared states of OH radical with the observed band. Seliger published his observations without interpretations in *Analytical Biochemistry* in 1960 (Vol. 1, No. 1). At that time we thought the observation of Seliger's was completely new. Later we found that L. Mallet had made the qualitative observation of the red-orange glow in 1927, P. Groh had published a wavelength in 1930, and P. Groh and K. A. Kirrmann in 1935 had associated the chemiluminescence emission with atmospheric "O_4" bands, using the wavelength coincidence alone as the assignment. G. Gattow and A. Schneider repeated the observations in a note in 1964, again believing them to be new (all of these sources are referenced in the Schaap reprint volume). Thus, Seliger simply uncovered what turned out to be in the quite dormant literature already, but he did present the first published spectrum of the single observed peak.

My reaction to Seliger's experiment was to encourage him to publish the observation in a physicochemically conspicuous place, and to buy a hypodermic syringe for my laboratory. Between 1960 and 1963, I demonstrated the red chemiluminescence glow to scores of laboratory visitors without a thought of a further spectral investigation. Then in 1963 Ahsan Ullah Khan came to my laboratory, and when I learned that at Oxford he had never operated a spectrograph, I immediately set him the task of calibrating himself on our Steinheil. But instead of using neon or other atomic line sources, evidently the red chemiluminescence was on my mind, as I said, "Why don't you photograph something interesting by measuring the peroxide–hypochlorite red chemiluminescence?" So Khan connected bottles of commercial bleach and 10% peroxide to a Y-tube and made photographs on the Steinheil on his first day in the laboratory. The next morning Khan brought in his first plate and I was astonished to see *two* very strong bands, and several other very weak ones. With considerable spectroscopic excitement the band interval was measured, and was found to be 1567 cm^{-1}, which was a satisfactory match for the ground state O_2 vibration frequency of 1580 cm^{-1} (the vibrational spacing observed in the transition expected to be slightly less). Thus, we were convinced we had identified the source of the red chemiluminescence as undoubtedly involving molecular O_2. Now a critical interpretation problem arose: (a) could the red glow which we measured at 6334 Å be due to a solvation shift of the 7620 Å band corresponding to the $^1\Sigma_g \rightarrow {}^3\Sigma_g^-$ oxygen transition, or (b) could the observed band be due to "O_4" molecules, as Groh and Kirrman had suggested earlier, an interpretation also favored by Stauff and Schmidkunz (1962), although both were on the basis of very incomplete spectral data. Khan and I did not understand the role of *simultaneous transitions* in colliding O_2 pairs at this point and favored the first interpretation in our first publication, over strong cautions of Robert Mulliken, Gerhard Herzberg, and others who were shown our observations.

Working in the completely different milieu of electrical discharges in gases, E. A. Ogryzlo and co-workers in Vancouver had observed strong IR emissions in O_2 gas from the $^1\Sigma_g$ and $^1\Delta_g$ excited state to ground state at 7619 and 12700 Å, but also weak bands at 6340 and 7030 Å, identical with the Khan and Kasha bands. Their publication in 1964 included a confirmation of the solution chemiluminescence observations of Khan and Kasha, but with the certain clarification that the red glow could not be due to solvation of single molecules, but instead involved oxygen–oxygen intermolecular interaction. However, they incorrectly referred to the emission as O_4 or O_2-dimer emission, stressing van der Waals interaction. Later experimental investigations (e.g., Ogryzlo *et al.*) and theoretical investigations (e.g., Wilse Robinson) showed that there is no stable dimer involved, and the unfortunate current use of "dimol" by many investigators as a term should be discouraged.

At this point the attention of organic chemists and photochemists was brought to bear strongly on the interaction of singlet molecular oxygen with organic molecules. Work by C. S. Foote and S. Wexler (1964), E. J. Corey and W. C. Taylor (1964), and E. McKeown and W. A. Waters (1966), in rapid succession opened up a gigantic new field of organic reaction chemistry, with especially illuminating comparisons with the empirically well-developed subject of dye-sensitized photooxidation. McKeown and Waters exposed a particularly broad panorama of the possibilities of the hypochlorite–peroxide reaction system as a source of singlet oxygen, and Foote and co-workers found parallels between yields and product distribution for the singlet oxygen reactions with organic substrate using the hypochlorite–peroxide system and using dye-sensitized photooxidation, convincingly invoking singlet oxygen mechanisms for the latter. The chemical period of singlet molecular oxygen fully came of age.

In retrospect, H. Kautsky stands out as a man before his time in relation to research on singlet molecular oxygen. I was attracted to Kautsky's writings through his interest in dye-sensitization experiments. One rainy week in Woods Hole in 1958, perusing more than a dozen of the long, prolix writings of Kautsky and co-workers, I was impressed by his very clever deductions concerning aggregation and sensitization of dyes in photooxidation reactions. These readings had a material effect on the interpretation of dye-aggregation phenomena which E. G. McRae and I gave using the molecular exciton model. I could not overlook the frequent mention of metastable single molecular oxygen in Kautsky's mechanisms of photosensitization. These did not impress me at the time for I knew that, in particular, H. Gaffron (my colleague at this Institute) had shown early that the energetics of $^1\Sigma_g$ oxygen excitation made the Kautsky mechanism improbable. The later recognition of the lower lying $^1\Delta_g$ oxygen state overcame the difficulty, but the Kautsky mechanism did not survive in the face of competing proposals by other more vigorous investigators. Nevertheless, Khan and I recognized that the Kautsky ideas should be revived once the chemical reality of metastable singlet oxygen in solution reactions was established.

Thus, this field of singlet molecular oxygen in its chemical period owes its renascence to the close interaction of detailed spectroscopic investigation and interpretation, coupled with the demonstration of the chemical consequences of the production of singlet molecular oxygen in solution systems. In retrospect, many older observations may be revived as having all contributed to the current understanding of the field. Yet the mere existence of information and even ideas in the literature may keep them dormant until the time of realization and justification of the importance of a topic has arrived through the stimulation of interaction between disciplines. In this case, the indispensable match was between the spectroscopists and the organic mechanisms chemists.

The chemical physics of singlet molecular oxygen has had a steady development, starting with the earliest work of Robert Mulliken, strongly developed by William Moffitt, and by Masao Kotani and others. There is an astonishing gap between even the most rudimentary aspects of the electronic theory and interactions of molecular oxygen and the interpretation of reactivities of singlet and triplet oxygen by the organic chemist.

We hope that this gap will be overcome in part by this volume and that a further enhancement of the mechanistic understanding of the interaction of oxygen with other molecules will ensue.

MICHAEL KASHA

REFERENCES

Ihde, Aaron J. (1961). "The Development of Modern Chemistry," pp. 40–54. Harper, New York.

Schaap, A. Paul (1976). "Singlet Molecular Oxygen." Dowden, Hutchinson, and Ross, Inc., Stroudsburg, Pennsylvania.

1

Singlet Oxygen Electronic Structure and Photosensitization

MICHAEL KASHA and DALE E. BRABHAM

I. Introduction 1
II. Primitive Representations of Oxygen Molecule States 2
A. Lewis Structures for O_2 2
B. Primitive Molecular Orbital Representations of O_2 States 3
III. Real and Complex Orbitals for O_2 5
A. The Real π Orbitals for the Oxygen Molecule 5
B. The Complex π Orbitals for the Oxygen Molecule 8
IV. The Molecular State Functions for O_2 12
A. The Molecular Orbital Configuration for Molecular Oxygen . . . 12
B. Spin and Orbital Angular Momenta for π Orbital Assignments . . 14
C. Construction of Determinantal Wave Functions for π Electron States of O_2 16
D. Electron Repulsion in the π Electron States of O_2 18
E. Degeneracies in the π Electron States of O_2 20
F. The Real π Electron State Functions for O_2 22
G. The Literature on Molecular Oxygen Structure 23
V. Sensitization Mechanisms for Singlet Oxygen 24
VI. Summary 31
References 32

I. INTRODUCTION

This chapter presents a discussion of the structure of molecular oxygen in its various lowest electronic states, and relates these structures to spectra, chemical reactivity, and sensitization mechanisms. The presentation is intended to serve as a bridge between the most primitive representations and the simplest adequately sophisticated ones.

Singlet Oxygen

ISBN 0-12-736650-4

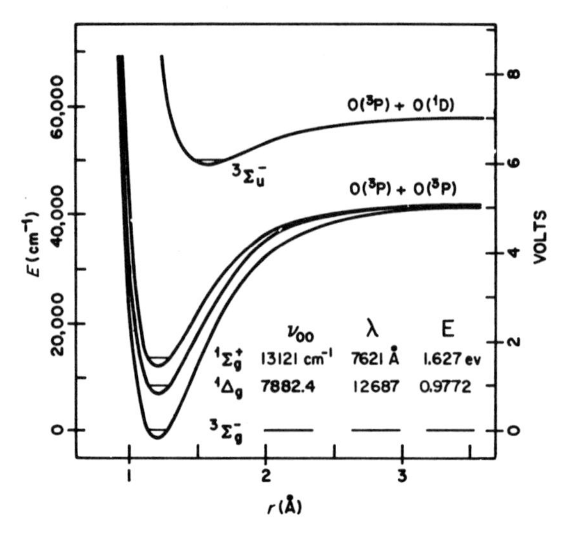

Fig. 1. Potential energy curves for molecular oxygen (after Herzberg, 1950).

Molecular oxygen, considered as a normal or ground state chemical species has been the subject of research for two centuries. As a physical species, research has centered on spectroscopic observations especially in reference to excited states which affect optical transmission of the earth's atmosphere. The last decade and a half has brought a new vigorous phase to oxygen research: the chemical reactivities of excited states of molecular oxygen, a phase of research promoted by the strong interaction of spectroscopic and chemical researches.

The potential energy diagram for the lowest electronic states of molecular oxygen is depicted in Fig. 1 (after Herzberg, 1950). The strongly allowed absorption ${}^3\Sigma_u^- \leftarrow {}^3\Sigma_g^-$ is known to spectroscopists as the Schumann–Runge band system with an origin at $\sim$2026 Å (49363 cm^{-1}), thereby giving the UV limit to atmospheric transmission. A forbidden transition ${}^3\Sigma_u^+ \leftarrow {}^3\Sigma_g^-$ (not shown in Fig. 1) is known as the Herzberg band system, and has an origin at $\sim$2800 Å (35713 cm^{-1}).

II. PRIMITIVE REPRESENTATIONS OF OXYGEN MOLECULE STATES

A. Lewis Structures for O_2

It is the low-lying singlet excited states of oxygen on which most of the new interest is focused (cf. Fig. 1). The first qualitative discussion of oxygen electronic structure was given by G. N. Lewis (1916) in his paper, "The Atom and the Molecule," in which electron-paired and diradical "structures" of oxygen and ethylene were discussed: "These two forms of oxygen (which, of course, may merge into one another by continuous gradations) can be represented by

:Ö::Ö: and :Ọ̈ : Ọ̈:

and the two forms of ethylene as

```
  H  H           H  H
H:C̈::C̈:H  and  H:C̣̈ : C̣̈:H    "
```

It is still generally assumed today that these Lewis structures correspond to "singlet" and "triplet" electron configurations. Lewis (1924a,b) later added further commentary to indicate localization of the "odd" electrons in the diradical form in which "one of the bonds is broken in such a way as to be equally divided between the two atoms, so that each atom possesses an odd electron," which recently was appraised as "a suggestion that in qualitative terms still adequately describes the ground state of molecular oxygen." In this chapter it will be shown that this is not an acceptable appraisal. Lewis (1924a,b) in this later paper modified his earlier concept of gradual tautomerization: "The enormous difference in magnetic properties between the oxygen molecule and other molecules to which we attribute double bonds seems to support the idea that the change from a nonmagnetic to a magnetic molecule is not a gradual process, but that the molecule must possess at least one unit of magnetic moment or no magnetic moment at all."

We have quoted extensively from Lewis' writings on molecular oxygen not to criticize their conceptual character for their time, but to point up their inadequacy for mechanistic deductions today. Yet in spite of a far deeper understanding of the rudiments of electronic structure of molecular oxygen now available, researchers in the forefront of triplet and singlet molecular oxygen reactions still rely on utterly inappropriate primitive representations for the problem. On the other hand, chemical physicists have not made it easy for the experimentalist to refine his models, so the need for a simple and fresh new exposé of the details of molecular electronic structures of various molecular oxygen states is quite pressing. We are devoting the first half of this chapter to this problem.

B. Primitive Molecular Orbital Representations of O_2 States

Primitive representations of the three lowest electronic states of molecular oxygen are reproduced in Fig. 2. The first column indicates the diatomic molecule state designations familiar now at least as labels to most researchers

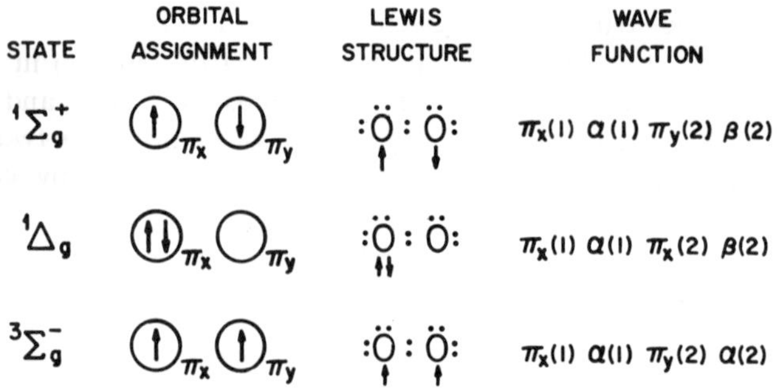

STATE	ORBITAL ASSIGNMENT	LEWIS STRUCTURE	WAVE FUNCTION
$^1\Sigma_g^+$	(↑)$_{\pi_x}$ (↓)$_{\pi_y}$	:Ö:Ö: (↑ ↓)	$\pi_x(1)\,\alpha(1)\,\pi_y(2)\,\beta(2)$
$^1\Delta_g$	(↑↓)$_{\pi_x}$ ( )$_{\pi_y}$	:Ö:Ö: (↑↓)	$\pi_x(1)\,\alpha(1)\,\pi_x(2)\,\beta(2)$
$^3\Sigma_g^-$	(↑)$_{\pi_x}$ (↑)$_{\pi_y}$	:Ö:Ö: (↑ ↑)	$\pi_x(1)\,\alpha(1)\,\pi_y(2)\,\alpha(2)$

Fig. 2. Primitive representations of molecular oxygen lowest singlet and triplet states.

dealing with molecular oxygen. We shall return to the significance of these state symbols later. Column 2 of Fig. 2 depicts what many chemists use and believe to represent the orbital assignment of electrons for the three lowest electronic states of molecular oxygen. The two circles represent "cells" of electron orbital occupancy, the orbital designations corresponding to the real molecular antibonding, degenerate (i.e., equal energy) π orbitals ($2p_x\pi_g$ and $2p_y\pi_g$, or serially, $1\pi_g{}^x$ and $1\pi_g{}^y$; or π_x, π_y) for diatomic molecules familiar to most chemists today (cf. Fig. 5, top line). These simple diagrammatic resolutions of Fig. 2, which seem so clearly to delineate pronounced differences in electronic orbital assignments for the three states of O_2, lend a false sense of security in that they simply do not correspond to the oxygen molecule states, even though they appear to do so with transparent clarity. Why are these simple pictures so inadequate? From one point of view, the diagrams seem to suggest three state configurations, one $^1\Sigma_g{}^+$, one $^1\Delta_g$, and one $^3\Sigma_g{}^-$ when the actual numbers are one, two, and three, respectively; we must, therefore, have a proper representation for these states and their correct number.

From another point of view, the wave functions (last column of Fig. 2) which describe these configurations are seen to be incorrect for the description of electron distribution functions. {In these wave functions, π_x and π_y represent the real degenerate pair for the configuration shown in Fig. 5 [top line: ... $(1\pi_g)^2$], (1) and (2) represent electron labeling, and α and β represent spin-up, spin-down spin functions with spin angular momentum eigenvalues $+(1/2)(h/2\pi)$ and $-(1/2)(h/2\pi)$}. These wave functions are inadequate because they fail to take into account electron indistinguishability (interchanges of indices 1 and 2) and also because the functions are symmetric under electron interchange, whereas proper electron wave functions are antisymmetric under electron interchange (the two-electron wave functions should change sign when 1 and 2 are interchanged). So these overall electron wave functions are incorrect, and cannot correspond to the apparent distinctions suggested by the cell diagrams of column 2, Fig. 2.

The (spin-designated) *Lewis structures* (column 3, Fig. 2) corresponding to the cell diagrams of column 2 can be criticized on even further grounds: (a) The Lewis structures designate electron localization on each atom, whereas the cell diagrams at least indicate occupancy of molecular orbitals delocalized with respect to the two oxygen atoms of the molecule. (b) The Lewis structures also incorporate the inadequacies of the cell diagrams, enumerated above. (c) The Lewis structures seem to impart a "nonbonding" electron role to four pairs of valency electrons; these assigned roles contrast to the bonding and antibonding roles of all the valency electrons designated by even the simplest molecular orbital picture (described later).

III. REAL AND COMPLEX ORBITALS FOR O_2

A. The Real π Orbitals for the Oxygen Molecule

In this section we shall discuss the form of the real and complex orbitals (Fig. 3) involved in building up the configuration of molecular oxygen. The diagrams of Fig. 3 represent schematic boundary surfaces for 90% of the electron probability, the contour lines merely indicating the three-dimensional shape of the boundary surface. As the well-known simple molecular orbital configuration of molecular oxygen indicates (top line of Fig. 5), all of the orbitals are fully occupied except the last pair, $1\pi_g$, each of which is half filled. This $1\pi_g$ orbital exists in two equal energy (or degenerate) forms, and only two electrons are available for populating the $1\pi_g$ pair.

The real molecular orbitals are shown on the left side of Fig. 3 and are universally known as the π-bonding molecular orbital ($2p_x\pi_u$) and the π-antibonding molecular orbital ($2p_x\pi_g$), totally familiar from examples in C=C bonds. These orbitals are expressible as first approximations by linear combinations of atomic orbitals for the component atoms.

$$(2p_x\pi_u) \equiv (1\pi_u{}^x) = \frac{1}{\sqrt{2}}(\phi_{2p_{x'}} + \phi_{2p_x})$$

$$(2p_x\pi_g) \equiv (1\pi_g{}^x) = \frac{1}{\sqrt{2}}(\phi_{2p_{x'}} - \phi_{2p_x})$$

It is clear that the signs of the linear combinations indicate constructive interference (+ combination) of the two atomic wave functions and thus bonding for the $1\pi_u{}^x$ orbital, and destructive interference (− combination)

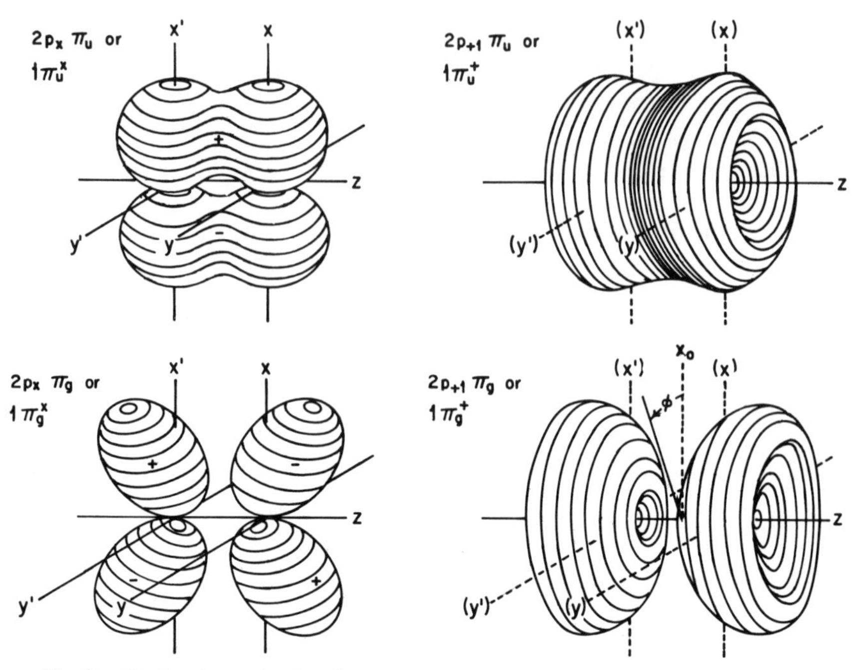

Fig. 3. Real and complex bonding and antibonding diatomic molecule π orbitals, showing boundary contour surface.

of the two atomic wave functions and thus antibonding between the two atoms (for the $1\pi_g{}^x$ orbital).

There is another set of y axis oriented π orbitals which may be expressed analogously.

$$(2p_y\pi_u) \equiv (1\pi_u{}^y) = \frac{1}{\sqrt{2}}(\phi_{2p_{y'}} + \phi_{2p_y})$$

$$(2p_y\pi_g) \equiv (1\pi_g{}^y) = \frac{1}{\sqrt{2}}(\phi_{2p_{y'}} - \phi_{2p_y})$$

Three simple observations can be made concerning the orbitals just described:

(a) The $1\pi_u{}^x$ and $1\pi_u{}^y$ orbitals and the $1\pi_g{}^x$ and $1\pi_g{}^y$ orbitals form exactly equivalent pairs (equal energy, degenerate). Thus, they each require double-electron occupancy, or a total of four electrons for closed-shell configurations.

(b) Each electron in a π_u or π_g orbital is fully delocalized with respect to the atomic centers, e.g., the Fig. 3 patterns are *one-electron wave functions*, or (mathematical, nonobservable) electron distribution functions; the square of these wave functions gives the point probability density for finding an electron (which is, in principle, a physical or observable quantity).

(c) The inversion operation î applied to the π_u function (point by point) clearly converts it into $-(\pi_u)$ (therefore it is designated u for *ungerade*, or odd, under inversion) whereas π_g under operator î goes to $+(\pi_g)$ (thus g for gerade, or even, under inversion).

There is another observation which now must be made concerning the real form of the orbitals. The chemist feels naturally at home in real Cartesian space, but does the oxygen molecule? The answer is a qualified no. In the free molecule there is certainly a physical operational significance to the interatomic axis (the z axis in Fig. 3), but in the free molecule there is no operational significance to the x or y axes, i.e., the xz and yz planes are not operationally defined. Therefore, it may be mathematically convenient for some purposes to use Cartesian coordinate orbital resolution, but we must remember that the π^x and π^y orbitals act as pairs and not as separate orbitals. This is clearly indicated by the symmetry operators. It was shown that inversion, î, transforms each π-type into itself or its negative, but this is not true of other symmetry operators. For example, the reflection operator in a plane containing the z axis ($\hat{\sigma}_v$) would at first sight send $(1\pi_u{}^x)$ into $-(1\pi_u{}^x)$ and $(1\pi_g{}^x)$ into $-(1\pi_g{}^x)$ (the common concept of "π-antisymmetry"). But this application of $\hat{\sigma}_v$, while perfectly applicable to ethylene π-bonds with a defined yz plane for the molecular skeleton, does not apply to a diatomic molecule. In a diatomic molecule the reflection plane $\hat{\sigma}_v$ may have any arbitrary orientation in space as long as it contains z (thus, the $D_{\infty h}$ point group table indicates $\infty\sigma_v$); if $\hat{\sigma}_v$ is chosen to bisect the xz and yz planes, and is applied to $\pi_u{}^x$, a combination of $\pi_u{}^x$ and $\pi_u{}^y$ results. This intertransforming characteristic is another manifestation of $\pi_u{}^x$ and $\pi_u{}^y$ orbital degeneracy. A similar result obtains even more irretrievably with the operator $\hat{C}_\phi$, rotation by an (arbitrary) angle about the z axis. If the orbital $\pi_u{}^x$ is rotated by any one of the infinite number (∞C_ϕ) of arbitrarily chosen angles ϕ, $\pi_u{}^x$ always goes into some combination of $\pi_u{}^x$ and $\pi_u{}^y$; only if ϕ is chosen to be 180° does $(1\pi_u{}^x)$ go into $-(1\pi_u{}^x)$; that exclusive choice, however, is not permitted for the cylindrically symmetrical ($D_{\infty h}$) molecule.

Similar deductions concerning the interdependence of the pair of real orbitals π_x and π_y are obtained if we consider angular momentum operators for the O_2 molecule. The net conclusion is that the real orbital resolution has only a mathematical and not an operational significance for the free molecule.

There is, however, a circumstance in which the real orbitals gain an operational significance and that is when the O_2 molecule is approached by an interacting or reacting molecule. Then the Cartesian axes and corresponding perpendicular planes gain physical meaning. Griffith (1964a,b) has utilized this resolution in predicting hydration effects in the O_2 molecule, and Kearns (1969) and others have used this resolution in discussion of reaction mechanisms, e.g., oxygen–ethylene correlation diagrams. So one could say that resolution of molecular oxygen orbitals into real Cartesian components "prepares" the molecule for perturbation effects in a Cartesian (cubic or octahedral) field.

B. The Complex π Orbitals for the Oxygen Molecule

The complex π orbitals (Fig. 3) are less familiar to the chemist, yet they are just as simply described as the real orbitals and have direct operational significance for the free molecule (uniaxial field defined by its own interatomic coordinate, the z axis). Actually the complex functions in atoms (full rotation group, with spherical polar coordinates r, θ, ϕ) and in homonuclear diatomic molecules ($D_{\infty h}$ infinite point group, cylindrical coordinates, r, z, ϕ) are derived naturally (in principle) from the wave mechanical solution, *and then transformed into real Cartesian functions* secondarily. The reason for the latter step in atoms is the same as that just described above; it is not that, e.g., p_x, p_y, and p_z are operationally defined in a free atom, it is that we *prepare* the atom in a triaxial Cartesian field for interactions with other atoms for molecule formation; it is the approaching perturbing atoms which may give p_x, p_y, and p_z operational significance.

Hence, since we have discussed the familiar real π orbitals first, we shall now reverse the normal process and essentially convert the real molecular wave functions back into the complex functions. The transformation is simple, and is mathematically identical to the resolution of circularly polarized light into plane (perpendicularly) polarized components P_x and P_y, or resolution of a plane polarized component, say P_x, into P_+ and P_- right and left circularly polarized components. The transformation equation is one of the Euler relations

$$e^{+i\phi} = \cos\phi + i\sin\phi$$

$$e^{-i\phi} = \cos\phi - i\sin\phi$$

We shall compare the transformation of real and complex orbitals for *atoms* with the transformation for *diatomic molecules*. The element in common for these is that for spherical polar coordinates in atoms the polar angle ϕ, for the 2π rotation about a z axis chosen as reference, is analogous to the axial 2π rotation angle ϕ about the z axis chosen as reference for diatomic molecules.

1. *Transformation of the Real 2p Orbitals to Complex 2p Orbitals for Atoms*

The real 2p atomic orbitals may be written

$$2p_x = xR(r) = r \sin\theta \textit{ cos } \phi \, R(r)$$

$$2p_y = yR(r) = r \sin\theta \textit{ sin } \phi \, R(r)$$

$$2p_z = zR(r) = r \cos\theta \cdot 1 \cdot R(r)$$

where the real $\Phi(\phi)$ functions are italicized for emphasis. If we consider an atom in a uniaxial field, such as a magnetic field, chosen to coincide with the z axis, we could apply the angular momentum operator

$$\hat{p}_\phi = \frac{h}{2\pi i} \frac{\partial}{\partial \phi}$$

to obtain the units of angular momentum (orbital angular momentum eigenvalue) corresponding to each 2p orbital. Taking the simple partial derivative with respect to ϕ

$$\frac{h}{2\pi i} \frac{\partial}{\partial \phi} (xR) = -\frac{h}{2\pi i} (r \sin\theta \sin\phi) R = (+i) \frac{h}{2\pi} (yR)$$

$$\frac{h}{2\pi i} \frac{\partial}{\partial \phi} (yR) = \frac{h}{2\pi i} (r \sin\theta \cos\phi) R = (-i) \frac{h}{2\pi} (xR)$$

$$\frac{h}{2\pi i} \frac{\partial}{\partial \phi} (zR) = \frac{h}{2\pi i} (r \cos\theta \cdot 0) R = (0) \frac{h}{2\pi} (zR)$$

we do not get eigenvalue equations, but find that the real x and y functions transform into each other instead of into themselves.

However, if we use the linear combinations* (the real functions being normalized)

$$p_+ = \frac{1}{\sqrt{2}} (p_x + ip_y)$$

$$p_- = \frac{1}{\sqrt{2}} (p_x - ip_y)$$

and then apply the Euler relations, we obtain complex atomic orbitals

* This particular linear combination ($1p_x + ip_y$) is chosen to permit the application of the Euler relation instead of, e.g., $1p_x + 1p_y$.

which are eigenfunctions of p_ϕ,

$$p_{+1} = r \sin\theta \, e^{+i\phi} \, R(r)$$
$$p_{-1} = r \sin\theta \, e^{-i\phi} \, R(r)$$
$$p_0 = r \cos\theta \, e^{0\phi} \, R(r)$$

with eigenvalues +1, −1, and 0 ($h/2\pi$), respectively.

2. *Transformation of Real π Orbitals to Complex π Orbitals for Diatomic Molecules*

Now we are in a position to compare the analogous result for diatomic molecular orbitals. As indicated earlier, the diatomic orbitals may in first approximation be written as linear combinations of atomic orbitals

$$\pi_g^{\,x} = \frac{1}{\sqrt{2}}(2p_{x'} - 2p_x)$$

$$\pi_g^{\,y} = \frac{1}{\sqrt{2}}(2p_{y'} - 2p_y)$$

The $\Phi(\phi)$ functions for diatomic molecular orbitals and for orbitals for the free atoms are identical, $\Phi(\phi) = (1/\sqrt{\pi})e^{im_l\phi}$.

Now the presence of a uniaxial field (z axis) is imposed by the molecular structure and does not depend on the imposition of an external field. Consequently, the π orbitals for the molecule should be eigenfunctions of the p_ϕ operator. However, as in the case of atoms, the real orbital components of the molecular orbitals will transform $x \rightarrow y$ and $y \rightarrow x$, so the real π orbitals separately do not behave properly as eigenfunctions.

Taking linear combinations of the real π orbitals, we construct the complex π orbitals as follows

$$\pi_g^{+1} = \frac{1}{\sqrt{2}}(\pi_g^{\,x} + i\pi_g^{\,y})$$

$$\pi_g^{-1} = \frac{1}{\sqrt{2}}(\pi_g^{\,x} - i\pi_g^{\,y})$$

These orbitals may then be written, by the Euler relation, as

$$\pi_g^{+1} = Z_{-}(z) \, R(r) \, e^{+i\phi}$$
$$\pi_g^{-1} = Z_{-}(z) \, R(r) \, e^{-i\phi}$$

The $2p_z$ and $2p_z$ orbitals, which we have not mentioned, were used to form the $2p_z\sigma_g$ and $2p_z\sigma_u$ sigma bonding and antibonding orbitals (labeled serially $3\sigma_g$ in Fig. 5, first line). It is now simply demonstrable that

$$\hat{p}_\phi(\pi_g^{+1}) = (+1)\frac{h}{2\pi}(\pi_g^{+1})$$

$$\hat{p}_\phi(\pi_g^{-1}) = (-1)\frac{h}{2\pi}(\pi_g^{-1})$$

so that the complex π orbital functions are indeed eigenfunctions of the appropriate angular momentum operator. We may attribute the $+1$ and $-1(h/2\pi)$ units of angular momenta to clockwise and counterclockwise electron currents about the z axis (analogous to the concept of circular polarization of light).

3. *Symmetry Properties of the Complex π Orbitals*

Now a question of interest is, how can we picture these complex orbitals. We may use the *modulus* of a complex function

$$|\Phi_{+1}^*\Phi_{+1}| = |e^{-i\phi}e^{i\phi}| = 1$$

which indicates that the function in real space has a constant numerical value as a function of ϕ, i.e., a unit circle gives the real spatial ϕ dependence of the orbitals π_g^{+1}, π_g^{-1}, π_u^{+1}, and π_u^{-1}. Thus, Fig. 3 pictures the resultant function in real space if the modulus of the complex function is used. It is evident that these are cylindrically symmetrical distribution patterns in real space. It can be regarded as a requirement of cylindrical symmetry that the proper orbitals for a diatomic molecule also conform to the symmetry operations of a cylindrical point group (e.g., $D_{\infty h}$). This section illustrates the fact that proper superposition of $\pi_g{}^x$ and $\pi_g{}^y$ real orbitals to yield complex ones leaves no "lumpiness" of the real orbitals. The complex π orbital functions in fact *have the appearance* of figures of revolution of the real π orbital functions about the z axis. Matters are more complicated if we consider the effect of the inversion operator, î, on these complex orbitals because of the actual complex function behavior under coordinate interchange.

It is evident at face value that the real orbitals $1\pi_g{}^x$ and $1\pi_g{}^y$ (Fig. 3) for a diatomic molecule are *gerade* (even) under inversion, and also antibonding. We should also expect the complex orbitals $1\pi_g{}^+$ and $1\pi_g{}^-$ to be *gerade* (even) and antibonding. But the node in Fig. 3 for $1\pi_g{}^+$ seems at first sight to be a contradiction. How can the node exist between the two atoms and still leave the $1\pi_g{}^+$ *gerade* under î? Of course in the real orbital $1\pi_g{}^x$ of Fig. 3 there are *two* nodes, allowing antibonding and *gerade* character to coexist. In the complex function $1\pi_g{}^+$ depicted as a modulus figure in real space, the nodes are concealed in the complex character of the wave function. (It will have occurred to the reader that the $1\pi_g{}^x$ resembles the real atomic orbital $3d_{xz}$ to which the molecular orbital should converge in the united-atom limit; and the $1\pi_g{}^+$ resembles the complex atomic orbital $3d_{+1}$, to which in turn it should converge in the united-atom correlation.)

To test the behavior of $1\pi_g{}^+$ under the inversion operator, î, we may write the orbital

$$1\pi_g{}^+ = [Z'(z) - Z(z)]\, R(r)\, e^{i\phi}$$

in which the linear combination of atomic orbitals is resolved in z-coordinate components, since the $\Phi(\phi)$ function is a common factor. Under the inversion operator, $(z) \rightarrow (-z)$, $(\phi) \rightarrow (\phi + \pi)$, and r (a positive number) remains unchanged in cylindrical coordinates. Therefore

$$\begin{aligned} \hat{\imath}(1\pi_g{}^+) &= [Z'(-z) - Z(-z)]\, R(r)\, e^{i\phi} e^{i\pi} \\ &= (-1)[Z'(z) - Z(z)]\, R(r) e^{i\phi}(-1) \\ \hat{\imath}(1\pi_g{}^+) &= (+1)(1\pi_g{}^+) \end{aligned}$$

using $e^{i\pi} = -1$; that is, the function is *gerade* (even). Thus, we cannot properly label the $1\pi_g{}^+$ orbital regions of Fig. 3 as to sign since there is no way in which to illustrate on the modulus figure the effect of inversion on the complex functions.

IV. THE MOLECULAR STATE FUNCTIONS FOR O_2

A. The Molecular Orbital Configuration for Molecular Oxygen

We shall now proceed to build up a set of molecular state functions corresponding to the lowest energy molecular orbital configuration. R. S. Mulliken (1928, 1932) gave the first detailed spectroscopic description of the electronic states of the oxygen molecule shown in Fig. 1, following Lennard-Jones (1929) and Hückel (1930).

The most general molecular orbital configuration for molecular oxygen is given in Fig. 4 and the first line of Fig. 5. Each oxygen atom ($Z = 8$) has the electron configuration $1s^2 2s^2 2p^4$. The inner atomic orbitals probably interact very weakly if at all, so that in most sources KK (atomic shells) is written to replace $(1\sigma_g)^2(1\sigma_u)^2$. The remaining twelve electrons may be considered to be the valency electrons, and according to the molecular orbital configuration, all of these are very busy acting in bonding or in antibonding roles. Thus, successively, there is a $2\sigma_g$ bonding pair, followed in energy by

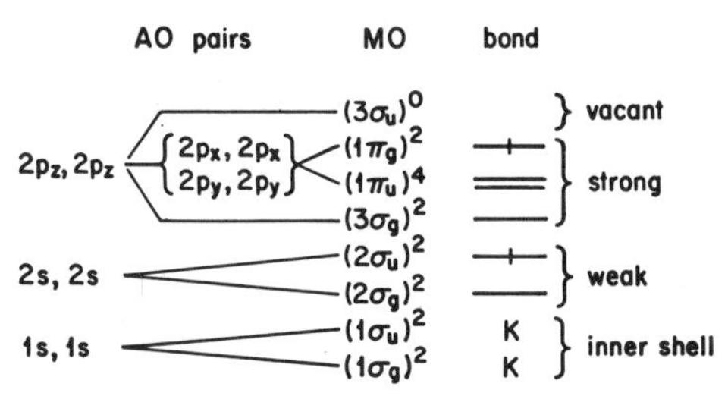

Fig. 4. Atomic orbital (AO) origin of molecular orbitals (MO) and bond type for homonuclear diatomic molecules.

a $2\sigma_u$ antibonding pair; these probably correspond to weak interactions arising from the 2s atomic orbitals.

Next in order on the energy scale comes a $3\sigma_g$ *bonding* pair, in an orbital arising from the $2p_z$ orbitals on each atom, thus giving the oxygen its fundamental σ bond. This is followed by the doubly-degenerate $1\pi_u$ *bonding orbital* which is pictured in part in Fig. 3, top row ($1\pi_u{}^x$ and $1\pi_u{}^y$ being the real pair, and alternatively, $1\pi_u{}^+$ and $1\pi_u{}^-$ constitute the complex pair, discussed in the last section). Finally, there are two electrons in the $1\pi_g{}^x$, $1\pi_g{}^y$ (or $1\pi_g{}^+$, $1\pi_g{}^-$) *antibonding orbital* pictured in part in Fig. 3, bottom row. All but the last pair of orbitals are fully occupied and constitute closed shell contributions, so do not determine overall symmetry, angular momentum, or spin.

How the last pair of electrons in the abstract configuration $\ldots(1\pi_g)^2$ leads to *six different electronic substates*, in three classes, is the subject of this section. Thus, it is not accurate to state that the ground state of molecular oxygen is represented by the configuration shown in Fig. 4 and the top line of Fig. 5, since this identical configuration gives rise to *three different electronic* states with subcomponents.

A word on the "number of bonds" in ground state molecular oxygen would be instructive. If we designate electron-pair bonds by the traditional link —, and now electron-pair *antibonds* by ⊹, we would arrive at the set of bonds shown in Fig. 4. Thus, the strong bonds holding the oxygen atoms together consist of one σ-bond and effectively one π-bond (the other π-bond is cancelled by the π-antibond; antibonding orbitals always more than cancel analogous bonding orbital effects). Thus, molecular oxygen in its ground state has *two effective (net) electron-pair bonds*, a picture entirely at variance with the Lewis structure representations of Fig. 2.

In particular we note that the two electrons in the π_g antibonding orbital effectively reduce the bonding strength of the four electrons in the π_u bonding orbitals, instead of being localized on atomic (essentially) nonbonding sites. If we were to translate the molecular orbital bonding information given in Fig. 4 into a chemical formula for molecular oxygen, the clumsy but informative result would be*

$$:\mathrm{O} \overset{\pi_g}{\underset{\sigma_g}{\equiv}} \mathrm{O}:$$

where the KK inner shell pairs are omitted since they are not valency electrons, and the "lone-pair" electrons shown are actually the weakly antibonding set $(2\sigma_g)^2(2\sigma_u)^2$. Because of the degeneracy of the π_g orbital, the above chemical formula corresponds to any of three different molecular states, which must now be examined.

B. Spin and Orbital Angular Momenta for π Orbital Assignments

Now we explore the building up of the proper molecular state functions. The configuration for O_2 given in Fig. 4 and the top line of Fig. 5 is unrevealing because of the varieties of electron assignment of that last pair of electrons, $1\pi_g^{\ 2}$. This is illustrated in the first column of Fig. 5. We choose the complex orbital set of Fig. 3 and the previous section, i.e., $\pi_g^{\ +}$ and $\pi_g^{\ -}$. Since electrons have spin, we must assign a spin wave function to each electron, either α (spin-up) with $+(1/2)(h/2\pi)$ units of spin angular momentum, or β (spin-down) with $-(1/2)(h/2\pi)$ units of angular momentum. In column 1 of Fig. 5, β spin function is designated by a bar over the orbital symbol, and an α spin function by an unbarred orbital symbol. For two electrons, with choices of + or − complex π orbital functions, and choices of α or β spin, we arrive at an array of six possible resulting orbital assignments (column 1, Fig. 5).

To classify the orbital configuration components we identify their orbital angular momentum and spin angular momentum characteristics. We can do this by inspection, or by analytically testing each orbital configuration under orbital angular momentum and spin angular momentum operators for eigenvalues, using motion about the z axis as reference (the general symbol $\hat{p}_\phi$ now being replaced by $\hat{L}_z$ and $\hat{S}_z$ for specific spectroscopic components). Inspection of $\pi_g^{\ +}\pi_g^{\ +}$ indicates that $+1(h/2\pi)$ plus $+1(h/2\pi)$ units of *orbital angular momentum* (characteristic of π^+) are involved), giving $+2(h/2\pi)$ total units, corresponding to a Δ state; $\pi_g^{\ +}\pi_g^{\ -}$ indicates $+1(h/2\pi)$ plus $-1(h/2\pi)$ units of *orbital angular momentum*, or 0 total for

* This formula is exactly analogous to Pauling's three-electron bond structure for oxygen (Pauling, 1960), where we perceive that the two extra electrons of the two three-electron bonds must be in the antibonding orbital.

$$O_2 \quad (1\sigma_g)^2\,(1\sigma_u)^2\,(2\sigma_g)^2\,(2\sigma_u)^2\,(3\sigma_g)^2\,(1\pi_u)^4\,(1\pi_g)^2$$

ORBITAL ASSIGNMENT	$h/2\pi$ EIGENVALUE UNDER $\hat{L}_z$	$\hat{S}_z$	COMPONENTS OF STATE
$\pi_g^+\ \bar{\pi}_g^+$	2	0	${}^1\Delta_g$
$\pi_g^+\ \pi_g^-$	0	1	${}^3\Sigma_g^-$
$\pi_g^+\ \bar{\pi}_g^-$	0	0	${}^3\Sigma_g^-, {}^1\Sigma_g^+$
$\bar{\pi}_g^+\ \pi_g^-$	0	0	${}^3\Sigma_g^-, {}^1\Sigma_g^+$
$\bar{\pi}_g^+\ \bar{\pi}_g^-$	0	-1	${}^3\Sigma_g^-$
$\pi_g^-\ \bar{\pi}_g^-$	-2	0	${}^1\Delta_g$

Fig. 5. Molecular oxygen configuration components.

a Σ state; and e.g., $\pi_g^-\pi_g^-$ indicates that $-1(h/2\pi)$ plus $-1(h/2\pi)$, or $-2(h/2\pi)$ total units, corresponding again to a Δ state (total orbital angular momentum units 0, 1, 2, 3, . . . , $(h/2\pi)$ corresponding to $\Sigma, \prod, \Delta, \Phi$ . . . states).

The spin angular momentum components are also discernible on inspection for the orbital configuration assignments of column 1, Fig. 5. The assignment $\pi_g^+\bar{\pi}_g^+$ indicates spin functions successively for the two electrons as $\alpha\beta$ or $+(1/2)(h/2\pi)$ plus $-(1/2)(h/2\pi)$ units, a net of 0 units of spin angular momentum, thus a *singlet* state component, ${}^1\Delta_g$; $\pi_g^+\pi_g^-$ indicates spin functions $\alpha\alpha$, $+(1/2)(h/2\pi)$ plus $+(1/2)(h/2\pi) = 1(h/2\pi)$ net units of spin angular momentum, thus a *triplet* state component, ${}^3\Sigma_g^-$. Rather unfamiliarly for chemists, orbital assignments like $\pi_g^+\bar{\pi}_g^-$, with spin function $\alpha\beta$, and $\bar{\pi}_g^+\pi_g^-$ with spin function $\beta\alpha$, therefore spin-component $M_S = 0$, can be *singlet state* components or a *triplet state* (zero spin) component, since $\alpha\beta - \beta\alpha$, and $\alpha\beta + \beta\alpha$ must later be formulated for the complete spin functions (see below).

Thus, the two ${}^1\Delta_g$ states are unambiguously or uniquely designated by the simple orbital configuration assignment; the $M_S = +1$ and $M_S = -1$ components of the ${}^3\Sigma_g^-$ state are also uniquely designated. The configuration for the ${}^1\Sigma_g^+$ state and the ${}^3\Sigma_g^-$ ($M_S = 0$) state must be resolved by taking linear combinations of the equivalent orbital assignments $\pi_g^+\bar{\pi}_g^-$ and $\bar{\pi}_g^+\pi_g^-$. The test of whether the linear combination is really ${}^3\Sigma_g^-$ ($M_S = 0$) or ${}^1\Sigma_g^+$ can be accomplished by factoring out the spin functions from the completed wave function, or by testing with the total spin angular momentum operator S^2 (distinguishing triplets from singlets).

C. Construction of Determinantal Wave Functions for π Electron States of O_2

The properties of an electron indicate that it behaves like a *fermion*, that is, it conforms to a Fermi–Dirac statistical distribution. Concomitant with this behavior is the requirement of an antisymmetric total wave function, i.e., one that changes sign upon electron interchange. It is convenient to express proper electron wave functions therefore in determinantal form (Slater determinant), since interchanging any two rows or columns of a determinant intrinsically changes its net sign, and such an interchange corresponds to interchange of electron numbering.

Determinantal wave functions for molecular oxygen based on complex orbital functions are tabulated in Fig. 6. For simplification, the g subscript has been dropped in this formulation, since all of the orbitals involved are the π_g orbitals, and the +, − superscripts have become subscripts. It is customary to simplify the writing of these determinants by writing only the principal diagonal. Thus for one of the ${}^1\Delta_g$ states

$$\Psi({}^1\Delta_g) = \frac{1}{\sqrt{2}}|\pi_+\bar{\pi}_+| = \frac{1}{\sqrt{2}}\begin{vmatrix} \pi_+(1)\alpha(1) & \cdots \\ \cdots & \pi_+(2)\beta(2) \end{vmatrix}$$

wherein the two-electron orbital assignment of column 1, Fig. 5 corresponds

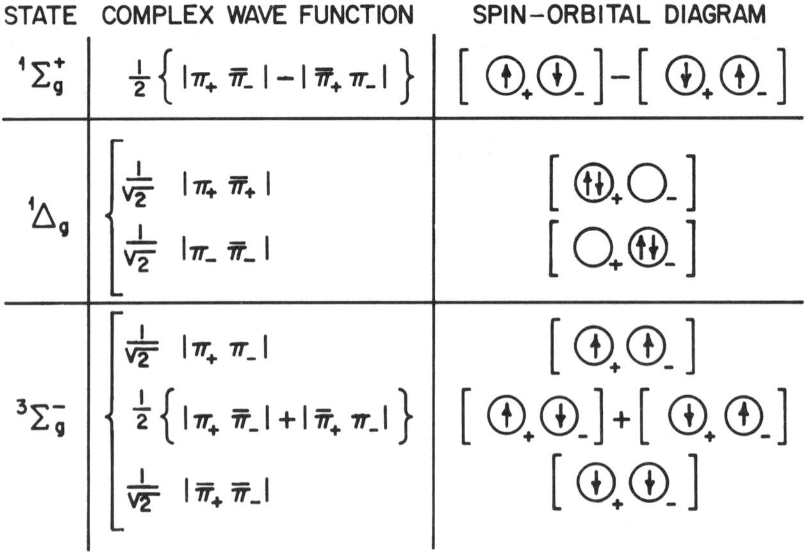

STATE	COMPLEX WAVE FUNCTION	SPIN–ORBITAL DIAGRAM
${}^1\Sigma_g^+$	$\frac{1}{2}\{|\pi_+\bar{\pi}_-| - |\bar{\pi}_+\pi_-|\}$	$[(\uparrow)_+(\downarrow)_-] - [(\downarrow)_+(\uparrow)_-]$
${}^1\Delta_g$	$\frac{1}{\sqrt{2}}|\pi_+\bar{\pi}_+|$	$[(\uparrow\downarrow)_+(\ )_-]$
	$\frac{1}{\sqrt{2}}|\pi_-\bar{\pi}_-|$	$[(\ )_+(\uparrow\downarrow)_-]$
${}^3\Sigma_g^-$	$\frac{1}{\sqrt{2}}|\pi_+\pi_-|$	$[(\uparrow)_+(\uparrow)_-]$
	$\frac{1}{2}\{|\pi_+\bar{\pi}_-| + |\bar{\pi}_+\pi_-|\}$	$[(\uparrow)_+(\downarrow)_-] + [(\downarrow)_+(\uparrow)_-]$
	$\frac{1}{\sqrt{2}}|\bar{\pi}_+\bar{\pi}_-|$	$[(\downarrow)_+(\downarrow)_-]$

Fig. 6. Complex wave functions for the lowest electronic states of molecular oxygen.

to the product obtained from the principal diagonal; the full determinant is then

$$\Psi(^1\Delta_g) = \frac{1}{\sqrt{2}} \begin{vmatrix} \pi_+(1)\alpha(1) & \pi_+(2)\alpha(2) \\ \pi_+(1)\beta(1) & \pi_+(2)\beta(2) \end{vmatrix}$$

This determinant may be expanded to yield

$$\Psi(^1\Delta_g) = \frac{1}{\sqrt{2}} [\pi_+(1)\alpha(1)\pi_+(2)\beta(2) - \pi_+(1)\beta(1)\pi_+(2)\alpha(2)]$$

Thus, in the determinant, columns label electrons 1 and 2, rows label orbital assignment $\pi_+\alpha$ and $\pi_+\beta$. For the determinantal form of the state wave function and in the expanded form of the wave function, it is evident that (a) interchange of 1 and 2 changes the sign of the overall wave function, as required for all proper (antisymmetrized) wave functions and (b) indistinguishability of electrons is now implicit in the formulation. The expanded expression can be factored into orbital and spin parts

$$\Psi(^1\Delta_g) = \frac{1}{\sqrt{2}} [\pi_+(1)\pi_+(2)][\alpha(1)\beta(2) - \beta(1)\alpha(2)]$$

where the spin function is clearly the antisymmetric two-electron spin function, i.e., singlet; and the orbital function is symmetric under electron interchange (for a net antisymmetry as required).

Using the linear combination of $(\pi_g{}^+\bar{\pi}_g{}^- - \bar{\pi}_g{}^+\pi_g{}^-)$ of Fig. 5, column 1, written in determinantal form for electron indistinguishability, we obtain the unique wave function for the $^1\Sigma_g{}^+$ state

$$\Psi(^1\Sigma_g{}^+) = \tfrac{1}{2}(|\pi_+\bar{\pi}_-| - |\bar{\pi}_+\pi_-|)$$

which factors after expansion of the determinants into

$$\Psi(^1\Sigma_g{}^+) = \tfrac{1}{2}\{[\pi_+(1)\pi_-(2)] + [\pi_-(1)\pi_+(2)]\}[\alpha(1)\beta(2) - \beta(1)\alpha(2)]$$

again revealing a singlet spin function, with a symmetric orbital function under electron interchange.

Finally, as a last example we resolve the $^3\Sigma_g{}^-$ ($M_S = 0$) component, for which Fig. 6 indicates the determinantal form of the linear combination $(\pi_g{}^+\bar{\pi}_g{}^- + \bar{\pi}_g{}^+\pi_g{}^-)$ of Fig. 5, column 1

$$\Psi(^3\Sigma_g{}^-, M_S = 0) = \tfrac{1}{2}(|\pi_+\bar{\pi}_-| + |\bar{\pi}_+\pi_-|)$$

which factors upon expansion of the determinants into

$$\Psi(^3\Sigma_g{}^-, M_S = 0) = \tfrac{1}{2}\{[\pi_+(1)\pi_-(2)] - [\pi_-(1)\pi_+(2)]\}[\alpha(1)\beta(2) + \beta(1)\alpha(2)]$$

revealing a *symmetric* or *triplet* spin function and an antisymmetric orbital function, under electron interchange. This last represents an electronic state with zero component of spin angular momentum with respect to the z axis. The remaining state functions of Fig. 6 are formulated analogously from the components of Fig. 5, column 1.

Thus, the compactness of the determinantal form as displayed in Fig. 6 for the lowest electronic states of molecular oxygen permits a full resolution into electronic state components and at the same time reveals their distinctive differences as well as features in common. We may summarize as follows.

From the single molecular orbital configuration $\ldots(1\pi_g)^2$ we have generated six different electronic states with distinctive electronic distribution patterns, energies, and magnetic characteristics. There is a unique $^1\Sigma_g{}^+$ state; there are in zeroth approximation two equal energy $^1\Delta_g$ states and three equal energy $^3\Sigma_g{}^-$ states. Arising as they do from a single molecular orbital configuration, we could anticipate the potential curve (Fig. 1) common features for the cluster of $^1\Sigma_g{}^+$, $^1\Delta_g$, and $^3\Sigma_g{}^-$ states: they have nearly coincident potential minima, indicating almost identical binding energies, and the states dissociate to a common limit.

D. Electron Repulsion in the π Electron States of O_2

The relative energy of the lowest molecular oxygen states was first discussed by Hückel (1930) in his paper: "Quantum Theory of the Double Bond." All of the states described by the wave functions of Fig. 6 would be degenerate in the absence of electron–electron interaction. Electron repulsion splits the states apart in energy, and Hückel showed that the energy of the $^1\Delta_g$ state should be approximately half of the sum of the energies of the $^1\Sigma_g{}^+$ and the $^3\Sigma_g{}^-$ states. The argument presented by Hückel is so elegant, revealing, and simple that it is worth presenting here, especially since it has escaped the attention it deserves. The idea is based essentially on correlation of motion of the two electrons in the π_g orbital with respect to the cylindrical axial angle ϕ. We may demonstrate Hückel's argument by starting with the spin-factored two-electron state functions just presented in Fig. 6. Factoring out the $Z(z)$ and $R(r)$ parts, and substituting the $\Phi(\phi)$ part of the function for the π orbitals using $\Phi(\phi) = (1/\sqrt{2\pi})e^{im_l\phi}$ with $m_l = +1$ for π_+ and -1 for π_-, we obtain the $\phi-$ dependence of the orbital part of the state functions ψ (primed)

$$\Psi'(^1\Sigma_g{}^+) = \tfrac{1}{2}\{[\pi_+(1)\pi_-(1)] + [\pi_-(1)\pi_+(2)]\}$$

$$= \frac{1}{4\pi}(e^{i\phi_1}e^{-i\phi_2} + e^{-i\phi_1}e^{i\phi_2}) \qquad = \frac{1}{4\pi}(e^{i(\phi_1-\phi_2)} + e^{-i(\phi_1-\phi_2)})$$

$$\Psi'(^1\Delta_g) = \frac{\sqrt{2}}{4\pi}[\pi_+(1)\pi_+(2)] = \frac{\sqrt{2}}{4\pi}[e^{i\phi_1}e^{i\phi_2}] \qquad = \frac{\sqrt{2}}{4\pi}e^{i(\phi_1+\phi_2)}$$

$$\Psi'(^1\Delta_g) = \frac{\sqrt{2}}{4\pi}[\pi_-(1)\pi_-(2)] = \frac{\sqrt{2}}{4\pi}[e^{-i\phi_1}e^{-i\phi_2}] = \frac{\sqrt{2}}{4\pi}e^{-i(\phi_1+\phi_2)}$$

$$\Psi'(^3\Sigma_g{}^-)_0 = \tfrac{1}{2}\{[\pi_+(1)\pi_-(2)] - [\pi_-(1)\pi_+(2)]\}$$

$$= \frac{1}{4\pi}(e^{i\phi_1}e^{-i\phi_2} - e^{-i\phi_1}e^{i\phi_2}) \qquad = \frac{1}{4\pi}(e^{i(\phi_1-\phi_2)} - e^{-i(\phi_1-\phi_2)})$$

so that for the $\Phi(\phi)$ parts of the orbital two-electron state functions

$$\Psi'(^1\Sigma_g{}^+) = \frac{1}{2\pi}\cos(\phi_1 - \phi_2)$$

$$\Psi'(^1\Delta_g) = \frac{\sqrt{2}}{4\pi}e^{i(\phi_1+\phi_2)}$$

$$\Psi'(^1\Delta_g) = \frac{\sqrt{2}}{4\pi}e^{-i(\phi_1+\phi_2)}$$

$$\Psi'(^3\Sigma_g{}^-)_0 = \frac{1}{2\pi}\sin(\phi_1 - \phi_2)$$

These two electron state functions show the dependence on the difference in angular position ϕ for the two π_g electrons in each state. The square (or complex modulus) of these wave functions corresponds to electron probability or point density. Since electron–electron repulsion is determined by the operator e^2/r_{12}, there is a great sensitivity of repulsive interaction to closeness of approach. The $\Psi(^1\Sigma_g{}^+)$ state shows a cosine squared dependence of the $(\phi_1 - \phi_2)$ angular difference (Fig. 7) for the probability density $(\Psi^*\Psi)$ for the two electrons, i.e., they tend to maximize $(\cos 0 = 1)$ probability density for zero angular difference: the electron repulsion interaction will be a maximum. The $\Psi(^3\Sigma_g{}^-)_0$ state shows a sine squared dependence of the $(\phi_1 - \phi_2)$ angular difference (Fig. 7) for a maximum probability density $(\Psi^*\Psi)$ for 90° angular difference $(\sin 90° = 1)$: the electrons avoid each other, and the electron repulsion interaction will be a minimum for the

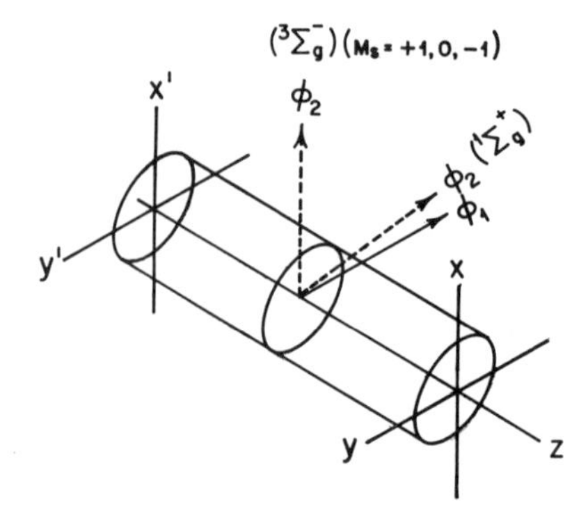

Fig. 7. Angular correlation for the two valency electrons in the lowest states of molecular oxygen.

triplet state. This is a well-known intuitive result, but Hückel's argument has the additional feature of an angular correlation analysis. The $\Psi(^1\Delta_g)$ state functions yield complex modulus $|\Psi^*\Psi| = 1$ so that the two electrons have an electron repulsion averaged over the entire orbital. Hückel (1930) used simple product orbitals for the two-electron functions. We have shown that his argument holds perfectly well for determinantal functions. Hückel then demonstrated by simple comparison of electron repulsion integrals that

$$E(^1\Delta) = (1/2)[E(^1\Sigma) + E(^3\Sigma)]$$

to first approximation.

A comment might be added on the consequence of the electron repulsion argument for triplet state lowering of energy versus singlet state raising of energy from a common or degenerate configuration. This is, of course, Hund's rule: the state of maximum multiplicity for a given configuration lies lowest. The role of spin in the Hückel argument is indirect: the antisymmetric and symmetric spin functions constrain the corresponding *orbital* occupancies to be, respectively, symmetric and antisymmetric; it is the form of the *orbital function* which then leads to cosine and sine dependence of angular correlation for singlet and triplet electron pairs.

E. Degeneracies in the π Electron States of O_2

Electron cell diagrams are given on the right side of Fig. 6 to offer some graphical comparisons of the various electron distributions corresponding to the different lowest electronic state components. The diagrams correspond exactly to the principal diagonal terms of the determinants. Thus, electron exchange is not shown in the cell diagrams because, as seen earlier, indexing of orbital occupation is given by the alternate diagonal product of the determinant. It is also clear from the cell diagrams that each state component has a qualitatively different arrangement of electrons, orbitals, and spins. The origin of the $^1\Delta_g$ state degeneracy stands out clearly. The $^1\Sigma_g{}^+$ state cell diagram has some apparent resemblance to that for the $^3\Sigma_g{}^-$ ($M_S = 0$) state, except for the sign of the linear combination of the determinants. Of course, the difference is crucial: when investigating electron probability density, the cross term or interference term appears with opposite sign.

The triplet state degeneracy can bear a little elaboration. We have already shown using the Hückel argument that

$$\Psi'(^3\Sigma_g{}^-)_{(M_S=0)} = \frac{1}{2\pi}\sin(\phi_1 - \phi_2)$$

leading to minimization of electron repulsion by angular correlation. The ϕ part of the orbital state wave function was given as

$$\Psi'(^3\Sigma_g{}^-)_{(M_S=0)} = \tfrac{1}{2}\{[\pi_+(1)\pi_-(2)] - [\pi_-(1)\pi_+(2)]\}_{\text{orb.}}$$

The determinantal forms of the state functions for the $M_S = +1$ and -1 triplet components are quite different in form (Fig. 6) from that for the $M_S = 0$ triplet component. If we expand the determinants, factor out the spin functions, and regroup the terms, we obtain for the $\Phi(\phi)$ part of the orbital state functions

$$\Psi'(^3\Sigma_g{}^-)_{(M_S=1)} = \tfrac{1}{2}\{[\pi_+(1)\pi_-(2)] - [\pi_-(1)\pi_+(2)]\}_{\text{orb.}}$$

$$\Psi'(^3\Sigma_g^-)_{(M_S=-1)} = \tfrac{1}{2}\{[\pi_+(1)\pi_-(2)] - [\pi_-(1)\pi_+(2)]\}_{\text{orb.}}$$

Thus, all of the orbital components of the three $^3\Sigma_g{}^-$ state components are identical and will have the same $\sin(\phi_1 - \phi_2)$ dependence (Fig. 7), with the same angular avoidance of electron repulsion. In other words, the three symmetric spin functions, even though differing from each other, constrain the orbital part of the triplet state components to be antisymmetric and identical, with energy identity (degeneracy).

It is interesting to point out that, at least in the free oxygen molecule, although the critical pair of electrons avoid each other, the correlation of motion is angle to angle, rather than end to end as in the primitive chemical picture for a triplet state. The other aspect which bears elaboration for the three triplet state components is the relation of the projection of spin angular momentum onto the z axis (the eigenvalues of the $\hat{S}_z$ operator) to the total spin for each component (the eigenvalues of the S^2 operator). The quantum numbers $M_S = +1$, 0, and -1 each correspond to eigenvalues of the $\hat{S}_z$ operator: they are successively the z components of spin angular momentum (in units $h/2\pi$). The total spin is given by the $\hat{S}^2 = \hat{S}_z^2 + \hat{S}_y^2 + \hat{S}_x^2$ operator, with eigenvalues $S(S+1)h^2/4\pi^2$. For each of the triplet state components the total spin quantum number S is 1; as triplet states, each component does have the same net spin. In this regard, the ${}^1\Sigma_g{}^+$ state differs from the ${}^3\Sigma_g{}^-$ ($M_S = 0$) component; for the singlet state, although $M_S = 0$, also $S = 0$, with zero net spin angular momentum.

F. The Real π Electron State Functions for O_2

Real state functions for molecular oxygen are presented in Fig. 8 for the three lowest states under discussion. A comparison of the determinantal form for these and for the complex state functions of Fig. 6 indicates an interesting juggling of terms and signs. The procedure for converting the complex forms is straightforward, and like the process of expanding the determinants and factoring out the spin functions summarized earlier, the reader will find it of interest to work out these conversions. The procedure is: (a) expand the determinants given in Fig. 6 into differences of products, with electron index labeling as exemplified earlier in this section; (b) substitute π_+ by $(1/\sqrt{2})(\pi_x + i\pi_y)$ and π_- by $(1/\sqrt{2})(\pi_x - i\pi_y)$; (c) cancel common terms, and regroup into new determinantal products; (d) condense to the

STATE	REAL WAVE FUNCTION	SPIN-ORBITAL DIAGRAM
${}^1\Sigma_g^+$	$\frac{1}{2}\{\lvert\pi_x\bar{\pi}_x\rvert + \lvert\pi_y\bar{\pi}_y\rvert\}$	$[(\uparrow\downarrow)_x(\ )_y] + [(\ )_x(\uparrow\downarrow)_y]$
${}^1\Delta_g$	$\frac{1}{2}\{\lvert\pi_x\bar{\pi}_x\rvert - \lvert\pi_y\bar{\pi}_y\rvert\}$	$[(\uparrow\downarrow)_x(\ )_y] - [(\ )_x(\uparrow\downarrow)_y]$
	$\frac{1}{2}\{\lvert\pi_x\bar{\pi}_y\rvert - \lvert\bar{\pi}_x\pi_y\rvert\}$	$[(\uparrow)_x(\downarrow)_y] - [(\downarrow)_x(\uparrow)_y]$
${}^3\Sigma_g^-$	$\frac{1}{\sqrt{2}}\lvert\pi_x\pi_y\rvert$	$[(\uparrow)_x(\uparrow)_y]$
	$\frac{1}{2}\{\lvert\pi_x\bar{\pi}_y\rvert + \lvert\bar{\pi}_x\pi_y\rvert\}$	$[(\uparrow)_x(\downarrow)_y] + [(\downarrow)_x(\uparrow)_y]$
	$\frac{1}{\sqrt{2}}\lvert\bar{\pi}_x\bar{\pi}_y\rvert$	$[(\downarrow)_x(\downarrow)_y]$

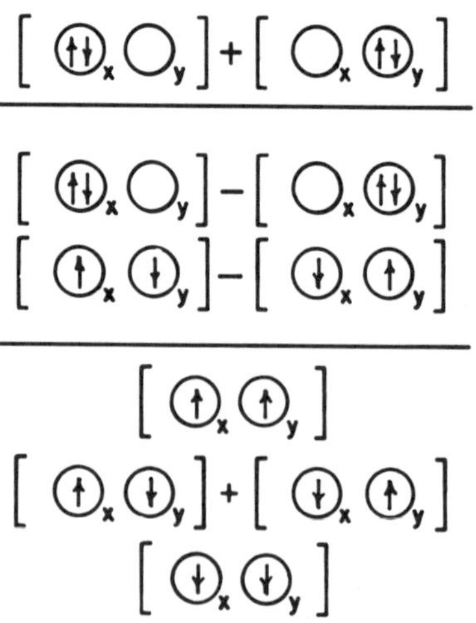

Fig. 8. Real wave functions for the lowest electronic states of molecular oxygen.

determinantal diagonal abbreviated form. The behavior of each transformation varies. For $^1\Sigma_g{}^+$, all imaginary terms cancel. For $^3\Sigma_g{}^-$ ($M_S = +1$) and $M_S = -1$ transformation, all real terms cancel; the common coefficient i for remaining terms is merely a phase factor and is eliminated by multiplying by $-i$. For the $^1\Delta_g$ transformations, one yields an expression in the form $A + iB$ and the other in the form $A - iB$. Since the goal is to obtain a real set of determinantal state functions, addition and subtraction yield $2A$ and $2iB$, which are simply renormalized and then appear as shown in Fig. 8.

The real state functions of Fig. 8 now confirm neatly what we learned about angular correlation for the $^1\Sigma_g{}^+$ and $^3\Sigma_g{}^-$ states. Each of the terms of $^1\Sigma_g{}^+$ is of the form $\pi_x\pi_x$ or $\pi_y\pi_y$, confirming the Fig. 7 correlation of electron pairs forced to move most probably in a common plane, with $\phi_1 - \phi_2$ a minimum. Each of the terms for the $^3\Sigma_g{}^-$ state components is of form $\pi_x\pi_y$, indicating that each electron of the pair moves most probably in perpendicular planes, minimizing repulsion with $\phi_1 - \phi_2$ equal to 90° as shown in Fig. 7. The $^1\Delta_g$ states represent xx, yy two-electron terms and also xy, xy terms: *taken together*, this degenerate pair of states averages the energy for the $^1\Sigma_g{}^+$ and the $^3\Sigma_g{}^-$ states, in the free molecule.

However, if a perturbing field is applied by an approaching molecule, the $^1\Delta_g$ state degeneracy will split. In such a case, xz (and yz) planes are defined by an approaching molecule. Thus, one could begin to elucidate reaction mechanisms on the basis of electron distribution in the real π orbital wave functions of the oxygen molecule. For example, the $^1\Sigma_g{}^+$ state would tend to have both of the π electrons in plane, so that concerted two-point addition reactions could be visualized (with the added step of polarization of the π electrons onto the atomic centers). The $^3\Sigma_g{}^-$ state(s) would tend to have each of the π electrons oriented in mutually perpendicular planes, so an intermediate with single point attachment might be anticipated. The $^1\Delta_g$ state, however, under perturbation by an approaching molecule in one plane would tend to show a distribution of products reflecting both concerted two-point addition and single-point attachment intermediates.

G. The Literature on Molecular Oxygen Structure

We have indulged in a somewhat detailed examination of the electronic structure of molecular oxygen in its three lowest states because of the absence of specific information in standard sources. Three classic reference works [Pauling and Wilson (1935); Eyring, Walter, and Kimball (1944); Kauzmann (1957)] fail to include *oxygen* in the index of the book. Under *diatomic molecules*, the last two references give the level of treatment corresponding to Fig. 4 with *ad hoc* state symbols, which is about the level of treatment of electronic structure of oxygen given in most contemporary valency theory and quantum theory books today.

Ballhausen and Gray (1964) presented an unlabeled set of state wave functions corresponding to those of Fig. 8, although the given atomic orbital composition indicates the functions to be real. Griffith (1964a,b) presented a set of real state wave functions for oxygen, with diverse forms of spin factoring. Kearns (1971) presented a set of complex state functions for oxygen, giving only the $M_S = 0$ component of the triplet. He also gave a corresponding set of real state functions. The normalization constants are inconsistent in the sources mentioned above; the present set agrees with those of Griffith. Kasha and Khan (1970) presented a complete set of state wave functions for molecular oxygen, unfortunately with an error in labeling: their Fig. 8 labels the orbitals as complex, whereas the form of the functions shows that they are the real set. Therefore, their Fig. 8 should have subscripts + and − replaced by x and y respectively, to agree with Fig. 8 of the present chapter.

Quantum mechanical calculations on the lowest electronic states of the oxygen molecule have been given by Moffitt (1951), Meckler (1953), and Kotani *et al.* (1957).

V. SENSITIZATION MECHANISMS FOR SINGLET OXYGEN

The experimental energy level diagram for the electronic transitions involving the lowest three electronic states of molecular oxygen (Fig. 9) is now firmly established by experimental and theoretical researches. The single molecule transitions (Fig. 1) ${}^1\Delta_g \leftarrow {}^3\Sigma_g^-$ and ${}^1\Sigma_g^+ \leftarrow {}^3\Sigma_g^-$ have been known as classical atmospheric absorption bands at low sun (Herzberg, 1950). These lowest two transitions are strictly forbidden for electric dipole radiation in the isolated molecule, and result in extraordinary metastability of the excited singlet states of oxygen. The lifetimes estimated by integrated absorption measurements are 45 min for the ${}^1\Delta_g$ state, and 7.1 sec for the ${}^1\Sigma_g^+$ state at *zero* pressure, but even at 1 atm pressure intermolecular collisions change the transition mechanism to electric dipole, with much shortened lifetimes. Figure 9 indicates intermolecular collision pairs with ground state molecular oxygen for the three lowest oxygen states, e.g., $({}^1\Delta_g)[{}^3\Sigma_g^-]$, as a mechanism for electronic relaxation of the multiplicity selection rule. The lifetimes in solution for the singlet molecular oxygen states become drastically shortened, with estimates of 10^{-3} sec for the ${}^1\Delta_g$ state and 10^{-9} sec for the ${}^1\Sigma_g^+$ state in water (Arnold *et al.*, 1968). The life-

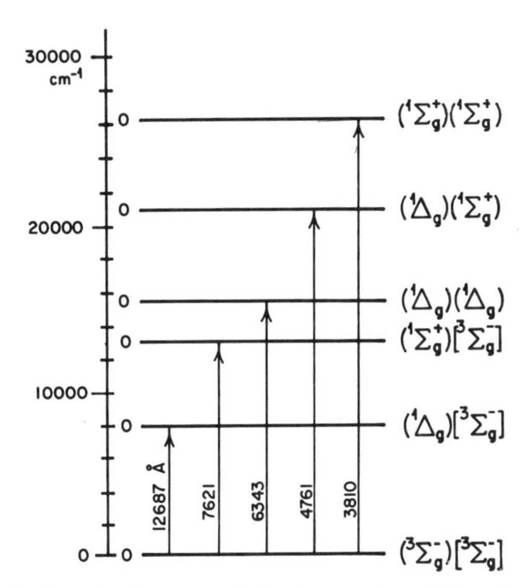

Fig. 9. Oxygen single molecule states and simultaneous transition molecular-pair states.

times of the ${}^1\Delta_g$ state of oxygen have been measured (Merkel and Kearns, 1972; Long and Kearns, 1975) to be in the range 10^{-6}–10^{-3} sec, with a strong solvent dependence. The first definitive observations on the ${}^1\Delta_g$ state were made by the Herzbergs (1934, 1947, 1948) after observation of this state in liquid oxygen absorption by Ellis and Kneser (1933).

Singlet molecular oxygen became a subject of intense laboratory study as a chemical reagent following the interpretation by Khan and Kasha (1963) of the chemiluminescence of the hypochlorite–oxygen reaction as due to liberated singlet oxygen. The complicated history of rediscoveries and reinterpretation has been carefully traced (Kasha and Khan, 1970; Khan and Kasha, 1970; Kearns, 1971). An indication of the vast attention that singlet oxygen is now receiving is given by the publication of a reprint volume summarizing the chemical history of this subject (Schaap, 1976).

The clue to spectroscopic and photochemical behavior of singlet oxygen excited states lies in understanding the *simultaneous transitions* which are possible for collision pairs of singlet oxygen molecules. The idea originated with Ellis and Kneser (1933), who sought to explain the blue color and absorption spectrum details of liquid oxygen on the basis of molecular-pair absorption. They were influenced by G. N. Lewis' concept of $(O_2)_2$ dimers as an explanation of the paramagnetic behavior of liquid oxygen diluted by liquid nitrogen. All subsequent research has indicated definitively that no dimers of molecular oxygen exist at ordinary conditions, and that the intermolecular perturbations, necessary to account for the collisional enhancement of the single-molecule transitions and the simultaneous transitions, are much smaller than kT. The term "dimol" which is commonly applied to singlet-molecular oxygen pairs is thus devoid of operational meaning and should not be used (Kearns, 1971; cf. Krupenie, 1972, esp. pp. 440–443; cf. Gray and Ogryzlo, 1969). Dimers of molecular oxygen are found in solid oxygen at very low temperatures.

Chemiluminescence observations of the hypochlorite–peroxide reaction were related to simultaneous transitions in molecular pairs by a series of investigators (Groh and Kirrmann, 1942; Stauff and Schmidkunz, 1962; Arnold *et al*., 1964; cf. Khan and Kasha, 1970), but it was the work of Ogryzlo's laboratory (Arnold *et al*., 1964; Browne and Ogryzlo, 1964) which related the observed chemiluminescence spectra of the hypochlorite–solution reaction to the electrical discharge molecular oxygen spectra in the vapor phase. Subsequently, Khan and Kasha (1970) gave a full spectroscopic correlation of absorption and emission spectra corresponding to simultaneous transition states shown in Fig. 9 for molecular pairs.

We may define the *simultaneous transition* as a *transition for a pair of atomic or molecular species to a composite excited state* (a one-photon two-molecule cooperative absorption), the energy of the new state arising by a nonresonance, superposition of two excited state energies of the component species; the converse simultaneous process may take place with the emission of a photon. We may neglect the shift of the new state relative to the sums of the component state energies on the basis of the very small interaction. The phenomenon of simultaneous transition is very widely observed in infrared spectroscopy and especially in electronic spectroscopy of forbidden transitions. The latter make the simultaneous transition easy to observe, both by the enhancement effects found, as well as by the natural "spectral window" they afford. In the case of singlet molecular oxygen, we may perceive a molecular pair state $(^1\Delta_g)(^1\Delta_g)$ as giving rise to a unique singlet multiplicity, S, whereas the ground state pair $(^3\Sigma_g{}^-)(^3\Sigma_g{}^-)$ would give rise to $S + T + Q$ multiplicity, permitting the multiplicity allowed singlet → singlet transition to occur.

The simultaneous transition for a pair of absorbing or emitting species does not require an actual "complex" to exist, but the two molecules (atoms, ions) must be within electron exchange "contact". In condensed phases, such as liquid oxygen, enough collision pairs exist for a weak but finite absorption to be observed. In the gas phase, 10 cm optical paths and oxygen pressures in excess of 100 atm are needed (cf. Khan and Kasha, 1970) to observe the simultaneous transition in absorption. In emission with singlet oxygen pairs, even at pressures below 1 atm, the simultaneous transition is readily observed. Blowing Cl_2 gas via a sintered glass filter into alkaline

15% H_2O_2, 1 *M* in OH^-, yields red emission easily seen in a half-darkened room [the $(^1\Delta_g)(^1\Delta_g) \rightarrow (^3\Sigma_g^-)(^3\Sigma_g^-)$ emission]. Obviously, the metastability of $^1\Delta_g$, even at 1 atm, is still great enough to permit two excited molecules to collide and produce a simultaneous transition state.

Sensitization mechanisms involving singlet molecular oxygen are of great chemical and biological importance. Khan and Kasha (1966) proposed that some organic molecule "chemiluminescences" could be the result of excitation by a *physical* energy transfer from singlet oxygen simultaneous transition states, such as the $(^1\Delta_g)(^1\Delta_g)$ state, or the $(^1\Sigma_g^+)(^1\Delta_g)$ state. The theory of such processes has been elaborated on in two papers from Kearns' laboratory (Kawaoka *et al.*, 1967; Khan and Kearns, 1968). We have been exploring energy resonances between the simultaneous transition states for singlet molecular oxygen (Fig. 9) and various dye molecules stable in the peroxide–hypochlorite reaction system. Here we shall review five possible mechanisms for such physical energy transfer excitation processes for the dye "chemiluminescences".

Mechanism a: Triplet Energy Pooling

A chemiluminescence mechanism suggested by Ogryzlo and Pearson (1968) and investigated by Wilson (1969) for rubrene and violanthrone is given in Fig. 10. This energy pooling mechanism requires two successive energy transfer excitations by a $^1\Delta_g$ state, first lifting the dye molecule to its lowest triplet state, then this triplet state being further promoted to or above its lowest singlet excited state. The energetic criteria are

$$\Delta E_{S_1} < 2(\Delta E_{^1\Delta_g})$$

$$\Delta E_{T_1} < \Delta E_{^1\Delta_g}$$

Molecules which have as low an energy for the lowest triplet state as is required by the energy pooling mechanism of Ogryzlo and Pearson are rare.

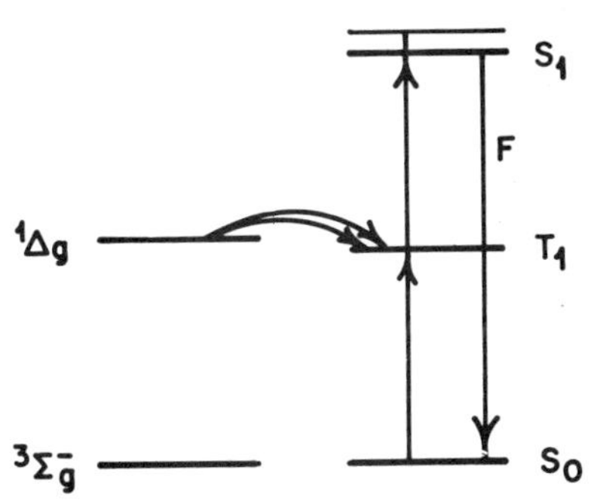

Fig. 10. Photosensitization by singlet oxygen. Mechanism a: Molecular triplet energy pooling.

Moreover, generally in dyes, singlet–triplet separations diminish as the lowest excited singlet state energy decreases (and molecular size increases). It is true that in aromatic hydrocarbons singlet–triplet splits can get as large as 12000 cm^{-1}. In the case of rubrene and also violanthrone, the lowest triplet state has not yet been located experimentally. We conclude that if the energy pooling mechanism is valid, it will apply to very few cases.

Mechanism b: Double-Molecule Sensitization

This is a form of the physical sensitization mechanism proposed by Khan and Kasha (1966). The possible criteria for the various simultaneous transition or double-molecule state (Fig. 9) are (cf. Fig. 11)

$$\Delta E_{S_1} < 2(\Delta E_{^1\Delta_g})$$

or

$$\Delta E_{S_1} < (\Delta E_{^1\Delta_g} + \Delta E_{^1\Sigma_g{}^+})$$

or

$$\Delta E_{S_1} < 2(\Delta E_{^1\Sigma_g{}^+})$$

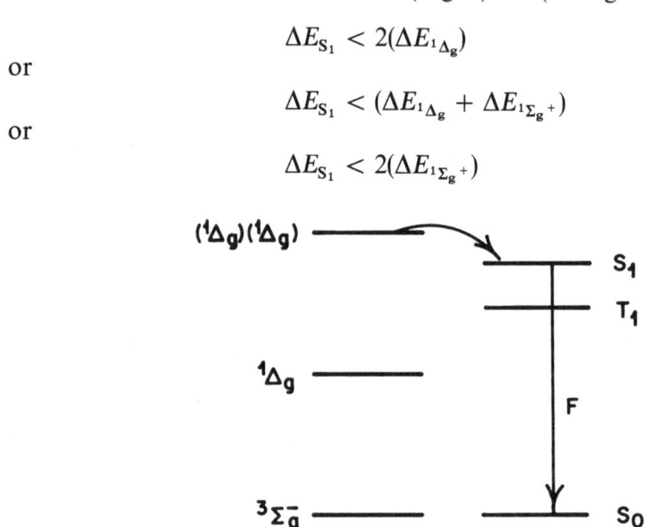

Fig. 11. Photosensitization by singlet oxygen. Mechanism b: Sensitization by simultaneous transition state.

We believe that this mechanism may apply to many red-luminescing dyes which satisfy the first energy criterion. As is commonly known, the $^1\Delta_g$ state seems to be the one predominantly produced in the hypochlorite–peroxide reaction, so the chance of $(^1\Delta_g)(^1\Delta_g)$ collision pairs existing is greatest, compared with the other two cases. It is impressive that the relative quantum yield of singlet-oxygen-sensitized dye luminescence is always much greater than the quantum yield of the oxygen simultaneous state direct emission. When the O_2–O_2 emission is observed, it is only from those molecules which have eluded quenching in the aqueous phase and escaped into the bubbles (Khan and Kasha, 1964), from which the characteristic red glow of $(^1\Delta_g)(^1\Delta_g)$ emission is seen. Obviously, energy transfer from molecular oxygen pairs to the acceptor molecule competes favorably in rate with the quenching. We believe the violanthrone and rubrene cases may prove to fit this present mechanism.

Mechanism c: Delayed Fluorescence Sensitization

The case where the lowest triplet level of an acceptor molecule lies below the simultaneous transition state of the singlet oxygen molecular pair may occur. Our study of the sensitized luminescence of fluorescein anion and Rose Bengal suggest that they may fit this mechanism. Then the dye fluorescence may be produced as a "delayed fluorescence" after Boltzmann activation from the sensitized triplet. The energy criteria are (cf. Fig. 12)

$$\Delta E_{T_1} < 2(\Delta E_{^1\Delta_g})$$

$$(\Delta E_{S_1} - \Delta E_{T_1}) \simeq kT$$

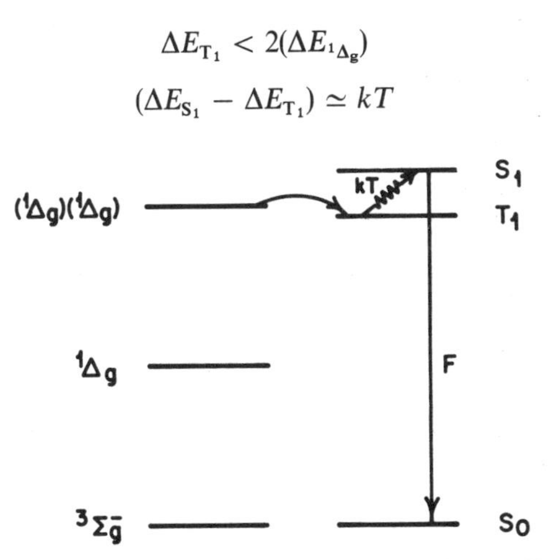

Fig. 12. Photosensitization by singlet oxygen. Mechanism c: Delayed fluorescence sensitization.

The mechanism would apply with restriction of quantum yield to the other simultaneous transition states of O_2 also. The temperature dependence may be complicated because the delayed fluorescence activation will compete with temperature-dependent quenching steps.

Mechanism d: Sensitization of Triplet–Triplet Annihilation

Triplet–triplet annihilation is a well-known phenomenon in molecular systems in condensed phases. If a large population of triplets is produced, this mechanism may be favored. We propose that the two independent triplets are excited by the simultaneous transition states of singlet oxygen, e.g., $(^1\Delta_g)(^1\Delta_g)$. Unlike the energy pooling (mechanism a) case, the energy criteria are not very restrictive (cf. Fig. 13)

$$\Delta E_{T_1} < 2(\Delta E_{^1\Delta_g})$$

$$\Delta E_{S_1} \gg 2(\Delta E_{^1\Delta_g}).$$

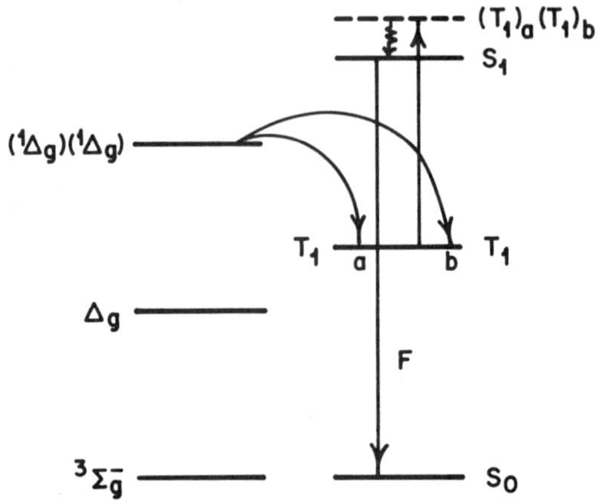

Fig. 13. Photosensitization by singlet oxygen. Mechanism d: Sensitization of triplet–triplet annihilation.

We have not yet found a case which exclusively requires this mechanism, but a careful search should reveal good examples.

Mechanism e: Induced Fluorescence

Recently we have uncovered an unusual case in the singlet-oxygen sensitized luminescence of methylene blue (Brabham and Kasha, 1974). Methylene blue has a fluorescence with a peak near 7000 Å, but in the presence of singlet oxygen in the hypochlorite–peroxide reaction, a new intense luminescence emission is observed with a broad band centered at 8130 Å. This emission occurs at the expected position for the triplet–singlet emission of methylene blue, although the intrinsic emission has eluded all attempts at its direct observation. In spite of this, methylene blue is known to be a good triplet sensitizer of photooxygenation. We believe this to be an example of the case described here by Fig. 14. The energetics of methylene blue fit the criterion

$$\Delta E_{T_1} < \Delta E_{^1\Sigma_g^+}$$

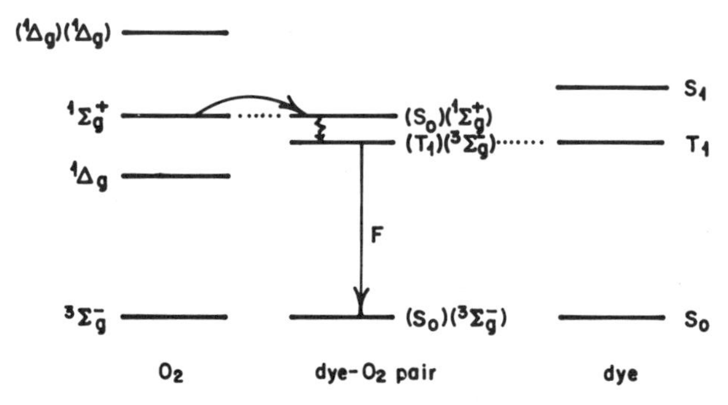

Fig. 14. Photosensitization by singlet oxygen. Mechanism e: Induced triplet–triplet fluorescence.

We believe that the $^1\Sigma_g{}^+$ state is the sensitizer, since the $(^1\Delta_g)(^1\Delta_g)$ state emission is not affected in first-order by the yield of the new luminescence. Other thiazine dyes are progressively less efficiently excited as they fail to satisfy this energy criterion. This sensitized "triplet" → "triplet" fluorescence in the complex [$(T)(T)$ yields $(S + T + Q)$ multiplicity in excited state; $(S)(T)$ yields (T) multiplicity in ground state] must occur in competition with the back excitation of triplet dye → normal oxygen (producing $^1\Delta_g$).

VI. SUMMARY

The electronic structure of molecular oxygen is examined at successive levels of sophistication from primitive models to antisymmetrized complex state functions and real state functions, for the $^3\Sigma_g^-$, $^1\Delta_g$, and $^1\Sigma_g{}^+$ states arising from the ... $(\pi_g)^2$ configuration. The treatment is at a level permitting an experimentalist, uninitiated into the chemical physics of electron quantum mechanics, to obtain an insight into a realistic understanding of the electronic states involved in oxygen excitation and interaction. The paper concludes with the discussion of five mechanisms, differentiated by energetic and spectroscopic criteria, for photosensitization of molecules by singlet oxygen.

ACKNOWLEDGMENTS

We express our thanks to W. H. Eberhardt of the Georgia Institute of Technology for calling our attention to the error in Fig. 8, Kasha and Khan (1970) noted above; to our colleagues W. Rhodes for informative discussions on complex wave function geometrical transformations, and J. Saltiel for valuable editorial critiques. We especially thank our colleague S. P. McGlynn of Louisiana State University for his penetrating technical and editorial comments. The preparation of this paper was suported by the Division of Biomedical and Environmental Research, U. S. Energy Research and Development Administration, under Contract No. EY-76-S-05-2690.

REFERENCES

Arnold, S. J., Ogryzlo, E. A., and Witzke, H. (1964). *J. Chem. Phys.* **40,** 1769.
Arnold, S. J., Kubo, M., and Ogryzlo, E. A. (1968). *Adv. Chem. Ser.* **77,** 133.
Ballhausen, C. J., and Gray, H. B. (1964). "Molecular Orbital Theory," p. 45. Benjamin, New York.
Brabham, D., and Kasha, M. (1974). *Chem. Phys. Lett.* **29,** 159.
Browne, R. J., and Ogryzlo, E. A. (1964). *Proc. Chem. Soc., London* p. 117.
Ellis, J. W., and Kneser, H. O. (1933). *Z. Phys.* **86,** 583.
Eyring, H., Walter, J., and Kimball, G. E. (1944). "Quantum Chemistry." Wiley, New York.
Gray, E. W., and Ogryzlo, E. A. (1969). *Chem. Phys. Lett.* **3,** 658.
Griffith, J. S. (1964a). *J. Chem. Phys.* **40,** 2899.
Griffith, J. S. (1964b). *In* "Oxygen in Animal Organisms" (F. Dickens and E. Neil, eds.), p. 481. Pergamon, Oxford.
Groh, P., and Kirrmann, K. A. (1942). *C. R. Hebd. Seances Acad. Sci.* **215,** 275.
Herzberg, G. (1934). *Nature (London)* **133,** 759.
Herzberg, G., and Herzberg, L. (1948). *Astrophys. J.* **108,** 167.
Herzberg, L., and Herzberg, G. (1947). *Astrophys. J.* **105,** 353.
Herzberg, G. (1950). "Molecular Spectra and Molecular Structure. I. Spectra of Diatomic Molecules," 2nd ed., p. 446. Van Nostrand-Reinhold, New York.
Hückel, E. (1930). *Z. Phys.* **60,** 423.
Kasha, M., and Khan, A. U. (1970). *Ann. N. Y. Acad. Sci.* **171,** 5.
Kauzmann, W. (1957). "Quantum Chemistry," Academic Press, New York.
Kawaoka, K., Khan, A. U., and Kearns, D. R. (1967). *J. Chem. Phys.* **46,** 1842.
Kearns, D. R. (1969). *J. Am. Chem. Soc.* **91,** 6554.
Kearns, D. R. (1971). *Chem. Rev.* **71,** 395.
Khan, A. U., and Kasha, M. (1963). *J. Chem. Phys.* **39,** 2105.
Khan, A. U., and Kasha, M. (1964). *Nature (London)* **204,** 241.
Khan, A. U., and Kasha, M. (1966). *J. Am. Chem. Soc.* **88,** 1574.
Khan, A. U., and Kasha, M. (1970). *J. Am. Chem. Soc.* **92,** 3293.
Khan, A. U., and Kearns, D. R. (1968). *J. Chem. Phys.* **48,** 3272.
Kotani, M., Mizuno, Y., and Kayama, K. (1957). *J. Phys. Soc. Jpn.* **12,** 707.
Krupenie, P. H. (1972). *J. Phys. Chem. Ref. Data* **1,** 423.
Lennard-Jones, J. E. (1929). *Trans. Faraday Soc.* **25,** 668.
Lewis, G. N. (1916). *J. Am. Chem. Soc.* **38,** 762.
Lewis, G. N. (1924a). *Chem. Rev.* **1,** 231.
Lewis, G. N. (1924b). *J. Am. Chem. Soc.* **46,** 2027.
Long, C. A., and Kearns, D. R. (1975). *J. Am. Chem. Soc.* **97,** 2018.
Meckler, A. (1953). *J. Chem. Phys.* **21,** 1750.
Merkel, P. B., and Kearns, D. R. (1972). *J. Am. Chem. Soc.* **94,** 7244.

Moffitt, W. (1951). *Proc. R. Soc. London* **210,** 224.
Mulliken, R. S. (1928). *Nature (London)* **122,** 505.
Mulliken, R. S. (1932). *Rev. Mod. Phys.* **4,** 1 (esp. pp. 54–56, and Fig. 48).
Ogryzlo, E. A., and Pearson, E. A. (1968). *J. Phys. Chem.* **72,** 2913.
Pauling, L. (1960). "Nature of the Chemical Bond," 2nd ed. Cornell Univ. Press, Ithaca, New York.
Pauling, L., and Wilson, E. B., Jr. (1935). "Introduction to Quantum Mechanics." McGraw-Hill, New York.
Schaap, A. P. (1976). "Singlet Molecular Oxygen." Dowden, Hutchinson, & Ross, Stroudsberg, Pennsylvania.
Stauff, J., and Schmidkunz, H. (1962). *Z. Phys. Chem. (Frankfurt am Main)* **35,** 295.
Wilson, T. (1969). *J. Am. Chem. Soc.* **91,** 2387.

2

Gaseous Singlet Oxygen

E. A. OGRYZLO

I. INTRODUCTION

In contrast to the condensed phase where singlet oxygen has come to mean only the low lying $O_2(^1\Delta_g)$ state, two singlet states of oxygen ($^1\Delta_g$ and $^1\Sigma_g$) have to be considered in the gas phase because of their interconvertibility (see Section V). Although the reactivity of $O_2(^1\Sigma_g^+)$ has never been observed, its relatively rapid relaxation to $O_2(^1\Delta_g)$ and the liberation of 15 kcal of electronic energy may be significant in some systems. We will therefore consider the interactions of both species in this chapter.

Several techniques for producing singlet oxygen in the gas phase can be found in the literature.

1. Kautsky and de Bruijn (1931) described a technique in which a solid sensitizer is deposited on silica gel, surrounded by gaseous oxygen and irradiated to produce singlet oxygen which can diffuse to another substrate via the gas phase. The experiment was studiously ignored for 33 years and

Singlet Oxygen

ISBN 0-12-736650-4

still has not been exploited for kinetic studies (Nilsson and Kearns, 1974).

2. Foner and Hudson (1956) presented mass spectrometric evidence which indicated that metastable oxygen molecules [possibly $O_2(^1\Delta_g)$] are present in an oxygen stream pumped through an electrical discharge. This was supported by calorimetric measurements (Elias *et al.*, 1959). However, it was not until 1964 (Arnold *et al.*, 1964; Bader and Ogryzlo, 1964) that direct spectroscopic measurements identified the two excited states present, opening up the possibility of using this technique for kinetic studies. Since then the discharge–flow system has been used to study both quenching processes and reactions of singlet delta ($^1\Delta_g$) and singlet sigma ($^1\Sigma_g{}^+$) oxygen. Because this source of singlet oxygen is useful in a variety of ways it will be described in some detail in Section II.

3. Snelling and co-workers have described two photochemical methods for producing singlet oxygen in the gas phase. The first direct spectroscopic identification of $O_2(^1\Delta_g)$ produced by energy transfer from a triplet state (benzene) was reported by Snelling (1968). The formation of $O_2(^1\Delta_g)$ when O_2 quenches other organic molecules in their triplet states has been confirmed by electron spin resonance (Kearns *et al.*, 1969; Wasserman *et al.*, 1969). The technique has not been used to study any reactions of singlet oxygen in the gas phase, but a number of $O_2(^1\Delta_g)$ quenching rate constants have been obtained with the technique (Findlay and Snelling, 1971). Snelling (1974) has also used the UV flash photolysis of ozone to produce $O_2(^1\Sigma_g{}^+)$ in the gas phase. Though some quenching and reactivity studies have been reported for this species (Gauthier and Snelling, 1975), the system is limited to molecules not affected by vacuum UV radiation, ozone, or oxygen atoms which are all present in the system. A similar technique has been used by Davidson *et al.* (1972) to obtain additional $O_2(^1\Sigma_g{}^+)$ quenching rates. These workers used photoexcited SO_2 to produce $O_2(^1\Sigma_g{}^+)$. The results of these studies will be presented in Section III in which we have collected all the available data for the physical quenching of $O_2(^1\Delta_g)$ and $O_2(^1\Sigma_g{}^+)$. These results are discussed in terms of current theories for such processes and also compared with the same reactions in the liquid phase.

II. SINGLET OXYGEN FROM ELECTRICAL DISCHARGE

The formation of metastable molecules in electrical discharges has its origins in the work of Warburg (1884), Lewis (1900a,b), and Strutt (1911) around the turn of the century. In this early work, nitrogen was subjected to a discharge in a static system and the decay of metastable species was observed after the cessation of the discharge. Several years later Wood

(1920) described a technique for removing metastable species from the discharge. Hydrogen was passed over electrodes in what is now called a "woods tube" and the metastable species (atoms in this case) were observed in the gas stream drawn out of the discharge by a vacuum pump. This technique remains the most effective method for producing metastables from N_2 and H_2.

With more corrosive species such as the halogens and oxygen, the technique was found unsatisfactory for the production of metastable species. Not only do the electrodes deteriorate with time, but the presence of metal electrodes in the stream leads to the premature destruction of the metastables. The elimination of internal electrodes from the discharge became possible with the development of powerful radio and microwave generators. Such radiation can penetrate glass vessels and maintain a discharge in a low pressure gas once it has been initiated with a Tesla-coil. One requires at least 50 watts of either radio (1–50 MHz) or microwave (2450 MHz) power and an efficient method of directing it into the gas stream. Although high power radio frequency generators are available, several factors have resulted in the almost exclusive use of microwave discharges for the formation of singlet oxygen. They are commercially available from several manufacturers; they operate at a frequency (2450 MHz) which does not interfere with other scientific equipment; most important, however, the power can be directed and localized in a short length of glass tubing through wave guides which couple the power very efficiently to the gas stream. Radiofrequency power is more difficult to localize and this creates problems when oxygen atoms are removed by the HgO technique described in Section I.

A. The Microwave Discharge

The electronic components used to create an electrodeless discharge in a low pressure stream of oxygen are shown within the dotted lines in Fig. 1. These include a microwave generator (an optional reflected power meter), a 6- to 8-ft coaxial transmission cable, and a tuned cavity (wave guide).

An increasing number of manufacturers produce microwave generators in North America and Europe. A 100 watt generator is, however, sufficient; somewhat higher concentrations can be obtained with a 200 watt generator. The addition of a reflector power meter to the unit makes it possible to tune the discharge cavity accurately. This not only maximizes the power absorbed by the gas stream, but more significantly it decreases the power reflected back to the magnetron. Such reflected power can overheat the magnetron causing its premature failure. However, magnetrons can handle a great deal of mismatching and some have survived many years of abuse in our laboratory without failure. In the absence of a reflector power meter the wave

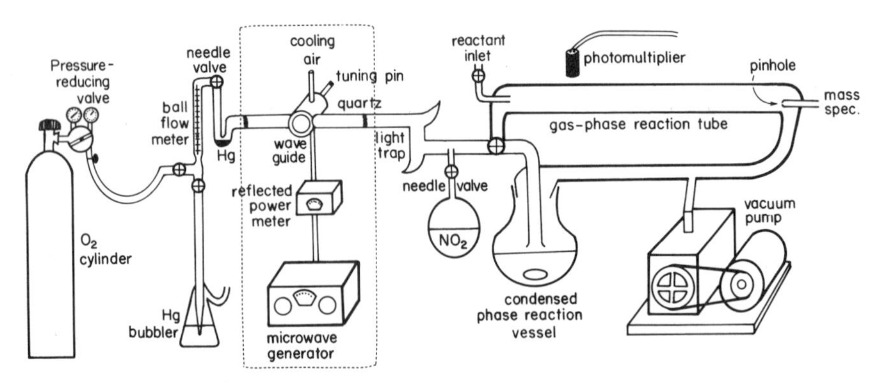

Fig. 1. Components of a discharge–flow system for the production of singlet molecular oxygen.

guide can be tuned to give the maximum $O_2(^1\Delta_g)$ concentration or the brightest discharge.

The power from the generator is coupled to the gas stream by a wave guide. For discharge–flow systems the "foreshortened $\frac{1}{4}$ wave cavity" (Fehrenfeld *et al.*, 1965) has proven most practical. It is available from several microwave generator manufacturers. This wave guide is normally constructed with 1 or 2 tuning screws and a cap which can be removed in order to fit the guide around the discharge tube (maximum diameter = 0.5 in.). The cavity requires cooling by compressed air. When the air stream is sufficiently strong, a Pyrex discharge tube is suitable; it will not collapse from overheating. However, somewhat higher concentrations of singlet oxygen can be obtained when the discharge operates hot. For this reason it is advisable to make the discharge tube of quartz.

B. The Flow System

The basic components of the conventional flow system used for the production of singlet oxygen are shown in Fig. 1. Standard grades of oxygen can be used directly from the cylinder if it is equipped with a pressure reducing valve. For kinetic studies, a somewhat steadier back pressure is usually necessary and this can be achieved by incorporating a simple mercury bubbler, as illustrated in Fig. 1. Since this portion of the flow system is at pressures in excess of atmospheric, rubber tubing can be used for connections.

The O_2 flow rate is most conveniently monitored by a commercial ball flow meter with a maximum gas flow of 100 cm/min at 1 atm. Fine control needle valves are required to regulate the O_2 and reactant flow. An all metal needle valve using flexible bellows is least prone to leak. Valves using Teflon needles do not maintain a constant or reproducible flow because of the elastic properties of the needle.

A small U-tube can be seen between the needle valve and the discharge area in Fig. 1. It serves two purposes: it prevents the radiation from the discharge from heating and thus affecting the needle valve and it, as well, serves as a retainer for a drop of mercury which can be warmed to saturate the oxygen stream with mercury vapor. We have found it convenient and advisable to heat the mercury electrically so that a permanent ring of mercuric oxide (about 1 in. wide) develops and then (with less heat) is maintained about 2 in. beyond the discharge. Without the continuous application of heat the mercuric oxide ring tends to move down toward the reaction tube. The ring removes the oxygen atoms from the stream but does not affect the $O_2(^1\Delta_g)$ or $O_2(^1\Sigma_g^+)$ coming from the discharge. Only if a very great excess of mercury is distilled through the discharge is the concentration of singlet oxygen reduced and sometimes briefly eliminated. A steady-state can be established so that the $O_2(^1\Delta_g)$ concentration remains constant over a period of hours or even days.

The complete removal of oxygen atoms can be determined by viewing the reaction tube in complete darkness. Because of trace amounts of N_2 in O_2 cylinders, small amounts of NO are formed in the discharge. The reaction

$$NO + O \rightarrow NO_2 + h\nu \qquad (1)$$

gives rise to a weak "white" glow which is easily seen in a completely darkened room. When atoms are absent, the reaction vessel appears red from the singlet oxygen emission. For most work such a "red" stream which has an oxygen atom concentration less than 10^{-4} of the $O_2(^1\Delta_g)$ concentration can be used without complications. However, for any experiment in which the presence of trace concentrations of atoms can lead to complications, very small amounts of NO_2 can be added (after the discharge) to remove the remaining atoms through the following rapid reaction (Kaufman, 1961)

$$NO_2 + O \rightarrow NO + O_2 \qquad (2)$$

If any atoms were present before the addition, a weak "white" glow appears at the inlet as a result of reaction (2) followed by reaction (1). When an excess of NO_2 is added, another "white" glow appears throughout the reaction vessel due to the slow energy transfer process (Arnold *et al.*, 1966).

$$NO_2 + 3O_2(^1\Delta_g) \rightarrow NO_2^* + 3O_2$$

Thus the point at which the visible emission from the reaction is a minimum corresponds to the complete removal of oxygen atoms from the system. We have found this procedure effective and essential in the study of quenching by hydrocarbons where a trace of oxygen atoms leads to anomalous results

(Furukawa and Ogryzlo, 1971). The amount of NO introduced in this way is normally less than that formed in the discharge from nitrogenous impurities.

A light trap is necessary between the discharge and the reaction vessel in order to avoid direct photolysis of the reactants by radiation from the discharge. It also serves to prevent the discharge radiation from interfering with the spectroscopic measurements in the reaction vessel. As a further precaution the entire discharge tube can be covered with black paint. This is essential if spectroscopic measurements are being made and also protects the experimenter's eyes from UV radiation which is transmitted by the quartz discharge tube.

The optimum size and shape of the reaction vessel depends on the rate of the reaction under study and the phase in which it is to be carried out. Two conventional vessels are illustrated in Fig. 1. The lower one is suitable for solid or liquid reactants. The singlet oxygen stream is made to impinge on the surface of the material. It is inadvisable to bubble the gas stream through the liquid since this creates too high and variable a pressure in the discharge region. However a variety of solid support materials can be used to increase the surface area of the substrate (Furukawa *et al.*, 1970). Though silica gel deactivates oxygen very rapidly, this causes no problem when the surface is covered with the substrate. Microcrystalline cellulose does not deactivate singlet oxygen as readily and is therefore somewhat more satisfactory as a support material. If a solution of the substrate is used, the reaction vessel must be cooled to reduce the solvent vapor pressure to less than 0.1 torr in order to prevent the rapid evaporation of the solvent. In a flow system the counterflow created by the evaporating solvent can prevent the oxygen stream from reaching the liquid surface. A bypass around the reaction vessel going directly to the pump is useful so that oxygen atoms can be completely eliminated before the gas stream is diverted over the substrate. In Fig. 1 the upper reaction vessel serves as such a bypass. However, it is included to illustrate the type of reaction vessel which has been used for kinetic studies of gas phase reactions. For these measurements the reaction tube is usually constructed from a straight piece of 20- to 50-mm diameter tubing about half a meter long. A reactant inlet is situated at the upstream end of the tube. A red sensitive photomultiplier can be used to monitor the relative $O_2(^1\Delta_g)$ concentration through its emission band at 6340 Å or the $O_2(^1\Sigma_g^+)$ concentration through its emission band at 7619 Å. The position of the photomultiplier along this reaction tube provides an accurate time axis for the reaction when the pressure in the system lies between 0.5 and 10 torr (Kaufman, 1961). The rate constant for the removal of $O_2(^1\Delta_g)$ by any species can therefore be determined by moving the photomultiplier along the tube. Since a rate constant determined in this way includes both reaction and quenching processes, a separate measurement of the rate of reactant disappearance is required to establish the reaction rate constant. This can be obtained with a mass spectrometer located at the end of the reaction tube, as shown in Fig. 1.

$O_2(^1\Sigma_g{}^+)$ quenching rate constants can also be determined with this type of discharge flow system, however a somewhat different technique is required because $O_2(^1\Sigma_g{}^+)$ is present along the reaction tube in a steady-state concentration which is determined by reactions (3)–(5).

$$O_2(^1\Delta_g) + O_2(^1\Delta_g) \rightarrow O_2(^1\Sigma_g{}^+) + O_2(^3\Sigma_g{}^-) \tag{3}$$

$$O_2(^1\Sigma_g{}^+) \xrightarrow{\text{wall}} O_2(^1\Delta_g) \tag{4}$$

$$O_2(^1\Sigma_g{}^+) + Q \rightarrow O_2(^1\Delta_g) + Q \tag{5}$$

Consequently a quenching constant can be obtained from s single measurement of the $O_2(^1\Sigma_g{}^+)$ emission intensity in the presence of a quencher (Q) from a knowledge of the rate constant for reaction (3) (1.2×10^3 liters mole^{-1} sec^{-1}) or reaction (4) which varies with the size of the container and condition of the walls).

We have found it unwise to use a liquid nitrogen trap to protect the pump from the reactants and products because explosive mixtures can accumulate. Dry ice traps are somewhat safer because ozone is not trapped, however unstable peroxides may still be concentrated in such traps. The safest procedure is simply to allow the products to enter the pump.

The pump must have a capacity of at least 50 liters per minute (at 1 atm) in order to draw the excited molecules out of the discharge before appreciable deactivation occurs. With such a pump the pressure in the reaction vessel can be set anywhere between 0.5 and 10 torr when the flow rate of O_2 is varied between about 25 and 500 cm/min. This O_2 flow rate is most easily measured by collecting the gas leaving the vacuum pump. The optimum pressure at which the $O_2(^1\Delta_g)$ concentration is highest varies with the construction of the flow system but usually lies near 5 torr. At higher flows oxygen atoms become more difficult to eliminate because they are swept past the mercuric oxide surface.

When the conventional flow system is operated in this pressure range the singlet oxygen has a half-life of about one second. In 10–20 mm tubing the gas stream takes 1 or 2 msec to travel a centimeter. Consequently the singlet oxygen can be transported a meter or two in such a flow system without great loss.

As noted above, a stream of $O_2(^1\Delta_g)$ is always accompanied by $O_2(^1\Sigma_g{}^+)$ because it is constantly being formed by reaction (3). However, its steady-state concentration is about 3 orders of magnitude lower than the $O_2(^1\Delta_g)$ concentration in the usual discharge–flow system. The addition of a good $O_2(^1\Sigma_g{}^+)$ quencher such as H_2O can be used to reduce this steady-state concentration in the flow system without affecting the $O_2(^1\Delta_g)$ concentration. However, it will be appreciated that this may leave unaffected any reaction of $O_2(^1\Sigma_g{}^+)$ which is faster than the quenching reaction with H_2O. Furthermore, if the water is condensed out of the stream at some point, the $O_2(^1\Sigma_g{}^+)$ concentration very quickly returns to its higher steady-state concentration. We emphasize these facts here because some workers have added H_2O and then condensed it out in the mistaken belief that $O_2(^1\Sigma_g{}^+)$ is thus permanently removed from the system.

C. Applications to Heterogeneous Systems

We will not attempt to discuss the details of condensed-phase reactions which have been studied with discharge-generated gaseous singlet oxygen since these will be discussed in other chapters of this book. Only the types of systems to which the technique has been applied will be described.

Historically, the exposure of discharge-generated singlet oxygen to a number of unsaturated compounds (Corey and Taylor, 1964) provided, simultaneously with the work of Foote and Wexler (1964), the first experimental support for Kautsky and de Bruijn's (1931) theory of photooxygenation. Since then the technique has been used to verify the involvement of singlet delta oxygen in the photooxygenation of a number of unsaturated cyclic compounds (Scheffer and Ouchi, 1970), fatty acids (Rawls and Van Santen, 1970), ribonucleosides (Clagett and Galen, 1971), and synthetic polymers (Kaplan and Kelleher, 1970). The singlet oxygen quenching abilities and consequent antioxidant properties of metal chelates have been tested extensively with discharge-generated singlet oxygen (Carlsson *et al.*, 1972 and 1974; Carlsson and Wiles, 1973; Breck *et al.*, 1974).

Absolute rate constants cannot be obtained in such heterogeneous systems, however, conventional "competitive techniques" can be used to obtain relative rate constants for both quenching and reaction processes.

A detailed discussion of a suitable flow system for such studies can be found in Section II,B. Care must be taken to eliminate all oxygen atoms from the system and we recommend that in addition to the mercuric oxide ring, a small flow of NO_2 be used to ensure the absence of this much more reactive species. When this is done, O_3 is also eliminated for reasons discussed in Section IV,D. $O_2(^1\Sigma_g{}^+)$ is the only other metastable species present in the system. For reasons discussed in Section V, the $O_2(^1\Sigma_g{}^+)$ concentration is usually only about 0.06% of the $O_2(^1\Delta_g)$ concentration, and it furthermore has not been found to contribute to the oxidation of any organic material in the condensed phase.

III. THE QUENCHING OF SINGLET OXYGEN

The two singlet oxygen species $O_2(^1\Sigma_g{}^+)$ and $O_2(^1\Delta_g)$ differ markedly in the ease with which they are deactivated. This difference can be attributed to the fact that the removal of the singlet delta state by a quencher Q which can be written as

$$O_2(^1\Delta_g) + Q \rightarrow O_2(^3\Sigma_g{}^-) + Q \tag{6}$$

is a "spin-forbidden" process unless Q has a multiplicity greater than one and furthermore liberates 22.54 kcal which must be accommodated in the products of reaction (6). On the other hand the relaxation of singlet-sigma oxygen, which can be written as

$$O_2(^1\Sigma_g{}^+) + Q \rightarrow O_2(^1\Delta_g) + Q \tag{7}$$

can occur to the singlet delta state which is not "spin-forbidden" and liberates only 15.0 kcal of electronic energy. These two factors combine to make most molecules quench $O_2(^1\Sigma_g{}^+)$ about 10^5 times more rapidly than $O_2(^1\Delta_g)$. However we shall see that there are molecules for which this difference is greatly reduced.

A. $O_2(^1\Sigma_g{}^+)$

Quenching rate constants for singlet sigma oxygen have been determined at 25°C by a number of workers and the results are summarized in Fig. 2. The short horizontal line beside each molecular formula gives the rate constant obtained by averaging values reported in the last few years (Davidson and Ogryzlo, 1973). To assist in the following discussion, the rate constants have been separated into seven somewhat arbitrary groups along the horizontal axis. Since an understanding of the quenching mechanism has come from theoretical studies of the relaxation of $O_2(^1\Sigma_g{}^+)$ by diatomic molecules, these will be considered first.

1. Diatomic Molecules

It can be seen in Fig. 2 that the quenching constants for diatomic molecules can differ by more than 4 orders of magnitude even though the interactions are strictly physical. The observation that the quenching rates for these species correlate reasonably well with their vibrational frequencies has led to the theory that the rate is determined by the ease with which the electronic energy from the transition in O_2 (15.0 kcal) can be distributed as vibrational and rotational motion in O_2 and the quencher. Such transitions can be induced by long range interactions (LRI) between the transition quadrupole for the ($^1\Sigma_g{}^+-{}^1\Delta_g$) transition on oxygen and the transition dipole or transition quadrupole on the quencher. Contributions from this quenching mechanism can be calculated with the Born–Bethe perturbation theory (Braithwaite *et al.*, 1976a). Alternatively, the transition can be induced by short range (repulsive) interactions (SRI). In this case the standard "distorted-wave approximation" can be used to calculate the transition probability. The results of both the SRI and LRI calculations are compared with the experimental rate constants in Table I. Unfortunately the early SRI

Table I Rate Constants for the Quenching of $O_2(^1\Sigma_g{}^+)$ by Diatomic Molecules at 300°K

	Rate constants (liter mole^{-1} sec^{-1})			
		Theoretical[k,l,m,n]		
Quencher	Experimental	SRI	LRI	Sum
H_2	8×10^8 [a]	3.6×10^8	1.8×10^8	5.4×10^8
	2.4×10^8 [b,c,d]			
	6.8×10^8 [e]			
	5.5×10^8 [f]			
HD	1.9×10^8 [a]	(3×10^6)	1.0×10^8	
	1.1×10^8 [c]			
D_2	1.2×10^7 [c,g]	(9.2×10^5)	1.2×10^7	
	1.1×10^7 [a]			
NO	3.3×10^7 [b]		9.3×10^6	
	2.4×10^7 [c]			
	2.5×10^7 [g]			
CO	1.8×10^6 [h]	(3.1×10^4)	2.5×10^6	
	2.6×10^6 [g]			
	2.0×10^6 [i]			
	1.5×10^6 [j]			
HBr	2.3×10^8 [k]	1.18×10^8	1.18×10^8	2.36×10^8

[a] Stuhl and Niki (1971).
[b] Becker *et al.* (1971).
[c] O'Brien and Myers (1972).
[d] Thomas and Thrush (1975).
[e] Davidson *et al.* (1972).
[f] Gauthier and Snelling (1975).
[g] Filseth *et al.* (1970).
[h] Noxon (1970).
[i] Stuhl and Welge (1969).
[j] Arnold *et al.* (1968).
[k] Braithwaite *et al.* (1976a).
[l] Braithwaite *et al.* (1976b).
[m] Kear and Abrahamson (1974).
[n] Numbers in parentheses are taken directly from Kear and Abrahamson (1974).

calculations (Kear and Abrahamson, 1974) did not include the possibility of rotational excitation of the products and hence yielded rather low results. (See the rate constants in parentheses in Table I.) In the case of H_2 and HBr where complete calculations have been carried out (Braithwaite *et al.*, 1976b) it has been shown that both LRI and SRI contribute to the quenching process which can be written as

$$O_2(^1\Sigma_g{}^+) + H_2\,(v = 0) \rightarrow O_2(^1\Delta_g) + H_2\,(v = 1) \tag{8}$$

$$O_2(^1\Sigma_g{}^+) + HBr\,(v = 0) \rightarrow O_2(^1\Delta_g) + HBr\,(v = 2) \tag{9}$$

i.e., H_2 and HBr are excited to the first and second vibrational level, respectively. These are the dominant processes because they are almost resonant, i.e., all the electronic energy can be taken up in a combination of easily excited vibrational modes and allowed rotational transitions. The poor quenchers like N_2 and especially O_2 have vibrational levels which poorly match the more favorable $O_2(^1\Sigma_g{}^+ - {}^1\Delta_g)$ transitions. Though such matching of electronic, vibrational, and rotational levels is somewhat "accidental" molecules with higher vibrational frequencies normally are more effective quenchers because Franck–Condon factors can make the process very unfavorable when a combination of too many vibrational energy levels must be excited in a single collision.

It is also worth noting that the theories described above do not require the transitions in the quencher to be optically active to induce energy transfer. H_2, for example, is a good quencher though it possesses no permanent dipole moment and hence does not absorb infrared radiation in the gas phase. This is in contrast to the theory proposed to account for the quenching in solution which relates the quenching rate correctly to the infrared absorption intensity (Merkel and Kearns, 1972).

2. *Polyatomic Molecules*

Theoretical calculations of quenching rates with polyatomic molecules have not yet been completed, however, it is not difficult to rationalize the relative quenching rates reported in Fig. 2.

Since the highest vibrational frequencies are found in molecules containing hydrogen atoms bound strongly to another atom, it is not surprising to find that H_2O and the alcohols are the best quenchers. NH_3 and other amines are also very good quenchers because of their high N–H frequencies. It is probably significant that trimethylamine which lacks an N–H bond has the lowest quenching constant of the amines in Fig. 2, and actually has a value comparable to a small alkane (like C_3H_8) where the quenching can be attributed to a combination of C–H bond frequencies.

In the group of hydrocarbons listed in Fig. 2 the quenching constant can be considered to rise with the number of C–H bonds in the molecule.

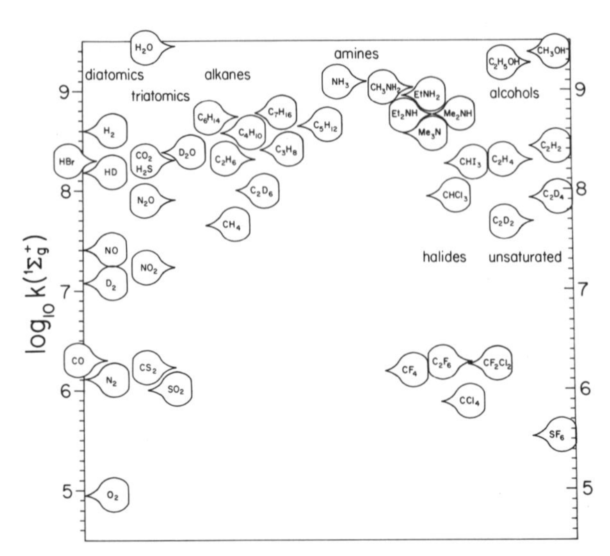

Fig. 2. Logarithm of the $O_2(^1\Sigma_g^+)$ quenching rate constant in the gas phase at 25°C. The values recommended by Davidson and Ogryzlo (1973) have been modified slightly by the recent work of Gauthier and Snelling (1975), Thomas and Thrush (1975), and Braithwaite *et al.* (1976b).

However, we would not expect such a simple statement to be very accurate or widely applicable since, as was pointed out earlier, the quenching rate depends on a somewhat accidental achievement of resonance in the change from electronic to nuclear motion in the system.

The alkyl halides provide an interesting example of how important the high frequency of the C–H bond is to the quenching. It can be seen in Fig. 2 that when the last hydrogen atom is removed from such a molecule the quenching constant drops by two orders of magnitude.

A few additional observations can be made about the rate constants in Fig. 2. (1) Unsaturated molecules have quenching constants which are not significantly different from structurally similar saturated molecules. (2) Deuteration results in lower rate constants. This can be attributed to the lower vibrational frequencies of the deuterated molecules. (3) Molecules like NO, O_2, and NO_2 with unpaired electrons have quenching rate constants consistent with their vibrational frequencies, i.e., their paramagnetism does not affect the quenching rates. (4) There is no relationship between quenching rate and the ionization energy of the quencher.

B. $O_2(^1\Delta_g)$

The quenching rate constants which have been reported for singlet delta oxygen are recorded in Fig. 3. The position of the short horizontal

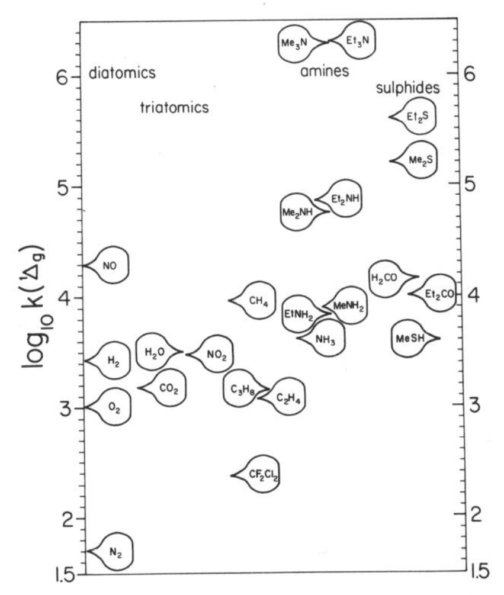

Fig. 3. Logarithm of the $O_2(^1\Delta_g)$ quenching rate constant in the gas phase at 25°C. The values recommended by Davidson and Ogryzlo (1973) have been modified slightly by the recent work of Yaron *et al.* (1976).

line next to each formula on the graph gives the current "best value" obtained by averaging a number of recent determinations (Davidson and Ogryzlo, 1973). Though the data is not as extensive as it is for $O_2(^1\Sigma_g^+)$ it is quite clear that some new factors influence the quenching rates.

1. The "Normal Quenchers"

Species such as H_2, H_2O, CO_2, C_3H_8, C_2H_4, NH_3, and a few others (which we call "normal quenchers") have $O_2(^1\Delta_g)$ quenching constants which are about 5 orders of magnitude smaller than the corresponding values for $O_2(^1\Sigma_g^+)$. As mentioned above, this can be attributed to the fact that the quenching reaction, given by equation (6) must (a) accommodate much more energy (22.54 kcal) in the products and (b) is spin forbidden. The difficulty that the system has in flipping the spin of one of the electrons is clear when the effect of "paramagnetic" quenchers such as NO, NO_2, and O_2 is noted. It can be seen by comparing the results in Figs. 2 and 3 that molecules with unpaired electrons have rate constants which are only 2 or 3 orders of magnitude lower than those for $O_2(^1\Sigma_g^+)$. This is easily understood since reaction (6) is formally spin allowed with quenchers possessing one or more unpaired electrons. In view of these observations the $O_2(^1\Delta_g)$ quenching rates can be considered to be reduced 2 or 3 orders of magnitude relative to the $O_2(^1\Sigma_g^+)$ rates because of spin conservation requirements and an additional 2 or 3 orders of magnitude due to the extra 6.5 kcal of electronic energy which must be accommodated in the products.

2. *The Ionization Energy and Spin Effects*

Another marked difference between the $O_2(^1\Sigma_g^+)$ and $O_2(^1\Delta_g)$ quenching rate constants can be seen when the amine values are compared. Whereas there is very little difference in the singlet sigma quenching abilities of the amines (see Fig. 2) the singlet delta quenching constants shown in Fig. 3 increase rapidly with methyl or ethyl substitution. This has been rationalized in terms of a charge-transfer interaction which occurs because of the high electron affinity of the singlet delta oxygen when the ionization energy of the quencher is low. Such an interaction could increase the quenching rate not only by bringing about a more intimate interaction between the two molecules but also by facilitating an electron spin change through "spin-orbit" interactions (Ogryzlo and Tang, 1970). It can be seen that a similar phenomenon probably occurs with the sulfides which also have low ionization energies.

No theoretical calculations of $O_2(^1\Delta_g)$ quenching rates in the gas phase have yet been described. It is clear that the multipolar interactions which dominate the $O_2(^1\Sigma_g^+)$ quenching will not be important in this case since the rather weak transition in O_2 is of the "magnetic dipole" type. Calculations based on the "distorted wave" approximation look promising in view of the relatively large quenching constant for H_2. However, it is clear that a somewhat different approach will be required to account for the amine, sulfide, and paramagnetic quenchers.

IV. REACTIONS OF SINGLET OXYGEN

With the possible exception of the interaction with O_3 (which can be classified as an energy transfer process) no reactions of $O_2(^1\Sigma_g^+)$ have been reported and the only reactions of $O_2(^1\Delta_g)$ which have been studied in the gas phase are additions to unsaturated hydrocarbons. Rate constants and Arrhenius parameters for these reactions are given in Table II. It is difficult to measure rate constants below 10^4 liter mole^{-1} sec^{-1} with a discharge flow system. This has limited the number of substrates which could be studied at room temperature. Values below 10^4 in Table II were obtained by extrapolating high temperature constants to room temperature. For comparison, values obtained in methanol solution by less direct techniques (Koch, 1968) are given in brackets. The reactions listed in Table II are classified into three types which will be discussed individually.

A. The Diels–Alder Reaction

The Diels–Alder reaction is observed with conjugated dienes and yields endoperoxides.

$$\mathrm{O{=}O} + \mathrm{{>}C{=}C{-}C{=}C{<}} \longrightarrow \mathrm{O{-}O{-}C{-}C{=}C{-}C} \text{ (cyclic)} \qquad (10)$$

The rate constants for these reactions of $O_2(^1\Delta_g)$ are very much higher than those for the corresponding "normal" Diels–Alder reactions (Wasserman, 1965), principally because of the much lower activation energies for the singlet oxygen reaction. However, when the rate constants in Table II are compared it can be seen that the values obtained in methanol are about two orders of magnitude greater than the values obtained in the gas phase. Furthermore, since the preexponential factor (A) is actually larger in the gas phase, this difference is due to a much lower activation energy for the reaction in solution. The most reasonable explanation for this difference is a somewhat polar transition state which is stabilized in the condensed phase relative to the nonpolar reactants.

The effect of substituents on the diene in these reactions is very similar to that reported for the same reactions in solution and for the normal Diels–Alder reaction (Wasserman, 1965). It has therefore been assumed by most workers that a "productlike" 6-center cyclic transition state occurs in all such reactions. The lowering of the activation energy for the reaction when electron donating groups are placed on the diene can then be accounted for by considering the interaction between the highest occupied molecular orbital (HOMO) on the diene and the lowest unoccupied molecular orbital (LUMO) on oxygen (Ashford and Ogryzlo, 1974). Such an interaction can give rise to a polar transition state and hence account for the difference between the gas phase and solution rates.

B. The Ene-Reaction

The ene-reaction in which hydroperoxides are formed is observed with olefins possessing an allylic hydrogen. It can be seen, in Table II, that the gas phase rate constants are also about two orders of magnitude smaller than the values obtained in methanol solution and once again this is due to a much lower activation energy for the condensed phase reaction. There is one exception (*cis*-2-butene) for which an activation energy of 10 kcal/mole

Table II Rate Constants and Arrhenius Parameter for Reactions of $O_2(^1\Delta_g)$ in the Gas Phase

Substrate	k (25°C) (liter mole^{-1} sec^{-1})	log A	E^* (kcal mole^{-1})
	Diels–Alder Reaction[a]		
	3.5×10^5	8.3	3.9
	1.6×10^4	8.2	5.5
O	2.1×10^4	8.2	5.3
O	3.5×10^5	8.3	3.9
	Ene-Reaction[b,c]		
	6×10^5 (2.4×10^7)	8.1 (7.7)	3.2 (0.5)
	3×10^4 (2.0×10^6)	8.1 (7.5)	4.9 (1.6)
	2.3×10^3	8.1 —	6.5 —
	9×10^2	8.2 —	7.3 —
	3×10^5 (1.3×10^7)	8.4 (8.00)	4.0 (1.3)
	7.3×10^2 (1.9×10^5)	8.4 —	7.5 —
	$<10^2$ (4.4×10^3)	8.4[e] (9.76)	>8.3 (5.4)
	3.2×10^5	8.5 —	4.0 —
	1.2×10^4 (1.7×10^6)	8.4 —	6.0 —
	9×10^2 (7×10^4)	8.4 —	7.4 —

Table II *(Continued)*

Substrate	k (25°C) (liter mole^{-1} sec^{-1})	log A	E^* (kcal mole^{-1})
	Dioxetane Reaction[d]		
$=$			21.0
$=\!\!/$ OCH_3			13.0
$=\!\!/$ OC_2H_5			12.3
$=\!\!/$ On-Bu			12.7

[a] Ashford and Ogryzlo (1974).
[b] Ashford and Ogryzlo (1975).
[c] Values in parentheses were obtained in solution.
[d] Bogan *et al.* (1975a,b; D. J. Bogan, R. S. Sheinson, R. G. Gann, and F. W. Williams, private communication).

has been reported for the solution reaction. Another study of that system would be useful since it is possible that this high value may be due to complications arising from the photochemical isomerization of this species.

Unlike the Diels–Alder reaction, the mechanism of the ene-reaction remains controversial. The uncertainty centers around the possible intermediacy of a peroxirane (perepoxide) as shown in the following equation.

$$O_2 + H\text{–C–C=C} \longrightarrow \left[\overset{\delta-}{O}\text{–}\overset{\delta+}{O}\text{(–C–C)}\;\text{peroxirane} \right] \longrightarrow \text{O–O–H ene product} \qquad (11)$$

The experimental evidence for the presence of a peroxirane intermediate in the ene-reaction is indirect (Hasty and Kearns, 1973; Schaap and Faler, 1973). In some theoretical CNDO/2 calculations, Inagaki and Fukui (1975) found that in the ethylene singlet oxygen interactions no stable peroxirane could be found and hence "the perepoxide structure cannot be a genuine intermediate." However, the MINDO/3 calculations of Dewar and Thiel (1975) did indicate a "stable" peroxirane as shown in Fig. 4. The theoretical path of the reaction of $O_2(^1\Delta_g)$ with both ethylene and propene is given by the solid line. This is somewhat discouraging since the results in Table II show that the activation energy is decreased 1.5 to 2 kcal by methyl substitution at the double bond. However, assuming the curves are qualitatively correct the rate-controlling step would be the formation of the peroxirane

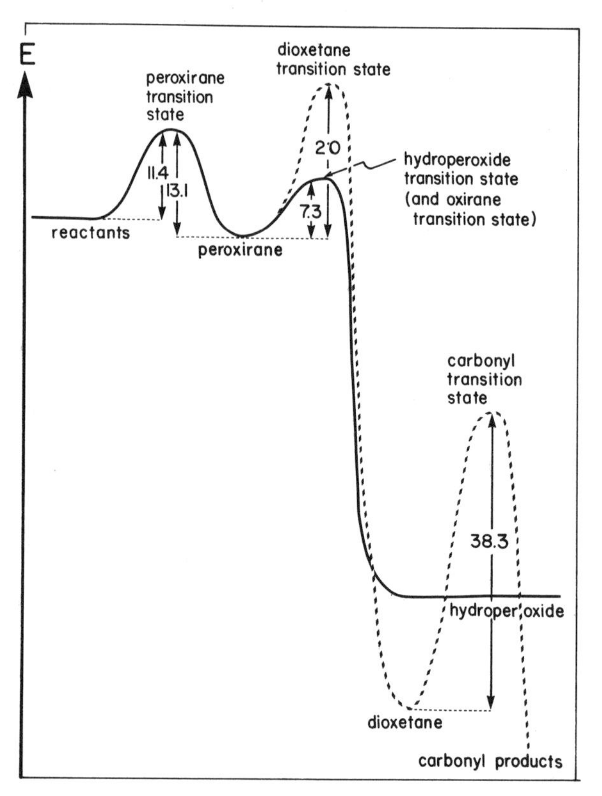

Fig. 4. Potential energy curves for the reaction of ethylene or propene with $O_2(^1\Delta_g)$ calculated by Dewar and Thiel (1975).

since the second step has a smaller activation energy. The fairly constant preexponential factor for these reactions are consistent with the peroxirane transition state. Despite the above evidence favoring the peroxirane intermediate doubts about its importance in the ene-reaction have been expressed (Kopecky and van de Sande, 1972), and by the more recent GVB-CI calculations of Harding and Goddard (1977). It may be worth noting that the gas phase data described above is consistent with a slightly polar "ene" intermediate and certainly does not require a peroxirane.

C. The Dioxetane Reaction

The dioxetane reaction has been reported to occur for olefins lacking an allylic hydrogen atom (Kearns, 1971). In the gas phase the reaction has only been observed by the appearance of electronically excited formaldehyde in the reaction.

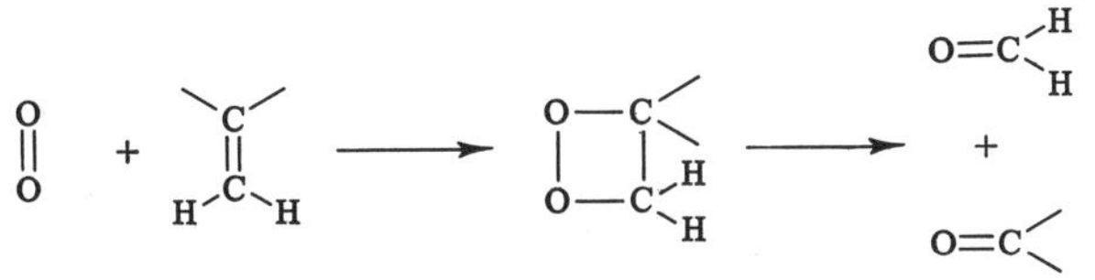

$$O_2 + H_2C{=}CH_2 \longrightarrow \text{(1,2-dioxetane)} \longrightarrow O{=}CH_2 + O{=}CH_2 \qquad (12)$$

The temperature dependence of the emission from formaldehyde was used (Bogan *et al.*, 1975a,b) to calculate the activation energies listed in Table II. A number of assumptions must be made to equate these activation energies to those for the formation of the dioxetane and it would therefore be useful to check these values by the rate of product disappearance. Nevertheless it is interesting to note that the MINDO/3 calculations shown in Fig. 4 (dotted line) predict an activation energy of about 18.3 kcal/mole for the formation of the dioxetane of ethylene. The experimental value is 21 kcal. However, the system is complicated by the fact that the reaction

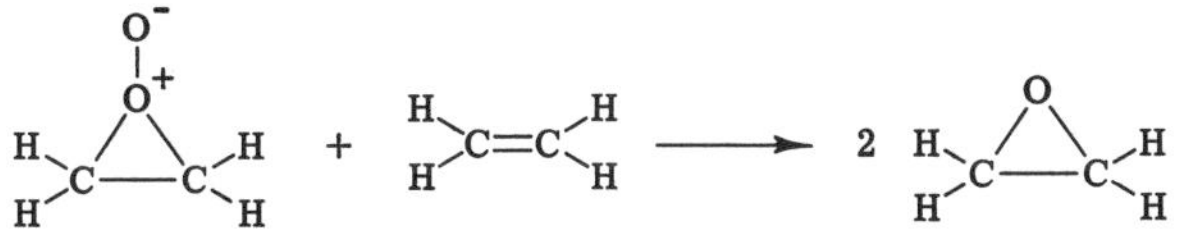

$$\text{(peroxirane, } H_2C{-}CH_2 \text{ with } O^+{-}O^-) + H_2C{=}CH_2 \longrightarrow 2\ \text{(oxirane, } H_2C{-}CH_2{-}O) \qquad (13)$$

is predicted to have a much lower activation energy than the rearrangement of the peroxirane to form the dioxetane. If this is true, the formaldehyde emission may be the result of the decomposition of the oxirane rather than the dioxetane. Work on these reactions has only been of an exploratory nature, and a more careful study of these systems is needed.

D. The Ozone Interaction

Because of its very low dissociation energy (25.36 kcal/mole) ozone is somewhat unique in its interactions with $O_2(^1\Delta_g)$ and $O_2(^1\Sigma_g^+)$. The rate constants for these interactions are the largest which have been reported for each species in the gas phase. In the case of $O_2(^1\Sigma_g^+)$ this interaction results in the dissociation of the ozone with a rate constant of 1.4×10^{10} liter mole^{-1} sec^{-1} (Gilpen *et al.*, 1971; Snelling, 1974). This large value is inconsistent with the relaxation of $O_2(^1\Sigma_g^+)$ to $O_2(^1\Delta_g)$ which occurs in the other energy-transfer processes discussed in Section II since this dissociation reaction would be 10.5 kcal endothermic. The reaction must therefore be

$$O_2(^1\Sigma_g^+) + O_3(^1A_1) \rightarrow 2O_2(^3\Sigma_g^-) + O(^3P) \qquad (14)$$

which is 12.1 kcal exothermic. Though the energy of the ozone triplet state has not been established, it is quite probable that the primary process can be written

$$O_2(^1\Sigma_g{}^+) + O(^1A_1) \rightarrow O_2(^3\Sigma_g{}^+) + O_3(^3B_1) \quad (15)$$

which is spin allowed.

The reaction of ozone with $O_2(^1\Delta_g)$ can be written

$$O_2(^1\Delta_g) + O_3(^1A_1) \rightarrow 2O_2(^3\Sigma_g{}^-) + O(^3P) \quad (16)$$

which is 2.8 kcal endothermic. The rate constant for this process is 2×10^6 (Wayne and Pitts, 1969; Yaron *et al.*, 1976) which is almost 4 orders of magnitude smaller than the $O_2(^1\Sigma_g{}^+)$ rate. This is consistent with the endothermicity of the reaction which reduces the rate by two orders of magnitude at room temperature. The additional two orders of magnitude difference can be attributed to either the less favorable electronic transition in O_2 or to a poorer "overlap" with the electronic transition in O_3. Any further analysis of these systems will require more information on the electronically excited states of ozone.

The relevance of these reactions to discharge–flow systems lies in the possible formation of ozone in the discharge products. When O_2 at 1 torr is passed through an electrical discharge, oxygen atoms are formed and these combine with molecular oxygen to form ozone in reaction (17)

$$O + O_2 + O_2 \rightarrow O_3 + O_2 \quad (17)$$

for which $k(25°C) = 2.2 \times 10^8$ liter mole^{-1} sec^{-1} (Hampson and Garvin, 1975). In the absence of any other species, ozone is removed by reaction (18)

$$O_3 + O \rightarrow 2\,O_2 \quad (18)$$

for which $k(25°C) = 5 \times 10^6$ liter mole^{-1} sec^{-1} (Hampson and Garvin, 1975). Consequently a steady-state ozone concentration is established in a flow system within a few milliseconds where

$$\frac{[O_3]}{[O_2]} = \frac{k_{17}}{k_{18}}[O_2] = \frac{2.2 \times 10^8}{5 \times 10^6}[O_2] = 44\,[O_2]$$

or at 1 torr

$$\frac{[O_3]}{[O_2]} = (44)(5 \times 10^{-5}) = 2.2 \times 10^{-3}$$

i.e., the $[O_3]$ is about 0.2% of the $[O_2]$ or, in the usual flow system, about 4% of the oxygen atom concentration. However, in the normal discharge-products there is also an $O_2(^1\Delta_g)$ concentration comparable to the atom concentration. In such a system the steady-state ozone concentration is given by

$$\frac{[O_3]}{[O_2]} = \frac{k_{17}}{k_{18}}[O_2] = \frac{2.2 \times 10^8}{5 \times 10^6}[O_2] = 44\,[O_2] \quad (19)$$

because of the destruction of ozone by both singlet delta ($^1\Delta$) and singlet sigma ($^1\Sigma$) oxygen. In the absence of any added quenchers, the [O] and [$^1\Delta$] are normally about 5% of the [O_2]. [$^1\Sigma$] assumes a steady-state value of about $6.5 \times 10^4[^1\Delta]$. Equation 19 then becomes

$$\frac{[O_3]}{[O_2]} = \frac{2.2 \times 10^8[O][O_2]}{5 \times 10^6[O] + 2 \times 10^6[^1\Delta] + 1.4 \times 10^{10}[^1\Sigma]}$$

i.e., the ozone concentration is reduced to 0.068% of the oxygen concentration. This is only a factor of 3 lower than the value it would have in the absence of the singlet molecular oxygen. Since most of this reduction is due to $^1\Sigma$, the addition of water to this system can result in an increase in the steady state ozone concentration. Such an increase may not be observed experimentally unless the oxygen is initially very dry.

When mercury is added to the discharge and most of the oxygen atoms are removed, the ozone destruction is due entirely to the singlet oxygen. If the [$^1\Delta_g$] remains essentially unchanged, a new steady-state ozone concentration is established and we obtain the interesting result that at 1 torr $[O_3]/[O] = 0.02$, i.e., the ozone concentration is about 2% of the oxygen atom concentration when the stream is mostly singlet oxygen. Since such steady-state concentrations are established in a few milliseconds, it should be clear that ozone need not be considered a "contaminant" in a discharge flow system when the oxygen atom concentration has been reduced to an insignificant value.

V. ENERGY-POOLING REACTIONS OF SINGLET OXYGEN

Though $O_2(^1\Delta_g)$ possess only 22.54 kcal/mole of electronic energy, it can give rise to photons with energies up to 75 kcal/mole through a number of novel energy-pooling processes.

At extremely low partial pressures of $O_2(^1\Delta_g)$ only the "normal" vibronic bands at 1.27, 1.58, and 2.1 μm are observed. However, at an $O_2(^1\Delta_g)$ concentration of 2×10^{-3} *M* emission from the following processes assume the same intensity:

$$O_2(^1\Delta_g) + O_2(^1\Delta_g) \xrightarrow{k=0.14} O_2(^3\Sigma_g^-) + O_2(^3\Sigma_g^-) + h\nu(6340\ \text{Å}) \quad (20)$$

$$O_2(^1\Delta_g) + O_2(^1\Delta_g) \xrightarrow{k=0.14} O_2(^3\Sigma_g^-) + O_2(^3\Sigma_g^-) + h\nu(7030\ \text{Å}) \quad (21)$$

Furthermore, under the same conditions the "disproportionation" reaction (Ogryzlo, 1970)

$$O_2(^1\Delta_g) + O_2(^1\Delta_g) \xrightarrow{k=1.3 \times 10^3} O_2(^1\Sigma_g^+) + O_2(^3\Sigma_g^-) \quad (22)$$

produces $O_2(^1\Sigma_g^+)$ which emits at 7619 Å. Since reaction (22) is about 10^3 times more rapid than reactions (20) and (21), the infrared emission at 7619 Å can be much more intense than the red emission (6340 and 7030 Å). Though the ratio of the two emission intensities varies with the quenching species present, it is not difficult to determine the emission intensity ratio in "pure" O_2. In the usual discharge flow system described in Section I steady-state $O_2(^1\Sigma_g^+)$ concentration is established within a few cm of the discharge by disproportionation reaction (22) and the quenching reaction which can be written as

$$O_2(^1\Sigma_g^+) + O_2(^3\Sigma_g^-) \rightarrow O_2(^1\Delta_g) + O_2(^3\Sigma_g^-) \tag{23}$$

The rate constant for this quenching reaction is normally found to be about 10^5 liter mole^{-1}sec^{-1} because of the presence of a small (unremovable) amount of water in the oxygen (Thomas and Thrush, 1975). Hence the $O_2(^1\Sigma_g^+)/O_2(^1\Delta_g)$ ratio under steady-state conditions is given by:

$$\frac{O_2(^1\Sigma_g^+)}{O_2(^1\Delta_g)} = \frac{k_{22}[O_2(^1\Delta_g)]}{k_{23}[O_2(^3\Sigma_g^-)]} = 1.3 \times 10^{-2} \frac{[O_2(^1\Delta_g)]}{[O_2(^3\Sigma_g^-)]}$$

Since the singlet delta oxygen is normally about 5% of the ground state O_2 in flow systems

$$\frac{O_2(^1\Sigma_g^+)}{O_2(^1\Delta_g)} = (1.3 \times 10^{-2})(0.05) = 6.5 \times 10^{-4}$$

However, because of the difference in radiative lifetimes of the two species [7 sec for $O_2(^1\Sigma_g^+)$ and 1 hr for $O_2(^1\Delta_g)$] the ratio of the emission intensities at 7619 and 12,690 Å is almost one (0.33).

With $O_2(^1\Sigma_g^+)$ present in the system as well as $O_2(^1\Delta_g)$ several additional energy pooling reactions can occur and have been recorded (Furukawa *et al.*, 1970):

$$O_2(^1\Delta_g) + O_2(^1\Sigma_g^+) \rightarrow 2\,O_2(^3\Sigma_g^-) + h\nu(4800\ \text{Å}) \tag{24}$$

$$O_2(^1\Delta_g) + O_2(^1\Sigma_g^+) \rightarrow O_2(^3\Sigma_g^-) + O_2(^3\Sigma_g^-\ v = 1) + h\nu(5200\ \text{Å}) \tag{25}$$

$$2\,O_2(^1\Sigma_g^+) \rightarrow 2\,O_2(^3\Sigma_g^-) + h\nu(3800\ \text{Å}) \tag{26}$$

$$2\,O_2(^1\Sigma_g^+) \rightarrow O_2(^3\Sigma_g^-) + O_2(^3\Sigma_g^-\ v = 1) + h\nu(4000\ \text{Å}) \tag{27}$$

The photons emitted in reaction (26) have an energy of 75.1 kcal/mole though the energy originally resided in $O_2(^1\Delta_g)$ molecules with an energy of 22.5 kcal/mole.

REFERENCES

Arnold, S. J., Ogryzlo, E. A., and Witzke, H. (1964). *J. Chem. Phys.* **40,** 1769.
Arnold, S. J., Finlayson, N., and Ogryzlo, E. A. (1966). *J. Chem. Phys.* **44,** 2529.

Arnold, S. J., Kubo, M., and Ogryzlo, E. A. (1968). *Adv. Chem. Ser.* **77,** 133.
Ashford, R. D., and Ogryzlo, E. A. (1974). *Can. J. Chem.* **52,** 3544.
Ashford, R. D., and Ogryzlo, E. A. (1975). *J. Am. Chem. Soc.* **97,** 3604.
Bader, L. W., and Ogryzlo, E. A. (1964). *Discuss. Faraday Soc.* **37,** 46.
Becker, K. H., Groth, W., and Schurath, U. (1971). *Chem. Phys. Lett.* **8,** 259.
Bogan, D. J., Sheinson, R. S., Gann, R. G., and Williams, F. W. (1975a). *J. Am. Chem. Soc.* **97,** 2560.
Bogan, D. J., Sheinson, R. S., Gann, R. G., and Williams, F. W. (1975b). *J. Am. Chem. Soc.* **98,** 1034.
Braithwaite, M., Davidson, J. A., and Ogryzlo, E. A. (1976a). *J. Chem. Phys.* **65,** 771.
Braithwaite, M., Ogryzlo, E. A., Davidson, J. A., and Schiff, H. I. (1976b). *Chem. Phys. Lett.* **42,** 158.
Breck, A. K., Taylor, C. L., Russel, K. E., and Wan, J. K. S. (1974). *J. Polym. Sci.* **12,** 1505.
Carlsson, D. J., Mendenhall, G. D., Suprunchuk, T., and Wiles, D. M. (1972). *J. Am. Chem. Soc.* **94,** 8960.
Carlsson, D. J., and Wiles, D. M. (1973). *J. Polym. Sci., Polym. Lett. Ed.* **11,** 759.
Carlsson, D. J., Suprunchuk, T., and Wiles, D. M. (1974). *Can. J. Chem.* **52,** 3728.
Clagett, D. C., and Galen, T. J. (1971). *Arch. Biochem. Biophys.* **146,** 196.
Corey, E. J., and Taylor, W. C. (1964). *J. Am. Chem. Soc.* **86,** 3881.
Davidson, J. A., and Ogryzlo, E. A. (1973). *In* "Chemiluminescence and Bioluminescence" (M. J. Cormier, D. M. Hercules, and J. Lee, eds.), p. 111. Plenum, New York.
Davidson, J. A., Kear, K. E., and Abrahamson, E. W. (1972–1973). *J. Photochem.* **1,** 307.
Dewar, M. J. S., and Thiel, W. (1975). *J. Am. Chem. Soc.* **97,** 3978.
Elias, L., Ogryzlo, E. A., and Schiff, H. I. (1959). *Can. J. Chem.* **37,** 1680.
Fehrenfeld, F. C., Evenson, K. M., and Broida, H. P. (1965). *Rev. Sci. Instrum.* **36,** 294.
Filseth, S. V., Zia, A., and Welge, K. H. (1970). *J. Chem. Phys.* **52,** 5502.
Findlay, F. D., and Snelling, D. R. (1971). *J. Chem. Phys.* **55,** 545 (1971).
Foner, S. N., and Hudson, R. L. (1956). *J. Chem. Phys.* **25,** 601.
Foote, C. S., and Wexler, S. (1964). *J. Am. Chem. Soc.* **86,** 3879.
Furukawa, K., and Ogryzlo, E. A. (1971). *Chem. Phys. Lett.* **12,** 370.
Furukawa, K., Gray, E. W., and Ogryzlo, E. A. (1970). *Ann. N.Y. Acad. Sci.* **171,** 175.
Gauthier, M. J. R., and Snelling, D. R. (1975). *J. Photochem.* **4,** 27.
Gilpen, R., Schiff, H. I., and Welge, K. H. (1971). *J. Chem. Phys.* **55,** 1087.
Hampson, R. F., Jr., and Garvin, D. (1975). *Natl. Bur. Stand. (U.S.), Tech. Note* **886.**
Harding, L. B., and Goddard, W. R. (1977). *J. Am. Chem. Soc.* **99,** 4521.
Hasty, N. M., and Kearns, D. R. (1973). *J. Am. Chem. Soc.* **95,** 3380.
Inagaki, S., and Fukui, K. (1975). *J. Am. Chem. Soc.* **97,** 7480.
Kaplan, M. L., and Kelleher, P. G. (1970). *Science* **169,** 1206.
Kaufman, F. (1961). *Prog. React. Kinet.* **1,** 1.
Kautsky, H., and de Bruijn, H. (1931). *Naturwissenschaften* **19,** 1043.
Kear, K., and Abrahamson, E. W. (1974). *J. Photochem.* **3,** 409.
Kearns, D. R. (1971). *Chem. Rev.* **71,** 395.
Kearns, D. R., Khan, A. U., Duncan, C. K., and Maki, A. H. (1969). *J. Am. Chem. Soc.* **91,** 1039.
Koch, E. (1968). *Tetrahedron* **24,** 6295.
Kopecky, K. R., and van de Sande, J. H. (1972). *Can. J. Chem.* **50,** 4034.
Lewis, P. (1900a). *Ann. Phys. (Leipzig)* [3] **2,** 459.
Lewis, P. (1900b). *Astrophysics* **12,** 8.
Merkel, P. B., and Kearns, D. R. (1972). *J. Am. Chem. Soc.* **94,** 7244.
Nilsson, R., and Kearns, D. R. (1974). *Photochem. Photobiol.* **19,** 181.
Noxon, J. F. (1970). *J. Chem. Phys.* **52,** 1852.
O'Brien, R. J., and Myers, G. H. (1972). *J. Chem. Phys.* **5,** 3832.

Ogryzlo, E. A. (1970). *Photophysiology* **5,** 35.
Ogryzlo, E. A., and Tang, C. W. (1970). *J. Am. Chem. Soc.* **92,** 5034.
Rawls, H. R., and Van Santen, P. J. (1970). *J. Am. Oil. Chem. Soc.* **47,** 121.
Schaap, A. P., and Faler, G. R. (1973). *J. Am. Chem. Soc.* **95,** 3381.
Scheffer, J. R., and Ouchi, M. D. (1970). *Tetrahedron Lett.* 223.
Snelling, D. R. (1968). *Chem. Phys. Lett.* **2,** 346.
Snelling, D. R. (1974). *Can. J. Chem.* **52,** 257.
Strutt, R. J. (1911). *Proc. R. Soc., London, Ser. A* **85,** 219.
Stuhl, F., and Niki, H. (1971). *Chem. Phys. Lett.* **7,** 473.
Stuhl, F., and Welge, K. H. (1969). *Can. J. Chem.* **47,** 1870.
Thomas, R. G. O., and Thrush, B. A. (1975). *J. Chem. Soc., Faraday Trans. 2* **71,** 664.
Warburg, E. (1884). *Arch. Sci. Phys. Nat.* **12,** 504.
Wasserman, A. (1965). *Diels-Alder Reactions*, Elsevier, Amsterdam, p. 50.
Wasserman, E., Kuck, V. J., Delvan, W. M., and Yager, W. A. (1965). *J. Am. Chem. Soc.* **91,** 1040.
Wayne, R. P., and Pitts, J. N., Jr. (1969). *J. Chem. Phys.* **50,** 3644.
Wood, R. W. (1920). *Proc. R. Soc. London, Ser. A* **97,** 455.
Yaron, M., von Engel, A., and Vidaud, P. H. (1976). *Chem. Phys. Lett.* **37,** 159.

3

Chemical Sources of Singlet Oxygen

ROBERT W. MURRAY

I. Introduction 59
II. Chemical Sources of Singlet Oxygen 60
A. The Hydrogen Peroxide–Hypochlorite System 60
B. The Reaction of Bromine with Hydrogen Peroxide 78
C. The Decomposition of Alkaline Solutions of Peracids 79
D. The Self-Reaction of *sec*-Butylperoxy Radicals 80
E. The Base-Induced Decomposition of Peroxyacetylnitrate 82
F. The Decomposition of Superoxide Ion 82
G. The Decomposition of Transition Metal Oxygen Complexes . . . 87
H. The Decomposition of Photoperoxides 91
I. The Reaction of Ozone with Organic Substrates 93
J. Other Sources 110
References 112

I. INTRODUCTION

Since the reawakening of interest in singlet molecular oxygen which began about fifteen years ago and the knowledge that it is probably involved in a variety of chemical, biochemical, and biomedical phenomena there has been a surge of interest in chemical methods of producing this interesting intermediate.

Efforts to produce singlet oxygen by chemical methods have sometimes been spurred by the hope that such methods could avoid some of the presumed disadvantages of the photosensitization method. In other cases the effort is made to verify a prediction based on the conservation of spin principle, that is, that the molecular oxygen produced in perhaps a well-known reaction ought to be in the excited singlet state. In other cases a particular chemical reaction is studied as a singlet oxygen source because the production of singlet oxygen in the observed reaction could have im-

Singlet Oxygen

ISBN 0-12-736650-4

portant consequences to pollution chemistry, a biomedical phenomenon, or some other area.

In many cases chemical oxygenation using singlet oxygen provides a useful synthetic path to desirable peroxidic materials or materials containing functional groups derived from peroxidic intermediates.

An attempt has been made in the following account to describe all reported chemical sources of singlet oxygen. In some cases the method is clearly not useful for preparative oxygenation. In other cases the method offers distinct advantages as an oxygenation method.

Other general methods of producing singlet oxygen include, besides photosensitization, the use of electrical or microwave discharges. The microwave discharge method is discussed in some detail by Ogryzlo (Chapter 2). The photosensitization technique is referred to by many contributors to this volume. Included is a discussion by Kearns of a method in which the sensitizing dye is excited by a ruby laser.

II. CHEMICAL SOURCES OF SINGLET OXYGEN

A. The Hydrogen Peroxide–Hypochlorite System

1. Spectroscopic Investigations

Beginning with Mallet (1927), a number of reports have appeared in the literature which describe a weak red chemiluminescence accompanying the reaction between hydrogen peroxide and hypochlorite ion. The nature of this chemiluminescence was clarified somewhat by Seliger (1960) who reported that it consisted primarily of a sharp band at 634 nm. In a highly significant step in the further understanding of the observed chemiluminescence Khan and Kasha (1963) attributed the emission to electronically excited oxygen. Two major peaks, at 633.4 and 703.2 nm, were observed and tentatively attributed to solvent-shifted (0,0) and (0,1) bands of the $^1\Sigma_g{}^+ \rightarrow {}^3\Sigma_g{}^-$ transition of molecular oxygen. In addition two other faint bands in the near infrared were observed. While it was ultimately shown that these assignments of the major bands were incorrect, the association of the chemiluminescence with excited oxygen was particularly important to the subsequent revival and exploitation of the original Kautsky (1933, 1937) postulate on the nature of the intermediate in photosensitized oxidation.

Arnold *et al.* (1964) provided the correct interpretation of the emissions bands reported by Khan and Kasha. Since the bands were observed in the gas phase as well as in solution they could not be solvent shifted bands from the $^1\Sigma \rightarrow {}^3\Sigma$ transition. Instead they proposed that the emissions at

634.3 and 703.8 nm were due to transitions from a $^1\Delta$ dimer to $^3\Sigma$ O_2 [(0,0) and (0,1) transitions, respectively].

Stauff *et al.* (1963) and Stauff and Schmidkunz (1962) had earlier suggested that the weak chemiluminescence which occurs in many oxidation systems, including aqueous hydrogen peroxide, could be due to molecular oxygen dimers which they termed van der Waals associates. Groh and Kirrmann (1942) observed that the emitting species must have a considerable lifetime since the emission required the energy of two molecules of excited oxygen which would have to collide before emitting. Arnold *et al.* (1964) also reported an extremely weak band at 760 nm corresponding to the single molecular $^1\Sigma_g{}^+ \rightarrow {}^3\Sigma_g{}^-$ emission. In their original proposal the $^1\Delta$ molecules were bound in a stabilized dimer [Eq. (1)].

$$2\,O_2(^1\Delta_g) \overset{M}{\rightleftharpoons} O_4 + h\nu \overset{M}{\rightleftharpoons} 2\,O_2(^3\Sigma_g{}^-) \tag{1}$$

A subsequent study (Arnold *et al.*, 1965) of the temperature dependence of the emission led to the conclusion that the emitting species is actually a colliding pair rather than a stabilized dimer [Eqs. (2) and (3)].

$$O_2(^1\Delta_g) + O_2(^1\Delta_g) \rightleftharpoons [O_2(^1\Delta_g)]_2 \tag{2}$$

$$[O_2(^1\Delta_g)]_2 \rightarrow [O_2(^1\Sigma_g{}^-)]_2 + h\nu \tag{3}$$

Browne and Ogryzlo (1964) also reported a definite assignment of the bands at 634 and 703 nm to the dimer emission, $2(^1\Delta_g) \rightarrow 2(^3\Sigma_g{}^-)$. Furthermore they identified the weak bands at 1070 and 1270 nm as being the (1,0) and (0,0) bands of the $^1\Delta_g \rightarrow {}^3\Sigma_g{}^-$ single molecule transition. The other single molecule transition, $^1\Sigma_g{}^+ \rightarrow {}^3\Sigma_g{}^-$ gave weak emission bands in the 760–870 nm region.

In 1970 Khan and Kasha provided a complete correlation of the chemiluminescence spectra observed in the aqueous hydrogen peroxide–hypochlorite system with high pressure gaseous oxygen absorption spectra. It was necessary to use high pressure for the absorption experiments since the dimer absorption requires interaction of a photon with an O_2–O_2 collision pair. The visible chemiluminescence bands observed include a red pair at 633.4 and 703.2 nm due to the $(^1\Delta_g)_2 \rightarrow 2(^3\Sigma_g{}^-)$ transition and a band in the green at 478 nm, due to the $(^1\Delta_g)(^1\Sigma_g{}^+) \rightarrow 2(^3\Sigma_g{}^-)$ transition. Emission due to the $(^1\Sigma_g{}^+)_2 \rightarrow (^3\Sigma_g{}^-)_2$ transition could not be observed although the absorption was seen in the high pressure gas phase oxygen system. Emission from the $^1\Sigma$ dimer is expected to be difficult to observe because of the efficient quenching of the $^1\Sigma$ state in aqueous media (Browne and Ogryzlo, 1964). Luminescence sensitization experiments using the peroxide–hypochlorite system do indicate that the $^1\Sigma_g{}^+$ excited dimer has some existence, however. The infrared chemiluminescence band at 762 nm due to the single molecule $^{-}\Sigma_g{}^{+} \rightarrow {}^{-}\Sigma_g{}^{-}$ emission was also observed. Balny and Douzou (1970) have also observed the red band at 633 nm due to the singlet delta dimol emission. Their observations were carried out using an apparatus in which singlet oxygen, from the reaction between hydrogen peroxide–hypochlorite, is produced in one chamber and then carried in a nitrogen stream to another chamber where singlet oxygen reactions can be observed. The apparatus was also used to carry out the conversion of 1,4-dimethoxy-9,10-diphenylanthracene to the 1,4-endoperoxide. Evidence for the reaction was obtained by monitoring the fluorescence emission of the starting material which was regenerated by thermal decomposition of the endoperoxide.

Khan and Kasha had earlier (1966) proposed that luminescence of any energetically favorable organic species may be produced as a result of energy transfer from excited molecular oxygen pairs. The proposal was based on the observation of the emissions described above and offered an alternative to other suggested explanations which were usually based on chemical mechanisms for excited state production.

Chemiluminescence attributed to the presence of O_2 ($^1\Delta$) and O_2 ($^1\Sigma$) has also been observed when chlorine instead of hypochlorite is used in the hydrogen peroxide reaction. The chemiluminescence was first reported by Seliger (1960). Assignment to the Σ and Δ states of singlet oxygen were made by Browne and Ogryzlo (1964). Using O_2 ($^1\Delta_g$), produced from an electrical discharge in the gas phase, as a calibration tool, the latter authors have estimated that the hydrogen peroxide–chlorine system gives a minimum yield of 2% O_2 ($^1\Delta_g$) (Browne and Ogryzlo, 1965).

The $^1\Delta_g$ oxygen dimol emission has been used in an elegant way by Kajiwara and Kearns (1973) who showed that the chemiluminescence due to this emission produced in the hypochlorite–hydrogen peroxide system is enhanced in D_2O solutions as compared to H_2O solutions. This observation provides direct spectroscopic evidence for the increased lifetime of $^1\Delta_g$ oxygen in D_2O as compared to H_2O, a characteristic which can be employed to diagnose for the involvement of singlet oxygen.

2. *Chemical Investigations*

The spectroscopic investigations described above demonstrated convincingly that the reaction between hypochlorite and hydrogen peroxide produced singlet oxygen. In 1964 a dramatic contribution was made to the recent resurgence of interest in singlet oxygen chemistry when Foote and Wexler (1964a) showed that singlet oxygen produced in the hypochlorite–hydrogen system could be used to oxygenate a number of mono- and diolefins. Furthermore, these oxygenations gave products which were identical to those obtained by photosensitized oxidation of the same substrates. Based on these observations and similar ones made by Corey and Taylor (1964), using a microwave discharge as a source of singlet oxygen, Foote and Wexler (1964b) concluded that the original proposal made by Kautsky (1933, 1937) that singlet oxygen was the active intermediate in photosensitized oxidation was correct.

Foote and Wexler (1964a) showed that oxygenation of 2,5-dimethylfuran in methanol using the hydrogen peroxide–hypochlorite system gave 2,5-dimethyl-2-hydroperoxy-5-methoxydihydrofuran [Eq. (4a)] in 84% yield. Likewise, 2,3-dimethyl-2-butene gave 2,3-dimethyl-3-hydroperoxy-1-butene in 63% yield [Eq. (4b)]. The diene, tetraphenylcyclopentadienone gave *cis*-dibenzoylstilbene in 50% yield presumably via an intermediate peroxide [Eq. (4c)]. Also, 1,3-cyclohexadiene was converted to 5,6-dioxabicyclo-[2,2,2]-2-octene [Eq. (4d)] under the same oxygenation conditions. In each case the product obtained is the same as that given by photosensitized oxidation of the corresponding substrate. The yields of the oxidation products obtained indicated that the procedure was a preparatively useful one.

$NaOCl/H_2O_2$, CH_3OH (4a)

(4b)

(4c)

(4d)

When hypochlorite reacts with hydrogen peroxide in aqueous media to liberate oxygen, the oxygen is derived from the hydrogen peroxide and not from the hypochlorite or water (Cahill and Taube, 1952). Evidence for this conclusion was obtained by using ^{18}O-enriched hydrogen peroxide or ^{18}O-enriched hypochlorite. Thus reactions of the type given in Eq. (5) are excluded while the reaction given in Eq. (6) appears to be responsible for O_2 formation (* indicates ^{18}O enrichment).

$$ClO^{-}* + H_2O_2 \rightarrow Cl^- + H_2O + O_2* \quad (5)$$

$$ClO^- + H_2O_2* \rightarrow H_2O + Cl^- + O_2* \quad (6)$$

If one adds to this information the intermediacy of a chloroperoxy ion then a mechanism for the production of electron-paired or singlet oxygen becomes available [Eq. (7)] (Kasha and Khan, 1970).

$$\begin{aligned} OCl^- + H_2O_2 &\rightarrow HOCl + {}^-O_2H \\ {}^-O_2H + HOCl &\rightarrow HOOCl + OH^- \\ HOOCl + OH^- &\rightarrow H_2O + {}^-OOCl \\ {}^-OOCl &\rightarrow {}^1O_2 + Cl^- \end{aligned} \quad (7)$$

Using a spin-state correlation diagram Kasha and Khan (1970) have shown that the ground state of the chloroperoxy anion correlates with the $^1\Delta_g$ state of molecular oxygen. Furthermore, they point out that the first excited state of the chloroperoxy anion also correlates with $^1\Delta_g$ oxygen and that one has to go to a higher excited state of the chloroperoxy anion to begin to get correlation with the $^1\Sigma_g^+$ state of molecular oxygen. They postulate that the $^1\Sigma_g^+$ oxygen which is observed to be present may arise from a reaction between two $^1\Delta_g$ molecules [Eq. (8)].

$$^1\Delta_g + {}^1\Delta_g \rightarrow {}^1\Sigma_g^+ + {}^3\Sigma_g^- \quad (8)$$

McKeown and Waters (1966) have also studied the use of the peroxide–hypochlorite system as an oxygenation method. One of the substrates they chose for study was the same as that used by Foote and Wexler, namely, 2,5-dimethylfuran. This material was converted in 10% yield to a mixture of 5-hydroxy-2,5-dimethyl-2-hydroperoxyfuran and 2-hydroperoxy-5-methoxy-2,5-dimethylfuran. The first compound was obtained presumably because the reaction medium used by these workers contained more water than that used by Foote and Wexler (1964a). While anthracene was too insoluble in the aqueous methanol medium used it was possible to observe the conversion of 9-methyl-10-phenylanthracene to its endoperoxide in 13% yield. These workers also used this method to oxygenate a number of steroid compounds. Thus, lumisteryl acetate was converted to its β-epidioxide in 13% yield and 2,4-chlolestadiene gave 5% of its epidioxide. Ergosterol itself was unreactive, but its acetate was oxidized to a mixture of three products in 10% yield.

The original Foote and Wexler communications were followed by a series of papers from this group devoted toward establishing further that singlet oxygen was the oxidant in dye-sensitized photooxidation and providing further examples of oxygenation using the hypochlorite–hydrogen peroxide systems. In 1965 Foote *et al.* demonstrated that the product selectivities observed in oxygenation using the hypochlorite–hydrogen peroxide system were the same as those obtained in dye-sensitized photooxidation. Results of additional studies of this type were also reported in 1968 (Foote, 1968). The chemical oxygenations were carried out by adding an aqueous hypochlorite solution dropwise to a methanolic or methanol/*tert*-butanol (1:1) solution of the substrate and excess hydrogen peroxide. Peroxides produced in the reactions were reduced by sodium borohydride before product distribution analysis. The results of the chemical and photooxidations using a number of simple olefins and (+)-limonene are given in Tables I and II, respectively. The product distributions obtained were essentially the same for the two methods and compared well with those previously reported for photooxidations of the same substrates. The product distribution results point to a common oxygenation reagent for the chemical and photochemical systems, namely, singlet oxygen.

Table I Products of Olefin Oxygenation[a,b]

			Photo-oxygenation		OCl^-/H_2O_2	
Olefin	Alcohol A	Alcohol B	%A	%B	%A	%B
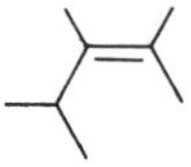	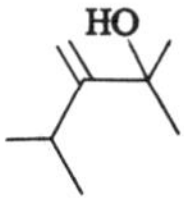 HO	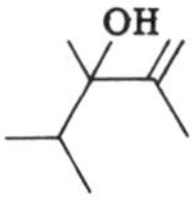 OH	40	60	39	61
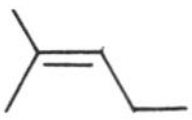	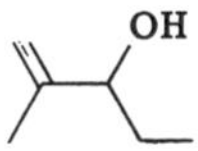 OH	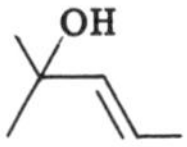 OH	49	51	51	49
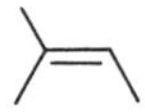	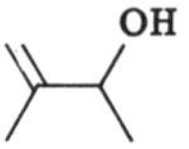 OH	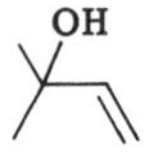 OH	48	52	49	51
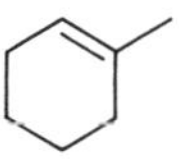	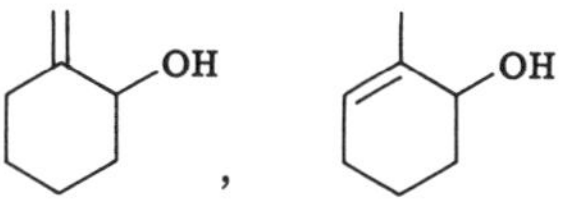 OH , OH	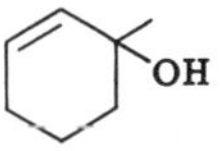 OH	44, 20	36	44, 20	36
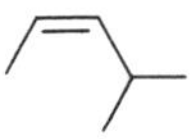	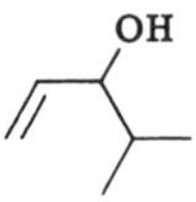 OH	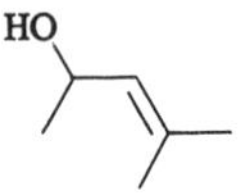 HO	96	4	94	6

[a] After reduction of the peroxides to the corresponding alcohols; mean of four or more analyses.

[b] Reprinted with permission from C. S. Foote, *Acc. Chem. Res.* **1**, 104 (1968a). Copyright by the American Chemical Society.

Table II Products from Oxidation of (+)-Limonene[a,b,f]

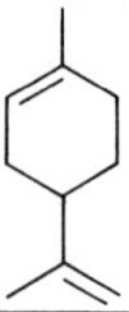

Product	Photosensitized autoxidation[c,d]	Singlet oxygen ($NaOCl + H_2O_2$)[d,e]
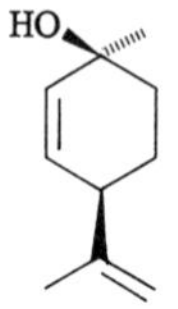 HO	31	34
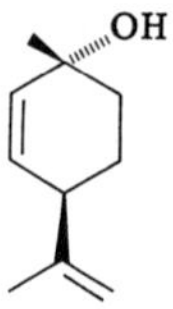 OH	11	9
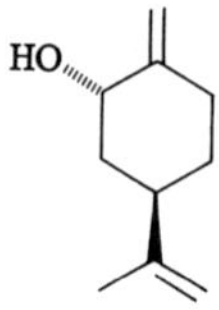 HO	21	18
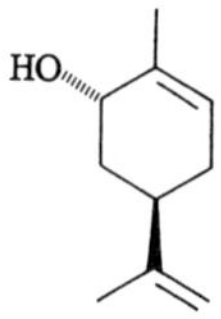 HO	10	9
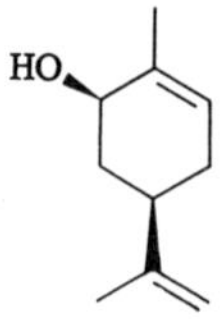 HO	3	7
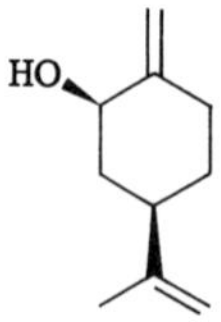 HO	25	24

The products obtained in the oxidation of limonene provided additional evidence concerning the mechanism of oxidation. The *trans*-carveol obtained in both the chemical and photochemical oxidations was found to display optical activity. A radical mechanism involving allylic hydrogen abstraction would have given an allylic free radical and, consequently, a racemic alcohol product. The results are more consistent with a nonradical, concerted process.

In the same study (1965) Foote and co-workers caution that use of the hypochlorite–hydrogen peroxide method with unreactive substrates can lead to serious side reactions which appear to be free radical in nature. Gollnick and Schenck (1964) had reported that oxidation of α-pinene with the chemical system gave a product distribution which was different from that obtained in photosensitized oxidations but which distribution was close to that given by free radical oxidation. Foote *et al.* (1965) repeated the chemical oxygenation of α-pinene in the presence of the free radical inhibitor, 2,6-di-*tert*-butyl phenol, and found that the product distribution obtained under these conditions was quite similar to that given by photosensitized oxidation. Since α-pinene is a relatively unreactive substrate it is apparent that the chemical oxygenation conditions permit competition from free radical side reactions.

Foote and co-workers (1968) expanded their work on the chemical oxygenation system by studying a number of reaction parameters. This same contribution provided further experimental details for the oxygenation reaction and reported that $\Delta^{9,10}$-octalin is converted to 10-hydroperoxy-$\Delta^{9,10}$-octalin when treated with sodium hypochlorite–hydrogen peroxide. Here again the product obtained is the same as that given by photooxygenation.

Attempts to optimize the yield of oxygenation product from tetramethylethylene led to reaction conditions giving an 88% yield. This yield was obtained by adding a 1.5 *M* aqueous sodium hypochlorite solution to a methanol solution which contained the acceptor (TME) at 0.21 *M*,

---

Table II—(*Continued*)

[a] In 1:1 *tert*-butanol–methanol.

[b] Per cent in reaction mixture after reduction.

[c] A very similar product distribution was reported from photosensitized autoxidation; free radical oxidation gives a drastically different product mixture.

[d] Products with unshifted double bonds were not found.

[e] $Ca(OCl)_2 + H_2O_2$ with added 2,6-di-*tert*-butyl phenol gave a virtually identical product distribution.

[f] Reprinted with permission from Foote, C. S., Wexler, S., and Ando, W. (1965). *Tetrahedron Lett.*, 4111–4118.

0.012 *M* NaOH, and 0.18 *M* H_2O_2. The reaction was carried out at $-20°C$. This yield could be obtained over the temperature range $-50°$ to $0°C$, but dropped to 40% to 50°C.

Foote *et al.* (1968) also introduced the concept of a β value as a measure of substrate reactivity. Kinetic studies had shown that the reactive species (RS) in the oxygenation had two major pathways open to it, that is, to react with the acceptor (A) to give the product AO_2, or to decay to ground-state oxygen [Eq. (9)]. Formally, RS could either be singlet oxygen or a sensitizer oxygen complex although later work suggests that the latter possibility is not likely.

$$\mathrm{A} + \mathrm{RS} \xrightarrow{k_1} \mathrm{AO_2}$$
$$\mathrm{RS} \xrightarrow{k_2} {}^3\mathrm{O_2} \tag{9}$$

Application of steady-state kinetics to this general scheme gives an expression for the quantum yield of product formation Φ_{AO_2} in which Φ_{RS} is the quantum yield of RS formation [Eq. (10)].

$$\Phi_{AO_2} = \Phi_{RS} \frac{k_1[A]}{k_1[A] + k_2} \tag{10}$$

If the quantity $\beta\ (= k_2/k_1)$ is introduced then the fraction of RS which gives product is given by

$$\frac{[A]}{\beta + [A]}$$

That is, β is the concentration of A at which one-half of the RS is converted to AO_2. A smaller β value thus indicates a more reactive substrate. For both photochemical and chemical oxygenation a plot of $1/AO_2$ versus $1/[A]$ should give a straight line and can be used to determine the β value for any acceptor. A series of β values determined by Foote *et al.* (1968) is given in Table III.

The β values are useful in setting acceptor concentrations for preparative oxygenation reactions, that is, the acceptor should be present at a concentration which is equal to or greater than its β value if a suitable amount of oxidized product is to be obtained. For olefins the general order of reactivity decreases as follows: tetraalkyl > trialkyl > dialkyl. Acyclic olefins are more reactive than compounds with double bonds in a six-membered ring. Substituted anthracenes, cyclohexadienes, and cyclopentadienes are all good acceptors. Further details on the measurement and use of β values as well as the determination of relative activities of acceptors were reported in 1968 by the Foote group (Higgins *et al.*, 1968).

Based on the observed chemistry, kinetics, and stereochemistry, Foote

Table III Relative Reactivities of Acceptors[a,h]

Acceptor	Photooxygenation		Hypochlorite/H_2O_2	
	k_6 (rel.)[g]	β (M)[b]	k_6 (rel.)[g]	β (M)[b]
2,5-Dimethylfuran[c]	2.4	0.001	3.8	0.001
Cyclopentadiene[c]	1.2	0.002	1.1	0.003
2,3-Dimethyl-2-butene[d]	(1.00)	0.0027	(1.00)	0.0030
1,3-Cyclohexadiene[c]	0.08[e]	0.03[e]	0.3	0.01
1-Methylcyclopentene[c]	0.05	0.05	0.03	0.1
trans-3-Methyl-2-pentene[c]	0.04	0.07	0.03	0.1
cis-3-Methyl-2-pentene[c]	0.03	0.1	0.03	0.1
2-Methyl-2-butene[d]	0.024	0.11	0.028	0.11
2-Methyl-2-pentene[d]	0.019	(0.14)	0.022	(0.14)
1-Methylcyclohexene[d]	0.0041	0.67	0.0056	0.55
cis-4-Methyl-2-pentene[d]	0.00026	10	0.00024	13
Cyclohexene[d]	0.000048	55	[f]	
trans-4-Methyl-2-pentene[d]	0.000047	57	0.000044	70

[a] Higgins *et al.*, 1968.

[b] Calculated, based on an average value for photosensitized and chemical oxygenation for 2-methyl-2-pentene of 0.14 M.

[c] By acceptor disappearance, usually $\pm 30\%$.

[d] By product appearance, usually $\pm 10\%$.

[e] Result of a single determination.

[f] Insufficiently reactive for determination.

[g] k_6 is for acceptor $+ {}^1O_2$.

[h] Reprinted with permission from Foote, C. S. (1968). *Acc. Chem. Res.* **1,** 104–110. Copyright by the American Chemical Society.

et al. (1968) again concluded that the reactive species in the chemical oxygenation as well as in dye-sensitized photooxygenation is singlet delta oxygen, O_2 (${}^1\Delta_g$).

In the case of tetramethylethylene oxygenation it was found that almost no allylic hydroperoxide was formed if the pH was less than 7. Instead the product was the methoxychloro adduct which is the addition product between the olefin and hypochlorite in methanol solvent. It was also found that the solvent has a major effect on product yield. The best solvents were methanol, ethanol, and methanol–*tert*-butyl alcohol (1:1). Poorer yields were obtained in isopropyl alcohol, pure or aqueous *tert*-butyl alcohol, tetrahydrofuran, dioxane, and acetonitrile.

Based on hypochlorite a 75% yield of singlet oxygen can be obtained in methanol. In the other water-miscible solvents, the yield was usually below 10%. The use of the hypochlorite–hydrogen peroxide system for chemical oxygenation thus has a solvent limitation.

Product distribution studies were extended to several more substrates. As seen in Table I, the chemical and photochemical methods gave very similar results. The relative reactivities of a series of olefins were determined using tetramethylethylene as a reference (Table III). The relative reactivity data are also in excellent agreement with those data available from photochemical studies.

A series of aryl-substituted trimethylstyrenes has been used to study the effect of substituents on rate and product distribution. In these compounds the steric effect is presumably constant while changing the substituent permits the electronic effect to be varied (Foote, 1968a; Foote and Denny, 1971a). Product distributions were found to be approximately the same whether photooxygenation or chemical oxygenation was used. The distribution was found to be in favor of product (**1**) [Eq. (11)]. Product distribution was 2.7/1.

$$Y\text{-}C_6H_4\text{-}C(CH_3)=C(CH_3)_2 \xrightarrow[\text{2. reduction}]{\text{1. } h\nu/O_2/\text{Sens. or } OCl^- - H_2O_2} \underset{(\mathbf{1})}{Y\text{-}C_6H_4\text{-}C(=CH_2)\text{-}C(CH_3)_2\text{-}OH} + Y\text{-}C_6H_4\text{-}C(CH_3)(OH)\text{-}C(=CH_2)CH_3 \quad (11)$$

The substituent (Y) was placed in both meta and para positions and was varied from methoxy to cyano. Electron-donating substituents were found to increase the rate of the reaction substantially. The Hammett ρ parameter was found to be -0.93 for the photooxygenation and -0.91 for the chemical oxygenation indicating a transition state which is assisted by electron-donating substituents. The electrophilicity found in this case is consistent with that found earlier for a series of alkyl-substituted ethylenes (Foote *et al.*, 1968).

The possible presence of charge separation in the transition state was further tested by determining the effect of solvent on the β value for the photooxygenation of 2-methyl-2-pentene. As shown in Table IV only a slight change in β value was observed over a great range of solvent polarities.

Table IV Solvent Effect on β $(k_5/k_6)^d$ for Photosensitized Oxygenation of 2-Methyl-2-pentene[e]

Solvent	β $(M)^a$
Ethyl acetate[b]	0.04
tert-Butyl alcohol	0.05
Bromobenzene[c]	0.06
Cyclohexanol[b]	0.075
Acetone[b]	0.08
Dimethyl sulfoxide[b]	0.08
Carbon disulfide[c]	0.09
Benzene[c]	0.10
Iodoethane	0.11
Methanol–*tert*-butyl alcohol (50:50)[b]	0.14
Methanol[b]	0.17

[a] Determined by plot of $1/\phi$ vs. 1/[A] [see Higgins *et al.* (1968)], maximum probable error around $\pm 30\%$.

[b] Rose Bengal sensitizer.

[c] Zinc tetraphenylporphine sensitizer.

[d] k_5 is for decay of $^1O_2$ to ground-state oxygen; k_6 is for acceptor $+ {}^1O_2$.

[e] Reprinted with permission from Foote, C. S. (1968a). *Acc. Chem. Res.* **1,** 104–110. Copyright by the American Chemical Society.

Several of the solvents used (bromobenzene, iodomethane) allow one to conclude that heavy atoms, which could have increased the rate of a decay to ground state oxygen via spin-orbit coupling have no appreciable effect on the β value (Foote and Denny, 1971b). The very slight change in β value observed suggests that there is very little charge separation in the transition state for this photooxygenation. This observation plus the previously cited lack of substituent directing effect in the trimethylstyrene system support a concerted mechanism for formation of the allylic hydroperoxide products.

On the basis of stereochemical considerations Gollnick and Schenck (1964) and Nickon (Nickon and Bagli, 1961) had suggested a cyclic transition state for the photooxygenation reaction [Eq. (12)].

O O H C C C   O O H C C C   (12)

According to this picture the oxygen is delivered on the same side of the molecule as the hydrogen is removed. The reaction should be favored when the hydrogen, which becomes the hydroperoxide hydrogen, is coplanar with the p orbitals of the double bond. On the basis of his systematic approach using a comparison between photosensitized and chemical oxygenations, Foote (1968, 1971) also favored the cyclic transition state with the oxygen free of sensitizer unlike the earlier picture (Gollnick and Schenck, 1964) which suggested that the sensitizer was still attached to the oxygen in the transition state. This concerted cycloaddition of oxygen to the olefin system is analogous to the mechanism generally accepted for the "ene" reaction (Berson *et al.*, 1966).

The singlet oxygen produced in the hypochlorite–hydrogen peroxide system is very effectively quenched by low concentrations of β-carotene (Foote *et al.*, 1970a,b). This conclusion was drawn from experiments using 2-methyl-2-pentene as the acceptor. In this case the solvent used was composed of methanol, benzene, and diglyme in a 1:1:1 ratio. This quenching by β-carotene was also observed in the photosensitized oxygenation of the same substrate and has important consequences regarding photodynamic action and the possible protective ability of β-carotene in chlorophyll-containing plants (Foote and Denny, 1968).

Investigations of the possible protective action of carotenoids led Foote and Brenner (1968) to study the oxidation of the triene, (**2**), using both photosensitized and chemical oxygenation methods. In both cases, four products, (**3**)–(**6**) were obtained and in approximately the same distribution [Eq. (13)].

(**2**)

1. $h\nu/O_2$/Sens
or
OCl^-/H_2O_2
2. reduction

(13)

OH + O O + OH + OH

(**3**) (**4**) (**5**) (**6**)

Products (**5**), (**6**), and (**4**) are those expected from ene and diene reactions, respectively. The allenic product, (**3**), appears to be the result of an ene-type reaction in which an sp^2-bonded hydrogen is transferred to the hydroperoxide moiety.

In further investigation of the role of carotenoids in chlorophyll-containing plants, Foote *et al.* (1970a) showed that singlet oxygen produced from hypochlorite–hydrogen peroxide can isomerize 15,15′-*cis*-β-carotene to the all-trans isomer. This observation supports the mechanism suggested by Foote and Denny (1968) for the protective action of β-carotene, namely that it quenches singlet oxygen with a resultant energy transfer to the carotene.

3. Additional Uses of the Hypochlorite–Hydrogen Peroxide System

The hypochlorite–hydrogen peroxide system has been used to oxygenate a wide variety of organic substrates in addition to those referred to above. Illustrative examples of this use are given here. In most cases, the products obtained are the same as those from sensitized photooxidations.

Bellin and Mahoney (1972) have oxidized 1,3,5-triphenylformazan (**7**) with the hypochlorite–hydrogen peroxide system. The formazan gives one mole of benzene and one mole of benzoic acid for each mole of formazan oxidized. The reaction is believed to proceed through the hydroperoxide (**8**) which can decompose to give the observed products [Eq. (14)]. The intermediate diazonium cation is presumably oxidized to give the trace amount of *p*-nitroaniline found.

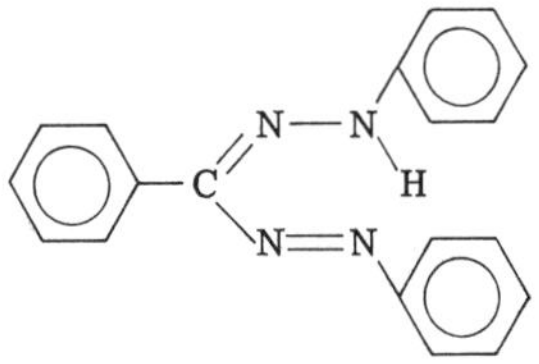

(**7**)

The photosensitized oxidation of heterocyclic compounds has been of great interest because of the possible relationship of such reactions to the photodynamic effect. Indoles have been of particular interest as models for tryptophyl residues in proteins. Evans (1971) has used hypochlorite–hydrogen peroxide to oxygenate 3-methylindole. The products, *o*-formamido-

(8)

$$\text{(8)} \rightarrow \left[C_6H_5-\overset{O}{\overset{\|}{C}}-N{=}N-C_6H_5 \right] + OH\cdot + \left[C_6H_5-\overset{+}{N}{\equiv}N \right]$$

$$\downarrow$$

$$C_6H_5-\overset{O}{\overset{\|}{C}}-O^- + C_6H_5-N{=}N-H \longrightarrow C_6H_6 + N_2 \qquad (14)$$

$$\left[C_6H_5-\overset{+}{N}{\equiv}N \right] \xrightarrow{OH^-} C_6H_5-OH$$

$$\left[C_6H_5-\overset{+}{N}{\equiv}N \right] \xrightarrow{OCl^-/H_2O_2} \text{oxidation products}$$

acetophenone and *o*-aminoacetophenone presumably arise from decomposition of an intermediate dioxetane [Eq. (15)].

$$\text{3-methylindole} \longrightarrow \text{dioxetane} \longrightarrow o\text{-}(HCONH)C_6H_4COCH_3 \longrightarrow o\text{-}(NH_2)C_6H_4COCH_3 \qquad (15)$$

Chemical oxygenation of 9,9′-bifluorenylidene to fluorenone presumably via an intermediate dioxetane has been reported (Richardson and Hodge, 1970). The reaction was carried out both photochemically and using hydrogen peroxide–hypochlorite [Eq. (16)].

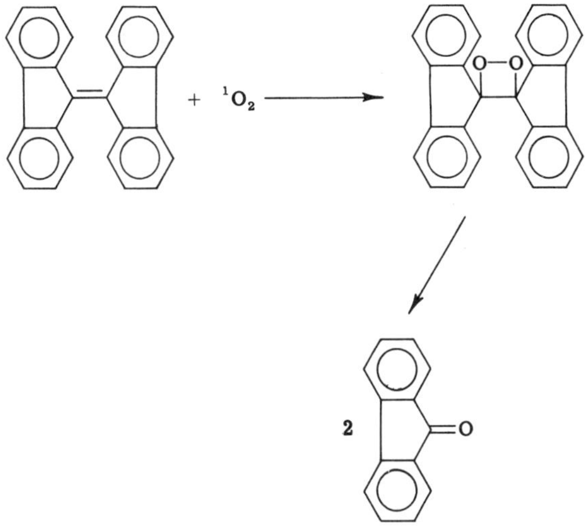

$$+ \; ^1O_2 \longrightarrow \qquad (16)$$

2

Bland and Craney (1974) have used the hydrogen peroxide–hypochlorite oxygenation system to show that vitamin D_3 is very susceptible to singlet oxygen attack. The dioxolane expected from a reaction of singlet oxygen with the diene system of the vitamin was not isolated, rather a product incorporating two methoxide groups was obtained. This product presumably arises from the intermediate dioxolane via reaction with the methanol solvent.

The possible role of singlet oxygen in the photodynamic oxidation of nucleic acid constituents has been studied using the hydrogen peroxide–hypochlorite system (Hallett *et al.*, 1970). Using suitable controls, various nucleic acid constituents were exposed to hydrogen peroxide–hypochlorite. Susceptibility to singlet oxygen attack was determined by analyzing for the percentage of the substrate remaining after a fixed period of time. The results suggest that the following constituents are attacked by singlet oxygen: guanine, uracil, thymine, guanosine, deoxyguanosine, xanthosine, guanylic acid, and deoxyguanylic acid. The susceptibility of these substrates to singlet oxygen attack parallels their photodynamic oxidation reactivities very closely.

Octalin (**9**) was converted by singlet oxygen to a mixture of allylic hydroperoxides which were reduced to the alcohols (**10**), (**11**), (**12**), and

(**13**) (Marshall and Hochstetler, 1966). The reaction was carried out by photosensitized oxidation or using hydrogen peroxide–hypochlorite. Both methods gave the same distribution of products [Eq. (17)].

(**9**) —— 1. OCl^-/H_2O_2; 2. $LiAlH_4$ ——> (**10**) 36% + (**11**) 18% + (**12**) + (**13**) cis + trans 34% (17)

Foster and Berchtold (1972) have reported a very interesting use of chemical oxygenation using hydrogen peroxide–hypochlorite. Benzene oxide (oxepin, (**14**)) was converted to the 1,4-endoperoxide, (**15**), when treated with singlet oxygen. The endoperoxide undergoes a facile and quantitative rearrangement to *trans*-benzene trioxide [*trans*-3,6,9-trioxatetracyclo[6.1.0.$0^{2,4}$.$0^{5,7}$]nonane, (**16**) Eq. (18)].

(**14**) —— OCl^-/H_2O_2 ——> (**15**) ——> (**16**) (18)

Chemically produced singlet oxygen using hypochlorite–hydrogen peroxide has been used to gain evidence for the involvement of singlet oxygen in the photosensitized oxidation of certain phenols (Saito *et al.*, 1970). Oxidation of the monomethyl ether of 4,6-di-*tert*-butylresorcinol (**17**) or the corresponding dimethyl ether (**18**) gave the hydroperoxide (**19**) (70%) and the epoxyketone (**20**) (75%), respectively [Eq. (19)]. The authors suggest

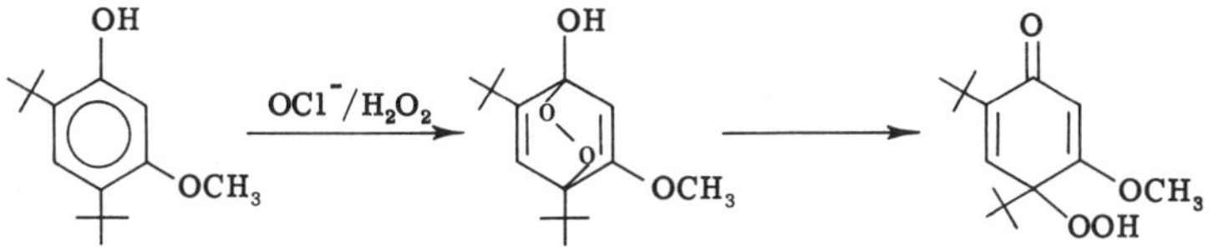

OCl^-/H_2O_2

(17) (21) (19)

(19)

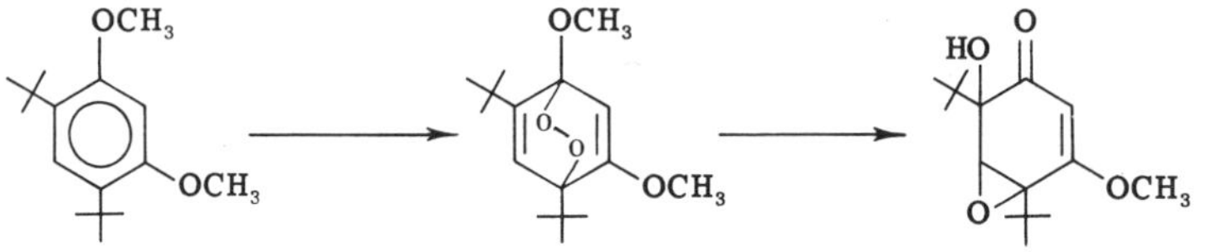

(18) (22) (20)

that the products may arise via the intermediate endoperoxides, (**21**) and (**22**). These products are the same as those obtained in photosensitized oxidations. Also the reactivities of (**17**) and (**18**) were found to be virtually the same in both photooxidation and chemical oxygenation conditions.

This same group (Matsuura *et al.*, 1971) also used chemical oxygenation with hypochlorite–hydrogen peroxide to argue for the involvement of singlet oxygen in the photosensitized oxidation of 1,5-cyclooctadiene, (**23**), to give the hydroperoxide, (**24**). While the hydroperoxide was formed under both sets of conditions, it is transformed to 4-hydroxy-5-cyclooctenone, (**25**), under the photosensitizing conditions [Eq. (20)].

A more complete discussion of the general reactions of singlet oxygen with these substrates is given by Saito and Matsuura in Chapter 10 of this volume.

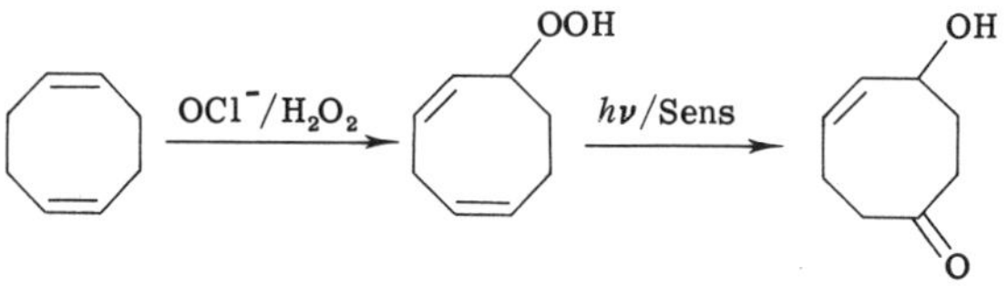

OCl^-/H_2O_2 $h\nu$/Sens (20)

(23) (24) (25)

4. *Summary*

The hydrogen peroxide–hypochlorite method of generating singlet oxygen has been of enormous importance in studies establishing the role of singlet oxygen in photosensitized oxidation. The spectroscopic evidence indicates that both $^1\Delta_g$ and $^1\Sigma_g^+$ states of singlet oxygen are produced in this reaction. Using methanol as solvent the efficiency of $^1\Delta_g$ singlet oxygen production can reach 80%.

The method has limitations. It clearly cannot be used with substrates which react with hydrogen peroxide. It requires high alkalinity which can be a problem. The reagent has a limited solubility in organic solvents. Its optimum use requires alcohol solvents which are not useful for all substrates.

B. The Reaction of Bromine with Hydrogen Peroxide

The decomposition of alkaline hydrogen peroxide with bromine was observed to be chemiluminescent by McKeon and Waters (1966). They have also shown that this system provides a good oxygenation method for anthracene-type aromatic hydrocarbons. The reactions involved are presumably similar to those which occur in the hydrogen peroxide–hypochlorite system [Eq. (21)–(23)].

$$HOO^- + Br{-}Br \rightarrow HOOBr + Br^- \tag{21}$$

$$HOOBr + OH^- \rightarrow H_2O + {}^-OOBr \tag{22}$$

$${}^-OOBr \rightarrow Br^- + {}^1O_2 \tag{23}$$

Using this system, McKeon and Waters (1966) were able to oxidize 9,10-dimethylanthracene, 9,10-diphenylanthracene, and 9-methyl-10-phenylanthracene to their corresponding endoperoxides in 27, 46, and 34% yields, respectively. Experimentally, two different approaches were used. In the first of these a solution of the hydrocarbon in alcoholic potassium hydroxide was stirred vigorously while hydrogen peroxide and bromine were added simultaneously and separately by burettes extending below the level of the solution.

The second approach uses two phases, the upper of which is a solvent (chlorobenzene) containing the hydrocarbon and the lower of which is an aqueous alkaline hydrogen peroxide solution. Bromine is then introduced into the lower phase in a rate controlled manner to permit oxygen to bubble through the top phase without carrying any bromine with it.

The hydrogen peroxide–bromine method is suitable only for 9,10-disubstituted anthracene compounds if endoperoxide products are desired. When 9-substituted anthracenes are used, 9-hydroxyanthrone derivatives rather than endoperoxides are obtained.

The authors have ruled out another possible route to endoperoxides, namely, prior 9,10-addition of bromine, followed by successive nucleophilic substitution reactions with hydroperoxy anions. Separate experiments showed that bromine rapidly undergoes the 1,4-addition reaction, but subsequent treatment of the dibromo adduct with alkaline hydrogen peroxide does not lead to endoperoxide.

The experimental conditions required impose a severe limitation on general use of this method for oxygenation purposes.

C. The Decomposition of Alkaline Solutions of Peracids

Organic peracids are unstable in alkaline solutions, decomposing to the corresponding acid with evolution of oxygen. McKeown and Waters (1966) have used a chemical trap to show that singlet oxygen is involved in these decompositions. Decomposition of diisoperoxyphthalic acid in ethanol at pH 8–9 in the presence of 9-methyl-10-phenylanthracene gave the endoperoxide in 21% yield based on the hydrocarbon. Peracetic acid failed to oxidize the hydrocarbon under these conditions, but when the pH was raised to 10, and a small amount of hydrogen peroxide added, 2% of the endoperoxide was formed.

A variation of the procedure is to permit decomposition of the peracid in alkaline solution containing hydrogen peroxide. Thus, when peroxytoluic acid was decomposed at pH 10 in the presence of hydrogen peroxide and 9-methyl-10-phenylanthracene, an 18% yield of the endoperoxide was obtained. In this latter procedure an earlier ^{18}O tracer study indicated that the molecular oxygen evolved originates mainly in the hydrogen peroxide (Akiba and Simamura, 1964).

In a more recent study Boyer *et al.* (1975) have shown that *m*-chloroperbenzoic acid reacts with 1,3-diphenylisobenzofuran and 2,5-diphenylfuran to give the typical singlet oxygen products, *o*-dibenzoylbenzene and *cis*-dibenzoylethylene, respectively. However, refluxing the peracid in dichloromethane with the singlet oxygen acceptor, 9,10-diphenylanthracene, failed to give any products.

A kinetic study revealed that the peracid was being consumed in a first-order process and not the bimolecular process which could lead to singlet oxygen production. Thus, the peracid produces the typical singlet oxygen products from the furan acceptors by a direct oxidation without involving singlet oxygen. The authors suggest that the oxidation involves oxygen atom transfer to give the epoxy compound which rearranges to the dicarbonyl product.

These results indicate that care must be exercised in using these furan acceptors as chemical tests for singlet oxygen. Such tests should be accompanied by further confirmatory experiments.

Based on the small number of reactions studied, the yields obtained, and competition from side reactions, it does not seem likely that the use of peracids as a source of singlet oxygen will compete very well with some of the other chemical methods available.

D. The Self-Reaction of *sec*-Butylperoxy Radicals

Russell has proposed (1957) that secondary peroxy radicals could undergo a self-reaction involving a cyclic transition state [Eq. (24)].

$$2\ \mathrm{\rangle C(H)-O-O\cdot} \rightleftharpoons [\text{cyclic transition state}] \longrightarrow \mathrm{\rangle C{=}O} + O_2 + \mathrm{\rangle C(H)OH} \quad (24)$$

Application of the conservation of spin principle suggests that there are two possibilities for this reaction with regard to the spin multiplicity of the products. In the first of these possibilities, oxygen would be eliminated in the triplet ground state. In this case the accompanying ketone would be found in the triplet excited state. While some chemiluminescence was observed in such reactions (Howard and Ingold, 1968), the quantum yield was so small that the chemiluminescence was not found to be useful in making a decision regarding this mode of decomposition.

In the second mode of decomposition, the molecular oxygen would be produced in an excited singlet state ($^1\Sigma$ or $^1\Delta$). Howard and Ingold have shown (1968) that when *sec*-butyl hydroperoxide is oxidized by ceric ion the molecular oxygen produced reacts with a singlet oxygen acceptor to give the expected singlet oxygen product. The acceptor used was 9,10-diphenylanthracene which is converted to 9,10-diphenylanthracene 9,10-endoperoxide [Eq. (25)].

$$\text{9,10-}(C_6H_5)_2\text{-anthracene} \xrightarrow[\text{sec-butyl hydroperoxide}]{\text{Ce(IV)}} \text{9,10-}(C_6H_5)_2\text{-anthracene 9,10-endoperoxide} \quad (25)$$

Based on the amount of endoperoxide isolated, it was concluded that singlet oxygen produced in this manner is trapped by the acceptor at an efficiency which is only one-third or one-fourth of that observed when the same acceptor is used and the $Br_2-H_2O_2$ system used as a source of singlet oxygen. Possible reasons for the apparently lower efficiency include (a) decomposition of the endoperoxide by ceric ion, (b) quenching of singlet oxygen by ceric ions, and (c) radical reactions which consume the acceptor. Howard and Ingold (1968) favor the last possibility as being the most probable since 9,10-diphenylanthracene is known to be quite reactive toward the radicals present in oxidizing hydrocarbons (Turner and Waters, 1956).

Comparable ceric ion oxidations of *tert*-butyl hydroperoxide or hydrogen peroxide gave no evidence for the involvement of singlet oxygen (Howard and Ingold, 1968).

Kellogg (1969) has also studied the disproportionation reaction of *sec*-peroxy radicals. In this work, the dominant chemiluminescence is believed to be due to phosphorescence from the excited triplet carbonyl compound. The photomultiplier used in these chemiluminescence studies was capable of detecting emission from $^1\Sigma$ singlet oxygen, but not $^1\Delta$. In addition, any $^1\Sigma$ emission would have overlapped with ketone phosphorescence. The authors conclude that there may be a small amount of $^1\Sigma$ emission, but that the major disproportionation path produces triplet carbonyl. The singlet oxygen present could be produced directly in the disproportionation reaction or by ground state oxygen quenching of excited carbonyl.

On the other hand, recent work by Nakano and co-workers (1976) adds strength to the view that the disproportionation reaction produces singlet oxygen. These workers have provided spectroscopic evidence for the production of both $^1\Sigma_g{}^+$ and $^1\Delta_g$ states of singlet oxygen in the aqueous ceric ion oxidation of *sec*-butyl hydroperoxide or linoleic acid hydroperoxide. The work was directed at adding to an understanding of the possible role of singlet oxygen in enzyme-mediated peroxidation reactions *in vivo*. In particular, the work relates to the role of superoxide dismutase which is known to inhibit the production of singlet oxygen in the dismutation of superoxide ion (see Section II,F). It had been shown that addition of the dismutase to the microsomal lipid peroxidation system does not lead to inhibition of light emission from singlet oxygen (Nakano *et al.*, 1975). This observation led to the suggestion that radical disproportionation reactions of lipid peroxides might be responsible for singlet oxygen production in the microsomal lipid peroxidation system rather than the superoxide dismutation reaction (Nakano *et al.*, 1976).

The reaction of *sec*-butyl hydroperoxide or linoleic acid hydroperoxide with ceric ion in aqueous media was shown to give emission from the ($^1\Sigma_g{}^+$, $^1\Delta_g$) and ($^1\Delta_g$, $^1\Delta_g$) dimers of singlet oxygen (Nakano *et al.*, 1976). Furthermore, the total light output of these systems was increased 1.5- to 3.0-fold when D_2O was used instead of H_2O. This observation is presumably due to the known increased lifetime of singlet oxygen in D_2O (Kajiwara and Kearns, 1973). The results support the suggestion that self-reactions of *sec*-peroxy radicals of unsaturated fatty acids are responsible for the production of singlet oxygen and the chemiluminescence observed in the microsomal lipoxygenation system (Nakano *et al.*, 1975).

E. The Base-Induced Decomposition of Peroxyacetylnitrate

The peroxyacyl nitrates (PAN) are a family of compounds which are produced by the action of sunlight on polluted atmospheres containing hydrocarbons and nitrogen oxides. They are of particular importance because of their ability to cause severe crop damage and eye irritation. It has been shown (Stephens, 1969) that a member of the PAN family, peroxyacetyl nitrate (**26**) decomposes in the presence of base to release molecular oxygen [Eq. (26)].

$$\underset{\textbf{(26)}}{CH_3\text{—}\overset{\overset{\displaystyle O}{\|}}{C}\text{—}OONO_2} + 2\,OH^- \rightarrow CH_3\text{—}\overset{\overset{\displaystyle O}{\|}}{C}\text{—}O^- + H_2O + O_2 + NO_2^- \quad (26)$$

The conservation of spin principle suggests that the oxygen so produced could be in an excited singlet state. It has been shown that this reaction does indeed lead to the production of singlet oxygen (Steer *et al.*, 1969). Evidence for the presence of the singlet delta state was observed by monitoring the near infrared emission at 1.27 μm which is due to the conversion of the excited delta state to ground-state oxygen [Eq. (27)]. The observation

$$O_2(^1\Delta_g) \rightarrow O_2(^3\Sigma_g^{\,-}) + h\nu \quad (27)$$

of O_2 ($^1\Delta_g$) does not preclude the possibility that the state which is initially produced is the ($^1\Sigma_g^{\,+}$) state since this latter state is known to be deactivated to the $^1\Delta_g$ state (Wayne, 1969).

Attempts to observe chemical reactions of the singlet oxygen produced in this fashion gave inconclusive results since peroxyacetyl nitrate itself is a powerful oxidant which competes with singlet oxygen for the acceptor molecules. This latter observation suggests that this method is of limited use as a source of singlet oxygen for chemical oxygenation.

F. The Decomposition of Superoxide Ion

Khan (1970) has reported that potassium superoxide evolves singlet oxygen in dimethyl sulfoxide solution. Evidence for the production of singlet oxygen was obtained in two ways. The decomposition was carried out in the presence of the singlet oxygen acceptor, 2,5-dimethylfuran. This reaction led after workup to a residue which gave a characteristic peroxide test with potassium iodide. Apparently no further work was done to characterize the product. Kearns (1971) has suggested that use of the peroxide test might not be a reliable indication of the presence of the expected product (2-methoxy-5-hydroperoxy-2,5-dihydrofuran) since hydrogen peroxide may have been generated by reaction of superoxide with 2,5-dimethylfuran. The hydrogen peroxide could have been responsible for the positive peroxide test.

More recently Khan (1977) has used a theoretical analysis to show that production of singlet oxygen from ion clusters of superoxide ion is critically dependent on the number of water molecules present. This analysis indicates that there is a narrow concentration range for the production of $^1\Delta$ singlet oxygen. Beyond that range, O_2 ($^1\Sigma_g{}^+$) becomes the predominant product. The analysis also predicts that singlet oxygen can be efficiently quenched by superoxide ion. This quenching has been observed experimentally by two groups (*vide infra*).

The fluorescence sensitization technique was also used to obtain evidence for the production of singlet oxygen. The potassium superoxide in dimethyl sulfoxide solution was permitted to decompose in the presence of a fluorescer (anthracene, esculin, eosin, and violanthrone) in a flow apparatus. In all cases fluorescence was observed. Based on the energy requirements for anthracene fluorescence and the observed fluorescence quenching by water, Khan further concluded that the singlet oxygen involved was in the Σ state. The mechanism for the formation of singlet oxygen given by Khan (1970) is shown in Eq. (28).

$$O_2{}^- \rightarrow O_2{}^* + e^- \tag{28}$$

Nillson and Kearns (1974) have reinvestigated the potassium superoxide–dimethyl sulfoxide system as a source of singlet oxygen. They found no luminescence when dry dimethyl sulfoxide was used, whether or not fluorescein was present. Addition of water did produce a weak light emission. They then added deuterium oxide instead of water but found no enhancement of the luminescence. They argue that if the luminescence were due to singlet oxygen it should have been enhanced since Kajiwara and Kearns (1973) had shown that the lifetime of singlet oxygen is greatly enhanced in deuterium oxide. Nilsson and Kearns (1974) also reported that they could observe no oxidation of tetramethylethylene or 1,3-diphenylisobenzofuran using the superoxide ion–dimethyl sulfoxide system whether or not water was present to aid the decomposition of the superoxide. On the basis of these results they feel that it is unlikely that appreciable amounts of singlet oxygen are produced in this system.

The production of singlet oxygen from a solution of superoxide anion in dimethyl sulfoxide has also been questioned by Poupko and Rosenthal (1973). These workers argue that release of singlet oxygen from superoxide ion should lead to a rapid decrease in time of the superoxide ion concentration. Using ESR at -196°C as a detection method, it was found that superoxide in dimethyl sulfoxide shows little change in the superoxide signal intensity after storage at room temperature for several hours.

In addition, the chemical substrates 2,3-dimethyl-2-butene, 2-methyl-2-pentene, and cyclohexene were all found to be unaffected by a dimethyl sulfoxide solution of superoxide ion after 24 hours storage at room temperature. When tetraphenylcyclopentadienone was used as substrate a reaction occurred, but the product obtained, while not identified, was not the singlet oxygen product, *cis*-dibenzoylstilbene.

Electrochemically produced superoxide ion has been shown to react with electrogenerated ferricenium (Ferr) ion to produce singlet oxygen [Eq. (29)] (Mayeda and Bard, 1974).

$$\mathrm{Ferr}^{\dot{+}} + O_2^{\dot{-}} \rightarrow \mathrm{Ferr} + {}^1O_2 \tag{29}$$

In this application of the electrogenerated chemiluminescence (ecl) technique, the goal was not to produce a luminescent product but to provide a method for generating singlet oxygen. Singlet oxygen was detected chemically using 1,3-diphenylisobenzofuran as the acceptor. A 55% yield of the oxidation product, *o*-dibenzoylbenzene, was obtained. In addition to ferrocene, several other sources of cation radicals were investigated (tetraanisylethylene, 1,1′,2,2′-tetramethylvinylene-3,3′-diindolizine, 10-methylphenothiazine, 9,10-diphenylanthracene) but none led to detectable amounts of singlet oxygen.

Mayeda and Bard (1974) have also shown that the dismutation of superoxide ion produces singlet oxygen [Eq. (30)]. Evidence for singlet

$$2\,O_2^{\dot{-}} + 2\,H^+ \rightarrow H_2O_2 + O_2 \tag{30}$$

oxygen production was again obtained by showing that 1,3-diphenylisobenzofuran could be converted to *o*-dibenzoylbenzene when electrogenerated superoxide was treated with water in the presence of the acceptor. Attempts to detect emissions from singlet oxygen dimers [$({}^1\Delta)_2$ and $({}^1\Delta, {}^1\Sigma)$] were unsuccessful.

The authors also present evidence that the production of singlet oxygen in the dismutation reaction is greatly decreased in the presence of the enzyme, superoxide dismutase (SOD). They suggest that in the enzyme-catalyzed dismutation reaction a number of lower energy elementary steps are involved which permit thermal dissipation of the reaction enthalpy and prevent singlet oxygen formation. The possibility that superoxide dismutase merely acts as a physical quencher was made less likely by showing that SOD has no effect on the rate of dye-sensitized oxidation of 1,3-diphenylisobenzofuran.

Schaap *et al.* (1974) have also presented evidence that SOD does not act as a physical quencher of singlet oxygen. They observed that SOD had no effect on the photosensitized oxidation of α-lipoic acid or 9,10-diphenylanthracene-2,3-dicarboxylic acid. Similarly, SOD was found not to affect the extent of reaction when α-lipoic acid is oxidized with 1-phospha-2,8,9-trioxaadamantane ozonide (**27**) as singlet oxygen source. Furthermore, these authors report that a chemiluminescence, which usually accompanies the decomposition of **27** and which they attribute to a singlet oxygen dimol emission, is not quenched by SOD.

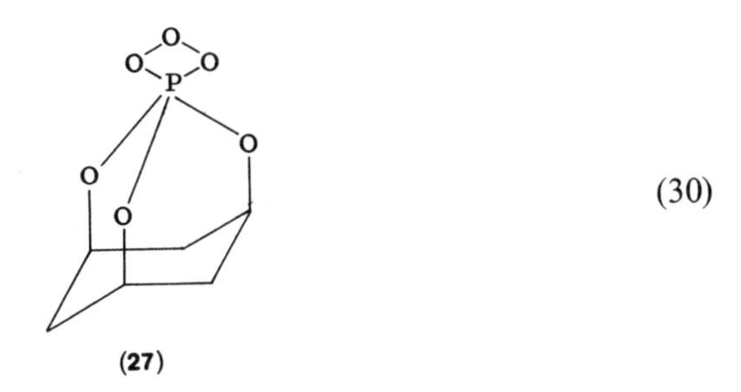

(30)

(27)

Goda *et al.* (1973) have described another singlet oxygen-producing biological system which involves superoxide as an intermediate. When the adrenal system, adrenodoxin reductase–adrenodoxin is used in the presence of nicotinamide adenine dinucleotide phosphate (NADPH), singlet oxygen is produced as evidenced by the chemiluminescence produced in a luminol system. Furthermore, it was found that singlet oxygen production was increased by about 200 times when cytochrome *c* was added to the reaction mixture. Addition of the singlet oxygen quencher, 1,4-diazabicyclo[2.2.2]-octane (DABCO), led to a pronounced decrease in the chemiluminescence.

The authors interpret their results in terms of the previously proposed (Khan, 1970) reaction sequence involving superoxide production [Eqs. (31) and (32)].

$$O_2 + e^- \rightarrow O_2^{\bar{\cdot}} \tag{31}$$

$$2\,O_2^{\bar{\cdot}} \rightarrow O_2^{2-} + {}^1O_2 \tag{32}$$

Addition of SOD to the reaction system led to a significant reduction in the chemiluminescence. This is presumably due to the ability of SOD to catalyze the dismutation reaction in such a way that triplet rather than singlet oxygen is produced.

The enhanced production of singlet oxygen in the presence of cytochrome *c* is significant because it suggests that the adrenal mitochondrial membranes could be damaged by singlet oxygen. The protective action of SOD would probably not be available since SOD does not appear to be present in the mitochondria.

It has been shown that the xanthine–xanthine oxidase system produces superoxide ion (Fridovich, 1970). Stauff *et al.* (1963) had earlier suggested that luminescence observed in this system may be due to the formation of excited states of oxygen. The luminescence observed consisted of a weak emission at 630 nm which is attributable to the $^1\Delta_g$ dimer. Stauff *et al.*

(1963) also proposed a reaction scheme to explain the luminescence [Eq.

$$\begin{aligned} O_2^{\cdot -} + H^+ &\rightarrow HO_2\cdot \\ HO_2\cdot + O_2^{\cdot -} &\rightarrow HO_2^- + {}^1O_2 \\ HO_2\cdot + HO_2\cdot &\rightarrow H_2O_2 + {}^1O_2 \\ {}^1O_2 + {}^1O_2 &\rightarrow ({}^1O_2)_2 \rightarrow 2\,O_2 + h\nu \end{aligned} \tag{33}$$

(33)]. Arneson (1970) has also investigated the chemiluminescence produced in the xanthine oxidase system. He has attributed the observed chemiluminescence to a singlet oxygen dimol emission. Arneson concurs that the singlet oxygen has its origin in superoxide ion, but proposes a different termination reaction involving hydroxyl radicals as being responsible for the production of the excited oxygen [Eq. (34)]. The required hydroxyl radicals arise from reaction of superoxide ion with hydrogen peroxide [Eq. (35)].

$$O_2^{\cdot -} + \cdot OH \rightarrow OH^- + O_2^* \tag{34}$$

$$O_2^{\cdot -} + H_2O_2 + H^+ \rightarrow O_2 + H_2O + \cdot OH \tag{35}$$

The apparently conflicting results on the question of singlet oxygen production in the dismutation of superoxide ion have been somewhat clarified by work of Guiraud and Foote (1976) and Rosenthal (1975). While studying the yield of singlet oxygen produced in the dismutation of tetramethylammonium superoxide using olefin traps, erratic results were obtained by Guiraud and Foote which raised the possibility that superoxide ion might itself act as a singlet oxygen quencher in an electron transfer process [Eq. (36)].

$${}^1O_2 + O_2^{\cdot -} \rightarrow O_2^{\cdot -} + {}^3O_2 + 22\ \text{kcal} \tag{36}$$

This possibility was examined by studying the kinetics of sensitized photooxygenation of diphenylisobenzofuran in the presence of superoxide ion. The results indicated that superoxide ion is an effective quencher of singlet oxygen with $k_q = 1.6 \times 10^9$ mole^{-1} sec^{-1}. In fact, its quenching rate constant is only a factor of ten below that of β-carotene which had been previously shown to be a very efficient singlet oxygen quencher (Foote and Denny, 1968). Using the quenching rate for superoxide ion derived from the diphenylisobenzofuran oxygenation studies it was possible to calculate the amount of singlet oxygen produced in the superoxide dismutation reaction. The initial indication was that the dismutation reaction produces a substantial amount of singlet oxygen. The authors point out that superoxide quenching of singlet oxygen in biological systems is not expected to be significant because of the expected low concentrations of superoxide ion.

Rosenthal (1975) also used tetramethylammonium superoxide to study its quenching effect on singlet oxygen. Superoxide was found to be an effective quencher of rubrene oxidation using either a self-sensitized photooxidation or a microwave discharge source of singlet oxygen. In this study, the quenching rate constant, k_q, was found to be $3.6 \pm 0.1 \times 10^7$ mole^{-1} sec^{-1}, a value lower than that reported by Guiraud and Foote (1976).

G. The Decomposition of Transition Metal Oxygen Complexes

1. Potassium Perchromate

Potassium perchromate was first synthesized and isolated by Riesenfeld *et al.* (1905). These workers also observed that the salt decomposes in aqueous solution with the stoichiometry shown in Eq. (37).

$$4\,CrO_8^{3-} + 2\,H_2O \rightarrow 4\,CrO_4^{2-} + 7\,O_2 + 4\,OH^- \quad (37)$$

An investigation of the use of this reaction as a singlet oxygen source was prompted in part by the speculation (Foote, 1968a) that certain biological oxidations could involve a metal-bound form of singlet oxygen. It has now been shown that this aqueous decomposition does indeed produce singlet oxygen (Peters *et al.*, 1972, 1975). Evidence for the production of singlet oxygen was obtained in several ways.

When a basic solution of potassium chromate was added to an aqueous methanol solution of hydrogen peroxide and 2,3-dimethyl-2-butene at 5°C, a red-brown precipate of potassium perchromate was observed to form. Simultaneous with its formation, the precipitate effervesced and the effervescence increased as the reaction mixture warmed to room temperature. It was shown that the typical singlet oxygen oxidation product of the olefin, 2,3-dimethyl-3-hydroperoxy-1-butene, had been formed in 35% yield. The yield can be increased by twofold if a buffer is used to lower the pH of the reaction mixture to 7.

Additional evidence for singlet oxygen production was obtained by using the product distribution method developed by Foote and co-workers (1965). As shown in Table V, the product distributions of reduced hydroperoxide products obtained when three different olefins are treated with aqueous perchromate are essentially the same as those obtained in dye-sensitized photooxidations of the same olefins.

The procedure described can only be used for substrates which are unreactive toward basic hydrogen peroxide. The use of these conditions can be avoided by isolating the dry perchromate salt. The salt can then be added to a cooled, 60% aqueous methanol solution of the substrate and the reaction mixture allowed to warm to room temperature. Using these

Table V Product Distributions from Olefin Oxygenations[c]

	Product alcohols		K_3CrO_8		Sensitized photooxygenation[a,b]	
Olefin	A	B	%A	%B	%A	%B
	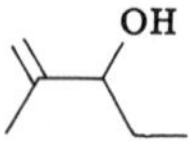 OH	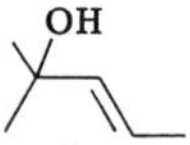 OH	46	53	49	51
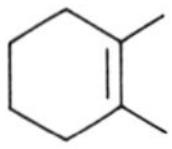	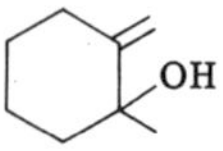 OH	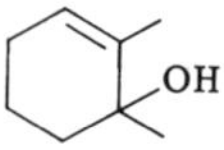 OH	82	18	89	11
	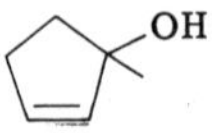 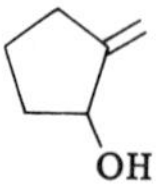 OH , OH	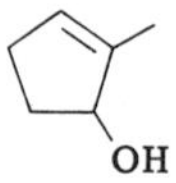 OH	54, 5	41	53, 4	43

[a] Foote, 1968a.
[b] Wasserman *et al.*, 1972.
[c] Reprinted with permission from Peters, J. W. *et al.* (1975). *J. Am. Chem. Soc.* **97,** 3299–3306. Copyright by the American Chemical Society.

latter conditions and two of the same three olefins singlet oxygen product distributions which were nearly the same as those shown in Table V were obtained (Peters *et al.*, 1972, 1975).

The aqueous decomposition of perchromate method was also used to compare relative reactivities of pairs of olefins with the reactivities obtained under dye-sensitized photooxidation conditions (Higgins *et al.*, 1968). As shown in Table VI, the relative reactivities compare quite well providing further evidence for the involvement of singlet oxygen in the perchromate decomposition (Peters *et al.*, 1972, 1975).

Table VI Relative Reactivities of Olefins by Product Analysis[b]

		k_a/k_b	
Olefin A	Olefin B	K_3CrO_8	Sensitized photooxygenation[a]
2-Methyl-2-pentene	1-Methylcyclopentene	2.7	2.0
2,3-Dimethyl-2-butene	2-Methyl-2-butene	41	41
2-Methyl-2-pentene	2-Methyl-2-butene	0.90	1.3

[a] Higgins *et al.*, 1968.
[b] Reprinted with permission from Peters, J. W. *et al.* (1975). *J. Am. Chem. Soc.* **97,** 3299–3306. Copyright by the American Chemical Society.

Spectral evidence was also obtained to support the observations indicating that perchromate decomposition produces singlet oxygen. Simultaneous with the gas evolution which accompanies the mixing of a basic potassium chromate solution with a dilute hydrogen peroxide solution at room temperature, an emission at 1.27 μm could be observed and monitored (Peters *et al.*, 1972, 1975). This emission is due to return of the singlet delta state of oxygen to the ground state as shown in Eq. (38).

$$O_2(^1\Delta_g) \rightarrow O^2(^3\Sigma_g^-) + h\nu(1.27\ \mu m) \quad (38)$$

When water was added to the dry perchromate salt at ~45°C a sudden and strong emission at 1.27 μm was observed which persisted for twenty minutes.

The method of Higgins *et al.* (1968) was also used to obtain an estimate of the yield of singlet oxygen using the perchromate decomposition method. The yield of singlet oxygen, based on potassium perchromate, was estimated to be 6 ± 2%.

The authors (Peters *et al.*, 1972, 1975) also used the well-known test for the involvement of singlet oxygen which is based on the physical quenching of the singlet oxygen by 1,4-diazabicyclo[2.2.2]octane (DABCO) a substrate which does not react chemically with the singlet oxygen (Ouannes and Wilson, 1968). Using 2,3-dimethyl-2-butene as the acceptor it was found, however, that DABCO inhibited the aqueous perchromate oxygenation at only 25% of the predicted efficiency. The reduced efficiency may be due to a different rate constant for the decay of singlet oxygen, k_d, in the solvent used, as well as a possible change in the value of the DABCO singlet oxygen quenching rate, k_q, as a result of solvent change from that used originally.

Because the perchromate decomposition method can be used in aqueous systems, it offered promise as a useful method for studying the photodynamic effect (Spikes and Livingston, 1969). It has been suggested (Foote, 1968b) that the photodynamic effect may involve singlet oxygen since the conditions used (substrate, sensitizing chromophore, O_2) are precisely those used to produce singlet oxygen in dye sensitized oxidations.

A series of purine and pyrimidine bases were oxidized by the perchromate decomposition method and the results compared with those obtained when the same bases are treated with singlet oxygen produced via the microwave discharge method (Rosenthal and Pitts, 1971). The results (Peters *et al.*, 1972, 1975) indicated that the perchromate decomposition method probably involves one or more oxidants, in addition to singlet oxygen, which can destroy the bases. Possibilities for the other oxidants include superoxide ion, perchromate ion, or protonated perchromate ion.

It was also found that oxidation of some olefinic substrates (e.g., 1,2-dimethylcyclopentene) by the perchromate method gave some nonsinglet oxygen products in addition to the expected singlet oxygen products. Thus, when perchromate is used as a singlet oxygen source caution must be exercised in interpreting the results strictly on the basis of singlet oxygen as the sole oxidant.

The use of EPR spectroscopy demonstrated that thermal decomposition of the neat potassium perchromate salt does not produce singlet delta oxygen in any significant quantity (Peters *et al.*, 1972, 1975).

2. *Chromium Pentoxide Etherate*

Chromium pentoxide etherate (**28**) is known to decompose at room temperature to give oxygen and a mixture of chromium salts (Connor and Ebsworth, 1964). It had been reported that stirring a solution of **28** and tetraphenylcyclopentadienone (tetracyclone) in the dark at room temperature for 12 hours led to the production of the endoperoxide (**29**) (Chan, 1970). Since **29** is the expected product of the reaction of tetracyclone and singlet oxygen, it was suggested that it may have been formed by decomposition of **28** to give singlet oxygen followed by reaction of the singlet oxygen with tetracyclone to give **29**. Subsequent to this report, a reinvestigation of the same reaction revealed that the material originally believed to be the endoperoxide is actually the known ketolactone (**30**) (Baldwin *et al.*, 1971). It was suggested that **30** may be formed by reaction of tetracyclone with chromic acid formed by degradation of **28**. At any rate it now appears that there is no evidence that decomposition of **28** gives singlet oxygen.

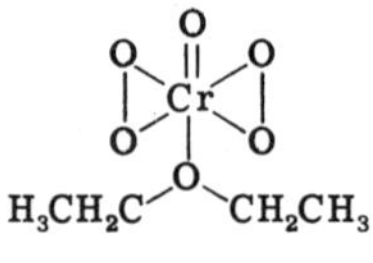

(**28**) 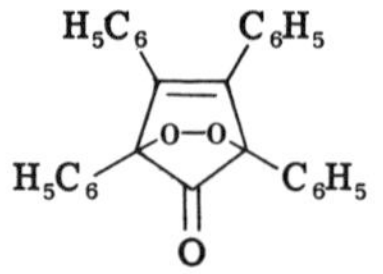 (**29**) 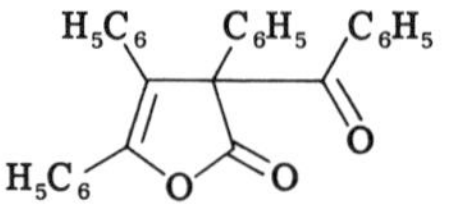 (**30**)

The results of this reinvestigation suggest that the report (Chan, 1971) indicating that singlet oxygen may be involved in oxidations by soybean lipoxidase will have to be reevaluated. This latter work rests heavily on the alleged conversion of tetracyclone to its endoperoxide (**29**). It now appears that the presumed endoperoxide was actually **30**.

3. *Others*

A dioxygen adduct of *meso*-tetraphenylporphyrin (pyridine) manganese(II), Mn(TPP)(py), was originally considered to be the singlet oxygen adduct, Mn(II)(TPP)($^1O_2$). This formulation was later abandoned in favor of a structure in which manganese is symmetrically bonded to a peroxide-like dioxygen fragment (Weschler *et al.*, 1975).

H. The Decomposition of Photoperoxides

It has been known for some time that many polycyclic aromatic photoperoxides will dissociate when heated to regenerate molecular oxygen and the parent hydrocarbon (Bergmann and McLean, 1941). The conservation of spin principle suggests that the molecular oxygen so generated should have singlet multiplicity. Some evidence that singlet oxygen may be involved in these decompositions has been available for some time. Thus Moreau *et al.* (1926) had observed that heating of rubrene endoperoxide, (**31**), regenerates the hydrocarbon quantitatively and gives a chemiluminescence due to singlet oxygen [Eq. (39)].

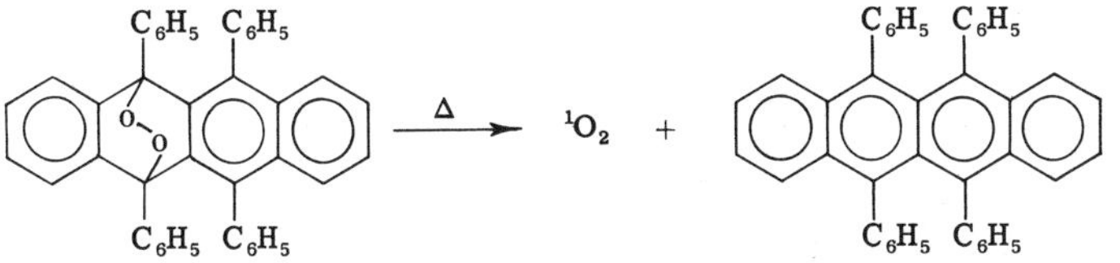

C_6H_5 C_6H_5 ... C_6H_5 C_6H_5 (**31**) $\xrightarrow{\Delta}$ $^1O_2$ + C_6H_5 C_6H_5 ... C_6H_5 C_6H_5 (39)

Wasserman and Scheffer (1967) and Wasserman *et al.* (1972) have shown that a similar decomposition of 9,10-diphenylanthracene peroxide produces singlet oxygen which can be used to oxidize a variety of singlet oxygen acceptors. The acceptors used and the products obtained are shown in Table VII.

The products obtained in this manner are the same as those produced when the same substrates are treated with singlet oxygen formed by other proven sources. The product yields are also comparable to those obtained when using other sources of singlet oxygen.

The reaction is usually carried out by decomposing the peroxide in refluxing benzene solution in the presence of the acceptor. In the general case the ratio of peroxide to acceptor used is 2:1. Other solvents (toluene, chloroform, dimethyl sulfoxide) may also be used.

Evidence that the reaction proceeds via decomposition of the peroxide to give singlet oxygen with subsequent reaction between substrate and singlet oxygen as opposed to a direct bimolecular reaction between substrate and peroxide was obtained by kinetic studies. By studying the oxygenation of tetracyclone at several temperatures, the activation energy for the decomposition of the 9,10-diphenylanthracene peroxide was determined to be 27.8 ± 0.2 kcal/mole.

The authors attempted similar oxidation studies using rubrene endoperoxide and found that this material was much less efficient as an oxygenation reagent than 9,10-diphenylanthracene peroxide. Thus, while the latter compound converts 1,3-diphenylisobenzofuran to *o*-dibenzoylbenzene in 99% yield, rubrene endoperoxide gave only a 36% yield.

Table VII Acceptors Studied and Products Obtained Using Singlet Oxygen from the Decomposition of 9,10-Diphenylanthracene Peroxide[d]

Acceptor	Product	Acceptor	Product
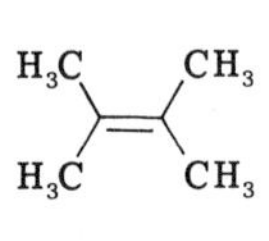	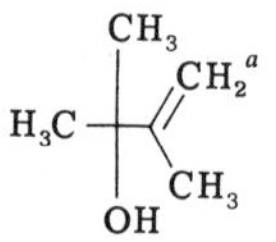	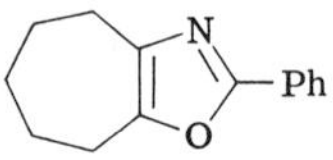	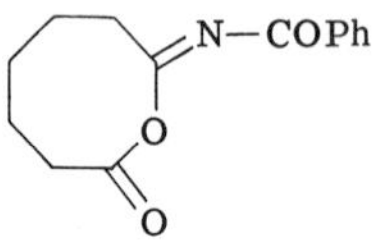
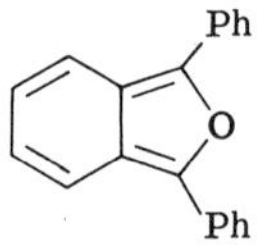	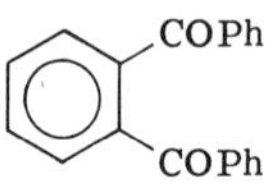	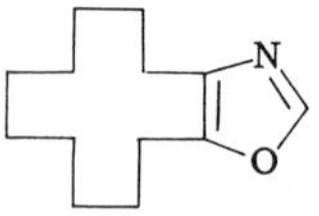	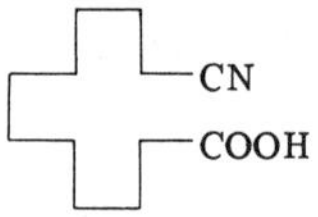
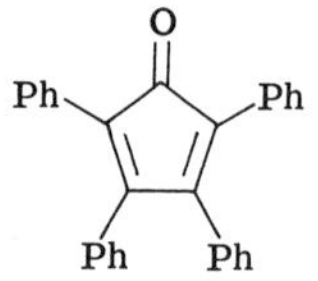	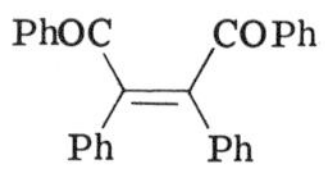	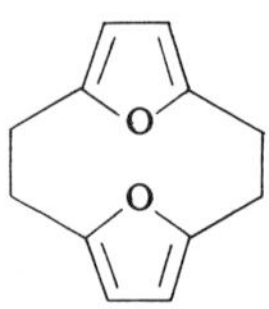	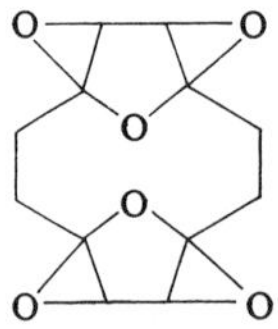
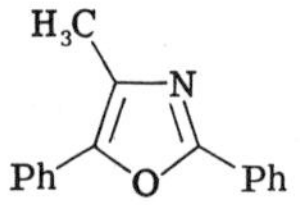	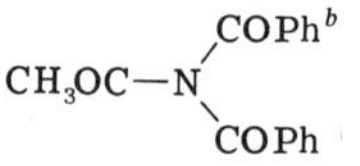	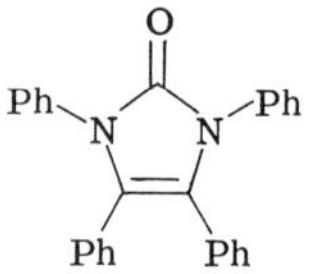	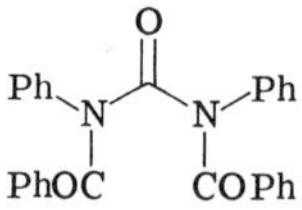
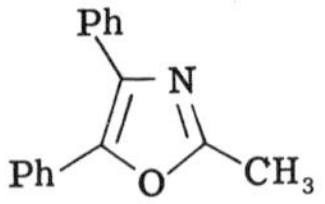	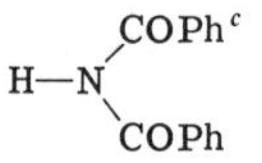	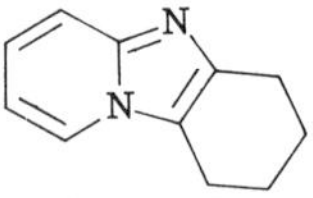	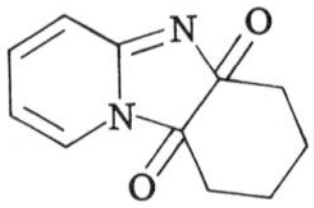

[a] After reduction of initially formed hydroperoxide.

[b] Plus dibenzamide (10%) formed during workup.

[c] Resulting from hydrolysis of *N*-acetyldibenzamide during workup.

[d] Reprinted with permission from Wasserman, H. H. *et al.* (1972). *J. Am. Chem. Soc.* **94,** 4991–4996. Copyright by the American Chemical Society.

Young and Hart (1967) have shown that an enolate anion can be oxygenated using singlet oxygen produced by thermal decomposition of 9,10-diphenylanthracene peroxide.

The decomposition of 9,10-diphenylanthracene endoperoxide as a source of singlet oxygen has been used by Matsuura *et al.* (1969, 1972) in their studies of the reactions of phenols with singlet oxygen. Thus, thermolysis of the endoperoxide in the presence of 2,6-di-*tert*-butylphenol using a nitrogen atmosphere led to the production of 3,5,3′,5′-tetra-*tert*-butyl-4,4′-diphenoquinone. When oxygen was used instead of nitrogen 2,5-di-*tert*-butyl-*p*-benzoquinone was produced in addition to the diphenoquinone.

The authors interpret their results as involving hydrogen abstraction from the phenol by singlet oxygen to give the phenoxy radical. Dimerization of the phenoxy radical, followed by dehydrogenation is suggested as the route to the diphenoquinone. The benzoquinone is postulated as arising from the reaction of the phenoxy radical with ground state oxygen followed by hydroperoxide formation and dehydration.

The decomposition of dibenzal diperoxide has been investigated by Abbott *et al.* (1970) who observed chemiluminescence from a variety of aromatic compounds which were present during the decomposition. These authors identified the emitting species as the lowest excited singlet state of the aromatic fluorescers used. They also identified an emission which may be due to a singlet oxygen collisional pair, $(^1\Delta_g)_2$. A mechanism was proposed which involves energy transfer from singlet oxygen to the aromatic fluorescer which then emits. The required singlet oxygen was seen as having two possible sources, namely, (1) direct formation in the decomposition of the diperoxide, or (2) by quenching of triplet benzaldehyde by ground state oxygen with the triplet benzaldehyde being formed in the peroxide decomposition.

Trozzolo and Fahrenholtz (1970) have described a photoperoxide whose decomposition occurs at room temperature to yield singlet oxygen which can be used to oxygenate typical singlet oxygen substrates. The photoperoxide used is the ozonide, **33**, which is produced when 2,5-diphenylfuran, **32**, is photooxidized at $-70°C$ with benzanthrone sensitization. The ozonide is warmed to room temperature in the presence of the acceptor molecules. Using this technique, 1,3-cyclohexadiene was converted to its endoperoxide in 70% yield, and tetramethylethylene to the allylic hydroperoxide in 50% yield. Decomposition of **33** in the presence of rubrene leads to the bleaching of the orange color of the hydrocarbon which accompanies formation of the colorless endoperoxide.

The possibility that the observed oxidations were the result of a bimolecular reaction between **33** and the substrates was ruled out by a kinetic study which showed that the decomposition is a first-order process [Eq. (40)].

$$\underset{(32)}{H_5C_6\text{-furan-}C_6H_5} \underset{\Delta,\ -O_2}{\overset{h\nu,\ \text{Sens.},\ O_2}{\rightleftharpoons}} \underset{(33)}{H_5C_6\text{-(O–O bridged furan)-}C_6H_5} \qquad (40)$$

I. The Reaction of Ozone with Organic Substrates

1. Phosphites

a. Triphenyl Phosphite. Ozone can react with certain organic substrates in such a way that the oxidized substrate incorporates only one oxygen atom of the three available in ozone. Examples include the oxidation of tertiary amines to amine oxides, phosphines to phosphine oxides, sulfides to sulfoxides, and sulfoxides to sulfones. Also included in this category are the ozonolyses of certain highly hindered terminal olefins where the major products are the corresponding epoxides (Bailey and Lane, 1967).

While investigating the course of some of these ozonizations, Knowles and Thompson (1959; Thompson, 1961) made some important observations concerning the stoichiometry of the reactions. In several cases, but particularly with triaryl phosphites, the stoichiometry of the reaction varied such that one, two, or all three of the oxygen atoms in the ozone could be utilized to produce the product phosphates. These results were correctly interpreted as requiring the formation of an ozone–phosphite adduct which could itself react with substrate phosphite to produce phosphate. In the case of triaryl phosphites, the 3:1 (phosphite: ozone) stoichiometry could be achieved by carrying out the reaction at high phosphite concentration or by first forming the adduct at low temperatures and then adding the same or another phosphite. With trialkyl phosphites, a 1:1 or 2:1, but no 3:1 stoichiometries could be observed. This latter result presumably reflects the difference in stability between the alkyl phosphite and aryl phosphite ozone adducts. In the cases of some phosphines and sulfides, greater than 1:1 stoichiometries could also be observed suggesting the formation of some kind of initial, unstable ozone–substrate product in these cases which is capable of carrying out further oxidations.

Further work was conducted on the nature of the ozone adduct in the triaryl phosphite cases. It was found that the adducts in these cases were stable in solution below about $-20°C$ and decomposed to the corresponding phosphate and molecular oxygen upon warming. The ozone adduct could also oxidize added phosphine, sulfide, or sulfoxide. On the basis of the ^{31}P magnetic resonance spectrum, the adduct was assigned the structure of a four-membered ring trioxide.

A consideration of the conservation of spin principle suggested that the molecular oxygen evolved in the decomposition of these adducts should have singlet multiplicity (Murray and Kaplan, 1968a; Corey and Taylor, 1964). In 1968, evidence was presented that this prediction was correct. Thus, the typical singlet oxygen acceptors, 2,3-dimethyl-2-butene and 1,3-cyclohexadiene were converted to their singlet oxygen products, 2,3-dimethyl-3-hydroperoxy-1-butene and 5,6-dioxabicyclo[2.2.2]-2-octene, respectively (Murray and Kaplan, 1968a). In subsequent work, the additional acceptors, tetraphenylcyclopentadienone, 9,10-diphenylanthracene, and α-terpinene were also smoothly converted to their singlet oxygen products, *cis*-dibenzoylstilbene, 9,10-diphenylanthracene-9,10-endoperoxide, and ascaridole, respectively (Murray and Kaplan, 1969).

A number of observations were made which bear on the question of whether these oxidations involved free singlet oxygen or whether a direct reaction was occurring between the phosphite–ozone adduct and the various substrates. When a 200-fold excess of the adduct is stored with rubrene at $-78°C$, the color of the rubrene is not discharged. However, when the solution is permitted to warm to $-25°C$, the rubrene color begins to fade and then disappears completely with an accompanying evolution of oxygen. The color discharge is presumably associated with the conversion of the bright orange color of the rubrene to its colorless endoperoxide by singlet oxygen.

The possibility of direct reaction between adduct and acceptors was further examined by attempting gas phase reactions between the oxygen evolved from decomposition of the solid adduct and several acceptors. All attempts to transfer the evolved oxygen to an adjacent flask containing solutions of acceptors failed. However, using a suitably designed apparatus, the oxygen evolved from the solid adduct was mixed with acceptor molecules in the gas phase. With the acceptors, 1,3-cyclohexadiene and α-terpinene, small yields ($\sim 0.01\%$, based on available oxygen) of the singlet oxygen products could be obtained (Murray and Kaplan, 1968b). The small yields are presumably due to collisional deactivation of singlet oxygen prior to encountering acceptor molecules.

Using the same apparatus, the oxygen evolved from the decomposition of the solid phosphite–ozone adduct was passed into the cavity of an EPR spectrometer in the gas phase. The characteristic absorptions of singlet delta oxygen were clearly observed (Wasserman *et al.*, 1968).

The combined weight of the evidence available at this point indicated that singlet oxygen was definitely produced in the decomposition of the phosphite–ozone adduct and that it appeared to be responsible for the solution oxygenations. Evidence has been presented subsequently, however, for a direct, bimolecular reaction between the adduct and the acceptor tetramethylethylene (Bartlett and Mendenhall, 1970). These workers showed that the allylic hydroperoxide can be produced from the reaction of the adduct and tetramethylethylene at temperatures considerably lower ($-60°C$) than those ($\sim -25°C$) at which the adduct gives off singlet oxygen at an appreciable rate. Furthermore, it was found that when the adduct is reacted with tetramethylethylene and 2,5-dimethylfuran competitively, the tetramethylene is found to be more reactive. Under conditions where singlet oxygen is very likely the oxidant, the furan is always found to be more reactive. The adduct does not appear to undergo corresponding bimolecular reactions with diene- or anthracene-type acceptors, or any such reactions are extremely slow.

Additional work on the nature of the bimolecular reaction indicated that it did not involve a radical chain process. Schaap and Bartlett (1970) subsequently showed that the triphenyl phosphite–ozone adduct can also perform another singlet oxygen-type oxidation. It had been shown that *cis*- and *trans*-diethoxyethylene, **34** and **35**, are converted stereospecifically to the dioxetanes (**36**) and (**37**) when treated with photochemically produced singlet oxygen (Bartlett and Schaap, 1970) [Eq. (41)]. The phos-

OCH_2CH_3 / OCH_2CH_3 (**34**) $\xrightarrow[\text{Sens}]{-78°C,\ h\nu,\ O_2,}$ H_3CH_2CO, H_3CH_2CO, H, H, O–O (**36**)

OCH_2CH_3 / H_3CH_2CO (**35**) $\xrightarrow[\text{Sens}]{-78°C,\ h\nu,\ O_2,}$ H_3CH_2CO, H, H, H_3CH_2CO, O–O (**37**) (41)

phite–ozone adduct also produces the dioxetanes from **34** and **35**. In this case, however, the reaction is not stereospecific. Both **34** and **35** give the same mixture of dioxetanes containing 19% of **36** and 81% of **37**. The lack of stereospecificity indicated that free singlet oxygen was not the oxidant. The authors proposed a mechanism [Eq. (42)] for the oxidation which incorporates a step permitting rotational equilibrium as suggested by the product distribution.

$(C_6H_5O)_3PO_3$ + OCH_2CH_3 / OCH_2CH_3 ⟶ $(C_6H_5O)_3P(O^-)$–O–O–, H_3CH_2CO, H, H, +, OCH_2CH_3 (42)

↓

37 + **36** ⟵ ^-O–O–, H, OCH_2CH_3, +, H, OCH_2CH_3 + $(C_6H_5O)_3PO$

The reaction also occurs, but at a slower rate, with *p*-dioxene which gives the corresponding dioxetane.

The possibility that the reaction of the phosphite–ozone adduct and ene-type acceptors at low temperatures might still be a singlet oxygen reaction has been considered and ruled out (Bartlett and Mendenhall, 1970). The possibility could exist if the adduct was dissociating and was in equilibrium with the phosphate and singlet oxygen. It was shown, however, that the formation of the adduct is not a reversible process, that is, it is not possible to produce the adduct by irradiating the phosphate in the presence of oxygen. Furthermore, the reaction between the adduct and the ene-type acceptor, 2-methyl-2-pentene, was shown to give the same yield of mixed allylic hydroperoxide products both in the presence and absence of the singlet oxygen quencher β-carotene. Presumably, if singlet oxygen were involved at the lower temperatures, then the β-carotene experiment would have given a lower yield of products (Bartlett *et al.*, 1970).

Attempts to assess the involvement of the direct reaction between the phosphite–ozone adduct and ene-type acceptors have been made by examining product distributions in several cases where the substrate gives more than one product (Murray *et al.*, 1970a). The first two substrates to be used for this purpose were 2-methyl-2-butene and 1,2-dimethyl-1-cyclohexene. As shown in Eq. (43) each of these substrates can give two allylic hydroperoxide products.

$$(CH_3)_2C{=}CHCH_3 + {}^1O_2 \longrightarrow CH_2{=}C(CH_3)CH(OOH)CH_3 + (CH_3)_2C(OOH)CH{=}CH_2$$

$$\text{1,2-dimethylcyclohexene} + {}^1O_2 \longrightarrow \text{2-methylenecyclohexyl}(CH_3)OOH + \text{cyclohex-2-enyl}(CH_3)(CH_3)OOH \qquad (43)$$

As shown in Table VIII for 2-methyl-2-butene oxidations, the distribution of hydroperoxides obtained using the phosphite–ozone adduct is not significantly different from that given by the photooxidation and hypochlorite–hydrogen peroxide methods of producing singlet oxygen. For 1,2-dimethyl-1-cyclohexene, the product distribution is again not significantly different than that obtained under conditions where singlet oxygen is believed to be the oxidant. In addition, the oxidation by the phosphite–ozone adduct in this case gave none of the radical oxidation product (Table IX). Bartlett *et al.* (1970) have also studied the oxidation of 1,2-dimethyl-1-cyclohexene with the phosphite–ozone adduct and again observed a product distribution which is almost the same as that given by singlet oxygen oxidations.

Table VIII Oxidation of 2-Methyl-2-butene Using Various Singlet Oxygen Sources[e]

Product	Photosensitized (%)	$Ca(OCl)_2 + H_2O_2$ (%)	$(\phi O)_3PO_3$ (%)
$(H_3C)_2C(OOH)-C(H)=CH_2$	54[a], 49[b]	52[b]	51[c], 56[d]
$H_2C=C(H_3C)-C(H)(OOH)-CH_3$	46, 51	48	49, 44

[a] Gollnick and Schenk, 1964.
[b] Foote *et al.*, 1965.
[c] CH_2Cl_2 solvent.
[d] CH_3OH solvent.
[e] Reprinted with permission from Murray, R. W. *et al.* (1970). *Ann. N.Y. Acad. Sci.* **171,** 121–129.

Table IX Oxidation of 1,2-Dimethyl-1-cyclohexene under Various Conditions[c]

Product	OCl^-/H_2O_2 (%)	Radical autoxidation (%)	Photooxidation (%)	$(\phi O)_3PO_3$ (%)
Cyclohexane with =CH_2, OOH, CH_3	91[a]	6[a]	89[a], 97[b]	96
Cyclohexene with CH_3, OOH, CH_3	9	39	11, 3	4
Cyclohexene with OOH, CH_3, CH_3	0	54	0	~0

[a] Foote and Wexler, unpublished, cited in Foote, 1968a.
[b] Murray *et al.*, 1970a.
[c] Reprinted with permission from Murray *et al.* (1970). *Ann. N.Y. Acad. Sci.* **171,** 121–129.

These results suggest that involvement of the direct reaction between the phosphite–ozone adduct and these ene-type acceptors must occur in such a way that the direct reaction mimics the singlet oxygen reaction very closely.

The possibility of bimolecular reactions between the phosphite–ozone adduct and olefin acceptors has also been examined by differential thermal analysis (DTA) (Koch, 1970). The DTA thermograms obtained were interpreted as indicating a bimolecular reaction between the adduct and a variety of substrates including both ene-type (2-methyl-2-butene, 2,3-dimethyl-2-butene) and diene-type (1,3-cyclohexadiene and 2,5-dimethylfuran). Activation energies for these bimolecular processes ranged from 7.2–11.1 kcal/mole. DTA was also used to obtain E_a for the unimolecular decomposition of the adduct. Using an adduct concentration of 0.05 moles/liter, E_a for the decomposition was determined to be 12.9 ± 0.7 kcal/mole. This value compares favorably with the one reported earlier (14.1 ± 1.8) by Murray and Kaplan (1969) who determined it by measuring oxygen evolution. Koch concluded, however, that his method did not provide clear evidence for or against the involvement of free singlet oxygen in the bimolecular processes.

The DTA method was also used to examine the reaction between the phosphite–ozone adduct and a variety of other reagents. In many cases reactions were observed with substrates which are unreactive or react only very slowly with singlet oxygen. Thus, the thermograms indicated bimolecular reactions with triphenyl phosphine, amines, hydrazine, and 1,4-diazabicyclo[2.2.2]octane (Koch, 1970).

In the earlier reports describing the use of the phosphite–ozone adduct as an oxygenating agent, the yields of typical singlet oxygen products obtained were in the range of 38–77% (Murray and Kaplan, 1969). In subsequent reports describing the use of the reagent, conditions were described which improved the yields to the 90–100% range (Murray *et al.*, 1970a; Bartlett and Mendenhall, 1970). In the earlier work, the adduct was formed by bubbling ozone into a solution of the triphenyl phosphite. To obtain the improved yields, a solution of the phosphite is added to a saturated solution of ozone at a rate such that ozone is always in excess. Under the latter conditions, reaction between free phosphite and the adduct is minimized giving an optimum yield of the adduct.

b. Mechanism of Decomposition of the Phosphite–Ozone Adduct. Details of the mechanism of decomposition of a variety of phosphite–ozone adducts have been examined further using decomposition kinetics and substituent effects (Stephenson and McClure, 1973). This work is based on an acceptance of the cyclic structure containing a pentacovalent phosphorous which was assigned earlier by Thompson (1961) on the basis of ^{31}P nuclear resonance data. It was further suggested that this cyclic intermediate is in equilibrium with a ring-opened form of the adduct [Eq. (44)]. The ring-

$$(RO)_3P + O_3 \longrightarrow (RO)_3P\langle O_3 \text{ ring} \rangle \rightleftharpoons (RO)_3P^{+}(O)\text{—O—O}^{-} \longrightarrow (RO)_3P{=}O + {}^1O_2 \quad (44)$$

opened intermediate then proceeds to the phosphate and singlet oxygen. Evidence cited to support the rapid preequilibration of open and closed forms of the adduct comes from a consideration of a thermochemical analysis indicating an oxygen-oxygen bond strength of 2 kcal/mole (Benson and Shaw, 1970) as compared to the reported activation energies for decomposition of 13–15 kcal/mole (Murray and Kaplan, 1969; Koch, 1970). By studying the rate of decomposition of a series of phosphite–ozone adducts, Stephenson and McClure came to the conclusion that the decomposition can involve two different mechanisms depending upon the structures of the phosphites. Thus, the process of decomposition is apparently more complex than is suggested by the simple scheme [Eq. (44)]. Based on the kinetic data, the phosphite–ozone adducts studied could be divided into two categories. The first category contains adducts whose phosphite structures indicate the presence of a steric constraint on the decomposition process. The phosphites in this first category lead to adducts whose steric constraints suggest that pseudorotation to other geometries is restricted. Such adducts are then proposed to decompose from a structure of the type represented by **38**, that is, decomposition occurs with simple extrusion of oxygen with no rearrangement about phosphorus. The second group of

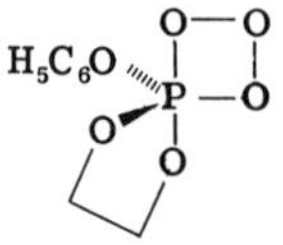

(38)

phosphites all have structures which suggest that a second mechanism is available for the decomposition. In these cases, the adducts are seen as undergoing a pseudorotation prior to loss of oxygen. Such pseudorotation leads to the formation of conformation **39** which requires little structural reorganization to give the phosphate and oxygen. The triphenyl phosphite–ozone adduct is an example of this second group [Eq. (45)].

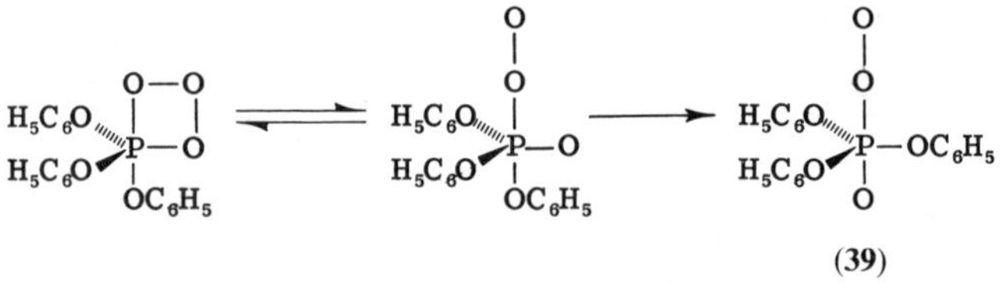

(45)

(39)

Support for this scheme comes from the observation that adducts in the first category have decomposition rates which are insensitive to substituent effects while those in the second category show substituent effects which reflect the affinity of the ester groups for equatorial positions in **39**.

c. Other Phosphites. A number of other phosphite–ozone adducts have been used to carry out oxygenation reactions. The original work of Thompson (1961) had indicated a greater stability of triarylphosphite–ozone adducts over the trialkyl compounds. This observation was confirmed by attempts to use the trialkylphosphite–ozone adducts as oxygenating reagents. Ozonization of triethylphosphite at $-78°C$, followed by addition of 1,3-cyclohexadiene, and warmup failed to give any oxygenation product. When the same procedure was followed at $-95°C$, however, cyclohexadiene endoperoxide could be formed in low yield ($\sim 10\%$) (Murray and Kaplan, 1969).

The bicyclic phosphite, 4-ethyl-2,6,7-trioxa-1-phosphabicyclo[2.2.2]-octane, which had been reported to be instantly oxidized by ozone (Chang, 1964), forms an ozone adduct of considerable stability. This adduct can be used as a preparative oxygenation reagent at temperatures above 0°C (Brennan, 1970). Thus, tetraphenylcyclopentadienone and 9,10-diphenylanthracene were converted to their oxidation products, *cis*-dibenzoylstilbene and 9,10-diphenylanthracene endoperoxide, in yields of 28 and 86%, respectively. Preliminary kinetic studies suggest that the adduct is greater than 100 times more stable than the triphenylphosphite–ozone adduct at 10°C.

The concept that restricted pseudorotation might lead to greater stability in the ozone adducts of bicyclic phosphites has been examined further by Schaap *et al.* (1975). The ozone adduct of 1-phospha-2,8,9-trioxaadamantane was found to give singlet oxygen quantitatively at 18°C. The decomposition of this adduct was found to have $E_a = 19.1 \pm 1.2$ kcal/mole indicating that it is about 106 times more stable than the triphenyl phosphite–ozone adduct at $-5°$. This new phosphite–ozone adduct was used to carry out all three types of singlet oxygen reactions (Table X). Good yields were obtained in all cases. When 1,2-dimethyl-1-cyclohexene is used as the substrate, the distribution of allylic hydroperoxides obtained is the same as that observed under singlet oxygen conditions. This bicyclic phosphite–ozone adduct is also water soluble and should permit studies of singlet oxygen reactions in aqueous media.

d. Further Uses of Phosphite–Ozone Adducts for Oxygenation. The ozone adducts of phosphites are convenient sources of singlet oxygen which can be used under extremely wild conditions and give excellent yields of the oxygenation products. Some further examples of the use of these reagents are given here.

Because use of the triphenylphosphite–ozone adduct permits generation of singlet oxygen without the use of light, it offers a potential advantage in studying the photodynamic effect or photodynamic action. These oxy-

Table X Oxidations with 1-Phospha-2,8,9-trioxaadamantane Ozonide[e]

Singlet oxygen acceptor	Products	Concentration (molar)		% Yield (isolated)[a,b]
		Acceptor	Ozonide	
	OOH	0.11	0.11	52
		0.11	0.22	77
O, O, C_6H_5, C_6H_5	O, COC_6H_5, O, COC_6H_5	0.11	0.22	82
	O, O	0.11	0.11	56
		0.11	0.33	89
		0.55	0.11	95[c]
	OOH [d]			95
		0.11	0.11	
	OOH [d]			5

[a] Products were identified by comparison with authentic samples.

[b] The isolated yields are based on starting alkene.

[c] Yield based on ozonide.

[d] Products from this reaction were analyzed by gas chromatography as the alcohols obtained by triphenylphosphine reduction of the hydroperoxides.

[e] Reprinted with permission from Schaap, A. P. *et al.* (1975). *J. Org. Chem.* **40,** 1185–1186. Copyright by the American Chemical Society.

genations can be carried out while avoiding further photochemical transformation of products.

The possibility that the disulfide bonds of cystine residues might be susceptible to the photodynamic effect was examined by using a series of dialkyl disulfides as models for the cystine residue (Murray *et al.*, 1971a). The disulfides were oxidized by the triphenyl phosphite–ozone adduct to mixtures of the corresponding thiolsulfinates and thiolsulfonates [Eq. (46)].

$$\mathrm{RSSR} + (C_6H_5O)_3P\langle O_3 \rangle \longrightarrow \mathrm{RS(=O)SR} + \mathrm{RS(=O)_2SR} \tag{46}$$

Using low temperature NMR spectroscopy, it was also shown that dimethyl disulfide can be converted to the corresponding thiolsulfinate at $-52°C$, that is, at a temperature well below that where the adduct has an appreciable rate of decomposition in the absence of acceptor. Thus, the reaction may involve singlet oxygen or a bimolecular reaction between the adduct and disulfide. The possibility that the disulfide might be inducing the decomposition of the adduct at the lower temperatures was also examined. When a solution of dimethyl disulfide was added to the adduct at $-78°C$, then 11% of the available oxygen in the adduct was evolved. In the absence of substrates, there is no observable evolution of oxygen from the adduct at this temperature. When they were subjected to photosensitized oxidation, the dialkyl disulfide substrates were also converted to the same products obtained by using the phosphite–ozone adduct.

An interesting example of the use of a phosphite–ozone adduct to accomplish dioxetane formation has been reported by Wieringa *et al.* (1972). While dioxetane formation is usually restricted to olefins without allylic hydrogens because of the availability of the ene reaction, it was found that adamantylidene adamantane is readily converted to the corresponding dioxetane in quantitative yield by using either photosensitized oxidation or the triphenylphosphite–ozone adduct. While the substrate used formerly bears allylic hydrogens, the adamantane structure makes these hydrogens unavailable for the ene reaction.

The observation of chemiluminescence when ketenes are reacted with the triphenylphosphite–ozone adduct in the presence of fluorescers has been interpreted as involving singlet oxygen oxidation of the substrates to dioxetanes which subsequently decompose (Bollyky, 1970). When a solution of the adduct was mixed with a solution of ketene or diphenylketene in the presence of 9,10-bis(phenylethynyl)anthracene or 9,10-diphenylanthracene and the mixed solution allowed to warm to room temperature, then a bright light emission was observed. When subjected to the same conditions diphenylketene glycol acetal gave only a weak emission and tetraphenylethylene was unreactive. The chemiluminescence observed is that of the added fluorescer (Flr) which is believed to receive its excitation via complex formation with intermediate dioxetane. Decomposition of the complex is postulated to lead to the corresponding carbonyl compounds and excited fluorescer [Eq. (47)].

$$R_2C{=}C{=}O + {}^1O_2 \longrightarrow \underset{R_2C-C=O}{\overset{O-O}{|\quad\;|}}$$

$$\underset{R_2C-C=O}{\overset{O-O}{|\quad\;|}} + Flr \longrightarrow [\text{Complex}] \tag{47}$$

$$[\text{Complex}] \longrightarrow Flr^1 + R_2C{=}O + CO_2$$

$$Flr^1 \longrightarrow h\nu + Flr$$

An intermediate dioxetane has also been postulated in the reaction between the tetrathioethylene, (**40**), and singlet oxygen from the triphenylphosphite–ozone adduct. The reaction gives the dithiooxalate, (**41**), and disulfide, (**42**), presumably from decomposition of the intermediate dioxetane (Adam and Lin, 1972) [Eq. (48)].

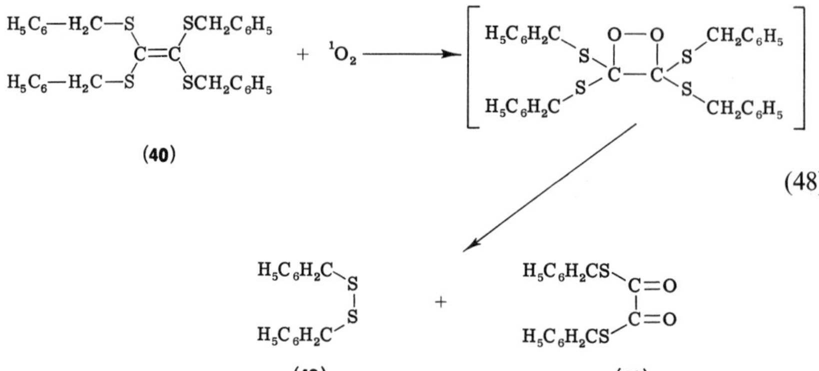

$$(H_5C_6H_2CS)_2C{=}C(SCH_2C_6H_5)_2 + {}^1O_2 \longrightarrow \left[(H_5C_6H_2CS)_2\underset{|}{C}\overset{O-O}{}\underset{|}{C}(SCH_2C_6H_5)_2 \right] \longrightarrow H_5C_6H_2C{-}S{-}S{-}CH_2C_6H_5 + H_5C_6H_2CS{-}C({=}O){-}C({=}O){-}SCH_2C_6H_5 \quad (48)$$

(**40**) (**42**) (**41**)

Observation of an ene-type reaction which probably involves a bimolecular reaction of the substrate with the triphenylphosphite–ozone adduct has been reported (Sam and Sutherland, 1972). Germacrene, (**43**), was found to react with the adduct at $-45°$ to give, after lithium aluminum hydride reduction, the alcohol, (**44**).

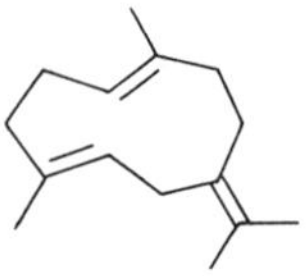

OH

(**43**) 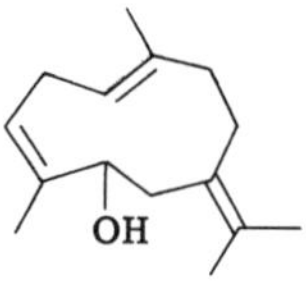 (**44**)

The temperature used for the reaction ($-70°$) suggests that a bimolecular process may be involved. In addition, photosensitized oxidation of (**43**) gives, after reduction, two different alcohols, (**45**) and (**46**). Thus, free

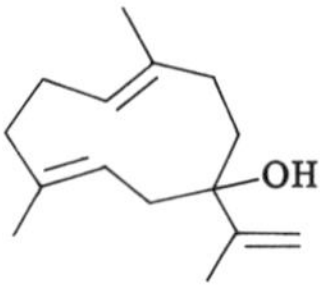

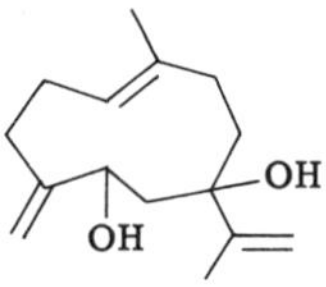

OH OH OH (49)

(**45**) (**46**)

singlet oxygen attacks the methyl groups while the adduct prefers to react at an endocyclic allylic methylene group.

The triphenylphosphite–ozone adduct was used in the very interesting synthesis of *trans*-benzene trioxide. The reaction involves the formation of the endoperoxide, (**47**), from oxepin, (**48**), and the adduct. The endoperoxide is subsequently thermolyzed at 45° to give the *trans*-trioxide, (**49**) (Foster and Berchtold, 1972).

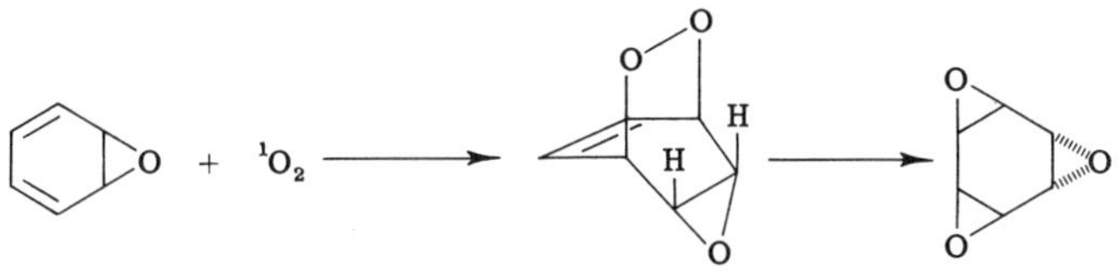

(**48**) + $^1O_2$ ⟶ (**47**) ⟶ (**49**)

The observation of quinone-type products in the photosensitized oxidation of certain phenols have prompted Matsuura and co-workers (1969, 1970) to suggest that singlet oxygen may be able to carry out a hydrogen abstraction reaction. In order to implicate singlet oxygen in these reactions, the substrates have been treated with chemical sources of singlet oxygen. Treatment of 2,6,-di-*tert*-butyl phenol with the triphenyl phosphite–ozone adduct leads to the formation of 2,6-di-*tert*-butyl-*p*-benzoquinone and 3,5,3′,5′-tetra-*tert*-butyl-4,4′-diphenoquinone. These products are the same as those found in the photosensitized oxidation and are postulated as arising from the corresponding phenoxy radical which is formed from the phenol in a hydrogen atom abstraction reaction by singlet oxygen.

Deuterium isotope effects have been used to examine in detail the mechanism involved in the conversion of 2,3-dimethyl-2-butene to its corresponding allylic hydroperoxide (Kopecky and van de Sande, 1972). In particular, an attempt was made to determine whether this conversion by various methods, including oxidation with the triphenylphosphite–ozone adduct, involved a dioxetane or peroxide intermediate. The partially deuterated isomers of 2,3-dimethyl-2-butene, 1,1,1-trideuterio-2-trideuteriomethyl-3-methyl-2-butene (**50**), and 2,3-bis(trideuteriomethyl)-2-butene, (**51**), were prepared and used to determine deuterium isotope effects in the hydroperoxide-forming reaction. When the phosphite–ozone adduct was used at $-28°C$, the isotope effect found for oxidation of both **50** and **51** was not significantly different that that observed in the oxidations of these compounds by established singlet oxygen conditions ($k_H/k_D = 1.45$ for **50** and $= 1.27$ for **51**). When the reaction with the adduct was carried

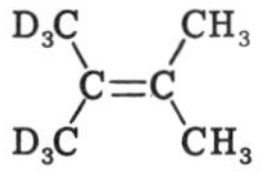

$(D_3C)_2C{=}C(CH_3)_2$ (**50**)

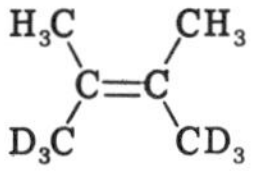

$(H_3C)(D_3C)C{=}C(CH_3)(CD_3)$ (**51**)

out at $-70°C$ the k_H/k_D observed for **50** was found to be 1.60 ± 0.04 and 1.65 ± 0.02 in duplicate runs. The authors felt that the difference between this value and the singlet oxygen value was significant. Oxidation of **51** at $-70°C$, however, gave an isotope effect which was essentially the same as that observed at $-28°C$. The authors conclude that at $-70°C$ the phosphite–ozone adduct oxidations of **50** and **51** do not proceed by way of a dioxetane or perepoxide intermediate or by way of singlet oxygen. The latter conclusion regarding singlet oxygen does not seem to be strongly supported by the isotope effect data, however.

2. *Other Substrates*

The successful use of phosphite–ozone adducts as oxygenation reagents has prompted the search for other cases where ozone might react with an organic substrate to give an oxygen-rich species which could decompose to give singlet oxygen. A number of classes of compounds react with ozone such that only one oxygen atom of the ozone is incorporated in the oxidized substrate. Included in this group are sulfides, phosphines, sulfoxides, tertiary amines, and certain highly hindered olefins. In their original work with phosphites Knowles and Thompson (1959, Thompson, 1961) did obtain evidence for a greater than 1:1 stoichiometry when phosphines and sulfides are ozonized suggesting the formation of an unstable ozone–substrate adduct. To date, however, none of these substrates has been shown to give a stable enough adduct with ozone to be useful as an oxygenation reagent.

Another compound which may have some potential as a source of singlet oxygen is the adduct between ozone and 9,10-dimethylanthracene. This adduct had been shown to have a trioxide structure (**52**). Furthermore, this material was observed to decompose at room temperature to give the endoperoxide **53** (Erickson *et al.*, 1962). The decomposition of the trioxide,

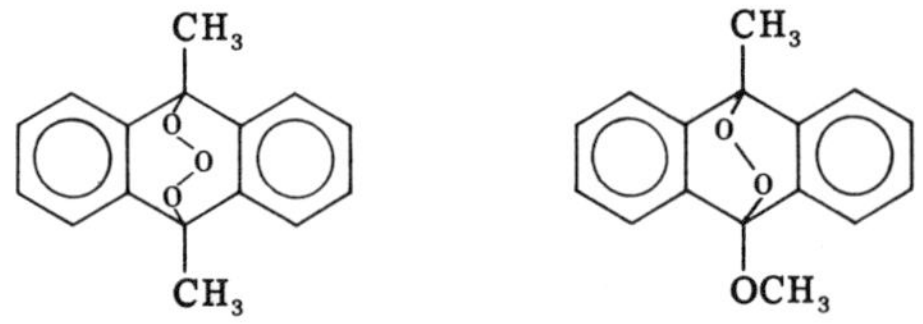

(52) (53)

(**52**), was reinvestigated in hopes that a decomposition pathway producing singlet oxygen might be uncovered. Decomposition of **52** in the presence of tetracyclone led to some bleaching of the purple tetracyclone color. Also, decomposition of the trioxide in the presence of 1,3-diphenylisobenzofuran led to the production of 1,2-dibenzoylbenzene which is the product expected from singlet oxygen oxidation of the furan (R. W. Murray, R. Lamberg, and S. Wassilak, unpublished results, 1973). Further work is needed before **52** can be regarded as a singlet oxygen source, however.

Another large group of organic compounds whose reactions with ozone have the potential of being singlet oxygen sources includes the following: hydrocarbons, silanes, ethers, alcohols, amines, aldehydes, and diazo compounds. In each of these classes of compounds the ozonization reaction has been postulated as involving a hydrotrioxide as an intermediate or unstable product. In many cases the final products of these ozonizations are seen as arising from decomposition of a hydrotrioxide with concomitant release of molecular oxygen.

Two early cases in which some evidence has been obtained that singlet oxygen is produced in the decomposition of an oxygen-rich intermediate, presumably a hydrotrioxide, are those of isopropyl alcohol and isopropyl ether (Murray *et al.*, 1970b).

When isopropyl ether was ozonized at low temperature and allowed to warm up in the presence of the highly colored singlet oxygen acceptors, rubrene and tetracyclone, the color of the substrates was immediately discharged. Color discharge and oxygen evolution were most evident at approximately $-10°C$. In the case of rubrene, the endoperoxide was isolated in 47% yield based on acceptor concentration. For tetracyclone, the singlet oxygen product, *cis*-dibenzoylstilbene, was isolated in 38% yield. Quantitative assays of bleaching of tetracyclone were made spectrophotometrically by measuring decreases in the absorbance maximum at 510 nm. Use of this technique indicated that for isopropyl ether 21.4% of absorbed ozone was available to bleach tetracyclone presumably as singlet oxygen. In the isopropyl alcohol case 3.13% of the absorbed ozone was available for bleaching. While tetracyclone is not one of the most reactive singlet oxygen acceptors the results of the bleaching experiments suggest that pathways other than those giving singlet oxygen may be available in the ozonization of these substrates. Nevertheless the tetracyclone bleaching technique has been used to gain a rough estimate of the potential for singlet oxygen availability in the ozonization of a wide variety of organic substrates (R. W. Murray and R. D. Smetana, unpublished results, 1971). Those substrates indicating the most promise were then examined further using additional diagnostic tests for singlet oxygen. (In all cases substrate ozonization was followed by a nitrogen purge prior to the addition of the singlet oxygen acceptors and warmup.)

When ozonized isopropyl ether was allowed to warm up in the presence of 2,3-dimethyl-2-butene, the singlet oxygen product, 2,3-dimethyl-3-hydroperoxy-1-butene, was obtained in 61% yield.

Oxidation of 2-methyl-2-butene with ozonized isopropyl ether gave, after reduction, the alcohols, 3-methyl-1-buten-3-ol and 2-methyl-1-buten-

3-ol in a ratio of 57:43. This ratio compares favorably with that of 52:48 which was obtained under typical singlet oxygen conditions (Foote *et al.*, 1965).

Products selectivity studies using ozonized isopropyl ether were also carried out on $^{3}\Delta$-carene (Murray *et al.*, 1971b). This substrate gives three allylic hydroperoxides in the ene reaction under typical singlet oxygen conditions. The products of the ozonized isopropyl ether oxidation of Δ^{3}-carene were first reduced to the corresponding alcohols and then analyzed by gas chromatography. Unfortunately, one of the products is a tertiary alcohol which undergoes some decomposition during the gas chromatographic analysis. This difficulty was partially overcome by isolating the pure tertiary alcohol and determining its susceptibility to decomposition during the analysis. The ozonized isopropyl ether reagent gave all three of the expected alcohol products, but their distribution, even after allowing for some decomposition of the tertiary alcohol, was somewhat different that that given under singlet oxygen conditions.

Control experiments have been run to show that other possible oxidants present in the system are not responsible for the observed reactions. Storage of ozonized isopropyl ether with tetracyclone for 24 hours at $-60°$ to $-80°$C led to no bleaching of the substrate. Warmup to 0°C led to complete bleaching of the substrate and evolution of oxygen, however. Also a fivefold excess of *tert*-butyl hydroperoxide did not bleach the tetracyclone after 15 min at room temperature. Similarly air and hydrogen peroxide did not bleach tetracyclone at room temperature. The results suggest the formation of an oxygen-rich intermediate at low temperature which can decompose upon warmup to give singlet oxygen. The overall efficiency of singlet oxygen production does suggest that other reactions are also occurring, however.

Experiments were run to determine whether the oxygen-rich intermediate could be a complex between ozone and the ether. When the oxygen-rich intermediate is formed, the reaction mixture purged with nitrogen, and warmup permitted in the presence of tetramethylethylene or diisopropylethylene, the typical ozonolysis products are not formed.

Subsequent work in the isopropyl ether case as well as several others indicated strongly that the oxygen-rich intermediate is a hydrotrioxide (Stary *et al.*, 1974, 1976). Low temperature NMR analysis of ozonized isopropyl ether showed the presence of a greatly deshielded proton ($\delta \approx 13.0$) which absorption decays upon warmup. Similar absorptions were found in ozonized benzaldehyde, 2-methyltetrahydrofuran and methyl isopropyl ether. In all cases this absorption disappears upon warmup. In most cases it is replaced by absorptions due to compounds presumably arising from decomposition of the intramolecularly hydrogen-bonded hydrotrioxides, (**54**), (**55**), and (**56**).

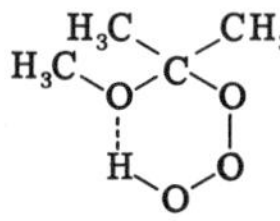

(54) 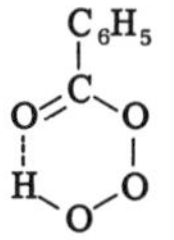 (55) 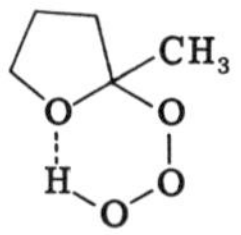 (56)

In each of these additional cases the warmup is accompanied by vigorous evolution of oxygen as was seen in the isopropyl ether case. Low temperature NMR was used to show that the absorptions assigned to the hydrotrioxides decomposed by a first-order kinetics process. In the cases of benzaldehyde, 2-methyltetrahydrofuran, and methyl isopropyl ether, the kinetics were examined over a series of temperatures to obtain the activation energies for decomposition of the hydrotrioxides (Table XI). These ozonized substrates were also used to oxidize the singlet oxygen acceptor, 1,3-diphenylisobenzofuran. The singlet oxygen product, *o*-dibenzoylbenzene, was obtained in 96, 80, and 30% yield from ozonized benzaldehyde, 2-methyltetrahydrofuran, and methyl isopropyl ether, respectively. For two of the substrates, benzaldehyde and 2-methyltetrahydrofuran, addition of the singlet oxygen acceptor, 1,2-dimethylcyclohexene, followed by warmup, led to a product distribution (Table XII) which is characteristic of singlet oxygen.

Table XI Summary of Activation Energy Data for Hydrotrioxides[c]

Hydrotrioxide	Solvent	E_a (kcal/mole)	Log A
	2-MeTHF	8.04 ± 0.2	4.1 ± 0.3
	2-MeTHF + Et_2O[a]	17.4 ± 0.4	11.7 ± 0.4
	C_6H_5CHO	10.7 ± 0.2	5.4 ± 0.4
	C_6H_5CHO + Et_2O[b]	16.4 ± 0.7	11.1 ± 0.6
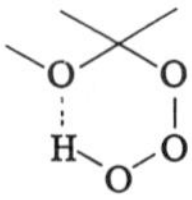	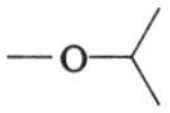	16.6 ± 0.6	11.5 ± 0.5

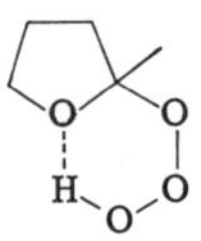

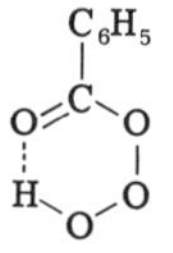

[a] 2-MeTHF, 15.4%; Et_2O, 84.6% (by weight).
[b] C_6H_5CHO, 21.3%; Et_2O, 78.7% (by weight).
[c] Reprinted with permission from Stary, F. E., *et al.* (1976). *J. Am. Chem. Soc.* **98,** 1880–1884. Copyright by the American Chemical Society.

Table XII Product Distributions in the Oxidation of 1,2-Dimethylcyclohexene[c]

	Method			
			Hydrotrioxide decomposition	
Product	OCl^{-}–H_2O_2 (%)	Photooxidation (%)	PhCHO (%)	2-MeTHF (%)
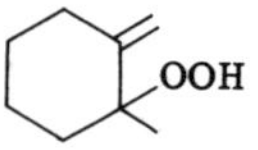 (2-methylene-1-methylcyclohexyl hydroperoxide) OOH	91[a]	89[a]	90[b]	92[b]
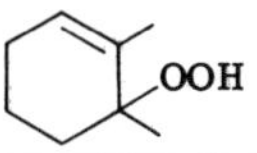 (1,2-dimethylcyclohex-2-enyl hydroperoxide) OOH	9[a]	11[a]	10[b]	8[b]

[a] Foote, 1968a.
[b] Stary *et al.*, 1976.
[c] Reprinted with permission from Stary, F. E., *et al.* (1976). *J. Am. Chem. Soc.* **98,** 1880–1884. Copyright by the American Chemical Society.

The combined evidence indicates that these substrates are ozonized to hydrotrioxides which decompose in the vicinity of 10°C to give singlet oxygen.

J. Other Sources

McKeown and Waters (1964) have reported chemiluminescence, attributed to singlet oxygen, when each of the following was treated with hydrogen peroxide: chloramine T, benzyl cyanide, acrylonitrile, and hydroxylamine. Attempts to observe chemical oxygenation using these systems have not been successful, however (McKeown and Waters, 1966).

Chemiluminescence accompanying the Trautz reaction has also been attributed to singlet oxygen (Bowen, 1964). When formaldehyde is oxidized by alkaline hydrogen peroxide in the presence of pyrogallol a bright orange-red chemiluminescence is observable, but these same conditions did not lead to chemical oxygenation (McKeown and Waters, 1966). When resorcinol was substituted for pyrogallol in the same reaction, a 5% yield of endoperoxide could be obtained from 9-methyl-10-phenylanthracene (McKeown and Waters, 1966).

It has been suggested (Smith and Kulig, 1975) that base-catalyzed disproportionation of hydrogen peroxide directly leads to the production of singlet oxygen. When cholesterol is treated with hydrogen peroxide in an aqueous sodium stearate dispersion, a variety of products are obtained. Some of the products are due to oxidation by hydrogen peroxide and ground state oxygen but a group of products is obtained which appear to be derived from 5α-hydroperoxycholest-6-en-3β-ol. While this latter compound was not isolated, its unique decomposition products were. The 5α-hydroperoxide is the expected allylic hydroperoxide from a singlet oxygen ene reaction of cholesterol. The total distribution of products suggests that disproportionation of hydrogen peroxide gives a ratio of $^1O_2/^3O_2$ of 1:3.

These findings may have a bearing on the reported formation of singlet oxygen in the dismutation of superoxide anion (Khan, 1970; Mayeda and Bard, 1974). Since the dismutation involves production of hydrogen peroxide, the role of the direct production of singlet oxygen in the base-catalyzed decomposition of hydrogen peroxide in the superoxide system needs to be considered.

Based on the observation of chemiluminescence due to excited oxygen, Stauff (1965) has indicated that a number of inorganic reactions may produce singlet oxygen. Treatment of Ce(IV) or Ti(III) ions with hydrogen peroxide, for example, leads to the referenced chemiluminescence. Stauff has suggested that singlet oxygen is produced in these systems via the intermediacy of hydroxy or hydroperoxy radicals. It is proposed, for example, that recombination of two hydrotrioxy radicals leads to hydrogen peroxide and an excited oxygen "associate" [Eq. (50)].

$$2 \cdot O_3H \rightarrow (O_2O_2)^* + H_2O_2 \tag{50}$$

Bowen and Lloyd (1963) have detected a chemiluminescence in the thermal decomposition of a number of organic hydroperoxides. It was suggested that this luminescence is probably due to oxygen molecules in the $^1\Sigma_g{}^+$ state which emit during transition to the ground state.

Turro *et al.* (1976) have suggested that the dark oxidation reaction of cyclic acetylene **57** by molecular oxygen involves the formation of an intermediate oxygen–acetylene complex which may decompose to produce singlet oxygen. If such a process occurs it would represent a catalytic thermal generation of $^1\Delta$ oxygen from $^3\Sigma$ oxygen.

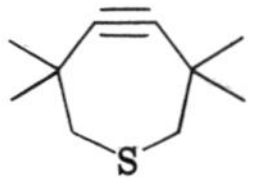

S

(57)

ACKNOWLEDGMENTS

The author wishes to acknowledge the efforts of his co-workers, whose names appear in the text, for their contributions to the work from the author's laboratory described in Section II,I. We are also grateful to the National Science Foundation for support of this work.

REFERENCES

Abbott, S. R., Ness, S., and Hercules, D. M. (1970). *J. Am. Chem. Soc.* **92,** 1128–1136.
Adam, W., and Lin, J.-C. (1972). *Chem. Commun.* pp. 73–74.
Akiba, K., and Simamura, O. (1964). *Chem. Ind.* (*London*) pp. 705–706.
Arneson, R. M. (1970). *Arch. Biochem. Biophys.* **136,** 352–360.
Arnold, J. S., Ogryzlo, E. A., and Witzke, H. (1964). *J. Chem. Phys.* **40,** 1769–1770.
Arnold, J. S., Browne, R. J., and Ogryzlo, E. A. (1965). *Photochem. Photobiol.* **4,** 963–969.
Bailey, P. S., and Lane, A. G. (1967). *J. Am. Chem. Soc.* **89,** 4473–4479.
Baldwin, J. E., Swallow, J. C., and Chan, H. W.-S. (1971). *Chem. Commun.* pp. 1407–1408.
Balny, C., and Douzou, P. (1970). *J. Chim. Phys.* **67,** 635–636.
Bartlett, P. D., and Mendenhall, G. D. (1970). *J. Am. Chem. Soc.* **92,** 210–211.
Bartlett, P. D., and Schaap, A. P. (1970). *J. Am. Chem. Soc.* **92,** 3223–3225.
Bartlett, P. D., Mendenhall, G. D., and Schaap, A. P. (1970). *Ann. N.Y. Acad. Sci.* **171,** 79–88.
Bellin, J. S., and Mahoney, R. D. (1972). *J. Am. Chem. Soc.* **94,** 7991–7995.
Benson, S. W., and Shaw, R. (1970). *Org. Peroxides* **1,** 105–139.
Bergman, W., and McLean, M. J. (1941). *Chem. Rev.* **28,** 367–394.
Berson, J. A., Wall, R. G., and Perlmutter, H. D. (1966). *J. Am. Chem. Soc.* **88,** 187–188.
Bland, J., and Craney, B. (1974). *Tetrahedron Lett.* pp. 4041–4044.
Bollyky, L. J. (1970). *J. Am. Chem. Soc.* **92,** 3230–3232.
Bowen E. J., and Lloyd, R. A. (1963). *Proc. R Soc. London, Ser. A* **275,** 465–468.
Bowen, E. J. (1964). *Pure Appl. Chem.* **9,** 473–479.
Boyer, R. F., Lindstrom, C. G., Darby, B., and Hylarides, M. (1975). *Tetrahedron Lett.* pp. 4111–4114.
Brennan, M. E. (1970). *Chem. Commun.* pp. 956–957.
Browne, R. J., and Ogryzlo, E. A. (1964). *Proc. Chem. Soc., London* p. 117.
Browne, R. J., and Ogryzlo, E. A. (1965). *Can. J. Chem.* **43,** 2915–2916.
Cahill, A. E., and Taube, H. (1952). *J. Am. Chem. Soc.* **74,** 2312–2318.
Chan, H. W.-S. (1970). *Chem. Commun.* pp. 1550–1551.
Chan, H. W.-S. (1971). *J. Am. Chem. Soc.* **93,** 2357–2358.
Chang, W.-H. (1964). *J. Org. Chem.* **29,** 3711–3712.
Connor, J. A., and Ebsworth, E. A. V. (1964). *Adv. Inorg. Chem. Radiochem.* **6,** 279–381.
Corey, E. J., and Taylor, W. C. (1964). *J. Am. Chem. Soc.* **86,** 3881–3882.
Erickson, R. E., Bailey, P. S., and Davis, J. C., Jr. (1962). *Tetrahedron* **18,** 388–395.
Evans, N. A. (1971). *Aust. J. Chem.* **24,** 1971–1973.
Foote, C. S. (1968a). *Acc. Chem. Res.* **1,** 104–110.
Foote, C. S. (1968b). *Science*, *162*, pp. 963–970.
Foote, C. S. (1971). *Pure Appl. Chem.* **27,** 635–645.
Foote, C. S., and Brenner, M. (1968). *Tetrahedron Lett.* pp. 6041–6044.
Foote, C. S., and Denny, R. W. (1968). *J. Am. Chem. Soc.* **90,** 6233–6235.
Foote, C. S., and Denny, R. W. (1971a). *J. Am. Chem. Soc.* **93,** 5162–5167.
Foote, C. S., and Denny, R. W. (1971b). *J. Am. Chem. Soc.* **93,** 5168–5171.
Foote, C. S., and Wexler, S. (1964a). *J. Am. Chem. Soc.* **86,** 3879–3880.
Foote, C. S., and Wexler, S. (1964b). *J. Am. Chem. Soc.* **86,** 3880–3881.
Foote, C. S., Wexler, S., and Ando, W. (1965). *Tetrahedron Lett.* pp. 4111–4118.
Foote, C. S., Wexler, S., Ando, W., and Higgins, R. (1968). *J. Am. Chem. Soc.* **90,** 975–981.
Foote, C. S., Chang, Y. C., and Denny, R. W. (1970a). *J. Am. Chem. Soc.* **92,** 5218–5219.
Foote, C. S., Denny, R. W., Weaver, L., Chang, Y., and Peters, J. (1970b). *Ann. N.Y. Acad. Sci.* **171,** 139–148.
Foster, C. H., and Berchtold, G. A. (1972). *J. Am. Chem. Soc.* **94,** 7939.

Fridovich, I. (1970). *J. Biol. Chem.* **245,** 4053–4057.
Goda, K., Chu, J.-W., Kimura, T., and Schaap, A. P. (1973). *Biochem. Biophys. Res. Commun.* **52,** 1300–1306.
Gollnick, K., and Schenck, G. O. (1964). *Pure Appl. Chem.* **9,** 507–525.
Groh, P., and Kirrmann, K. A. (1942). *C.R. Hebd. Seances Acad. Sci., Ser. C* **215,** 275–276.
Guiraud, H. J., and Foote, C. S. (1976). *J. Am. Chem. Soc.* **98,** 1984–1986.
Hallett, F. R., Hallett, B. P., and Snipes, W. (1970). *Biophys. J.* **10,** 305–315.
Higgins, R., Foote, C. S., and Cheng, H. (1968). *Adv. Chem. Ser.* **77,** 102–117.
Howard, J. A., and Ingold, K. U. (1968). *J. Am. Chem. Soc.* **90,** 1056–1058.
Kajiwara, T., and Kearns, D. R. (1973). *J. Am. Chem. Soc.* **95,** 5886–5890.
Kasha, M., and Khan, A. U. (1970). *Ann. N.Y. Acad. Sci.* **171,** 5–23.
Kautsky, H. (1933). *Chem. Ber.* **66,** 1588–1600.
Kautsky, H. (1937). *Biochem. Z.* **291,** 271–284.
Kearns, D. R. (1971). *Chem. Rev.* **71,** 395–427.
Kellogg, R. E. (1969). *J. Am. Chem. Soc.* **91,** 5433–5436.
Khan, A. U. (1970). *Science* **168,** 476–477.
Khan, A. U. (1977). *J. Am. Chem. Soc.* **99,** 370–371.
Khan, A. U., and Kasha, M. (1963). *J. Chem. Phys.* **39,** 2105–2106.
Khan, A. U., and Kasha, M. (1966). *J. Am. Chem. Soc.* **88,** 1574–1576.
Khan, A. U., and Kasha, M. (1970). *J. Am. Chem. Soc.* **92,** 3293–3300.
Knowles, W. S., and Thompson, Q. E. (1959). *Chem. Ind.* (*London*) p. 121.
Koch, E. (1970). *Tetrahedron* **26,** 3503–3519.
Kopecky, K. R., and van de Sande, J. H. (1972). *Can. J. Chem.* **50,** 4034–4049.
McKeown, E., and Waters, W. A. (1964). *Nature* (*London*) **203,** 1063.
McKeown, E., and Waters, W. A. (1966). *J. Chem. Soc. B* pp. 1040–1046.
Mallet, L. (1927). *C.R. Hebd. Seances Acad. Sci., Ser. C* **185,** 352–354.
Marshall, J. A., and Hochstetler, A. R. (1966). *J. Org. Chem.* **31,** 1020–1025.
Matsuura, T., Yoshimura, N., Nishinaga, A., and Saito, I. (1969). *Tetrahedron Lett.* pp. 1669–1671.
Matsuura, T., Horinaka, A., Yoshida, H., and Butsugan, Y. (1971). *Tetrahedron* **27,** 3095–3100.
Matsuura, T., Yoshimura, N., Nishinaga, A., and Saito, I. (1972). *Tetrahedron* **28,** 4933–4938.
Mayeda, E. A., and Bard, A. J. (1973). *J. Am. Chem. Soc.* **95,** 6223–6226.
Mayeda, E. A., and Bard, A. J. (1974). *J. Am. Chem. Soc.* **96,** 4023–4024.
Moreau, C., Dufraisse, C., and Dean, P. M. (1926). *Compt. Rend.* **182,** 1440.
Murray, R. W., and Kaplan, M. (1968a). *J. Am. Chem. Soc.* **90,** 537–538.
Murray, R. W., and Kaplan, M. L. (1968b). *J. Am. Chem. Soc.* **90,** 4161–4162.
Murray, R. W., and Kaplan, M. L. (1969). *J. Am. Chem. Soc.* **91,** 5358–5364.
Murray, R. W., Lin, J. W.-P., and Kaplan, M. L. (1970a). *Ann. N.Y. Acad. Sci.* **171,** 121–129.
Murray, R. W., Lumma, W. C., Jr., and Lin, J. W.-P. (1970b). *J. Am. Chem. Soc.* **92,** 3205–3207.
Murray, R. W., Smetana, R. D., and Block, E. (1971a). *Tetrahedron Lett.* pp. 299–302.
Murray, R. W., Kaplan, M. L., Lin, J. W.-P., Lumma, W. C., Jr., and Smetana, R. D. (1971b). *Proc. Clean Air Congr. 2nd, 1970* 324–330.
Nakano, M., Noguchi, T., Sugioka, K., Fukuyama, H., Sato, M., Shimizu, Y., Tsuji, Y., and Inaba, H. (1975). *J. Biol. Chem.* **250,** 2404–2406.
Nakano, M., Takayama, K., Shimizu, Y., Tsuji, Y., Inaba, H., and Migita, T. (1976). *J. Am. Chem. Soc.* **98,** 1974–1975.
Nickon, A., and Bagli, J. F. (1961). *J. Am. Chem. Soc.* **83,** 1498–1508.
Nilsson, R., and Kearns, D. R. (1974). *J. Phys. Chem.* **78,** 1681–1683.
Ouannes, C., and Wilson, T. (1968). *J. Am. Chem. Soc.* **90,** 6527–6528.
Peters, J. W., Pitts, J. N., Jr., Rosenthal, I., and Fuhr, H. (1972). *J. Am. Chem. Soc.* **94,** 4348–4350.

Peters, J. W., Bekowies, P. J., Winer, A. M., and Pitts, J. N., Jr. (1975). *J. Am. Chem. Soc.* **97,** 3299–3306.
Poupko, R., and Rosenthal, I. (1973). *J. Phys. Chem.* **77,** 1722–1724.
Richardson, W. H., and Hodge, V. (1970). *J. Org. Chem.* **35,** 1216–1217.
Riesenfeld, E. H., Wohlers, H. E., and Kutsch, W. A. (1905). *Ber. Dtsch. Chem. Ges.* **38,** 1885–1898.
Rosenthal, I. (1975). *Israel J. Chem.* **13,** pp. 86–90.
Rosenthal, I., and Pitts, J. N., Jr. (1971). *Biophys. J.* **11,** 963–966.
Russell, G. A. (1957). *J. Am. Chem. Soc.* **79,** 3871–3877.
Saito, I., Kato, S., and Matsuura, T. (1970). *Tetrahedron Lett.* pp. 239–242.
Sam, T. W., and Sutherland, J. K. (1972). *Chem. Commun.* pp. 424–425.
Schaap, A. P., and Bartlett, P. D. (1970). *J. Am. Chem. Soc.* **92,** 6055–6057.
Schaap, A. P., Thayer, A. L., Faler, G. R., Goda, K., and Kimura, T. (1974). *J. Am. Chem. Soc.* **96,** 4025–4026.
Schaap, A. P., Kees, K., and Thayer, A. L. (1975). *J. Org. Chem.* **40,** 1185–1186.
Seliger, H. H. (1960). *Anal. Biochem.* **1,** 60–65.
Smith, L. L., and Kulig, M. J. (1975). *J. Am. Chem. Soc.* **98,** 1027–1029.
Spikes, J. D., and Livingston, R. (1969). *Adv. Radiat. Biol.* **3,** 29–121.
Stary, F. E., Emge, D. E., and Murray, R. W. (1974). *J. Am. Chem. Soc.* **96,** 5671–5672.
Stary, F. E., Emge, D. E., and Murray, R. W. (1976). *J. Am. Chem. Soc.* **98,** 1880–1884.
Stauff, J. (1965). *Photochem. Photobiol.* **4,** 1199–1205.
Stauff, J., and Schmidkunz, H. (1962). *Z. Phys. Chem.* [N.S.] **35,** 295–313.
Stauff, J., Schmidkunz, H., and Hartmann, G. (1963). *Nature* (*London*) **198,** 281–282.
Steer, R. P., Darnall, K. R., and Pitts, J. N., Jr. (1969). *Tetrahedron Lett.* pp. 3765–3767.
Stephens, E. R. (1969). *Adv. Environ. Sci.* **1,** 119–146.
Stephenson, L. M., and McClure, D. E. (1973). *J. Am. Chem. Soc.* **95,** 3074–3076.
Thompson, Q. E. (1961). *J. Am. Chem. Soc.* **83,** 845–851.
Trozzolo, A. M., and Fahrenholtz, S. R. (1970). *Ann. N.Y. Acad. Sci.* **171,** 61–66.
Turner, A. H., and Waters, W. A. (1956). *J. Chem. Soc.* pp. 879–888.
Turro, N. J., Ramamurthy, V., Liu, K.-C., Krebs, A., and Kemper, R. (1976). *J. Am. Chem. Soc.* **98,** 6758–6761.
Wasserman, E., Murray, R. W., Kaplan, M. L., and Yager, W. A. (1968). *J. Am. Chem. Soc.* **90,** 4160–4161.
Wasserman, H. H., and Scheffer, J. R. (1967). *J. Am. Chem. Soc.* **89,** 3073–3075.
Wasserman, H. H., Scheffer, J. R., and Cooper, J. L. (1972). *J. Am. Chem. Soc.* **94,** 4991–4996.
Wayne, R. P. (1969). *Advan. Photochem.* **7,** 311–371.
Weschler, C. J., Hoffman, B. M., and Basolo, F. (1975). *J. Am. Chem. Soc.* **97,** 5278–5280.
Wieringa, J. H., Strating, J., Wynberg, H., and Adam, W. (1972). *Tetrahedron. Lett.* pp. 169–172.
Young, R. H., and Hart, H. (1967). *Chem. Commun.* pp. 827–828.

4

Solvent and Solvent Isotope Effects on the Lifetime of Singlet Oxygen

DAVID R. KEARNS

I. Introduction 115
II. Experimental Studies of Solvent-Induced Quenching of $^1\Delta$ 116
A. Spectroscopic Measurements of τ_Δ in Solution 116
B. Independent Test of the Solvent Deuteration Effect on τ_Δ Lifetime 122
C. Temperature Dependence of the Lifetime of Singlet Oxygen . . . 123
D. Anomalous Behavior of Benzene–CS_2 Mixtures 123
E. Comparison of Solution and Gas Phase Quenching of $^1\Delta$ 124
III. Theoretical Analysis of the Radiationless Decay of Singlet Oxygen . . 125
A. General Formulation 127
B. Relation between τ_Δ and Solvent Infrared Band Intensities 128
C. Quantitative Test of Theory 130
D. The Effect of Solvent Deuteration on τ_Δ 131
E. Temperature Effect 131
F. Lack of a Heavy-Atom Effect 132
G. Search for "Super Solvents" 132
IV. Use of Solvent Deuteration as a Diagnostic Test for Singlet Oxygen . 133
V. Interaction of Singlet Oxygen with Ground and Excited State Dyes . . 134
VI. Summary 135
References 136

I. INTRODUCTION

One reason electronically excited singlet molecular oxygen ($^1\Delta$) is such an important intermediate in various photophysical and photooxygenation reactions is that it has a relatively long lifetime in solution (from microseconds to milliseconds). Early studies of the β-carotene quenching of singlet oxygen indicated that the lifetime of $^1\Delta$ (τ_Δ) was longer than 10 μsec

Singlet Oxygen

ISBN 0-12-736650-4

in typical organic solvents (Foote and Denny, 1968), and for some time it was generally believed among investigators in the field that τ_Δ was more or less independent of the nature of the solvent (Foote, 1968; Kearns, 1971). Part of the reason for the early incorrect notions regarding solvent effects of τ_Δ can be attributed to the fact that the singlet oxygen lifetime in many of the common solvent organic solvents is about the same. The other reason is that prior to the introduction of laser and flash photolysis techniques, there was no "direct measurement" of τ_Δ. When these measurements were carried out, extraordinary solvent effects of τ_Δ were observed (Merkel and Kearns, 1971, 1972a). Thus, a knowledge of τ_Δ is crucial to understanding solvent effects of photooxygenation reactions. In the course of this work a general theoretical framework has been developed for interpreting the data and for predicting τ_Δ in solvent systems yet to be studied (Merkel and Kearns, 1972a). The solvent deuterium isotope effect on singlet oxygen discovered during this work has provided a simple, virtually unambiguous test for the involvement of singlet oxygen in any chemical, enzymatic, or photochemical process.

In this chapter, we summarize the experimental results on the solvent induced deactivation of singlet oxygen and present a theoretical analysis of the quenching of $^1\Delta$ in solution. From this analysis, the important physical properties of the solvent responsible for the quenching are identified and it is shown how the lifetime of singlet oxygen in different solvents can be directly correlated with the infrared absorption properties of the solvent. The theory accounts for the observed solvent deuterium effect and the lack of a pronounced temperature effect on the decay rate. The possibility of finding special solvents in which the lifetime of singlet oxygen is extremely long or short is discussed.

II. EXPERIMENTAL STUDIES OF SOLVENT-INDUCED QUENCHING OF $^1\Delta$

A. Spectroscopic Measurements of τ_Δ in Solution

The true lifetime of $^1\Delta$ in solution has been obtained by comparing the rate at which it reacts with a good acceptor, A, with the rate, $1/\tau_\Delta$, at which it is quenched to its ground state by the solvent. Under usual photooxygenation conditions, the two important competitive decay modes of $^1\Delta$ are

$$\text{Radiationless decay:} \quad {}^1\Delta \xrightarrow{k_q = 1/\tau_\Delta} {}^3\Sigma \tag{1}$$

$$\text{Chemical reaction:} \quad {}^1\Delta + \text{A} \xrightarrow{k_A} \text{AO}_2 \tag{2}$$

where k_A includes both chemical and physical quenching of $^1\Delta$ by A. With high concentrations of oxygen the quenching of $^1\Delta$ by ground state oxygen molecules could become important (Matheson and Lee, 1972) and in some cases quenching of $^1\Delta$ by excited state sensitizer molecules may have to be included (Duncan and Kearns, 1971). However, in most cases, the above scheme is adequate. Thus, by monitoring the rate of disappearance of A in a kinetic experiment where the singlet oxygen decays both by reaction and by solvent-induced relaxation, it is possible to determine the overall lifetime of singlet oxygen $\tau = (k_q + k_A[A])^{-1}$, as well as the unquenched lifetime. $\tau_\Delta = 1/k_q$. To accomplish this, it is necessary to have methods for generating relatively large concentrations of singlet oxygen in a short time and for following the decay of the singlet oxygen after its generation. In our experiments a ruby laser was used to excite the singlet oxygen sensitizer, methylene blue (MB), in an oxygenated solution containing a singlet oxygen acceptor, 1,3-diphenylisobenzofuran (Merkel and Kearns, 1971, 1972a,b).

A schematic diagram of the experimental apparatus is given in Fig. 1. A Q-switched (cryptocyanine passive dye cell) ruby laser was used to generate 694 nm pulses of ~20 nsec duration. Transient absorption changes produced in a 1-cm^2 cell (C in Fig. 1) were monitored at 90° to the exciting laser beam. The oscilloscope (O) was triggered synchronously with the laser pulse by a second photomultiplier (P1) which viewed a portion of the laser light. The laser pulse excites methylene blue molecules (MB) from their ground state, ^1M, to an excited singlet state, $^1M^*$, which either rapidly return to their ground state or intersystem cross to a lower lying triplet state, $^3M^*$. Under our experimental conditions, the laser pulse was sufficiently intense to cause virtually all methylene blue molecules in a 5×10^{-5} M solution to be rapidly

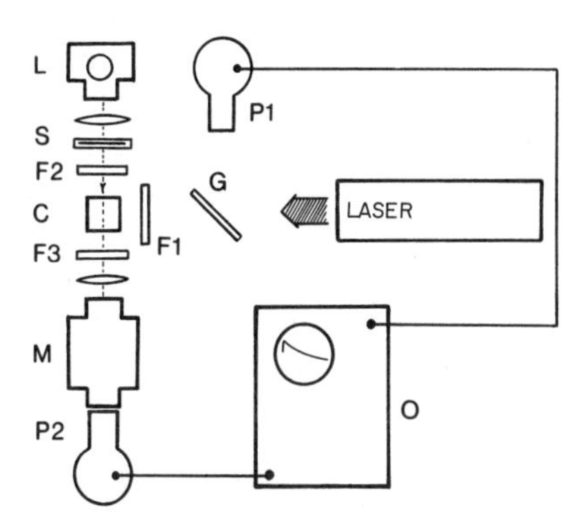

Fig. 1. Schematic diagram of laser photolysis apparatus; P1 (triggering photomultiplier), L (analyzing lamp), M (monochromator), P2 (monitoring photomultiplier), C (sample cell), S (shutter), G (glass plate), F1 (corning 2-64 filter), F2 (5-60 filter), F3 (4-96 filter), O (oscilloscope).

converted to their triplet state. Because of the relatively high concentration of ground state oxygen in the solution, the methylene blue triplet state molecules are rapidly quenched to their ground state by energy transfer to molecular oxygen and this results in the generation of approximately 5×10^{-5} M of $^1\Delta$ in less than 10^{-7} sec. The sequence of events in the photogeneration of singlet oxygen is outlined below.

$^1M + h\nu \rightarrow {}^1M^*$	Excitation to lowest singlet state
$^1M^* \rightarrow {}^3M^*$	Intersystem crossing to triplet
$^3M^* \rightarrow {}^1M$	Intersystem crossing to ground state
$^3M^* + O_2(^3\Sigma) \rightarrow {}^1M + O_2(^1\Delta)$	Energy transfer to oxygen

1,3-Diphenylisobenzofuran (DPBF) was chosen as the acceptor because it

DPBF

reacts rapidly with singlet oxygen and its disappearance in solution can easily be monitored by a loss of absorption at 410 nm. Furthermore, this dye–acceptor combination is well suited for the laser kinetic experiments since MB exhibits strong absorption at the fundamental ruby wavelength (694 nm), but is nearly transparent where absorption by DPBF is strong ($\sim$420 nm). A highly reactive acceptor is required for these experiments since the maximum fractional bleaching can be shown to be equal $[^1\Delta]_0 k_A/k_q$, where $[^1\Delta]_0$ is the concentration of singlet oxygen produced by the light pulse (Merkel and Kearns, 1972b). Finally, to investigate the effect of adding a second acceptor on the decay of singlet oxygen, it is desirable that the acceptor concentration remain nearly constant during the bleaching of DPBF. This condition requires that k_q/k_A and hence $1/k_A$ for DPBF be substantially smaller than that of the second acceptor.

Under conditions where the concentration of A does not change substantially during the experiment, solution of the kinetic equations describing processes (1) and (2) yields

$$[AO_2]_\infty - [AO_2] = \frac{k_A[A][^1O_2]_0}{(k_q + k_A[A])} e^{-(k_q + k_A[A])t} \tag{3}$$

where $[AO_2]_\infty$ is the concentration of peroxide after all the singlet oxygen has decayed. From this, it follows that $\log(\Delta OD) = 2.3(k_q + k_A[A])t + \text{const.}$,

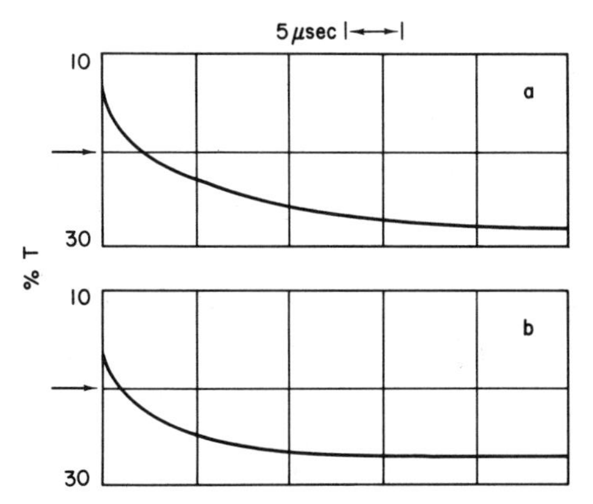

Fig. 2. Bleaching in an oxygenated solution of methylene blue (10^{-4} *M*) in methanol of (a) 1.5×10^{-5} *M*, 1,3-diphenylisobenzofuran monitored at 410 nm in a 1-cm cell and (b) 1.0×10^{-4} *M*, 1,3-diphenylisobenzofuran monitored at 435 nm in a 2-mm cell. Arrows indicate %T before the pulse (Merkel and Kearns, 1972a,b).

where ΔOD is the optical density of the acceptor at time *t* minus the asymptotic value. Plots of ΔOD vs. *t* at two acceptor concentrations provide values for both k_A and $\tau_\Delta = 1/k_q$.

Photobleaching of 1,3-diphenylisobenzofuran in methanol under two sets of conditions is illustrated in Fig. 2. In (a), a low acceptor concentration (1.5×10^{-5} *M*) results in a decay rate limited primarily by τ_Δ. In deoxygenated solution, no bleaching is observed and the methylene blue triplet decays at the characteristic rate. This eliminates the possibility of a direct reaction between the acceptor and triplet state methylene blue molecules. Increasing the acceptor concentration to (1.0×10^{-4} *M*) accelerates the decay rate by consumption of a significant portion of $^1\Delta$ (Fig. 2b). Under these conditions, the decay rate is primarily determined by reaction with DPBF. The addition of a second oxidizable acceptor (tetramethylethylene, 5×10^{-4} *M*) decreases both the duration and yield of 1,3-diphenylisobenzofuran photooxidation in accordance with expectations.

The slight curvature in the first-order plots of the decay curves in Fig. 3 reflects the consumption of acceptor as photooxidation progresses. From the slopes, we obtain values of $\tau_\Delta = 7 \pm 1$ μsec and $k_A = 7 \pm 1 \times 10^8$ moles^{-1} sec^{-1}. This yields a value for $\beta = (\tau_\Delta k_A)^{-1}$ of 2×10^{-4} for 1,3-diphenylisobenzofuran, which is reasonably close to that observed in 1,1,2-trichlorotrifluroethane (Matheson and Lee, 1970). The absolute reaction rate constant for tetramethylethylene (TME) is calculated to be $k_A = 4 \pm 1 \times 10$ moles^{-1} sec^{-1}, and the β value computed from these data is in agreement with the previously determined value (Foote, 1968).

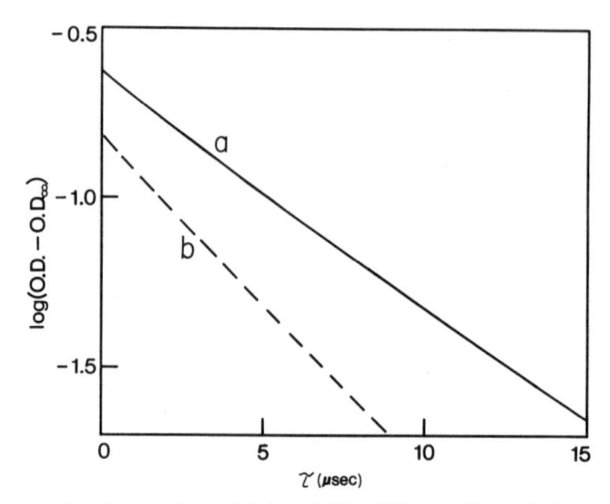

Fig. 3. First-order decay plots of (a) and (b) of Fig. 2. Note slight curvature in (a).

The same procedure has been used to measure the $^1\Delta$ lifetimes in a number of other solvents and these results are summarized in Table I and II (Merkel and Kearns, 1972a,b). In each case, concentrations of methylene blue were such that fractional bleaching was small enough to prevent substantial departures from first-order linearity.

The lifetime of $^1\Delta$ in water was extrapolated from data on pure methanol and a 50:50 mixture of methanol and water (DPBF was not soluble in pure water and apparently aggregates in solutions containing more than about

Table I Lifetimes of Singlet Oxygen in Various Solvents

Solvent	$\tau_{1\Delta}$ (μsec)	OD_{7880} (cm^{-1}) (1 cm)	OD_{6280} (cm^{-1}) (1 cm)
H_2O	2[a]	0.47	3.4
CH_3OH	7	0.18	3.9
C_2H_5OH	12	0.14	2.0
C_6H_{12}	17	0.09	0.08
C_6H_6	24	0.009	0.11
$CH_3C(=O)CH_3$	26	0.015	0.08
CH_3CN	30	0.016	0.14
$CHCl_3$	60	0.002	0.013
CS_2	200	<0.0005	—
C_6F_6	600	<0.0005	—
CCl_4	700	<0.0005	—
Freon 11	1,000	<0.0005	—

[a] Extrapolated from data obtained using 1:1 mixture of H_2O and CH_3OH.

70% H_2O). The strong dependence of the lifetime of singlet oxygen upon solvent is illustrated by a comparison of the results for 50% H_2O–50% methanol and for CS_2, shown in Fig. 4. For other solvents, the lifetimes range from 2 μsec (in water) to approximately 1 msec (in Freon). Most of the measured values are estimated to be accurate to within 20%, except for CS_2 (~30%) and CCl_4 (~50%). For reasons which will become evident later, we have also listed the optical densities (1-cm cell) of the pure solvents at 7880 cm^{-1} and 6280 cm^{-1} in Table I.

Adams and Wilkinson (1972) have also used a laser flash photolysis technique to measure the decay of singlet oxygen in various solvents. By frequency doubling the ruby laser light, they were able to use a number of different sensitizers and thus determine the influence of the sensitizer on the singlet oxygen decay. A number of different aromatic hydrocarbons and methylene blue were used as sensitizers and they found that the decay of singlet oxygen was virtually independent of sensitizer used.

Contrary to initial expectations, the results presented in Table I indicate the lifetime of singlet oxygen in solution is extremely sensitive to the nature of the solvent. The shortest lifetimes (2 μsec) are observed in aqueous solution or alcohols, whereas the longest lifetimes (1,000 μsec) are observed with the perfluorinated hydrocarbons. One of the most significant observations (see Table II) is that the lifetime of singlet oxygen in many solvents is increased by a factor of 10 when the solvent is deuterated (Merkel *et al.*, 1972).

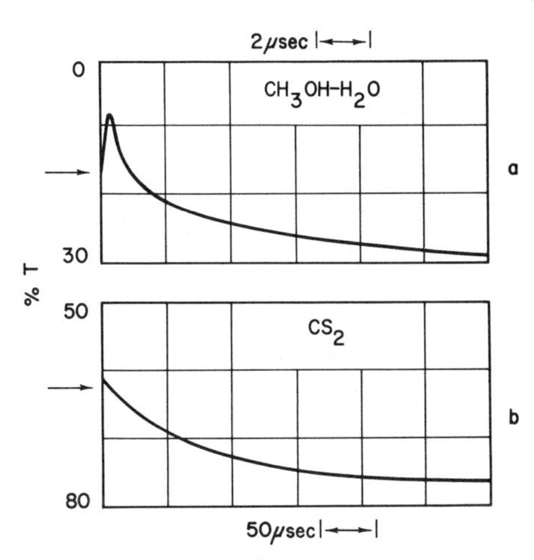

Fig. 4. A comparison of the decay of $^1\Delta$ in water–methanol (1:1) and CS_2. The decay of $^1\Delta$ was monitored by DPBF in (a) oxygenated 50% H_2O–50% CH_3OH with 1.5×10^{-5} *M* DPBF and 5×10^{-5} *M* MB (at 410 nm), and in (b) air-saturated CS_2 (containing 1% CH_3OH) with 0.5×10^{-5} *M* DPBF and 2×10^{-5} *M* DB (at 415 nm). Note the different time scales for (a) and (b) (Merkel and Kearns, 1972a,b).

B. Independent Test of the Solvent Deuteration Effect on τ_Δ Lifetime

Since the decay of singlet oxygen in the laser photolysis experiments could only be followed indirectly by monitoring the disappearance of some acceptor, an independent means of testing some of the results given in Tables I and II was desirable. Ordinarily, the decay of singlet oxygen in solution is so rapid that it is impossible to observe emission from excited oxygen molecules which are actually dissolved in solution (Kasha and Khan, 1970). However, when large amounts of singlet oxygen are generated, as in the decomposition of hydrogen peroxide, dimol emission may be observed from bubbles which form (Kasha and Khan, 1970), and this system was used to provide an independent test of the effect of solvent deuteration on the lifetime of singlet oxygen (Kajiwara and Kearns, 1973). The sequence of events in the hypochlorite-induced decomposition of H_2O_2 in basic solution is presumably as follows:

$$HO_2^- + HOCl \rightarrow H_2O + {}^1O_2(\ell) + Cl^-$$

$$H_2O_2 + OCl^- \rightarrow H_2O + {}^1O_2(\ell) + Cl^-$$

$${}^1O_2(\ell) \rightarrow {}^1O_2(g)$$

$${}^1O_2(g) + {}^1O_2(g) \rightarrow 2\,O_2(g) + h\nu$$

where O_2 (g) and O_2 (ℓ) refer to oxygen dissolved in the liquid and gas phases, respectively. If the decomposition is carried out under conditions where most of the singlet oxygen decays in solution before escaping into a gas bubble

Table II Solvent Deuteration Effects on the Lifetime of Singlet Oxygen

Solvent	τ_Δ (μsec)	Infrared absorption	
		$OD_{7880\ cm^{-1}}$ (1 cm cell)	$OD_{6280\ cm^{-1}}$ (1 cm cell)
H_2O	2	0.47	3.4
D_2O	20	0.06	0.27
H_2O/CH_3OH, 1:1	3.5	0.33	3.7
D_2O/CH_3OH, 1:1	11	0.10	1.0
D_2O/CD_3OD, 1:1	35	0.03	0.30
$CH_3C(=O)CH_3$	26	0.015	0.08
$CD_3C(=O)CD_3$	26	0.002	0.13
$CHCl_3$	30	0.002	0.013
$CDCl_3$	300	<0.005	—

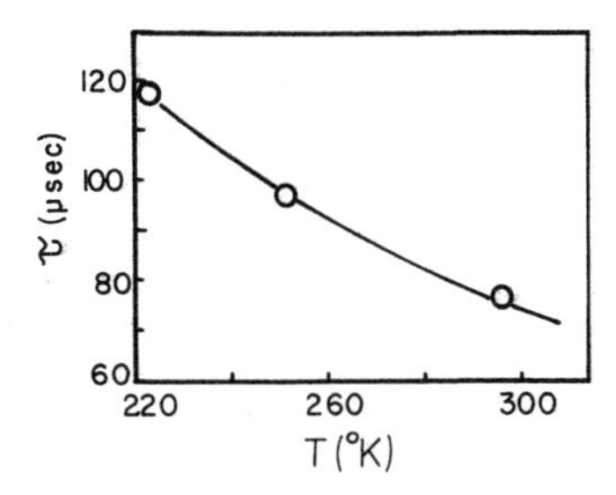

Fig. 5. Effect of temperature on the lifetime of singlet oxygen in $CHCl_3$ (Long and Kearns, 1975).

(dilute solution), then the intensity of the chemiluminescence resulting from the reaction will be very sensitive to the lifetime of singlet oxygen in solution. Such experiments have been carried out, and as expected, when the decomposition of H_2O_2 is carried out, D_2O chemiluminescence intensity is 50–100 times brighter than when it is carried out in H_2O solution (Kajiwara and Kearns, 1973). This confirms our other experiments which indicated that the lifetime of $^1\Delta$ is 10 times longer in D_2O than H_2O.

C. Temperature Dependence of the Lifetime of Singlet Oxygen

The variation of the lifetime of singlet oxygen in $CHCl_3$ with temperature is shown in Fig. 5. While the data are rather limited, there is only a 50% decrease in the lifetime on going from $-50°$ to $+25°C$. Adams and Wilkinson (1972) studied the effect of temperature on τ_Δ over the limited range from 0° to 23°C and found little change in the decay rate. They also found that the decay constant obtained in experiments using methylene blue as the sensitizer was the same using either the doubled ruby frequency (347 nm) or the primary frequency (694 nm). The relatively small temperature effect on the $^1\Delta$ lifetime indicates that large temperature effects observed in the photooxygenation of an organic substrate are probably due to a temperature effect on the addition of the excited oxygen to the acceptor (Koch, 1968).

D. Anomalous Behavior of Benzene–CS_2 Mixtures

Foote *et al.* (1972) studied the sensitized photooxygenation of anthracene in a solvent system consisting of a mixture of benzene and CS_2 and obtained the results shown in Fig. 6 where β, the ratio of singlet oxygen decay rate (k_D) to the rate of reaction of singlet oxygen with the acceptor (k_A), is plotted as a function of the percent of benzene in the solution. Unexpectedly,

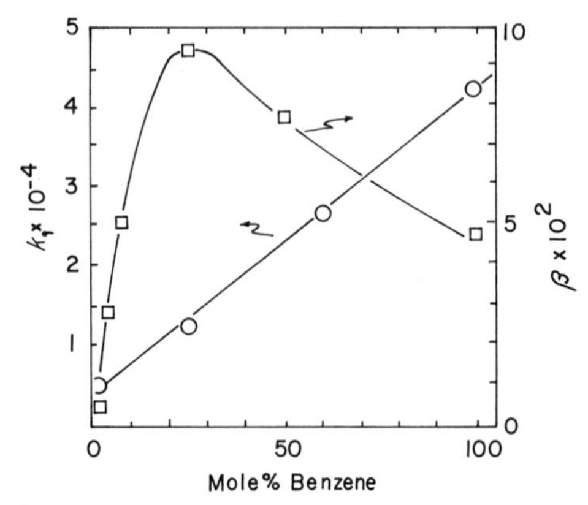

Fig. 6. Variation in β, the rate of the sensitized photooxidation of anthracene in mixtures of CS_2 and benzene, and k_q, the rate constant for decay of singlet oxygen in these same mixtures (Long and Kearns, 1975).

they observed a pronounced maximum in β at about 25 mole% benzene and this raised the intriguing possibility that the lifetime of singlet oxygen might be abnormally short in the mixed solvent system. To test this possibility, we measured the singlet oxygen lifetime in this mixed solvent system with the results shown in Fig. 6 (Long and Kearns, 1975). The data show that the lifetime of singlet oxygen varies in a very simple fashion with the percent of benzene and the lifetime for any mixture can be computed simply from the solvent composition and the lifetimes in each of the pure solvents. The anomalous behavior of the sensitized photooxygenation of anthracene in this solvent system cannot be attributed to anomalies in the lifetime of the singlet oxygen. Other factors are apparently responsible.

E. Comparison of Solution and Gas Phase Quenching of $^1\Delta$

It is interesting to compare the quenching rates obtained in solution with those obtained in the gas phase. In the gas phase, the quenching process can be described by the following scheme:

$$^1\Delta + \mathrm{Q} \underset{k_{-1}}{\overset{k_1}{\rightleftharpoons}} [^1\Delta \cdots Q] \xrightarrow{k_2} {}^3\Sigma + \mathrm{Q}$$

where Q is the quencher, k_1 is the rate constant for formation of a collision complex, k_{-1} is the rate constant for dissociation, and k_2 is the first-order quenching constant of the $[^1\Delta + \mathrm{Q}]$ collision complex. A kinetic analysis assuming steady state for $[^1\Delta + \mathrm{Q}]$ and $k_2 \ll k_{-1}$ gives

$$\frac{-d[^1\Delta]}{dt} = \frac{k_1 k_2}{k_{-1}}[^1\Delta][\mathrm{Q}] = k_\mathrm{Q}[^1\Delta][\mathrm{Q}]$$

where k_Q is the experimentally observed second-order quenching constant. In the gas phase where the collision frequency is 10^{11} sec^{-1} at 1 M and the estimated lifetime of the collision complex is 10^{-12} sec, we find $k_2 \sim 10\ k_Q$. From the observed gas phase values for k_Q, we calculate $k_2 \simeq 3 \times 10^4$ sec^{-1} for both water and benzene (Findlay *et al.*, 1969). To the extent that gas phase and solution state collision complexes are the same, k_2 should be the same for both phases.

To relate the gas phase k_2 values to the k_q values in solution, we assume that each singlet oxygen molecule is in constant contact with N nearest neighbor solvent molecules in solution so that $k_q \sim Nk_2$. If we assume $N = 5$ for water and 2 for benzene, then k_2 is calculated from the solution state data to be 10^5 sec^{-1} and 2×10^4 sec^{-1} for water and benzene, respectively, in reasonable agreement with the values derived from gas phase data.

It is interesting to apply the same analysis to the quenching of $^1\Sigma$. From the vapor phase rate constant for the quenching of $^1\Sigma$ oxygen by water (Filseth *et al.*, 1970; Stuhl and Niki, 1970), we estimate the lifetime of $^1\Sigma$ in liquid water to be on the order of 10^{-11} seconds ($k_2 \simeq 2 \times 10^{10}$). The ease with which it is quenched is the main reason why $^1\Sigma$ oxygen is not important in most solution state photooxygenation reactions, except perhaps as a precursor to the formation of $^1\Delta$.

This summarizes much of what is known experimentally about the lifetime of singlet oxygen in different solvent systems. We now turn to the development of a theoretical model which is capable of qualitatively and, in some cases, quantitatively accounting for the various observations.

III. THEORETICAL ANALYSIS OF THE RADIATIONLESS DECAY OF SINGLET OXYGEN

The salient experimental observations which we would like any model to account for are (i) the lifetime of singlet oxygen varies over almost a factor of 1,000, depending upon the solvent, (ii) there is a substantial solvent deuterium isotope effect on the singlet oxygen lifetime in many solvents, and (iii) there is relatively little temperature effect on the singlet oxygen lifetime. In addition, it would be desirable if we could account for the absolute magnitude of the solvent quenching rate in some solvents.

A brief consideration of the data presented in Tables I and II reveals that the lifetime of singlet oxygen is uncorrelated with most of the usual solvent properties. For example, there is no correlation between τ_Δ and solvent polarity. The lifetime of $^1\Delta$ in nonpolar cyclohexane is an order of magnitude greater than the lifetime in water, but approximately half that is highly polar acetonitrile. Likewise, viscosity does not appear to play a prominent role in the radiationless decay of singlet oxygen. Carbon disulfide has approximately one-third the viscosity of water yet the $^1\Delta$ lifetime is two orders of magnitude *greater* in the former solvent. Solvent polarizability, ionization potential, and oxygen solubility also appear to be unimportant factors in the radiationless decay of singlet oxygen.

However, one solvent property which does correlate well with the $^1\Delta$ lifetime is the intensity of the infrared absorptions near 7880 cm^{-1} and 6280 cm^{-1}, resonant with $0' \rightarrow 0$ and $0' \rightarrow 1$, $^1\Delta \rightarrow {}^3\Sigma$ transitions (see Fig. 7). This correlation is reminiscent of the electric dipole resonant electronic energy transfer treated by Förster (1948, 1959) and expanded by Dexter (1953). In the present case, however, the energy transfer requires the conversion of electronic excitation of oxygen into vibrational excitation of the solvent. The substantial deuterium effects observed suggests that solvent vibrations play a prominent role in the relaxation of the singlet oxygen. We now outline the theoretical framework which permits us to account for these observations.

With large molecules, radiationless decay primarily involves intramolecular mechanisms (Henry and Kasha, 1968; Bixon and Jortner, 1968; Schlag *et al.*, 1971) in which only small amounts of energy are taken up by indirectly coupled low frequency vibrations of the medium. In the absence of heavy atom effects, relaxation is thus expected to be essentially independent

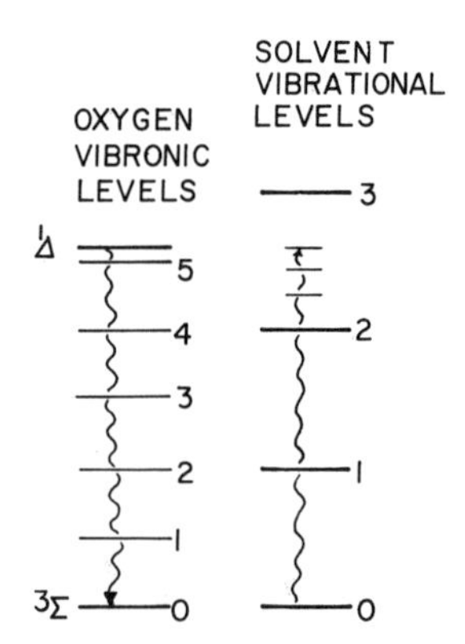

Fig. 7. A schematic diagram showing the vibronic levels of molecular oxygen and selected vibrational levels of a solvent molecule. The C–H vibrational levels of the solvent are indicated with bold lines, and other vibrational modes are indicated by thin lines. In this diagram we have indicated a quenching process in which the electronically excited singlet oxygen molecule is deactivated to the ground vibrational state of the ground electronic state ($0' \rightarrow 0$ transition) and the solvent molecule is excited to some high vibrational level involving 2 quanta of C–H vibrations and several quanta of some other mode(s).

of the nature of the solvent. In contrast to the relaxation of large molecules where the solvent plays only a minor role, the solvent may have a major role in the radiationless deactivation of small molecules, such as oxygen, where internal vibrational modes cannot provide the required set of vibrational states nearly degenerate with the initial state.

A. General Formulation

The radiationless decay of singlet oxygen can be treated theoretically using the general formalism developed earlier by Robinson and Frosch (1962, 1963) in which radiationless transitions are considered to arise from interaction of zero-order nonstationary Born–Oppenheimer initial (ψ_i) and final (ψ_f) states under the influence of an interaction Hamiltonian $\mathscr{H}'$. In this approach the rate constant for the transition from the initial state i to a particular final state f may be written as

$$k_{if} = \frac{2\pi}{\hbar\alpha} |\beta_{if}|^2 \simeq \frac{2\pi\tau_{\text{vib}}}{\hbar^2} |\beta_{if}|^2 \tag{4}$$

where α is the energy of interaction between the final state and the molecules of the medium, τ_{vib} is the vibrational relaxation time of the solvent, and $\beta_{if} = \langle\psi_i|\mathscr{H}'|\psi_f\rangle$. If transitions occur to a number of different final states, then the net quenching rate constant will involve a sum over all possible final states.

If intermolecular electron exchange is neglected, the interaction between O_2 and some solvent molecule with a dipole moment $\vec{\mu}$ can be approximated by

$$\mathscr{H}' = \sum_i \frac{q_i \mu \cos\theta_i}{R_i^2} \tag{5}$$

where q_i is the charge of an electron on oxygen nucleus located at a distance R_i from the center of the solvent and θ_i is the angle formed between R_i and $\vec{\mu}$. If we express wavefunctions as products of oxygen (ψ) and solvent (Ω) wavefunctions and use the above expression for $\mathscr{H}'$ the following expression for β_{if} is obtained:

$$\beta_{if} = \sum_{m,n} \left\langle \psi^0(^1\Delta)\Omega_0 \left| \sum_i \frac{q_i\mu\cos\theta_i}{R_i^2} \right| \psi^m(^3\Sigma)\Omega_n \right\rangle = \sum_{m,n} \beta_{mn} \tag{6}$$

where the superscripts on ψ and subscripts on Ω index the vibrational levels. The final state indices, m and n, are of course related by conservation of energy.

Making use of the Born–Oppenheimer separability, the matrix elements can be further factored to obtain the following expression (Merkel and Kearns, 1972b):

$$\beta_{mn} = \beta_{el} F_m^{1/2} \langle \Omega_0 | \mu | \Omega_n \rangle \tag{7}$$

in which

$$\beta_{el} = \left\langle \psi(^1\Delta) \middle| \sum_i \frac{e \cos \theta_i}{R_i^2} \middle| \psi(^3\Sigma) \right\rangle \tag{8}$$

The ψ's denote purely electronic oxygen wavefunctions, Ω_0 and Ω_n are vibrational wavefunctions for the ground electronic state of the *solvent* molecule and $F_m = |\langle \chi_o(^1\Delta) | \chi_m(^3\Sigma) \rangle|^2$ is the Franck–Condon overlap between the vibrational wavefunctions, χ, for the specific initial and final electronic states of oxygen indicated. Since $\psi(^1\Delta)$ and $\psi(^3\Sigma)$ are orthogonal and not explicit functions of nuclear coordinates all terms in β_{el} involving the nuclear coordinates of oxygen vanish.

From the nature of the electronic wavefunctions for oxygen (Kearns, 1971), we immediately conclude that in the absence spin-dependent perturbation $^1\Delta$ cannot couple with $^3\Sigma$. Consequently, in nonheavy atom solvents there is no solvent induced direct mixing of the $^3\Sigma$ and $^1\Delta$ states. However, the $^1\Sigma$ and $^3\Sigma$ states are strongly coupled by a spin-orbit coupling matrix element of $\sim 140\ \text{cm}^{-1}$ and, therefore, nonheavy atom solvents can cause *indirect* mixing of $^1\Delta$ and $^3\Sigma$ by induced mixing of $^1\Sigma$ and $^1\Delta$ states. Application of second-order perturbation theory (Kauzmann, 1957) to this mixing scheme yields the following modified expression for β_{el}:

$$\beta_{el} = \beta'_{el} \frac{\beta_{so}}{\Delta E} \tag{9}$$

in which

$$\beta'_{el} = \left\langle \psi(^1\Delta) \middle| \sum_i \frac{e \cos \theta_i}{R_i^2} \middle| \psi(^1\Sigma) \right\rangle \tag{10}$$

$$\beta_{so} = \langle \psi(^1\Sigma) | \mathscr{H}_{so} | \psi(^3\Sigma) \rangle = 140\ \text{cm}^{-1} \tag{11}$$

and ΔE in Eq. (9) is the difference in energy between the ground $^3\Sigma$ state and $^1\Delta$ (i.e., $7880\ \text{cm}^{-1}$).

B. Relation between τ_Δ and Solvent Infrared Band Intensities

For reasons which will become evident, it is desirable to expand μ as is often done in treatments of infrared vibrational selection rules (Pitzer, 1953) and write

$$\vec{\mu} = \vec{\mu}_o + \sum_j \left(\frac{\partial \mu}{\partial Q_j} \right)_{Q_j = 0} \vec{Q}_j + \cdots \tag{12}$$

where μ_o is the static state dipole moment of the solvent molecule and Q_j is some normal coordinate. Substituting Eqs. (12), (9), (10), into Eq. (7) we obtain

$$\beta_{mn} = \beta'_{el} \frac{\beta_{so}}{\Delta E} F_m^{1/2} M_n \tag{13}$$

where

$$M_n = \left\langle \Omega_o \middle| \mu_o + \sum_j \left(\frac{\partial \mu}{\partial Q_j} \right)_{Q_j = 0} Q_j \middle| \Omega_n \right\rangle \tag{14}$$

When electronic energy of oxygen is converted into vibrational excitation of the solvent, $n \neq 0$ and thus the term containing μ_o vanishes due to the orthogonality of the solvent vibrational wavefunctions ($\langle \Omega_o | \Omega_n \rangle = 0$).

Using the above results, the following expression for the quenching rate constant is obtained:

$$k_{if} \simeq \frac{2\pi\tau_{\text{vib}}}{\hbar^2} \left| \beta'_{el} \frac{\beta_{so}}{\Delta E} \right|^2 \sum_{m,n} F_m |M_n|^2 \tag{15}$$

The quantity M_n which appears in the above equation is the same factor which determines the intensity of $o \to n$ transitions of the solvent in the infrared region (Pitzer, 1953) and, consequently, M_n can be calculated directly from the near IR spectral data on the solvent.

The quantity F_m is the Franck–Condon factor for a transition from the zeroth vibrational level of the $^1\Delta$ state to the mth vibrational level of the ground $^3\Sigma$ electronic state. Because the potential energy curves for oxygen in its $^1\Delta$ and $^3\Sigma$ state are so similar, both with regard to average internuclear separation (1.207 versus 1.215 Å) and force constants ($\omega = 1580$ versus $\omega = 1509$ cm^{-1}), the Franck–Condon factors decrease extremely rapidly as m increases. For the isolated molecule, the values decrease as $\sim 1{:}10^{-2}{:}10^{-4}{:}2 \times 10^{-7}$ as m increases from 0 to 3 (Herzberg, 1950). Complete intramolecular conversion of energy from $^1\Delta$ into vibration of $^3\Sigma$ requires excitation to $m = 5$. Because of the rapid decrease in F_m, relaxation of $^1\Delta$ oxygen is expected to be most rapid when *large* amounts of electronic energy can be transferred to vibrational excitation of the solvent. For example, if all of the $^1\Delta$ excitation energy is transferred to the solvent, theory predicts that the $^1\Delta$ decay rate will be directly proportional to the solvent IR absorption intensity at $\sim$7880 cm^{-1}. For many solvents the absorption bands at this energy are not particularly strong. In this case, the oxygen may be quenched to its $m = 1$ vibrational state of the ground electronic state provided the solvent picks up (7880–1600) $\sim$6280 cm^{-1} of vibrational excitation. When this occurs, the solvent absorption in the region of 6280 cm^{-1} becomes important.

C. Quantitative Test of Theory

It is evident the theory can at least qualitatively account for the dramatic variation of the lifetime of singlet oxygen in various solvents and for the correlation between infrared absorption intensity and quenching rate constants. As a quantitative test of the theory, we have computed the lifetime of $^1\Delta$ in water (Merkel and Kearns, 1972b). Although several additional assumptions go into the calculation ($\tau_{vib} \simeq 10^{-12}$ sec, $R_{H_2O\text{-}O_2} \simeq 5$ Å, $F_o \sim 1$), the computed value ($k_{if} = 1/\tau_\Delta = 6 \times 10^4\ \text{sec}^{-1}$) is in reasonably good agreement with the experimental value of $5 \times 10^5\ \text{sec}^{-1}$.

In the above estimate of k_q we only considered the $^1\Delta(0) \rightarrow {}^3\Sigma(0)$ transition. In order to properly include the effects of the $0 \rightarrow 1$, $0 \rightarrow 2$ and other nonadiabatic transitions, we need the corresponding Franck–Condon factors. Isolated molecule Franck–Condon factors are probably not appropriate for oxygen molecules in solution since there is experimental evidence that these may be rather sensitive to solvent perturbations. For example, Evans (1969) has observed $0 \leftarrow 0 : 0 \leftarrow 1$ intensity ratio of $\sim 1:0.3$ for the $^1\Delta \leftarrow {}^3\Sigma$ absorption of oxygen at high pressure in 1,1,2-trichlorotrifluroethane. If we assume that $F_1 \simeq 0.1$ (intermediate between the isolated molecule and high pressure cases), then using the data in Tables I and II we obtain the following approximate expression relating the $^1\Delta$ lifetime to solvent IR absorption:

$$k_q = \frac{1}{\tau_\Delta} (\mu\text{sec})^{-1} \simeq 0.5(\text{OD}_{7880}) + 0.005(\text{OD}_{6280}) + \text{higher terms} \quad (16)$$

where OD_{7880} and OD_{6280} are optical densities of 1 cm of solvent at 7880 cm^{-1} and 6280 cm^{-1}, respectively.

Examination of the data in Table I indicates that for most hydrocarbon solvents, higher terms in Eq. (16) can be neglected. The lifetime of $^1\Delta$ in chloroform, carbon disulfide, and carbon tetrachloride, however, is substantially shorter than that calculated using only the 0–0 and 0–1 terms, but the correlation can be improved by inclusion of terms involving transitions to higher vibrational levels of $^3\Sigma$. The $0 \rightarrow 2$ $^1\Delta \rightarrow {}^3\Sigma$ transition energy is at 4700 cm^{-1}. Benzene has a very strong absorption in this region (OD of 1 cm $\simeq 7.0$). A Franck–Condon factor of $F_2 \simeq 5 \times 10^{-3}$ (versus 10^{-4} for an isolated oxygen molecule) would bring the predicted and observed $^1\Delta$ lifetimes in benzene into good agreement. Chloroform and carbon disulfide exhibit moderate absorption within $\sim kT$ of 4700 cm^{-2} and thus radiationless decay of $^1\Delta$ to the $m = 2$ level of $^3\Sigma$ is expected to be important in these solvents also. Of the remaining solvents studied only acetone shows an absorption at 4700 cm^{-1} which is strong (OD $\simeq 3.0$ in a 1 cm cell) relative to that at 7880 cm^{-1} and 6280 cm^{-1}. With an F_2 of 5×10^{-3}, the value of

τ_Δ in acetone calculated from Eq. (16) is closer to the measured lifetime. Transitions further into the infrared evidently must be included to account for the $^1\Delta$ lifetime in carbon tetrachloride.

When Franck–Condon factors for oxygen in various solvents become available, more accurate calculations of radiationless decay rates will be permitted. Singlet oxygen lifetimes are expected to be longest in solvents such as nitrogen and argon which lack intensity in the infrared region. It is worth noting that if the solvent oxygen interaction is especially strong, transitions to the $m = 5$ vibrational level of $^3\Sigma$ may become more allowed. Hence, transfer of excitation energy to the solvent would be less important and the static dipole moment of the solvent could become significant in radiationless decay.

D. The Effect of Solvent Deuteration on τ_Δ

Since the infrared overtone bands of common solvents in the regions of interest are usually due to C–H or O–H vibrations, the theory immediately permits us to understand the large solvent deuteration effects on the lifetime of singlet oxygen. A comparison of the infrared absorption intensities of H_2O and D_2O (Table II) suggests, for example, that the lifetime in the latter solvent should increase by approximately a factor of 9 and experimental studies of the photooxidation efficiencies in H_2O and D_2O indicate that the $^1\Delta$ lifetime is actually ten times longer in D_2O (Merkel *et al.*, 1972). A similar 10-fold increase in the singlet oxygen lifetime is predicted, and observed, in going from a 1:1 mixture of $H_2O{:}CH_3OH$ to $D_2O{:}CD_3OD$. Order of magnitude increases in the $^1\Delta$ lifetime upon deuteration predicted for methanol and chloroform are also consistent with the experimental observations shown in Table II.

Solvent deuteration will not always lead to an increase in the lifetime of singlet oxygen. This is well illustrated in the case of acetone where a reduction in absorption intensity at 7880 cm^{-1} upon deuteration of acetone is almost completely compensated for by an increase in intensity at 6280 cm^{-1} (Merkel and Kearns, 1972b). Thus, while deuteration will usually lead to a decrease in near IR absorption intensities (and, hence, increased $^1\Delta$ lifetime) a situation could arise in which deuteration would create new resonances that would actually enhance $^1\Delta$ relaxation.

E. Temperature Effect

According to the simple quenching theory no significant temperature dependence is expected so in this sense there is agreement between theory and experiment. The small temperature dependence of τ_Δ which is observed could be due to any one or a combination of different effects including subtle changes in the intensity of some of the IR overtone bands of the solvent and an increase in β_{el} due to harder O_2–solvent collisions at the higher temperatures. Because of the low intensity of absorption intensity in the appropriate spectral region it was not possible to determine whether or not there are changes in the near IR spectrum of $CHCl_3$ with temperature.

F. Lack of a Heavy-Atom Effect

Since the $^1\Delta \rightarrow {}^3\Sigma$ transition is formally spin forbidden, one might expect to observe a large external heavy atom solvent effect on the relaxation of $^1\Delta$. Our theoretical analysis indicates that this may not be the case, however, for the following reason. There is a strong spin–orbit mixing of the $^1\Sigma$ state with the $^3\Sigma$ ground state, and mixing of $^1\Delta$ with $^3\Sigma$ is *indirectly* achieved by an electrostatically-induced mixing of $^1\Delta$ with $^1\Sigma$. In order for heavy atom solvents to influence the relaxation, the heavy atoms would have to provide some new route for mixing $^1\Delta$ and $^3\Sigma$. Since a similar process is responsible for the external heavy atom enhancement of the radiationless relaxation of an excited triplet state molecule to its ground state, we can use triplet data to place an upper limit on the importance of this mechanism in the oxygen case. For many aromatic hydrocarbons in a bromine containing solvent (tetrabromobenzene) the radiative lifetimes are on the order of 50 msec, and nonradiative transition rates are estimated to be much smaller than 10^2 sec^{-1} (Giachino and Kearns, 1970). If we assume that the electronic matrix elements are similar in the case of oxygen, then heavy atom solvents are not expected to enhance the relaxation of $^1\Delta$ since nonheavy atom solvents already give relaxation rates of $\sim 10^5$ sec^{-1}.

G. Search for "Super Solvents"

One strong prediction of the simple theory is that the lifetime of singlet oxygen should be longest in solvents which have only low frequency molecular vibrations. Although the approximate two-term expression which we used to account for the variation of a singlet oxygen lifetime in the hydrocarbon solvents would predict extraordinarily long lifetimes for singlet oxygen in solvents such as CCl_4, Freon 11, and C_6F_6, experimentally the lifetime is not longer than about 1 msec. This may in part be due to quenching of singlet oxygen by the sensitizer. It also may be that relaxation of the electronically excited oxygen can proceed via transitions to very high vibrational states of the ground electronic state oxygen, and terms representing these processes were purposely omitted in our initial simplified theoretical treatment. Furthermore, these higher order terms are not needed to account for

τ_Δ in the hydrocarbon solvents and there is difficulty in obtaining Franck–Condon factors for the appropriate oxygen transitions. A measurement of the emission spectrum of oxygen *dissolved* in solution could provide the needed data, but such experiments have not yèt been successful. As an alternative we measured the absorption spectrum of oxygen dissolved under high pressure in various solvents, but again were unable to detect the $3 \leftarrow 0$ or $4' \leftarrow 0$ transitions (Long and Kearns, 1975). We therefore do not have an independent measure of the appropriate oxygen Franck–Condon factors needed to carry out a more rigorous test of the theory at this time. This aspect notwithstanding, the results presented here have shown that there are some solvent systems in which the lifetime of singlet oxygen can approach 1 msec and these may prove to be useful in other studies of the physical and chemical properties of singlet oxygen.

IV. USE OF SOLVENT DEUTERATION AS A DIAGNOSTIC TEST FOR SINGLET OXYGEN

Because deuteration involves a very minor perturbation of the solvent, the solvent deuteration effect on the lifetime of singlet oxygen can be used as a powerful diagnostic tool for investigating the role of singlet oxygen in various photochemical, photophysical, and photobiological processes.

The solvent deuteration effect was first used to establish that the methylene blue sensitized photooxygenation of three amino acids, tryptophan, methionine, and histidine, involves singlet oxygen (Nilsson *et al.*, 1972). In subsequent studies, comparison of the rate of the dye-sensitized photodynamic inactivation of various enzymes [alcohol dehydrogenase, trypsin (Nilsson and Kearns, 1973), lysozyme (Schmidt and Rosenkranz, 1972; Kepka and Grossweiner, 1973)] in H_2O and D_2O was used to demonstrate that singlet oxygen is involved in the photoinactivation process. Application of some of the other procedures which have been used to test for singlet oxygen reactions (quenching by β-carotene, special dyes, and azides) are all subject to numerous complications in the enzyme systems and in the case of more complex biological systems, the problems with these tests are even greater. However, the H_2O/D_2O comparison is an extremely simple method for evaluating the role of singlet oxygen in a photodynamic process. Recently, Kobayashi and Ito (1976) used this test to demonstrate the participation of singlet oxygen in the *in vivo* photodynamic inactivation and induction of genetic changes in yeast (*Saccaromyces cerevisiae*). In this connection, it is interesting to note that Epstein and Taylor (1966) developed an assay for the presence of small amounts of carcinogenic compounds in water based on their photodynamic effect on *Paramecium caudatum*. Since it is very likely that singlet oxygen is involved, we may predict that the sensitivity of the test might be significantly improved (perhaps a factor of 10) if the reactions were carried out in D_2O rather than H_2O.

The H_2O/D_2O comparison has also been used to explore the role of singlet oxygen in some nonphotochemical enzymatic reactions (Nilsson and Kearns, 1974). For example, it has been suggested that the oxidation of organic substrates by the enzyme soybean lipoxidase involves singlet oxygen. However, we found no enhancement of the rate of oxidation when the reaction was run D_2O as compared with H_2O, and concluded that there was no evidence for the participation of singlet oxygen in the reaction. We anticipate that the solvent deuteration test will find wide applications in future studies of other biological systems where singlet oxygen reactions are believed to be involved.

V. INTERACTION OF SINGLET OXYGEN WITH GROUND AND EXCITED STATE DYES

Since dye molecules are commonly used as singlet oxygen sensitizers in photochemical reactions, it is important to realize that there are some situations in which the dye molecules themselves may cause quenching of $^1\Delta$. Excited triplet state molecules can be efficient sensitizers of singlet oxygen but they are also extraordinarily efficient quenchers of $^1\Delta$ once it is formed. Evidence for this quenching was discovered during an investigation of the photosensitized formation of $^1\Delta$ in the gas phase (Duncan and Kearns, 1971). In those experiments, sensitizers (naphthalene or quinoxaline) were introduced into an oxygen flow downstream from a microwave discharge cavity. EPR spectroscopy was then used to monitor the $^1\Delta$ concentration even further downstream; in this way it was possible to separately study the formation of $^1\Delta$ by energy transfer from photoexcited sensitizers, or by the microwave discharge, or from both sources operating together. Surprisingly, it was found that when both lamp and the discharge were turned on, the $^1\Delta$ concentration was smaller than when the microwave discharge was operated alone. Ground state sensitizer (quinoxaline) had no effect on the $^1\Delta$ generated by the microwave discharge alone, and in the combined microwave–photoexcitation experiments, the $^1\Delta$ concentration immediately returned to the steady-state dark value when the lamp was turned off. The lifetime of the excited singlet state of quinoxaline is so short (probably less than 10^{-9} sec) that quenching of $^1\Delta$ cannot be due to an interaction with excited singlet state quinoxaline. The quenching must, therefore, involve the interaction of triplet state quinoxaline molecules with $^1\Delta$ which reaches concentrations as high as 10% in the flow experiment discharge.

$$^3\text{Quinoxaline} + {}^1\Delta \rightarrow [\text{Quinoxaline} \cdots O_2] \rightarrow ?$$

Energetically, the product of the quenching could be quinoxaline in its second excited triplet state and ground state oxygen, or perhaps some charge transfer state intermediate. An alternative mechanism discovered by Khan and his co-workers (Bolton *et al.*, 1972; Kenner and Khan, 1975, 1976) is depicted below.

$$O_2(^1\Delta) + {}^3M^* \rightarrow {}^1M^* + O_2(^3\Sigma)$$

In this rather unusual process, the excited state of the sensitizer ($^1M^*$) is regenerated as the result of annihilation of $^1\Delta$ and triplet state sensitizer ($^3\Sigma^*$).

Since the quenching of $^1\Delta$ by excited triplet state sensitizers is so rapid in the gas phase, in solution the quenching will undoubtedly be diffusion controlled. Because the rate constant for the triplet state sensitized *formation* of $^1\Delta$ by energy transfer to ground state oxygen molecules is only 1/9th the diffusion controlled value (Patterson *et al.*, 1970; Geacintov *et al.*, 1972), this means that the maximum concentration of singlet oxygen which can be photochemically generated in solution is limited to about 10% of the total oxygen concentration in solution, and this may have important consequences in high intensity flash photolysis experiments.

Any dye molecule which has a triplet state which lies below $^1\Delta$ will be incapable of sensitizing the formation of $^1\Delta$. However, such dyes should be good quenchers of singlet oxygen and this has been shown for the case of a dye which is used in Q-switching neodymium lasers (Merkel and Kearns, 1972b). This dye has bimolecular rate constant for quenching of 3×10^{10} M^{-1} which is slightly higher than the value for β-carotene quenching of singlet oxygen which is discussed elsewhere in this book. Another group of dyes which have been found to efficiently quench singlet oxygen have recently been described by Smith *et al.* (1975).

VI. SUMMARY

Laser flash photolysis studies have demonstrated that the lifetime of singlet oxygen is very sensitive to the nature of the solvent, ranging from 2 to 1,000 μsec. Deuteration of the solvent can easily lead to a 10-fold increase in the lifetime of $^1\Delta$, and this provides one of the most unambiguous diagnostic tests for the involvement of singlet oxygen in a reaction, photooxygenation, enzymatic oxygenation, chemiluminescence and other photophysical processes. Various applications of the solvent deuterium isotope effect are described. The lifetime of $^1\Delta$ is relatively temperature independent. A theory has been developed which qualitatively, and in some cases even quantitatively, accounts for the various experimental observations regarding solvent and temperature effects on τ_Δ. Finally, the molecular properties required of solvent systems in which the lifetime of singlet oxygen is either very long or very short are also described.

ACKNOWLEDGMENTS

The support of the U.S. Public Health Service and the National Science Foundation is most gratefully acknowledged. I also want to express my appreciation to Dr. Ahsan Khan who stimulated our initial studies on singlet oxygen, and my co-workers in subsequent studies: C. Duncan, T. Kajiwara, K. Kawaoka, C. Long, P. Merkel, and R. Nilsson.

REFERENCES

Adams, D. R., and Wilkinson, F. (1972). *J. Chem. Soc., Faraday Trans.* **2,** 586–593.
Bixon, M., and Jortner, J. (1968). *J. Chem. Phys.* **48,** 715–726.
Bolton, P. H., Kenner, R. D., and Khan, A. U. (1972). *J. Chem. Phys.* **57,** 5604–5605.
Dexter, D. L. (1953). *J. Chem. Phys.* **21,** 836.
Duncan, C. K., and Kearns, D. R. (1971). *Chem. Phys. Lett.* **12,** 306–309.
Epstein, S. S., and Taylor, F. B. (1966). *Science* **154,** 261–263.
Evans, D. F. (1969). *Chem. Commun.* pp. 367–368.
Filseth, S. V., Zia, A., and Welge, K. H. (1970). *J. Chem. Phys.* **52,** 5502–5510.
Findlay, F. D., Fortin, C. J., and Snelling, D. R. (1969). *Chem. Phys. Lett.* **3,** 204–206.
Foote, C. S. (1968). *Acc. Chem. Res.* **1,** 104–110.
Foote, C. S., and Denny, R. W. (1968). *J. Am. Chem. Soc.* **90,** 6233–6235.
Foote, C. S., Peterson, E. R., and Lee, K.-W. (1972). *J. Am. Chem. Soc.* **94,** 1032–1033.
Förster, T. (1948). *Ann. Phys. (Leipzig)* [6] **2,** 55–75.
Förster, T. (1959). *Discuss. Faraday Soc.* **27,** 7–17.
Geacintov, N. E., Benson, R., and Pomeranz, S. B. (1972). *Chem. Phys. Lett.* **17,** 280–282.
Giachino, G. G., and Kearns, D. R. (1970). *J. Chem. Phys.* **52,** 2964–2974.
Henry, B. R., and Kasha, M. (1968). *Annu. Rev. Phys. Chem.* **19,** 161–192.
Herzberg, G. (1950). "Spectra of Diatomic Molecules." Van Nostrand-Reinhold, New York.
Kajiwara, T., and Kearns, D. R. (1973). *J. Am. Chem. Soc.* **95,** 5886–5890.
Kasha, M., and Khan, A. U. (1970). *Ann. N.Y. Acad. Sci.* **171,** 5–23.
Kauzmann, W. (1957). "Quantum Chemistry," p. 175. Academic Press, New York.
Kearns, D. R. (1971). *Chem. Rev.* **71,** 395–427.
Kenner, R. D., and Khan, A. U. (1975). *Chem. Phys. Lett.* **36,** 643–644.
Kenner, R. D., and Khan, A. U. (1976). *J. Chem. Phys.* **64,** 1877–1882.
Kepka, A. G., and Grossweiner, L. I. (1973). *Photochem. Photobiol.* **18,** 49–61.
Kobayashi, K., and Ito, T. (1976). *Photochem. Photobiol.* **23,** 21–28.
Koch, E. (1968). *Tetrahedron* **24,** 6295–6318.

Long, C. A., and Kearns, D. R. (1975). *J. Am. Chem. Soc.* **97,** 2018–2020.
Matheson, I. B. C., and Lee, J. (1970). *Chem. Phys. Lett.* **7,** 475–476.
Matheson, I. B. C., and Lee, J. (1972). *Chem. Phys. Lett.* **14,** 350–351.
Merkel, P. B., and Kearns, D. R. (1971). *Chem. Phys. Lett.* **12,** 120–122.
Merkel, P. B., and Kearns, D. R. (1972a). *J. Am. Chem. Soc.* **94,** 1029–1030.
Merkel, P. B., and Kearns, D. R. (1972b). *J. Am. Chem. Soc.* **94,** 7244–7253.
Merkel, P. B., Nilsson, R., and Kearns, D. R. (1972). *J. Am. Chem. Soc.* **94,** 1030–1031.
Nilsson, R., and Kearns, D. R. (1973). *Photochem. Photobiol.* **17,** 65–68.
Nilsson, R., and Kearns, D. R. (1974). *J. Phys. Chem.* **78,** 1681–1683.
Nilsson, R., Merkel, P. B., and Kearns, D. R. (1972). *Photochem. Photobiol.* **16,** 117–124.
Patterson, L. K., Porter, G., and Topp, M. R. (1970). *Chem. Phys. Lett.* **7,** 612–614.
Pitzer, K. S. (1953). "Quantum Chemistry," p. 372. Prentice-Hall, Englewood Cliffs, New Jersey.
Robinson, G. W., and Frosch, R. P. (1962). *J. Chem. Phys.* **37,** 1962–1973.
Robinson, G. W., and Frosch, R. P. (1963). *J. Chem. Phys.* **38,** 1187–1203.
Schlag, E. W., Schneider, S., and Fischer, S. F. (1971). *Annu. Rev. Phys. Chem.* **22,** 465–526.
Schmidt, H., and Rosenkranz, P. (1972). *Z. Naturforsch., Teil B* **27,** 1436–1437.
Smith, W. F., Jr., Herkstroeter, W. G., and Eddy, K. L. (1975). *J. Am. Chem. Soc.* **97,** 2764–2770.
Stuhl, F., and Niki, H. (1970). *Chem. Phys. Lett.* **7,** 473–474.

5

Quenching of Singlet Oxygen

CHRISTOPHER S. FOOTE

I. Introduction . . . 139
II. Methodology . . . 141
A. Distinction of Singlet Oxygen Quenching from Other Modes of Inhibition . . . 141
B. Steady-State Techniques for Determining Rates . . . 143
C. Other Steady-State Kinetic Techniques . . . 147
D. Nondifferential Technique . . . 148
E. Separation of k_Q from k_R . . . 149
F. Time-Resolved Techniques for Determining Rates . . . 150
G. Gas Phase Methods . . . 151
III. Mechanisms of Quenching $^1O_2$ . . . 151
A. Energy-Transfer Quenching . . . 151
B. Charge-Transfer Quenching . . . 152
C. Other Mechanisms of $^1O_2$ Quenching . . . 153
IV. Types of Quencher . . . 154
A. Carotenes . . . 154
B. Other Systems with Extensive Conjugation . . . 155
C. Amines . . . 157
D. Phenols . . . 159
E. Metal Complexes . . . 160
F. Sulfides . . . 161
G. Bilirubin and Biliverdin . . . 162
H. Miscellaneous Compounds . . . 162
V. Biological Applications . . . 164
VI. Rate Constants . . . 165
References . . . 167

I. INTRODUCTION

The term "quenching of singlet oxygen" can be used to encompass both "chemical" quenching, in which singlet oxygen reacts with a "quencher," Q, to give a product, QO_2 (reaction 1, rate constant k_R), and "physical" quenching, in which the interaction leads only to deactivation of singlet oxygen to the ground state, with no O_2 consumption or product formation

Singlet Oxygen

ISBN 0-12-736650-4

(reaction 2, rate constant k_Q) However, in this chapter, the former process will be called "reaction," and the latter "quenching"; where both processes occur together, their sum ($k_R + k_Q$) will be called "total quenching."

$$^1O_2 + Q \xrightarrow[k_R]{} QO_2 \tag{1}$$

$$^1O_2 + Q \xrightarrow[k_Q]{} {}^3O_2 + Q \tag{2}$$

In this chapter, I will concentrate on the quenching reaction, and will discuss chemical reaction only in cases where it occurs along with quenching. The emphasis will be on the types of quenchers which have been identified and on their rates and mechanisms of quenching, where known. Methods of determining quenching rates and of demonstrating singlet oxygen quenching will also be discussed. Emphasis will be on solution chemistry throughout. Quenching of singlet oxygen by solvent is covered in another chapter in this volume (Kearns, Chapter 4).

The first clear demonstration of singlet oxygen quenching (although it was not so interpreted at the time) was by Schenck and Gollnick in 1958. These workers carried out a careful kinetic analysis of the photooxygenation of nicotine. In this pioneering (and overlooked) work, the reactive intermediate in the photooxygenation was shown to undergo two competitive reactions with nicotine, one of which led to oxygen consumption, and another which led only to deactivation of the reactive intermediate. Since the reactive intermediate was not recognized to be singlet oxygen at the time, the possible generality and significance of the observation was not recognized.

The first reports of the quenching of singlet oxygen as such appeared almost simultaneously in 1968: quenching by β-carotene (Foote and Denny, 1968) was suggested to involve energy transfer and that by 1,4-diazabicyclo-[2.2.2]octane (DABCO) (Ouannès and Wilson, 1968), to involve a charge-transfer mechanism. These papers stimulated a large volume of subsequent work which has confirmed and greatly extended the initial observations. Singlet oxygen quenching has become a subject of practical interest in numerous biological systems, and in the protection of materials against oxidation in the environment.

Several reviews have appeared in which singlet oxygen quenching has been the principal subject (Foote *et al.*, 1970b; Shlyapintokh and Ivanov, 1976; Belluš, 1978) or part of a more comprehensive survey of singlet oxygen reactions (Kearns, 1971; Ranby and Rabek, 1975; Foote, 1976). In order to avoid duplication with these sources, this review will concentrate on methodology and will give illustrative examples only, with enough rate constants to give a representative treatment (see Table I in Section VI, and text). Where several different rate constants for a given compound are available, I have chosen the one that I consider to be most reliable. Usually the basis for the choice is that the technique used by the investigator is a reliable one, or that one investigator has published a large number of rate constants using the same technique. The review of Belluš (1978) should be consulted for nearly complete tables of rate constants through mid-1976.

II. METHODOLOGY

In this section, I discuss kinetic and other techniques used to establish the mechanism of inhibition of a photoreaction (whether by quenching of singlet oxygen or inhibition at a previous stage of reaction), for measuring the rate constant of total quenching, and for separating physical from chemical quenching. These techniques are classified into steady-state and time-resolved methods.

A. Distinction of Singlet Oxygen Quenching from Other Modes of Inhibition

The kinetics of photooxidation reactions have been studied in great detail. The following general scheme holds for photooxidation of a substrate (A) which is known to react *only with singlet oxygen** (at rate constant k_A) in the presence of a quencher (Q). The quencher can (in principle) inhibit the reaction of A by quenching singlet (^{1}Sens) or triplet (^{3}Sens) sensitizer (rate constant $k_Q{}^S$ or $k_Q{}^T$, respectively) or by quenching or reacting with singlet oxygen (k_Q or k_R). The "trivial" mechanism of competitive light absorption by the quencher can easily be ruled out by suitable choice of sensitizer and excitation wavelength. In Scheme I, the terms so far undefined are fluorescence (k_F), radiationless decay (k_S), and intersystem crossing (k_{isc}), all from sensitizer singlet; energy transfer from ^{3}Sens to O_2 (k_O); and decay of $^1O_2$ (k_d) (Gollnick, 1968; Foote *et al.*, 1970a,b; Foote, 1976).

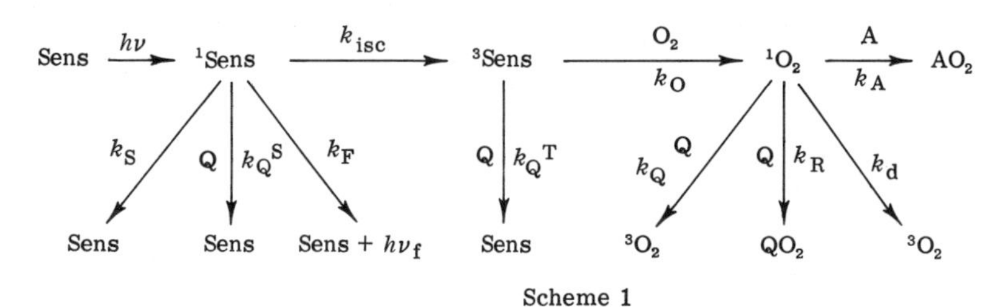

Scheme 1

* If the substrate can also react with a sensitizer excited state, the analysis is complicated considerably; methods of demonstrating the mode of reaction of A are similar to those used for Q and are discussed in Foote (1976); see also Davidson and Trethewey (1977).

The steady-state kinetic equation for the quantum yield of formation of product AO_2 (ϕ_{AO_2}) is given by

$$\phi_{AO_2} = \underbrace{\left(\frac{k_{isc}}{k_S + k_F + k_{isc} + k_Q{}^S[Q]}\right)}_{A}\underbrace{\left(\frac{k_O[O_2]}{k_O[O_2] + k_Q{}^T[Q]}\right)}_{B}$$

$$\times\underbrace{\left(\frac{k_A[A]}{k_A[A] + (k_R + k_Q)[Q] + k_d}\right)}_{C} \tag{3}$$

In this expression, terms A, B, and C reflect the partitioning of singlet sensitizer, triplet sensitizer,* and singlet oxygen toward product AO_2. Although this expression appears very complex, each term can be studied separately.

Singlet sensitizer quenching (term A) is usually negligible because the short lifetime of the singlet requires a very high concentration of quencher to intercept a substantial fraction. This process has recently been studied by Davidson and Trethewey (1976, 1977), who pointed out that the singlet states of rose bengal and methylene blue can be quenched by amines and β-carotene if sufficiently high concentrations are used. For example, singlet Rose Bengal is 50% quenched at a DABCO concentration of 0.2 *M*, or of β-carotene of 0.02 *M*. However, these concentrations are several orders of magnitude higher than those necessary for 50% quenching of $^1O_2$. Thus, singlet dye quenching is too inefficient to be of concern with most quenchers, and will not be discussed further in this chapter.

Under normal conditions, term A of Eq. (3) is a constant, equal to the quantum yield of intersystem crossing (ϕ_{isc}) of the sensitizer. Under these conditions, there are two extreme cases, one in which *only* triplet sensitizer quenching occurs $\{(k_Q + k_R)[Q] \ll k_A[A] + k_d\}$, and one in which *only* singlet oxygen quenching occurs ($k_Q[Q] \ll k_O[O_2]$). For these extreme cases, Eq. (3) can be inverted and reduced to Eqs. (4) and (5), respectively.†

* Term B does not include radiationless deactivation of triplet sensitizer, since this is small relative to $k_0[O_2]$ under normal conditions.

† In the discussion below, $[AO_2]$, the amount of product formed in a given time of irradiation, is used as it is proportional to ϕ_{AO_2} and can usually be directly measured by spectroscopy, gas chromatography, etc. Loss of A or O_2 uptake can be used interchangeably with AO_2. It is often not necessary to determine the proportionality constant. The following discussion is illustrative of various kinetic techniques which have been used, but many other methods of plotting, etc., have been devised and are useful in various cases.

$$\phi_{AO_2}^{-1} = \phi_{isc}^{-1}\overbrace{\left(1 + \frac{k_Q{}^T[Q]}{k_O[O_2]}\right)}^{B^{-1}}\overbrace{\left(1 + \frac{k_d}{k_A}[A]^{-1}\right)}^{C^{-1}} \tag{4}$$

(no singlet oxygen quenching)

$$\phi_{AO_2}^{-1} = \phi_{isc}^{-1}\left\{\overbrace{1 + \frac{k_d + (k_R + k_Q)[Q]}{k_A}[A]^{-1}}^{C^{-1}}\right\} \tag{5}$$

(no triplet quenching; term B = 1)

Equations (4) and (5) have consequences which allow triplet sensitizer quenching to be distinguished in several ways from singlet oxygen quenching. Many of the techniques used below have been used in "one point" quenching experiments, i.e., some inhibition of reaction is shown when a certain amount of quencher is added to a given reaction. However, because most singlet oxygen quenchers will also quench sensitizer excited states, or react with other strong oxidizing agents besides singlet oxygen, considerably more confidence can be attached to the results of a complete kinetic analysis; in doubtful cases, several different techniques should be used and their numerical results compared with the results calculated from known rate constants. In general, triplet dye quenching (like singlet dye quenching) requires reasonably high concentrations of quenchers, since the quencher must compete with oxygen, whose rate constant (k_O) is on the order of $2 \times 10^9\ M^{-1}\ \text{sec}^{-1}$ (Gollnick, 1968; Patterson *et al.*, 1970; Nilsson *et al.*, 1972).

B. Steady-State Techniques for Determining Rates

1. *Intercepts of Plots*

Plots of $[AO_2]^{-1}$ (the amount of product formed in a given time) vs. $[A]^{-1}$, under conditions where [A] does not change appreciably, are straight lines. Where there is no triplet quenching [Eq. (5)] such plots at various [Q] have *constant* intercepts. An example is the quenching of 2-methyl-2-pentene oxidation by azide ion, shown in Fig. 1 (Foote *et al.*, 1972b). The ratio of slope to intercept of these plots is $\{k_d + (k_Q + k_R)[Q]\}/k_A$. A plot of slope/intercept versus [Q] allows determination of both k_d/k_A (from the intercept) and $(k_Q + k_R)/k_A$ (from the slope). If k_d is known, which it is for many solvents (Wilkinson, 1978; Adams and Wilkinson, 1972; Young *et al.*, 1973b; Farmilo and Wilkinson, 1973; Merkel and Kearns, 1971, 1972a,b)

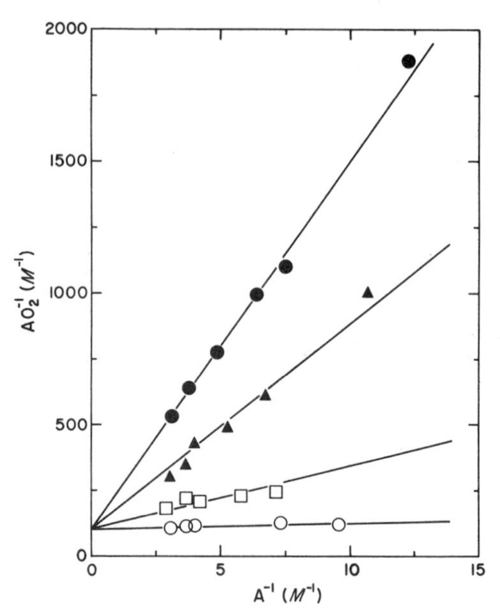

Fig. 1. Quenching of methylene blue sensitized photooxidation of 2-methyl-2-pentene (A) by NaN_3 (Foote *et al.*, 1972b). $[N_3^-]$ = ○, 0.0 *M*; □, 5×10^{-3} *M*; ▲, 10^{-2} *M*; ●, 5×10^{-2} *M*. Constant intercept implies $^1O_2$ quenching.

k_A and $(k_Q + k_R)$ can be determined in this way. *Constancy of intercept in plots of* $[AO_2]^{-1}$ *versus* $[A]^{-1}$ *at different* [Q] *is diagnostic for singlet* O_2 *quenching* (*if* A *reacts only with* $^1O_2$) *and implies that* A *and* Q *compete for a common intermediate.* Unfortunately, intercepts are often difficult to measure with accuracy, especially with substrates of limited solubility.

A second diagnostic tool is the fact that plots of the above type do not depend on O_2 concentration, in contrast to the case where there is triplet quenching (see below). Thus slopes and slope/intercept ratios are the same under air and O_2.

In the case where triplet quenching but no singlet oxygen quenching occurs, Eq. (4) applies. and plots of $[AO_2]^{-1}$ versus $[A]^{-1}$ at different [Q] have an intercept equal to

$$\phi_{\mathrm{isc}}^{-1}\left(1 + \frac{k_Q{}^{\mathrm{T}}[\mathrm{Q}]}{k_{\mathrm{O}}[\mathrm{O}_2]}\right)$$

This intercept depends on both quencher and oxygen concentrations. The ratio $k_Q{}^{\mathrm{T}}/k_{\mathrm{O}}$ can be determined from plots of the above intercepts versus either [Q] or $[O_2]^{-1}$. Since k_{O} is known for many sensitizers (Gollnick,

1968; Patterson *et al.*, 1970; Nilsson *et al.*, 1972), k_Q^T can be determined. An example of a case of this sort is the quenching of 2-methyl-2-pentene photooxygenation by benzoquinone (Fig. 2) (Weaver, 1970; Foote *et al.*, 1970b). *Variation in intercept of plots of* $[AO_2]^{-1}$ *versus* $[A]^{-1}$ *with different* [Q] *is diagnostic for quenching of sensitizer excited states, and implies that* A *and* Q *do not compete for a common intermediate.* The ratio of slope to intercept of plots in this case is k_d/k_A and is independent of [Q] or $[O_2]$.

In the case where *both* triplet sensitizer and $^1O_2$ quenching are present, both the intercept and the ratio of slope to intercept depends on [Q]; both quenching rate constants can be determined from these plots (Foote *et al.*, 1970a).

2. *Oxygen Dependence*

As Eq. (3) shows, and the analysis of the previous section suggests, quenching of sensitizer triplet is competitive with its reaction with oxygen. Therefore, triplet quenching shows an oxygen pressure dependence, and is larger at low O_2 pressures than at high; the rate of quenching can be determined from this dependence (see Section II,B,1). Singlet O_2 quenching shows no such dependence, since the trapping of sensitizer triplet by oxygen is complete except at extremely low $[O_2]$ (Gollnick, 1968).

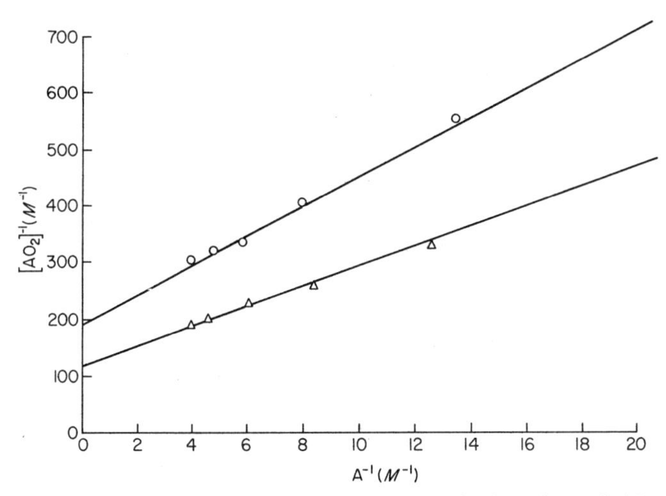

Fig. 2. Quenching of Rose Bengal sensitized photooxidation of 2-methyl-2-pentene (A) by benzoquinone (Weaver, 1970). [Benzoquinone] = △, 0.0 *M*; ○, 1.27×10^{-3} *M*. Slope/intercept: △, 0.148 ± 0.01 *M*; ○, 0.138 ± 0.008 *M*. Increase in intercept with constant slope/intercept ratio implies pure triplet sensitizer quenching.

3. $^1O_2$ *Decay Rate Dependence*

From Eqs. (4) and (5), it follows that the degree of quenching of triplet sensitizer should not depend on k_d, the decay of singlet O_2 in the solvent used; singlet oxygen quenching, however, depends on the extent to which $^1O_2$ is undergoing radiationless decay, and thus often will be larger in solvents of long $^1O_2$ lifetime. A particularly useful technique is to compare the amount of quenching of reaction of a $^1O_2$ acceptor in CH_3OD versus CH_3OH, or in D_2O versus H_2O, where large differences in $^1O_2$ lifetime occur without large changes in other solvent properties (Nilsson *et al.*, 1972; Merkel and Kearns, 1972b; Merkel *et al.*, 1972). If singlet oxygen quenching is involved, slope/intercept ratios of $[AO_2]^{-1}$ versus $[A]^{-1}$ plots are $\{k_d + (k_Q + k_R)[Q]\}/k_A$; if inhibition is by triplet sensitizer quenching of the reaction of a $^1O_2$ acceptor, they are k_d/k_A; thus, *in both cases they depend on* k_d. However, the dependence is less in the singlet oxygen case, especially at high [Q]. The effect should be quantitatively compared with the expected results. Most previous uses of this technique have assumed that triplet sensitizer behavior and the other rate constants in the system are more or less independent of whether deuterated or protiated solvent is used, an assumption which has not been tested.

This technique is frequently used in "one point" experiments, to establish the intermediacy of singlet oxygen in the oxidation of an acceptor. The rate of reaction of the acceptor in deuterated versus protiated solvent are compared under a single set of conditions. The above discussion and Eq. (5) show clearly that the size of the effect (reaction rate in D_2O/rate in H_2O) measured in this way is complex, and can vary from roughly 10* (when $k_d \gg \{k_A[A] + (k_Q + k_R)[Q]\}$, and the radiationless decay limits the $^1O_2$ lifetime) to 1.0, when the reverse inequality is true, and $^1O_2$ lifetime is limited by reaction or quenching.

4. *Dependence on* $k_A[A]$

The dependence of the amount of quenching on the rate constant k_A for reaction of $^1O_2$ with A, and on the A concentration is similar to its dependence on k_d: none for triplet quenching, but substantial for singlet oxygen quenching. From Eqs. (3) or (5) it is evident that for $^1O_2$ quenching, the amount of quenching will be largest when the term $(k_d + k_A[A])$ is smallest; this is obvious, since A and Q compete for $^1O_2$.

5. *Dependence on Source of* $^1O_2$

An additional check on the quenching mechanism can be obtained by comparing the amount of inhibition of the reaction of A with nonphoto-

* The value of $k_d^{H_2O}/k_d^{D_2O}$ (Merkel and Kearns, 1972a).

chemically generated $^1O_2$ to that for photooxygenation. Such checks have been successfully carried out in several systems, for example, with β-carotene quenching of the $NaOCl/H_2O_2$ oxygenation of 2-methyl-2-pentene (Foote and Denny, 1968). The results were identical, within the limited accuracy of the experimental data, ruling out triplet quenching as a possible mechanism for the inhibition.

C. Other Steady-State Kinetic Techniques

If the photooxygenation is known to go by way of a singlet oxygen mechanism, and if it is known that no quenching of triplet sensitizer occurs under the conditions, (for example by checking for $[O_2]$ dependence of quenching) there are several other techniques which can be used to measure quenching rates. One such technique, developed by Young *et al.* (1971), uses a fluorescent compound, such as diphenylisobenzofuran (DPBF) or diphenylfuran, as a photooxygenation acceptor, with its reaction being followed by loss of fluorescence. Because of the sensitivity of this technique, it can be used at extremely low concentration (where $k_A[A] \ll k_d$); under these conditions, the loss of acceptor is first-order, and the kinetic expression, Eq. (6), which results is integrable.

$$\frac{-dA}{[A]} = K\left\{\frac{k_A}{(k_R + k_Q)[Q] + k_d}\right\} dt \tag{6}$$

where K is a constant, the rate of $^1O_2$ generation.

Under conditions where Q concentration is constant during a run (which requires that $k_R \ll k_A$), this technique gives linear plots of log fluorescence versus time, whose slopes (S) are

$$S = K\left\{\frac{k_A}{(k_R + k_Q)[Q] + k_d}\right\}$$

Plotting S_0/S_Q (slope in absence/slope in presence of quencher) versus [Q] gives a straight line of slope $(k_R + k_Q)/k_d$ and an intercept of 1.

Another technique is to determine the amount of product AO_2 formed in a given time at various concentrations of quencher. From Eq. (5),

$$[AO_2^{-1}] = \text{const.}\left\{1 + \frac{k_d + (k_R + k_Q)[Q]}{k_A}[A]^{-1}\right\} \tag{7}$$

and a plot of $[AO_2]^{-1}$ versus [Q] at constant [A] gives a straight line; the ratio of slope to intercept of these plots is

$$S/I = \frac{k_R + k_Q}{k_A[A] + k_d}$$

Substitution of the value of [A] and an independently determined value of k_d/k_A allows calculation of the ratio $(k_R + k_Q)/k_d$. Both of the above methods measure total inhibition of product formation, and cannot be used without modification if the quencher inhibits an excited state of the sensitizer. However, they are much faster and easier to use than the more complete method, and often yield more precise values of rate constants if their limitations are understood.

A technique related to those above has been developed by Matheson and Lee (1972a,b,c). Using $^1O_2$, directly generated by laser photolysis at 1.065 μm at high O_2 pressure in Freon 113, they measured the inhibition of tetracyclone photooxidation by amines. Unfortunately, the quenching rates they obtain are relative to the quenching rate of $^1O_2$ by ground state O_2, which is difficult to measure accurately. The absolute rates reported by these investigators are low relative to those of other workers, probably for this reason. [See Foote and Ching (1975) and Merkel and Kearns (1975) for further discussion of problems with this technique.]

D. Nondifferential Technique

All the techniques in Sections II,A–C (with the exception of Young's method) are based on the use of the differential form of the kinetic equations, and are used for convenience of plotting. These techniques require that [A] be nearly constant during a run. In practice, allowing conversions of up to about 15% gives satisfactory results. At higher conversions, it is better to use $[A]_{mean}$, the average of initial and final [A].

An integral form of the equation can be used, although it is limited to a series of "one point" experiments at different [Q] (Carlsson *et al.*, 1972, Monroe, 1977). In this technique, if singlet oxygen quenching is the only important inhibition, for two solutions, both containing the same initial concentration of acceptor, one with (Q superscript) and one without (0 superscript) quencher, the following equation holds. (Subscripts I and F refer to initial and final concentrations).

$$k_Q + k_R = \frac{k_A([A]_F{}^Q - [A]_F{}^0) + k_d \ln([A]_F{}^Q/[A]_F{}^0)}{[Q]\ln([A]_I/[A]_F{}^Q)} \tag{8}$$

It is particularly important to establish the absence of sensitizer quenching when using this technique, because it responds to total inhibition of the reaction, and there are no internal checks such as plot linearity or constancy of intercept. One useful check would be to use several quencher concentrations and find consistent values. This technique should be used at the lowest feasible [Q] to avoid sensitizer quenching.

E. Separation of k_Q from k_R

All the above techniques yield values of the rate constant for total quenching, $(k_R + k_Q)$. To determine one of these, the other or their ratio must be measured independently.

This can be done in several different ways, most of which involve comparison of the amount of [Q] which reacts (or the amount of product formed) with the amount of singlet oxygen which is removed. These quantities are equal if $k_R \gg k_Q$ (as it is for many singlet oxygen substrates, Foote and Ching, 1975; Merkel and Kearns, 1975). If $k_R \ll k_Q$, the amount of quencher which reacts is very small relative to the amount of singlet oxygen quenched, as in the case of β-carotene* (Foote and Denny, 1968). The most interesting cases are the few where both k_Q and k_R are significant. With amines (Ouannès and Wilson, 1968; Smith, 1972; Gollnick and Lindner, 1973), sulfides (Foote and Peters, 1971a,b), and tocopherols (Foote *et al.*, 1974; Fahrenholtz *et al.*, 1974; Stevens *et al.*, 1974b) this has been found to be the case, and the ratio appears to be solvent dependent, although systematic studies have not yet been carried out. Bilirubin also shows comparable values of k_R and k_Q (Foote and Ching, 1975).

The techniques used for separation of k_R and k_Q will not be discussed at length; original references discussed above should be consulted for details. The methods range from measurement of the loss of quencher by absorption spectroscopy in comparison with the absolute amount of singlet oxygen quenched, calculated from gas chromatographic measurements of $[AO_2]$ at various [A] and [Q], to determination of k_R by competition experiments, comparing the amount of product formed from A and from Q. Many variations are possible.

Slightly different techniques can be used to separate k_Q and k_R in experiments where the photooxidation of quencher [Q] is studied separately, rather than the quenching of the reaction of an acceptor [A] by [Q]. In this case, Eq. (3) must be modified slightly to give Eq. (9) (P. B. Merkel and W. F. Smith, personal communication, 1977).

$$\phi_{QO_2} = \phi_{^1O_2}\left\{\frac{k_R[Q]}{(k_R + k_Q)[Q] + k_d}\right\} \tag{9}$$

In this equation, the *maximum* value of QO_2 {at [Q] high enough so that $(k_R + k_Q)[Q] \gg k_d$} is $\phi_{^1O_2}k_R/(k_R + k_Q)$. A relative value of $\phi_{^1O_2}$ can be measured by determining the maximum rate of photooxidation of an

* The fact that $k_R \ll k_Q$ need not imply indefinite or even very great stability toward photooxidation for a quencher, however. With β-carotene, even though $k_Q \sim 10^4 k_R$, k_R is $\sim 10^6\ M^{-1}\ \text{sec}^{-1}$, and carotene disappears rapidly on extended irradiation (Foote and Denny, 1968; Foote *et al.*, 1970b).

acceptor (such as dimethylfuran) which does not quench $^1O_2$, or an absolute value can be determined by actinometry in a similar experiment (Foote and Ching, 1975).

An alternative technique is to use Eq. (9) at limiting *low* [Q], where $(k_R + k_Q)[Q] \ll k_d$. Here $\phi_{QO_2} = \phi_{^1O_2}(k_R/k_d)[Q]$. With $\phi_{^1O_2}$ measured as above, and a known k_d value, k_R is determined (P. B. Merkel and W. F. Smith, personal communication, 1977).

F. Time-Resolved Techniques for Determining Rates

Several techniques for mèasuring rate constants using pulsed laser light sources have been developed. The original method made use of the elegant technique of Wilkinson and Kearns (Adams and Wilkinson, 1972; Merkel and Kearns, 1971, 1972a,b) where the lifetime of $^1O_2$ is measured indirectly by observing the loss of acceptor (usually diphenylisobenzofuran, DPBF) as a function of time after pulsing a photosensitizing system. Under conditions where $^1O_2$ lifetime is governed by k_d (i.e., low DPBF concentration), the lifetime can be measured from the DPBF loss curve (see Wilkinson, 1978, for a discussion of how this is best done). The rates of quenchers are measured by their effect on the lifetime of $^1O_2$. This technique has been modified by using triplet β-carotene (which has a λ_{max} in the 515 nm region) as a probe for $^1O_2$ (Farmilo and Wilkinson, 1973; Garner and Wilkinson, 1978). The triplet carotene (^{3}Car) is formed by energy transfer from $^1O_2$ as shown below:

$$
\begin{array}{ccc}
^1O_2 + \text{Car} & \rightarrow & ^3O_2 + {}^3\text{Car} \\
\downarrow \text{Q} & & \downarrow \\
^3O_2 & & \text{Car}
\end{array}
$$

This reaction takes place so rapidly under suitable conditions of carotene concentration and the ^{3}Car lifetime is short enough that the ^{3}Car concentration follows the $^1O_2$ concentration closely, and its lifetime is a direct measure of the $^1O_2$ lifetime; added $^1O_2$ quenchers cause ^{3}Car to decay more rapidly because they deactivate $^1O_2$.

Matheson and co-workers (1974a) have adapted a pulsed-laser system of direct $^1O_2$ generation to studies of $^1O_2$ quenching. As in their steady-state technique, a knowledge of the rate constant of ground state oxygen quenching is required; this technique also requires knowledge of beam width and absorption parameters, and has given rate constants substantially lower than those reported by others. It is not clear whether the difference is caused by systematic errors from the different equipment used by Matheson *et al.*, or whether it is caused by a solvent effect, since to avoid explosions, most of their work has been done in Freon-113, which has not been used much by other investigators.

G. Gas Phase Methods

Several methods have been developed for measuring rates of deactivation of $^1O_2$ in the gas phase, where its concentration is more readily monitored than in solution. These techniques use the emission of $^1\Delta_gO_2$ (either the direct emission at 1270 nm or the "dimol" emission at 634 nm, the paramagnetic resonance absorption of $O_2(^1\Delta_g)$, or its reaction with a cobalt surface, which produces heat. Effects of quenchers on the concentration of $^1O_2$ are used to determine the rates of deactivation. These techniques will not be discussed extensively here because they are useful only in the gas phase, and are thus outside the scope of this review (see, for example, Furukawa *et al.*, 1970).

III. MECHANISMS OF QUENCHING $^1O_2$

Two major mechanisms of singlet oxygen quenching have been established: charge transfer and energy transfer.* Both mechanisms are important; the energy-transfer mechanism appears to be more efficient in many cases, having a maximum rate of about 2×10^{10} M^{-1} sec^{-1}, at (or only slightly below) the diffusion-controlled rate, whereas the rates for charge-transfer quenching so far found are usually below 10^9 M^{-1} sec^{-1}, and often far lower.

A. Energy-Transfer Quenching

This mechanism was first suggested for carotene quenching of $^1O_2$ (Foote and Denny, 1968); it is the reverse of the reaction by which singlet oxygen is formed (by energy transfer from triplet sensitizer to triplet oxygen) (Gollnick, 1968). The mechanism involves formation of triplet quencher and ground state oxygen, and to be efficient, requires that the triplet state of the quencher be very near or below the energy of $^1\Delta_gO_2$, 22 kcal.

$$^1O_2 + Q \rightarrow {}^3O_2 + {}^3Q$$

Although the energy transfer mechanism has been well documented for β-carotene (Farmilo and Wilkinson, 1973; Mathis and Vermeglio, 1972; Mathis and Kleo, 1973; Herkstroeter, 1975a,b; Bensasson *et al.*, 1976), it has not yet been clearly demonstrated for any other case, but is likely to be the mechanism of quenching by certain dyes (Merkel and Kearns, 1972a;

* As mentioned in the introduction, electronic → vibrational energy transfer, which appears to be the mechanism of the (relatively slow) quenching by solvents, will not be discussed (Merkel and Kearns, 1972a).

Herkstroeter, 1975b; Smith *et al.*, 1975), metal complexes (Farmilo and Wilkinson, 1973; Wilkinson, 1978; Carlsson *et al.*, 1972, 1974; Guillory and Cook, 1973; Flood *et al.*, 1973) and other compounds with very extensive conjugated systems which would permit the triplet energy of the quencher to be below 22 kcal.

Quenching of singlet oxygen by triplet dyes has also been reported (Kenner and Khan, 1975, 1976; Duncan and Kearns, 1971); this reaction is also probably an energy transfer reaction, but it is unlikely to be important except at extremely high light intensities, for example, under flash photolysis conditions. This will not be discussed further.

$$^3\text{Dye} + {}^1O_2 \rightarrow {}^3O_2 + {}^1\text{Dye}$$

It should be noted that compounds with low triplet energies and extended conjugated systems also have low oxidation potentials, so that charge-transfer would also be a feasible mechanism of singlet oxygen quenching in these systems.

B. Charge-Transfer Quenching

The second mechanism, which is more general than the energy-transfer mechanism, is charge-transfer quenching. This reaction involves interaction of the very electron-poor singlet oxygen molecule with electron donors to give a charge-transfer complex (or perhaps even complete electron-transfer in some cases); intersystem crossing restrictions are relaxed in the complex, which can then dissociate to donor and ground-state oxygen (Ouannès and Wilson, 1968; Furukawa and Ogryzlo, 1972; Young and Martin, 1972; Young and Brewer, 1978; Ogryzlo, 1978).

$$D + {}^1O_2 \rightleftharpoons [D^+ \cdots O_2^-]^1 \rightleftharpoons [D^+ \cdots O_2^-]^3 \rightleftharpoons D + {}^3O_2$$

The *reverse* of this process has been suggested recently to account for chemistry which was interpreted as coming from thermally activated generation of singlet oxygen from ground-state oxygen in the presence of certain donors (Turro *et al.*, 1976). This reaction is a special case of a general mechanism of quenching of excited molecules by ground state donors or acceptors (Weller, 1967; Rehm and Weller, 1969). The rate of the general reaction has been shown to be a simple function of the oxidation potential of the donor, the reduction potential of the acceptor and the excitation energy of the excited component. Since singlet oxygen acts as the electron accepting component, the most easily oxidized compounds are the best quenchers. Young and Martin (1972) have shown that the rate of $^1O_2$ quenching by DABCO fits a plot of the Weller function for hydrocarbon fluorescence quenching, and also that the rate of singlet oxygen quenching by amines is correlated with their ionization potentials; a similar relationship was reported for gas phase quenching data (Furukawa and Ogryzlo, 1972). Substituted dimethylanilines give a Hammett ρ for $^1O_2$ quenching of -1.71 (Young *et al.*, 1974). Substituted phenols and methoxybenzenes give a linear correlation of the logarithm of their rate of total quenching of singlet oxygen with their half-wave potential for oxidation (Foote *et al.*, 1976; Thomas and Foote, 1978; Thomas, 1973). Paquette *et al.* (1976) have used a very similar frontier-electron treatment to arrive at the conclusion that substituted hydrazides can quench singlet oxygen by a charge-transfer mechanism, although definite evidence for such quenching was not presented.

Types of compound which probably quench or react by the charge transfer mechanism include amines, phenols, some metal complexes, and perhaps such compounds as sulfides, iodide, azide, superoxide ion, and similar electron-rich compounds. (See discussion of each compound type for references.) Some compounds (such as amines, sulfides, and phenols) which react by this mechanism not only quench but also react. The proportion of reaction versus quenching is dependent on structure and probably on solvent. It is likely that the charge-transfer complex which results can either transfer the electron back, giving ground state oxygen, or combine to give product (DO_2)

$$D + {}^1O_2 \rightarrow (D^+ \cdots O_2^-) \begin{cases} \rightarrow D + {}^3O_2 \\ \rightarrow DO_2 \end{cases}$$

The ratio of reaction to quenching varies with structure; DABCO quenches, but does not react (Ouannès and Wilson, 1968; Smith, 1972) whereas triethylamine has a quenching/reaction ratio of 8.4 in pyridine (Smith, 1972). Similarly, triphenyl phenol quenches without reaction in methanol or benzene (Thomas, 1973), whereas α-tocopherol gives a quenching/reaction ratio of 13.5 in methanol (Foote *et al.*, 1974). The ratio of quenching/reaction also seems to depend on solvent, but systematic investigations of this problem have not yet been carried out.

C. Other Mechanisms of $^1O_2$ Quenching

No other quenching mechanisms have been clearly identified. External heavy-atom quenching by halogenated solvents does not seem to be important (Foote and Denny, 1971; Young *et al.*, 1973b; Wilkinson, 1978). The quenching rate of $^1O_2$ by ethyl iodide is $2 \times 10^4\ M^{-1}\ sec^{-1}$, which is comparable to that of most organic solvents without heavy atoms (Wilkinson, 1978). The lifetime of singlet oxygen in bromobenzene actually seems to be slightly longer than that in benzene (Merkel and Kearns, 1972a; Farmilo and Wilkinson, 1973). Formation of a metal–O_2 complex which decays to ground-state oxygen has been discussed as a possible mechanism for quenching by metal complexes (Carlsson *et al.*, 1972). The "energy pooling" process, by which two singlet oxygen molecules interact to give to ground-state molecules and chemiluminescence is a special case of energy transfer, and is probably important only in the gas phase (Furukawa *et al.*, 1970).

$$2\,{}^1O_2 \rightarrow 2\,{}^3O_2 + h\nu$$

IV. TYPES OF QUENCHER

A. Carotenes

Carotenes were among the first compounds to be conclusively identified as singlet oxygen quenchers (Foote and Denny, 1968). They were investigated because it was suspected that part of their biological role might be to prevent biological damage by quenching singlet oxygen (see Section V). Carotenes with eleven or more conjugated double bonds quench at a rate which is close to the diffusion-controlled limit. Although $3 \times 10^{10}\ M^{-1}\ \text{sec}^{-1}$ was originally suggested for the rate of quenching by β-carotene, this was an upper limit, set by the diffusion rate of oxygen (Foote *et al.*, 1970a). The most reliable value for this quenching constant, as determined by pulsed laser techniques is about $1.3 \times 10^{10}\ M^{-1}\ \text{sec}^{-1}$ in benzene (Wilkinson, 1978). Several other carotenes in the 9–15 double-bond range quench at a comparable rate (Foote *et al.*, 1970a).

As discussed above, carotenes have been shown to quench singlet oxygen by an energy transfer mechanism. This mechanism was suggested originally because (1) quenching of singlet O_2 by 15,15′-*cis*-β-carotene is accompanied by isomerization to the trans isomer (Foote *et al.*, 1970c) and (2) the rate of quenching depends on the number of conjugated double bonds (Fig. 3) (Foote *et al.*, 1970a). From the dependence of rate on length of the conjugated chain, it was suggested that energy transfer is exothermic to carotenes of 11 or more conjugated double bonds, and endothermic to those with fewer. Mathews-Roth *et al.* (1974) found a similar relationship to obtain for a variety of naturally occurring carotenoids; they found the rate begins to drop off with an 8-double bond carotenoid.

Direct measurements of triplet energies of carotenes are not available. However, triplet energies estimated by indirect means (extrapolation from shorter polyenes of known triplet energy) (Mathis, 1970; Mathis and Kleo, 1973; Bensasson *et al.*, 1976) or ability to quench sensitizers of known triplet energy (Bensasson *et al.*, 1976; Herkstroeter, 1975a; Mathis and Kleo, 1973) are consistent with the conclusions reached by singlet oxygen

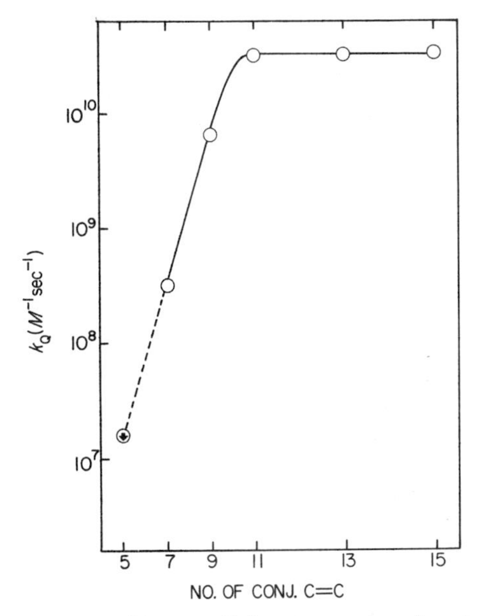

Fig. 3. Singlet oxygen quenching rate (k_Q) versus number of conjugated double bonds of five isoprenologous carotenoids (7–15 C=C) and retinol (5 C=C) (Foote *et al.*, 1970a). Value for retinol is an upper limit.

quenching; carotenes with nine or more double bonds have triplet energies ≤ 22 kcal; those with fewer have higher energies, and energy transfer from singlet oxygen is correspondingly inefficient.

The values of the quenching rate constant for β-carotene determined by Matheson and Lee (1972a,b) are an order of magnitude lower than those found by other investigators. The values obtained by these workers for other systems are also consistently lower than those of others for reasons which are not clear, but which may reside in a systematic error in their technique, already discussed.

Carotenes, although they quench $^1O_2$ very efficiently, are not very stable to oxidation, especially in the presence of light. This factor limits their usefulness as photostabilizers. The nature of the oxidation products is not completely certain, but Schenck and Schade (1970) reported the formation of β-ionone; that and several conventional photooxidation products were reported by Tsukida *et al.* (1966).

B. Other Systems with Extensive Conjugation

Certain other heavily conjugated systems also quench singlet oxygen efficiently. The mechanism of quenching by these compounds is probably also energy-transfer to a low-lying triplet. Two such systems are compounds **1** (Lee, 1972; Foote, 1976) and **2** (Merkel and Kearns, 1972a) whose structures and quenching rate (M^{-1} sec^{-1}) are shown below. Like the closely related **1**, several stilbene quinones such as **3** also quench $^1O_2$ efficiently (Taimr and Pospišil, 1976).

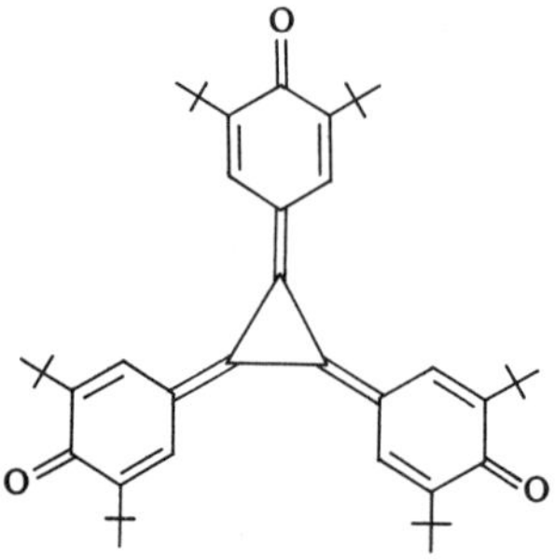

3×10^{10} (C_6H_6)

(**1**)

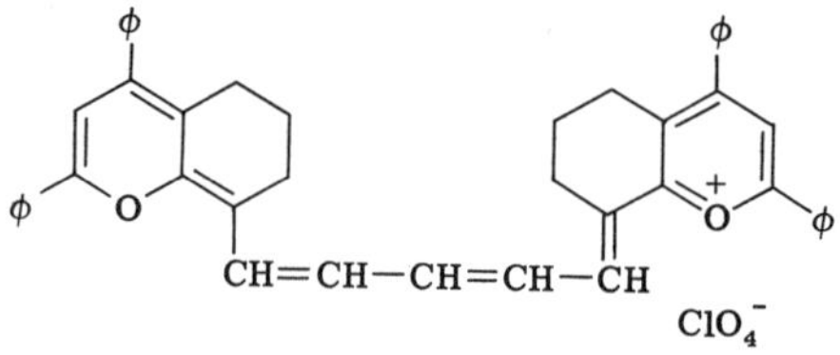

3×10^{10} (CH_3CN)

(**2**)

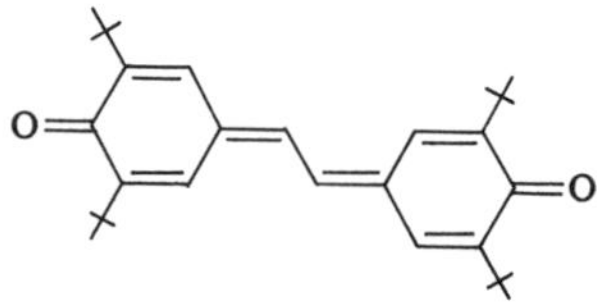

(**3**)

It is not certain whether these compounds quench with or without reaction; a large number of cyanine dyes similar to **2** have been found to react with $^1O_2$ at a high rate (Byers *et al.*, 1976). However, both **1** and **2** absorb in the near infrared, so that they probably have low triplet levels.

Many azomethine dyes were studied by Smith *et al.* (1975) and found to be efficient quenchers of $^1O_2$. Some of the dyes were better quenchers than predicted from their energy-transfer ability alone. Evidence has been presented (Herkstroeter, 1975b) that at least some of these compounds have triplet energies below 22 kcal and thus quench by the energy transfer mechanism. A combination of the energy- and charge-transfer mechanisms was found to be most likely. Rates of $^1O_2$ quenching ranging up to 10^9 M^{-1} sec^{-1} in pyridine, and over 10^9 M^{-1} sec^{-1} in CH_3CN were reported. Examples of dyes and their quenching rates are shown below. These dyes do not react appreciably with singlet oxygen.

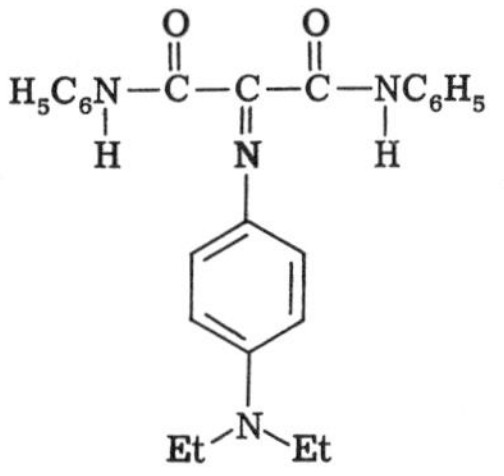

1.6×10^7
(pyridine)

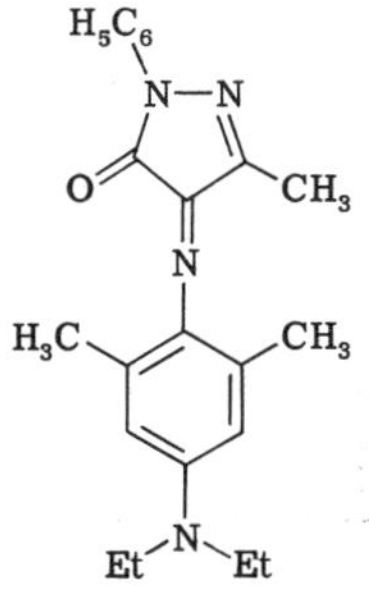

4×10^9
(CH_3CN)

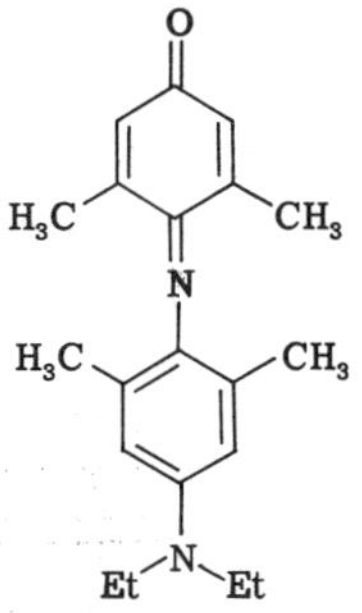

10^9
(pyridine)

C. Amines

The evidence for the charge transfer mechanism of $^1O_2$ quenching by amines was discussed in Section III,B. Primary aliphatic amines do not quench $^1O_2$ very efficiently; secondary and tertiary aliphatic amines quench at rates which vary from about $1\text{–}100 \times 10^6$ M^{-1} $\sec^{-1}$ (Young *et al.*, 1973a; Young and Brewer, 1978; Monroe, 1977). The rates appear to be somewhat lower in methanol than in other solvents, probably because of hydrogen bonding to the basic nitrogen. The rates of aromatic amines depend on substituent; the most electron-rich compounds have rates which approach 10^9 M^{-1} $\sec^{-1}$ (Young *et al.*, 1973a; Carlsson *et al.*, 1974; Dalle *et al.*, 1972).

DABCO quenches singlet oxygen without any reaction at all in all solvents that have been investigated (Ouannès and Wilson, 1968; Foote *et al.*, 1970b; Smith, 1972; Gollnick and Lindner, 1973). Other amines react to a greater or lesser extent; the products are complex and often result from secondary reactions. A principal process is oxidation of alkyl groups at the position α to the nitrogen to the aldehyde level. Abstractable hydrogen atoms α to the amine are a requirement for reaction to occur; in their absence, amines are pure quenchers for singlet oxygen (Gollnick and Lindner, 1973). *N*-Methyl groups, in particular are readily oxidized, leading to demethylation (Lindner *et al.*, 1972; Fisch *et al.*, 1970, 1971; Belluš *et al.*, 1972).

Davidson and Trethewey (1975, 1977) studied the photooxidation of triethylamine and other amines in various solvents. They fully explored the remarkable complexity of these reactions, which can involve singlet and triplet excited states of the sensitizer as well as singlet oxygen, and showed conditions under which each reaction occurs.

They found that a change of mechanism occurs as a function of amine concentration. At high concentrations, the dye triplet (and singlet, to some extent) is quenched by the amine; at lower concentrations, the reaction goes by a singlet oxygen mechanism. The use of water in the solvent promotes the triplet quenching mechanism for two reasons: the concentration of oxygen in water is an order of magnitude lower in water than it is in organic solvents, and the singlet oxygen lifetime is minimal in water. One can calculate (using Davidson's data for the rate constant for Rose Bengal triplet quenching) that half the triplet would be quenched at 0.25 M triethylamine in pure methanol under pure oxygen. At concentrations below 0.02 M amine, negligible triplet quenching would be expected. However, triplet quenching would still be appreciable at 0.02 M amine under air or in water, and can be more efficient in other solvents. In addition, Davidson and Trethewey showed that the amine oxidation rate can either be promoted or inhibited by an increase in oxygen or amine concentration, depending on the conditions.

Matheson and Lee (1972c) studied the quenching of singlet oxygen by amines in Freon 113 by a laser technique; their rate constants appear to be lower than those found by other workers by 1–2 orders of magnitude. Since charge transfer is important in the quenching, it might be expected that rates in the nonpolar Freon-113 would be lower than in more polar solvents. For example, rates are somewhat lower in the gas phase than in solution (Furukawa and Ogryzlo, 1972), but this does not seem to account for all the difference, since rates for carotene (Matheson and Lee, 1972a,b), bilirubin (Matheson *et al.*, 1974b), and biliverdin (Matheson and Toledo, 1977), measured by these investigators are also an order of magnitude lower, compared to those of other groups.

Monroe (1977) has recently determined rates of quenching of rubrene photooxidation by many amines, using the "integrated" technique (see Section II,D). The values he determines for quenching rates are somewhat higher than those of other workers (e.g., DABCO, $5.2 \times 10^7\ M^{-1}\ \mathrm{sec}^{-1}$; triethylamine $6.5 \times 10^7\ M^{-1}\ \mathrm{sec}^{-1}$ in $CHCl_3$) and may reside in the fact that these amines can also quench sensitizer excited states (see Section II,D). However, Monroe showed that rubrene fluorescence was not being quenched under the conditions, and that the rate of photooxidation was independent of $[O_2]$, which tends to rule out singlet or triplet sensitizer quenching.

Atkinson *et al.* (1973) showed that amino groups were effective in protecting various reactive groups in a molecule against singlet oxygen attack. Paquette *et al.* (1976) showed a similar effect for hydrazide groups, although they did not demonstrate that quenching was actually occurring.

Deneke and Krinsky (1976, 1977) found that DABCO and other cyclic tertiary diamines added to the reaction of NaOCl and H_2O_2 (a chemical singlet oxygen source) gave *enhanced* $^1O_2$ dimol emission; an acyclic diamine inhibited the emission, as did NaN_3. The effect did not appear to be due to an acceleration of the rate of the hypochlorite–H_2O_2 reaction. This puzzling phenomenon has not yet been satisfactorily explained.

D. Phenols

Phenols are capable of both reacting with and quenching singlet oxygen, depending both on their substitution and reaction conditions. The rate of total quenching ranges up to $7 \times 10^8\ M^{-1}\ \mathrm{sec}^{-1}$ (for tocopherol in methanol, Foote *et al.*, 1974). Thomas (1973; Foote *et al.*, 1976; Thomas and Foote, 1978) made a detailed study of the reaction rate of a number of 2,4,6-trisubstituted phenols, and found the log of the total quenching rate for singlet oxygen to be a linear function of the halfwave oxidation potential of the phenol. Phenols and phenol methyl ethers (Saito *et al.*, 1972) fit on the same line, suggesting that the phenolic hydrogen does not take part in the rate-determining step. They also found evidence for the intermediacy of the phenoxy radicals in one case by an absorption spectroscopic technique. The ratio of quenching to reaction varied from very large with triphenylphenol to about 4 for 2,6-di-*tert*-butyl phenol in benzene.* There seems to be a solvent effect in the case of the tocopherols, where k_Q/k_R is 13.5 in methanol (Foote *et al.*, 1974) but 120 in pyridine (Fahrenholtz *et al.*, 1974) and about 100 in benzene (Stevens *et al.*, 1974a). The trend is similar to that observed with sulfides (Foote and Peters, 1971a,b). The effect of solvent on total rate has not been studied extensively, but the rate seems to be somewhat faster in methanol than in nonpolar solvents.

Based on the rate effects and the detection of the phenoxy radical, Foote *et al.* (1976) suggested that the mechanism of phenol oxidation is that shown below. Other authors have also suggested a similar mechanism (Stevens *et al.*, 1974a; Saito *et al.*, 1972). The mechanism appears to involve charge-transfer but not complete electron transfer in the rate-determining step, since the slope of a plot of a log k versus $E_{1/2}$ is -3.2, whereas it would have to be -16.9 for complete electron-transfer (Thomas and Foote, 1978).

* 2,6-Di-*tert*-butyl phenol has a rate constant of $10^6\ M^{-1}\ \mathrm{sec}^{-1}$ in CH_3OH; the value of $k_d/(k_R + k_Q)$ (equal to the concentration at which it quenches half the singlet oxygen) is 0.11 M in CH_3OH (Thomas, 1973). Thus, this compound, frequently used to inhibit free radical side reactions, is not completely inert to singlet oxygen; however, at concentrations below 0.01 M, its influence on the $^1O_2$ reaction is small in methanol, although it will be larger for solvents in which $^1O_2$ has a longer lifetime, and users should be aware of this.

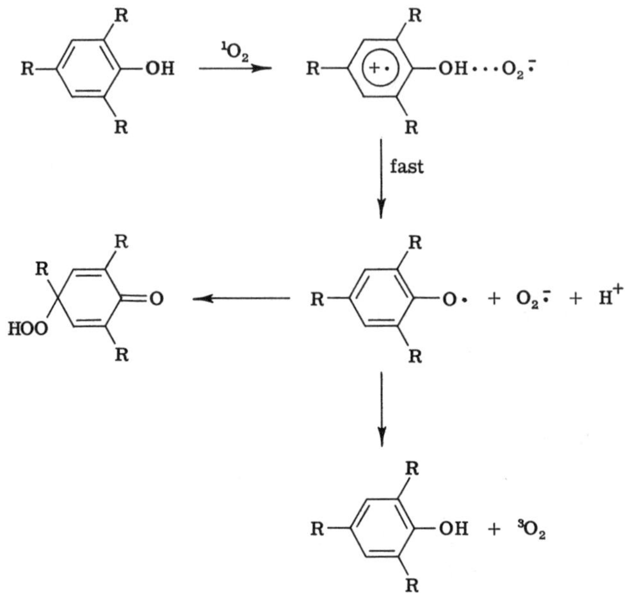

However, it is important to note that, like amines, phenols also react rapidly with sensitizer triplets. Tyrosine, which would not be expected to react very rapidly with singlet oxygen (Foote *et al.*, 1976) appears to interact readily with dye triplets (Zwicker and Grossweiner, 1963; Crysochoos and Grossweiner, 1968). Muszkat and Weinstein (1975) also obtained CIDNP results which they interpreted as showing a Rose Bengal triplet–phenol interaction as the primary step in the photooxygenation of phenols. However, a singlet oxygen mechanism has also been demonstrated by competition kinetics (Foote *et al.*, 1976; Thomas, 1973). It is probable that the mechanism can switch from reaction with triplet sensitizer to reaction with singlet oxygen as the concentration of the phenol is lowered, as with amines.

Grams and Eskins (1972) were the first to measure the reactivity of singlet oxygen toward tocopherols. They showed that the total quenching rate of different tocopherols parallels their biological activity, but did not separate k_R and k_Q.

Brabham and Lee (1976) have also used their laser technique to measure a quenching rate for α-tocopherol in Freon 113; as with other substrates, their rate is lower than that found by other workers by an order of magnitude, presumably for similar reasons.

E. Metal Complexes

A large number of metal complexes, primarily Ni(II) chelates, have been screened for singlet oxygen quenching ability. Some complexes are extremely reactive; rates ranging up to diffusion-controlled have been observed. Many of these compounds are used commercially as photostabilizers for polymers, and their ability to quench singlet oxygen has been suggested to be important to their role as photostabilizers (Guillory and Cook, 1973; Carlsson and Wiles, 1973; Breck *et al.*, 1974; Felder and Schumacher, 1973). However, there are a number of good photostabilizers which do not quench singlet oxygen efficiently, and a number of good singlet oxygen quenchers which are not good stabilizers (Zweig and Henderson, 1975a,b; Wiles, 1978; Carlsson *et al.*, 1973). Some of the stabilizers also retard the photooxidative degradation of unsaturated food oils (Carlsson *et al.*, 1976).

There appear to be several features which contribute to the ability of a compound to protect polymers against photodegradation: stability toward photooxidation, ability to inhibit free radical chain reaction, selective light absorption (leading to a screening effect) without sensitizing ability, ability to quench excited states other than oxygen, ability to decompose peroxides without initiating radical chain reactions, and ability to quench singlet oxygen. The "ideal" photostabilizer would combine all these attributes. It seems likely that singlet oxygen quenching ability is certainly not the only, and perhaps not even the major function of photostabilizers, but it may be important under some conditions, particularly in the early stages of photodegradation. Certain amine and phenol antioxidants probably act in a similar multiple way.

The mechanism of $^1O_2$ quenching by metal complexes is not certain. However, the most effective quenchers of $^1O_2$ have triplet energies estimated to be near or below the excitation energy of $^1O_2$ (Farmilo and Wilkinson, 1973; Carlsson *et al.*, 1972; Guillory and Cook, 1973), and a correlation between $^1O_2$ quenching ability and ability to quench triplet pentacene (which is nearly isoenergetic with singlet oxygen) has been made (Furue and Russell, 1978). It thus seems likely that an energy-transfer mechanism is involved, at least for some of the complexes. However, a charge-transfer mechanism would also be possible for some.

In general, the quenching ability of complexes decreases in the order Ni(II) > Co > Cu, Pt, Pd, Zn; there does not seem to be much correlation with the magnetic properties of the complex. Even simple salts such as $NiCl_2 \cdot 6H_2O$ can quench with a reasonably high rate (Carlsson *et al.*, 1972, 1974). Sulfur substituents seem to be the best, especially the bisdithio-α-diketone ligands, whose Ni(II) chelates have $^1O_2$ quenching rates above 10^{10} M^{-1} $\sec^{-1}$, similar to that of β-carotene (Zweig and Henderson, 1975a); many other complexes quench at only slightly lower rates (Wiles, 1978; Furue and Russell, 1978; Wilkinson, 1978).

F. Sulfides

Sulfides both react with singlet oxygen and quench it. The quenching/reaction ratio is solvent dependent (Foote and Peters, 1971a,b). In protic solvents, the process involves mainly reaction; in aprotic solvents, largely quenching occurs; the total quenching rate does not vary much with solvent. Although this solvent effect was suggested to be caused by hydrogen bonding to an intermediate persulfoxide, it is quite likely that the actual explanation is more complex, since very recent work has shown the mechanism involves more steps than previously believed (Kacher, 1977; M. L. Kacher and C. S. Foote, unpublished). The role of solvent is not clear, but it may conceivably influence the partitioning of a charge-transfer complex between reaction and quenching.

$$R_2S \xrightarrow{^1O_2} \begin{cases} \xrightarrow{\text{protic solvent}} R_2SO \\ \xrightarrow{\text{aprotic solvent}} R_2S + {}^3O_2 \end{cases}$$

Disulfides behave very similarly to sulfides (Murray and Jindal, 1972; Stevens *et al.*, 1974a). Again, both quenching and reaction are observed.

G. Bilirubin and Biliverdin

Because of its possible importance in the phototherapy of neonatal jaundice (Bergsma *et al.*, 1970; Odell *et al.*, 1972) the interaction of bilirubin and biliverdin with singlet oxygen has received considerable scrutiny (Foote, 1976). Bilirubin reacts with singlet oxygen (McDonagh, 1971, 1972; Bonnett and Stewart, 1972a,b, 1975). The total process involves both reaction and quenching (Foote and Ching, 1975; Stevens and Small, 1976); the rate of quenching is on the order of $2 \times 10^9\ M^{-1}\ \text{sec}^{-1}$ and k_R is an order of magnitude lower. Matheson *et al.* (1974) also studied the photooxidation of bilirubin using a laser technique, and again found a reaction rate in Freon which is considerably lower. However, they also found a rate in D_2O in basic solution which was very high. Biliverdin was shown to be a quencher by McDonagh (1972); its quenching rate is $3.3 \times 10^9\ M^{-1}\ \text{sec}^{-1}$ and its reaction rate is immeasurably slow (Stevens and Small, 1976).

Matheson and Toledo (1977) studied the biliverdin quenching; as usual their rates were somewhat low, although not by as much as in other cases (a factor of 4). Interestingly, they found rates of biliverdin quenching in basic D_2O ranging up to $6 \times 10^{10}\ M^{-1}\ \text{sec}^{-1}$, which is in excess of the diffusion-controlled rate and suggests that something peculiar is going on in this system.

H. Miscellaneous Compounds

A large number of miscellaneous compounds have been found to quench singlet oxygen. It is likely that most of these quench by the charge-transfer mechanism, but no mechanistic studies have been carried out. Several inorganic anions are quenchers for $^1O_2$; among these are I^- (Rosenthal and Frimer, 1976) N_3^- (Foote *et al.*, 1972b; Hasty *et al.*, 1972; Gollnick *et al.*, 1972), and O_2^- (Guiraud and Foote, 1976; Rosenthal, 1975). Although charge transfer is the most probable mechanism of quenching for these compounds, nucleophilic adduct formation is also conceivable.

$$N:^- + {}^1O_2 \rightarrow N{-}O{-}O^- \rightarrow N^- + {}^3O_2$$

Nitroso compounds quench singlet oxygen efficiently; their rate approaches diffusion-controlled (Singh and Ullman, 1975, 1976). Only a small fraction of the quenching proceeds through reaction. An energy-transfer mechanism was suggested. Closely related to nitroso compounds are azodioxides **4**, which are nitroso dimers. Singh and Ullman (1976) found several cyclic azodioxides to be excellent quenchers of triplet sensitizers, but only **5** quenched singlet oxygen at all. The authors suggested that the singlet oxygen quenching was occurring by way of the dinitroso compound **6**, which is present in equilibrium with **5** at low concentration.

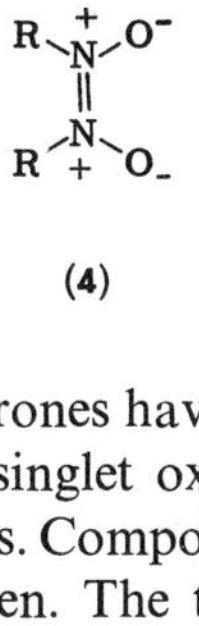

(4) 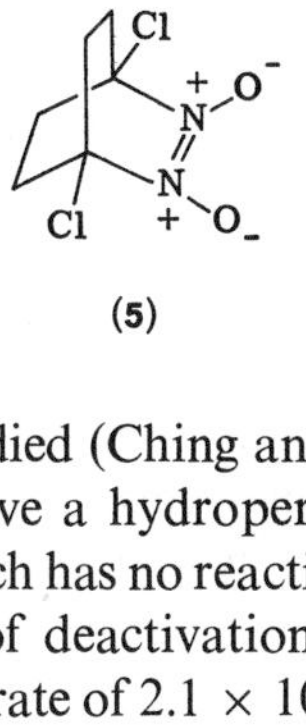 (5) 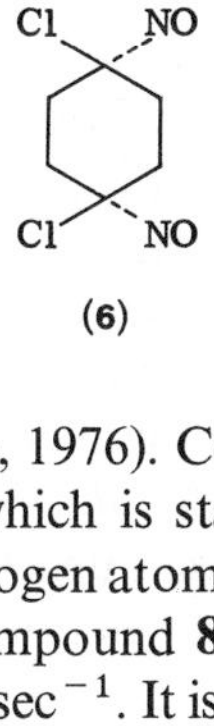 (6)

Two nitrones have been studied (Ching and Foote, 1976). Compound **7** reacts with singlet oxygen to give a hydroperoxide which is stable at low temperatures. Compound **8**, which has no reactive hydrogen atoms, quenches singlet oxygen. The total rate of deactivation for compound **8** is 5×10^7 M^{-1} sec^{-1} whereas **7** reacts at a rate of 2.1×10^7 M^{-1} sec^{-1}. It is not certain what the mechanism of quenching by **8** is, but it is probably charge-transfer.

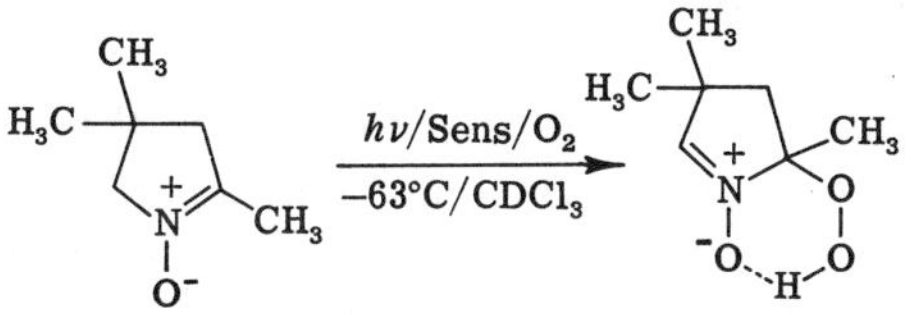

$h\nu$/Sens/O_2, $-63°C/CDCl_3$

(7) 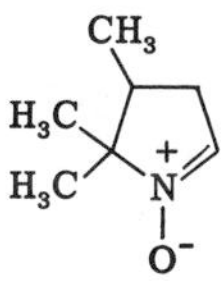 (8)

Several piperidine-*N*-oxyl derivatives (**8**) have been shown to be moderately efficient singlet oxygen quenchers (Felder and Schumacher, 1973;

Belluš *et al.*, 1972). The *N*-oxyls are much better quenchers than the corresponding secondary amines, but about as good as the N–CH_3 derivatives.

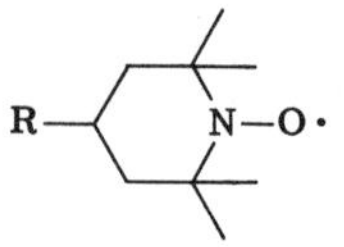

R—[piperidine ring with gem-dimethyl groups]—N—O·

(9)

Nilsson *et al.* (1972), as part of their definitive study of amino acid photooxidation, reported that tryptophan, histidine, and methionine quench singlet oxygen about 10 times as fast as they react with it. However, the total quenching rates and reaction rates were gotten from two distinctly different types of experiment. Merkel and Kearns (1975) reported that diphenylisobenzofuran did not quench singlet oxygen, and that its only reaction with singlet oxygen was chemical. Foote and Ching (1975) came to similar conclusions regarding most olefins and other single oxygen acceptors, and, in general, there does not seem to be any evidence that ordinary singlet oxygen acceptors (hydrocarbons, heterocycles) quench singlet oxygen at all. Thus, the conclusion of Nilsson *et al.* seems suspect, and should be restudied.

Several authors have suggested that singlet oxygen can be quenched by superoxide dismutase, the enzyme which is responsible for the dismutation and detoxification of superoxide ion *in vivo* (Weser and Paschen, 1972; Paschen and Weser, 1973; Weser *et al.*, 1975; Finazzi Agró *et al.*, 1972). However, careful investigation failed to show any singlet oxygen quenching activity for superoxide dismutase (Hodgson and Fridovich, 1974; Michelson, 1974; Goda *et al.*, 1974) and it seems likely that what was suggested to be singlet oxygen quenching actually derives from inhibition of superoxide ion rather than singlet oxygen.

V. BIOLOGICAL APPLICATIONS

Singlet oxygen quenchers have been used to test for the presence of singlet oxygen in a number of biological systems. Because of lack of space, this area will not be covered in detail here; only illustrative examples will be given. The papers by Foote (1976) and Krinsky (this volume, Chapter 12) have a more detailed survey of this area. Many of the biological uses of singlet oxygen quenchers have not taken account of the pitfalls mentioned in Section II. It is important that the biological papers be read very carefully to establish that singlet oxygen quenching (and not sensitizer quenching or some other interaction) is actually responsible for the observations.

Perhaps the most dramatic use of singlet oxygen quenchers in a biological system has been the use of β-carotene to treat erythropoetic protoprophyria (EPP), a human photosensitivity disease. The photosensitivity of patients with this disease seems to be caused by deposition of photosensitizing porphyrins in the skin. Blood from such patients also has elevated protoporphyrin levels in the red cells. Dramatic reports have appeared recently that the photosensitivity of patients with EPP can be largely relieved by orally administering β-carotene (Mathews–Roth *et al.*, 1970; DeLeo *et al.*, 1976). The mechanism of action is not known, but singlet oxygen quenching is an attractive hypothesis.

Nilsson *et al.* (1975) showed tht photosensitized hemolysis of red blood cells by kynurenic acid (related to the photooxidation product of tryptophan) probably involves singlet oxygen. Competitive quenching by β-carotene and histidine, and the deuterium isotope effect on the lifetime of singlet oxygen were used in this study. In the same study, it was shown that chlorpromazine and demethylchlorotetracycline, which are photosensitizing drugs, appear to react by a nonsinglet oxygen mechanism, while hematoporphyrin appears to involve mixed type 1 and type 2 mechanisms.

The observation of a D_2O effect and inhibition by sodium azide led to the conclusion that singlet oxygen was responsible for a substantial proportion of the gene conversion which occurs on sensitized photodynamic action in yeast cells (Kobayashi and Ito, 1976, 1977; Ito, 1977).

Michelson and Durosay (1977) showed that photolysis of human erythrocytes at 255 nm does not appear to involve singlet oxygen, but that the treatment with a reduced riboflavin system known to generate superoxide apparently did. The D_2O effect was the major test used by these authors, and should probably be supplemented with other tests of a more quantitative nature.

VI. RATE CONSTANTS

Table I is a representative selection of quenching rate constants from the literature selected by the criteria discussed in Section I. Where several values are given by an author, the most recent is listed. The rate constants are listed as $k_Q + k_R$ (total quenching); where it was determined separately, k_R (reaction) is also listed. More complete tables may be found in Belluš (1978). Where rates for a given compound type are described adequately in the text, they are not repeated in Table I.

Table I Rates of Total Quenching ($k_Q + k_R$) or Reaction (k_R) for Various Compounds

Compound	$(k_Q + k_R) \times 10^{-6}$ $(M^{-1}\ sec^{-1})$[a]	$k_R \times 10^{-6}$ $(M^{-1}\ sec^{-1})$[a]	Solvent	Ref.
Carotenes				
β-Carotene (11 C=C)	1.2×10^4		C_6H_6	b
Isozeaxanthine (11 C=C)	1.2×10^4		$C_6H_6:CH_3OH$ (3:2)	c
Lutein (10 C=C)	1.4×10^4		$C_6H_6:CH_3OH$ (3:2)	c
p-438 (9 C=C)	2.1×10^4		$C_6H_6:CH_3OH$ (3:2)	c
C_{35}-Carotenoid (9 C=C)	2.4×10^3		$C_6H_6:CH_3OH$ (4:1)	d
p-422 (8 C=C)	8.2×10^3		$C_6H_6:CH_3OH$ (3:2)	c
C_{30}-Carotenoid (7 C=C)	68		$C_6H_6:CH_3OH$ (4:1)	d
Systems of extended conjugation				
See Section IV,B for examples				
Amines				
Et_2NH	1.9		CH_3OH	e
	15		$CHCl_3$	f
Et_3N	14		CH_3OH	e
	65		$CHCl_3$	f
DABCO	8.9		CH_3OH	e
	17		CH_3OH	g
	52		$CHCl_3$	f
	29		CS_2	g
	40		C_6H_6	g
	34		Isooctane	b
Phenols				
α-Tocopherol	530	36	CH_3OH	h
	148	~1.5	C_6H_6	i
	250	2.1	Pyridine	j
	90		Cyclohexane	i
	120		Isooctane	j
2,4,6-Triphenyl phenol	239	Very small	CH_3OH	k
	21.3		C_6H_6	k
2,6-Di-*tert*-butyl-4-methyl phenol	4.0	0.20	CH_3OH	k
	0.64	0.12	C_6H_6	k
Ni(II) complexes				
dithiobenzil	2.2×10^4		Toluene	l
bis(diisopropyldiethiophosphato)	7.6×10^3		CCl_4	m
"Negopex A"	3.1×10^3		C_6H_6	b
Sulfides				
Diethyl sulfide	20	0.61	C_6H_6	n,o
Diethyl sulfide	17	17	CH_3OH	n,o
Thioanisole	0.8		C_6H_6	o
Bilirubin and biliverdin				
See Section IV,G				

Table I (*Continued*)

Compound	$(k_Q + k_R) \times 10^{-6}$ $(M^{-1}\ sec^{-1})^a$	$k_R \times 10^{-6}$ $(M^{-1}\ sec^{-1})^a$	Solvent	Ref.
Miscellaneous compounds				
$Bu_4N^+I^-$	280		ϕBr : acetone (2:1)	p
NaN_3	220		CH_3OH	q
NaN_3	280		$CH_3OH:H_2O$ (8:1)	r
O_2^-	1.25×10^3		DMSO	s
O_2^-	36		DMSO	t
2-Nitroso-2-methylpropane	9.3×10^3		$CH_2Cl_2:CH_3OH$ (15:1)	u

[a] Calculated using the following k_d values: CH_3OH, 1.1×10^5 [b]; C_6H_6, 4.0×10^4 [b]; $CHCl_3$, 1.7×10^4 [b]; CS_2, 5×10^3 [b]; C_6H_5/CH_3OH, 4:1 (v:v), 5.9×10^4, (Young and Brewer, 1978).
[b] Wilkinson, 1978.
[c] Mathews-Roth *et al.*, 1974.
[d] Foote *et al.*, 1970a.
[e] Young *et al.*, 1973a.
[f] Monroe, 1977.
[g] Foote *et al.*, 1972a.
[h] Foote *et al.*, 1974.
[i] Stevens *et al.*, 1974b.
[j] Fahrenholtz *et al.*, 1974.
[k] Thomas and Foote, 1978.
[l] Zweig and Henderson, 1975a.
[m] Furue and Russell, 1978.
[n] Foote and Peters, 1971a.
[o] Kacher, 1977.
[p] Rosenthal and Frimer, 1976.
[q] Hasty *et al.*, 1972.
[r] Foote *et al.*, 1972b.
[s] Guiraud and Foote, 1976.
[t] Rosenthal, 1975.
[u] Singh and Ullman, 1976.

ACKNOWLEDGMENT

Contribution No. 3869 from UCLA Department of Chemistry. Work of the Author's group and preparation of this manuscript were supported by NSF and NIH grants.

REFERENCES

Adams, D. R., and Wilkinson, F. (1972). *J. Chem. Soc., Faraday Trans. 2* **68,** 586.

Atkinson, R. S., Brimage, D. R. G., Davidson, R. S., and Gray, E. (1973). *J. Chem. Soc., Perkin Trans. 1* p. 960.

Belluš, D. (1978). *In* "Singlet Oxygen—Reactions with Organic Compounds and Polymers" (B. Rånby and J. F. Rabek, eds.), p. 61. Wiley, New York.

Belluš, D., Lind, H., and Wyatt, J. F. (1972). *Chem. Commun.* p. 1199.

Bensasson, R., Land, E. J., and Maudinas, B. (1976). *Photochem. Photobiol.* **23,** 189.

Bergsma, D., Hsia, Y.-Y., and Jackson, C., eds. (1970). "Bilirubin Metabolism in the Newborn." Williams & Wilkins, Baltimore, Maryland.

Bonnett, R., and Stewart, J. C. M. (1972a). *Arch. Int. Physiol. Biochim.* **80,** 951.

Bonnett, R., and Stewart, J. C. M. (1972b). *Biochem. J.* **130,** 895.

Bonnett, R., and Stewart, J. C. M. (1975). *J. Chem. Soc., Perkin Trans. 1* p. 224.
Brabham, D. E., and Lee, J. (1976). *J. Phys. Chem.* **80,** 2292.
Breck, A. K., Taylor, C. L., Russell, K. E., and Wan, J. K. S. (1974). *J. Polym. Sci.* **12,** 1505.
Byers, G. W., Gross, S., and Henrichs, P. M. (1976). *Photochem. Photobiol.* **23,** 37.
Carlsson, D. J., and Wiles, D. M. (1973). *Polym. Lett. Ed.* **11,** 759.
Carlsson, D. J., Mendenhall, G. D., Suprunchuk, T., and Wiles, D. M. (1972). *J. Am. Chem. Soc.* **94,** 8960.
Carlsson, D. J., Suprunchuk, T., and Wiles, D. M. (1973). *Polym. Lett. Ed.* **11,** 61.
Carlsson, D. J., Suprunchuk, T., and Wiles, D. M. (1974). *Can. J. Chem.* **52,** 3728.
Carlsson, D. J., Suprunchuk, T., and Wiles, D. M. (1976). *J. Am. Oil Chem. Soc.* **53,** 656.
Ching, T.-Y., and Foote, C. S. (1976). *Tetrahedron Lett.* p. 3771.
Chrysochoos, J., and Grossweiner, L. I. (1968). *Photochem. Photobiol.* **8,** 193.
Dalle, J. P., Magous, R., and Mousseron-Canet, M. (1972). *Photochem. Photobiol.* **15,** 411.
Davidson, R. S., and Trethewey, K. R. (1975). *J. Chem. Soc., Chem. Commun.* p. 674.
Davidson, R. S., and Trethewey, K. R. (1976). *J. Am. Chem. Soc.* **98,** 4008.
Davidson, R. S., and Trethewey, K. R. (1977). *J. Chem. Soc., Perkin Trans. 2* **169,** 173 and 178.
DeLeo, V. A., Poh-Fitzpatrick, M., Mathews-Roth, M. M., and Harber, L. C. (1976). *Am. J. Med.* **60,** 8.
Deneke, C. F., and Krinsky, N. I. (1976). *J. Am. Chem. Soc.* **98,** 4041.
Deneke, C. F., and Krinsky, N. I. (1977). *Photochem. Photobiol.* **25,** 299.
Duncan, C. K., and Kearns, D. R. (1971). *Chem. Phys. Lett.* **12,** 306.
Fahrenholtz, S. R., Doleiden, F. H., Trozzolo, A. M., and Lamola, A. A. (1974). *Photochem. Photobiol.* **20,** 519.
Farmilo, A., and Wilkinson, F. (1973). *Photochem. Photobiol.* **18,** 447.
Felder, B., and Schumacher, R. (1973). *Angew. Makromol. Chem.* **31,** 35.
Finazzi Agró, A., Giovagnoli, C., De Sole, P., Calabrese, L., Rotilio, G., and Mondovi, B. (1972). *FEBS Lett.* **21,** 183.
Fisch, M. H., Gramain, J. C., and Olesen, J. A. (1970). *Chem. Commun.* p. 13.
Fisch, M. H., Gramain, J. C., and Olesen, J. A. (1971). *Chem. Commun.* p. 663.
Flood, J., Russell, K. F., and Wan, J. K. S. (1973). *Macromolecules* **6,** 669.
Foote, C. S. (1976). *In* "Free Radicals in Biology" (W. A. Pryor, ed.), Vol. 2, p. 85. Academic Press, New York.
Foote, C. S., and Ching, T.-Y. (1975). *J. Am. Chem. Soc.* **97,** 6209.
Foote, C. S., and Denny, R. W. (1968). *J. Am. Chem. Soc.* **90,** 6233.
Foote, C. S., and Denny, R. W. (1971). *J. Am. Chem. Soc.* **93,** 3795.
Foote, C. S., and Peters, J. W. (1971a). 23rd IUPAC Congr., *Special Lect.* **4,** 129.
Foote, C. S., and Peters, J. W. (1971b). *J. Am. Chem. Soc.* **93,** 3795.
Foote, C. S., Chang, Y. C., and Denny, R. W. (1970a). *J. Am. Chem. Soc.* **92,** 5216.
Foote, C. S., Denny, R. W., Weaver, L., Chang, Y., and Peters, J. (1970b). *Ann. N. Y. Acad. Sci.* **171,** 139.
Foote, C. S., Chang, Y. C., and Denny, R. W. (1970c). *J. Am. Chem. Soc.* **92,** 5218.
Foote, C. S., Peterson, E. R., and Lee, K.-W. (1972a). *J. Am. Chem. Soc.* **94,** 1032.
Foote, C. S., Fujimoto, T. T., and Chang, Y. C. (1972b). *Tetrahedron Lett.* p. 45.
Foote, C. S., Ching, T.-Y., and Geller, G. G. (1974). *Photochem. Photobiol.* **20,** 511.
Foote, C. S., Thomas, M., and Ching, T.-Y. (1976). *J. Photochem.* **5,** 172; see Thomas and Foote, 1978.
Furue, H., and Russell, K. E. (1978). *In* "Singlet Oxygen—Reactions with Organic Compounds and Polymers" (B. Rånby and J. F. Rabek, eds.), p. 316. Wiley, New York.

Furukawa, K., and Ogryzlo, E. A. (1972). *J. Photochem.* **1,** 163.
Furukawa, K., Gray, E. W., and Ogryzlo, E. A. (1970). *Ann. N. Y. Acad. Sci.* **171,** 175.
Garner, A., and Wilkinson, F. (1978). *In* "Singlet Oxygen—Reactions with Organic Compounds and Polymers" (B. Ranby and J. F. Rabek, eds.), p. 48. Wiley, New York.
Goda, K., Kimura, T., Thayer, A. L., Kees, K., and Schaap, A. P. (1974). *Biochem. Biophys. Res. Commun.* **58,** 660.
Gollnick, K. (1968). *Adv. Photochem.* **6,** 1.
Gollnick, K., and Lindner, J. H. E. (1973). *Tetrahedron Lett.* p. 1903.
Gollnick, K., Haisch, D., and Schade, G. (1972). *J. Am. Chem. Soc.* **94,** 1747.
Grams, G. W., and Eskins, K. (1972). *Biochemistry* **11,** 606.
Guillory, J. P., and Cook, C. F. (1973). *J. Polym. Sci., Polym. Chem. Ed.* **11,** 1927.
Guiraud, H. J., and Foote, C. S. (1976). *J. Am. Chem. Soc.* **98,** 1984.
Hasty, N., Merkel, P. B., Radlick, P., and Kearns, D. R. (1972). *Tetrahedron Lett.* p. 49.
Herkstroeter, W. G. (1975a). *J. Am. Chem. Soc.* **97,** 3090.
Herkstroeter, W. G. (1975b). *J. Am. Chem. Soc.* **97,** 4161.
Hodgson, E. K., and Fridovich, I. (1974). *Biochemistry* **13,** 3811.
Ito, T. (1977). *Photochem. Photobiol.* **25,** 47.
Kacher, M. L. (1977). Ph.D. Dissertation, University of California, Los Angeles.
Kearns, D. R. (1971). *Chem. Rev.* **71,** 395.
Kenner, R. D., and Khan, A. U. (1975). *Chem. Phys. Lett.* **36,** 643.
Kenner, R. D., and Khan, A. U. (1976). *J. Chem. Phys.* **64,** 1877.
Kobayashi, K., and Ito, T. (1976). *Photochem. Photobiol.* **23,** 21.
Kobayashi, K., and Ito, T. (1977). *Photochem. Photobiol.* **25,** 385.
Lee, K.-W. (1972). M.S. Thesis, University of California, Los Angeles.
Lindner, J. E., Kuhn, H. J., and Gollnick, K. (1972). *Tetrahedron Lett.* p. 1705.
McDonagh, A. F. (1971). *Biochem. Biophys. Res. Commun.* **44,** 1306.
McDonagh, A. F. (1972). *Biochem. Biophys. Res. Commun.* **48,** 408.
Matheson, I. B. C., and Lee, J. (1972a). *Chem. Phys. Lett.* **14,** 350.
Matheson, I. B. C., and Lee, J. (1972b). *Bull. Am. Phys. Soc.* [2] **17,** 329.
Matheson, I. B. C., and Lee, J. (1972c). *J. Am. Chem. Soc.* **94,** 3310.
Matheson, I. B. C., and Toledo, M. M. (1977). *Photochem. Photobiol.* **25,** 243.
Matheson, I. B. C., Lee, J., Yamanashi, B. S., and Wolbarsht, M. L. (1974a). *J. Am. Chem. Soc.* **96,** 3343.
Matheson, I. B. C., Curry, N. U., and Lee, J. (1974b). *J. Am. Chem. Soc.* **96,** 3348.
Mathews-Roth, M. M., Pathak, M. A., Fitzpatrick, T. B., Harber, L. C., and Kass, E. H. (1970). *N. Engl. J. Med.* **282,** 1231.
Mathews-Roth, M. M., Wilson, T., Fujimori, E., and Krinsky, N. (1974). *Photochem. Photobiol.* **19,** 217.
Mathis, P. (1970). Ph.D. Thesis, University of Paris XI.
Mathis, P., and Kleo, J. (1973). *Photochem. Photobiol.* **18,** 343.
Mathis, P., and Vermeglio, A. (1972). *Photochem. Photobiol.* **15,** 157.
Merkel, P. B., and Kearns, D. R. (1971). *Chem. Phys. Lett.* **12,** 120.
Merkel, P. B., and Kearns, D. R. (1972a). *J. Am. Chem. Soc.* **94,** 1029.
Merkel, P. B., and Kearns, D. R. (1972b). *J. Am. Chem. Soc.* **94,** 7244.
Merkel, P. B., and Kearns, D. R. (1975). *J. Am. Chem. Soc.* **97,** 462.
Merkel, P. B., Nilsson, R., and Kearns, D. R. (1972). *J. Am. Chem. Soc.* **94,** 1030.
Michelson, A. M. (1974). *FEBS Lett.* **44,** 97.

Michelson, A. M., and Durosay, P. (1977). *Photochem. Photobiol.* **25,** 55.
Monroe, B. M. (1977). *J. Phys. Chem.* **81,** 1861. I thank Dr. Monroe for a prepublication copy of his manuscript and discussions.
Murray, R. W., and Jindal, S. L. (1972). *Photochem. Photobiol.* **16,** 147.
Muszkat, K. A., and Weinstein, M. (1975). *J. Chem. Soc., Chem. Commun.* p. 143.
Nilsson, R., Merkel, P. B., and Kearns, D. R. (1972). *Photochem. Photobiol.* **16,** 117.
Nilsson, R., Swanbeck, G., and Wennersten, G. (1975). *Photochem. Photobiol.* **22,** 183.
Odell, G. B., Schaffer, R., and Simopoulos, A. P., eds. (1972). "Phototherapy in the Newborn: An Overview." Natl. Acad. Sci., Washington, D.C.
Ogryzlo, E. A. (1978). *In* "Singlet Oxygen—Reactions with Organic Compounds and Polymers" (B. Rånby and J. F. Rabek, eds.), p. 17. Wiley, New York.
Ouannès, C., and Wilson, T. (1968). *J. Am. Chem. Soc.* **90,** 6528.
Paquette, L. A., Liotta, D. C., Liao, C. C., Wallis, T. G., Eickman, N., Clardy, J., and Gleiter, R. (1976). *J. Am. Chem. Soc.* **98,** 6413.
Paschen, W., and Weser, U. (1973). *Biochim. Biophys. Acta* **327,** 217.
Patterson, L. K., Porter, G., and Topp, M. R. (1970). *Chem. Phys. Lett.* **7,** 612.
Rånby, B., and Rabek, J. F. (1975). "Photodegradation, Photooxidation, and Photostabilization of Polymers." Wiley, New York.
Rehm, D., and Weller, A. (1969). *Ber. Bunsenges. Phys. Chem.* **73,** 834.
Rosenthal, I. (1975). *Isr. J. Chem.* **13,** 86.
Rosenthal, I., and Frimer, A. (1976). *Photochem. Photobiol.* **23,** 209.
Saito, I., Imuta, M., and Matsuura, T. (1972). *Tetrahedron* **28,** 5307.
Schenck, G. O., and Gollnick, K. (1958). *J. Chim. Phys.* **55,** 892.
Schenck, G. O., and Schade, G. (1970). *Chimia* **24,** 13.
Shlyapintokh, V. Ya., and Ivanov, V. B. (1976). *Usp. Khim.* **45,** 202.
Singh, P., and Ullman, E. F. (1975). *Abstr. Pap., 169th Natl. Meet., Am. Chem. Soc.* p. 0–11.
Singh, P., and Ullman, E. F. (1976). *J. Am. Chem. Soc.* **98,** 3018.
Smith, W. F., Jr. (1972). *J. Am. Chem. Soc.* **94,** 186.
Smith, W. F., Jr., Herkstroeter, W. G., and Eddy, K. L. (1975). *J. Am. Chem. Soc.* **97,** 2764.
Stevens, B., and Small, R. D., Jr. (1976). *Photochem. Photobiol.* **23,** 33.
Stevens, B., Perez, S. R., and Small, R. D. (1974a). *Photochem. Photobiol.* **19,** 315.
Stevens, B., Small, R. D., and Perez, S. R. (1974b). *Photochem. Photobiol.* **20,** 515.
Taimr, L., and Pospišil, J. (1976). *Angew. Makromol. Chem.* **52,** 31.
Thomas, M. (1973). Ph.D. Dissertation, University of California, Los Angeles; see Thomas and Foote, 1978.
Thomas, M. J., and Foote, C. S. (1978). *Photochem. Photobiol.* **27,** 683.
Tsukida, K., Yokota, M., and Ikeuchi, K. (1966). *Bitamin* **33,** 179.
Turro, N. J., Ramamurthy, V., Liu, K.-C., Krebs, A., and Kemper, R. (1976). *J. Am. Chem. Soc.* **98,** 6758.
Weaver, L. (1970). M.S. Thesis, University of California, Los Angeles.
Weller, A. (1967). *Fast React. Primary Processes Chem. Kinet., Proc. Nobel Symp., 5th, 1967* p. 413.
Weser, U., and Paschen, W. (1972). *FEBS Lett.* **27,** 248.
Weser, U., Paschen, W., and Younes, M. (1975). *Biochem. Biophys. Res. Commun.* **66,** 769.
Wiles, D. M. (1978). *In* "Singlet Oxygen—Reactions with Organic Compounds and Polymers" (B. Rånby and J. F. Rabek, eds.), p. 320. Wiley, New York.
Wilkinson, F. (1978). *In* "Singlet Oxygen—Reactions with Organic Compounds and Polymers" (B. Rånby and J. F. Rabek, eds.), p. 27. Wiley, New York.
Young, R. H., and Brewer, D. R. (1978). *In* "Singlet Oxygen—Reactions with Organic Compounds and Polymers" (B. Rånby and J. F. Rabek, eds.), p. 36. Wiley, New York.

Young, R. H., and Martin, R. L. (1972). *J. Am. Chem. Soc.* **94,** 5183.
Young, R. H., Wehrly, K., and Martin, R. (1971). *J. Am. Chem. Soc.* **93,** 5774.
Young, R. H., Martin, R. L., Feriozi, D., Brewer, D., and Kayser, R. (1973a). *Photochem. Photobiol.* **17,** 233.
Young, R. H., Brewer, D., and Keller, R. A. (1973b). *J. Am. Chem. Soc.* **95,** 375.
Young, R. H., Brewer, D., Kayser, R., Martin, R., Feriozi, D., and Keller, R. A. (1974). *Can. J. Chem.* **52,** 2889.
Zweig, A., and Henderson, W. A. (1975a). *J. Polym. Sci., Polym. Chem. Ed.* **13,** 717.
Zweig, A., and Henderson, W. A. (1975b). *J. Polym. Sci., Polym. Chem. Ed.* **13,** 993.
Zwicker, E. F., and Grossweiner, L. I. (1963). *J. Phys. Chem.* **67,** 549.

6

1,2-Cycloaddition Reactions of Singlet Oxygen

A. PAUL SCHAAP and K. A. ZAKLIKA

Singlet Oxygen

ISBN 0-12-736650-4

I. INTRODUCTION

A. Photooxidative Cleavage Reactions

The first sensitized photooxygenation was discovered by Windaus and Brunken (1928) with the isolation of a stable peroxide, later shown to be endoperoxide **2** (Fieser, 1936), from the irradiation of ergosterol (**1**) in the presence of a dye and oxygen. Schenck and Ziegler (1944) subsequently found that this reaction could be extended to simple 1,3-dienes with the synthesis of ascaridole (**4**) from α-terpinene (**3**). We now recognize these reactions as examples of the 1,4-cycloaddition of singlet oxygen to *cisoid* conjugated dienes.

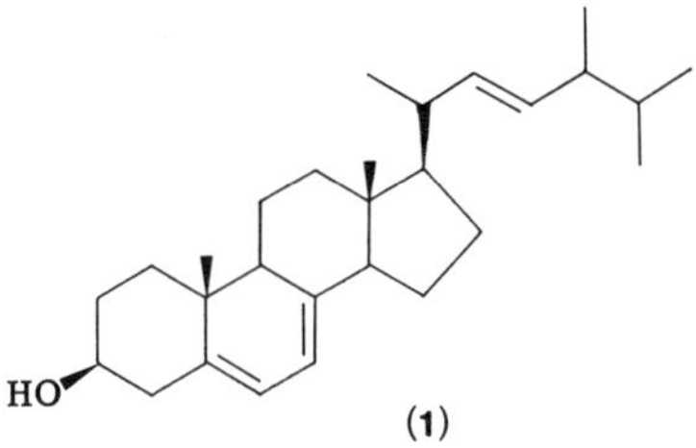

HO (**1**) Sens, $h\nu$ O_2 (1)

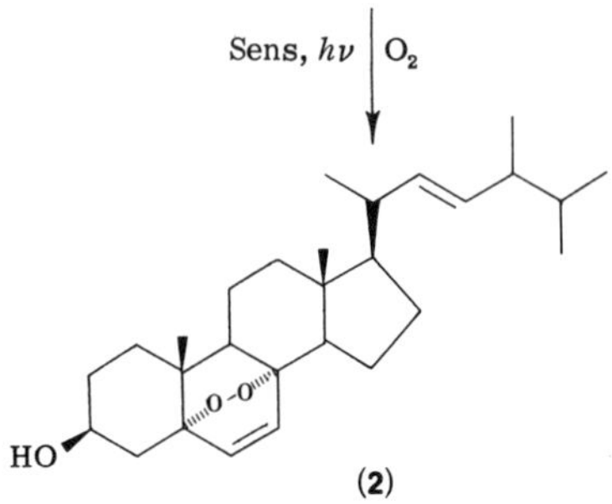

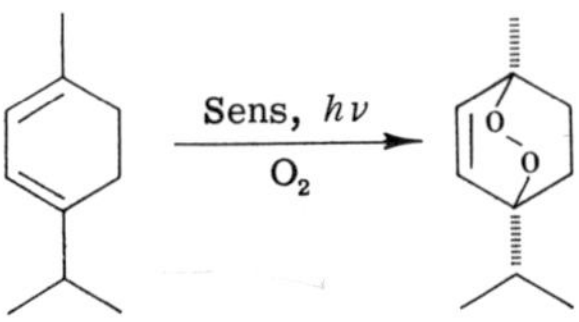

HO (**2**)

Sens, $h\nu$ O_2 (2)

(**3**) (**4**)

Schenck also observed that photooxygenation of olefins possessing allylic hydrogen atoms gave allylic hydroperoxides in which the double bond is shifted, e.g., (+)-α-pinene (**5**) yields hydroperoxides **6** and **7** (Gollnick and Schenck, 1964). Subsequent investigations have shown that the allylic hydroperoxides are formed by an "ene" type reaction of singlet oxygen with olefins (see Chapter 8 by Gollnick and Kuhn).

Sens, $h\nu$, O_2 — OOH + OOH (3)

(5) (>99%) (6) (<1%) (7)

While the products from most photooxygenations could be accommodated by these two modes of reaction, there were, however, several examples of photooxygenation in the literature which resulted in the oxidative cleavage of carbon–carbon bonds [Eq. (4)]. In general, these carbonyl products were

$R_1R_2C{=}CR_3R_4 \xrightarrow[O_2]{\text{Sens, } h\nu} R_1COR_2 + R_3COR_4$ (4)

attributed to secondary reactions of the initially formed allylic hydroperoxides. For example, the oxidative cleavage of **8** and **12** has been proposed to occur via a normal ene reaction with subsequent rearrangement of the corresponding allylic hydroperoxide under the photooxygenation conditions (Gollnick, 1968). Precedent for this mechanism was provided by the known Hock-type cleavage of allylic hydroperoxides such as **15** at elevated tem-

(8) Sens, $h\nu$, O_2 → [(9) OOH] → (10) CHO + CH_3CHO (11) (5)

(12) Sens, $h\nu$, O_2 → [(13) HOO] → (14) + CH_3CHO (11) (6)

OOH $\xrightarrow{H^+}$ [$\overset{+}{OOH_2}$ $\xrightarrow{-H_2O}$ $\overset{+}{O}$] $\xrightarrow{H_2O}$ 2 O (7)

(15) (16)

peratures or in the presence of a trace of acid (Schenck and Schulte-Elte, 1958). As allylic hydroperoxides are isolable in most cases, the reason for the cleavage of **9** and **13** under the mild conditions of a photooxygenation is not clear. However, the intervention of allylic hydroperoxides in photo-oxidative cleavage reactions still deserves consideration despite the recent work on the intermediacy of 1,2-dioxetanes in singlet oxygen reactions.

In 1968 and 1969 several groups reported that olefins without allylic atoms were also oxidatively cleaved by singlet oxygen. Obviously in such cases the earlier explanation of carbonyl formation via secondary reactions of allylic hydroperoxides does not apply. Foote and Lin (1968) found that photooxygenation of enamines such as **17** produced the carbonyl cleavage products quantitatively. Similarly, the photooxygenation of tetraphenyl-

Ph H Ph N $\xrightarrow{^1O_2}$ O Ph Ph + O H N (8)

(17) (18) (19)

Ph Ph Ph Ph $\xrightarrow{^1O_2}$ 2 O Ph Ph (9)

(20) (18)

Me N N Me $\xrightarrow{^1O_2}$ 2 O N Me + $h\nu$ (10)

(21) (22)

ethylene (**20**) and other aryl-substituted ethylenes resulted in the formation of ketones and aldehydes (Rio and Berthelot, 1969). Singlet oxygen generated by the thermal decomposition of triphenyl phosphite ozonide or by an electric discharge was found to cleave 10,10′-dimethyl-9,9′-biacridylidene (**21**) to *N*-methylacridone (**22**) (McCapra and Hann, 1969).

B. Isolation of 1,2-Dioxetanes

Kohler was the first to claim the isolation of 1,2-dioxetanes **24**, obtained from the autoxidation of hindered enols **23** (Kohler and Thompson, 1937; see also Doering and Haines, 1954). However, subsequent spectral characterization of these crystalline peroxides has shown them to be α-hydroperoxyketones **25** (Rigaudy, 1948; Fuson and Jackson, 1950; Pinkus *et al.*, 1970). Similarly, a dioxetane structure **27** was originally pro-

(24)

(23) + $^3O_2$ (11)

(25)

posed for an autoxidation product of cyclohexene (Stephens, 1928); but this was later identified as allylic hydroperoxide **28** (Criegee, 1936).

(27)

(26) + $^3O_2$ (12)

(28)

Interest in 1,2-dioxetanes was intensified with the suggestion by McCapra (1968) that these strained, high-energy peroxides might decompose concertedly to form electronically excited products which would emit light, such as is observed in various chemiluminescent and bioluminescent reactions. Kopecky and Mumford described the first successful isolation of a 1,2-dioxetane (Kopecky and Mumford, 1969; Kopecky *et al.*, 1968, 1975). The base-catalyzed cyclization of α-bromohydroperoxide **29** gave 3,3,4-trimethyl-1,2-dioxetane (**30**) as a yellow crystalline material (mp 5°–7°C) which decomposed in solution at 60°C with a half-life of ~25 min. The decomposition is accompanied by chemiluminescence in support of McCapra's earlier proposal. The stability of 1,2-dioxetane **30** indicated that 1,2-dioxetanes formed in the addition of singlet oxygen to olefins should be isolable compounds (see Chapter 7 by Bartlett and Landis).

HOO Br; Me, Me; Me H **(29)** $\xrightarrow{HO^-}$ O—O; Me, Me, H, Me **(30)** $\xrightarrow{\Delta}$ Me(C=O)Me + Me(C=O)H + $h\nu$ (13)

Shortly after Kopecky's initial report, Bartlett and Schaap were successful in isolating a 1,2-dioxetane from the 1,2-cycloaddition of singlet oxygen to an olefin (Bartlett *et al.*, 1970; Bartlett and Schaap, 1970). Sensitized photooxygenation of *cis*-diethoxyethylene (**31**) at −78°C in $CFCl_3$ gave crystalline **32** (mp −4°C), which decomposed violently at ambient temperature but formed ethyl formate smoothly when heated to 60°C in solution. The thermolysis of **32** in the presence of fluorescent hydrocarbons

EtO OEt **(31)** $\xrightarrow[-78°C]{^1O_2}$ O—O; EtO OEt **(32)** $\xrightarrow{60°C}$ 2 EtOC(=O)H (14)

is accompanied by chemiluminescence (Wilson and Schaap, 1971). Mazur and Foote (1970) independently observed that 1,2-dioxetane **33** is obtained from the addition of singlet oxygen to tetramethoxyethylene.

O—O; MeO, MeO, OMe, OMe

(33)

In the following sections, the reactions of singlet oxygen with various types of olefins to give, in general, carbonyl cleavage products are considered. Although in several cases the intermediate 1,2-dioxetanes have been isolated or characterized spectroscopically, in most reactions the intermediacy of 1,2-dioxetanes is inferred from the structures of the final products. Caution should be exercised in making the assumption that all singlet oxygen cleavage reactions proceed via 1,2-dioxetanes.

This point is well illustrated by the recent study of Turner and Herz (1977) of the reaction of singlet oxygen with dihydrohexamethyl(Dewar benzene) (**34**). Photooxygenation in CH_2Cl_2 at ambient temperature gave the carbonyl product **35** and allylic hydroperoxide **36**; the reaction at $-78°C$, however, formed only **36**. However, **36** was converted quantitatively to **35** in $CHCl_3$ on standing at ambient temperature. That the rearrangement is probably an acid-catalyzed Hock-type cleavage was illustrated by the exothermic conversion of **36** to **35** on the addition of a drop of aqueous HCl.

(**34**) $\xrightarrow{^1O_2}$ (**35**) + (**36**); (**36**) $\xrightarrow{H^+}$ (**35**) (15)

Even when 1,2-dioxetanes are isolated from photooxygenation solutions, they need not be primary cycloaddition products. Rearrangement of initial 1,4-cycloadducts has been observed to lead to subsequent formation of dioxetanes (Section VIII).

Additionally, the reader should be aware of the possible formation of other reactive species besides singlet oxygen under photooxidation conditions. Sensitizer triplets are efficiently quenched by oxygen to generate $^1O_2$; however, with sensitizers such as fluorescein and eosin a small fraction of the encounters results in electron transfer and the formation of $O_2^{\cdot -}$ and the radical cation of the sensitizer. Reactions of these species could be significant in those cases where the substrate is relatively unreactive toward $^1O_2$, or the sensitizer is present in high concentration (see Foote, 1968b, Ericksen

$$^3\text{Sens} + {}^3O_2 \longrightarrow \text{Sens} + {}^1O_2$$
$$^3\text{Sens} + {}^3O_2 \longrightarrow \text{Sens}^{\cdot +} + O_2^{\cdot -} \qquad (16)$$

et al., 1977, and references therein).

II. REACTIONS OF SINGLET OXYGEN WITH NITROGEN-ACTIVATED OLEFINS

A. Enamines

In 1968 three papers appeared describing the photooxidative cleavage of enamines to carbonyl fragments. Huber (1968) illustrated the synthetic utility of the reaction by the preparation of progesterone (**38**) [Eq. (17)].

Sens, $h\nu$ / O_2

CHO

(**37**) (**38**) (17)

Pfoertner and Bernauer (1968) showed that several indole-derived enamines such as **39** underwent a similar reaction [Eq. (18)]. Mechanistically more

Me Sens, $h\nu$ / O_2 Me CHO

Me Me

(**39**) (**40**) (18)

informative, however, was the work of Foote and Lin (1968) who reported that photooxygenation of enamines **41** proceeded by way of an intermediate stable at low temperatures, which formed carbonyl products **42** and **43** in high yield upon warming. Detailed examination by low temperature

Sens, $h\nu$ / O_2

(**42**) (**43**) (19)

(**41a**) R, R = $(CH_2)_5$; $R_1 = R_2 = Me$
(**41b**) R, R = $(CH_2)_5$; $R_1 = R_2 = Ph$
(**41c**) R, R = $(CH_2)_5$; $R_1 = Ph$; $R_2 = H$
(**41d**) $R = R_1 = R_2 = Me$

^{1}H-NMR spectroscopy of the intermediate derived from **41d** revealed that it had an oligomeric structure such as **44** rather than the expected structure

45 or **46**. The ready photooxidative cleavage of **41b** and **41c** showed that

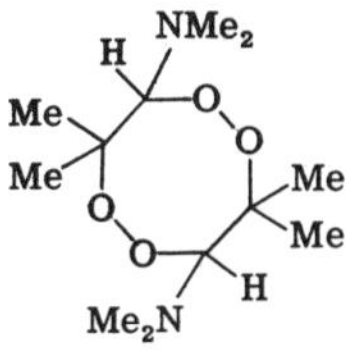

(**44**) 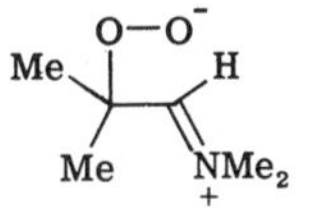 (**45**) 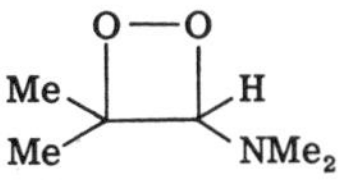 (**46**)

allylic hydroperoxides were not obligatory intermediates in the formation of **42** and **43**. These amino-substituted hydroperoxides **47** could have resulted from an ene reaction, and their subsequent Hock cleavage should be particularly facile [Eq. (20)]. The failure to observe any ene product from

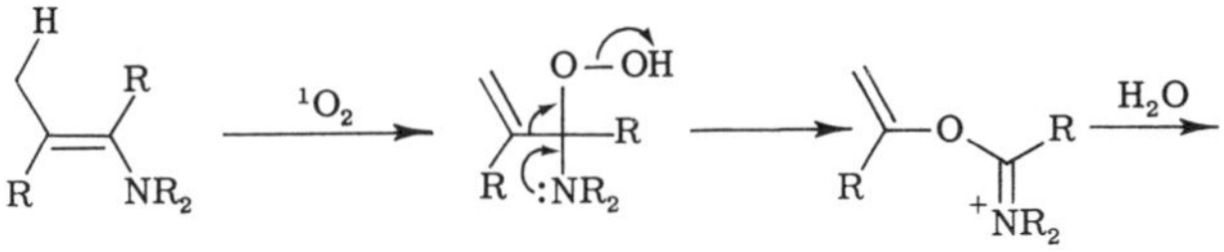

$\xrightarrow{^1O_2}$ (**47**) $\longrightarrow$ $\xrightarrow{H_2O}$ (**42**) + (**43**) (20)

41a and **41b** may be understood in terms of the extremely high reactivity of the enamine double bond. For instance, under conditions where photooxidation of **41c** was rapid, β-methoxystyrene failed to react. It is also relevant to note that whereas **48** reacts with singlet but not triplet oxygen to yield a dioxetane (Mazur and Foote, 1970), **49** reacts spontaneously in air (with chemiluminescence) forming tetramethylurea as the major product (Urry and Sheeto, 1965).

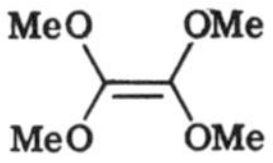

(**48**) 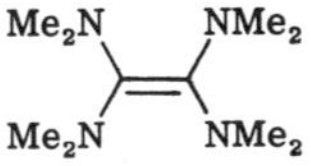 (**49**)

Foote continued his investigations and showed by competition experiments using tetramethylethylene that the cleavage was an authentic singlet oxygen reaction (Foote *et al.*, 1970). The unstable intermediate appeared to be chemiluminescent and was trapped by reduction in the case of **41d** to give amino alcohol **50**, suggesting that zwitterion **45** might have been a

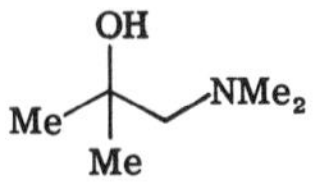

(**50**)

precursor to **44**. Subsequently, photooxygenation of **41d** in dilute solution led to the spectroscopic detection of monomeric dioxetane **46** as a primary product (Foote *et al.*, 1975). Similar behavior was found for 1,2-diphenyl-1-(4-morpholino)-ethylene, from which could be isolated a solid that decomposed explosively at $-6°C$ giving *N*-benzoylmorpholine and benzaldehyde.

The absolute rate of reaction of **41d** with singlet oxygen was determined to be $4.9 \times 10^8\ M^{-1}\ s^{-1}$, only two orders of magnitude less than diffusion controlled. Solvent polarity was without effect on this rate, with the exception of a diminished reactivity in methanol, ascribed to lowered electron availability resulting from hydrogen bonding to the enamine. Foote proposed that addition of singlet oxygen occurred by an electron transfer or charge transfer mechanism (see also Mazur, 1971). Subsequent investigations of

$^1O_2$ (21)

the photosensitized oxygenation of alkenes via a nonsinglet oxygen mechanism (Ericksen *et al.*, 1977) have shown that the mechanism outlined in Eq. (21) is indeed feasible.

Shortly after Foote's observation of dioxetane **46**, Wasserman and Terao (1975) reported the photooxygenation of a series of cyclic enamines **51** in methanol to yield diketones **52** as the major products [Eq. (22)]. Wasserman explained the products as arising from intermediate dioxetane

Sens, $h\nu$ / O_2, MeOH (22)

(52)

(51a) $n = 0$ **(51d)** $n = 3$
(51b) $n = 1$ **(51e)** $n = 7$
(51c) $n = 2$

53 by β-elimination [Eq. (23)]. Support for this mechanism came from the low temperature isolation of crystalline materials having molecular weights,

51 $\xrightarrow{^1O_2}$ **(53)** ⟶ **(54)** ⟶ **(52)** (23)

infrared spectra, and microanalyses consistent with the structures **53b–53d**. It was observed that olefin **55a** for which the β-elimination is precluded, underwent solely cleavage of the double bond, while **55b** yielded 28% of the diketone in addition to the usual cleavage products.

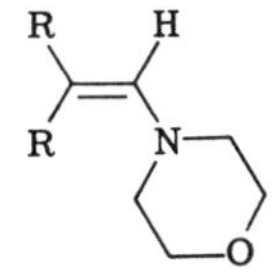

(**55a**) R = Ph
(**55b**) R = Et

Ando *et al.* (1975a) have also reported photooxygenation of enamines in pyridine to give products other than those of normal double bond cleavage. A typical example is outlined in Eq. (24) together with the product yields.

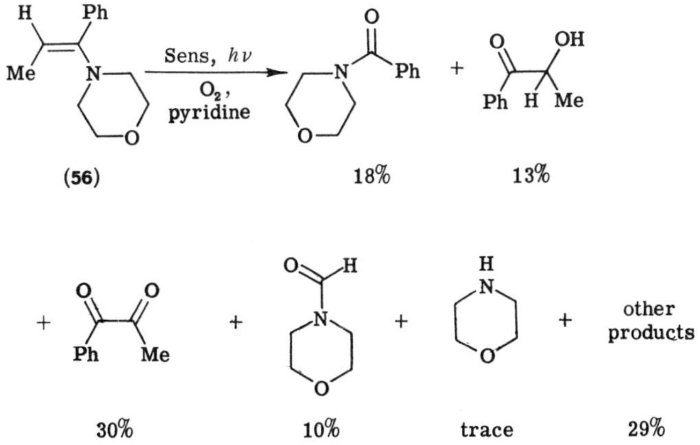

Sens, $h\nu$; O_2, pyridine

(**56**) → 18% + 13% + 30% + 10% + trace + other products 29% (24)

A number of routes to these products may be envisaged although Ando favors biradical cleavage of a dioxetane intermediate (compare with Section IV).

Recently McCapra *et al.* (1976) have shown that the photooxygenation of **57** at −70°C gives rise to an intermediate whose decomposition at 0°C yields the characteristic products of dioxetane cleavage and is accompanied by light emission [Eq. (25)].

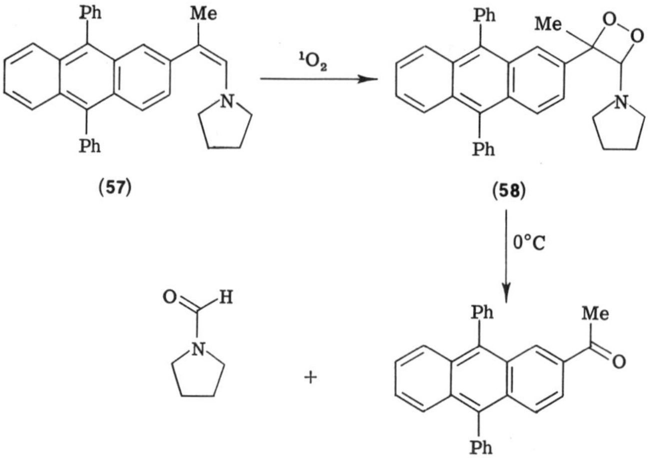

(57) (58) (25)

(60) (59)

The cautionary note sounded by McCapra against the sometimes common assumption that light emission implies a dioxetane structure is a timely one. The claimed (Akutagawa *et al.*, 1976) stable aminodioxetane **61** has subsequently been shown (McCapra *et al.*, 1976; Goto and Nakamura, 1976) to have the structure **62**.

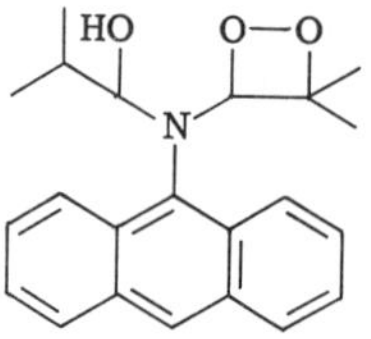

(61) 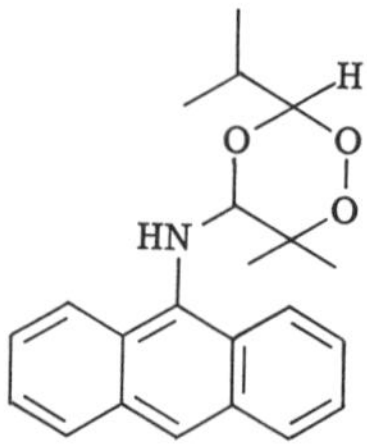 (62)

B. Enaminoketones

The reaction of singlet oxygen with enamines was first extended to enaminoketones by Orito *et al.* (1974). In the course of the synthesis of the alkaloids *cis*-alpinine and *cis*-alpinigenine, enaminoketone **63** was converted to ketolactone **67** through presumed dioxetane **64**. The alternative intervention of a zwitterion has been suggested by Dewar and Thiel (1975).

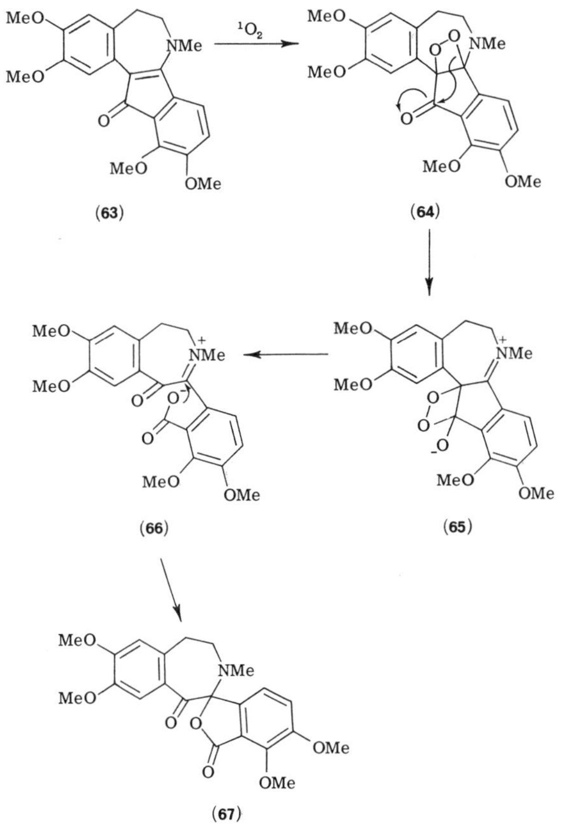

MeO
MeO
NMe
$^1O_2$
O
OMe
+

(63) (64) (65) (66) (67)

(26)

Subsequently, Wasserman and Ives (1976) developed a mild, convenient synthetic procedure for the conversion of ketones (**68**) to α-diketones (**70**) in high overall yield, utilizing the photooxidative cleavage of enaminoketones (**69**). The sequence is outlined in Eq. (27). The procedure is additionally

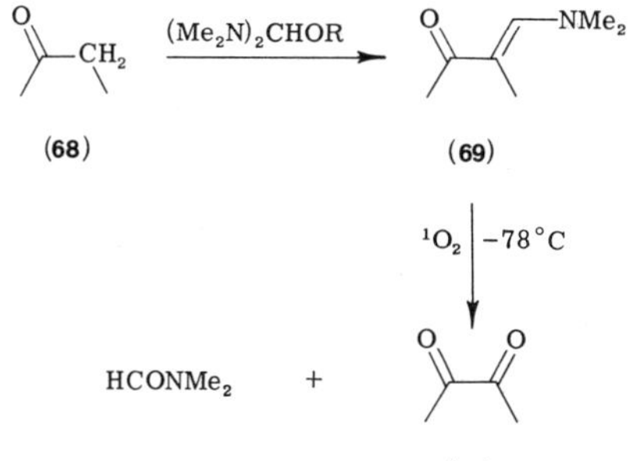

CH_2
$(Me_2N)_2CHOR$
NMe_2
$^1O_2$
−78°C
$HCONMe_2$
+

(68) (69) (70)

(27)

straightforward since **69** need not be isolated, and can be further converted to **70** by photooxygenation using a polymer-bound sensitizer (Schaap *et al.*, 1975) which facilitates isolation of the diketone. The utility of the method is illustrated by the conversion of menthone (**71**) to **72** in 81% yield. Machin

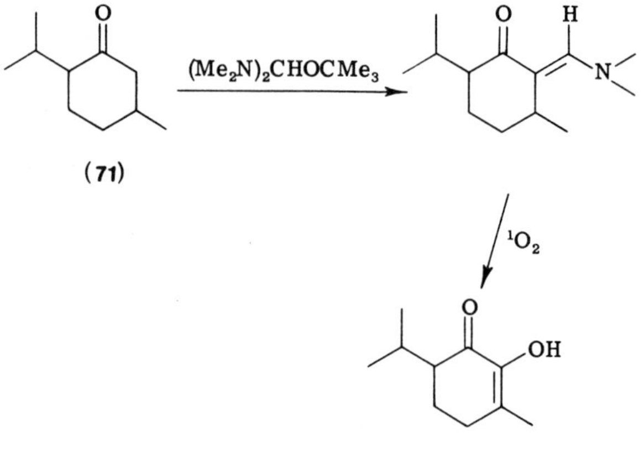

$(Me_2N)_2CHOCMe_3$

(**71**)

$^1O_2$

(28)

(**72**)

and Sammes (1976) have described the photooxidative cleavage of benzylidenepiperazinediones **73** to give the corresponding piperazinetriones and benzaldehyde. NMR evidence was presented for the intermediate 1,2-dioxetanes.

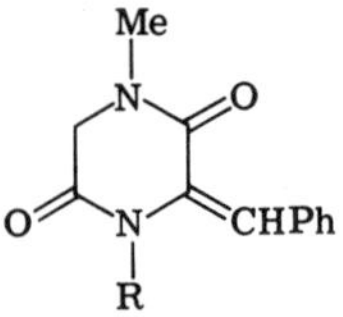

(**73a**) R = H
(**73b**) R = Me

The successful cleavage of enaminoketones is undoubtedly due to the low π ionization potential of amino-substituted olefins. The ready cleavage of **69** may be contrasted with the behavior of keto olefin **74** which does not cleave photooxidatively even though the bond is doubly oxygen-activated (Matsuura *et al.*, 1970).

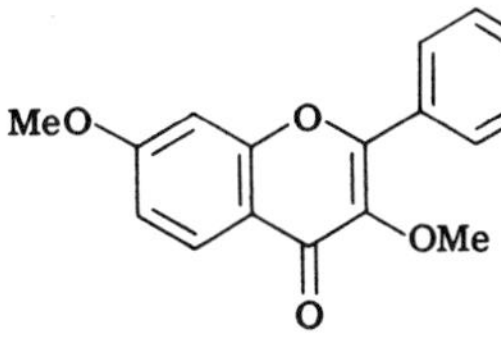

(74)

C. Vinylogous Enamines

McCapra and Hann (1969) were the first to recognize that, like enamines, vinylogous enamines should react with singlet oxygen. Treatment of olefin **75** with triphenyl phosphite ozonide, with bromine and alkaline hydrogen peroxide, or with singlet oxygen generated by radiofrequency discharge, gave an intermediate presumed to be **76**. The thermolysis of **76** formed *N*-methylacridone (**77**), and was accompanied by chemiluminescence. Singer (Lee *et al.*, 1976) later showed that **75** also underwent photooxygena-

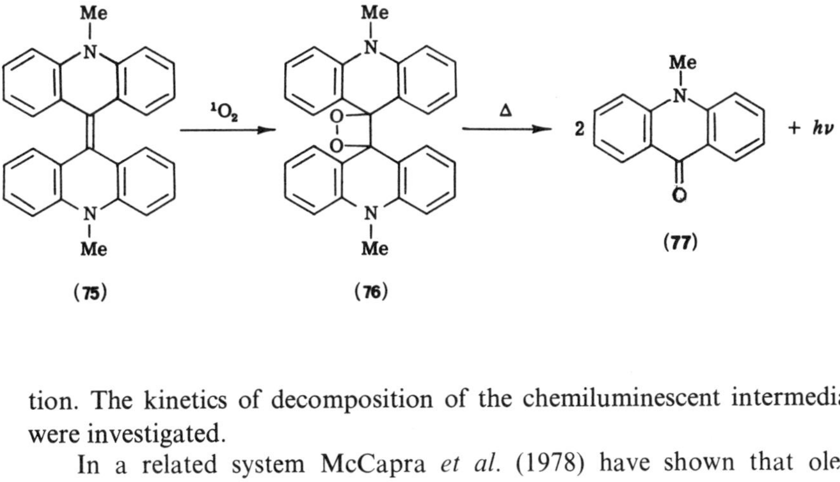

(75) $\xrightarrow{^1O_2}$ (76) $\xrightarrow{\Delta}$ 2 (77) + $h\nu$ (29)

tion. The kinetics of decomposition of the chemiluminescent intermediate were investigated.

In a related system McCapra *et al.* (1978) have shown that olefin **78** possesses a low ionization potential (6.98 eV) comparable to tetraethoxyethylene (6.90 eV) and adds singlet oxygen readily. Low temperature spectroscopic studies were consistent with the formation of dioxetane **79**. Cleavage of **79** to *N*-methylacridone (**77**) and benzaldehyde occurred rapidly at room temperature, and the singlet excited state of **77** was populated with high efficiency (25%), resulting in unusually intense chemiluminescence [Eq. (30)].

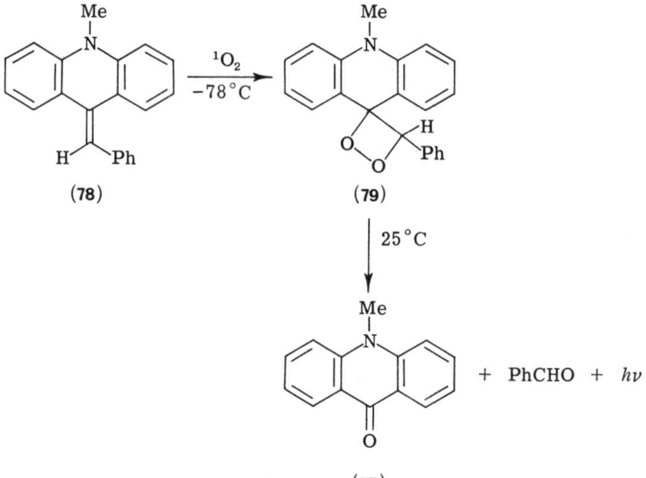

Me N H Ph (78) $\xrightarrow[-78°C]{^1O_2}$ Me N O O H Ph (79) $\xrightarrow{25°C}$ Me N O (77) + PhCHO + $h\nu$ (30)

III. REACTIONS OF SINGLET OXYGEN WITH OXYGEN-ACTIVATED OLEFINS

A. Photooxidative Cleavage of Vinyl Ethers

Bartlett and Schaap observed that several electron-rich vinylene diethers react readily with singlet oxygen to afford carbonyl cleavage products in essentially quantitative yield (Bartlett *et al.*, 1970; Schaap, 1970) [Eqs. (31)–(34)]. Kearns *et al.* (1970) also described the photooxygenation of 1,4-dioxene [Eq. (31)]. The cleavage of β-ethoxystyrene to benzaldehyde and ethyl formate was reported by Rio and Berthelot (1969) [Eq. (35)]. As these

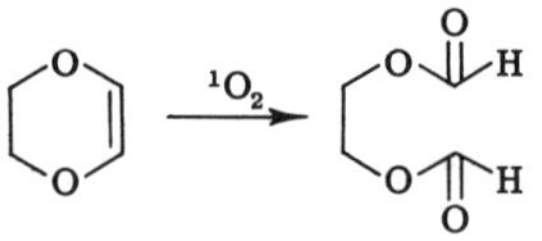

$\xrightarrow{^1O_2}$ (31)

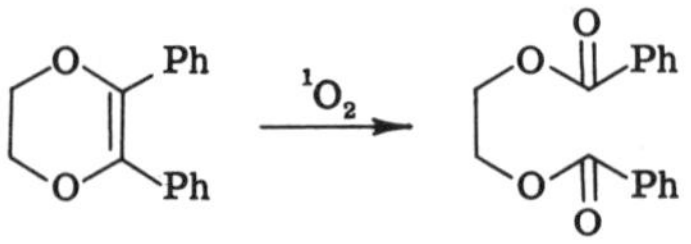

$\xrightarrow{^1O_2}$ (32)

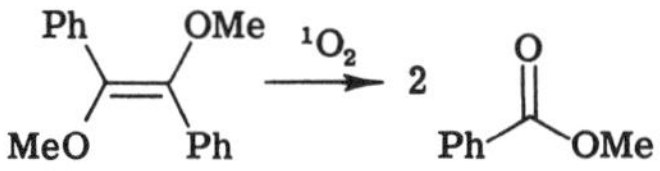

(33)

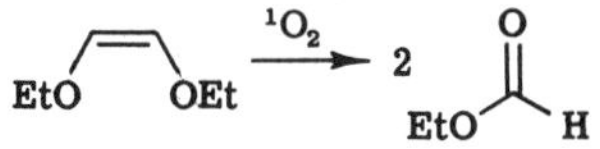

(34)

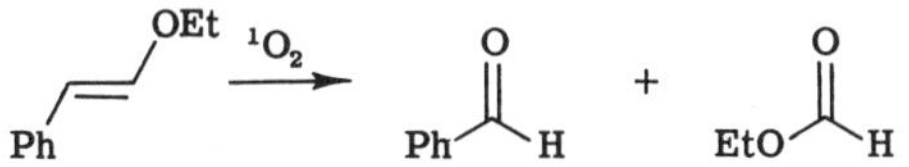

(35)

olefins do not possess allylic hydrogen atoms, the results again indicated that carbonyl cleavage products can be formed in singlet oxygen reactions by a mechanism other than the Hock-type cleavage of an allylic hydroperoxide.

The photooxygenation of 6-(*p*-anisyl)-3,4-dihydro-2*H*-pyran (**80a**) and the 4,4-dimethyl analog **80b** gave only the cleavage products **81** (Atkinson, 1970, 1971). The reaction of singlet oxygen with **82** has been reported to give

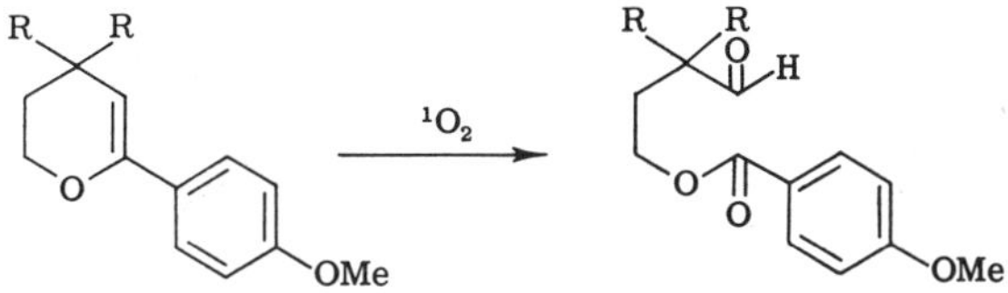

(36)

(**80a**) R = H
(**80b**) R = Me
(**81**)

benzophenone and carbonate **83** (Bollyky, 1970).

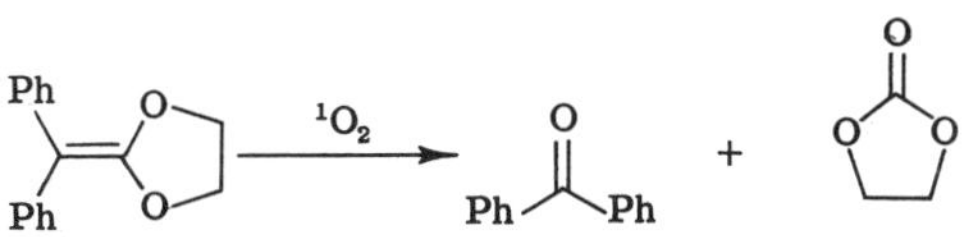

(37)

(**82**) (**83**)

Photooxygenation of **84** in benzene yielded **85** as the major product and 8% of **86** which was shown not to be formed by photoextrusion of dimethylacetylene from **84** (Font *et al.*, 1970). The reaction has been viewed in terms of a competition between cleavage paths *a* and *b* in the presumed dioxetane.

MeO OMe Cl Cl Cl Cl (84) $\xrightarrow[O_2]{h\nu}$ [MeO (a) O—O (b) OMe] (a) (b) (38)

MeO_2C Cl Cl Cl MeO_2C Cl (85)   Cl Cl Cl Cl (86) + $[MeO_2CCO_2Me]$

No reaction was observed upon direct irradiation of 9,10-dimethoxyphenanthrene (**87**) in chloroform or methanol (Rio and Berthelot, 1972). However, if the photooxygenation was carried out in the presence of acetic acid or triphenylphosphine, the products were diester **88** (58%) or quinone **89** (90%), respectively. These results were claimed as evidence for reversible formation of the dioxetane.

MeO OMe (87) $\underset{(-O_2)}{\overset{^1O_2}{\rightleftharpoons}}$ [MeO OMe O—O] $\xrightarrow{H^+}$ MeO_2C CO_2Me (88) (39)

$\downarrow Ph_3P$

O O (89)

B. Stereochemistry of 1,2-Cycloaddition

Photoxygenation of *cis*- and *trans*-diethoxyethylenes (**90**) and (**92**) at $-78°C$ in acetone-d_6 resulted in the stereospecific formation of 1,2-dioxetanes **91** and **93** (Bartlett and Schaap, 1970). While the addition of singlet

EtO OEt (**90**) —Sens, $h\nu$, O_2 / $(CD_3)_2CO$, -78°C→ O—O EtO OEt (**91**) —Δ→ 2 EtOCHO (40)

EtO OEt (**92**) —Sens, $h\nu$, O_2 / $(CD_3)_2CO$, -78°C→ O—O EtO OEt (**93**) —Δ→ 2 EtOCHO (41)

oxygen to **90** in $CFCl_3$ with tetraphenylporphine gave only **91**, the reaction with **92** in $CFCl_3$ led to the formation of a mixture of **93** and **91** that varied from 81% to 40% **93**. Using Rose Bengal solubilized in $CFCl_3$ with a crown ether, the photooxygenation of **92** resulted in 93% of **93**.

With the symmetrically substituted olefins **90** and **92**, one could not unequivocally rule out the cycloaddition of singlet oxygen with the olefin as the antarafacial partner to form **93** from **90**, and **91** from **92**. Therefore, a study of the addition of singlet oxygen to *cis*- and *trans*-ethoxyphenoxy-ethylene (**94**) and (**96**) was undertaken (Schaap and Tontapanish, 1971, 1972). The configurations of the 1,2-dioxetanes could be assigned in this case on the basis of ^{1}H-NMR vicinal coupling constants. Photooxygenation of **94** in $CFCl_3$ formed exclusively *cis*-1,2-dioxetane **95**, while mixtures of **97** (90–75%) and **95** (10–25%) were obtained from the trans isomer **96**. The lack of stereospecificity in the photooxygenation of trans olefins **92**

PhO OEt (**94**) —Sens, $h\nu$, O_2 / $CFCl_3$, -78°C→ O—O PhO OEt (**95**) —Δ→ PhOCHO + EtOCHO (42)

PhO OEt (**96**) —Sens, $h\nu$, O_2 / $CFCl_3$, -78°C→ O—O PhO OEt (**97**) —Δ→ PhOCHO + EtOCHO (43)

and **96** in $CFCl_3$ is as yet not well understood. In view of the higher reactivity of cis-olefins **90** and **94** relative to the trans isomer, it could not be established whether the stereochemical integrity of the trans-olefins was lost during the photooxygenations.

Additionally, Rio and Berthelot (1971b) reported that singlet oxygen adds stereospecifically to *cis*- and *trans*-α-dimethoxystilbene (**98**) and (**100**).

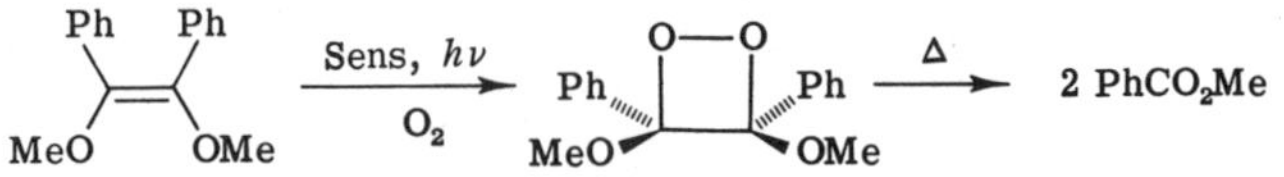

(44)

(**98**) (**99**)

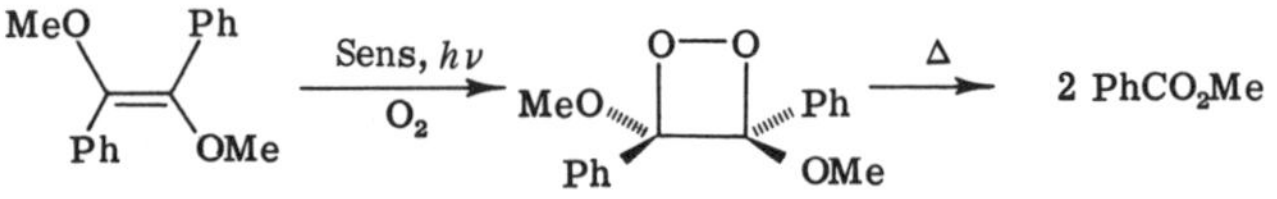

(45)

(**100**) (**101**)

The stereospecific addition of singlet oxygen to **90** and **92** in acetone-d_6 is in sharp contrast to the stereochemical course of the *direct* reaction of triphenyl phosphite ozonide (Schaap and Bartlett, 1970). The ozonide formed at −78°C by the addition of ozone to triphenyl phosphite is stable at that temperature. It reacts with **90** and **92** in $CFCl_3$ at −78°C to yield a virtually identical mixture of dioxetanes. A plausible mechanism for this reaction is shown in Eq. (47).

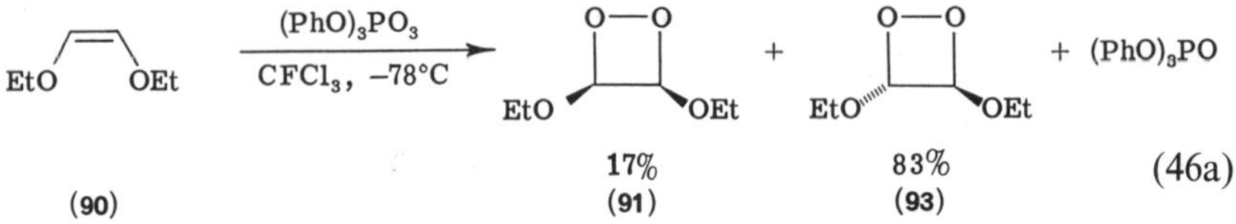

(46a)

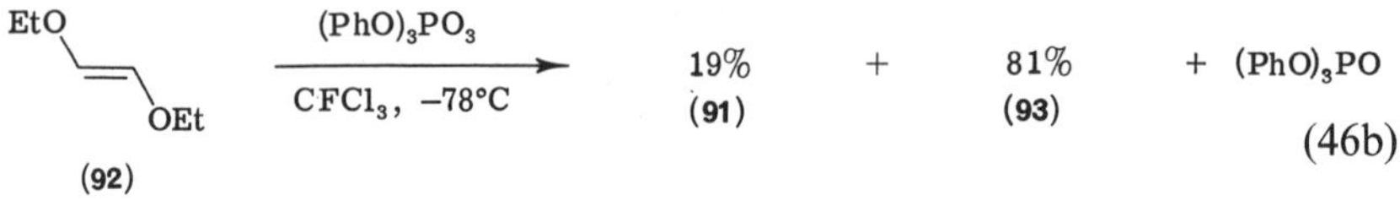

(46b)

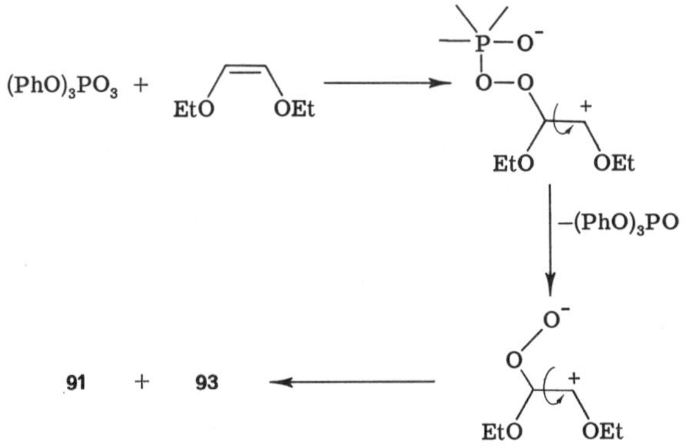

(47)

C. Isolated 1,2-Dioxetanes

A relatively stable 1,2-dioxetane **103** ($t_{1/2}$ = 102 min at 56°C) was isolated by Mazur and Foote (1970) from the photooxygenation of tetramethoxyethylene (**102**). The isolated peroxide was carefully shown by vapor pressure osmometry and cryoscopic methods to be a monomeric dioxetane.

Addition of singlet oxygen at −78°C to 1,4-dioxene (**104**) and 1,3-dioxole (**106**) was found to yield 1,2-dioxetanes **105** and **107** which were characterized spectroscopically (Schaap, 1971). 1,2-Dioxetanes **105** and **107** are of comparable stability to **91** (Schaap, 1971; Wilson *et al.*, 1976). The formation of dioxetanes from 2,3-diaryl-1,4-dioxenes has also been reported (Zaklika *et al.*, 1978a,b).

MeO OMe / MeO OMe (102) —Sens, $h\nu$, O_2 / Et_2O, −70°C→ O—O MeO OMe MeO OMe (103) —Δ→ 2 MeO(C=O)OMe (48)

(104) —Sens, $h\nu$, O_2 / $(CD_3)_2CO$, −78°C→ (105) —Δ→ O-CHO / O-CHO (49)

(106) —Sens, $h\nu$, O_2 / $(CD_3)_2CO$, −78°C→ (107) —Δ→ O-CHO / O-CHO (50)

Photooxygenation of **108** gave an 80% yield of a mixture of the two isomers of **109**, an unexpectedly stable 1,2-dioxetane with a half-life of 4 days at 37°C (Basselier and LeRoux, 1971). Evidence for the structure of the crystalline peroxide is provided principally by the formation of **110** and benzophenone from the thermolysis of **109** and its reduction with thiourea to **111**. Similar results were obtained where methyl was replaced by hydrogen in **108**.

Ph Ph Ph Ph Me Ph (108) —Sens, $h\nu$, O_2 / Me_2CO, −70°C→ Ph Ph Ph Me O O Ph Ph (two isomers) (109) (51)

Ph Ph
Ph
Me O O
(110)
+
O
Ph Ph
Δ
109
thiourea
Ph Ph
Ph OH
Me O OH
Ph Ph
(two isomers)
(111)

(52)

The 1,2-dioxetane **113a** was obtained as a crystalline compound (mp 110°C dec.) from the photooxygenation of **112a** (Basselier *et al.*, 1971). 1,2-Dioxetanes **113b** and **113c** were also identified by NMR spectroscopy. Substitution with deactivating acetyl or benzoyl groups rendered benzofurans **112d** and **112e** unreactive toward singlet oxygen (see also Rio and Berthelot, 1971a). In light of the results of Burns *et al.* (1976) on the reactions of singlet oxygen with indenes (Section V,C), consideration of other possible structures for **113** should not be neglected.

R_3
R_2
R_1 O
Sens, $h\nu$, O_2
−78°C
R_3 O O
R_1 O R_2
(113)

(53)

(**112a**) R_1 = H ; R_2 = R_3 = Me
(**112b**) R_1 = OMe; R_2 = Ph ; R_3 = H
(**112c**) R_1 = OMe; R_2 = Me; R_3 = Ph
(**112d**) R_1 = H ; R_2 = COMe; R_3 = H
(**112e**) R_1 = H ; R_2 = Et ; R_3 = COPh

O
R_3
R_1 O R_2
O
(114)
Δ
113
thiourea
R_3 OH
OH
R_2
R_1 O
(115)

(54)

D. Solvent Effects and Reactivities

Dihydropyran (**116**) has an activated double bond and allylic hydrogen atoms and is therefore able to undergo both 1,2-cycloaddition and the ene reaction with singlet oxygen. Photooxygenation in acetone gives two products, dioxetane cleavage product **117** and unsaturated lactone **119**, derived from dehydration of allylic hydroperoxide **118**. The formation of **118** from **116** and its ready dehydration to **119** had been noted earlier by Schenck (1952). The ratio of **117** and **119** has been taken as a measure of the relative rates of the two competing processes (Bartlett *et al.*, 1970;

$^1O_2$ $^1O_2$ CHO CHO OOH H $-H_2O$

(116) (117) (118) (119) (55)

Schaap, 1970). The ratio varies over a 58-fold range from benzene as the solvent to acetonitrile; i.e., polar solvents favor the 1,2-cycloaddition over the ene reaction (Table I). See also Frimer *et al.* (1977).

The moderate solvent effect is not observed for the reaction of singlet oxygen with tetraethoxyethylene (**120**). The rate constant for the addition

EtO OEt EtO OEt —Sens, $h\nu$, O_2→ 2 EtO–C(=O)–OEt

(120) (121) (56)

Table I Solvent Effect on the Products from Dihydropyran (**116**)

Solvent	Product ratio, **117**/**119**
Benzene	0.094
Acetone	0.80
Dichloromethane	2.67
Acetonitrile	5.51

Table II Solvent Effect on the Rate of Addition of Singlet Oxygen to Tetraethoxyethylene (**120**)

Solvent	k_r (M^{-1} sec^{-1})
Benzene	7.3×10^7
Chloroform	7.8×10^7
Acetone	4.5×10^7
Acetonitrile	4.3×10^7

of singlet oxygen to **120** has been determined in four solvents (Faler, 1977) (Table II). Competitive rate experiments of **120** with tetramethylethylene, and a comparison of the β values of the two olefins, indicate that **120** does not appreciably quench singlet oxygen. The implications of these results for the mechanism of the reaction are discussed in Section IX,B.

Electronic effects on the 1,2-cycloaddition of singlet oxygen to vinyl ethers have been probed by an investigation of the effect of substitution in **122** on its rate of reaction (A. P. Schaap, K. A. Zaklika, and B. Kaskar, unpublished observations). Relative rates are found to correlate better with

$$\mathbf{(122)} \xrightarrow{^1O_2} \mathbf{(123)} \tag{57}$$

σ^+ ($r = 0.989$) to give a ρ value of -0.64, than with σ ($r = 0.858$) with a ρ value of -1.02. The better correlation with σ^+ indicates some resonance stabilization of the rate-limiting transition state.

E. 1,4-Cycloaddition Reactions of Singlet Oxygen with Vinyl Ethers

Foote *et al.* (1973) have observed a novel 1,4-cycloaddition of singlet oxygen to 1,1-diaryl-2-methoxyethylenes instead of the expected addition to yield 1,2-dioxetanes. Photooxygenation of **124** in benzene gave **126**, **127**, and benzophenone (10%). Intermediate **125** could be obtained as an unstable solid by carrying out the reaction at -78°C in dimethyl ether. Whether the benzophenone is the result of a rearrangement of **125** to the dioxetane or of a direct 1,2-cycloaddition of singlet oxygen to **124** is not yet established. Similar results were found with 1,1-di-*p*-anisyl-2-methoxyethylene.

Photooxygenation of 2-(2-anthryl)-1,4-dioxene (**128**) at -78°C in CH_2Cl_2 gave a mixture of endoperoxide **129** (31%), 1,2-dioxetane **130**

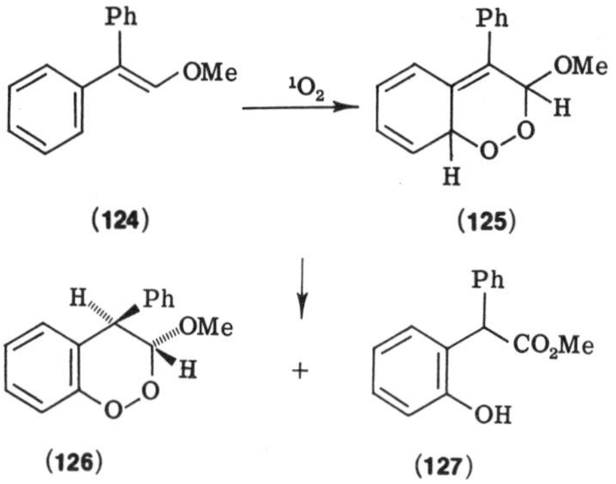

(58)

(124) (125) (126) (127)

(16%), diester **131** (23%), and 30% of unknown material (Schaap *et al.*, 1977). Endoperoxide **129** was isolated from this mixture in 26% overall yield by column chromatography at −55°C. Unlike endoperoxide **125** which undergoes spontaneous rearomatization to **126**, **129** is relatively stable (see Section VIII).

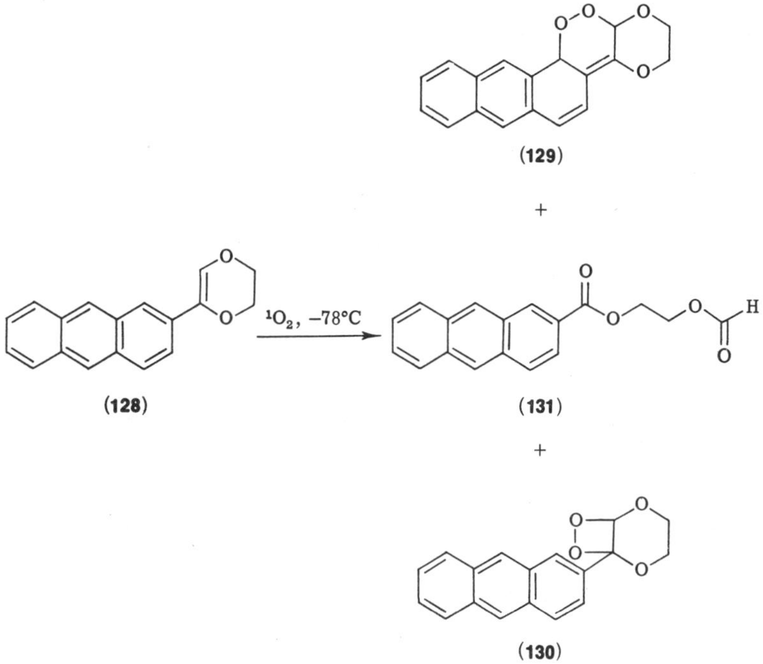

(59)

(128) (129) (131) (130)

In contrast to mono-anthryl-substituted 1,4-dioxene **128**, 2,3-di(2-anthryl)-1,4-dioxene reacts with singlet oxygen to principally yield the corresponding 1,2-dioxetane (Zaklika *et al.*, 1978a). The reduced reactivity of the latter olefin towards 1,4-cycloaddition results presumably from steric hindrance to coplanarity of the anthryl groups with the carbon–carbon double bond of the 1,4-dioxene moiety.

IV. REACTIONS OF SINGLET OXYGEN WITH SULFUR-ACTIVATED OLEFINS

The original report of the photooxygenation of thioethylenes was made by Adam and Liu (1971). They observed that singlet oxygen cleaves 1,1-dithioethylenes **132** to the carbonyl products **133** and **134** expected from a 1,2-cycloaddition [Eq. (60)] (Adam and Liu, 1972a). In contrast, these

Sens, $h\nu$, O_2, −78°C; CH_2Cl_2, workup at 25°C (60)

(**133**) (**134**)

(**132a**) R = Ph
(**132b**) R = $(CH_2)_5$
(**132c**) R =

authors (Adam and Liu, 1972b) have shown that the photooxygenation of tetrathioethylenes **135** gives rise to dithiooxalates **136** and disulfides **137**. Olefin **135e** yields a single product **138**, which is unaccompanied by any detectable amount of **134**. The reaction did not involve triplet oxygen.

Sens, $h\nu$, O_2, - 78°C; CH_2Cl_2, workup at 25°C; + RSSR (61)

(**135a**) R = Me
(**135b**) R = $PhCH_2$
(**135c**) R = *p*-anisyl
(**135d**) R = Ph

(**136**) (**137**)

Relative photooxygenation rates for **135a–135e** were as expected for reaction with singlet oxygen, and the same products were obtained using chemically generated singlet oxygen.

Sens, $h\nu$, O_2, - 78°C; CH_2Cl_2, workup at 25°C (62)

(**135e**) (**138**)

These unusual products were suggested to arise (Adam and Liu, 1972b) by radical fragmentation of dioxetane **139**, dominated by the low strength of the C–S bond (65 kcal/mole) compared to that of the C–C bond (83 kcal/mole), as outlined in Eq. (63). Interestingly, the authors were unable to observe the light emission that might be expected from **139**, though the multistep nature of the cleavage could account for this.

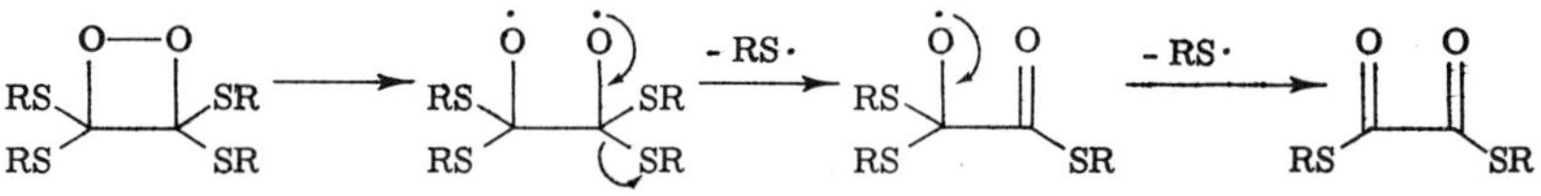

(63)

(139)

Ando *et al.* (1972, 1973, 1974, 1975b) have also photooxidized a variety of mono-, di-, tri-, and tetrathioethylenes. A multiplicity of products was found, among them **136**, **137**, and traces of dithiocarbonates. The rather vigorous conditions employed (Ando *et al.*, 1972, 1973) may leave some doubt whether all the products detected were the result of an initial 1,2-cycloaddition, though the intermediacy of dioxetanes was postulated to account for them.

At present, the initial formation of dioxetanes is a convenient and not implausible rationalization of the results. As recognized by Adam and Liu (1972b), considerable further low temperature studies are required to establish the nature of the primary photooxygenation products.

The reaction of singlet oxygen with vinylogs of thioethylenes has also been reported. Schoenberg and Ardenne (1968) found that 2,6-diphenyl-4-diphenylmethylene-4*H*-thiopyran (**140**) undergoes cleavage of the exocyclic double bond upon photooxidation [Eq. (64)]. A later publication by Ishibe

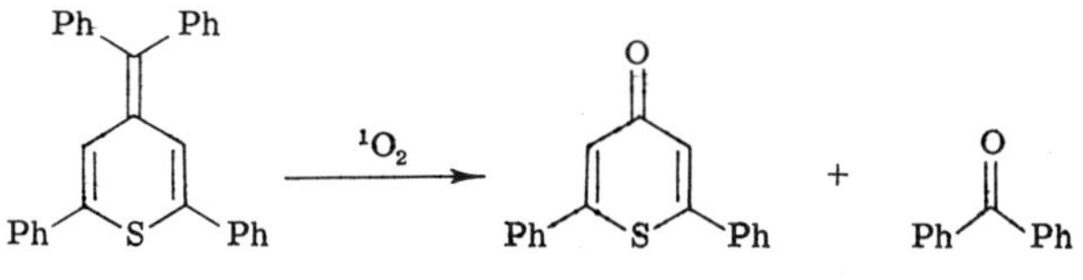

(64)

(140)

et al. (1971) appears to indicate that related olefin **141** is stable to singlet oxygen, but this may be unintentional.

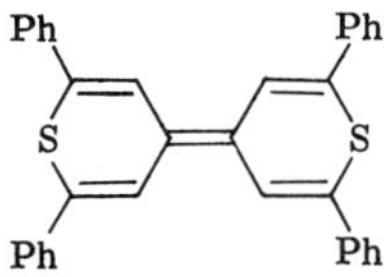

(141)

V. REACTIONS OF SINGLET OXYGEN WITH OLEFINS LACKING HETEROATOM ACTIVATION

In most cases, 1,2-cycloaddition of singlet oxygen to olefins requires the presence of electron-donating groups in the alkene, such as $-NR_2$, $-OR$, and $-SR$. Photooxidation of alkyl-substituted olefins generally results in the formation of allylic hydroperoxides. However, if the ene reaction is precluded by the substitution pattern or for steric reasons, 1,2-dioxetanes or their cleavage products may be formed. These reactions have been grouped into several rather general categories with some unavoidable overlap.

A. Hindered Olefins

The successful addition of singlet oxygen to the hindered olefin adamantylideneadamantane (**142**) by Wynberg and co-workers in 1972 proved to be of considerable significance (Wieringa *et al.*, 1972). Treatment of **142** with singlet oxygen generated photochemically or from triphenyl phosphite ozonide resulted in the formation of **143**, a solid dioxetane of unprecedented stability (mp 163°C). Dioxetane **143** was stable to reagents such as borohydride, but could be reduced with zinc and acetic acid to the corresponding diol. Thermolysis yielded adamantanone by a chemiluminescent process. These authors noted that under the conditions required to oxygenate **142**,

HO OH

Zn

HOAc

$^1O_2$

85%

O—O

(65)

(**142**) (**143**)

Δ

2 =O + $h\nu$

related olefins **144** and **145** were inert. A later collaborative effort (Schuster

R

R

(**144**)

(**145a**) R = H
(**145b**) R = Ph
(**145c**) R = *t*-Bu

et al., 1975) by the groups of Turro, Schaap and Adam found that the cleavage of **143** required an activation energy of ca 35 kcal/mole, some 10 kcal/mole higher than that of simple dioxetanes. X-ray crystal data obtained recently by Numan *et al.* (1977) not only confirm the structure of **143** but also provide some insight into the reasons for the unusual stability of this 1,2-dioxetane. Wynberg and Numan (1977) have also reported the photooxygenation of **146** to give optically active dioxetane **147**, whose cleavage is accompanied by circularly polarized chemiluminescence (Wynberg *et al.*, 1977).

$^1O_2$ (66)

(**146**) (**147**)

The initial suggestion by Sharp (1960) and later by Kearns (1969) that reactions of singlet oxygen might occur via a perepoxide intermediate (see Section IX) prompted Schaap and Faler (1973) to reinvestigate the photooxygenation of **142**. Examination of molecular models and the known stability of **143** indicated that the rearrangement of perepoxide **148** to **143** might be sufficiently slowed by steric restrictions to permit trapping of **148**. It was found that photooxidation of **142** in pinacolone gave rise to the expected dioxetane **143** (81%), but also to epoxide **149** (19%). *tert*-Butyl acetate was also identified as a trace product. Epoxide **149** was an authentic singlet oxygen product, as evidenced by the ineffectiveness of radical scavengers, by inhibition of formation with the singlet oxygen quencher 1,4-

Ad Ad $^1O_2$ (**142**) (**148**) (67)

Ad = (**143**) + (**149**)

diazabicyclo[2.2.2]octane (DABCO), and by competitive inhibition by *cis*-dimethoxyethylene, a reactive singlet oxygen acceptor. In contrast, photooxidation of **142** in dichloromethane yielded less than 0.1% of **149**. Epoxide **149** was proposed to arise from perepoxide **148** by a Baeyer–Villiger type mechanism.

A later study by Bartlett and Ho (1974) of related hindered olefin 7,7′-binorbornylidene (**150**), showed that the epoxide corresponding to **149** was a coproduct with the dioxetane. However, solvents other than

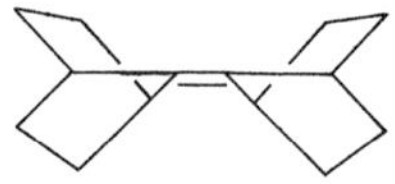

(**150**)

pinacolone proved to afford larger amounts of epoxide than pinacolone itself. For instance, benzene gave as much as 95% epoxide as compared to 74% in pinacolone. Corresponding oxygenation of the solvent was not detected. The experiments involving **150** and additional results on **142** have been summarized by Bartlett (1976) in a review article, and it is clear that Baeyer–Villiger trapping is not an important pathway to epoxide. On the other hand, tetracyanoethylene, and to a lesser extent cyanide ion, increase the ratio of epoxide to dioxetane formed from **150**—a result felt by Bartlett to be compatible with interception of a perepoxide. Oxygenated products from either of the trapping agents have, however, not yet been reported.

Two mechanisms for the formation of epoxide during singlet oxygen addition to olefins have been suggested by Dewar *et al.* (1975a) [Eqs. (68), (69)]. Examination of Bartlett's results provides no evidence for pathway

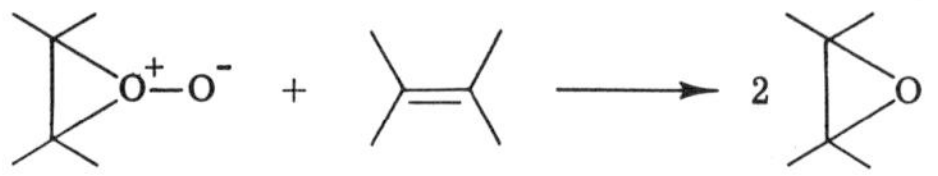

$$\text{perepoxide } (\overset{+}{O}-O^-) + \text{olefin} \longrightarrow 2\ \text{epoxide} \quad (68)$$

$$\text{perepoxide } (\overset{+}{O}-O^-) + {}^1O_2 \longrightarrow \text{epoxide} + O_3 \quad (69)$$

(68). Indeed, decreased concentrations of **150** lead to enhanced yields of epoxide. In view of the sterically hindered double bond in **150**, however, the possibility exists that this mechanism is important for other olefins. In a general sense, the data are consistent with pathway (69), especially if it is assumed that the ozone formed epoxidizes a second molecule of olefin with concomitant regeneration of singlet oxygen. Nevertheless, the suggestion by Jefford and Boschung (1976) that epoxide yields depend on the sensitizer used for photooxygenation allows only tentative conclusions at present.

In connection with the appearance of epoxide in the photooxygenation of olefins, the report by Shimizu and Bartlett (1976) of an efficient photosensitized epoxidation should be noted. In this case, however, singlet oxygen is not involved, as witnessed by the lack of most of the characteristic products of photooxygenation.

McCapra and Beheshti (1977) have obtained evidence for the intermediacy of perepoxide **152** or a closely related species with strong carbonium ion character in the photooxygenation of camphenylideneadamantane (**151**). In addition to the expected 1,2-dioxetane **153**, 1,2-dioxolane **154**, with a rearranged carbon skeleton, was isolated. Consistent with the proposed mechanism are the observations that the ratio of the two products is independent of sensitizer, and addition of the singlet oxygen quencher DABCO inhibited the formation of both **153** and **154**. The ratio **153**/**154** is sensitive to solvent (CH_2Cl_2, 7:3; MeCN, 3:7; Me_2CO, 1:9; MeOH, 1:9), but seems not to relate directly to polarity. The authors also report the isolation of chiral **153** utilizing optically active camphenilone in the synthesis of **151**.

$^1O_2$

(151) (152)

(70)

152

(153)

(154)

B. Aryl-Substituted Olefins

The photooxidative cleavage of aryl olefins is generally a slow process. Rio and Berthelot (1969) have examined the reactions of a variety of examples **155a–g** for which photooxygenation times of 24 hours were typical [Eq. (71)]. The cleavage of olefins (**156**) with electron-donating aryl substituents

$$\mathrm{Ph(R_1)C{=}C(R_2)(R_3)} \xrightarrow[O_2]{\text{Sens, } h\nu} \mathrm{Ph(R_1)C{=}O} + \mathrm{O{=}C(R_2)(R_3)} \tag{71}$$

(**155a**) $R_1 = R_2 = R_3 = H$
(**155b**) $R_1 = H$; $R_2 = Ph$; $R_3 = H$
(**155c**) $R_1 = R_2 = H$; $R_3 = Ph$
(**155d**) $R_1 = Ph$; $R_2 = R_3 = H$
(**155e**) $R_1 = R_2 = Ph$; $R_3 = H$
(**155f**) $R_1 = R_2 = R_3 = Ph$
(**155g**) $R_1 = R_2 = H$; $R_3 = OEt$

[Eq. (72)] was reported not to be significantly more facile, though the direct substitution of electron donors at the double bond (e.g., **155g**) markedly enhanced reactivity. Olefins with electron-withdrawing substituents at the double bond were unreactive, as were acenaphthylene and phenanthrene. The rate of reaction depended on the solvent, apparently as a function of

$$\mathrm{4\text{-}R_1C_6H_4(R_3)C{=}CH(R_2)} \xrightarrow{^1O_2} \mathrm{4\text{-}R_1C_6H_4(R_3)C{=}O} + \mathrm{O{=}CH(R_2)} \tag{72}$$

(**156a**) $R_1 = OMe$; $R_2 = R_3 = H$
(**156b**) $R_1 = OMe$; $R_2 = Ph$; $R_3 = p$-anisyl
(**156c**) $R_1 = NMe_2$; $R_2 = H$; $R_3 = Ph$
(**156d**) $R_1 = NMe_2$; $R_2 = H$; $R_3 = p$-dimethylanilino

singlet oxygen lifetime. Methylene blue, the sensitizer with the lowest triplet energy [34 kcal/mole (Denny and Nickon, 1973)] was found to be most effective. The studies of Saito and Matsuura (1972) point up the delicate balance between different reaction pathways for related oleffns. Whereas stilbenes are slowly cleaved by singlet oxygen to benzaldehyde [Eq. (71)], α,α'-dimethylstilbenes yield the ene products virtually quantitatively [Eq. (73)]. The addition of electron donors to the aryl ring is, however, sufficient

$$\mathrm{Me(Ph)C{=}C(Ph)Me} \text{ (or cis)} \xrightarrow{^1O_2} \mathrm{Me{-}C(Ph)(OOH){-}C(Ph){=}CH_2} \tag{73}$$

to redirect the reaction towards cleavage product (7%) at the expense of ene product (80%), as indicated in Eq. (74). The photooxidative cleavages of

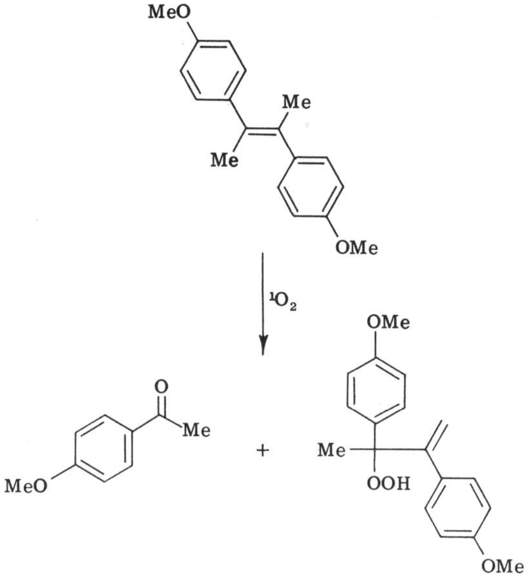

(74)

9,9′-bifluorenylidene (Richardson and Hodge, 1970) and of 9,9′-fluorenylideneanthrone (Ismail and El-Shafei, 1957) have also been observed.

Zimmerman and Keck's (1975) reported photooxygenation of the 1-methylene-4,4-diphenyl-2,5-cyclohexadienes **157** is important because it represents the first successful isolation of dioxetane products **158** from unactivated aryl olefins. The dioxetanes **158** could subsequently be thermolyzed to cleavage products of **159** and **160** together with **161**, formed by di-π-methane rearrangement of electronically excited **159**. The observation

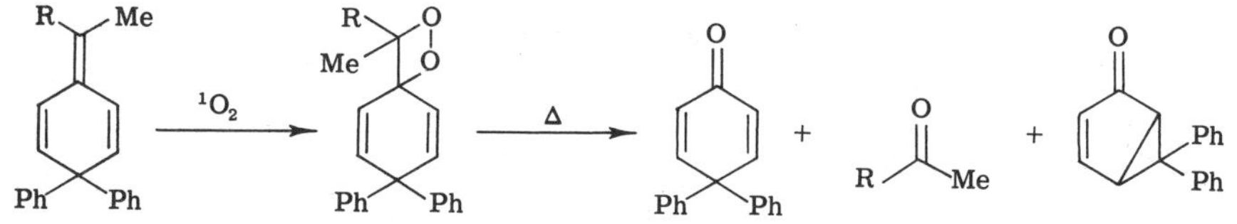

(157a) R = Ph
(157b) R = *m*-anisyl
(157c) R = β-naphthyl
(158) **(159)** **(160)** **(161)** (75)

that the relative yields of **159** and **161** were insensitive to the nature of R led Zimmerman *et al.* (1976) to conclude that cleavage of 1,2-dioxetanes populates n,π^* triplet states selectively even when lower-lying π,π^* states are available, as in **160c**. However, an investigation of several structurally related 1,2-dioxetanes by McCapra *et al.* (1978) has shown that this preference for the formation of n,π^* triplet states is not universal.

A number of photooxygenations have been investigated in which the aryl-substituted double bond is part of a ring. 1,2-Diphenylcyclopentene (**162**) gives solely the ene product under a variety of conditions (Rio and Bricout, 1971; Rio and Charifi, 1970). In acetone and methanol photo-

Ph Ph $\xrightarrow{^1O_2}$ HOO Ph Ph (76)

(**162**) (**163**)

oxygenation of 1,2-diphenylcyclobutene (**164**) also leads to ene product **165**, with a trace of **166** (Schultz and Schlessinger, 1970). However, in benzene, dichloromethane, or dichloromethane containing HCl gas **165** is not found. Instead, scission product **166** occurs in enhanced yield together with a polymer. Since acid-catalyzed rearrangement of **165** leads quantitatively to furan **167**, the authors considered **166** to arise from a dioxetane. Politzer and Griffin (1973) have reported the reaction of the arylcyclopropenes **169**

HOO Ph Ph $\xrightarrow{H^+}$ Ph O Ph

(**165**) (**167**)

Ph Ph $^1O_2$ (77)

$^1O_2$

(**164**)

[O—O Ph Ph] → Ph O O Ph

(**168**) (**166**)

and **170** with singlet oxygen. The products obtained in dichloromethane were strongly dependent on substitution as outlined in Eq. (78). The photooxygenation of **169** in acetone resulted in polymerization, while in methanol–

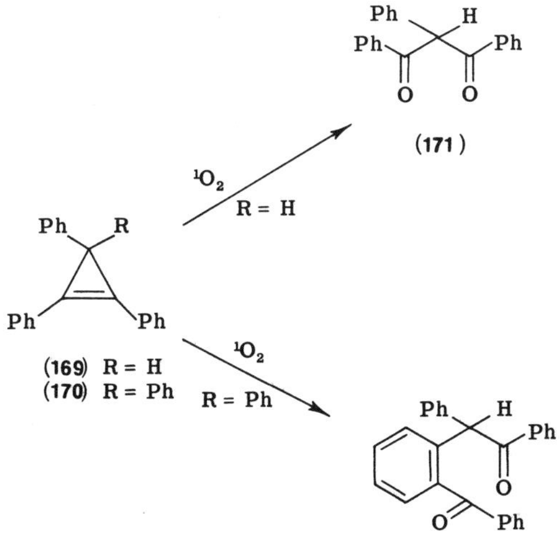

(78)

(171)

(169) R = H
(170) R = Ph

(172)

benzene, an unidentified isomer of **171** is formed together with another product, also unidentified. At low temperature (Griffin *et al.*, 1977) both olefins gave rise to unisolated intermediates, which were formulated as dioxetanes on the basis of the chemiluminescence attending their cleavage. Olefin **169** was proposed to give **173**, while the similar chemiluminescence behavior (principally the activation energy) of the intermediate formed from **170** or indene **174** led to structure **175**. Though **173** and **175** were not other-

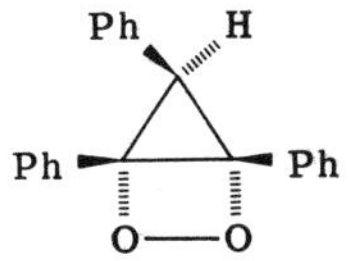

(173) 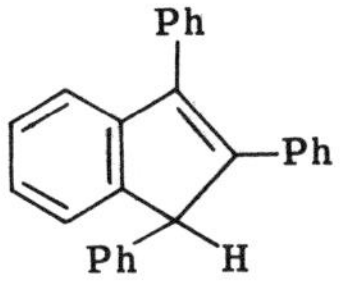 (174) 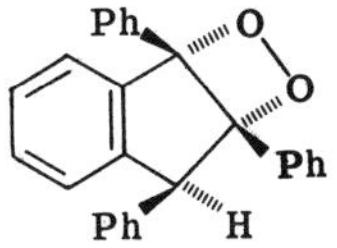 (175)

wise characterized, the dioxetane structures are plausible. The authors noted this, believing the primary singlet oxygen product of both olefins to be unrearranged dioxetane **173**, though the possibility exists that **173** and **175** are secondary products of initial ene or 1,4-cycloaddition reactions.

The related monoaryl cycloalkenes **176**, **179**, and **181** have been photooxygenated by Jefford and Rimbault (1976) to yield the products outlined in Eqs. (79)–(81). In the case of **181**, there is some solvent dependence of

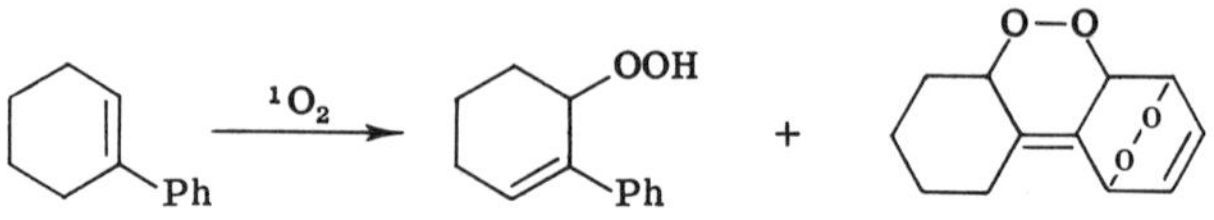

(79)

(176) (177) (178)

(80)

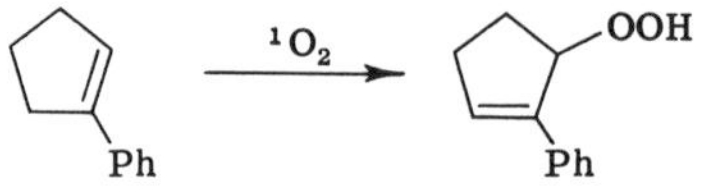

(179) (180)

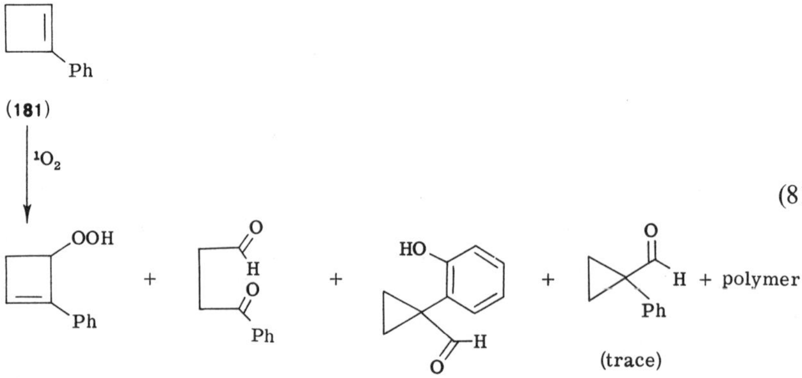

(181)

(81)

(182) (183) (184) (185)

yields. In carbon tetrachloride, **183** is the only nonpolymeric product, while **184** is found only in the more polar solvents. It is not known whether the products are interconvertible under the photooxygenation conditions, or whether radical reactions may have occurred. The authors speculate that a perepoxide intermediate could account for their results.

C. Indenes and Related Olefins

Gollnick (1968) made the first reference to the photooxygenation of indene (**186**) which gave homophthalaldehyde (**188**), suggested by him to arise from allylic hydroperoxide **187**. The formation of **188** in the reaction

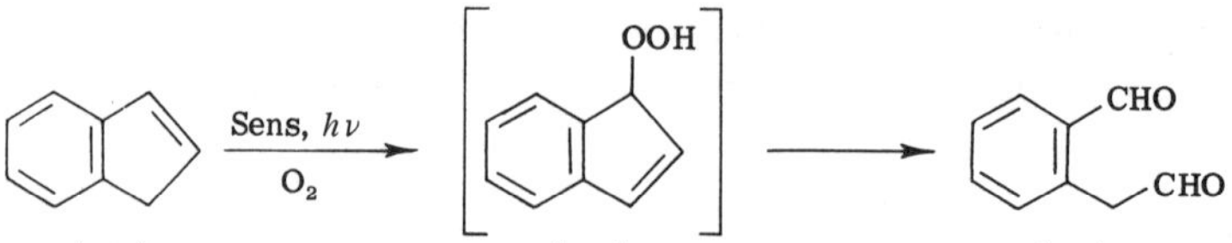

(82)

(186) (187) (188)

of singlet oxygen with indene was confirmed by Kearns (Fenical *et al.*, 1969a), who generated singlet oxygen both photochemically and by microwave discharge. To rule out a mechanism involving an ene reaction to afford **187** followed by scission to **188**, these authors prepared **187** and showed it to be stable under the photooxidation conditions. Subsequently, Mazur (1971) indicated that Kearns and co-workers may have synthesized the alcohol corresponding to **187** rather than the hydroperoxide itself, thereby invalidating this control. However, by careful experiments using authentic **187**, Mazur confirmed the original conclusion of Kearns that the allylic hydroperoxide was not an intermediate.

In methanol at ambient temperature, Kearns reported that indene was oxidized to **188** (6%), **189** (74%), **190** (16%), and **191** (4%) (Fenical *et al.*, 1969a). Bisacetal **189** was shown to be rapidly formed from **188** under the

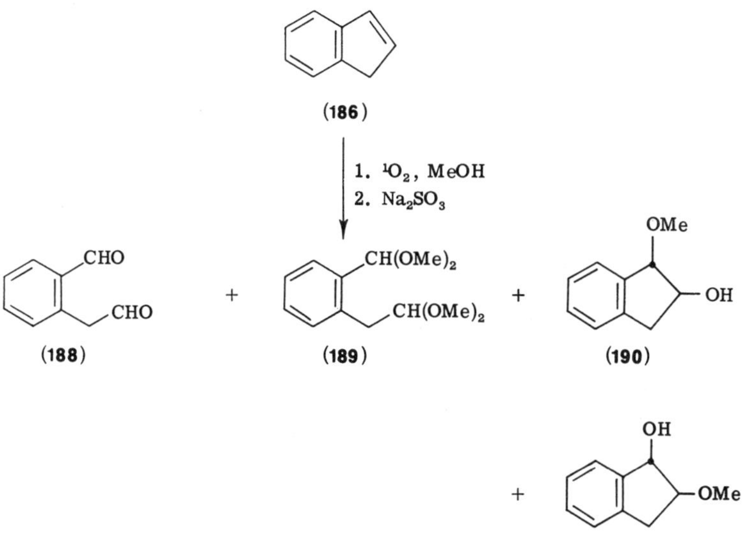

(186)

1. $^1O_2$, MeOH
2. Na_2SO_3

CHO, CHO (188) + CH(OMe)$_2$, CH(OMe)$_2$ (189) + OMe, OH (190) + OH, OMe (191) (83)

photooxidation conditions. Alcohols **190** and **191** were attributed to the corresponding hydroperoxides derived from solvent attack on the presumed dioxetane intermediate **192**. When dioxetanes were shown to be stable to

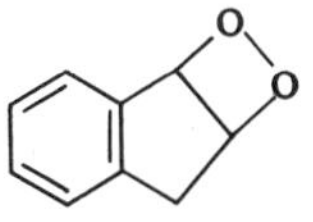

(192) 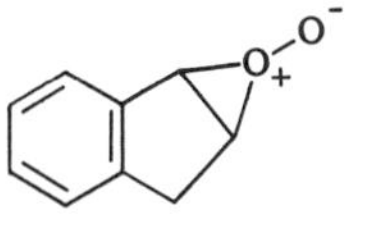 (193)

attack by methanol (Richardson *et al.*, 1975), Kearns postulated trapping of perepoxide **193** (Hasty *et al.*, 1972). A second report by Fenical *et al.* (1969b) claimed interception of **192** by azide ion, but this was later shown by other authors (Foote *et al.*, 1972; Gollnick *et al.*, 1972) and by Kearns himself (Hasty *et al.*, 1972) to be erroneous.

The observation of 1,4-cycloaddition of singlet oxygen to 1,1-diaryl-2-methoxyethylenes rather than the expected 1,2-addition (see Section III,E), led Foote and co-workers (Foote *et al.*, 1973) to reinvestigate the photooxygenation of indene (a styrene analog). A distinction was found between the oxidation of indenes in aprotic solvents (Burns *et al.*, 1976) as compared to protic solvents (Burns and Foote, 1974). Though slow at ambient temperature in aprotic solvents, the photooxidation of indenes **194** proceeds readily in acetone at $-78°C$ to give dioxygenated products **195** (Burns *et al.*, 1976). These authors suggest that the mechanism for formation of these

Sens, $h\nu$ / O_2 → (40–90%) (**195**) (84)

(**194a**) $R_1 = R_2 = R_3 = R_4 = H$
(**194b**) $R_1 = R_4 = H$; $R_2 = R_3 = Ph$
(**194c**) $R_1 = R_3 = R_4 = H$; $R_2 = i\text{-}Pr$
(**194d**) $R_1 = R_4 = H$; $R_2 = Me$; $R_3 = Ph$
(**194e**) $R_1 = R_2 = R_4 = H$; $R_3 = Ph$
(**194f**) $R_1 = R_3 = H$; $R_2 = t\text{-}Bu$; $R_4 = Me$
(**194g**) $R_1 = Me$; $R_2 = i\text{-}Pr$; $R_3 = R_4 = H$

unusual products involves initial 1,4-addition of singlet oxygen to indene **194** followed by rearrangement of endoperoxide **196** to bisepoxide **197**, with subsequent addition of a second molecule of singlet oxygen.

(**194**) $\xrightarrow{^1O_2}$ (**196**) → (**197**) $\xrightarrow{^1O_2}$ (**195**) (85)

In contrast to the reaction in aprotic solvents, photooxygenation of indenes occurs rapidly in methanol with the formation of the scission products noted previously for **186**. Burns and Foote (1974) investigated this reaction at low temperature and were able to isolate several 1,2-dioxetanes **198** [Eq. (86)]. It remains to be established whether the 1,2-dioxetanes **198** are formed directly in a 1,2-cycloaddition facilitated by the polar,

$^1O_2$, −78°C, MeOH

(194b)–(194d) and (194h) $R_1 = R_3 = H$; $R_2 = Me$; $R_4 = t$-Bu

31–55% (198)

(86)

protic solvent or by a methanol-assisted rearrangement of initially formed endoperoxide **196** of the type described in Section VIII.

$^1O_2$

(194) (196) (192)

(87)

Burns and Foote (1976) extended their study to include the 1,2-dihydronaphthalenes **199** which, unlike the indenes, also undergo the ene reaction. Photooxygenation proceeded similarly both in acetone and in methanol according to the general scheme outlined in Eq. (88). As shown,

$^1O_2$, −78°C, Me_2CO

(199a) $R_1 = H$; $R_2 = Ph$
(199b) $R_1 = Ph$; $R_2 = H$
(199c) $R_1 = R_2 = Ph$
(199d) $R_1 = Me$; $R_2 = Ph$
(199e) $R_1 = Me$; $R_2 = H$
(199f) $R_1 = R_2 = H$

(88)

(2 stereoisomers) (200b-f) + (201a, c-f) + (202d, e)

the course of the oxidation is strongly dependent on substitution; **199a** yields only **201a**, **199b** only **200b**, and **199c–f**, a mixture. Efforts to identify 1,2-dioxetanes as products of the reaction were unsuccessful, though at room temperature in methanol, photooxygenation of **199b** and **199e** gave 23% and 12%, respectively, of the aldehyde **203**.

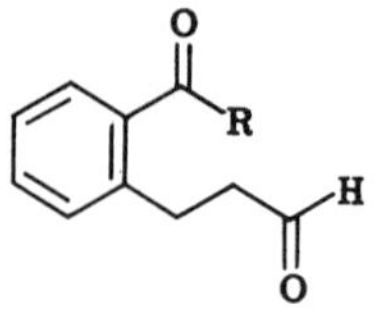

(203)

Several other systems are also related to the indenes. Early reports (Rigaudy *et al.*, 1972) that **204** forms a stable bis-1,2-dioxetane **205** when treated with singlet oxygen were shown to be in error, the correct dioxygenated product being **206** (Rigaudy *et al.*, 1974), presumably formed via opening of an intermediate cyclobutene to a *cis*-diene [Eq. (89)]. Photooxygenation of the structurally similar **207** provided no evidence for ene product

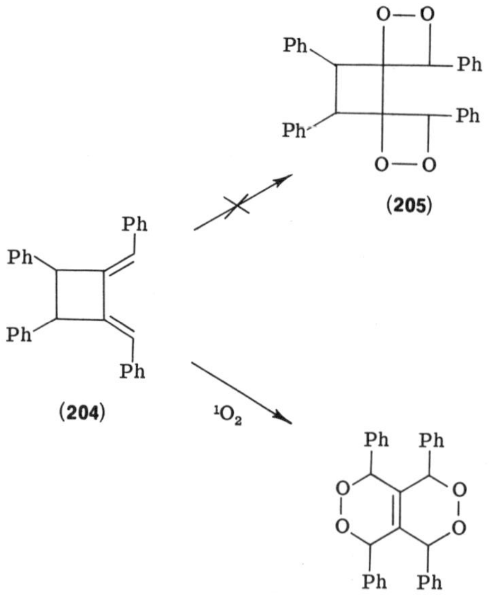

(205)

(204) $^1O_2$ (89)

(206)

208 or dioxetane **209**, leading only to the isolation of bisendoperoxide **210** in 63% yield (Rio *et al.*, 1973). The intermediate monooxygenated 1,4-

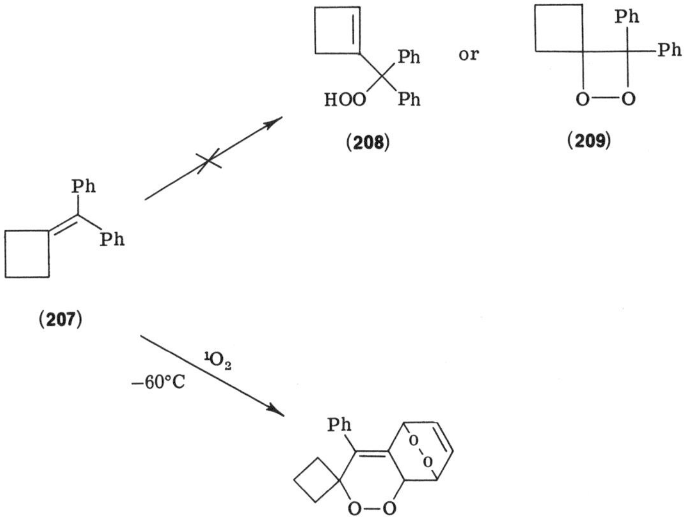

Ph, Ph, HOO, or, Ph, Ph, O—O

(208) (209)

Ph, Ph

(207)

$^1O_2$, −60°C

Ph, O, O, O—O

(210)

(90)

adduct that is the expected precursor of **210** could not be isolated as indeed was the case with the indenes. Matsumoto and Kondo (1975b), however, have described the first example of a stable 1,4-adduct of singlet oxygen in which an aromatic nucleus functions as part of the diene. Photooxygenation of the vinylnaphthalenes **211** resulted in stereospecific addition of singlet oxygen to give the endoperoxides **212**. In contrast α-substituted vinylnaphthalenes **213** did not give such endoperoxides, but underwent

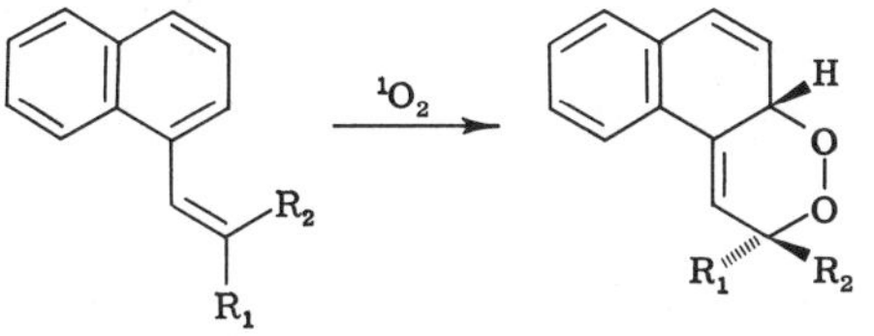

R_2, R_1, $^1O_2$, H, O, O, R_1, R_2

(91)

(**211a**) R_1 = Ph; R_2 = H
(**211b**) R_1 = R_2 = H
(**211c**) R_1 = R_2 = Me
(**211d**) R_1 = Me; R_2 = H
(**211e**) R_1 = H; R_2 = Me

(**212**)

the ene reaction with singlet oxygen to the allylic hydroperoxides **214** [Eq. (92)]. This may be understood in terms of a *peri* interaction which does not allow the required *cisoid* arrangement of the vinyl moiety and the naphthalene ring. 1,4-Adducts **216** are similarly formed with singlet oxygen

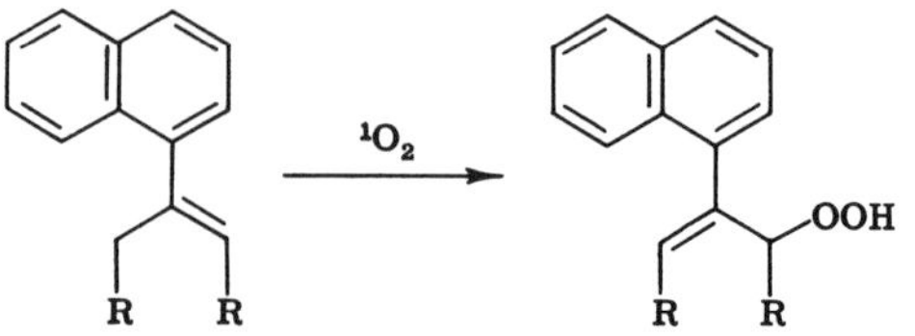

$^1O_2$ (92)

(**213a**) R = H
(**213b**) R = $(CH_2)_2$
(**213c**) R = $(CH_2)_3$

(**214**)

and 2-vinylthiophenes **215** (Matsumoto *et al.*, 1975), the reaction occurring more readily with the heteroaromatic nucleus than with the benzenoid one. Examples are given in Eq. (93).

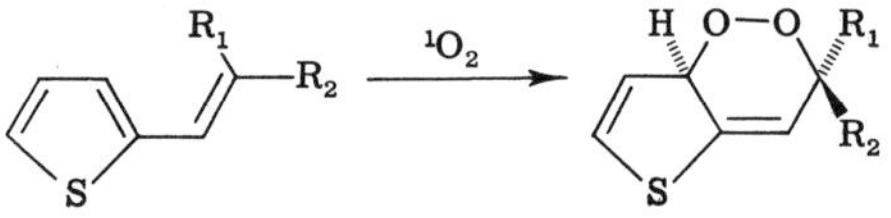

$^1O_2$ (93)

(**215a**) $R_1 = R_2 =$ Me
(**215b**) $R_1 = R_2 =$ H
(**215c**) $R_1 =$ H; $R_2 =$ Me
(**215d**) $R_1 =$ H; $R_2 =$ Ph
(**215e**) $R_1 =$ H; $R_2 =$ 1-naphthyl

(**216**)

D. Alkyl-Substituted Olefins

The photooxidative scission of dihydrohexamethyl(Dewar benzene) (Turner and Herz, 1977) has already been mentioned in the introduction as resulting from Hock cleavage of an initially formed allylic hydroperoxide rather than from cleavage of a dioxetane. Since photooxygenation of **217** and **218** gives no cyclopentane ketoaldehydes (Jefford and Boschung, 1974), relief of ring strain appears to be a dominant factor in the rearrangement described by Herz. Norbornene itself (Rigaudy, 1968) is precluded from undergoing the ene reaction, which would result in an anti-Bredt olefin. Its

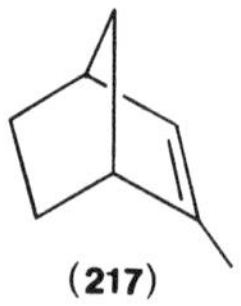

(**217**) 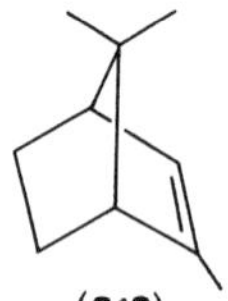 (**218**)

photooxygenation is therefore of some interest in regard to 1,2-cycloaddition of oxygen, but contradictory reports have appeared concerning its reactivity. Litt and Nickon (1968) found that norbornene had undergone no reaction after one month of attempted photooxygenation as evidenced by

IR, NMR, and GC analysis in conjunction with negative chemical tests for peroxide. Under their conditions, the poorly reactive olefin, cyclohexene (Foote, 1968a), had a half-life of one day. This result receives confirmation from the work of Matsuura *et al.* (1973) on the relative reactivities of a number of olefins. A calculation based on these reactivities, and the assumption that the lowest relative reactivity determined by these authors represented a detection limit, indicates that **219** is at least 44 times less reactive towards singlet oxygen than is cyclohexene. In contrast, Jefford and Boschung (1974) have reported that photooxidation of 2 M solutions of **219** led to 5% yields of oxygenated products within 5 hours. The formation of these products was inhibited by DABCO. The major components isolated from them were epoxide **220** and dialdehyde **221**. It is not entirely certain that formation of **220** is a consequence of singlet oxygen reaction. Competing reactions are often a problem with poorly reactive olefins. Both radical-induced (Hiatt, 1971; Mayo, 1968; Filippova *et al.*, 1973) and sensitizer-initiated epoxidations (Shimizu and Bartlett, 1976; Bartlett and Landis, 1977) are established phenomena. The danger of the assumption that inhibition of reaction by DABCO necessarily implies involvement of singlet oxygen has also been emphasized (Davidson and Trethewey, 1976).

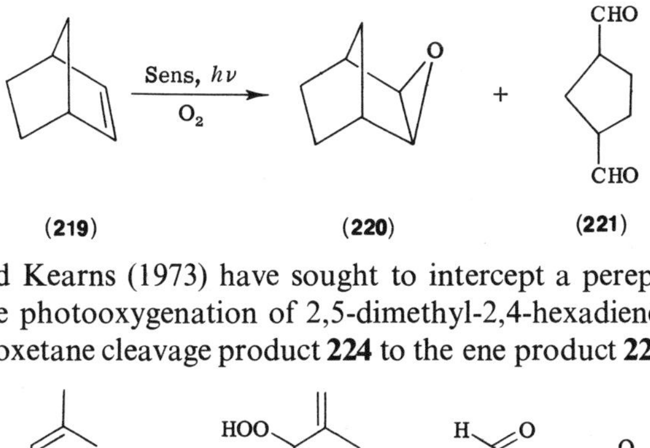

(94)

(219) **(220)** **(221)**

Hasty and Kearns (1973) have sought to intercept a perepoxide intermediate in the photooxygenation of 2,5-dimethyl-2,4-hexadiene (**222**). The ratio of the dioxetane cleavage product **224** to the ene product **223** was found

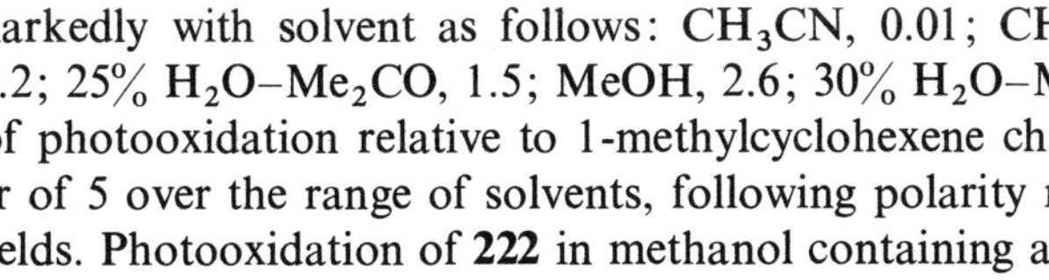

(95)

(222) **(223)** **(224)**

to vary markedly with solvent as follows: CH_3CN, 0.01; CH_2Cl_2, 0.1; Me_2CO, 0.2; 25% H_2O–Me_2CO, 1.5; MeOH, 2.6; 30% H_2O–MeOH, 5.5. The rate of photooxidation relative to 1-methylcyclohexene changed only by a factor of 5 over the range of solvents, following polarity rather than product yields. Photooxidation of **222** in methanol containing acid or base led to primarily products of solvent incorporation as outlined in Eq. (96).

Dioxetane **225** was prepared by low temperature photooxygenation of **222**

in methanol, and was found not to give significant amounts of the solvent incorporation products when subjected to the normal reaction conditions. Kearns concluded that these results were best explained by the interception of a perepoxide which is a precursor to **225**. Hasty and Kearns, however,

(225)

also reported in footnote 19 of their paper that *trans,trans*-2,4-hexadiene (admittedly less hindered) was photooxidized to the endoperoxide, which was accompanied by solvent incorporation products in neutral nucleophilic solvents. There is other precedent for formation of endoperoxides by reaction of singlet oxygen with linear conjugated dienes (Kondo and Matsumoto, 1972; Matsumoto and Kondo, 1975a), and the propensity of endoperoxides for rearrangement is well known (Section VIII). In view of this, the possibility that the behavior of **222** can be explained by the reactions of an initially formed endoperoxide may need investigation.

The involvement of 1,2-dioxetanes has been suggested in three other systems. Photooxygenation of thujopsene (**226**) results in the formation of the ketoaldehyde (Ito *et al.*, 1971). Deuterium incorporation studies were believed to exclude a Hock cleavage mechanism, and the intermediacy of a dioxetane was invoked. In an investigation of the ene reaction, Kellogg

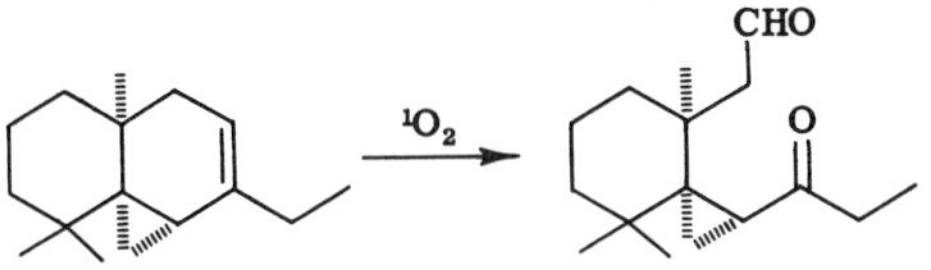

CHO

$^1O_2$ (97)

(226)

and Kaiser (1975) have described the reaction of olefins **227** and **228** with singlet oxygen. The major products were those of the ene reaction, but the

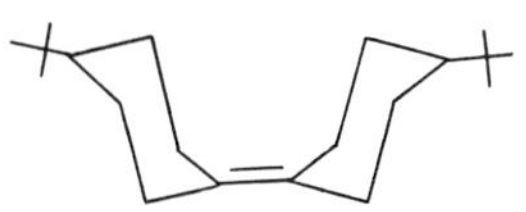

(227) 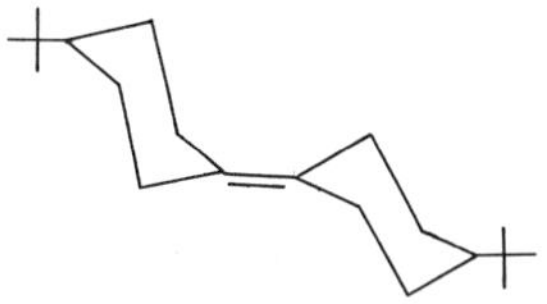 (228)

crude photooxidation mixtures were chemiluminescent. In the case of **227** borohydride reduction afforded 5–10% yields of diol **229**. Similar reduction

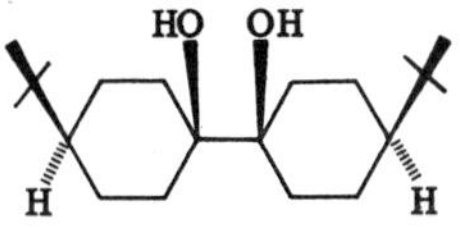

HO OH

H H H

(229) 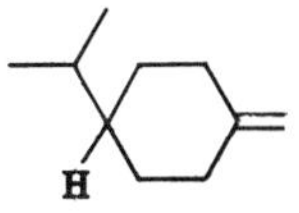 (230)

of photooxidation mixtures of cyclohexylidenecyclohexane resulted in 1% yields of both the corresponding diol and the epoxide. It was suggested that 1,2-cycloaddition to a dioxetane might be feebly competitive with the ene reaction for these olefins. In contrast, **230** is completely unreactive towards singlet oxygen (Klein and Rojahn, 1965). An unusual reaction of ground state oxygen with relatively stable cyclobutadiene **231** to give a 1,2-dioxetane has been reported (Maier, 1974).

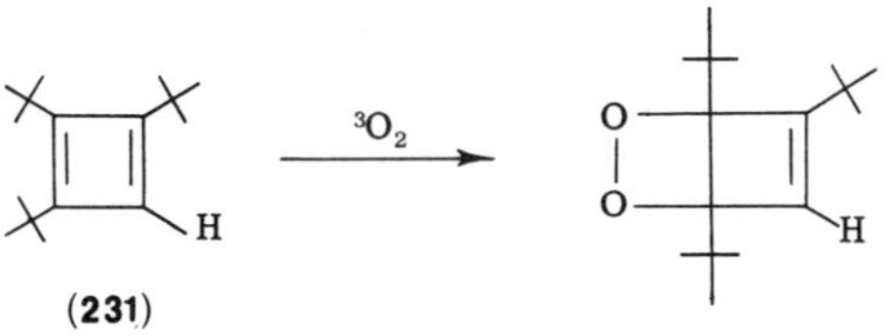

$^3O_2$ (98)

H O O H

(231)

VI. REACTIONS OF SINGLET OXYGEN WITH OLEFINS IN THE GAS PHASE

Bogan *et al.* (1975, 1976) have recently described experiments on the reaction of singlet oxygen with ethylene and several vinyl ethers in the gas phase which, unlike the chemiluminescence studies in solution, provide detailed spectral information about the primary excited products. The reactions were carried out at low pressure in a flow system using singlet oxygen produced by the electric discharge method. The observation of the high resolution emission spectrum of excited formaldehyde (HCHO*) in the reaction of ethylene with singlet oxygen has been taken as evidence for the transient existence of the parent 1,2-dioxetane **233**, the synthesis of which in the condensed phase has yet to be accomplished. The chemiluminescence intensity exhibited first-order dependence on both singlet oxygen and

$$CH_2{=}CH_2 \xrightarrow[\text{electric discharge}]{O_2 \text{ gas phase}} \underset{\text{H}_2\text{C}-\text{CH}_2}{\text{O}-\text{O}} \longrightarrow \overset{O^*}{\underset{CH_2}{\|}} + \overset{O}{\underset{CH_2}{\|}} \quad (99)$$

(232) **(233)** **(234)**

ethylene with an activation energy over the temperature range 500°–650°K of 21.0 ± 1.3 kcal/mole. On the basis of thermochemical calculations, these authors suggest that the gas phase reaction is a two-step process involving the formation of a vibrationally excited dioxetane with subsequent fragmentation to the products before collisional stabilization. They indicate that the rate-limiting step under these conditions is the addition of $^1O_2$ to $CH_2{=}CH_2$. In similar studies with the electron-rich olefin, ethyl vinyl ether, a significantly lower activation energy for the addition was obtained (9.8 ± 1.0 kcal/mole).

Relatedly, Ashford and Ogryzlo (1974, 1975) have determined Arrhenius parameters for some gas phase 1,4-cycloaddition and ene reactions of singlet oxygen.

VII. PHOTOOXYGENATION OF VARIOUS UNSATURATED FUNCTIONAL GROUPS

In the course of their study of enamine photooxygenation, Foote and co-workers reported that the ynamines **235** undergo a ready reaction with singlet oxygen with high yields of the corresponding α-ketoamides (Foote and Lin, 1968; Foote *et al.*, 1970). The possibility of dioxetene intermediates remains to be investigated.

$$R_1-C{\equiv}C-N{<}^{R_2}_{R_3} \xrightarrow{^1O_2} R_1-C(=O)-C(=O)-N(R_2)R_3 \quad (100)$$

(235a) R_1 = Me; $R_2 = R_3$ = Et
(235b) R_1 = Ph; $R_2 = R_3 = (CH_2)_5$

(236)

The strained acetylene **237** reacts with $^1O_2$ generated photochemically with polymer-bound Rose Bengal at $-90°C$ to yield an intermediate which is stable at that temperature, but yields **239** on warming to $-30°C$, with chemiluminescence that is similar to the fluorescence of **239** (Turro *et al.*, 1976). The structure of the intermediate is formulated as dioxetene **238**. Acetylenes **240** and **241** show similar results upon reaction with singlet oxygen.

$^1O_2$, $-90°C$; $-30°C$; $+ h\nu$

(237) **(238)** **(239)**

(101)

(240) **(241)**

Allenes are relatively unreactive towards singlet oxygen. However, Greibrokk (1973) observed that upon prolonged (2–4 days) photooxygenation, tetraphenylallene (**242**) was converted to benzophenone and CO_2. 1,3-Diphenylallene and triphenylallene gave the expected products. Tetraoxaspirocycloheptanes **243** were proposed as intermediates; but given their likely thermal and photochemical instability, it is doubtful that any appreciable amount of these intermediates would be found in solution. A

Sens, $h\nu$, O_2; 2 PhCOPh $+ CO_2$ (102)

(242) **(243)** **(244)**

more plausible mechanism would involve photooxidative cleavage of the allene to the corresponding ketene with subsequent oxidation to the ketone or aldehyde, and CO_2.

In 1970 Bollyky reported that reaction of diphenylketene (**245**) with singlet oxygen affords benzophenone (**244**) (see also Adam, 1975). Turro *et al.* (1977) subsequently characterized the intermediate α-peroxylactones by IR and NMR spectroscopy. The spontaneous decomposition of triphenyl phosphite ozonide at −20°C was used as the source of singlet oxygen in these experiments.

$$\mathrm{Ph_2C{=}C{=}O}\ (\mathbf{245}) \xrightarrow{^1O_2} \left[\text{dioxetanone}\right] \longrightarrow \mathrm{Ph_2C{=}O}\ (\mathbf{244}) + CO_2 \quad (103)$$

$R_2C{=}C{=}O$

(**246**) R = CF_3
(**247**) R = *t*-Bu

Photooxygenation of thiones **248** (Ishibe *et al.*, 1971), sulfines **250** (Zwanenburg *et al.*, 1970), and thione **251** (Worman *et al.*, 1972) yields ketones.

$$(\mathbf{248}) \xrightarrow[O_2]{\text{Sens}, h\nu} (\mathbf{249}) + [SO] \quad (104)$$

X = O, S

$$R_2C{=}S{=}O\ (\mathbf{250}) \xrightarrow{\text{Sens}, h\nu} R_2C{=}O + SO_2 \quad (105)$$

$$(\mathbf{251}) \xrightarrow{h\nu, O_2} \text{tetramethylcyclobutanedione} \quad (106)$$

Adam and Liu (1972b) have described the photooxygenation of ylide **252** which gave triphenylphosphine oxide and diphenyldithiocarbonate in 90 and 98% isolated yields, respectively. The reaction was rapid and did not

$$\mathrm{Ph_3P{=}C(SPh)_2}\ (\mathbf{252}) \xrightarrow[-78^\circ C,\ CH_2Cl_2]{h\nu,\ O_2} \mathrm{Ph_3P{=}O} + \mathrm{O{=}C(SPh)_2} \quad (107)$$

require added sensitizer. Control experiments indicated that **252** does not react with ground state oxygen.

The photooxidative cleavage of the C=N bond in benzophenone derivatives **253** has been observed by Wamser and Herring (1976) [Eq. (108)].

$$Ph_2C{=}N{-}OR \xrightarrow[O_2]{\text{Sens, } h\nu} Ph_2C{=}O + O{=}N{-}OR \tag{108}$$

(**253**)
R = H, Me, :$^-$

VIII. REARRANGEMENT OF ENDOPEROXIDES TO 1,2-DIOXETANES

During the course of their investigations of the photooxygenation of aromatic hydrocarbons such as rubrene, Moureu *et al.* (1926) noted that thermolysis of corresponding photoperoxide **254** regenerated the hydrocarbon with concomitant evolution of molecular oxygen and weak chemiluminescence. Endoperoxides of 1,4-dialkoxy-substituted anthracenes with the oxygen bridge across the 1,4-positions decomposed at lower temperatures

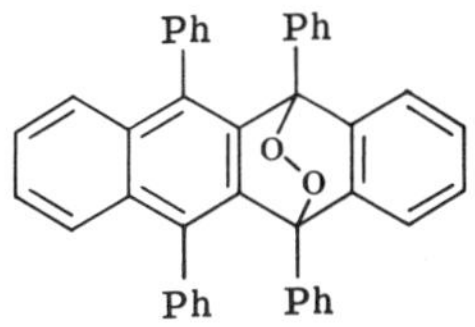

Ph Ph
O–O
Ph Ph

(**254**)

and with more intense light emission (Dufraisse and Velluz, 1942). Rigaudy (1968) has shown that in benzene at ambient temperature endoperoxide **255** yields only 20% of acene **256** with the major product being aldehyde ester **259**. The mechanism proposed by Rigaudy for the formation of **259** involves the rearrangement of **255** to **257**. However, Baldwin *et al.* (1968), who found that **255** is converted to **259** with HCl in benzene or ether, suggested that a plausible mechanism for the reaction involves an acid-catalyzed rearrangement of endoperoxide **255** to 1,2-dioxetane **258** with subsequent cleavage to **259**. Investigations by Wilson (1969), and Lundeen and Adelman (1970) of the chemiluminescence which accompanies the decomposition of **255** have shown that the light emission is the result of an acid-induced rearrangement of **255** to an intermediate, which cleaves to yield **259** in an elec-

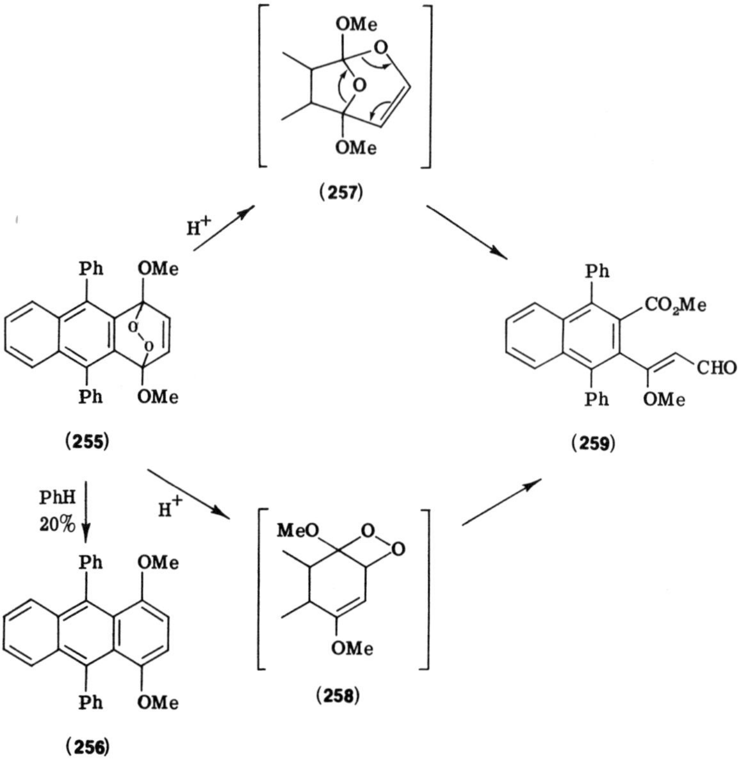

(255) (257) (258) (259) (256) (109)

tronically excited state. Wilson has suggested that the chemiluminescence results are best interpreted in terms of the intermediacy of 1,2-dioxetane **258**. However, calculations with standard bond energies indicate that the decomposition of the Rigaudy intermediate would also be sufficiently exothermic to produce excited **259** (Lundeen and Adelman, 1970). Further study of this reaction would seem to be warranted.

LeRoux and Goasdoue (1975) have isolated 1,2-dioxetanes from the acid-catalyzed rearrangement of polyarylfulvene endoperoxides. Treatment of endoperoxide **261** with *p*-nitrobenzoic acid or zinc chloride in benzene–methanol gave 1,2-dioxetane **263** in 85% yield. The authors proposed a mechanism for the rearrangement involving **262**.

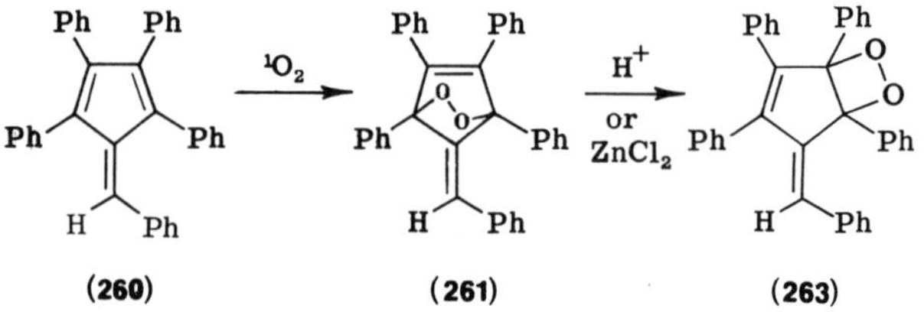

(260) (261) (263) (110)

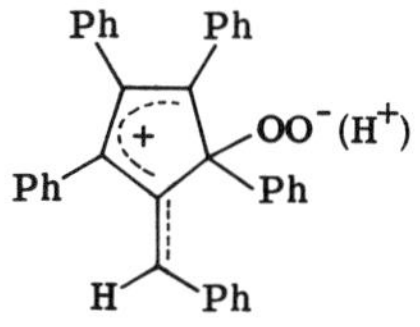

Ph Ph
OO⁻ (H⁺)
Ph Ph
H Ph

(**262**)

Schaap *et al.* (1977) have recently described the silica gel catalyzed rearrangement of an endoperoxide to a 1,2-dioxetane. The use of a heterogeneous catalyst for the rearrangement facilitated the separation of the catalyst from the sensitive 1,2-dioxetane. The endoperoxide is obtained by the Diels–Alder reaction of singlet oxygen with a vinyl-substituted anthracene, in which the aromatic system functions as part of the diene (see Section III,E). Treatment of endoperoxide **264** in *o*-xylene with silica gel at ambient temperature results in the rearrangement of **264** to the 1,2-dioxetane **265** in

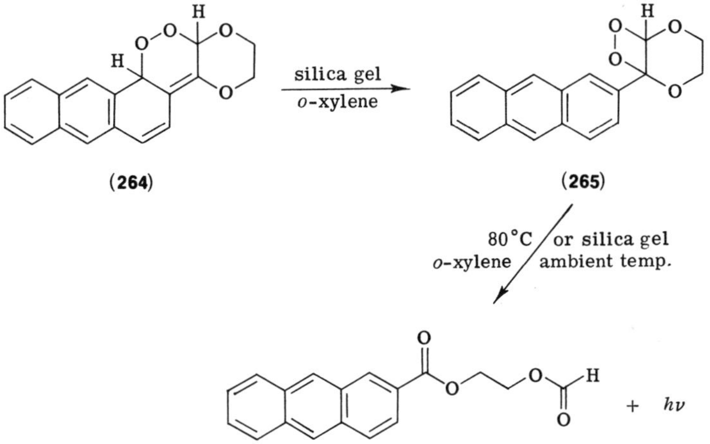

silica gel / *o*-xylene (111)

(**264**) (**265**)

80 °C *o*-xylene / or silica gel ambient temp.

+ $h\nu$

(**266**)

quantitative yield. A plausible mechanism for the rearrangement involves an acid-catalyzed opening of **264** to give **267** with subsequent cyclization to yield **265**. The behavior of **264** contrasts with that of the endoperoxide obtained from the photooxygenation of 1,1-diphenyl-2-methoxyethylene which undergoes spontaneous rearomatization (Section III,E).

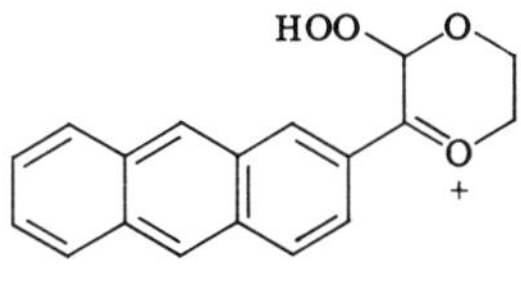

HOO

(**267**)

It had been observed that protracted contact of 1,2-dioxetane **265** with silica gel resulted in decomposition to **266** (Schaap *et al.*, 1977). Subsequent investigations have shown that silica gel also catalyzes the cleavage of **265** to **266** at ambient temperature (Zaklika *et al.*, 1978a). However, in contrast to the nonluminescent, catalyzed decomposition of 1,2-dioxetanes with transition metals (Bartlett, 1976), amines (Lee and Wilson, 1973), and electron-rich olefins (Lee and Wilson, 1973), the conversion of **265** to **266** in the presence of silica gel is accompanied by a greatly enhanced singlet chemiexcitation quantum yield. The chemiexcitation quantum yield (singlet excited **266**) for the thermolysis at 80°C in *o*-xylene is 0.9%. However, in the presence of silica gel the quantum yield is increased to a minimum of 12%.

IX. THEORETICAL AND MECHANISTIC CONSIDERATIONS

The enhanced reactivity of π electron-rich olefins in 1,2-cycloadditions with oxygen in the $^1\Delta_g$ excited state is in accord with the formulation of this species as an electrophilic reagent—a conclusion in harmony with its electronic structure. Bond formation thus requires transfer of electron density from the highest occupied molecular orbital (HOMO) of the olefin to the lowest unoccupied molecular orbital (LUMO) of singlet oxygen. It is the extent of this charge transfer and the molecular geometry attending it that is the basis of the mechanistic description of singlet oxygen reactivity. A number of the mechanisms that have been considered are outlined here, but it should be recognized that distinctions between some of them may be at times rather artificial.

(a) A biradical mechanism (Harding and Goddard, 1977).

$^1O_2$ •O—O O—O (112)

(b) A symmetry-allowed concerted [2s + 2a] path with singlet oxygen as the antarafacial partner (Bartlett and Schaap, 1970)

$^1O_2$ O O O—O (113)

(c) A symmetry-allowed initial formation of a perepoxide or peroxirane (Sharp, 1960; Kearns, 1971)

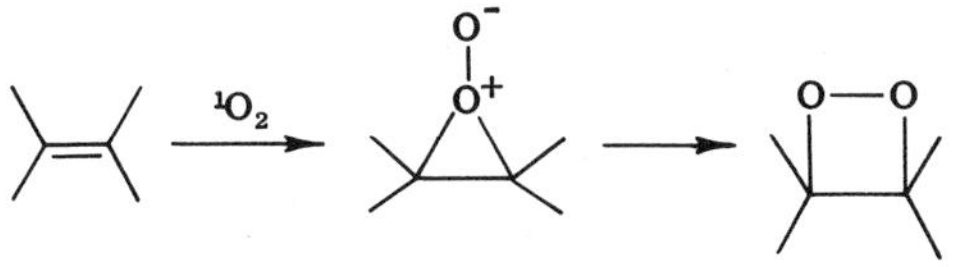

(114)

(d) A zwitterionic mechanism (Foote *et al.*, 1970)

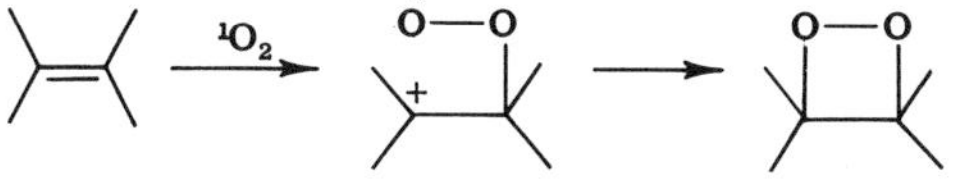

(115)

(e) One electron transfer to give an initial radical cation and superoxide radical anion pair (Mazur, 1971; Foote *et al.*, 1975; Ericksen *et al.*, 1977)

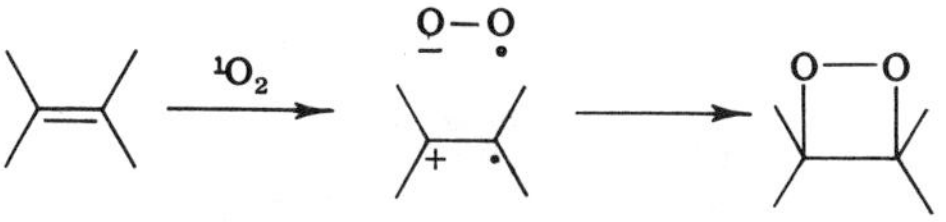

(116)

(f) A charge transfer mechanism, proposed to be possible when HOMO of the olefin lies higher in energy than LUMO of $^1O_2$. Under these conditions a [2s + 2s] addition appears to be allowed (Kearns, 1971; Foote, 1971). This mechanism is related to the "effectively allowed" type of [2s + 2s] cycloaddition proposed by Epiotis *et al.* (1976).

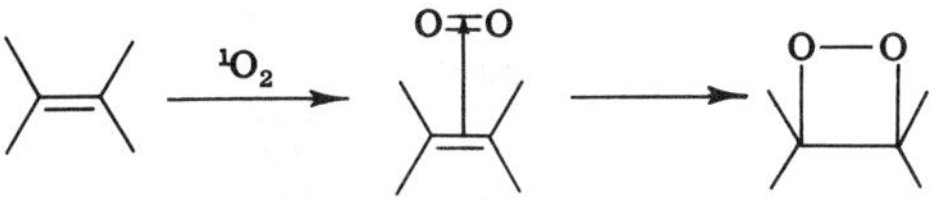

(117)

A. Theoretical Perspectives

Kearns (1969) was the first to investigate 1,2-cycloadditions of singlet oxygen from the standpoint of orbital symmetry conservation. He constructed orbital and state correlation diagrams for the [2s + 2s] cycloaddition (Fig. 1b). These led to the conclusion that the reaction was forbidden, insofar that it exhibits an activation energy. Since the magnitude of the olefin HOMO–$^1O_2$ LUMO energy separation relates to the size of the activation energy (Klopman and Hudson, 1967; Fukui and Fujimoto, 1968, 1969; Devaquet, 1969; Sustmann and Binsch, 1971), Kearns speculated that the reaction could occur when the gap was small. Indeed, if HOMO were higher in energy that LUMO, the addition would be allowed. In this case, the mechanism was recognized as one of charge transfer (Foote, 1971; Kearns, 1971). Examination of similar correlation diagrams for the concerted formation of perepoxide (Fig. 1a) and [2s + 2a] (Fig. 1c) adducts (Kearns, 1971) showed that both processes were allowed. These conclusions were supple-

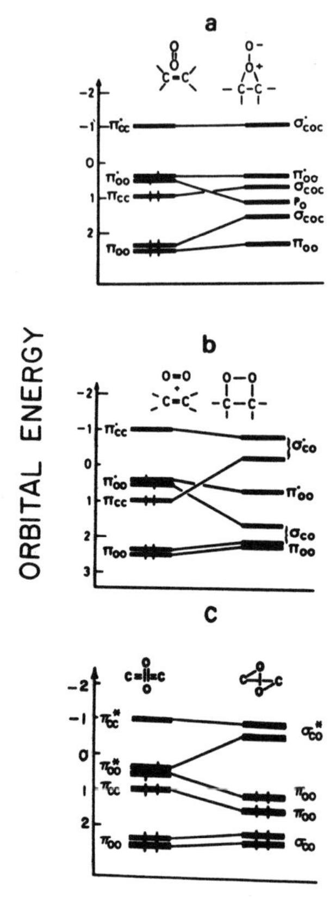

Fig. 1. Orbital correlation diagrams for the concerted addition of molecular oxygen to an olefin. Three different modes of addition are depicted with transition states which are assumed to have C_s (a), C_{2r} (b), and C_2 (c) symmetries, respectively. The reader should note that the orbital occupation shown is that of ground state oxygen; $^1\Delta_g$ oxygen has two electrons in the same π^*_{oo} orbital. Reproduced from Kearns (1971).

mented by CNDO/2 calculations. The perepoxide was found to bear a charge of $-0.4\,e$ on the exocyclic oxygen and to be capable of rearrangement to ene products. The dioxetane was also found to undergo this path in competition with cleavage to carbonyl products, a result at variance with experimental observations (Kopecky and Mumford, 1969). In general, perepoxide formation was energetically preferred over direct formation of dioxetane, but the rearrangement of one to the other was not explicitly studied.

Fueno (Yamaguchi *et al.*, 1973a) approached the problem of reaction of singlet oxygen with both ethylene and 2-aminopropylene, using spin- and space-symmetry criteria (Yamaguchi *et al.*, 1973b) in conjunction with CNDO/2 calculations. Geometries for the $^1O_2 + CH_2{=}CH_2$ reaction ranged from the parallel [2s + 2s] to the orthogonal [2s + 2a] mode. The former was found to be a forbidden process leading to a two-step path, while the latter was allowed. In the case of 2-aminopropylene, a model involving a declined attitude of $^1O_2$, with closer approach at the unsubstituted terminus of the double bond, led to a zwitterionic transition state for this reaction.

An initial study of the $^1O_2 + CH_2{=}CH_2$ reaction by Fukui (Inagaki *et al.*, 1972) based on HOMO–LUMO interactions and CNDO calculations suggested that the geometries **268** or **269** leading to a perepoxide were favored. The [2s + 2a], the [2s + 2s], and intermediate geometries of ap-

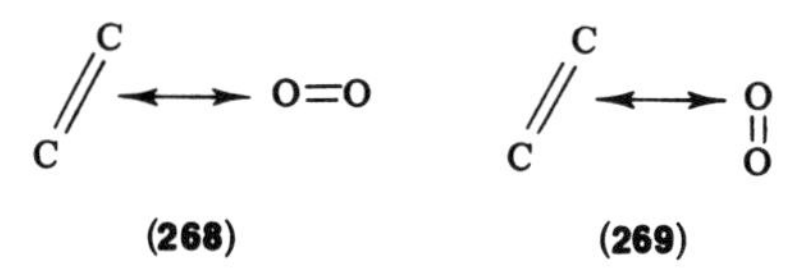

proach were found to be of higher energy, as was an unsymmetrical path with greater bonding at one end of the C=C linkage than the other. The similarity of $^1O_2$ and benzyne was also discussed (Inagaki and Fukui, 1973). A subsequent calculation (SCF CNDO/2) of the reaction coordinate of the $^1O_2 + CH_2{=}CH_2$ cycloaddition (Inagaki and Fukui, 1975) described the perepoxide as a "quasi" intermediate. This description follows from the existence of a flat region on the potential energy surface corresponding to a perepoxide, which may undergo rearrangement to the dioxetane with negligible activation energy. The perepoxide thus exists, but is too transient to exhibit the chemistry of an intermediate—a vexing conclusion for the experimentalist.

To date, the most extensive theoretical study of the addition of singlet oxygen to olefins is that of Dewar and Thiel (1975, 1977) using the MINDO/3 method. Their results are summarized in Figs. 2–4 (* denotes a transition state structure). Perepoxide and zwitterion intermediates are involved. The perepoxide (Fig. 2) was calculated to be a symmetrical, polar species with a dipole moment of 4.26 D and a charge of $-0.4\ e$ on the exocyclic oxygen atom (cf. Inagaki and Fukui, 1975; Kearns, 1971).

The following characteristics were found for the various transition states.

1. An olefin–perepoxide transition state is an early, nonpolar one. The C–C and C–O bond lengths are similar to those in the starting materials, and a charge of only $-0.09\ e$ develops on the exocyclic oxygen (Fig. 2). For this reason *cis*- and *trans*-alkyl isomers do not differ in stability (Fig. 3).

2. A perepoxide–dioxetane transition state is unsymmetrical and resembles the perepoxide in polarity. Charges of $-0.53\ e$ and $+0.49\ e$, respectively are developed on the oxygen and carbon atoms forming the new bond (Fig. 2).

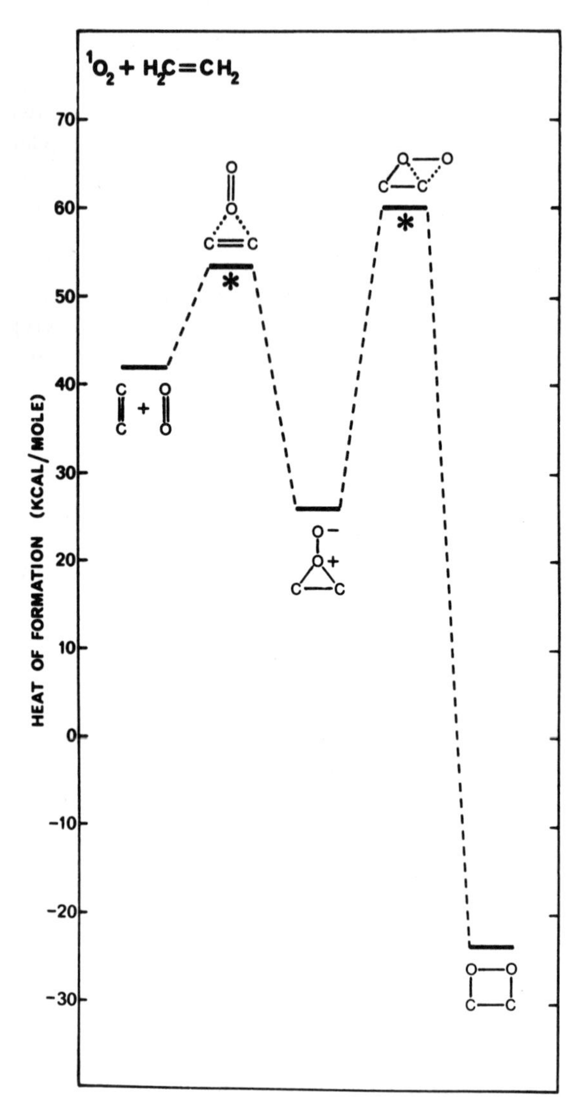

Fig. 2. MINDO/3 calculation of the energetics for the reaction: $O_2 + CH_2{=}CH_2$. MINDO/3 also predicts cleavage of the dioxetane to triplet formaldehyde, with an activation energy of 38.3 kcal/mole (Dewar and Kirschner, 1974).

3. A perepoxide–ene transition state is again reactant like, the migrating hydrogen atom being closer to carbon than to oxygen (Fig. 3). The polarity is, however, lower, since the charge is $-0.10\ e$ on carbon and $-0.36\ e$ on oxygen, while the dipole moment is 2.94 D. The requirement for neutrality appears to indicate a charge of $+0.12\ e$ on the migrating hydrogen atom, suggesting the reaction is an atom (not proton) transfer.

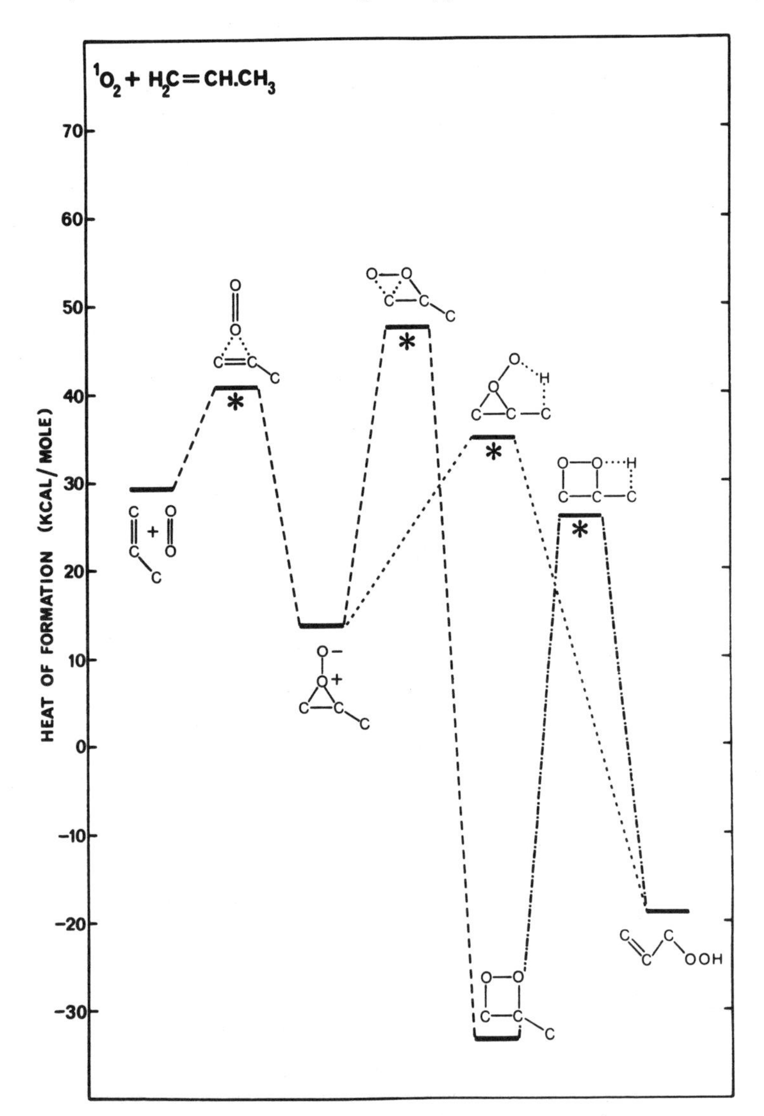

Fig. 3. MINDO/3 calculation of the energetics for the reaction: $^1O_2 + CH_2{=}CH.CH_3$. ---, Formation of dioxetane; ···, formation of ene product from perepoxide; -·-·, formation of ene product from dioxetane.

4. An olefin–zwitterion transition state is also early, as evidenced by the O–O bond length, which is similar to that in $^1O_2$, and a geometry not consistent with any 1,4-interaction (Fig. 4).

5. A zwitterion–dioxetane transition state resembles the open zwitterion, and the dipole moment of 8.59 D shows it to be highly polar (Fig. 4).

Some general conclusions may be made. Cycloadditions to alkyl olefins

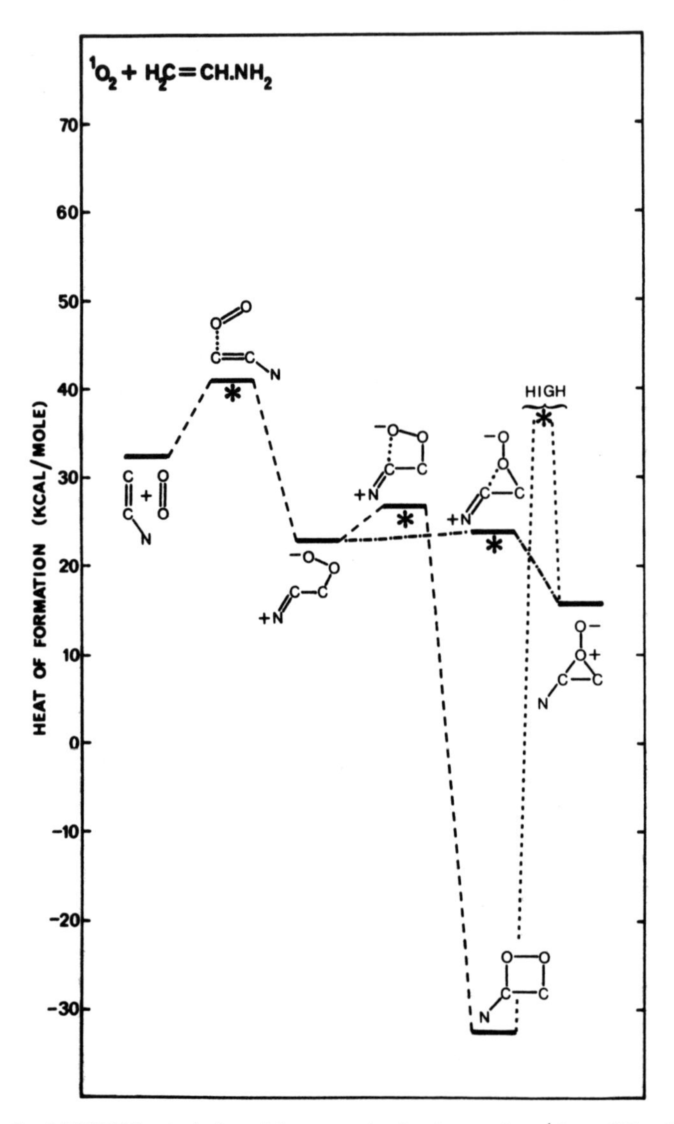

Fig. 4. MINDO/3 calculation of the energetics for the reaction: $^1O_2 + CH_2{=}CH.NH_2$. ---, Formation of dioxetane; -·-·-, formation of perepoxide; ···, rearrangement of perepoxide to dioxetane.

(Figs. 2–4) are predicted to form dioxetanes via perepoxide intermediates, which can also undergo reaction to ene products. Rearrangement of the perepoxide (polar transition state) should be rate-limiting for dioxetane products, while its rate of formation (nonpolar transition state) should be determining for ene products. When the ene reaction is precluded, it is possible that a significant steady state concentration of perepoxide may be formed, in view of the depth of the potential well containing this intermediate. This seems not to have been recognized in the original paper. Cycloadditions of $^1O_2$ to 1-heterosubstituted olefins are predicted to occur exclusively by zwitterionic intermediates (Fig. 4). These may close directly to dioxetanes, or to perepoxides. It may be noted that the activation energies for these processes run contrary to expectations based on Hammond's postulate (Leffler, 1953; Hammond, 1955). Whether or not ene products are possible, initial addition of $^1O_2$ to give a zwitterion is predicted to be rate limiting. The subsequent formation of a perepoxide should be very facile; however, its further rearrangement to ene products appears to require significantly more activation energy than does cyclization of the zwitterion to dioxetane. The general conclusions appear at present to be in reasonable accord with experiment, as discussed by Dewar and Thiel (1975).

Nevertheless, the merits of the MINDO/3 method have been the subject of a lively exchange in the literature (Pople, 1975; Hehre, 1975; Dewar, 1975; Dewar *et al.*, 1975b; Harding and Goddard, 1977). The principal difficulties relevant here are: the overestimation by MINDO/3 of oxirane stabilities (by 14 kcal/mole) (Dewar *et al.*, 1975a); the suggestion that C=C bond energies are underestimated (Pople, 1975); and an 8 kcal/mole excess in the calculated heat of atomization of O_2 ($^1\Delta_g$) (Dewar *et al.*, 1975b). Recently, for example, on the basis of *ab initio* and thermochemical calculations Goddard (Harding and Goddard, 1977) has found that perepoxide formation from ethylene and $^1O_2$ is 16.5 kcal/mole endothermic. While there is a general consensus among theoretical chemists that perepoxide formation is a feasible mechanism, the experimentalist has yet to act as the final arbiter.

B. Experimental Observations

The study of the 1,2-cycloadditions of singlet oxygen shares common features with that of [2 + 2] cycloadditions of olefins. Several factors, however, cause difficulty and lead to ambiguous results. The reagent itself is unstable, and its concentration is seldom known. It is incapable of structural modification. More importantly, because the oxygen atom has no stereochemistry, valuable mechanistic clues are lost. The possibility that under the reaction conditions singlet oxygen is not the only oxidizing species present, and the frequent instability of the reaction products are further complications. Nevertheless, a number of results are available, though they are not always definitive.

1. Stereochemistry

The stereochemistry of a number of polyalkoxyolefins is retained during cycloaddition of singlet oxygen (Section III,B). This result provides an indication that the reaction may be concerted in these cases. The slight steric requirement of the oxygen molecule would facilitate a [2s + 2a] pathway, though the electrophilic center found in ketenes is lacking (Woodward and Hoffman, 1970; Brady and Hoff, 1970). If this view is correct, experiments with suitably substituted olefins require that singlet oxygen be the antarafacial partner. However, the alternative of a stepwise pathway needs considerations. It is found that for 1,2-cycloadditions via biradicals, bond rotation is faster than ring closure (Bartlett *et al.*, 1969; Bartlett and Wallbillich, 1969) so that a biradical pathway is unlikely. Zwitterionic intermediates are, however, stereochemically more stable as a result of favorable 1,4-coulombic interactions (Huisgen, 1977). The lack of a solvent effect on the rate of photooxidation of *cis*-diethoxyethylene (Bartlett and Schaap, 1970) and the absence of stereochemical leakage during oxygenation in the polar solvent acetone as compared to Freon is evidence, albeit not compelling, against a zwitterionic pathway.

2. Epoxide Products from Photooxygenation

The formation of a perepoxide intermediate has been suggested as a rather general mode of singlet oxygen cycloaddition (Dewar and Thiel, 1975). There is, however, some disagreement concerning the expected stability of this species (see Inagaki and Fukui, 1975; Dewar and Thiel, 1975; Harding and Goddard, 1977). The related episulfoxides **270** and aziridine *N*-oxides **271** have been synthesized (Baldwin *et al.*, 1971a,b). They are found to undergo a partially stereospecific extrusion to yield the olefin and a more facile ene-type reaction [Eq. (118)]. No evidence has been found for

X—OH

H

X^{+}—O^{-} (118)

(270) X = S
(271) X = N-R

\+ OX

their rearrangement to 1,2-oxathietanes or 1,2-oxazetidines. The observation by Schaap and Faler (1973) that adamantylideneadamantane affords the corresponding epoxide as an authentic product of singlet oxygen reaction provides the major evidence for the involvement of perepoxides in the cycloaddition of singlet oxygen. Though their original conclusions have required some modification, their results have been substantively confirmed in related systems (Bartlett and Ho, 1974). The epoxide has been suggested to arise by trapping of the perepoxide by singlet oxygen (Dewar *et al.*, 1975a) and presently available evidence seems generally consistent with this mechanism (Bartlett, 1976). Authenticated formation of epoxides is restricted to hindered olefins, perhaps as a result of special stability conferred on the perepoxide. In this regard, the unusual reaction of adamantylideneadamantane with ozone (Bartlett, 1976) and with halogens (Wieringa *et al.*, 1970) should be noted. It is not known whether such a perepoxide is merely in equilibrium with the reactants, or lies along the potential surface to products. Recent work by McCapra and Beheshti (1977) on camphenylideneadamantane strongly suggests a perepoxide can undergo solvent depending partitioning to 1,2-dioxetane and 1,2-dioxolane. However, additional kinetic data are needed to exclude the possibility that 1,2-dioxolane formation via perepoxide is competing with direct addition of singlet oxygen to the olefin to give 1,2-dioxetane.

3. *Solvent Effects*

Negligible solvent effects are found for reaction of singlet oxygen by the ene mode (Foote, 1968a) and in [4 + 2] cycloaddition (Stevens and Perez, 1974). With the exception of dihydropyran (Section III,D) for which an explanation has been proposed (Dewar and Thiel, 1975), and camphenylideneadamantane (Section V,A and above), there is presently no well defined instance of a solvent effect on rates of uncomplicated 1,2-cycloadditions of $^1O_2$. For example, *N,N*-dimethylisobutenyl amine (Foote *et al.*, 1975), tetraethoxyethylene (Section III,D), *cis*-diethoxyethylene (Schaap, 1970), and adamantylideneadamantane (G. R. Faler and A. P. Schaap, unpublished) all exhibit rates that are essentially independent of solvent. While this may be considered to exclude transition states or intermediates with significant charge separation, it has been pointed out (Foote *et al.*, 1975) that no solvent effect need be observed even for a quite polar transition state in case of olefins of high reactivity. This is reasonable in view of the finding that activation energies for cycloaddition reactions of singlet oxygen lie typically in the range 0.1 to 5.0 kcal/mole (Koch, 1968; Stevens *et al.*, 1974). The reader will recall that MINDO/3 calculations in general predict nonpolar reactantlike transition states for the 1,2-cycloadditions of singlet oxygen.

4. *Relative Reactivities*

The rate constants for the reaction of singlet oxygen with a number of olefins are collected in Table III, together with those for a typical ene and [4 + 2] reaction. For an electrophilic reaction involving some transfer of

Table III Reactivities toward Singlet Oxygen

No.	Olefin	Solvent	k_r (M^{-1} sec^{-1})	Ref.
1	Me, Me, NMe_2	PhH	4.9×10^8	[a]
2	OEt	Me_2CO	2.9×10^4	[b]
3	EtO, OEt	Me_2CO	4.7×10^7	[b]
4	EtO, OEt, EtO, OEt	Me_2CO	4.5×10^7	[b]
5	O, O	Me_2CO	2.2×10^5	[b]
6	O, Ph, O	Me_2CO	1.0×10^7	[c]
7	O, Ph, O, Ph	Me_2CO	1.5×10^7	[c]
8	Me, Me, Me, Me	MeOH	4.8×10^7	[d]
9	Me, Me	MeOH	2.4×10^7	[e]

[a] Foote *et al.* (1975).
[b] Faler (1977).
[c] A. P. Schaap, K. A. Zaklika, and B. Kaskar (unpublished).
[d] Foote and Ching (1975).
[e] Stevens and Perez (1974).

charge to oxygen, one might expect a rough correlation of rates with "π electron richness" or ionization potential of the olefin. This expectation is realized to a certain extent, for example, in the ordering of the rates $1 > 3 > 2$ and $7 > 6 > 5$. It is noteworthy that the tabulated rates of [2 + 2] and [4 + 2] addition of singlet oxygen are similar, whereas marked differences appear for addition of polycyanoolefins to similar substrates by these two modes. Evidence has been adduced that the cyanoolefins undergo [2 + 2] addition by a late, zwitterionic transition state (Huisgen, 1977), while charge transfer interactions occur in the Diels–Alder reaction (see Kiselev and Miller, 1975, and compare Ashford and Ogryzlo, 1974). The trends in the reactivity toward $^1O_2$ are not, however, without discrepancies. The difference between 5 and 3 may be ascribable to a conformational effect, but the identical rates for 3 and 4 are striking indeed. A similar "saturation" effect may be evidenced by the small difference in rates for 6 and 7, though other explanations are possible. This type of behavior is presently not well understood. A change in mechanism may be one interpretation.

5. *Linear Free Energy Relationships*

The photooxygenation of the 2-aryl-1,4-dioxenes proceeds cleanly and efficiently ($> 90\%$) to the diesters formed by oxidative scission of the double bond (Section III,D). Dioxetanes have been isolated in high yield both from the unsubstituted 1,4-dioxene, and from its 2,3-diaryl derivatives. As NMR evidence for dioxetanes is available in this case also, this reaction appears to be an uncomplicated 1,2-cycloaddition. Substituents in the aryl ring affect the rate. Relative rate experiments (A. P. Schaap, K. A. Zaklika, and B. Kaskar, unpublished observations) lead to a Hammett plot. A better correlation is found using σ^+ values ($\rho = -0.64$, $r = 0.989$) than with σ ($\rho = -1.02$, $r = 0.858$). This conclusion was tested with the *p*-methylthio substituent which has $\sigma = 0.0$ but $\sigma^+ = -0.60$ (Leffler and Grunwald, 1963). A correlation with σ thus implies a rate ratio for this substituent relative to the unsubstituted parent that is close to unity. In fact, a ratio of 2.6 was found, indicating that σ^+ should be used. Comparison of the ρ values for this and related reactions (Table IV) shows that the 1,2-cycloaddition has only a modest electron demand. The correlation with σ^+ is consistent with an unsymmetrical transition state involving development of positive charge α to the phenyl substituent. Nevertheless, the possibility is not excluded that destabilizing effects on the olefin itself are an adequate explanation.

6. *Other Effects of Substitution*

Foote *et al.* (1972) have pointed out that a very delicate electron density requirement appears to exist for dioxetane formation. For instance, these

Table IV Hammett Correlations

Substrate	Reaction	ρ	Ref.
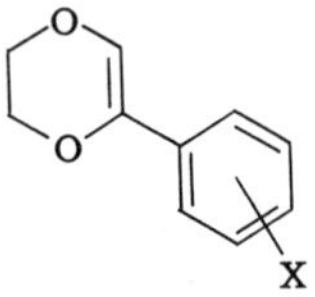	$^1O_2$, 1,2-addition (Me_2CO)	-0.64 [σ^+]	a
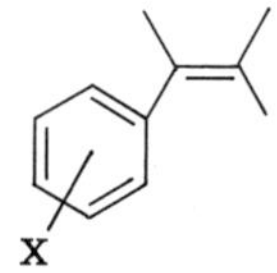	$^1O_2$, ene mode (MeOH–2% C_5H_5N)	-0.92 [σ]	b
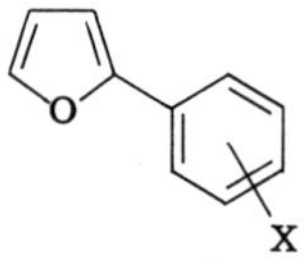	$^1O_2$, 1,4-addition (MeOH)	-0.84 [σ]	c
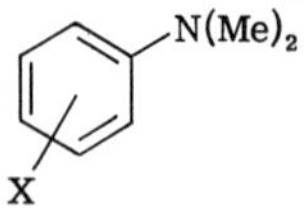 N(Me)$_2$, X	$^1O_2$, quenching only (MeOH)	-1.72 [σ]	d
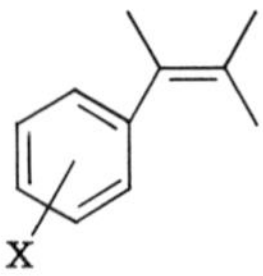	Peracid epoxidation (PhH)	-0.87 [$\sigma - 0.37(\sigma^+ - \sigma)$]	b
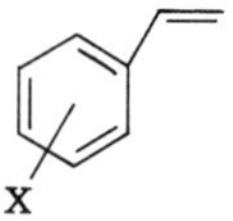	Peracid epoxidation (PhH)	-1.3 [$\sigma - 0.48(\sigma^+ - \sigma)$]	e
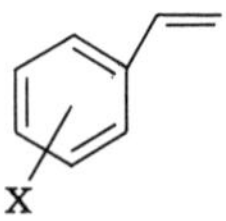	Molybdenum-catalyzed epoxidation (PhH)	-1.4 [σ or σ^+]	f

[a] A. P. Schaap, K. A. Zaklika, and B. Kaskar (unpublished).
[b] Foote and Denny (1971).
[c] Young *et al.* (1972).
[d] Brewer (1974).
[e] Ishii and Inamoto (1960).
[f] Howe and Hiatt (1971).

authors found that olefin **124** undergoes 1,4-addition, while the α,α′-dimethoxystilbenes **100** add singlet oxygen in a 1,2-fashion (Rio and Berthelot, 1971b). A related observation has been made by Saito and Matsuura

Ph OMe / Ph H — (**124**)

Ph OMe / MeO H — (**100**) and also cis

(1972) on the competition between ene and 1,2-addition modes (Section V,B). In contrast, Foote and Denny (1971) have shown that this effect seems absent in competitive formation of two different ene products. There is an apparent relationship to symmetry of substitution. *cis*- and *trans*-1,2-Diethoxyethylenes are of roughly comparable reactivity toward singlet oxygen, yet for 1,1-diethoxyethylene only quenching is observed (Faler, 1977), implying a reactivity difference of at least three orders of magnitude. These results stand in contrast to the dipolar addition of tetracyanoethylene to these enol ethers, where the 1,1-diethoxyolefin is much more reactive (Huisgen and Steiner, 1973). Further examination of these effects is likely to offer considerable insight into the mechanism of cycloaddition.

Presently available results are in accord with mechanisms of cycloaddition involving small amounts of charge transfer in the transition state. A larger extent of charge transfer is not, however, precluded for highly activated olefins (Foote *et al.*, 1975; Mazur, 1971). A concerted reaction or one involving short-lived, stereochemically-stable intermediates seems favored. The limited correlation of substitution with reactivity, and the fact that epoxide formation is observed only with hindered, unactivated olefins suggest that a single cycloaddition mechanism may not adequately account for all the results. Although definitive mechanistic conclusions are as yet not possible, considerable progress has been made, particularly in view of the fact that the reagent itself was unambiguously identified only eleven years ago.

ACKNOWLEDGMENTS

The authors' research in this area has been supported by the U.S. Army Research Office, the National Science Foundation, and the Petroleum Research Fund as administered by the American Chemical Society. An Alfred P. Sloan Fellowship to A. Paul Schaap is gratefully acknowledged. We also express our appreciation to Dr. Paul Burns for many helpful discussions.

REFERENCES

Adam, W. (1975). *Chem. Ztg.* **99,** 142.
Adam, W., and Liu, J.-C. (1971). *Abstr. 162nd Natl. Meet., Am. Chem. Soc.* No. 167.
Adam, W., and Liu, J.-C. (1972a). *J. Chem. Soc., Chem. Commun.* p. 73.
Adam, W., and Liu, J.-C. (1972b). *J. Am. Chem. Soc.* **94,** 1206.
Akutagawa, M., Aoyama, H., Omote, Y., and Yamamoto, H. (1976). *J. Chem. Soc., Chem. Commun.* p. 180.
Ando, W., Suzuki, J., Arai, T., and Migita, T. (1972). *J. Chem. Soc., Chem. Commun.* p. 477.
Ando, W., Suzuki, J., Arai, T., and Migita, T. (1973). *Tetrahedron* **29,** 1507.
Ando, W., Watanabe, K., Suzuki, J., and Migita, T. (1974). *J. Am. Chem. Soc.* **96,** 6766.
Ando, W., Saiki, T., and Migita, T. (1975a). *J. Am. Chem. Soc.* **97,** 5028.
Ando, W., Watanabe, K., and Migita, T. (1975b). *Tetrahedron Lett.* p. 4127.
Ashford, R. D., and Ogryzlo, E. A. (1974). *Can. J. Chem.* **52,** 3544.
Ashford, R. D., and Ogryzlo, E. A. (1975). *J. Am. Chem. Soc.* **97,** 3604.
Atkinson, R. S. (1970). *Chem. Commun.* p. 177.
Atkinson, R. S. (1971). *J. Chem. Soc. C* p. 784.
Baldwin, J. E., Basson, H. H., and Krauss, H., Jr. (1968). *Chem. Commun.* p. 984.
Baldwin, J. E., Höfle, G., and Choi, S. C. (1971a). *J. Am. Chem. Soc.* **93,** 2811.
Baldwin, J. E., Bhatnagar, A. K., Choi, S. C., and Shortridge, T. J. (1971b). *J. Am. Chem. Soc.* **93,** 4082.
Bartlett, P. D. (1976). *Chem. Soc. Rev.* **5,** 149.
Bartlett, P. D., and Ho, M. S. (1974). *J. Am. Chem. Soc.* **96,** 627.
Bartlett, P. D., and Landis, M. E. (1977). *J. Am. Chem. Soc.* **99,** 3033.
Bartlett, P. D., and Schaap, A. P. (1970). *J. Am. Chem. Soc.* **92,** 3223.
Bartlett, P. D., and Wallbillich, G. E. H. (1969). *J. Am. Chem. Soc.* **91,** 409.
Bartlett, P. D., Dempster, C. J., Montgomery, L. K., Schueller, K. E., and Wallbillich, G. E. H. (1969). *J. Am. Chem. Soc.* **91,** 405.
Bartlett, P. D., Mendenhall, G. D., and Schaap, A. P. (1970). *Ann. N. Y. Acad. Sci.* **171,** 79.
Basselier, J.-J., and LeRoux, J. P. (1971). *Bull. Soc. Chim. Fr.* p. 4443.
Basselier, J.-J., Cherton, J.-C., and Caille, J. (1971). *C. R. Hebd. Seances Acad. Sci.* **273,** 514.
Bogan, D. J., Sheinson, R. S., Gann, R. G., and Williams, F. W. (1975). *J. Am. Chem. Soc.* **97,** 2560.
Bogan, D. J., Sheinson, R. S., and Williams, F. W. (1976). *J. Am. Chem. Soc.* **98,** 1034.
Bollyky, J. (1970). *J. Am. Chem. Soc.* **92,** 3230.
Brady, W. T., and Hoff, E. F., Jr. (1970). *J. Org. Chem.* **35,** 3733.
Brewer, D. R., III. (1974). Ph.D. Dissertation, Georgetown University, Washington, D.C.
Burns, P. A., and Foote, C. S. (1974). *J. Am. Chem. Soc.* **96,** 4339.
Burns, P. A., and Foote, C. S. (1976). *J. Org. Chem.* **41,** 908.
Burns, P. A., Foote, C. S., and Mazur, S. (1976). *J. Org. Chem.* **41,** 899.
Criegee, R. (1936). *Justus Liebigs Ann. Chem.* **522,** 75.
Davidson, R. S., and Trethewey, K. R. (1976). *J. Am. Chem. Soc.* **98,** 4008.
Denny, R. W., and Nickon, A. (1973). *Org. React.* **20,** 133.
Devaquet, A. (1969). *Mol. Phys.* **18,** 233.
Dewar, M. J. S. (1975). *J. Am. Chem. Soc.* **97,** 6591.
Dewar, M. J. S., and Kirschner, S. (1974). *J. Am. Chem. Soc.* **96,** 7578.
Dewar, M. J. S., and Thiel, W. (1975). *J. Am. Chem. Soc.* **97,** 3978.
Dewar, M. J. S., and Thiel, W. (1977). *J. Am. Chem. Soc.* **99,** 2338.
Dewar, M. J. S., Griffin, A. C., Thiel, W., and Turchi, I. J. (1975a). *J. Am. Chem. Soc.* **97,** 4439.
Dewar, M. J. S., Haddon, R. C., Li, W.-K., Thiel, W., and Weiner, P. K. (1975b). *J. Am. Chem. Soc.* **97,** 4540.

Doering, W. von E., and Haines, R. M. (1954). *J. Am. Chem. Soc.* **76,** 482.
Dufraisse, C., and Velluz, L. (1942). *Bull. Soc. Chim. Fr.* **9,** 171.
Epiotis, N. D., Yates, R. L., Carlberg, D., and Bernardi, F. (1976). *J. Am. Chem. Soc.* **98,** 453.
Ericksen, J., Foote, C. S., and Parker, T. L. (1977). *J. Am. Chem. Soc.* **99,** 6455.
Faler, G. R. (1977). Ph.D. Dissertation, Wayne State University, Detroit, Michigan.
Fenical, W., Kearns, D. R., and Radlick, P. (1969a). *J. Am. Chem. Soc.* **91,** 3396.
Fenical, W., Kearns, D. R., and Radlick, P. (1969b). *J. Am. Chem. Soc.* **91,** 7771.
Fieser, L. F. (1936). "The Chemistry of Natural Products Related to Phenanthrene." Van Nostrand-Reinhold, New York.
Filippova, T. V., Blyumberg, E. A., and Kas'yan, L. F. (1973). *Dokl. Akad. Nauk SSSR* **210,** 664.
Font, J., Serratosa, F., and Villarasa, L. (1970). *Tetrahedron Lett.* p. 4105.
Foote, C. S. (1968a). *Acc. Chem. Res.* **1,** 104.
Foote, C. S. (1968b). *Science* **162,** 963.
Foote, C. S. (1971). *Pure Appl. Chem.* **27,** 635.
Foote, C. S., and Ching, T.-Y. (1975). *J. Am. Chem. Soc.* **97,** 6209.
Foote, C. S., and Denny, R. W. (1971). *J. Am. Chem. Soc.* **93,** 5162.
Foote, C. S., and Lin, J. W.-P. (1968). *Tetrahedron Lett.* p. 3267.
Foote, C. S., Lin, J. W.-P., and Wong, S.-Y. (1970). *Prepr., Div. Pet. Chem., Am. Chem. Soc.* **15,** E89.
Foote, C. S., Fujimoto, T. T., and Chang, Y. C. (1972). *Tetrahedron Lett.* p. 45.
Foote, C. S., Mazur, S., Burns, P. A., and Lerdal, D. (1973). *J. Am. Chem. Soc.* **95,** 586.
Foote, C. S., Dzakpasu, A. A., and Lin, J. W.-P. (1975). *Tetrahedron Lett.* p. 1247.
Frimer, A. A., Bartlett, P. D., Boschung, A. F., and Jewett, J. G. (1977). *J. Am. Chem. Soc.* **99,** 7977.
Fukui, K., and Fujimoto, H. (1968). *Bull. Chem. Soc. Jpn.* **41,** 1989.
Fukui, K., and Fujimoto, H. (1969). *Bull. Chem. Soc. Jpn.* **42,** 3399.
Fuson, R. C., and Jackson, H. L. (1950). *J. Am. Chem. Soc.* **72,** 1637.
Gollnick, K. (1968). *Adv. Photochem.* **6,** 1.
Gollnick, K., and Schenck, G. O. (1964). *Pure Appl. Chem.* **9,** 507.
Gollnick, K., Haisch, D., and Schade, G. (1972). *J. Am. Chem. Soc.* **94,** 1747.
Goto, T., and Nakamura, H. (1976). *Tetrahedron Lett.* p. 4627.
Greibrokk, T. (1973). *Tetrahedron Lett.* p. 1663.
Griffin, G. W., Politzer, I. R., Ishikawa, K., Turro, N. J., and Chou, M.-F. (1977). *Tetrahedron Lett.* p. 1287.
Hammond, G. S. (1955). *J. Am. Chem. Soc.* **77,** 334.
Harding, L. B., and Goddard, W. A., III. (1977). *J. Am. Chem. Soc.* **99,** 4520.
Hasty, N. M., and Kearns, D. R. (1973). *J. Am. Chem. Soc.* **95,** 3380.
Hasty, N., Merkel, P. B., Radlick, P., and Kearns, D. R. (1972). *Tetrahedron Lett.* p. 49.
Hehre, W. J. (1975). *J. Am. Chem. Soc.* **97,** 5308.
Hiatt, R. (1971). *In* "Oxidation" (R. L. Augustine and D. J. Trecker, eds.), Vol. 2, p. 113. Decker, New York.
Howe, G. R., and Hiatt, R. R. (1971). *J. Org. Chem.* **36,** 2493.
Huber, J. E. (1968). *Tetrahedron Lett.* p. 3271.
Huisgen, R. (1977). *Acc. Chem. Res.* **10,** 117.
Huisgen, R., and Steiner, G. (1973). *Tetrahedron Lett.* p. 3763.
Inagaki, S., and Fukui, K. (1973). *Bull. Chem. Soc. Jpn.* **46,** 2240.
Inagaki, S., and Fukui, K. (1975). *J. Am. Chem. Soc.* **97,** 7480.
Inagaki, S., Yamabe, S., Fujimoto, H., and Fukui, K. (1972). *Bull. Chem. Soc. Jpn.* **45,** 3510.
Ishibe, N., Odani, M., and Sunami, M. (1971). *Chem. Commun.* p. 118.
Ishii, Y., and Inamoto, Y. (1960). *J. Chem. Soc. Jpn., Ind. Chem. Sect.* **63,** 765.

Ismail, A. F. A., and El-Shafei, Z. M. (1957). *J. Chem. Soc.* p. 3393.
Ito, S., Takeshita, H., and Hirama, M. (1971). *Tetrahedron Lett.* p. 1181.
Jefford, C. W., and Boschung, A. F. (1974). *Helv. Chim. Acta* **57,** 2257.
Jefford, C. W., and Boschung, A. F. (1976). *Tetrahedron Lett.* p. 4771.
Jefford, C. W., and Rimbault, C. G. (1976). *Tetrahedron Lett.* p. 2479.
Kearns, D. R. (1969). *J. Am. Chem. Soc.* **91,** 6554.
Kearns, D. R. (1971). *Chem. Rev.* **71,** 395.
Kearns, D. R., Fenical, W., and Radlick, P. (1970). *Ann. N. Y. Acad. Sci.* **171,** 34.
Kellogg, R. M., and Kaiser, J. K. (1975). *J. Org. Chem.* **40,** 2575.
Kiselev, V. D., and Miller, J. G. (1975). *J. Am. Chem. Soc.* **97,** 4036.
Klein, E., and Rojahn, W. (1965). *Tetrahedron* **21,** 2173.
Klopman, G., and Hudson, R. F. (1967). *Theor. Chim. Acta* **8,** 165.
Koch, E. (1968). *Tetrahedron* **24,** 6295.
Kohler, E. P., and Thompson, R. B. (1937). *J. Am. Chem. Soc.* **59,** 887.
Kondo, K., and Matsumoto, M. (1972). *J. Chem. Soc., Chem. Commun.* p. 1332.
Kopecky, K. R., and Mumford, C. (1969). *Can. J. Chem.* **47,** 709.
Kopecky, K. R., Vande Sande, J. H., and Mumford, C. (1968). *Can. J. Chem.* **45,** 25.
Kopecky, K. R., Filby, J. E., Mumford, C., Lockwood, P. A., and Ding, J.-Y. (1975). *Can. J. Chem.* **53,** 1103.
Lee, D. C.-S., and Wilson, T. (1973). *In* "Chemiluminescence and Bioluminescence" (M. J. Cormier, D. M. Hercules, and J. Lee, eds.), p. 265. Plenum, New York.
Lee, K.-W., Singer, L. A., and Legg, K. D. (1976). *J. Org. Chem.* **41,** 2685.
Leffler, J. E. (1953). *Science* **117,** 340.
Leffler, J. E., and Grunwald, E. (1963). "Rates and Equilibria of Organic Reactions," p. 204. Wiley, New York.
LeRoux, J. P., and Goasdoue, C. (1975). *Tetrahedron* **31,** 2761.
Litt, F. A., and Nickon, A. (1968). *Adv. Chem. Ser.* **77,** p. 8.
Lundeen, G. W., and Adelman, A. H. (1970). *J. Am. Chem. Soc.* **92,** 3914.
McCapra, F. (1968). *Chem. Commun.* p. 155.
McCapra, F., and Beheshti, I. (1977). *J. Chem. Soc., Chem. Commun.* p. 517.
McCapra, F., and Hann, R. A. (1969). *Chem. Commun.* p. 442.
McCapra, F., Chang, Y. C., and Burford, A. (1976). *J. Chem. Soc., Chem. Commun.* p. 608.
McCapra, F., Beheshti, I., Burford, A., Hann, R. A., and Zaklika, K. A. (1978). *J. Chem. Soc., Chem. Commun.* p. 944.
Machin, P. J., and Sammes, P. G. (1976). *J. Chem. Soc., Perkin I* p. 628.
Maier, G. (1974). *Angew. Chem., Int. Ed. Engl.* **13,** 425.
Matsumoto, M., and Kondo, K. (1975a). *J. Org. Chem.* **40,** 2259.
Matsumoto, M., and Kondo, K. (1975b). *Tetrahedron Lett.* p. 3935.
Matsumoto, M., Dobashi, S., and Kondo, K. (1975). *Tetrahedron Lett.* p. 4471.
Matsuura, T., Matsushima, H., and Nakashima, R. (1970). *Tetrahedron* **26,** 435.
Matsuura, T., Honnaka, A., and Nakashima, R. (1973). *Chem. Lett.* p. 887.
Mayo, F. R. (1968). *Acc. Chem. Res.* **1,** 193.
Mazur, S. (1971). Ph.D. Dissertation, University of California, Los Angeles.
Mazur, S., and Foote, C. S. (1970). *J. Am. Chem. Soc.* **92,** 3225.
Moureu, C., Dufraisse, C., and Dean, P. M. (1926). *C. R. Hebd. Seances Acad. Sci.* **182,** 1584.
Numan, H., Wieringa, J. H., Wynberg, H., Hess, J., and Vos, A. (1977). *J. Chem. Soc., Chem. Commun.* p. 591.
Orito, K., Manske, R. H., and Rodrigo, R. (1974). *J. Am. Chem. Soc.* **96,** 1944.
Pfoertner, K., and Bernauer, K. (1968). *Helv. Chim. Acta* **51,** 1787.

Pinkus, A. G., Haq, M. Z., and Lindberg, J. G. (1970). *J. Org. Chem.* **35,** 2555.
Politzer, I. R., and Griffin, G. W. (1973). *Tetrahedron Lett.* p. 4775.
Pople, J. A. (1975). *J. Am. Chem. Soc.* **97,** 5306.
Richardson, W. H., and Hodge, V. (1970). *J. Org. Chem.* **35,** 1216.
Richardson, W. H., Montgomery, F. C., Slusser, P., and Yelvington, M. B. (1975). *J. Am. Chem. Soc.* **97,** 2819.
Rigaudy, J. (1948). *C. R. Hebd. Seances Acad. Sci.* **226,** 1993.
Rigaudy, J. (1968). *Pure Appl. Chem.* **16,** 169.
Rigaudy, J., Capdevielle, P., and Maumy, M. (1972). *Tetrahedron Lett.* p. 4997.
Rigaudy, J., Capdevielle, P., Combrisson, S., and Maumy, M. (1974). *Tetrahedron Lett.* p. 2757.
Rio, G., and Berthelot, J. (1969). *Bull. Soc. Chim. Fr.* p. 3609.
Rio, G., and Berthelot, J. (1971a). *Bull. Soc. Chim. Fr.* p. 1705.
Rio, G., and Berthelot, J. (1971b). *Bull. Soc. Chim. Fr.* p. 3555.
Rio, G., and Berthelot, J. (1972). *Bull. Soc. Chim. Fr.* p. 822.
Rio, G., and Bricout, D. (1971). *Bull. Soc. Chim. Fr.* p. 3557.
Rio, G., and Charifi, M. (1970). *Bull. Soc. Chim. Fr.* p. 3598.
Rio, G., Bricout, D., and Lacombe, L. (1972). *Tetrahedron Lett.* p. 3583.
Rio, G., Bricout, D., and Lacombe, L. (1973). *Tetrahedron* **29,** 3553.
Saito, J., and Matsuura, T. (1972). *Chem. Lett.* p. 1169.
Schaap, A. P. (1970). Ph.D. Dissertation, Harvard University, Cambridge, Massachusetts.
Schaap, A. P. (1971). *Tetrahedron Lett.* p. 1757.
Schaap, A. P., and Bartlett, P. D. (1970). *J. Am. Chem. Soc.* **92,** 6055.
Schaap, A. P., and Faler, G. R. (1973). *J. Am. Chem. Soc.* **95,** 3381.
Schaap, A. P., and Tontapanish, N. (1971). *Prepr., Div. Pet. Chem., Am. Chem. Soc.* **16,** A78.
Schaap, A. P., and Tontapanish, N. (1972). *J. Chem. Soc., Chem. Commun.* p. 490.
Schaap, A. P., Thayer, A. L., Blossey, E. C., and Neckers, D. C. (1975). *J. Am. Chem. Soc.* **97,** 3741.
Schaap, A. P., Burns, P. A., and Zaklika, K. A. (1977). *J. Am. Chem. Soc.* **99,** 1270.
Schenck, G. O. (1952). *Angew. Chem.* **64,** 12.
Schenck, G. O., and Schulte-Elte, K. H. (1958). *Justus Liebigs Ann. Chem.* **618,** 185.
Schenck, G. O., and Ziegler, K. (1944). *Naturwissenschaften* **32,** 157.
Schoenberg, A., and Ardenne, R. (1968). *Chem. Ber.* **101,** 346.
Schultz, A. G., and Schlessinger, R. H. (1970). *Tetrahedron Lett.* p. 2731.
Schuster, G. B., Turro, N. J., Steinmetzer, H.-C., Schaap, A. P., Faler, G., Adam, W., and Liu, J.-C. (1975). *J. Am. Chem. Soc.* **97,** 7110.
Sharp, D. B. (1960). *Abstr., 138th Natl. Meet., Am. Chem. Soc.* No. 79P.
Shimizu, N., and Bartlett, P. D. (1976). *J. Am. Chem. Soc.* **98,** 4193.
Stephens, H. N. (1928). *J. Am. Chem. Soc.* **50,** 568.
Stevens, B., and Perez, S. R. (1974). *Mol. Photochem.* **6,** 1.
Stevens, B., Perez, S. R., and Ors, J. A. (1974). *J. Am. Chem. Soc.* **96,** 6846.
Sustmann, R., and Binsch, G. (1971). *Mol. Phys.* **20,** 1.
Turner, J. A., and Herz, W. (1977). *J. Org. Chem.* **42,** 1657.
Turro, N. J., Ramamurthy, V., Liu, K.-C., Krebs, A., and Kemper, R. (1976). *J. Am. Chem. Soc.* **98,** 6758.
Turro, N. J., Ito, Y., Chow, M.-F., Adam, W., Rodriguez, O., and Yany, F. (1977). *J. Am. Chem. Soc.* **99,** 5838.
Urry, W. H., and Sheeto, J. (1965). *Photochem. Photobiol.* **4,** 1067.
Wamser, C. C., and Herring, J. W. (1976). *J. Org. Chem.* **41,** 1476.
Wasserman, H. H., and Ives, J. L. (1976). *J. Am. Chem. Soc.* **98,** 7868.

Wasserman, H. H., and Terao, S. (1975). *Tetrahedron Lett.* p. 1735.
Wieringa, J. H., Strating, J., and Wynberg, H. (1970). *Tetrahedron Lett.* p. 4579.
Wieringa, J. H., Strating, J., Wynberg, H., and Adam, W. (1972). *Tetrahedron Lett.* p. 169.
Wilson, T. (1969). *Photochem. Photobiol.* **10,** 441.
Wilson, T., and Schaap, A. P. (1971). *J. Am. Chem. Soc.* **93,** 4126.
Wilson, T., Golan, D. E., Harris, M. S., and Baumstark, A. L. (1976). *J. Am. Chem. Soc.* **98,** 1086.
Windaus, A., and Brunken, J. (1928). *Justus Liebigs Ann. Chem.* **460,** 225.
Woodward, R. B., and Hoffman, R. (1970). "The Conservation of Orbital Symmetry." Verlag Chemie, Weinheim.
Worman, J. J., Shen, M., and Nichols, P. C. (1972). *Can. J. Chem.* **50,** 3923.
Wynberg, H., and Numan, H. (1977). *J. Am. Chem. Soc.* **99,** 603.
Wynberg, H., Numan, H., and Dekkers, H. P. J. M. (1977). *J. Am. Chem. Soc.* **99,** 3870.
Yamaguchi, K., Fueno, T., and Fukutome, H. (1973a). *Chem. Phys. Lett.* **22,** 461.
Yamaguchi, K., Fueno, T., and Fukutome, H. (1973b). *Chem. Phys. Lett.* **22,** 466.
Young, R. H., Martin, R. L., Chinh, N., Mallon, C., and Kayser, R. H. (1972). *Can. J. Chem.* **50,** 932.
Zaklika, K. A., Burns, P. A., and Schaap, A. P. (1978a). *J. Am. Chem. Soc.* **100,** 318.
Zaklika, K. A., Thayer, A. L., and Schaap, A. P. (1978b). *J. Am. Chem. Soc.* **100,** 4916.
Zimmerman, H. E., and Keck, G. E. (1975). *J. Am. Chem. Soc.* **97,** 3527.
Zimmerman, H. E., Keck, G. E., and Pflederer, J. L. (1976). *J. Am. Chem. Soc.* **98,** 5574.
Zwanenburg, B., Wagenaar, A., and Strating, J. (1970). *Tetrahedron Lett.* p. 4683.

Note Added in Proof: The following are papers recently published in this area.

Ando, W., and Kohmoto, S. (1978). *J. Chem. Soc., Chem. Commun.* p. 120.
Bartlett, P. D., and Becherer, J. (1978). *Tetrahedron Lett.* p. 2983.
Friedrich, E., Lutz, W., Eichenauer, H., and Enders, D. (1977). *Synthesis.* p. 893.
Harding, L. B., and Goddard, W. A., III. (1978). *Tetrahedron Lett.* p. 747.
Inagaki, S., Fujimoto, H., and Fukui, K. (1976). *Chem. Lett.* p. 749.
Jefford, C. W., and Rimbault, C. G. (1978). *J. Am. Chem. Soc.* **100,** 6437.
Lerdal, D., and Foote, C. S. (1978). *Tetrahedron Lett.* p. 3227.
Matsumoto, M., Dobashi, S., and Kondo, K. (1977). *Tetrahedron Lett.* p. 2329, 3361.
Mirbach, M. J., Henne, A., and Schaffner, K. (1978). *J. Am. Chem. Soc.* **100,** 7127.
Okada, K. O., and Mukai, T. (1978). *J. Am. Chem. Soc.* **100,** 6509.
Paquette, L. A., Hertel, L. W., Gleiter, R., and Böhm, M. (1978). *J. Am. Chem. Soc.* **100,** 6510.
Rajee, R., and Ramamurthy, V. (1978). *Tetrahedron Lett.* p. 5127.
Rousseau, G., Lechevallier, A., Huet, F., and Conia, J. M. (1978). *Tetrahedron Lett.* p. 3287.
Rousseau, G., LePerchec, P., and Conia, J. M. (1977). *Tetrahedron Lett.* p. 2517.
Saito, I., Matsuura, T., Nakagawa, M., and Hino, T. (1977). *Acc. Chem. Res.* **10,** 346.
Takeshita, H., Hatsui, T., and Shimooda, I. (1978). *Tetrahedron Lett.* p. 2889.
Takeshita, H., and Hatsui, T. (1978). *J. Org. Chem.* **43,** 3080.
Turro, N. J., Chow, M.-F., and Ito, Y. (1978). *J. Am. Chem. Soc.* **100,** 5580.

7

The 1,2-Dioxetanes

PAUL D. BARTLETT and MICHAEL E. LANDIS

I. Physical Properties and Synthetic Methods 244
II. Decomposition of 1,2-Dioxetanes 255
A. Chemiluminescence. 255
B. "Titration" of Singlet and Triplet Products 260
C. TMD-Initiated Reactions 261
D. Other Dioxetanes 262
III. The Mechanism of Dioxetane Decomposition 264
IV. Rearrangements and Alternative Cleavages 266
V. Catalysts for the Decomposition of 1,2-Dioxetanes 269
A. Quantum Chain Processes. 269
B. Amine Catalysis 270
C. Metal Ion Catalysis 270
VI. Nonluminescent Reactions of 1,2-Dioxetanes 274
A. Reactions with Trivalent Phosphorus Compounds 274
B. Reactions with Divalent Sulfur Compounds 278
C. Reactions with Divalent Tin 280
D. Reactions with Boron Trifluoride. 280
References 283

Despite extensive investigation of photooxidations of conjugated and other unsaturated systems, and despite proposals of 1,2-dioxetanes as intermediates, these four-membered ring peroxides remained hypothetical intermediates until 1968. The failure to observe them earlier resulted from several unfavorable factors. (1) Most of the alkenes investigated had some hydrogen vicinal to the double bond, and such alkenes readily undergo the ene reaction. Singlet oxygen thus yields allylic hydroperoxides rather than any cyclic product. (2) Many alkenes not containing allylic hydrogen are too unreactive to compete with the rapid quenching of the excited oxygen. (3) Although it would be conceivable that triplet, ground-state oxygen should enter into cycloaddition with olefins by way of biradicals, the formation of the intermediate species is so endothermic that the process does not in fact occur at an appreciable rate at ordinary temperatures.

Singlet Oxygen

ISBN 0-12-736650-4

It is interesting in this connection that tri-*tert*-butylcyclobutadiene has recently been observed to react with ground state oxygen to yield the doubly tertiary dioxetane

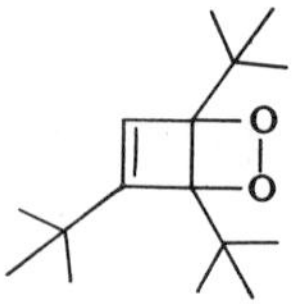

either because the high destabilization inherent in the cyclobutadiene ring makes the initiation much less endothermic, or perhaps because this anti-aromatic, four-electron system exists, at least transiently, in a triplet ground state, and therefore reacts concertedly with triplet oxygen (Maier, 1974).

Although enamines and enol ethers react with singlet oxygen to yield isolable dioxetanes (Section I), the discovery (Kopecky and Mumford, 1968) that 1,2-dioxetanes can be made generally from bromohydroperoxides made these compounds available for study whether they could be prepared from singlet oxygen or not.

The concept of a four-membered ring containing a peroxide link has been often entertained in the past, especially as an intermediate in oxidative cleavage of the carbon–carbon double bond to carbonyl groups. However, some of the best known cleavages of this sort have been shown to proceed by way of a free radical copolymerization chain (Mayo, 1958). There was little surprise at the failure to isolate compounds in which the combination of the weak O–O bond and the strain of the four-membered ring offered the possibility of very exothermic cleavage to two carbonyl compounds. Indeed some of the interest in dioxetanes since their discovery has centered on the question of why such compounds survive as long as they do.

I. PHYSICAL PROPERTIES AND SYNTHETIC METHODS

The first dioxetane, trimethyl-1,2-dioxetane (**1**), was prepared by the bromohydroperoxide method (Kopecky and Mumford, 1968, 1969). Fluorescence from an excited carbonyl compound (acetone or acetaldehyde) accompanied the thermal decomposition of **1**. Since this initial work of

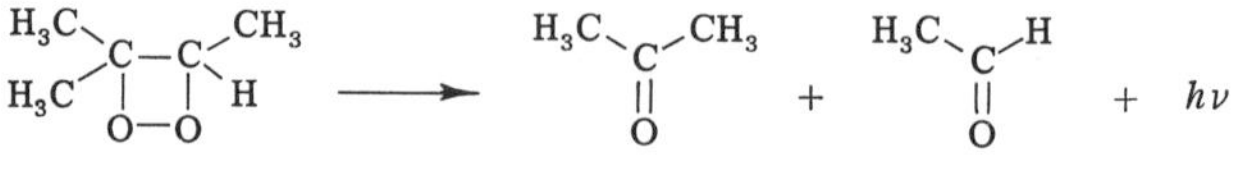

$$(CH_3)_2C{-}CH(CH_3)\ (\text{ring O{-}O}) \longrightarrow H_3C{-}C(=O){-}CH_3 + H_3C{-}C(=O){-}H + h\nu$$

(**1**)

Kopecky and Mumford, many 1,2-dioxetanes have been synthesized and characterized.

Most 1,2-dioxetanes are yellow; for tetramethyl-1,2-dioxetane the color has been attributed to a long wavelength tail of an absorption at 240 nm, $\varepsilon = 24$ (Turro *et al.*, 1973). The color of dioxetanes represents a continuation of the trend noted (Criegee and Paulig, 1955) in comparing the tertiary cyclic peroxide tetramethyl-1,2-dioxacyclopentane with the corresponding dioxacyclohexane. The shift of the UV absorption to longer wavelength, in this case by about 30 nm, and its further extension in the dioxetane, is a normal result of forcing the unfavorable eclipsed conformation upon the nonbonding electrons of the peroxide oxygen atoms. This effect is seen earlier in the spectra of the cyclic disulfides (Bullock *et al.*, 1954); the related conformational strain is responsible for S_8 rather than S_6 being the stablest form of elemental sulfur (Pauling, 1960). It is likely that this effect makes the 1,2-dioxetanes more strained than the oxetanes that are sometimes used as thermochemical models for them (O'Neal and Richardson, 1970).

Infrared spectra of several dioxetanes show absorptions typical of peroxide stretching modes in the region of 870–915 cm^{-1} (Schuster *et al.*, 1975; Wasserman and Terao, 1975). ^{13}C-NMR spectra of tetramethyl-1,2-dioxetane (TMD) and the dioxetane from biadamantylidene have been recorded (Schuster *et al.*, 1975; Wieringa *et al.*, 1972); ring carbons absorb at 89.4δ and 95δ for TMD and biadamantylidene dioxetane, respectively.

There exists quite a range of thermal stabilities within the family of dioxetanes. On the short-lived end of the scale are the α-peroxylactones (Adam and Liu, 1972c; Adam and Steinmetzer, 1972) with half-lives in carbon tetrachloride at room temperature of 5–10 minutes. Most simple di-, tri-, and tetra-substituted dioxetanes must be heated to effect decomposition; for example, the half-life of tetramethyl-1,2-dioxetane at 70°C in benzene is approximately 30 min. On the long-lived side of the family are the dioxetanes from biadamantylidene (Wieringa *et al.*, 1972) and binorbornylidene (Bartlett and Ho, 1974), with half-lives at 160°C of several hours.

Simple alkyl-substituted dioxetanes have been synthesized from β-halohydroperoxides, products of the reaction of alkenes with an electrophilic halogen donor (e.g., 1,3-dibromo-5,5-dimethylhydantoin) in the presence of concentrated hydrogen peroxide (Kopecky and Mumford, 1968, 1969). Ring formation is effected by treatment of the halohydroperoxide either with a silver salt (for tertiary halides) or with base (for primary or secondary halides).

Br / OOH $\xrightarrow{Ag^+}$ O—O (Kopecky *et al.*, 1971)

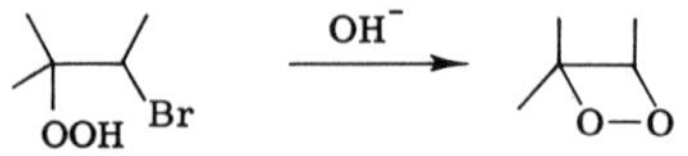

(Kopecky and Mamford, 1968, 1969)

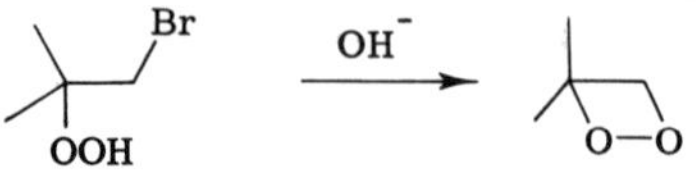

(Richardson and Hodge, 1971)

The addition of singlet oxygen to enol ethers, enamines (Foote and Liu, 1968; Foote *et al.*, 1975; Huber, 1968), pyrroles (Lightner *et al.*, 1976), and alkenes with unreactive allylic hydrogen generates the corresponding dioxetanes.

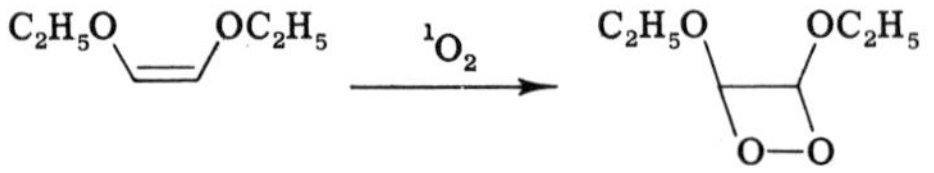

(Bartlett and Schaap, 1970)

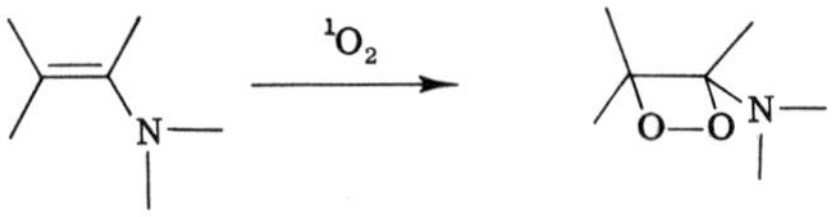

(Foote *et al.*, 1975)

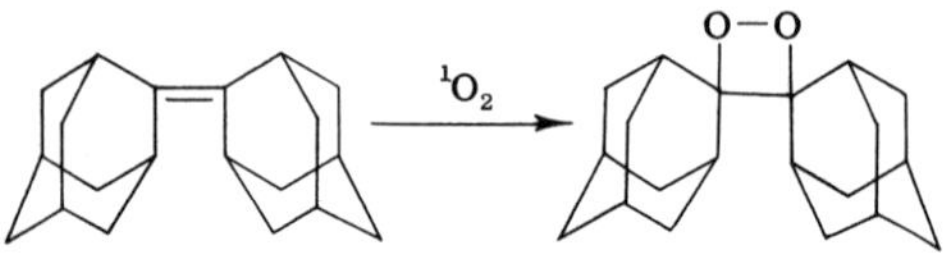

(Wieringa *et al.*, 1972)

Several mechanistic alternatives have been advanced to account for this reaction. A concerted [2s + 2a] (antarafacial on oxygen) cycloaddition (Woodward and Hoffman, 1969) is consistent with the observed retention of configuration in the reaction of singlet oxygen with *cis*- and *trans*-diethoxyethylene (Bartlett and Schaap, 1970).

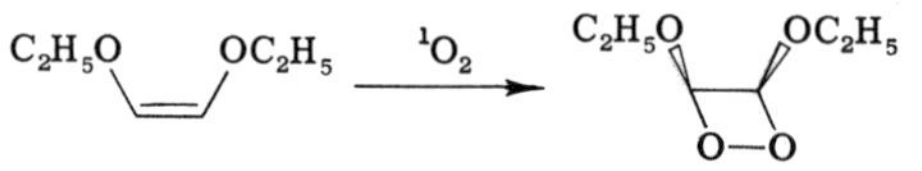

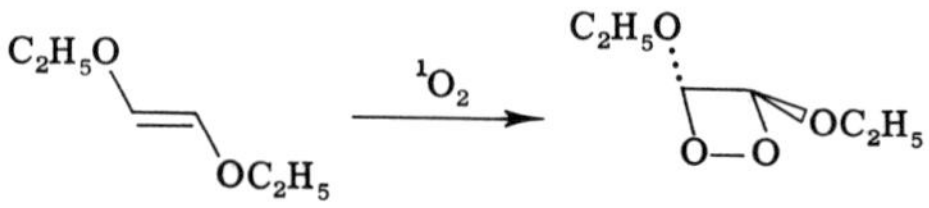

The closely related reaction with enamines has been rationalized as proceeding via initial electron transfer, generating a pair of radical ions that collapse to product (Foote *et al.*, 1975).

In the cases of biadamantylidene (P. D. Bartlett and M. J. Shapiro, unpublished results, 1975), 7,7-binorbornylidene (Bartlett and Ho, 1974), and norbornene (Jefford and Boschung, 1974, 1977), dioxetane formation by singlet oxygen in every solvent used is accompanied by epoxidation, and there is no evidence that any solvent becomes oxidized in the process. The first example of this to be observed was with biadamantylidene in pinacolone (Schaap and Faler, 1973), but this solvent has proved to have no unique properties with respect to singlet oxygen reactions.

Recent MINDO/3 calculations (Dewar and Thiel, 1975) show 1,4-dipolar ions as favored intermediates in the reaction of enamines and enol ethers with singlet oxygen; reactions of alkenes without reactive allylic hydrogens are expected to proceed via peroxiranes. Deoxygenation of peroxiranes (to epoxides) by ethylene (yielding ethylene oxide) or by singlet oxygen (yielding ozone) are also suggested (Dewar *et al.*, 1975; cf. Ho, 1974; Sharp, 1960).

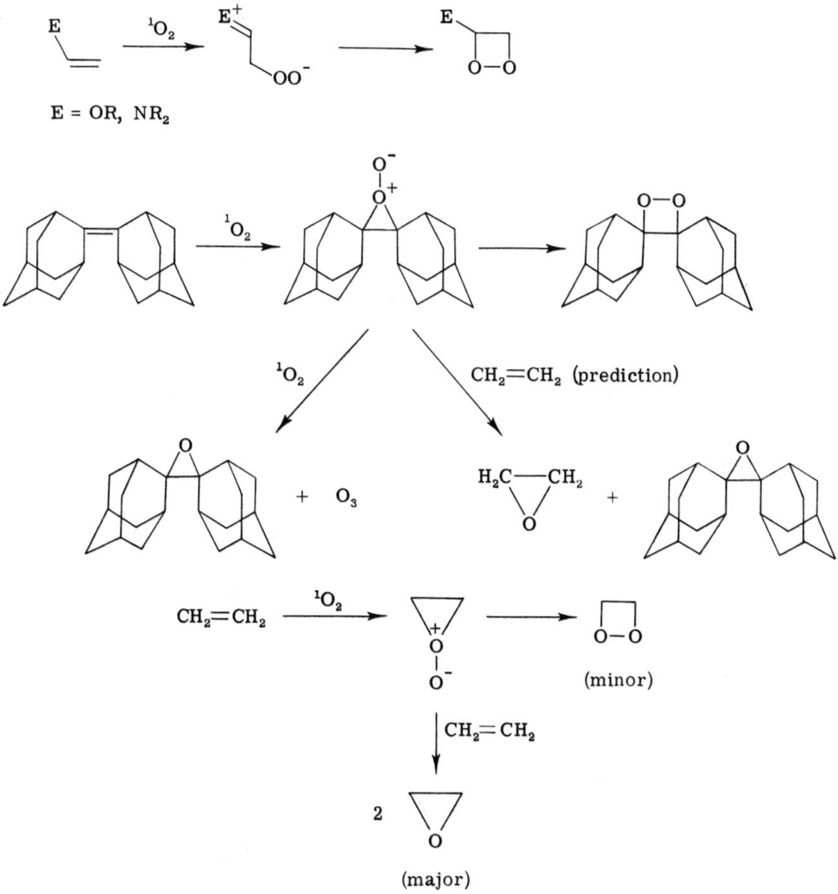
E
$^1O_2$
E^+
OO^-
E
O—O
E = OR, NR_2
O^-
O^+
$^1O_2$
O—O
$^1O_2$
CH_2=CH_2 (prediction)
O
+ O_3
H_2C—CH_2
O
+
O
CH_2=CH_2
$^1O_2$
+
O
O^-
O—O
(minor)
CH_2=CH_2
2
O
(major)

Table I Characterized 1,2-Dioxetanes

Dioxetane	Thermal stability	Synthetic method[a]	Ref.
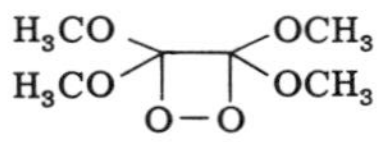 H_3CO, H_3CO, OCH_3, OCH_3, O–O	E_a = 28.6 (benzene or xylene)	B	Mazur and Foote, 1970 Wilson *et al.*, 1976
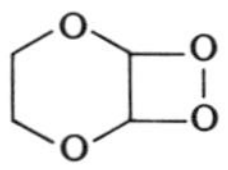	E_a = 24.5 (benzene)	B	Schaap, 1971 Wilson *et al.*, 1976
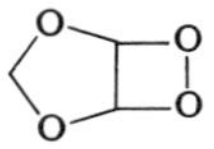	Complete decomposition at 60°C in 30 min	B	Schaap, 1971
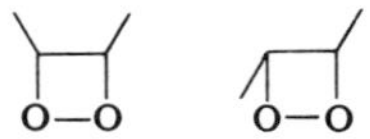	—	C (base)	White *et al.*, 1973
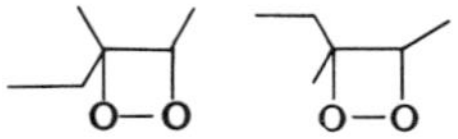			
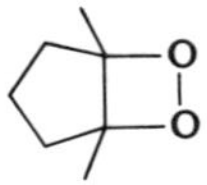	—	C (Ag^+)	Turro *et al.*, 1974a
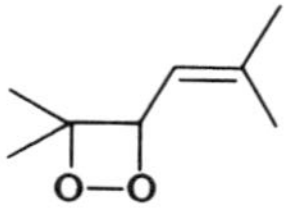	$\tau_{1/2}^{44°}$ = 37 min (CCl_4)	B (CH_3OH)	Hasty and Kearns, 1973

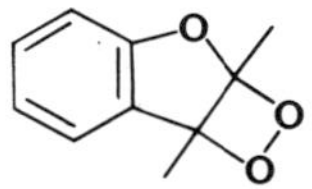	—	B	Basselier *et al.*, 1971
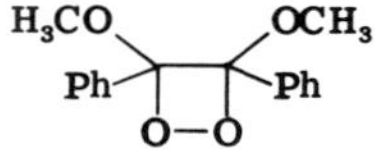 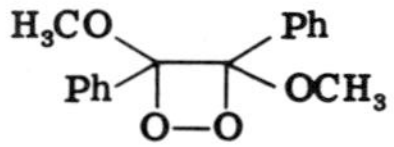	Complete decomposition in ether, methanol, and ethanol in 2 hr at 67°–68°C; dichloromethane or chloroform, 5 min	B	Rio and Berthelot, 1971
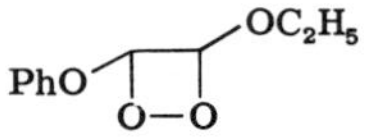 	—	B	Schaap and Tontapanish, 1972
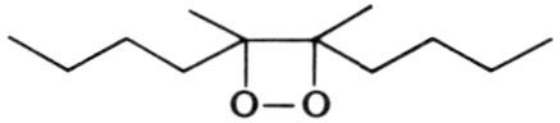	$E_a = 25.4$ (decalin)	C (Ag^+)	Darling and Foote, 1974
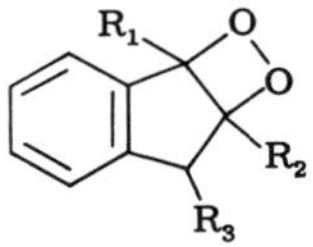 $R_1 = R_2 = \phi$, $R_3 = H$	$\tau_{1/2}^{35°} = 6$ hr ($CHCl_3$)	B (CH_3OH)	Burns and Foote, 1974
$R_1 = CH_3$, $R_2 = \phi$, $R_3 = H$	m.p. 69°–70°C (dec.)		
$R_1 =$ *tert*-butyl, $R_2 = R_3 = H$	m.p. 49.5°–51°C (dec.)		
$R_1 = CH_3$, $R_2 = H$, $R_3 =$ *tert*-butyl	m.p. 56°–58° C (dec.)		
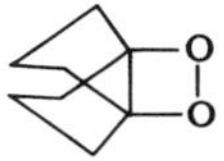	—	C (Ag^+)	Ding and Li, 1972

(*Continued*)

Table I *(Continued)*

Dioxetane	Thermal stability	Synthetic method[a]	Ref.
H_3C, R, O—O, Ph, Ph; R = Ph—			
R = (3-methoxyphenyl, OCH_3)	$\tau_{1/2}^{71.4} = 25$ min (CCl_4)	B (CH_3OH)	Zimmerman and Keck, 1975
	—		
R = (2-naphthyl)	m.p. 89°–94°C (dec.)		
O, N, O—O, $(CH_2)_n$; *n* = 1	Violent decomposition at room temperature	B	Wasserman and Terao, 1975
n = 2	—		
n = 3	—		
O—O, N	—	B	Foote *et al.*, 1975
Ph, Ph, O—O, N, O	Solid sample exploded at −6°C		

	$\tau_{1/2}^{33^\circ} = 17$ min (CCl_4)	A	Adam and Steinmetzer, 1972; Adam and Liu, 1972 ; Adam *et al.*, 1974
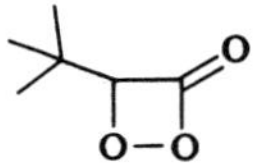	$E_a = 19.4$ (CCl_4)		
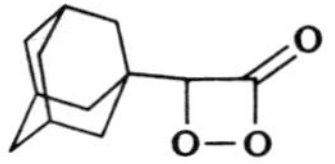	$E_a = 14.3$ (CCl_4)		
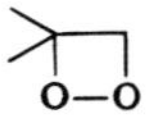	$E_a = 23.0$ (CCl_4)	C (base)	Richardson *et al.*, 1974b
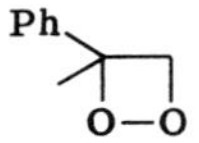 Ph	$E_a = 22.9$ (CCl_4)		
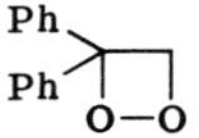 Ph Ph	$E_a = 22.7$ (benzene)		
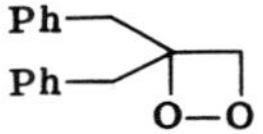 Ph Ph	$E_a = 24.3$ (benzene)		
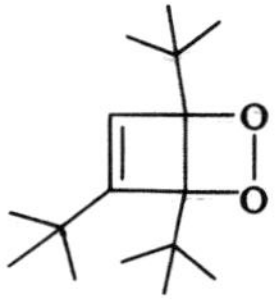	—	O_2 + 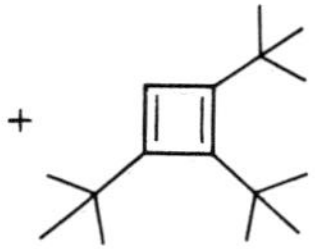	Maier, 1974

(*Continued*)

Table I (*Continued*)

Dioxetane	Thermal stability	Synthetic method[a]	Ref.
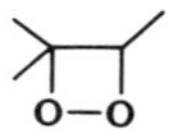	E_a = 23.5 (CCl_4)	C (base)	Kopecky and Mumford, 1969; Kopecky *et al.*, 1971, 1973, 1975
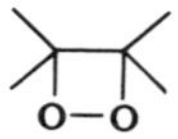	E_a = 25.8 (CCl_4)	C (Ag^+)	
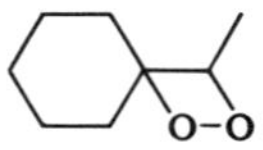	$\tau_{1/2}^{50°}$ = 15 min (benzene)	C (base)	
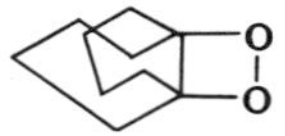	E_a = 22.7 (benzene)	C (Ag^+)	
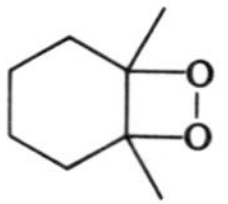	E_a = 25.7 (benzene)	C (Ag^+)	
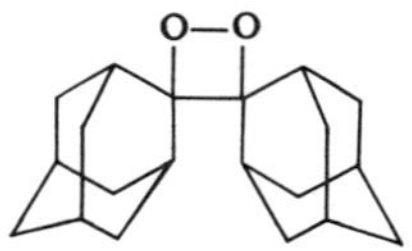	E_a = 34.6 (*o*-xylene) m.p. = 164°–165°C	B	Wieringa *et al.*, 1972; Schuster *et al.*, 1975
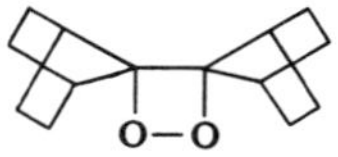	m.p. 129°–130°C, 200°C (dec.)	B	Bartlett and Ho, 1974

Compound	Data	Method[a]	Reference
 H_5C_2O, OC_2H_5 (O–O)	E_a = 23.6 (benzene)	B	Bartlett and Schaap, 1970; Schaap and Bartlett, 1970; Wilson and Schaap, 1971
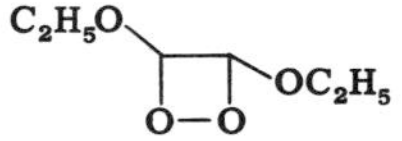 C_2H_5O, OC_2H_5 (O–O)	$[k_{trans}/k_{cis}]^{50°} = 1.4$		
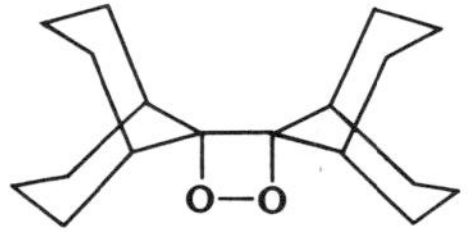 O–O	E_a = 37 (*o*-xylene) m.p. = 139°–140°C	B	P. D. Bartlett, A. L. Baumstark, and M. J. Shapiro, unpublished results, 1975
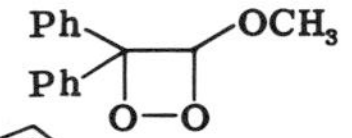 Ph, Ph, OCH_3 (O–O)	E_a = 25.5 (benzene)	C (Ag^+)	Baumstark *et al.*, 1976
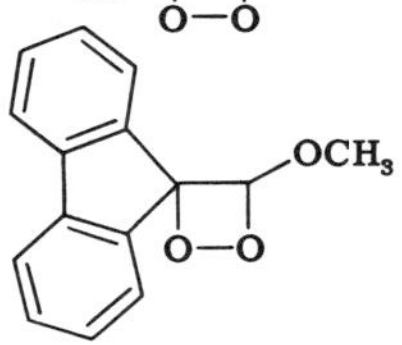 OCH_3 (O–O)	E_a = 21.0 (benzene)		Bartlett and Landis, 1977

[a] A, Dehydration with *N*,*N*-dicyclohexylcarbodiimide; B, singlet oxygen with alkene; and C, from halohydroperoxide with method for ring closure (base or Ag^+).

The hypothetical process represented by the bottom equation applies to the most unhindered olefin (ethylene), which is also unactivated and has not been observed to react in solution with singlet oxygen in competition with quenching. It is an interesting question whether this can be the mechanism of the slow thermal formation of ethylene oxide from ethylene and oxygen, representing leakage through the spin-inversion barrier (Meadus and Ingold, 1964; cf. Barton *et al.*, 1975).

Photooxidation (in aprotic solvents) of certain alkenes with aromatic substituents generates [2 + 4] cycloaddition products involving a double bond from the aromatic ring (Foote *et al.*, 1973); the reaction in methanol affords 1,2-dioxetanes (Burns and Foote, 1974).

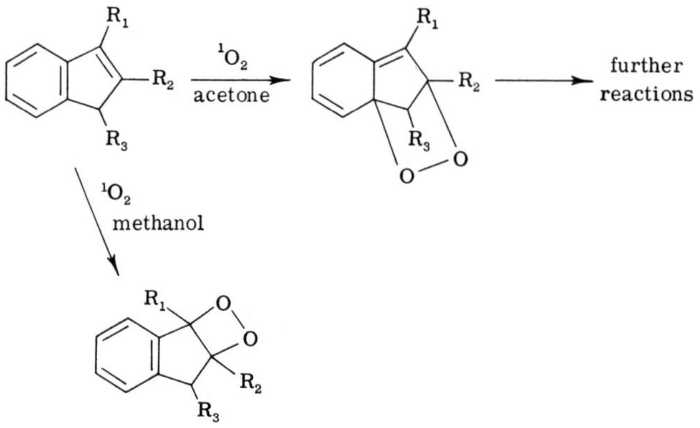

R_1 R_2 R_3 $^1O_2$ acetone further reactions $^1O_2$ methanol

The nature of this unusual solvent effect has not yet been resolved, but it is consistent with the view that dioxetane formation is a more polar reaction than either 1,4-cycloaddition or ene reaction (see Frimer *et al.*, 1977; McCapra and Beheshti, 1977).

Reports of the preparation of dioxetanes by ozonolysis (Story *et al.*, 1972; Yang and Carr, 1973) were later shown to be incorrect (Bailey *et al.*, 1973; Kopecky *et al.*, 1973) or were retracted (Yang and Libman, 1974).

α-Peroxylactones (dioxetanones) have been synthesized from α-hydroperoxycarboxylic acids by dehydration with *N*,*N*-dicyclohexylcarbodiimide (Adam and Liu, 1972c; Adam and Steinmetzer, 1972).

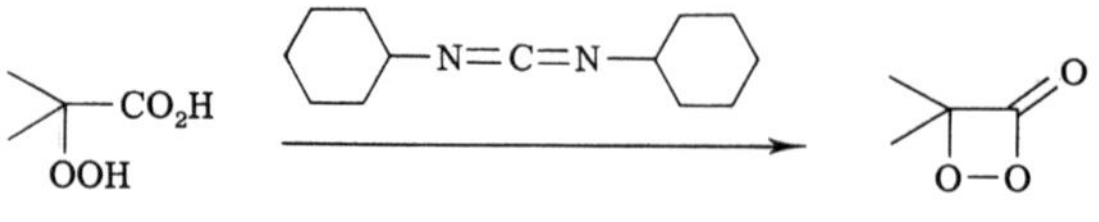

CO_2H OOH —N=C=N— O O—O

All known isolated and characterized 1,2-dioxetanes and α-peroxylactones with method of synthesis are listed in Table I.

II. DECOMPOSITION OF 1,2-DIOXETANES

All of the 1,2-dioxetanes listed in Table I luminesce upon thermal decomposition. A good deal of research has been devoted to the determination of the nature of the electronically excited products derived from dioxetanes as well as to the mechanism by which these products are generated.

The efficiency of production of excited states generated during the thermolysis of 1,2-dioxetanes has been determined by the following general methods: (1) measurement of the yield of light emitted as fluorescence from an excited carbonyl product; (2) measurement of the yield of light emitted when an external fluorescer is added to accept the excitation energy of a product; and (3) measurement of the extent of some chemical reaction induced by an excited product.

A. Chemiluminescence

Wilson and Schaap (1971) found that *cis*-diethoxy-1,2-dioxetane decomposed thermally with nearly unit efficiency to the triplet excited state of ethyl formate. The analytical method involved the investigation of the chemiluminescence generated when the dioxetane was thermolyzed in the presence of added fluorescers, 9,10-dibromoanthracene (DBA) and 9,10-diphenylanthracene (DPA). The intensity of chemiluminescence was greater in the presence of DBA (than in the presence of an equivalent amount of DPA), although the *fluorescence efficiency* of DPA on direct excitation was much greater than that of DBA. Since it had been demonstrated that DBA was a more efficient acceptor of triplet energy than was DPA (Belyakov and Vassil'ev, 1970), it was concluded that the predominant excited species generated from the dioxetane was triplet rather than singlet ethyl formate. The yield of triplet formate (ϕ^{*3}) was determined by evaluating the yield of chemiluminescence (ϕ_{chem}) at infinite concentration of DBA.

$$\text{cis-}(H_5C_2O)_2\text{-dioxetane} \xrightarrow[\phi^{*3}]{\Delta} [HC(=O)OC_2H_5]^{*3}\ (P^{*3}) + HC(=O)OC_2H_5\ (P)$$

$$P^{*3} \xrightarrow[\phi_{e.t.}]{DBA} DBA^{*1} + P$$

$$DBA^{*1} \xrightarrow{\phi_{Fl}^{DBA}} DBA + h\nu$$

$$\phi_{chem} = \phi^{*3} \times \phi_{e.t.} \times \phi_{Fl}^{DBA}$$

With Belyakov and Vasil'ev's value (1970) for the efficiency of the spin-forbidden transfer of energy from triplet formate to DBA ($\phi_{e.t.}$) and with the fluorescence efficiency (ϕ_{Fl}^{DBA}) of DBA, a value of 1.0 was determined for ϕ^{*3}.

White *et al.* (1969) demonstrated for the first time that a dioxetane could be used instead of light to generate an excited ketone. The thermolysis of trimethyl-1,2-dioxetane (**1**) in the presence of acenaphthylene (**2**) resulted in the formation of both the cis and the trans photodimers of **2** in approximately the same cis to trans ratio as that of the acetone-photosensitized dimerization. Thermolysis of **1** in the presence of *trans*-stilbene (**3**) (White *et al.*, 1970) afforded the stationary state of 41% *cis*- and 59% *trans*-stilbene also observed in acetone photosensitization. When the thermal decomposition of **1** was carried out in the presence of 4,4-diphenyl-2,5-cyclohexadienone (**4**) (White *et al.*, 1969) or santonin (**5**) (White *et al.*, 1973), photorearrangement of reactants was observed.

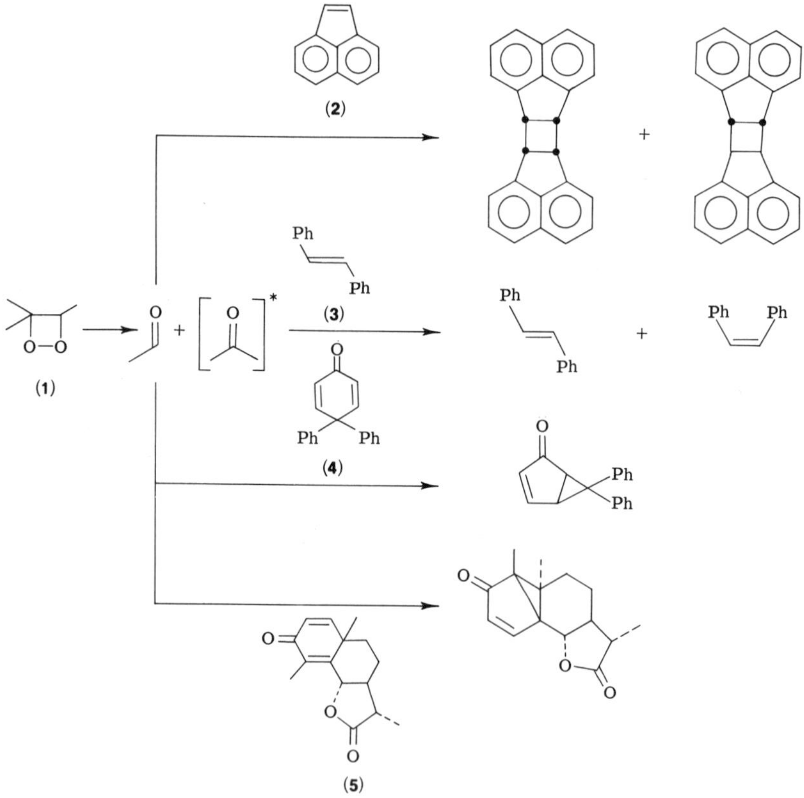

The yields of photoproducts based on dioxetane were 0.4% (for **4**), 1.5% (for **2**), 2.9% (for **5**), and 4% (for **3**). A larger yield (14%) of excited product from **1** was determined by evaluating the yield of chemiluminescence generated during the thermolysis of **1** in the presence of added fluorescer, europium tris(thenoyltrifluoroacetonate)-1,10-phenanthroline (**6**) (Wildes and White, 1971; White *et al.*, 1973).

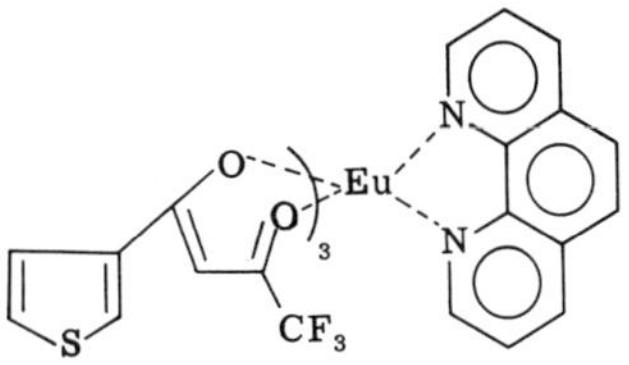

(**6**)

Norrish type I cleavage products from excited dibenzylketone were generated during the thermal decomposition of 3,3-dibenzyl-1,2-dioxetane (Richardson *et al.*, 1972b).

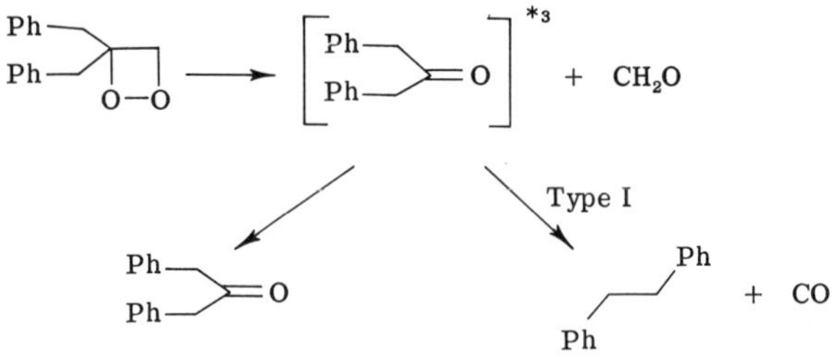

Type I

+ CH_2O

+ CO

With the yield of cleavage products (~2%) and with the quantum efficiency (0.7) for bibenzyl formation from excited dibenzyl ketone (Engel, 1970; Robbins and Eastman, 1970), a value of 3% was determined for the yield of triplet dibenzyl ketone from dioxetane.

The yield of excited 2-hexanone generated during the decomposition of 3,4-dimethyl-3,4-di-*n*-butyl-1,2-dioxetane has been determined by evaluating the yields of Norrish type II photoproducts, acetone and 1,2-dimethylcyclobutanol (Darling and Foote, 1974). Irradiation of 2-hexanone in the presence of high concentrations of one isomer of pentadiene (a triplet quencher) allows the determination of products derived from singlet ketone (ϕ_a^1 and ϕ_{cb}^1) as well as the efficiency of intersystem crossing ϕ_{isc}, based on the extent of isomerization of the diene.

$\Phi_a^1 = 0.1$ + $H_2C{=}CHCH_3$

α_1 *1

Δ $\Phi_{cb}^1 = 0.008$ OH

$\Phi_{isc} = 0.2$

α_3 *3

$\Phi_a^3 = 0.57$ + $H_2C{=}CHCH_3$

O—O

$\Phi_{cb} = 0.25$ OH

Here α_1 = yield of singlet excited 2-hexanone from dioxetane; α_3 = yield of triplet excited 2-hexanone from dioxetane; ϕ_{isc} = quantum yield for intersystem crossing from excited singlet to excited triplet 2-hexanone; ϕ_a^1, quantum yield of acetone formation from singlet excited 2-hexanone; ϕ_a^3 = quantum yield of acetone formation from triplet excited 2-hexanone; ϕ_{cb}^1 = quantum yield of 1,2-dimethylcyclobutanol formation from singlet excited 2-hexanone; and ϕ_{cb}^3 = quantum yield of 1,2-dimethylcyclobutanol formation from triplet excited 2-hexanone. Irradiations of 2-hexanone in the absence of triplet quencher gives quantum efficiencies for product formation from triplet ketone (ϕ_a^3 and ϕ_{cb}^3) by subtracting the quantum yields for product formation from singlet ketone from the total observed yields and dividing that number by the intersystem crossing yield. The yields of excited products (α_1 and α_3) generated during the thermolysis of dioxetane are then determined by the observed yields of photoproducts (ϕ_{obs}^{cb}) and the known quantum efficiencies via Eqs. (1) and (2). In this manner, α_1 and α_3 were determined as 5 and 3.5%, respectively.

$$\phi_{obs}^{a} = \alpha_1(\phi_a^1 + \phi_{isc}\phi_a^3) + \alpha_3\phi_a^3 \tag{1}$$

$$\phi_{obs}^{cb} = \alpha_1(\phi_{cb}^1 + \phi_{isc}\phi_{cb}^3) + \alpha_3\phi_{cb}^3 \tag{2}$$

The yield of excited benzophenone generated during the thermolysis of 3,3-diphenyl-1,2-dioxetane has recently been determined (Richardson *et al.*, 1974a). Thermal decomposition of the dioxetane in the presence of *trans*-stilbene affords a low yield of *cis*-stilbene (1.9%); thermolysis in the presence of 2-methyl-2-butene generates the diphenyloxetane in 0.5% yield.

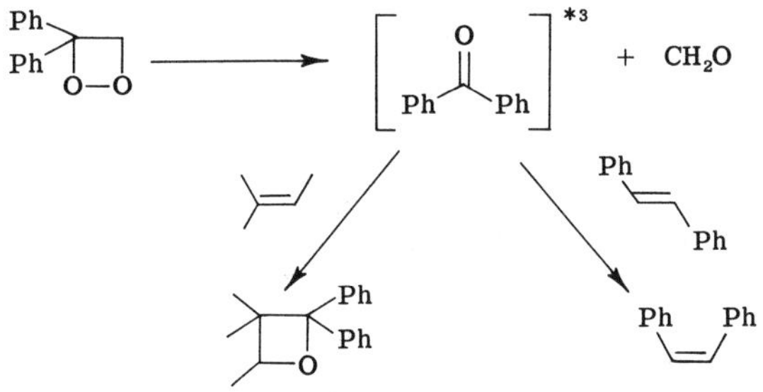

$+ \ CH_2O$

With the quantum yield of the benzophenone-sensitized isomerization of *trans*-stilbene (0.55, Valentine and Hammond, 1972) and with the quantum efficiency of oxetane formation (0.22), a value in the range of 2–3.5% was determined for the yield of triplet benzophenone from this dioxetane.

Thermal decomposition of dioxetanes **7a–7c** (Zimmerman and Keck, 1975) generated 4,4-diphenyl-2,5-cyclohexadienone (**4**), and **8**, the product of the di-π-methane rearrangement of triplet (or singlet) **4**. The yield of photoproduct **8** was rather insensitive to the nature of dioxetane substituent R. Thus, the yield of triplet **4** (E_T, 68 kcal/mole) was not substantially reduced when the second ketonic product of the thermolysis was β-acetonaphthone (E_T, 59 kcal/mole).

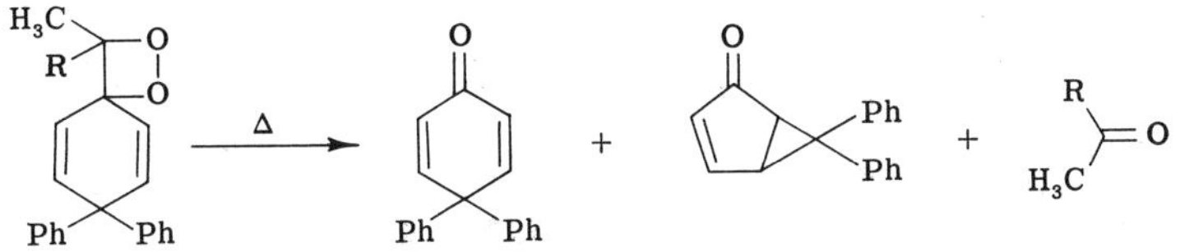

(7)	(4)	(8)
(7a) R = C_6H_5	82.9%	17.1%
(7b) R = m-$CH_3OC_6H_4$	86.0%	14.0%
(7c) R = β-naphthyl	88.0%	12.0%

Interestingly, the thermal decomposition of **7c** preferentially generated the excited $n \rightarrow \pi^*$ carbonyl fragment even when a $\pi \rightarrow \pi^*$ fragment (β-acetonaphthone) was energetically more accessible. Little or no "inter- or intramolecular drainage" from $n \rightarrow \pi^*$ to $\pi \rightarrow \pi^*$ ketonic products was observed. With the quantum yield for rearrangement of triplet **4** (0.68, Zimmerman and Swenton, 1964), the yield of excited **4** from dioxetane was determined as 20% (for **7a**).

The yields of excited products generated during the thermolysis of dioxetanes **9** and **10** from p-dioxene and tetramethoxyethylene have been determined by evaluating chemiluminescent yields in the presence of added fluorescers, 9,10-dibromoanthracene and 9,10-diphenylanthracene (Wilson *et al.*, 1976).

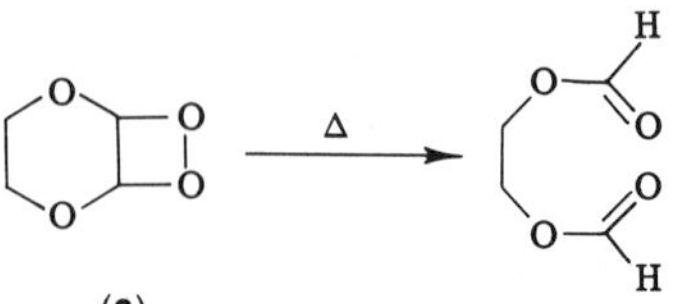

(9)

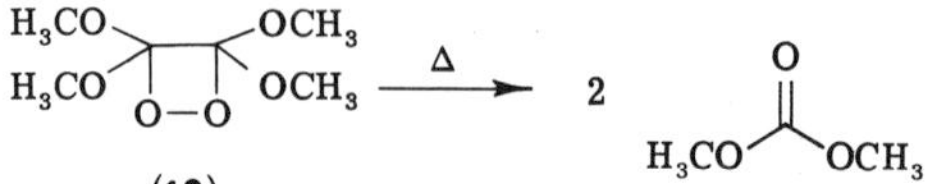

(10)

The yields of singlet (ϕ^{*1}) and triplet (ϕ^{*3}) diformate from **9** were 0.01 and 30%, respectively; the yield of singlet and triplet dimethyl carbonate from **10** were 1% and 10%, respectively.

B. "Titration" of Singlet and Triplet Products

The most extensively investigated of the dioxetanes is tetramethyl-1,2-dioxetane (TMD). A useful tool in determining the nature of the excited acetone from TMD decomposition has been its interception with *trans*-dicyanoethylene. In calibrating experiments it is observed that irradiation of acetone in the presence of *trans*-dicyanoethylene (*trans*-DCE) affords an oxetane (**11**) and isomerization to *cis*-DCE (Dalton *et al.*, 1970).

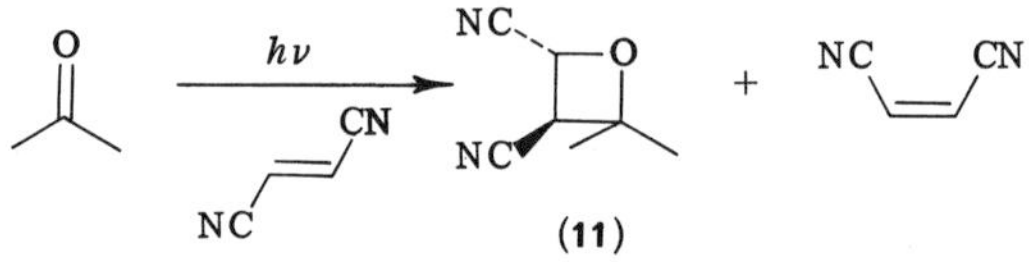

(11)

A triplet quencher, 1,3-pentadiene, added to the solution to be irradiated inhibits the cis, trans isomerization without appreciably affecting the oxetane formation; this serves as evidence that these two processes arise from triplet and singlet acetone, respectively.

Because of competitive intersystem crossing by singlet acetone, the fraction of DCE isomerization passes through a maximum with increasing concentration of DCE. The Stern–Volmer plot of $(\text{oxetane})^{-1}$ versus $(\text{DCE})_0^{-1}$ is linear throughout and enables extrapolation of oxetane yield to infinite DCE concentration. It is then possible to use *trans*-DCE to distinguish quantitatively between triplet acetone resulting from intersystem crossing of singlet and that formed directly in thermal dioxetane decomposition. Unlike the case in the irradiation experiments, the yield of *cis*-dicyanoethylene was not depressed at high initial concentrations of *trans*-DCE when TMD was used as the source of excited acetone, and the yield of oxetane was unexpectedly low (Turro and Lechtken, 1972). These results indicate that TMD generates predominantly triplet rather than singlet acetone. The observed yields of oxetane (0.022%) and *cis*-alkene (4.8%) from TMD initiation and the quantum yields of oxetane formation (0.08) and isomerization (0.1) allow the calculation of the yield of singlet and triplet acetone from TMD as 0.28 and 48%, respectively. The result of this "chemical titration" with *trans*-dicyanoethylene was confirmed by Wilson *et al.* (1976), by evaluating chemiluminescent yields from TMD in the presence of added fluorescers, DPA and DBA (singlet, 0.15%, triplet, 30%).

C. TMD-Initiated Reactions

Thermolysis of TMD in the presence of acenaphthylene or 1-(*o*-tolyl)-1,2-propanedione produces corresponding photoproducts in low yields (Filby, 1973). The biacetyl-sensitized decomposition of TMD

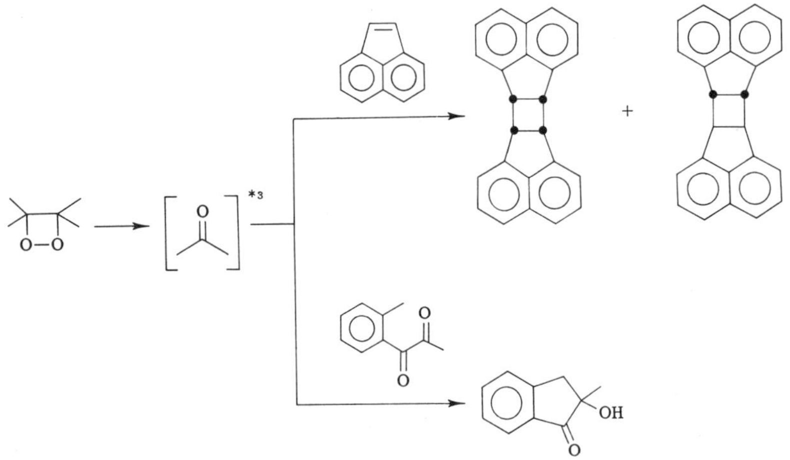

generates excited acetone that can sensitize Norrish type II reactions of butyrophenone (Turro and Lechtken, 1973c).

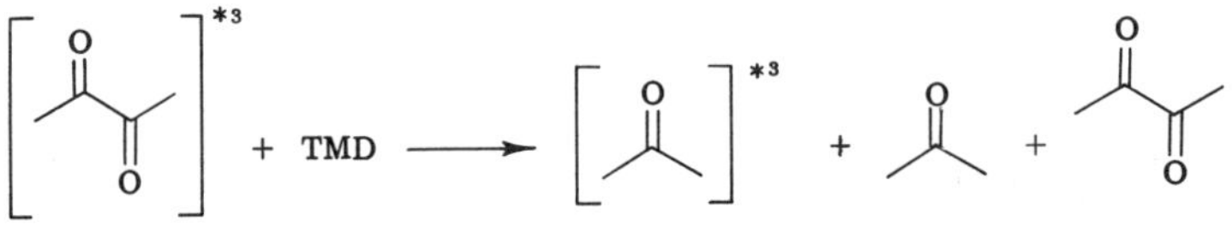

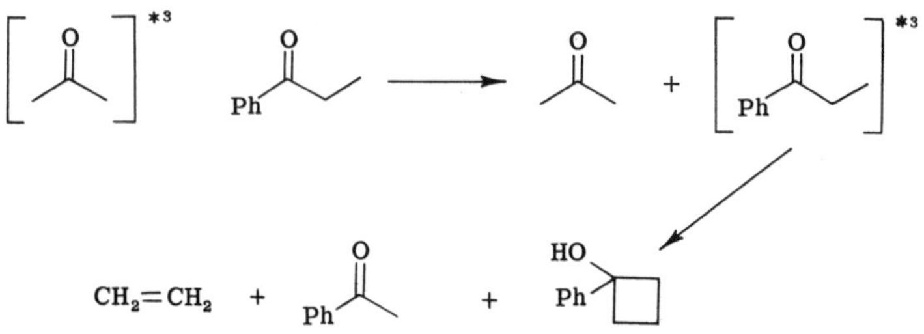

$CH_2{=}CH_2$ + Ph + HO Ph

The excitation energy of biacetyl is insufficient to sensitize butyrophenone photoreactions without TMD; thus, the biacetyl–TMD initiated system has been called an example of "anti-Stokes sensitization." Chemically induced dynamic nuclear polarization (CIDNP) has been observed in the TMD initiated decomposition of benzoyl peroxide (Bartlett and Shimizu, 1975), and polarized spectra of the products indicate triplet acetone as the sensitizer for the decomposition.

Thermolysis of TMD in 1,4-cyclohexadiene affords products attributable to hydrogen abstraction from solvent by excited acetone (Wilson *et al.*, 1973; Landis, 1974). Products of this reaction include benzene, isopropyl alcohol, 1,4-dihydrocumyl alcohol, and bicyclohexadienyl. The yield of acetone-derived products is 10%,

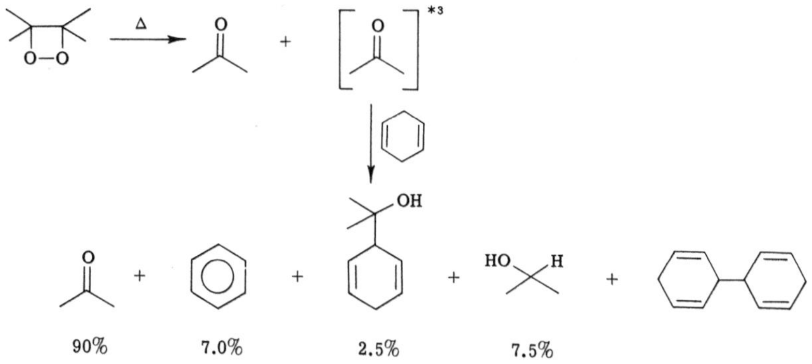

90% 7.0% 2.5% 7.5%

giving a minimum yield of excited acetone from TMD of 20%. These products are all normal results of the ability of triplet excited acetone to abstract hydrogen from a doubly allylic site. It is known that singlet acetone is only ~1/1000 as reactive toward hydrogen abstraction as the triplet state (Wagner, 1967).

D. Other Dioxetanes

As with acetone (see Section II,B) adamantanone singlets form an oxetane with *trans*-dicyanoethylene whereas the triplet isomerizes the nitrile.

The yields of excited singlet and triplet adamantanone generated from the dioxetane (**12**) of biadamantylidene have been reported as 2% and 15%, respectively (Schuster *et al.*, 1975).

(**12**)

From the yields of oxetane and *cis*-dicyanoethylene (values extrapolated to infinite initial concentration of *trans*-dicyanoethylene) and with the quantum efficiencies for oxetane formation and isomerization, the yields of excited singlet and triplet adamantanone were evaluated. The ratio of triplet to singlet products (7.5) as determined by this "chemical titration" was comparable to that (10.0) determined by comparisons of intensities of chemiluminescence generated from **12** in the presence of added fluorescers, 9,10-diphenylanthracene (DPA, singlet acceptor) and 9,10-dibromoanthracene (DBA, triplet acceptor).

The yields of excited singlet (ϕ_S) and triplet (ϕ_T) carbonyl fragments from several dioxetanes (TMD, **1**, **13**, **14**, **15**) have been determined (Turro *et al.*, 1974a) by evaluating the yields of chemiluminescence generated during the thermolysis of the dioxetanes in the presence of added fluorescers, DBA and DPA.

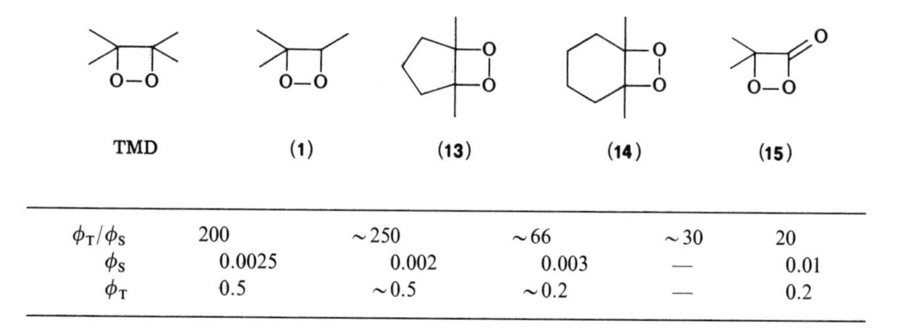

	TMD	(**1**)	(**13**)	(**14**)	(**15**)
ϕ_T/ϕ_S	200	~250	~66	~30	20
ϕ_S	0.0025	0.002	0.003	—	0.01
ϕ_T	0.5	~0.5	~0.2	—	0.2

The yields of excited singlet and triplet acetone from TMD were determined independently by "chemical titration" (see Section II,B) with *trans*-dicyanoethylene (Turro and Lechtken, 1972). The yields of excited products from TMD as determined by luminescence intensity measurements were set equal to those determined by "chemical titration"; the singlet and triplet yields from **1**, **13**, **14**, and **15** were then determined relative to those for TMD.

III. THE MECHANISM OF DIOXETANE DECOMPOSITION

The simplest view of dioxetane decomposition as a suprafacial [2 + 2] cycloreversion yielding one product in its ground state and the other in an excited singlet state (McCapra, 1968; Kopecky and Mumford, 1969) has had to be modified in view of the demonstration (see above, Section II) that the excited product primarily produced is not in the singlet but in the triplet state. Turro and Lechtken (1973a) cut this knot by proposing that the normally forbidden spin inversion from singlet to triplet becomes an integral part of a concerted cleavage mechanism. This comes about through a torque imposed on the electron spin by the twisting motion of the ring components and their orbitals. (The triplet in any case is of lower energy than the singlet and, except for the spin barrier, naturally a more favored product of the thermal dioxetane decomposition.) Recent MINDO/3 calculations (Dewar and Kirschner, 1974) support the possibility of such a concerted triplet mechanism for the decomposition of the model substance 1,2-dioxetane. Activation parameters for its decomposition to two molecules of formaldehyde were calculated based on: (a) a $[2\pi + 2\pi]$ cycloreversion; (b) a two-step process involving an intermediate biradicaloid; and (c) a concerted ring fragmentation generating an excited triplet product. The activation energies for processes (a) and (b) were calculated as 65 and 45 kcal/mole, respectively; whereas the intersection of the triplet and singlet dioxetane surfaces [process (c)] was calculated to occur at 38.3 kcal/mole above the ground state of dioxetane. Thus, a concerted decomposition generating a triplet product required the least activation and was permitted as the preferred mode of ring fragmentation.

A two-step ring cleavage involving the intermediacy of a 1,4-biradical has also been proposed to account for the luminescent decomposition of 1,2-dioxetanes (Richardson *et al.*, 1974b). Initial rupture of the O–O bond will generate a singlet biradical that either closes (to regenerate dioxetane), cleaves (to singlet carbonyl products), or undergoes a spin inversion (generating a triplet biradical). The triplet biradical either regenerates the singlet

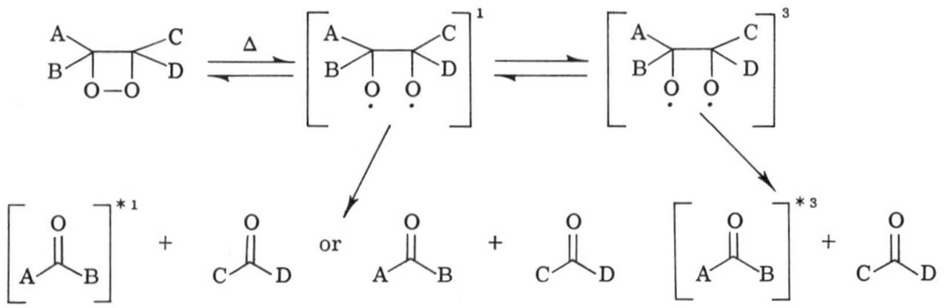

Table II Experimental and Calculated[a] Activation Parameters for 1,2-Dioxetanes

Dioxetane	Experimental E_a	Experimental log A	Calculated E_a	Calculated log A	Ref.
	—	—	16.7	12.6	Richardson and O'Neal, 1972
	—	—	20.9	12.8	Richardson and O'Neal, 1972
	19.4 (CCl_4)	—	22.0	12.6	Adam and Steinmetzer, 1972; Richardson and O'Neal, 1972
	23.0 (CCl_4)	12.2	22.9	12.9	Richardson *et al.*, 1972a
Ph	22.9 (CCl_4)	12.1	22.9	12.65	Richardson *et al.*, 1972a
Ph, Ph	22.7 (CCl_4)	12.36	22.6	12.71	Richardson *et al.*, 1974b
Ph, Ph	24.3 (benzene)	12.83	24.6	12.71	Richardson *et al.*, 1974b
	—	—	21.5	13.1	O'Neal and Richardson, 1970
	—	—	21.7	12.7	O'Neal and Richardson, 1970
	—	—	22.7	12.6	O'Neal and Richardson, 1970
	—	—	21.7	12.4	O'Neal and Richardson, 1970
	25.8 (CCl_4)	—	24.7	12.3	Kopecky *et al.*, 1975; O'Neal and Richardson, 1970
	25.7 (benzene)	—	22.2	—	Kopecky *et al.*, 1975
	22.7 (benzene)	—	22.2	—	Kopecky *et al.*, 1975

[a] Thermochemical calculations for stepwise decomposition.

or cleaves to one triplet and one unexcited product. Thermochemical calculations (cf. Benson, 1968) of the activation parameters for the decomposition of 1,2-dioxetanes based on this biradical mechanism (Richardson *et al.*, 1972a,b; O'Neal and Richardson, 1970) and α-peroxylactones (Richardson and O'Neal, 1972) are in agreement with those determined experimentally (Table II). The activation parameters of a series of dioxetanes (as tabulated below) are reasonably insensitive to the nature of the ring substituent (Richardson *et al.*, 1974b), suggesting a transition state for decomposition

Dioxetane	E_a
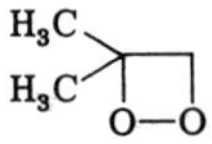 3,3-dimethyl-1,2-dioxetane (H_3C, H_3C)	23.0
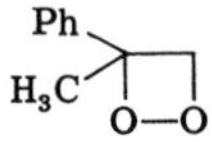 3-methyl-3-phenyl-1,2-dioxetane (Ph, H_3C)	22.9
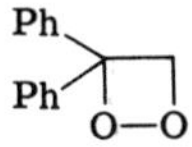 3,3-diphenyl-1,2-dioxetane (Ph, Ph)	22.7

involving little development of the incipient carbonyl function. A two-step mechanism, with initial O–O rupture, is consistent with this experimental observation. Recent CNDO/2 (Evleth and Feler, 1973), CNDO/S (Roberts, 1974), and molecular orbital calculations (Barnett, 1974) predict a two-step biradical mechanism as the lowest energy pathway in the decomposition of 1,2-dioxetane. An attractive point of the biradical mechanism is that the biradical in reversible equilibrium with the dioxetane is in a state close to that of the intimate radical pairs in which the rapid spin relaxation is observed that is responsible for chemically induced dynamic nuclear polarization.

IV. REARRANGEMENTS AND ALTERNATIVE CLEAVAGES

The reaction of singlet oxygen with alkenyl thioethers and certain enamines affords oxidation products not involving cleavage of the original ethylenic C–C bond. These products have been rationalized as arising from 1,2-dioxetane intermediates, and in one instance a dioxetane has been isolated and shown to produce these products (Wasserman and Terao, 1975). The photooxidation of certain enamines gives products attributable to

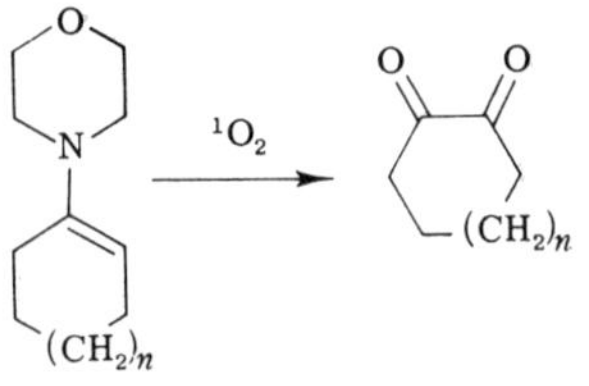

$^1O_2$; $(CH_2)_n$; $(CH_2)_n$

n = 0, 1, 2, 3, 7

$^1O_2$ −78°C $n = 2$

warm to room temperature in methanol

silica gel

reflux benzene

O N OH O $(CH_2)_2$

competitive C–C and C–N bond cleavage from a dioxetane intermediate (Ando *et al.*, 1975a), while others give only the anticipated ring fragmentation

C–C cleavage

Ph N O + H_3C CH_3

49%

H_3C Ph H_3C N O $^1O_2$

C–N cleavage

Ph OH CH_3 CH_3 + Ph CH_3

22% trace

products (Ando *et al.*, 1975a; Foote *et al.*, 1975). The photooxidation of an enaminoketone (see below) reportedly generates a dioxetane that produces

H_3C H_3C N O $^1O_2$ → H_3C CH_3 + H N O

Ph Ph N O $^1O_2$ → Ph Ph O–O N O → Ph H + Ph N O

rearranged products in preference to those predicted for ring fragmentation (Orito *et al.*, 1974).

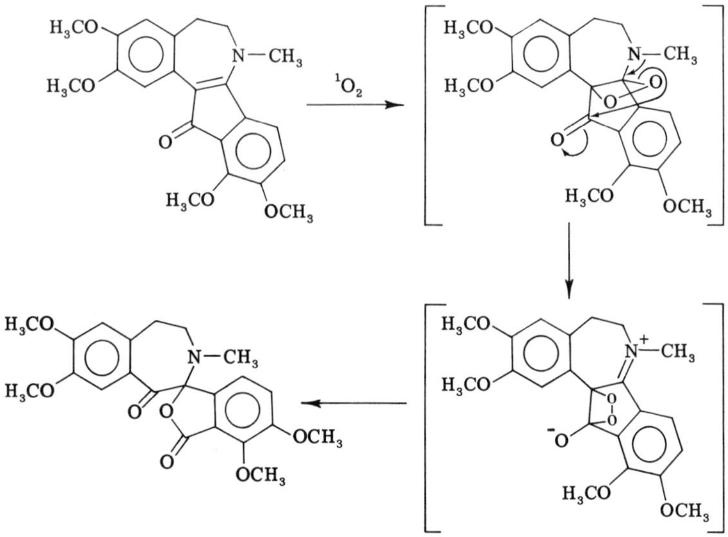

As with enamines, the photooxidation of thioethylenes leads to products without cleavage of the original double bond (Ando *et al.*, 1972, 1973, 1974; Ando *et al.*, 1975b; Adam and Liu, 1972a,b). For example, the oxidation of tetrakis(ethylthio)ethylene generates diethylthiooxalate and diethyldisulfide (Ando *et al.*, 1972, 1973; Adam and Liu, 1972a,b). Other alkenyl thioethers afford mixtures of products attributable to competitive C–C and C–S bond cleavage from intermediate dioxetanes (Ando *et al.*, 1974, 1975b).

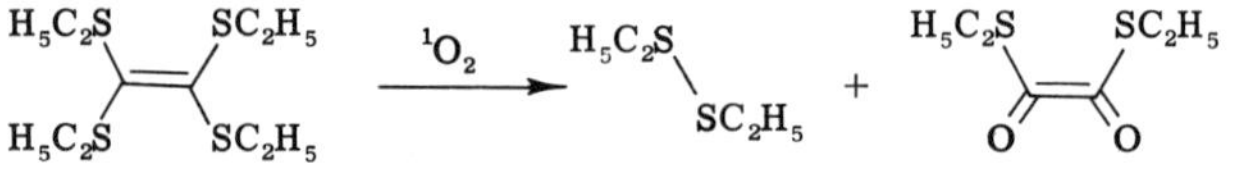

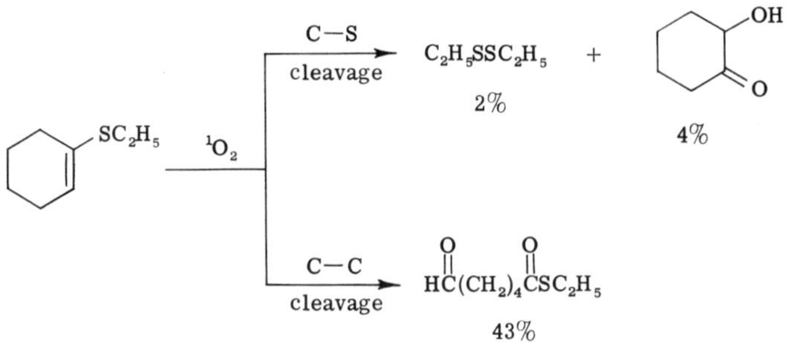

H_5C_2S $\xrightarrow{^1O_2}$ [H_5C_2S … O—O] + ene products

C—C cleavage / C—S cleavage

65% / $C_2H_5SSC_2H_5$ + 4% / OHC, HO 26%

Oxidation of 2-(1,3-dithia-2-cyclohexylidenyl)-1,3-dithiane affords 1,2,5,9-tetrathiadodecane-7,8-dione as the sole product (Adam and Liu, 1972a,b).

$\xrightarrow{^1O_2}$ O O S S $(CH_2)_3$ $(CH_2)_3$ S—S

V. CATALYSTS FOR THE DECOMPOSITION OF 1,2-DIOXETANES

A. Quantum Chain Processes

Both *cis*-diethoxy-1,2-dioxetane (Wilson and Shaap, 1971) and tetramethyl-1,2-dioxetane (Lechtken *et al.*, 1973) exhibit chain kinetics in their thermal decompositions indicative of the process

$$D + {}^3K^* \rightarrow 2K + {}^3K^*$$

where D is dioxetane and K is ketone or ester cleavage product. The excited carbonyl compound on the right is in each such step one of the cleavage fragments of the dioxetane molecule whose decomposition is being induced. This chain step efficiently replaces the inducing $^3K^*$ with a new one without giving rise to any light emission (Adam *et al.*, 1975). Therefore, the higher the concentration of the dioxetane the more induced decomposition accompanies the spontaneous cleavage, and the lower the total light yield in the decomposition of a given sample.

A high concentration of an efficient triplet quencher, such as 1,3-pentadiene, can eliminate the induced decomposition and make the observed rate cleanly first-order. It can also quench the luminescence coming from $^3K^*$, but not that coming from the smaller amount of $^1K^*$. Saturation with air generally produces only partial quenching, since at low oxygen concentrations energy transfer from triplet ketone to the fluorescer 9,10-dibromoanthracene (DBA) competes with quenching of the ketone triplet by oxygen. Because the excited DBA in the singlet state is not quenched by oxygen, some complicated effects are seen (Lechtken *et al.*, 1973). This "quantum chain" decomposition of TMD in degassed solution has also been initiated by direct irradiation of the dioxetane or by singlet sensitization with pyrene (Turro and Waddell, 1975).

Dioxetanes undergo efficient photolytic decomposition, the decomposition of TMD yielding excited acetone with nearly unit quantum efficiency. In cyclohexane solution at 77°K, the direct irradiation (Turro *et al.*, 1973) or the pyrene-sensitized (Baron and Turro, 1974) decomposition of TMD, like the thermal decomposition, produces high yields of triplet acetone; however, the yield of excited singlet acetone increases steadily as the wavelength of the irradiation is decreased. At 366 nm, at 6°C, the yields of singlet and triplet acetone are 10 and 43%, respectively; at 297 nm, the yield of singlet acetone is 30%.

B. Amine Catalysis

It has been demonstrated that *cis*-diethoxy-1,2-dioxetane is subject to amine-catalyzed decomposition to ethyl formate via a nonluminescent mechanism (Lee and Wilson, 1973). Triethylamine, diethylamine, and 1,4-diazabicyclo[2.2.2]octane induce greatly accelerated decomposition (activation energy for the catalyzed reaction, ~12 kcal/mole). A correlation of the catalytic efficiency of the amine with its ionization potential indicates the possibility of a charge transfer complex between amine (donor) and dioxetane (acceptor). The complex, with a weakened O–O bond, might then be rapidly cleaved to unexcited carbonyl products. Several other 1,2-dioxetanes, including 3,3-diphenyl-1,2-dioxetane (Landis, 1974) are subject to a similar catalysis not seen with tetramethyl-1,2-dioxetane (T. Wilson, private communication, 1974).

C. Metal Ion Catalysis

Tetramethyldioxetane decomposes in ordinary methanol and ethanol at rates that are both much faster (Turro and Lechtken, 1973a,b) and less reproducible (Wilson *et al.*, 1973) than in benzene. This proved to be the result of traces of powerful catalysts in the alcoholic solvents that could be removed only by treatment with ethylenediaminetetraacetic acid (EDTA) or chelating resins (Richardson, 1973; Richardson *et al.*, 1975; Wilson *et al.*, 1973). Wilson *et al.* (1973) established that (a) in rigorously deionized methanol or ethanol the rate of TMD decomposition was essentially the same as in benzene, (b) the superposed catalytic reaction in the unpurified solvents was nonchemiluminescent, and (c) the luminescent portion of the reaction in all solvents had the rate and activation parameters characteristic of the reaction in benzene (Table III). The spontaneous, uncatalyzed reaction could be distinguished from the catalytic one not only by its chemiluminescence but also by the hydrogen abstraction from 1,4-cyclohexadiene performed by the excited triplet acetone in degassed solution, and by the chain character seen in degassed solutions in the absence of a triplet quencher. The catalytic effects of the following series of transition metal ions (as the dichloride) were investigated in purified methanol: Cu^{2+}, Ni^{2+}, Co^{2+}, Zn^{2+}, Mn^{2+}, and Cd^{2+} (Bartlett *et al.*, 1974b; Baumstark, 1974). Treatment of TMD in CD_3OD with one equivalent of each salt at room temperature resulted in the quantitative formation of acetone. The rate of decay of the chemiluminescent intensity from TMD in treated methanol with 9,10-dibromoanthracene as fluorescer was measured at 57°C before and after the addition of known amounts of transition metal salts. The second-order rate constants for the catalyzed decomposition, k_2, were calculated from

$$k_2 = \frac{k_{obsd} - k_u}{\text{metal salt}} = \frac{k_{cat}}{\text{metal salt}}$$

where k_u is the rate constant for the uncatalyzed decomposition and k_{obsd} is the observed rate constant for the catalyzed decomposition. All catalyzed decompositions were nonchemiluminescent.

Cupric ion was found to be the most effective of the metal ions in catalyzing the decomposition of TMD. Addition of equimolar $CuCl_2 \cdot 2H_2O$ to solutions of TMD in acetonitrile-d_3 produced a yellow-green solution in contrast to the pale blue solution produced upon addition of $CuCl_2 \cdot 2H_2O$

Table III Rates of Decomposition of TMD in Treated Solvents

Solvent	Treatment	$k_1 \times 10^5$ (sec^{-1})[a]
Benzene	Shaken with saturated aqueous Na_2EDTA	3.5
	None	6–7
Methanol[b]	With 5×10^{-4} M Na_2EDTA	3.3
	None	200–500
Ethanol[c]	With 1×10^{-4} M Na_2EDTA	4.5
	None	110–190

[a] Temperature, 57°C; [TMD], 7.2×10^{-3} M; [DBA], 8×10^{-4} M.
[b] Solvent system, 50 μl benzene + 550 μl methanol.
[c] Solvent system, 50 μl benzene + 550 μl ethanol.

Table IV Cupric Ion[a] Catalyzed Decomposition of TMD[b] in Treated Methanol

Anion	Anion/Cu^{2+}	k_{cat} (sec^{-1} at 57°C)
None	—	8.6×10^{-2}
Acetate	2.1	2.6×10^{-2}
Troponoxy[c]	2.0	1.4×10^{-3}
Citrate	2.0	4.0×10^{-4}
EDTA	1.2	0

[a] [$CuCl_2$], 6.6×10^{-5} *M*; [DBA] = 7.8×10^{-4} *M*.
[b] [TMD] ~ 10^{-3} *M*.
[c] No Na^+ or Cl^- present.

to methanol-d_4. The NMR spectrum of the acetonitrile-d_3 solution showed slow decomposition of TMD to acetone (~20% after 60 min) whereas the reaction in methanol-d_4 was essentially instantaneous. In methanol, cupric chloride dissociates to the solvated ion $Cu^{2+}(CH_3OH)_x$; but in acetonitrile, the salt forms a complex $[CuCl_n(CH_3CN)_{4-n}]^{(n-2)-}$ ($n = 0$–4) (Massey, 1973). The cupric ion catalysis was subject to a counterion effect (Table IV), with catalytic efficiency decreasing with increasing chelation of the metal ion. These results indicate that access to one or perhaps two coordination sites on the metal is essential for the catalysis.

The second-order rate constants, k_2, for the catalyzed decomposition of TMD by the metal salts show a linear free energy correlation (Fig. 1) with the Lewis acidities of the metal ions, as measured by the negative logarithms of the dissociation constants (K_{mal}) of the metal malonates (Prue, 1952; Bender, 1971).

A mechanism [Eq. (3)] for the catalysis involving initial coordination of TMD by the metal ion is consistent with the experimental observations.

TMD + M^{2+} ⇌ [O—O ··· M complex]$^{2+}$ ⟶ [O···M···O complex]$^{2+}$ (3)

Other mechanisms involving electron transfer [Eq. (4)] or oxidative addition [Eq. (5)] may also be written.

TMD + M^{2+} ⟶ (O• O⁻ species) + M^{3+} ⟶ 2 (acetone, O) + M^{2+} (4)

TMD + M^{2+} ⟶ [O—M—O complex]$^{2+}$ ⟶ 2 (acetone, O) + M^{2+} (5)

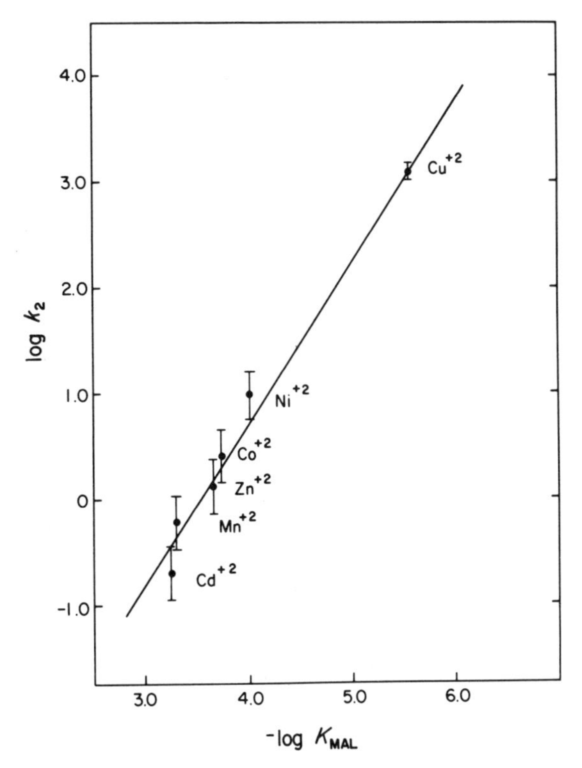

Fig. 1. Linear free energy relationship observed for the metal ion catalyzed decomposition of TMD.

Full electron transfer [Eq. (4)] seems unlikely since the logarithms of the second-order rate constants (k_2) do not correlate with the third ionization potentials of the neutral metals (Cotton and Wilkinson, 1972). Complexation of TMD by a metal ion [as in Eq. (3)] might facilitate decomposition by removing orbital symmetry restrictions against a concerted fragmentation to two unexcited molecules of acetone (Woodward and Hoffmann, 1969). Similar catalytic effects, with metal ions acting as Lewis acids, have been observed in many other reactions, including decarboxylations of dimethyloxaloacetic acid (Steinberger and Westheimer, 1951) and of acetonedicarboxylic acid (Prue, 1952), solvolysis of *tert*-butyl halides (Rudakov and Kozhevnikov, 1971), bromination of 2-carbethoxycyclopentanone, and hydrolysis of glycine methyl ester (Tanaka and Ozaki, 1967).

This particular series of divalent metal salts are alike in their inability to undergo facile valence expansion. Complexes of rhodium(I) and iridium(I), on the other hand, have been shown to undergo oxidative addition reactions with reactive carbon–carbon σ or π bonds (Fraser *et al.*, 1973; Cassar and Halpern, 1970). Many of these complexes also facilitate the decomposition of TMD (Bartlett and McKennis, 1977); the relative catalytic efficiencies of these complexes (Table V) are consistent with an oxidative addition mechanism, in which the metallocycle fragments to acetone and the metal in its initial valency.

Table V Catalytic Efficiencies of Complexes of Rhodium and Iridium in Tetramethyl-1,2-dioxetane Decomposition

Complex	Relative rate
$Rh(Ph_3P)_3Cl$	1.0
$Rh(CO)(Ph_3P)_2Cl$	1.48
$Rh(CO)(Ph_3P)_2I$	3.0
$Rh(CO)(Ph_3As)_2Cl$	6.1
$[Rh(CO)_2Cl]_2$	31.5
$Ir(CO)(Ph_3P)_2Cl$	61
$[Rh(Nbd)Cl]_2$[a]	185
$CuCl_2\ 2H_2O$ (in methanol)	76.8

[a] Nbd, norbornadiene.

An interesting extension of metal-catalyzed decompositions of TMD is the fact that lanthanide shift reagents [europium(III) tris-6,6,7,7,8,8,8-heptafluoro-2,2-dimethyl-3,4-octanedione and europium(III)-tris-2,2,6,6-tetramethyl-3,5-heptanedione] could not be used in the NMR study of TMD because of rapid catalytic destruction of the dioxetane (P. D. Bartlett and M. J. Shapiro, unpublished results, 1975).

VI. NONLUMINESCENT REACTIONS OF 1,2-DIOXETANES

A. Reactions with Trivalent Phosphorus Compounds

Reactions of triphenylphosphine with several dioxetanes (in carbon tetrachloride) lead to the formation of allylic alcohols, an epoxide and ring-cleavage products (Kopecky *et al.*, 1975).

Ph_3P

O–O O OH O

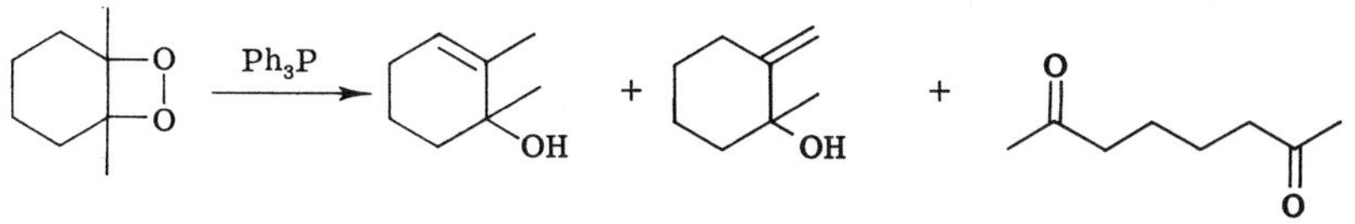

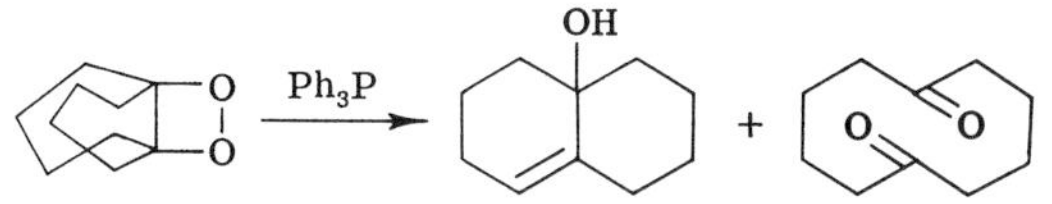

If the temperature is kept from rising, the reaction of triphenylphosphine with tetramethyl-1,2-dioxetane (TMD) reveals the intermediacy of a pentacoordinate phosphorus species, a phosphorane, in the course of deoxygenation to an epoxide (Bartlett *et al.*, 1973).

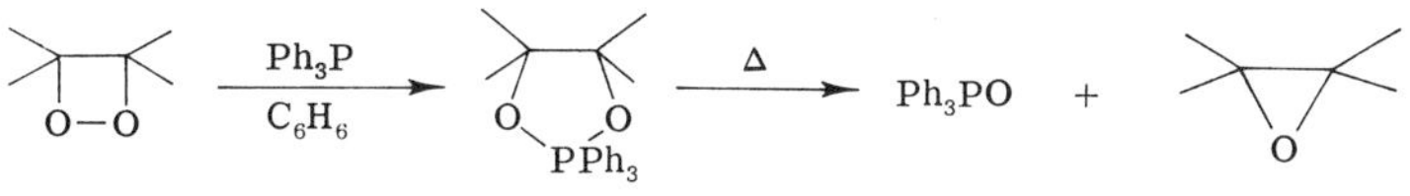

(**16**)

The nature of phosphorane formation and the thermal decomposition of phosphorane **16** to triphenylphosphine oxide and tetramethylethylene oxide have been examined (Bartlett *et al.*, 1974a; Baumstark, 1974).

The reactions of triphenyl phosphine, trimethyl phosphite, triethyl phosphite, and methyl diphenyl phosphinite with TMD produce phosphoranes **16**, **17**, **18**, and **19**, respectively, as identified by their characteristic

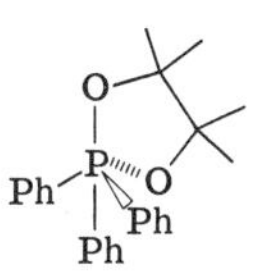

(**16**) 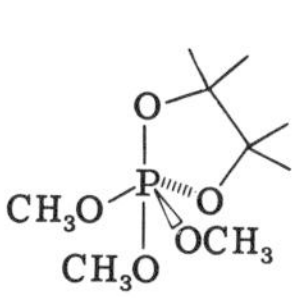 (**17**) 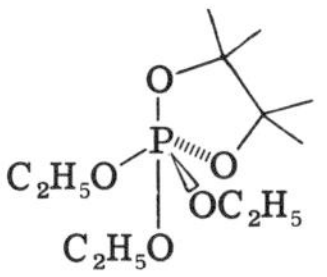 (**18**) 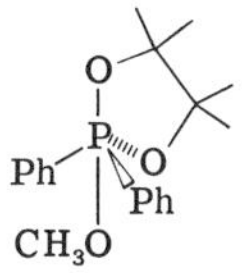 (**19**)

(Hellwinkel, 1972) ^{31}P chemical shifts (Table VI). Second-order rate constants for the reaction of each phosphorus compound with TMD are shown in Table VII.

D. B. Denney (Denney and Jones, 1969; Denney *et al.*, 1972) has discussed two extreme mechanistic possibilities for phosphorane formation during the reaction of trivalent phosphorus compounds with dialkyl peroxides. Corresponding concerted and stepwise mechanisms for reactions with TMD are shown below.

$$\text{TMD} + R_3P \longrightarrow [\text{concerted TS}]\ PR_3 \longrightarrow \text{phosphorane}\ PR_3 \quad (6)$$

$$\text{TMD} + R_3P \longrightarrow \text{dipolar ion } (O^-, \overset{+}{P}R_3) \longrightarrow \text{phosphorane}\ PR_3 \quad (7)$$

As with reactions with acyclic peroxides (Denney *et al.*, 1961, 1968, 1972; Chang *et al.*, 1971), the reactions with TMD show little or no dependence on solvent polarity (Table VII). Unlike its reaction with TMD, triphenyl phosphine generates a dipolar ion when treated with S_8, and the reaction rate increases greatly with increasing solvent polarity (Bartlett and Meguerian, 1956). An ionic mechanism is therefore unlikely for the dioxetane reaction, so the reaction pathway must be concerted [Eq. (6)] or homolytic [Eq. (8)]. Since **16** can be formed by neither mechanism with both oxy

$$\text{TMD} + R_3P \longrightarrow \text{biradical } (O\cdot, \dot{P}R_3) \longrightarrow \text{phosphorane}\ PR_3 \quad (8)$$

substituents apical, an attractive reason is at hand for the tenfold jump in the rate of formation from **16** to **19**, namely the unfavorable entry of a phenyl group into an apical position of the phosphorane (Gorenstein and Westheimer, 1970; Gorenstein, 1970; Muetterties and Schunn, 1966). That **17** and **18** form more slowly than **16** is consistent with an opposing depression of the general reactivity of the phosphorus atom as phenyl groups are re-

Table VI ^{31}P-NMR Data[a] for Phosphoranes (**16**–**19**) and Related Compounds

δ for X_3P	X_3
+48.4	Ph_3
+54.3	$(OCH_3)_3$
+60[b]	$(OEt)_3$
+37.4	Ph_2, OMe

[a] All reported chemical shifts relative to 85% phosphoric acid. Spectra were recorded at normal probe temperature.

[b] Denney and Jones, 1969.

Table VII Rate Constants for Reaction of Trivalent Phosphorus Compounds with TMD

Reagent	TMD[a] ($M \times 10^3$)	X_3P ($M \times 10^3$)	k_2 (M^{-1} sec^{-1})[c] Benzene	CH_3CN/Benzene[b]
Ph_2POMe	2.6	18	12 ± 2	
Ph_2POMe	3.3	35		10 ± 1
Ph_3P	6.6	70	1.0 ± 0.1	
Ph_3P	6.1	65		1.0 ± 0.1
$(MeO)_3P$	6.6	70	0.28 ± 0.03	
$(MeO)_3P$	6.6	70–140		0.16 ± 0.05
$(EtO)_3P$	2.6–6.6	35–280	0.30 ± 0.03	
$(EtO)_3P$	6.6	70–140		0.18 ± 0.05

[a] All runs contained 7.8×10^{-4} M 9,10-dibromoanthracene; 23 ± 1°C.
[b] 550 μl acetonitrile to 100 μl benzene.
[c] Disappearance of TMD was pseudo first-order through at least three half-lives. All rate constants are the average of at least two experiments.

placed by alkoxy groups (Kirby and Warren, 1967). None of the experimental observations, however, discriminates between a radical and a concerted process [Eq. (8) vs. Eq. (6)]. Probably the best reason for preferring Eq. (6) is that the intermediate in Eq. (8) is likely to be formed by attack of P on one O with inversion, which makes it likely that if this were the mechanism some or most of the intermediate would take the exothermic course of forming epoxide and R_3PO directly. Actually these products are formed in a later reaction of **16** which shows a polar solvent effect.

The rate of decomposition of phosphorane **16** to triphenyl phosphine oxide and tetramethylethylene oxide increases with increasing solvent polarity (a factor of 100 in acetonitrile versus benzene, as solvent. Like the decomposition of other oxyphosphoranes (Ramirez *et al.*, 1968; Denney and Jones, 1969) and unlike phosphorane *formation*, the decomposition of **16** probably proceeds with initial heterolytic cleavage of a P–O bond. Rotation about the C–C bond and backside displacement by the alkoxide (with inversion at carbon) of triphenyl phosphine oxide generates the epoxide. Unlike **16**, phosphoranes **17–19** do not undergo facile thermolytic decom-

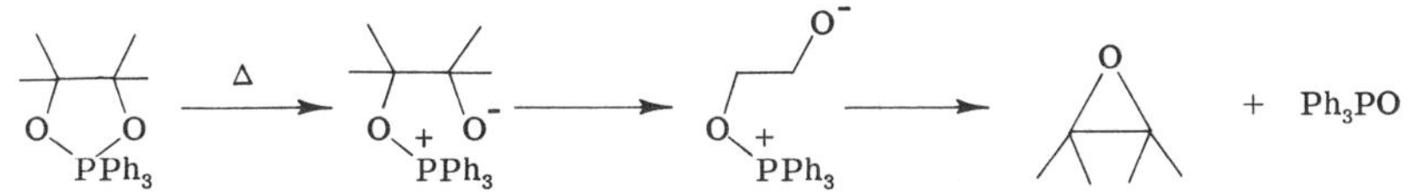

position; elevated temperatures (~150°C) are required and complex mixtures of products are observed (Baumstark, 1974). The inversion in this step has recently been verified in the case of 3,4-diphenyl-1,2-dioxetane (Bartlett *et al.*, 1977b).

B. Reactions with Divalent Sulfur Compounds

Reactions of dioxetanes with diphenyl sulfide (Wasserman and Saito, 1975) and with dialkyl sulfides and sulfoxylates (Campbell *et al.*, 1975) have been reported. Tetravalent sulfur species, sulfuranes, have been isolated from the low temperature reaction of 3,3-dimethyl- and trimethyl-1,2-dioxetane with dimethyl- or di-*n*-propylsulfoxylate (Campbell *et al.*, 1975).

$(R'O)_2S$

R = H, CH_3
R′ = CH_3, C_3H_7

At room temperature, these sulfuranes cleave to dialkyl sulfates and deoxygenated dioxetane-derived products.

polymer + CH_3OSOCH_3 ; —CHO + $(C_3H_7O)_2SO$

Although no intermediate sulfurane was isolated, the reaction of diphenyl sulfide with 3,3-dimethyl- and trimethyl-1,2-dioxetane (Campbell *et al.*, 1975) generates epoxides, and the reaction of trimethyl-1,2-dioxetane with dimethyl sulfide and tetrahydrothiophene produces 3-hydroxy-3-methyl-2-butanone in a base-catalyzed rearrangement.

Ph_2S → + Ph_2SO

R = H, CH_3

Ph_2SO → + Ph_2SO_2

$(CH_3)_2S$ or (tetrahydrothiophene) → OH

The deoxygenation of the dioxetane derived from the addition of singlet oxygen to *cis*-dimethoxystilbene has been accomplished with diphenyl sulfide (Wasserman and Saito, 1975). The nature of the photooxidation of 2,3-diphenyl-*p*-dioxene and imidiazoles is perturbed by the addition of

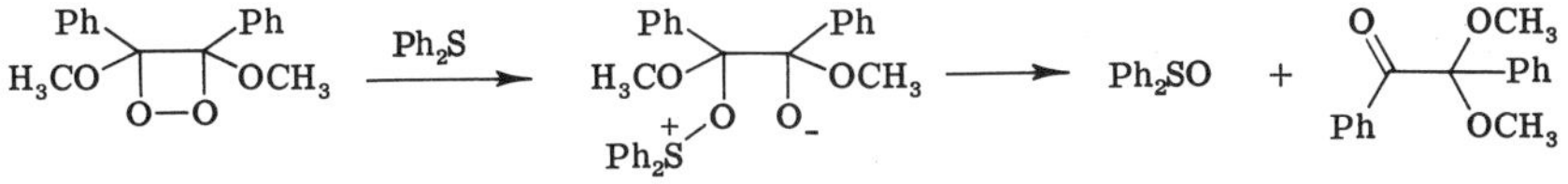

Ph_2S ⟶ Ph_2SO +

diphenyl sulfide, the products of the diphenyl sulfide reaction being rationalized as arising via deoxygenation of dioxetane intermediates (Wasserman and Saito, 1975).

$^1O_2$ ⟶ Ph_2S

$^1O_2$ ⟶ Ph_2S

R = Ph, CH_2Ph

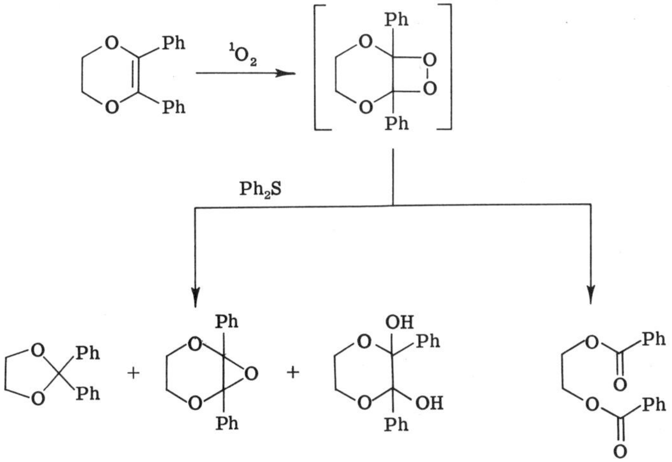

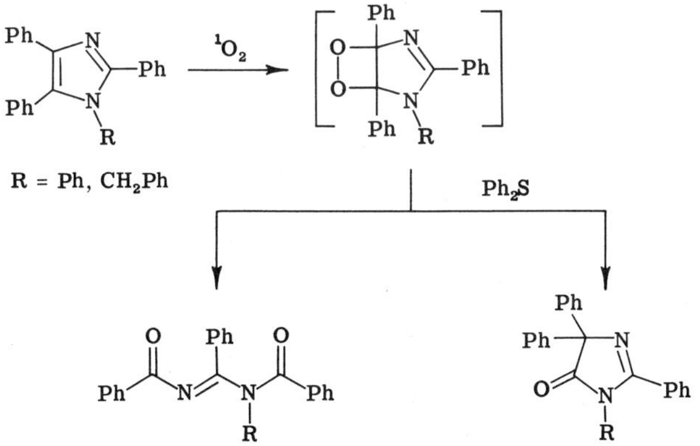

C. Reactions with Divalent Tin

Reactions of tetramethyl-1,2-dioxetane (TMD) with stannous chloride or stannous acetate in methanol-d_4 produce a mixture of pinacol and acetone (Baumstark, 1974). The reaction with stannous chloride yields 100% acetone

$SnCl_2$: 63% acetone, 37% pinacol

$Sn(O_2CCH_3)_2$: 67% acetone, 33% pinacol

only in *p*-dioxane; the two most aqueous media investigated, 8:1 water–methanol and 8:1 water–acetonitrile, still yielding acetone and pinacol in a 32:68 ratio. Presumably the stannous salt undergoes insertion to the intermediate stannic pinacolate (**20**), which may fragment to acetone and stannous compound, or may be hydrolyzed to pinacol and an Sn(IV) product. The

$(CH_3)_2C$–O–SnX_2–O–$C(CH_3)_2$ (ring)

(20)

low sensitivity of the product composition to solvent may mean that the fragmentation to acetone occurs by Lewis acid catalysis before **20** is formed, as with the catalysts of Fig. 1, and does not occur after formation of the pinacolate.

A model for the formation of **20** is the insertion of stannous chloride into benzoyl peroxide to yield stannic benzoate (Razuvaev *et al.*, 1954). This view is supported by the stability of dibutyltin pinacolate (Mehrotra and Gupta, 1965).

D. Reactions with Boron Trifluoride

While most transition metal salts catalyze ring fragmentation of dioxetanes, boron trifluoride induces the rearrangement of TMD to pinacolone, formally a reduction product (Baumstark, 1974; Bartlett *et al.*, 1977a).

Treatment of TMD in carbon tetrachloride at room temperature with gaseous boron trifluoride for a few seconds resulted in the conversion of TMD to pinacolone (69%) and acetone (9%). No molecular oxygen was evolved during the reaction characterized by low product recovery (78%); but a small amount of a nonorganic peroxide (hydrogen peroxide; up to

Table VIII Product Yields and Total Peroxide Titer for the Low Temperature Reaction of TMD with Boron Trifluoride

TMD								
Initial (mmoles)	% Recovered	**21** (%)	Pinacolone (%)	Acetone (%)	Total product recovery (%)	Titratable peroxide[a] (mmoles)	Peroxide (mmoles) − recovered TMD (mmoles) = "H_2O_2" (mmoles)	"H_2O_2" (mmoles) / Pinacolone (mmoles)
13.3 mg (0.1146)	4 (0.0046 mmole)	28	32 (0.0367 mmole)	4	67	0.0250	0.0214	0.58
8.3 mg (0.0716)	7 (0.005 mmole)	23	55 (0.0394 mmole)	3	88	0.0301	0.0251	0.64
7.3 mg (0.063)	7 (0.0044 mmole)	20	58 (0.0365 mmole)	3	88	0.0353	0.0309	0.85

[a] By iodometric titration, under conditions such that pinacolone diperoxide does not liberate iodine.

10% based on dioxetane) remained after complete conversion of TMD. The low temperature reaction of boron trifluoride with TMD in dichloromethane allowed more complete characterization of the reaction pathway. Under conditions where 27% of initial TMD was recovered, the product composition was 70% pinacolone, 5% acetone, and 25% of the dimeric peroxide (**21**) of pinacolone; the average product recovery was 90%.

BF_3, −78°C

70% 5% 25%

(**21**)

Iodometric titration of the reaction mixture (under conditions of high conversion of TMD to products) indicated the presence of a peroxide (hydrogen peroxide) in substantial excess of recovered TMD (with conditions so that **21** does not liberate iodine) as shown in Table VIII.

The formation of dimeric peroxide **21** indicates the intermediacy of the carbonyl oxide of pinacolone (Criegee *et al.*, 1955) during the course of the reaction of TMD with boron trifluoride. Pinacolone formation can be accounted for by competitive acid-catalyzed hydrolysis of the carbonyl oxide by trace amounts of moisture, as indicated by the presence of 0.6–0.8 equivalents of hydrogen peroxide per equivalent of pinacolone. Acetone may be generated by ring fragmentation, concomitant with rearrangement from the initial monooxygen complex of TMD with the Lewis acid. Monooxygen coordination of TMD thus results predominantly in rearrangement, whereas coordination by divalent transition metal salts generates acetone

+ BF_3 → major → (**22**); minor →

22 + H_2O → H_2O_2 + BF_3 +

22 + **22** → **21** + 2 BF_3

as the sole product (Bartlett *et al.*, 1974b). An attractive explanation for this difference in reaction mode involves the nature of the TMD–Lewis acid complex; namely, that coordination of TMD by transition metal ions involves both oxygen atoms of the dioxetane.

$$\text{TMD} + M^{2+} \longrightarrow [\text{TMD}\cdot M]^{2+} \longrightarrow [(\text{Me}_2\text{C}{=}\text{O})_2 M]^{2+}$$

REFERENCES

Adam, W., and Liu, J.-C. (1972a). *J. Chem. Soc. D* pp. 73–74.
Adam, W., and Liu, J.-C. (1972b). *J. Am. Chem. Soc.* **94,** 1206–1209.
Adam, W., and Liu, J.-C. (1972c). *J. Am. Chem. Soc.* **94,** 2894–2895.
Adam, W., and Steinmetzer, H.-C. (1972). *Angew. Chem., Int. Ed. Engl.* **11,** 540–542.
Adam, W., Simpson, G. A., and Yany, F. (1974). *J. Phys. Chem.* **78,** 2559–2569.
Adam, W., Duran, N., and Simpson, G. A. (1975). *J. Am. Chem. Soc.* **97,** 5464–5467.
Ando, W., Suzuki, J., Arai, T., and Migita, T. (1972). *J. Chem. Soc. D* pp. 477–478.
Ando, W., Suzuki, J., Arai, T., and Migita, T. (1973). *Tetrahedron* **29,** 1507–1513.
Ando, W., Watanabe, K., Suzuki, J., and Migita, T. (1974). *J. Am. Chem. Soc.* **96,** 6766–6768.
Ando, W., Saiki, T., and Migita, T. (1975a). *J. Am. Chem. Soc.* **97,** 5028–5029.
Ando, W., Watanabe, K., and Migita, T. (1975b). *Tetrahedron Lett.* pp. 4127–4130.
Bailey, P. S., Carter, T. P., Jr., Fischer, C. M., and Thompson, J. A. (1973). *Can. J. Chem.* **51,** 1278–1283.
Barnett, G., *Can. J. Chem.* **52,** 3837–3843.
Baron, W. J., and Turro, N. J. (1974). *Tetrahedron Lett.* pp. 3515–3518.
Bartlett, P. D., and Ho, M. S. (1974). *J. Am. Chem. Soc.* **96,** 627–629.
Bartlett, P. D., and Landis, M. E. (1977). *J. Am. Chem. Soc.* **99,** 3033.
Bartlett, P. D., and McKennis, J. S. (1977). *J. Am. Chem. Soc.* **99,** 5334.
Bartlett, P. D., and Meguerian, G. (1956). *J. Am. Chem. Soc.* **78,** 3710–3715.
Bartlett, P. D., and Schaap, A. P. (1970). *J. Am. Chem. Soc.* **92,** 3223–3225.
Bartlett, P. D., and Shimizu, N. (1975). *J. Am. Chem. Soc.* **97,** 6253–6254.
Bartlett, P. D., Baumstark, A. L., and Landis, M. E. (1973). *J. Am. Chem. Soc.* **95,** 6486–6487.
Bartlett, P. D., Baumstark, A. L., Landis, M. E., and Lerman, C. L. (1974a). *J. Am. Chem. Soc.* **96,** 5267–5268.
Bartlett, P. D., Baumstark, A. L., and Landis, M. E. (1974b). *J. Am. Chem. Soc.* **96,** 5557–5558.
Bartlett, P. D., Baumstark, A. L., and Landis, M. E. (1977a). *J. Am. Chem. Soc.* **99,** 1890.
Bartlett, P. D., Landis, M. E., and Shapiro, M. J. (1977b). *J. Org. Chem.* **42,** 1661.
Barton, D. H. R., Haynes, R. K., Leclerc, G., Magnus, P. D., and Menzies, I. D. (1975). *J. Chem. Soc., Perkin Trans. 1,* pp. 2055–2065.
Basselier, J. J., Cherton, J.-C., and Caille, J. (1971). *C. R. Hebd. Seances Acad. Sci., Ser. C* **273,** 514–517.
Baumstark, A. L., (1974). Thesis, Harvard University, Cambridge, Massachusetts.
Baumstark, A. L., Wilson, T., Landis, M. E., and Bartlett, P. D., *Tet. Lett., 1976,* 2397–2400.
Belyakov, V. A., and Vassil'ev, R. V. (1970). *Photochem. Photobiol.* **11,** 179–192.
Bender, M. L. (1971). "Mechanisms of Homogeneous Catalysis from Protons to Proteins," p. 218. Wiley, New York.
Benson, S. W. (1968). "Thermodynamical Kinetics." Wiley, New York.

Bullock, M. W., Brockman, J. A., Jr., Patterson, E. L., Pierce, J. V., von Saltza, M. H., Sanders, F., and Stokstad, E. L. R. (1954). *J. Am. Chem. Soc.* **76,** 1828–1832.
Burns, P. A., and Foote, C. S. (1974). *J. Am. Chem. Soc.* **96,** 4339–4340.
Campbell, B. S., Denney, D. B., Denney, D. Z., and Shih, L. S. (1975). *J. Am. Chem. Soc.* **97,** 3850–3851.
Cassar, L., and Halpern, J. (1970). *J. Chem. Soc. D* pp. 1082–1083.
Chang, B. C., Conrad, W. E., Denney, D. B., Denney, D. Z., Edelman, R., Powell, R. L., and White, D. W. (1971). *J. Am. Chem. Soc.* **93,** 4004–4009.
Cotton, F. A., and Wilkinson, G. (1972). "Advanced Inorganic Chemistry," 3rd ed., p. 801. Wiley, New York.
Criegee, R., and Paulig, G. (1955). *Chem. Ber.* **88,** 712–716.
Criegee, R., Kerckow, A., and Zincke, H. (1955). *Chem. Ber.* **88,** 1878.
Dalton, J. C., Wriede, P. A., and Turro, N. J. (1970). *J. Am. Chem. Soc.* **92,** 1318–1326.
Darling, T. R., and Foote, C. S. (1974). *J. Am. Chem. Soc.* **96,** 1625–1627.
Denney, D. B., and Jones, D. H. (1969). *J. Am. Chem. Soc.* **91,** 5821–5825.
Denney, D. B., Goodyear, W. F., and Goldstein, B. (1961). *J. Am. Chem. Soc.* **83,** 1726–1733.
Denney, D. B., Denney, D. Z., and Wilson, L. A. (1968). *Tetrahedron Lett.* pp. 85–89.
Denney, D. B., Denney, D. Z., Hall, C. D., and Marsi, K. L. (1972). *J. Am. Chem. Soc.* **94,** 245–249.
Dewar, M. J. S., and Kirschner, S. (1974). *J. Am. Chem. Soc.* **96,** 7578–7579.
Dewar, M. J. S., and Thiel, W. (1975). *J. Am. Chem. Soc.* **97,** 3978–3986.
Dewar, M. J. S., Griffin, A. C., Thiel, W., and Turchi, I. J. (1975). *J. Am. Chem. Soc.* **97,** 4439–4440.
Ding, J. Y., and Li, K. (1972). *Tui-Wan Ta Hsueh Nang Hsueh Yuan Shih Yen Lin, Yen Chiu Paoka* No. 108.
Engel, P. S. (1970). *J. Am. Chem. Soc.* **92,** 6074–6076.
Evleth, E. M., and Feler, G. (1973). *Chem. Phys. Lett.* **22,** 499–502.
Filby, J. E. (1973). Thesis, University of Alberta, Edmonton.
Foote, C. S., and Liu, J. W.-P. (1968). *Tetrahedron Lett.* pp. 3267–3270.
Foote, C. S., Mazur, S., Burns, P. A., and Lerdal, D. (1973). *J. Am. Chem. Soc.* **95,** 586–588.
Foote, C. S., Dzakpasu, A. A., and Liu, J. W.-P. (1975). *Tetrahedron Lett.* pp. 1247–1250.
Fraser, A. R., Bird, P. H., Bezman, S. Z., Shaply, J. R., White, R., and Osborn, J. A. (1973). *J. Am. Chem. Soc.* **95,** 597–598.
Frimer, A. A., Bartlett, P. D., Boschung, A. F., and Jewett, J. G. (1977). *J. Am. Chem. Soc.* **99,** 7977.
Gorenstein, D. (1970). *J. Am. Chem. Soc.* **92,** 644–650.
Gorenstein, D., and Westheimer, F. H. (1970). *J. Am. Chem. Soc.* **92,** 634–644.
Hasty, N. M., and Kearns, D. R. (1973). *J. Am. Chem. Soc.* **95,** 3380–3381.
Hellwinkel, D. (1972). *Org. Phosphorus Compd.* **3,** 185–339.
Ho, M. (1974). Thesis, Harvard University, Cambridge, Massachusetts.
Huber, J. E. (1968). *Tetrahedron Lett.* pp. 3271–3272.
Jefford, C. W., and Boschung, A. F. (1974). *Helv. Chim. Acta* **57,** 2257–2261.
Jefford, C. W., and Boschung, A. F. (1977). *Helv. Chim. Acta* **60,** 2673.
Kirby, A. J., and Warren, S. G. (1967). "The Organic Chemistry of Phosphorus." Am. Elsevier, New York.
Kopecky, K. R., and Mumford, C. (1968). *Abstr., 51st Annu. Conf., Chem. Inst. Can., 1968* p. 41.
Kopecky, K. R., and Mumford, C. (1969). *Can. J. Chem.* **47,** 709–711.
Kopecky, K. R., Van de Sande, J. H., and Mumford, C. (1971). *In* "Symposium on Oxidations by Singlet Oxygen Presented Before the Division of Petroleum Chemistry, Inc.," Petr 027 (abstr.). Am. Chem. Soc., Washington, D.C.
Kopecky, K. R., Lockwood, P. A., Filby, J. E., and Reid, R. W. (1973). *Can. J. Chem.* **51,** 468–470.

Kopecky, K. R., Filby, J. E., Mumford, C., Lockwood, P. A., and Ding, J.-Y. (1975). *Can. J. Chem.* **53,** 1103–1122.
Landis, M. E. (1974). Thesis, Harvard University, Cambridge, Massachusetts.
Lechtken, P., Yekta, A., and Turro, N. J. (1973). *J. Am. Chem. Soc.* **95,** 3027–3028.
Lee, D. C. S., and Wilson, T. (1973). *In* "Chemiluminescence and Bioluminescence" (M. J. Cormier, D. M. Hercules, and J. Lee, eds.), pp. 265–281. Plenum, New York.
Lightner, D. A., Bisacchi, G. S., and Norris, R. D. (1976). *J. Am. Chem. Soc.* **98,** 802–807.
McCapra, F. (1968). *J. Chem. Soc. D* p. 155.
McCapra, F., and Beheshti, I. (1977). *J. Chem. Soc. Chem. Comm.* 517.
Maier, G. (1974). *Angew. Chem., Int. Ed. Engl.* **13,** 425–438, esp. p. 432.
Massey, A. G. (1973). *In* "Comprehensive Inorganic Chemistry" (A. F. Trotman-Dickenson, ed.), Vol. 3, Chapter 27, p. 62. Pergamon, Oxford.
Mayo, F. R. (1958). *J. Am. Chem. Soc.* **80,** 2465–2480 and 6701.
Mayo, F. R. (1968). *Acc. Chem. Res.* **1,** 193–201.
Mazur, S., and Foote, C. S. (1970). *J. Am. Chem. Soc.* **92,** 3225–3226.
Meadus, F. W., and Ingold, K. U. (1964). *Can. J. Chem. Eng.* p. 86.
Mehrotra, R. C., and Gupta, V. D. (1965). *J. Organomet. Chem.* **4,** 145–150.
Muetterties, E. L., and Schunn, R. A. (1966). *Q. Rev., Chem. Soc.* **20,** 245–299.
O'Neal, H. E., and Richardson, A. H. (1970). *J. Am. Chem. Soc.* **92,** 6553–6557.
Orito, K., Manske, R. H., and Rodrigo, R. (1974). *J. Am. Chem. Soc.* **96,** 1944–1945.
Pauling, L. (1960). "The Nature of the Chemical Bond," 3rd ed., pp. 134–135. Cornell Univ. Press, Ithaca, New York.
Prue, J. E. (1952). *J. Chem. Soc.* p. 2337.
Ramirez, F., Gulati, A. S., and Smith, C. P. (1968). *J. Org. Chem.* **33,** 13–19.
Razuvaev, G. A., Moryzanov, B. N., Dlin, E. P., and Oldekop, Yu. A. (1954). *J. Gen. Chem. USSR* (*Engl. Transl.*) **24,** 265–267.
Richardson, W. H. (1973). Personal communication to PDB.
Richardson, W. H., and Hodge, V. F. (1971). *J. Am. Chem. Soc.* **93,** 3996–4004.
Richardson, W. H., and O'Neal, H. E. (1972). *J. Am. Chem. Soc.* **94,** 8665–8668.
Richardson, W. H., Yelvington, M. B., and O'Neal, H. E. (1972a). *J. Am. Chem. Soc.* **94,** 1619–1623.
Richardson, W. H., Montgomery, F. C., and Yelvington, M. B. (1972b). *J. Am. Chem. Soc.* **94,** 9277–9278.
Richardson, W. H., Montgomery, F. C., Yelvington, M. B., and Ranney, G. (1974a). *J. Am. Chem. Soc.* **96,** 4045–4046.
Richardson, W. H., Montgomery, F. C., Yelvington, M. B., and O'Neal, H. E. (1974b). *J. Am. Chem. Soc.* **96,** 7525–7532.
Richardson, W. H., Montgomery, F. C., Slusser, P., and Yelvington, M. B. (1975). *J. Am. Chem. Soc.* **97,** 2819–2821.
Rio, G., and Berthelot, J. (1971). *Bull. Soc. Chim. Fr.* pp. 3555–3557.
Robbins, W. K., and Eastman, R. H. (1970). *J. Am. Chem. Soc.* **92,** 6076–6077.
Roberts, D. R. (1974). *J. Chem. Soc. D* p. 683.
Rudakov, E. S., and Kozhevnikov, I. V. (1971). *Tetrahedron Lett.* pp. 1333–1336.
Schaap, A. P. (1971). *Tetrahedron Lett.* pp. 1757–1760.
Schaap, A. P., and Bartlett, P. D. (1970). *J. Am. Chem. Soc.* **92,** 6055–6057.
Schaap, A. P., and Faler, G. R. (1973). *J. Am. Chem. Soc.* **95,** 3381–3382.
Schaap, A. P., and Tontapanish, N. (1972). *J. Chem. Soc. D.* p. 490.
Sharp, D. B. (1960). Abstracts 138th National Meeting, Am. Chem. Soc. New York, No. 79 P.
Schuster, G. B., Turro, N. J., Steinmetzer, H.-C., Schaap, A. P., Faler, G., Adam, W., and Liu, J. C. (1975). *J. Am. Chem. Soc.* **97,** 7110–7118.
Steinberger, R., and Westheimer, F. H. (1951). *J. Am. Chem. Soc.* **73,** 429–435.
Steinmetzer, H.-C., Yekta, A., and Turro, N. J. (1974). *J. Am. Chem. Soc.* **96,** 282–284.
Story, P. R., Whited, E. A., and Alford, J. A. (1972). *J. Am. Chem. Soc.* **94,** 2143–2144.

Tanaka, K., and Ozaki, A. (1967). *Bull. Chem. Soc. Jpn.* **40,** 1728–1730.
Turro, N. J., and Lechtken, P. (1972). *J. Am. Chem. Soc.* **94,** 2886–2888.
Turro, N. J., and Lechtken, P. (1973a). *J. Am. Chem. Soc.* **95,** 264–266.
Turro, N. J., and Lechtken, P. (1973b). *Pure Appl. Chem.* **33,** 363–388.
Turro, N. J., and Lechtken, P. (1973c). *Tetrahedron Lett.* pp. 565–568.
Turro, N. J., and Waddell, W. H. (1975). *Tetrahedron Lett.* pp. 2069–2072.
Turro, N. J., Lechtken, P., Lyons, A., Hautala, R. R., Carnahan, E., and Katz, T. J. (1973). *J. Am. Chem. Soc.* **95,** 2035–2037.
Turro, N. J., Lechtken. P., Schuster, G., Orell, J., Steinmetzer, H.-C., and Adam, W. (1974a). *J. Am. Chem. Soc.* **96,** 1627–1629.
Turro, N. J., Schore, N. E., Steinmetzer, H.-C., and Yekta, A. (1974b). *J. Am. Chem. Soc.* **96,** 1936–1938.
Turro, N. J., Lechtken, P., Schore, N. E., Schuster, G., Steinmetzer, H.-C., and Yekta, A. (1974c). *Acc. Chem. Res.* **7,** 97–105.
Valentine, D., Jr., and Hammond, G. S. (1972). *J. Am. Chem. Soc.* **94,** 3449–3454.
Wagner, P. J. (1967). *J. Am. Chem. Soc.* **89,** 2503–2505.
Wasserman, H. H., and Saito, I. (1975). *J. Am. Chem. Soc.* **97,** 905–906.
Wasserman, H. H., and Terao, S. (1975). *Tetrahedron Lett.* pp. 1735–1738.
White, E. H., Wiecko, J., and Roswell, D. F. (1969). *J. Am. Chem. Soc.* **91,** 5194–5196.
White, E. H., Wiecko, J., and Wei, C. C. (1970). *J. Am. Chem. Soc.* **92,** 2167–2168.
White, E. H., Wildes, P. D., Wiecko, J., Doshan, H., and Wei, C. C. (1973). *J. Am. Chem. Soc.* **95,** 7050–7058.
Wieringa, J. H., Strating, J., Wynberg, H., and Adam, W. (1972). *Tetrahedron Lett.* pp. 169–172.
Wildes, P. D., and White, E. H. (1971). *J. Am. Chem. Soc.* **93,** 6286–6288.
Wilson, T., and Schaap, A. P. (1971). *J. Am. Chem. Soc.* **93,** 4126–4136.
Wilson, T., Landis, M. E., Baumstark, A. L., and Bartlett, P. D. (1973). *J. Am. Chem. Soc.* **95,** 4765–4766.
Wilson, T., Golan, D. E., Harris, M. S., and Baumstark, A. L. (1976). *J. Am. Chem. Soc.* **98,** 1086–1091.
Woodward, R. B., and Hoffmann, R. (1969). *Angew. Chem., Int. Ed. Engl.* **8,** 781–853.
Yang, N. C., and Carr, R. V. (1973). *Tetrahedron Lett.* p. 5143.
Yang, N. C., and Libman, J. (1974). *J. Org. Chem.* **39,** 1782–1784.
Zimmerman, H. E., and Keck, G. E. (1975). *J. Am. Chem. Soc.* **97,** 3527–3528.
Zimmerman, H. E., and Swenton, J. S. (1964). *J. Am. Chem. Soc.* **86,** 1436–1437.
Zimmerman, H. E., Crumrine, D. S., Döpp, D., and Huyffer, P. S. (1969). *J. Am. Chem. Soc.* **91,** 434–445.

8

Ene-Reactions with Singlet Oxygen*

K. GOLLNICK and H. J. KUHN

I. General Aspects 287
II. Liquid Phase Ene-Reactions 288
A. Determination of Rate Constants 288
B. Solvent Effects 290
C. Electronic Effects 297
D. Steric Hindrance 309
E. Conformational Effects 318
F. Secondary Products 328
G. Mechanism of the Ene-Reaction 329
III. Tabular Survey of the Ene-Reaction of Various Olefins with 342
Singlet Oxygen 342
A. Olefins with Acyclic Double Bonds 360
B. Olefins with Semicyclic Double Bonds 368
C. Olefins with Endocyclic Double Bonds . . . .
D. Miscellaneous Heterocompounds Which May Also 413
Undergo Ene-Type Reactions 419
References . . . .

I. GENERAL ASPECTS

Interaction of singlet oxygen, $^1\Delta_g\ O_2$, with alkyl-substituted olefins may result in the formation of allylic hydroperoxides with a shifted double bond as required by an ene-type reaction [6, 42, 55, 83, 90, 121]:

* Dedicated to Professor Dr. Günther O. Schenck for his 65th birthday on May 14, 1978.

Singlet Oxygen

ISBN 0-12-736650-4

The reaction was discovered by Schenck [230, 231] in 1943 and named the "Schenck reaction" by Schönberg [253] because of its usefulness in synthetic organic chemistry. Allylic hydroperoxides are easily transformed into allylic alcohols and α,β-unsaturated carbonyl compounds.

The inevitably shifted double bond clearly distinguishes the ene-type reaction from the well-known autoxidation reaction. The latter proceeds by a primary dehydrogenation of the olefin and a subsequent addition of triplet ground-state oxygen, $^3\Sigma_g^- \ O_2$, to the allylic radical.

$+R^\bullet$, $-RH$; (1) $+{}^3O_2$, (2) $+[H]$

allylic hydroperoxides with rearranged and unrearranged C=C bonds

The difference between the two types of reaction is illustrated beyond doubt, for example with (+)-limonene, α-pinene, and cholesterol, as will be discussed below.

Investigations on $^1O_2$ reactions with olefins have been carried out predominantly in the liquid phase. As a result, ene-product formation seems to be affected little by the nature of the solvent but is strongly dependent on the stereoelectronic and steric effects exerted by the olefin on the attacking electrophilic singlet oxygen.

II. LIQUID PHASE ENE-REACTIONS

A. Determination of Rate Constants

Once the singlet oxygen molecule is produced by photosensitization or any other method with an efficiency α,* it may undergo (1) solvent-induced relaxation to its triplet ground state, (2) reaction with an olefin to products, and in some particular cases (3) deactivating collisions with olefins leading to its triplet ground state.

$$^1O_2 \rightarrow {}^3O_2 \qquad k_d = 1/\tau_{^1O_2}\ (\text{sec}^{-1}) \qquad (1)$$

$$^1O_2 + \text{olefin A} \rightarrow \text{products AO}_2 \qquad k_r\ (\text{liter/mole sec}) \qquad (2)$$

$$^1O_2 + \text{olefin A} \rightarrow {}^3O_2 + \text{olefin A} \qquad k_q\ (\text{liter/mole sec}) \qquad (3)$$

The rates with which oxygen and the olefin A are consumed and the

* In the case of photosensitization, $\alpha = I_{Abs}\Phi_{^1O_2}$, with I_{Abs} = moles of photons absorbed per unit volume and unit time by the sensitizer, and $\Phi_{^1O_2}$ = the quantum yield of singlet oxygen formation [76, 83].

products AO_2 are formed are accordingly given as

$$-\frac{d[O_2]}{dt} = -\frac{d[A]}{dt} = \frac{d[AO_2]}{dt} = v = \alpha \frac{k_r[A]}{(k_r + k_q)[A] + k_d} \tag{4}$$

However, reaction (3) does not appear to play a significant role. Mono-olefins such as 2-methyl-2-butene, *cis*- and *trans*-2-butenes, limonene, α- and β-pinene, Δ^3-carene, etc., do not quench the rates of the more efficient singlet oxygen reactions with 2,5-dimethylfuran and 2,3-dimethyl-2-butene [78, 90]. Furthermore, if the concentration of 2-methyl-2-butene, e.g., is increased, the rate of oxygen consumption approaches the maximum rate possible (which is observed with 2,5-dimethylfuran at relatively low concentrations of this furan acceptor). These results show that simple olefins either do not act at all according to reaction (3) or quench singlet oxygen with efficiencies that are similar to or smaller than those obtained with solvents, i.e., $k_q[A] \leq k_d$.

The rate expression thus simplifies to

$$v = \alpha \frac{k_r[A]}{k_r[A] + k_d} = \alpha \frac{[A]}{[A] + \beta} \tag{5}$$

The "acceptor half-value concentration" [76, 83, 134, 232, 248]

$$\beta = k_d/k_r \text{ (mole/liter)} \tag{6}$$

may principally be determined by plotting $1/v$ vs. $1/[A]$ according to

$$\frac{1}{v} = \frac{1}{\alpha} + \frac{\beta}{\alpha} \times \frac{1}{[A]} \tag{7}$$

by which $1/v$ is linearly related to $1/[A]$ so that $1/\alpha$ and β/α are easily obtained from the intercept of the straight-line curve with the ordinate and from the slope of the straight line, respectively.

By use of 2,5-dimethylfuran or 2,3-dimethyl-2-butene as standard substrates, for which zero-order reactions are observed as long as their concentrations exceed about 5×10^{-3} mole/liter (i.e., $\beta \ll [A]$), α-values may precisely be determined as

$$v(\text{2,5-dimethylfuran}) = v_0 = \alpha \tag{8}$$

Thus, from a plot of v_0/v vs. $1/[A]$, β of a substrate A may directly be obtained as the slope of the straight-line curve, since

$$v_0/v = 1 + \beta/[A] \tag{9}$$

β-Values of various olefins evaluated under similar experimental con-

Table I Singlet Oxygen ($^1\Delta_gO_2$) Lifetimes in Solution at Room Temperature [148, 167, 168].

Solvent	$10^6 \times \tau_{^1O_2}$ (sec)	Solvent	$10^6 \times \tau_{^1O_2}$ (sec)
H_2O	2	Acetone	26
H_2O/CH_3OH (1:1)	3.5	Acetone-d_6	26
CH_3OH	5[a]	Acetonitrile	30
CH_3OH	7	Pyridine	33[b]
D_2O/CH_3OH (1:1)	11	t-C_4H_9OH	34[b]
C_2H_5OH	12	D_2O/CD_3OD (1:1)	35
Cyclohexane	17	Chloroform	60
n-C_4H_9OH	19[b]	CS_2	200
D_2O	20	$CDCl_3$	300
D_2O	32[c]	Hexafluorobenzene	600
Benzene	12.5[a]	CCl_4	700
Benzene	24	CF_3Cl (Freon 11)	1000

[a] From Adams and Wilkinson [4].
[b] From Young *et al.* [285].
[c] From Matheson *et al.* [154a]; cf. also Gorman *et al.* [91a].

ditions (especially for similar or identical solvents) allow the determination of relative rates, which may otherwise be obtained experimentally by competition reactions. Until recently, only relative rates were accessible. With the advent of lasers, however, it became possible to determine singlet oxygen lifetimes in solution (for selected values see Table I), which permit us to calculate absolute rate constants from β-values:

$$k_r = 1/(\beta \times \tau_{^1O_2})(\text{liter/mole sec}) \tag{10}$$

B. Solvent Effects

Ene-reactions appear to be only slightly solvent dependent as may be judged from the few examples available (see Table II) i.e., p,p'-disubstituted α,α'-dimethylstilbenes [77], 2-methyl-2-pentene [62], and citronellol [248].

The same conclusion may be drawn from the results obtained with 2,5-dimethyl-2,4-hexadiene (**1**), which is reported to give a 1,2-dioxetane (**3**) in addition to an ene-product (**2**), the ratio of which depends rather strongly on the solvent [95] (see Table III). Changing the solvent from acetonitrile to aqueous methanol increased this ratio by a factor larger than 500, whereas the relative rate, k_r(2,5-dimethyl-2,4-hexadiene)/k_R(1-methylcyclohexene), was enhanced by a factor of about five. If we assume that 1-methylcyclohexene, which affords only ene-products, reacts with singlet oxygen with an approximately solvent independent rate as do the other

Table II Rate Constants[j] of Singlet Oxygen ($^1\Delta_g$ O_2) Reactions with Olefinic Substrates in Methanol at Room Temperature

Substrate	I.P.[a] (eV)	k_r (liter/mole sec)	Calc. from Ref.
1. POLYENES			
β-Carotene		$\leq 10^7$	
		$k_q = 1.4 \times 10^{10}$	56
2. TETRASUBSTITUTED OLEFINS			
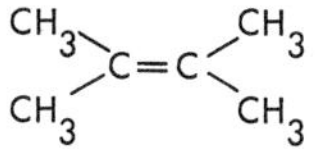 $(CH_3)_2C{=}C(CH_3)_2$ (TME)	8.30	4.7×10^7	82
	8.42	4.0×10^7	168
	8.53	1.5×10^7 [b]	78
	8.34*		
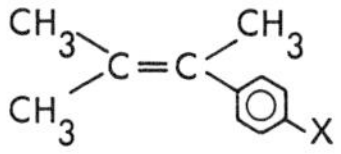 $(CH_3)_2C{=}C(CH_3)(C_6H_4X)$			
$\rho = -0.93$ e.g. X = $N(CH_3)_2$		1.6×10^7	61
= H		4.0×10^6 [c]	61
= CN		1.0×10^6 [c]	61
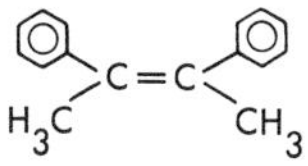 $(C_6H_5)(H_3C)C{=}C(C_6H_5)(CH_3)$		2.1×10^6	
		from $\frac{k_{cis}}{k_{trans}} = 10$	141
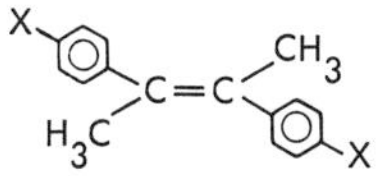 $(X{-}C_6H_4)(H_3C)C{=}C(CH_3)(C_6H_4{-}X)$			
$\rho = -1.35$ X = OCH_3		6.8×10^5	77
		2.8×10^6 [d]	77
= CH_3		2.9×10^5	77
		1.2×10^6 [d]	77
= H		2.1×10^5	77
		8.6×10^5 [d]	77
= NO_2		1.9×10^4	77
		7.6×10^4 [d]	77
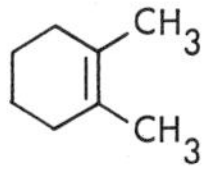 1,2-dimethylcyclohexene (CH_3, CH_3)	8.2*	1.0×10^7	168
		4.7×10^6	82
3. TRISUBSTITUTED OLEFINS			
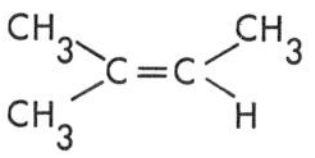 $(CH_3)_2C{=}C(CH_3)H$	8.60		
	8.68		
	8.80	2.5×10^6	82
	8.83	2.4×10^6	99
	8.85		
	8.95		
	8.82*		

(Continued)

Table II *(Continued)*

Substrate	I.P.[a] (eV)	k_r (liter/mole sec)	Calc. from Ref.
$CH_3(C_2H_5)C{=}C(CH_3)H$ (CH₃ and CH₃ on the same side)	8.69* cis-trans mixture	1.9×10^6	99
$CH_3(C_2H_5)C{=}C(H)CH_3$ (C₂H₅ and CH₃ on the same side)		1.4×10^6	99
$(CH_3)_2C{=}C(C_2H_5)H$	8.7*	7.7×10^5	82
		8.8×10^5	62
in CH_3OH/benzene (1:4)		7.7×10^5	62
$t\text{-}C_4H_9OH$		5.8×10^5	60
$CH_3OH/t\text{-}C_4H_9OH$ (1:1)		3.7×10^5	60
Pyridine		6.2×10^5	62
CS_2		6.3×10^5	62
Acetone		4.8×10^5	62
Benzene		4.2×10^5	62
Average Value:		$(6.1 \pm 1.7) \times 10^5$	
$(CH_3)_2C{=}C(CH(CH_3)_2)H$		1.1×10^5	82
$(CH_3)_2C{=}C(C(CH_3)_3)H$	8.62*	3.3×10^4	82
$(CH_3)_2C{=}C(CH_2\text{-}R)H$			
R = phenyl		1.1×10^6 [g]	245, 248
= $-CH_2-CH(CH_3)-CH_2CH_3$		7.9×10^5 [g]	245
Citronellol = $-CH_2-CH(CH_3)-CH_2-CH_2OH$	8.63*	8.8×10^5	248
		3.3×10^6 [h]	248
		4.4×10^5 [i]	248
Linalool = $-CH_2-C(CH_3)(OH)-CH{=}CH_2$		7.9×10^5 [g]	245, 248
1-Methylcyclopentene (structure)	8.65*	7.3×10^5	59

Table II (*Continued*)

Substrate	I.P.[a] (eV)	k_r (liter/mole sec)	Calc. from Ref.
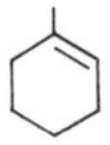	8.59*	1.2×10^5	59, 82
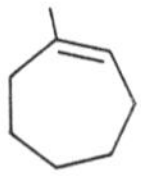	8.57*	1.1×10^6	59
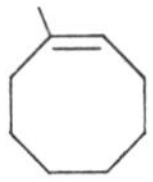	8.56*	1.8×10^5	59
 Δ^4-Carene		6.8×10^5	82
 Δ^3-Carene	8.51*	3.5×10^5	82
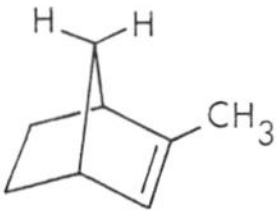 2-Methylnorbornene		7.4×10^4 [d]	112
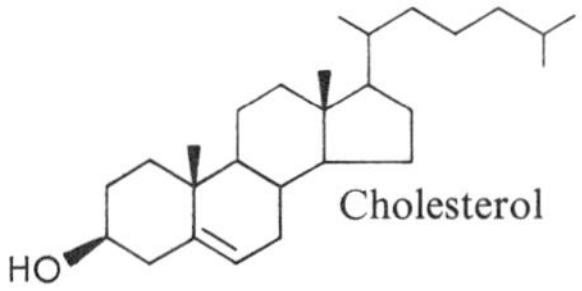 Cholesterol		3.4×10^4 [e]	236
 α-Pinene	8.31*	1.7×10^4	82
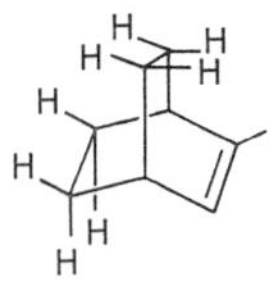		4.6×10^3	68
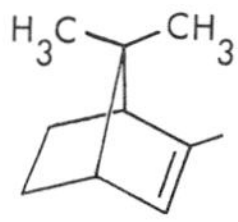 2,7,7-Trimethylnorbornene		1.6×10^3 [d]	114

(*Continued*)

Table II (*Continued*)

Substrate	I.P.[a] (eV)	k_r (liter/mole sec)	Calc. from Ref.
4. DISUBSTITUTED OLEFINS			
 cis-Δ^2-Carene		8.4×10^5	82
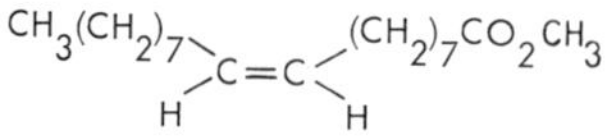 $CH_3(CH_2)_7$(H)C=C(H)$(CH_2)_7CO_2CH_3$ Methyl Oleate		7.4×10^{4} [e]	44
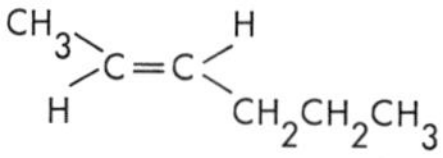 CH_3(H)C=C(H)$CH_2CH_2CH_3$	9.16	7.9×10^4	138
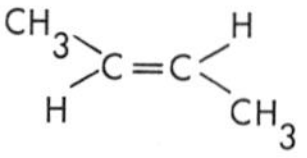 CH_3(H)C=C(H)CH_3	9.13 9.27	1.2×10^4	134
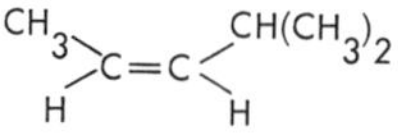 CH_3(H)C=C(H)$CH(CH_3)_2$	9.13	1.2×10^{4} [f]	99
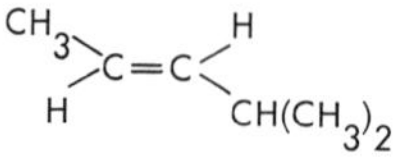 CH_3(H)C=C(H)$CH(CH_3)_2$	9.05* 9.13	2.2×10^{3} [f]	99
	9.0 9.01; 9.10 9.12* 9.20; 9.27 10.20	8.7×10^4	161
	8.72 8.95 9.01* 9.11	5.4×10^3 2.2×10^{3} [f]	82 99
	9.0*	5.5×10^3	161
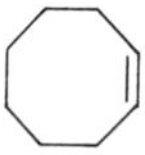	9.0*	7.6×10^2	161
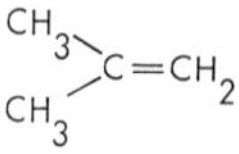 $(CH_3)_2C=CH_2$	8.95 9.23 9.35 9.41 9.58	8.9×10^4	138

Table II (*Continued*)

Substrate	I.P.[a] (eV)	k_r (liter/mole sec)	Calc. from Ref.
$\Delta^{4(10)}$-Carene		7.0×10^4	83
(methylenecyclopentane)	9.13 9.14*	$4.8 \times 10^{4\,d}$	112
2-Methylenenorbornane		$2.1 \times 10^{4\,d}$	112
7,7-Dimethyl-2-methylenenorbornane		$5.4 \times 10^{3\,d}$	114
β-Pinene	8.7*	3.7×10^3	248
5. MONOSUBSTITUTED OLEFINS			
$H_2C{=}CH{-}(CH_2)_6{-}CH_3$	9.53*	8.7×10^2	138

[a] I.P.s taken from Robin [216], Turner [272], and Vedeneyev *et al.* [273] and own unpublished (*) results (PS 16 Photoelectron Spectrometer, gas phase, vertical values given) [207]. Adiabatic I.P.s were found to be generally about 0.2 eV lower than the vertical values in our measurements. Reference: Xe = 12.13; 13.44; Ar = 15.76 eV.

[b] In methanol/H_2O 2:3.

[c] In methanol containing 2% pyridine.

[d] In acetonitrile.

[e] In pyridine.

[f] In methanol/*tert*-butanol 1:1.

[g] In methanol/isopropanol 1:1 (Assumption: $\tau\ ^1O_2 \sim 7 \times 10^{-6}$ sec).

[h] In methanol/H_2O 7:3.

[i] In *n*-butanol.

[j] For rate constants in $CHCl_3$, see [168a].

Table III Solvent Effects on the Product Distribution, Relative Rates, and Partial Rate Constants of the Singlet Oxygen ($^1\Delta_g O_2$) Reaction with 2,5-Dimethyl-2,4-hexadiene (**1**) at Room Temperature

Solvent	Product ratio 1,2-dioxetane[a]/ ene-product	k_r/k_R[a,b]	k_r[c] (liter/mole sec) (calc.)	k_{ene} (liter/mole sec) (calc.)	$k_{dioxetane}$ (liter/mole sec) (calc.)
Acetonitrile	0.01	6.3	7.6×10^5	7.6×10^5	0.08×10^5
Dichloromethane	0.1	5.0	6.0×10^5	5.4×10^5	0.6×10^5
Acetone	0.2	3.2	3.8×10^5	3.2×10^5	0.6×10^5
Acetone/water (3:1)	1.5	13	1.6×10^6	6.4×10^5	9.6×10^5
Methanol	2.6	28	3.4×10^6	9.4×10^5	24.6×10^5
Methanol/water (7:3)	5.5	29	3.5×10^6	5.4×10^5	29.6×10^5

[a] Exp. values from Hasty and Kearns [95].

[b] k_R for reaction of 1-methylcyclohexene (CH_3) $+ {}^1O_2$.

[c] Calculated by assuming $k_R = 1.2 \times 10^5$ liter/mole sec in all solvents.

olefins cited above, and if we furthermore assume that the ene-product and the dioxetane are formed via two independent pathways (Mechanism 1) rather than via a common intermediate such as a perepoxide (**4**) (Mechanism 2), partial rate constants k_{ene} and $k_{dioxetane}$ may be calculated with the result that k_{ene} varies by a factor of less than three, whereas $k_{dioxetane}$ varies by a factor of about 400. The 1,2-cycloaddition reaction thus seems to be rather dependent on the nature of the solvent, in contrast to the ene-reaction.

Mechanism 1:

1 + $^1O_2$ $\xrightarrow{k_{ene}}$ HOO– **2**

1 + $^1O_2$ $\xrightarrow{k_{dioxetane}}$ O–O **3**

$$k_r = k_{ene} + k_{dioxetane}$$

Mechanism 2:

k'_{ene} HOO 2

1 + $^1O_2$ $\xrightarrow{k_r}$ 4

$k'_{dioxetane}$ 3

It is, of course, not possible to decide which of the two mechanisms prevails. If a common intermediate were involved and if its formation would be rate-determining, the results with 2,5-dimethyl-2,4-hexadiene had to be explained, however, by assuming (1) a nearly solvent independent formation of the zwitterionic intermediate perepoxide from two neutral reactants, and (2) two rearrangement modes of the perepoxide, one of which is little (or not) solvent dependent (k'_{ene}), whereas the other ($k'_{dioxetane}$) depends rather strongly on the solvent. The latter assumption was made in order to explain dioxetane/ene-product ratios with 1-ethylthio-2-ethyl-1-hexene, EtS—CH=C(Et)C_4H_9, which vary from 1.8 in CCl_4 to 4.5 in acetonitrile and to >100 in methanol [7].

C. Electronic Effects

Hammett plots of log k_r vs. σ-values of para and meta substituents for $^1O_2$ reactions with α,β,β-trimethylstyrenes (**5**) [61] and α,α'-dimethylstilbenes (**8**) [77] in solution revealed ρ-values of -0.93 and -1.35, respectively, indicating that singlet oxygen behaves like an electrophilic species.

$$X\text{-}C_6H_4\text{-}C(CH_3)=C(CH_3)_2 \xrightarrow{^1O_2} X\text{-}C_6H_4\text{-}C(=CH_2)\text{-}C(CH_3)_2\text{-}OOH + CH_3\text{-}C(C_6H_4X)(OOH)\text{-}C(CH_3)=CH_2$$

5 **6** **7**

X = *m*-CH_3, *m*-OCH_3, *m*-Cl, *m*-CN, *p*-$N(CH_3)_2$, *p*-OCH_3, *p*-CH_3, *p*-Cl, *p*-CN, H

2.7 : 1 (ratio independent of X)

8 $\xrightarrow{+^{1}O_2}$ **9**

$X = OCH_3, CH_3, H, NO_2$

In agreement with this result, substitution of olefinic hydrogens by electron-donating alkyl groups enhances the rate of ene-reactions with singlet oxygen. Table IV is a compilation of rate parameters of gas phase and liquid phase $^{1}O_2$ reactions with olefins, for which steric factors may be neglected and the number and conformations of allylic hydrogens be ignored.

Substitution of an olefinic hydrogen by a methyl group increases the rate by a factor of about 20 in both phases, and another increase of about one order of magnitude of the rate constant is observed if we compare the most reactive olefin, 2,3-dimethyl-2-butene, with 2,5-dimethylfuran (which undergoes a 1,4-cycloaddition, i.e., a Diels–Alder reaction). Thus, the relative rates are similar in both phases.

However, the absolute rates in methanolic solution are about two orders of magnitude larger than those obtained for the corresponding gas phase reactions, which is entirely due to a larger activation energy by about 3 kcal/mole for the latter reactions [8]. For the entropy of activation, a practically constant value of -23 e.u. is observed in gas phase reactions whereas an average value of about -26 e.u. may be estimated from the few solution data, indicating a highly ordered transition state to be involved in both phases. The decrease of the activation energy when changing from the gas phase to the methanolic solution may further indicate that this transition state is somewhat polar [8]. However, whatever the mechanism of the ene-reaction with singlet oxygen finally may be, the transition state involved should resemble the starting material (olefin + $^{1}O_2$) rather than the reaction product (allylic hydroperoxide). The activation energies amount to only about 1 to 5 kcal/mole whereas the product formation is exothermic by about 45 kcal/mole.* Consequently, we shall consider the stereochemistry of the starting olefins rather than the stereochemistry of the resulting products in order to evaluate steric and conformational effects involved in ene-reactions with singlet oxygen.

Kearns [119] pointed out that the reactivity of olefins toward singlet

* Loss in bond energy (bond energies taken from Gordon and Ford [91]): $H—CH_2CH{=}CH_2 = 85$; $O{=}O = 119–23(^{1}\Delta_gO_2$ energy$) - 44 = 52$. ($HO—OH = 51$; $CH_3O—OCH_3 = 36$; thus $HO—OCH_3 \approx 44$ kcal/mole). Gain in bond energy: $HO—C_2H_5 = -92$; $H—OOH = -90$. Overall gain in bond energy of $(+137 - 182) = -45$ kcal/mole.

oxygen (measured as log k_r) increases linearly with decreasing $\pi_{C=C}$-ionization potentials of the olefins, i.e., with higher occupied $\pi_{C=C}$-orbitals. Figure 1 shows that an excellent linear correlation is observed between log k_r and the lowest ionization potentials (I.P.) (Table II) for open chain olefins and that this correlation holds also approximately for five- and seven-membered cyclic olefins. (The rate of *trans*-2-hexene appears to be too large and should be reexamined.) Cyclohexenes, however, possess distinctly lower reaction rates than would be expected from this correlation, but their log k_r values are also linearly correlated with their ionization potentials as is indicated by the dashed line in Fig. 1. The reason for this behavior as well as for the rather low reaction rates observed with α- and β-pinene will be discussed in the following paragraphs on steric and conformational effects.

A linear correlation is also observed between log k_r and the charge-transfer absorption maxima of olefins with tetracyanoethylene since the positions of these maxima parallel the olefins' π-ionization potentials [220]. Lowering of the π-ionization potential, however, may gradually open up another reaction channel, namely the 1,2-cycloaddition of $^1O_2$ to give dioxetanes (see Chapter 6). Whether the 1,2-cycloaddition does compete with the ene-reaction or not, however, seems to depend rather strongly on the solvent polarity as was discussed in the preceding paragraph.

If a molecule possesses two (or more) isolated C=C bonds of different degrees of alkyl substitution, the higher substituted double bond will prefer-

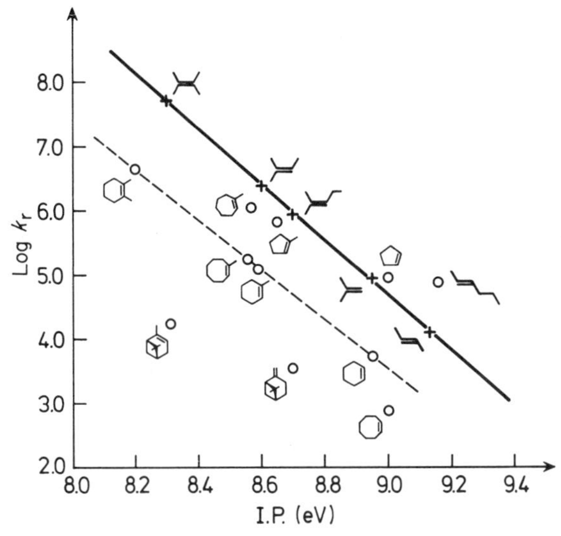

Fig. 1. Correlation between the rate of the ene-reaction of simple olefins with singlet oxygen in methanol at 20°C and the lowest ionization potentials of the olefins.

Table IV Comparison of Rate Constants of $^1O_2$ Reactions with Several Olefins in the Gas Phase and in Methanolic Solution at Room Temperature

	Gas phase[a]				Methanolic solution				
Substrate	k_r (liter/mole sec)	E_A (kcal/mole)	$\Delta S^\ddagger$ (e.u.)	k_r/k_{TME}	k_r (liter/mole sec)	E_A[b] (kcal/mole)	$\Delta S^\ddagger$ (e.u.)	k_r/k_{TME}	k_r calc. from Ref.
[c]	1.6×10^7 [d]			26	4.0×10^8	~ 0[e]	-19.1[h]	8.51	168
	7.3×10^7 [g]	1.1[g]	-21.1		1.0×10^8				99
	1.0×10^6 [d]				4.7×10^7	0.5	-23.7	$\equiv 1.00$	82
	0.6×10^6	3.2	-23.6	$\equiv 1.00$	4.0×10^7				168
	3.2×10^5	4.0	-22.4	0.53					
	4×10^5 [d]			0.67	1.0×10^7	1.3	-24.5	0.21	168
	3×10^5	4.0	-22.5	0.58	4.7×10^6		-26.0	0.10	82
	3×10^4	4.9	-24.1	0.05	2.5×10^6	1.6	-26.2	0.05	82
					2.4×10^6			0.05	99
	1.5×10^4 [d]			0.015	7.3×10^5			0.015	59
	1.2×10^4	6.0	-22.3	0.020					

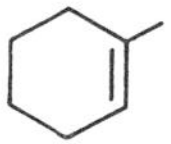	7.3×10^2	7.5	−22.9	0.001	1.2×10^5	$(2.0)^{e,f}$	(−29.0)	0.003	59, 82
	2.3×10^3	6.5	−23.9	0.004					
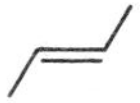	9×10^2	7.3	−23.1	0.001					
	9×10^2	7.4	−22.8	0.001	8.7×10^4			0.002	161
	$<10^2$	>8.3	(−24.2)	<0.0001	5.4×10^3	5.4	−25.8	0.0001	82

[a] Ashford and Ogryzlo [8].

[b] Koch [134].

[c] 1,4-Cycloaddition of $^1\Delta_g$ O_2.

[d] Herron and Huie [98].

[e] Schenck and Koch [243].

[f] For limonene.

[g] For furan after Hollinden and Timmons [100].

[h] The diffusion-controlled rate in methanol at 25°C is given by $k_{Diff} = 8RT/3000\eta$ or $8RT/2000\eta$. Since $\eta_{MeOH}^{25°} = 1.005 \times 10^{-2}$ Poise, $k_{Diff}^{MeOH} = 6.6 \times 10^9$ or 9.9×10^9 liter/mole sec, respectively. If these values and $\Delta H^‡ = 0$ are introduced into the expression for k_r of the Transition State Theory, a limitation value of $\Delta S^‡ = -13.7$ or -12.9 e.u. is obtained. A more realistic value of k_{Diff} in methanol for small molecules as $^1O_2$ may be 2×10^{10} liter/mole sec; the limitation value of $\Delta S^‡ = -11.5$ e.u. in this case.

entially or even exclusively be involved in the singlet oxygen reaction provided that no steric hindrance or conformational effect will counteract such a preference. A regioselective or regiospecific ene-reaction with respect to the various double bond systems available may thus be due to electronic effects.

Regiospecific ene-reactions with respect to the double bond systems* available in certain non-conjugated dienes were shown to occur with (+)-limonene (**10**) [83, 249], (+)-isosylvestrene [85], (−)-caryophyllene and (−)-isocaryophyllene [84, 85, 256], *trans,trans*-1-methyl-cyclodeca-1,6-diene (**11**) [34], α-terpinolene [132] (**12**), and germacrene [222].† Again, in each case the preference of singlet oxygen attack on the higher substituted double bond correlates with the lower $\pi_{C=C}$-ionization potential of this double bond [202].

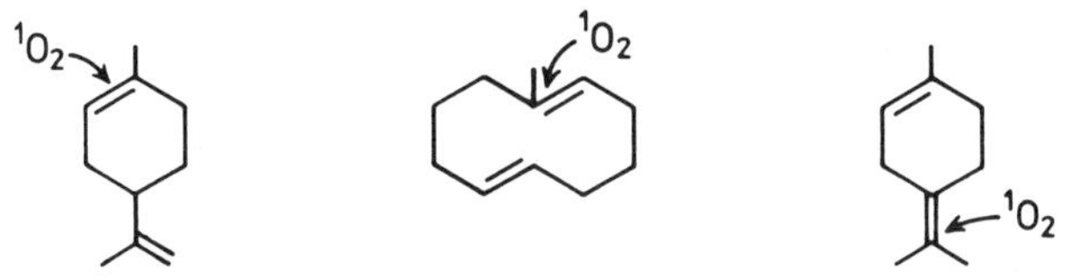

$^1O_2$ $^1O_2$ $^1O_2$

10 **11** **12**

β-Myrcene (**13**), which contains a trisubstituted isolated double bond and a conjugated diene moiety, reacts with $^1O_2$ to give two allylic hydroperoxides (**14** and **15**); the latter slowly react further with singlet oxygen via a 1,4-cycloaddition reaction to peroxides **16** and **17**, respectively.

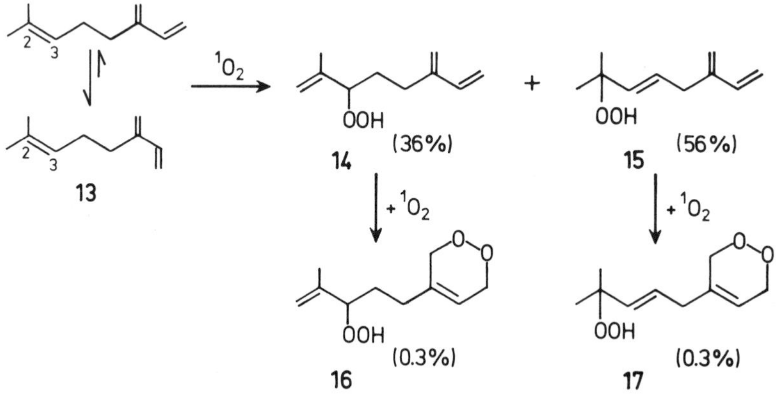

2 3 2 3 **13** $^1O_2$ OOH (36%) **14** + OOH (56%) **15** $+{}^1O_2$ $+{}^1O_2$ O O OOH (0.3%) **16** O O OOH (0.3%) **17**

α-Mycrene (**18**), however, yields only the Diels–Alder product (**19**) [155].

* See footnote * on page 303.

† Products and product distributions are given for all ene-reactions listed by us in Section III of this chapter.

$+{}^1O_2$

18 19

With respect to the double bond systems, the ${}^1O_2$ attack occurs regiospecifically on both myrcenes.* Generally, if the 1,3-diene system is forced into a cis arrangement as in cyclohexadienes, cyclopentadienes, furans, etc., it is more reactive toward an attack of singlet oxygen than an isolated double bond. The preferred attack on the trisubstituted double bond in β-myrcene is certainly mainly due to the fact that the equilibrium conformation of the diene system is practically that of a *s-trans*-1,3-diene. If, however, the isolated double bond becomes disubstituted as in α-myrcene, the rate of reaction with this double bond is so much reduced that it cannot compete with the 1,4-cycloaddition even though the equilibrium concentration of the *s-cis*-1,3-diene system is rather small.† A correlation between the reactivities and $\pi_{C=C}$-ionization potentials holds for both myrcenes in that the ionization potential of the trisubstituted double bond in β-myrcene is slightly lower by about 0.2 eV, that of the terminal double bond in α-myrcene is fairly larger by about 0.5 eV than the ionization potential of the diene moiety [202].

It is also mainly due to electronic effects that monoolefins are, in general, not overoxidized, i.e., that only one molecule of oxygen is consumed per molecule of olefin. From the preceding discussion it is apparent that this should be the case if the shifted double bond in the allylic hydroperoxide has a lower degree of alkyl substitution than the double bond of the start-

* However, with respect to the attack of ${}^1O_2$ on C-2 and C-3 of the isolated double bond in β-myrcene, the ene-reaction itself is only slightly regioselective. In order to distinguish the two regiospecificities (or regioselectivities) the term "locospecificity" ("locoselectivity") [88] may be introduced. It should replace the term "regiospecificity (regioselectivity) with respect to the double bond systems" and it will be used in this sense in Table XII.

† Note that this statement does not violate the Curtin–Hammett principle. The rates of ene-production by ${}^1O_2$ reaction with the isolated double bonds of **13** and **18** are given by k_{ene} [13] and by k'_{ene} [18]. Assume now that only 1% of **13** and **18** exists in the *s-cis*-1,3-diene form; the rates of Diels–Alder product formation by 1,4-addition of ${}^1O_2$ to the *s-cis*-diene systems of **13** and **18** are then given by $k_{DA} \times 0.01 \times$ [**13**] and $k_{DA} \times 0.01 \times$ [**18**], respectively. Thus, the ratios of [ene product]/[Diels–Alder product] are given by $100k_{ene}/k_{DA}$ in the case of **13** and by $100k'_{ene}/k_{DA}$ in the case of **18**. If k_{DA} is of the order of 10^7, $k_{ene} \approx 10^6$ and $k'_{ene} \approx 10^4$ liters/mole sec, **13** should interact with ${}^1O_2$ to give more than 90% of the allylic hydroperoxides **14** and **15**, whereas **18** should give more than 90% of the peroxide **19**.

ing olefin. In favor of a further lowering of the reactivity of the shifted double bond in the allylic hydroperoxide is the fact that an allylic OR group (R = H, OH, Ac, alkyl, etc.) reduces the electron density of the double bond (and thus enhances the $\pi_{C=C}$-ionization potential) by its negative inductive effect. This is demonstrated quite informatively with Δ^5-cholestenes (**20**) (see

C_8H_{17} $^1O_2$ Pyridine C_8H_{17} R^1 R^2 H R^3 **20** R^1 R^2 OOH R^3 **21**

Table V). Table V also shows the reduced influence of a homoallylic OR and halogen substituent. If the substituent is even further removed from the reacting double bond as, for example, in 3β-substituted $\Delta^{8(14)}$-ergostenes **22** (see Table VI) the influence is practically nil. An allylic OR group at C-7 in the α-position drastically reduces the reactivity which is certainly

Table V Influence of Allylic and Homoallylic Substituents on the Rate of Singlet Oxygen Reactions with Δ^5-Cholestenes in Pyridine at Room Temperature

20 → 21			k_r (liter/mole sec) calc. from Eisfeld [48]
R^1	R^2	R^3	
H	H	H	3.9×10^4
OH	H	H	3.4×10^4 [a]
OH	OH	H	1.4×10^3
OH	H	OH	3.0×10^3
OAc	H	H	1.3×10^4
OAc	OAc	H	No reaction
OAc	H	OAc	No reaction
OBz	H	H	1.1×10^4
OBz	OBz	H	No reaction
OBz	H	OBz	No reaction
OCH_3	H	H	2.0×10^4
F	H	H	8.9×10^3
Cl	H	H	7.6×10^3
Br	H	H	6.6×10^3

[a] From Schenck *et al.* [236].

Table VI Influence of Allylic OR Substituents on the Rate of Singlet Oxygen Reactions with $\Delta^{8(14)}$-Ergostenes in Pyridine at Room Temperature

Steroid	Substituent	k_r (liter/mole sec) calc. from Eisfeld [48]
22	$R^1 = H$	2.0×10^6
	$R^1 = Ac$	1.8×10^6
26	$R^2 = H$	1.1×10^4
	$R^2 = Ac$	2.3×10^3

in part due to steric hindrance toward the attacking singlet oxygen, exerted by the 7α-OR group on the α-side and by the angular methyl group at C-13 on the β-side of **26**.

C_9H_{19} R^1O **22** $\xrightarrow[\text{Pyridine}]{+\,^1O_2}$ C_9H_{19} RO OOH **23** R^1 = H or Ac: 76% + C_9H_{19} OOH **24** 20% + C_9H_{19} OOH **25** 4%

C_9H_{19} 17 13 3 7 AcO OR^2 **26** $\xrightarrow[\text{Pyridine}]{+\,^1O_2}$ C_9H_{19} OOH **27** R^2 = H or Ac: 50% + C_9H_{19} OOH **28** 50%

On the other hand, if the shifted double bond in the allylic hydroperoxide is alkylated to a higher degree than the double bond of the starting olefin as is the case with 3β-acetoxy-Δ^7-cholestene (**29**) [252], or if the shifted

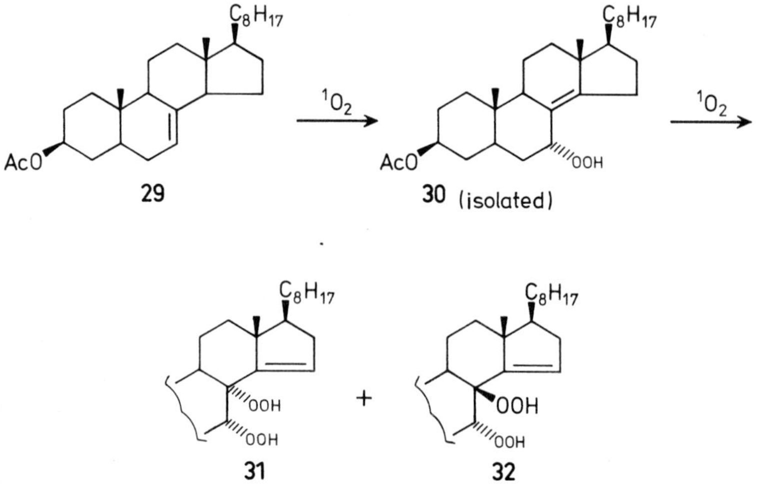

C_8H_{17} C_8H_{17}

AcO 29 $\xrightarrow{^1O_2}$ AcO OOH 30 (isolated) $\xrightarrow{^1O_2}$

C_8H_{17} C_8H_{17}

OOH OOH 31 + OOH OOH 32

33 $\xrightarrow{^1O_2}$ [OOH 34] $\xrightarrow{^1O_2}$ OOH O O 35

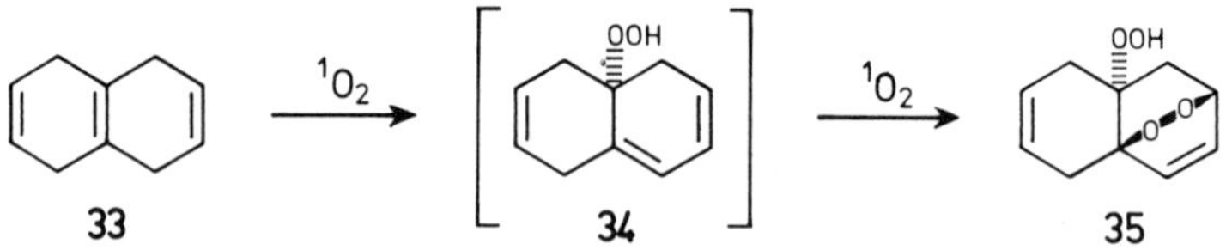

36 CO_2H $\xrightarrow{^1O_2}$ [OOH CO_2H 37] $\xrightarrow{^1O_2}$

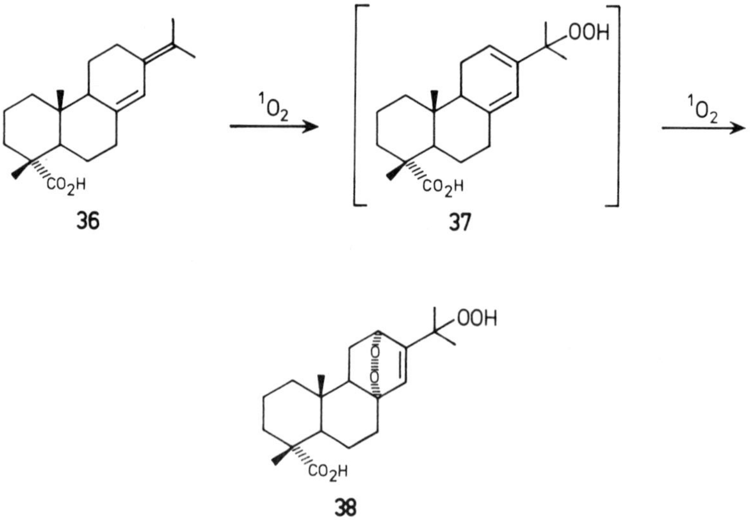

OOH O O CO_2H 38

double bond becomes part of a *cis*-1,3-diene system as occurs with isotetralin (**33**) [83, 240] and neoabietic acid (**36**) [254], the reactivity decreasing effect of the allylic hydroperoxy group may be compensated or overcompensated by the electronic effect of the newly formed double bond system so that the primary oxygenation product may not be isolable.

Another aspect of electronic effects on singlet oxygen reactions shall finally be discussed. Generally, alkyl-substituted *cis*-1,3-diene systems such as α-terpinene (**39**) undergo only 1,4-cycloaddition reactions with $^1O_2$, although allylic hydrogens as required for an ene-reaction should be readily available [81].* This result is due to an activation energy which is almost zero for the 1,4-cycloaddition and an activation entropy which appears to be reduced to a smaller negative value as compared to the activation entropies observed for ene-reactions.

OOH + Further allylic hydroperoxides ← $^1O_2$ (not observed) — **39** + $^1O_2$ → **40**

41 39 40

If there exists an equilibrium between a *s-cis*-1,3-diene system and a *s-trans*-1,3-diene system, as for example in (+)-nopadiene (**42**) [81, 246],

42 —$^1O_2$→ **43** (35%) + [**44** (HOO)] → **45** (23%) + CH_3CHO

β-ionone [170], β-carotene and carotenoids [58, 105–107, 171, 173, 271, 274], competition between 1,4-cycloaddition and the ene-reaction may occur.

Under dissolution of aromaticity, vinyl-substituted benzenes [215], indene [26], 1-vinylnaphthalenes [156], and 2-vinylthiophenes [157] may undergo 1,4-cycloaddition reactions with singlet oxygen. Substitution of the olefinic double bond may, however, alter the reaction pathway drastically. Table VII compares the competition between ene-reactions and cycloaddition reactions which occur with 1,2-dihydronaphthalenes **46** and β-methylstyrenes **51**. The most impressive effects are executed by a phenyl

* An interesting exception to the rule is discussed in the following paragraph.

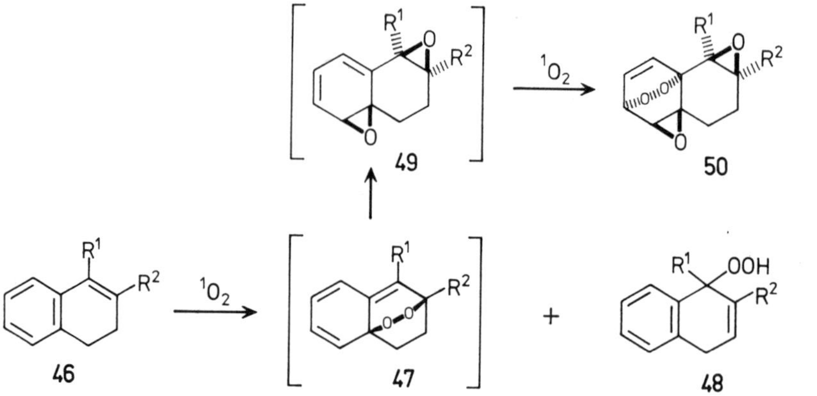

Table VII Competition of Ene-Reactions with Cycloaddition Reactions during $^1O_2$ Reactions with 1,2-Dihydronaphthalenes (**46**) and β-Methylstyrenes (**51**)

R^1	R^2	Compound	**47** (%)[a]	**48** (%)[a]	Compound	**52** (%)	**53** (%)
H	H	**46a**	75	25	**51a**	100[b,c]	0
Ph	H	**46b**	100	0	**51b**	100[c,d]	0
H	Ph	**46c**	0	100			
Ph	Ph	**46d**	65	35			
CH_3	H	**46e**	60	40	**51e**	0	100[a,e]
CH_3	Ph	**46f**	10	90	**51f**	0	100[f]

[a] Burns and Foote [27].
[b] Schenck and Gollnick [248].
[c] Gollnick [83].
[d] Schenck and Köller [240].
[e] Compared with α,β,β-trimethylstyrene **5** [61], which preferably forms the β-hydroperoxide (**6**) with the conjugated double bond.
[f] Gollnick and Kotsonis [77].

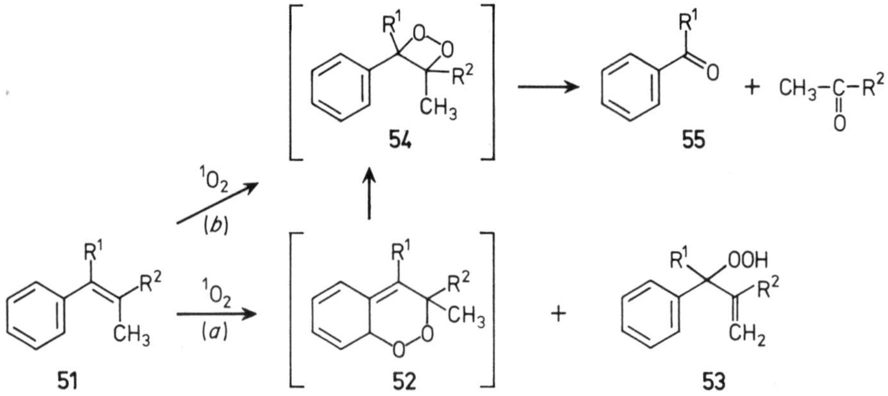

group which directs the $^1O_2$ attack toward a cycloaddition reaction,* when it is in the α-position, but toward an ene-reaction, when it is in the β-position.

Burns and Foote [27] explain their results with 1,2-dihydronaphthalenes by assuming that preferentially products are formed which possess conjugated olefinic moieties, an argument that seems to hold also for the preferred formation of the β-hydroperoxides from α,β,β-trimethylstyrenes (**5**). This assumption does not explain why there is practically no formation of α-methylene β-hydroperoxides (**56**) from **46e** and **46f**.

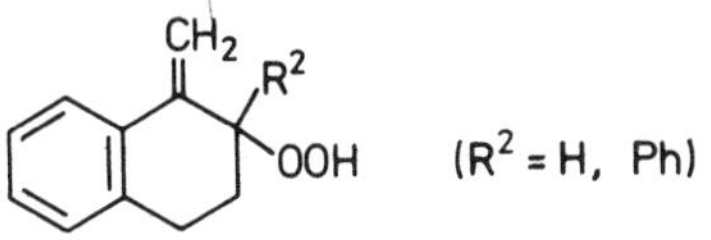

CH_2 R^2 OOH (R^2 = H, Ph)

56

If one assumes, however, that during the approach of the electrophilic singlet oxygen toward one of the double-bond carbons there is a slight electron deficiency created at the other and that the effect of a phenyl group is to stabilize this electron deficiency, the results obtained with compounds **46** as well as with compounds **51** may be rationalized. However, one would expect some influence of para substituents on the product distribution with α,β,β-trimethylstyrenes (**5**), but the ratio of **6**:**7** remains unchanged as 2.7 (see above). Furthermore, alkyl groups should be able to stabilize developing electron deficiencies, too, so that a certain Markovnikov directing effect in sterically unhindered trialkylated olefins such as 2-methyl-2-butenes, $(CH_3)_2C{=}CH{-}CH_2R$ (R = H, CH_3, Ph, etc., see Table II), should be observed; this, however, is not the case (see Table IX). The effect of aryl groups on the mode of $^1O_2$ reactions certainly deserves further attention.

D. Steric Hindrance

Bulky groups in the neighborhood of the olefinic double bond may prevent singlet oxygen from approaching this double bond and/or the allylic hydrogen that is involved in the ene-reaction. Bicyclic monoterpenes and norbornenes are appropriate substrates to demonstrate such steric effects.

* It is not known whether the cycloaddition of $^1O_2$ to β-methylstyrenes occurs as a 1,4-addition followed by a rearrangement to a dioxetane [path (*a*)] or as a 1,2-addition [path (*b*)] affording the dioxetane directly. A Hock cleavage of a primarily formed allylic hydroperoxide as assumed earlier [83], which could also account for the observed cleavage products, appears to be rather unlikely.

Δ^3-Carene (**57**) is stereospecifically attacked on the α-face of the molecule, i.e., trans to the isopropylidene bridge [79, 83]. Attack on C-3 and C-4 on the α-side occurs regio-nonselectively in agreement with the observations made with simple trialkylated olefins such as 2-methyl-2-butene and 2-methyl-2-pentene (see Table IX) and in agreement with the assumption that on the α-face of Δ^3-carene the allylic hydrogens in positions 2α, 5α, and 10 should be perfectly accessible.

57 (93%) ⇌ (7%) at 20°C [1]

$+ ^1O_2$ ↓

58 50% + 59 27% + 60 23%

k_P^α: 17×10^4 — 9.5×10^4 — $8 \cdot 10^4$ l/mole sec

α-Pinene (**61**), which may be used as a probe for the occurrence of singlet oxygen [76], is stereospecifically attacked on the α-face; in addition, this attack proceeds regiospecifically at C-3 [233, 246].

2,7,7-Trimethylnorbornene (**70b**) is stereoselectively attacked on its α-(or endo)-face, whereas 2-methylnorbornene (**70a**) preferentially experiences $^1O_2$ attack on its β-(or exo)-face [112, 114] (see Table VIII).

Let us assume that the enthalpies of activation, $\Delta H^\ddagger$, are the same for the substrates **57**, **61**, **70a**, and **70b** mentioned above and for 1-methylcycloalkenes of Table X (which, of course, may only approximately be so). Then let us compare the partial rate constants k_P^α and k_P^{endo} for the production of allylic hydroperoxides with exocyclic double bonds, i.e., 8×10^4 for **60**, 1.6×10^4 for **62**, 0.13×10^4 for **72b**, 0.10×10^4 for **72a**, and 2×10^4 for

+ $^1O_2$ → 62 (94%) + 63 (<1%)

$k_p^\alpha = 1.6 \times 10^4$ l/mole sec

61

+ R•/$^3O_2$ → 62 (14%) + 64 (33%) + 65 (9%)

+ 66 + 67 (13%)

+ 63 + 68 + 69 (31%)

70 + $^1O_2$ → 71 exo-product + 72 endo-product

$$\mathbf{73} + {}^1O_2 \longrightarrow \mathbf{74}$$

Table VIII Product Distributions and Partial Rate Constants of Singlet Oxygen Reactions with 2-Methylnorbornenes and 2-Methylenenorbornanes in Acetonitrile at Room Temperature

Compound	R,R	Product ratio[a] (exo/endo)	Partial rate constants (liter/mole sec) k_P^{exo}	k_P^{endo}
70a	H	66	6.6×10^4	0.10×10^4
70b	CH_3	0.19	0.026×10^4	0.13×10^4
73a	H	28	2.0×10^4	0.07×10^4
73b	CH_3	0.67	0.2×10^4	0.32×10^4

[a] From Jefford *et al.* [112, 114].

118*. Obviously, only the two 2-methylnorbornenes, **70a** and **70b**, exercise some steric hindrance toward a reaction with singlet oxygen on their α-faces, and this steric hindrance should be executed by the two endo hydrogens at C-5 and C-6. If we now consider the partial rate constants k_P^{exo}, i.e., 0.026×10^4 for **71b** and 6.6×10^4 for **71a**, the latter appears to have a similar value as the k_P^{α} rate constants discussed above, which indicates that a hydrogen at C-7 syn to the double bond in 2-methylnorbornene (marked in the stereochemical formula **70c**) has no significant steric effect on the approaching singlet oxygen, as is true for the hydrogen at C-6 of α-pinene (marked in the stereochemical formula **61a**). On the other hand, if the 2-methylnorbornene carries a syn methyl group at C-7, k_P^{exo} is reduced by a factor of about 250.

* The (average) value of 4×10^4 liter/mole sec given in Table X has to be divided by 2 in order to account for the fact that with 1-methylcycloalkenes, which presumably do not exert any significant steric hindrance towards an attack of singlet oxygen, both faces of these molecules are equivalent with respect to the formation of allylic hydroperoxides with exocyclic double bonds.

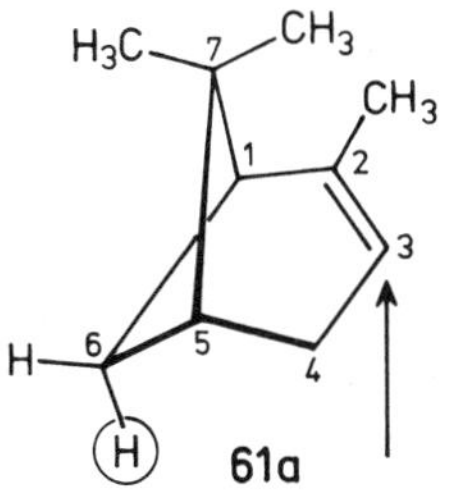

61a 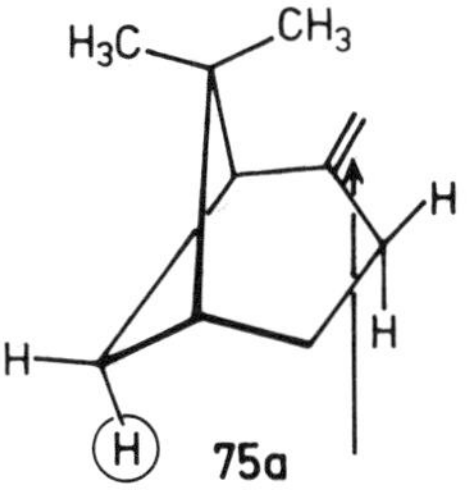 75a

α-Pinene: $k_P^\alpha = 1.6 \times 10^4$ l/mole sec β-Pinene: $k_P^\alpha = 0.37 \times 10^4$ l/mole sec

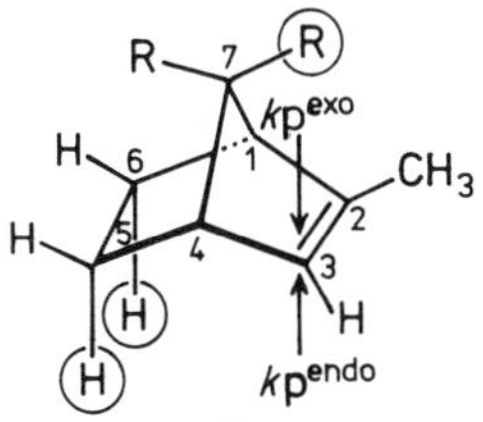

70c 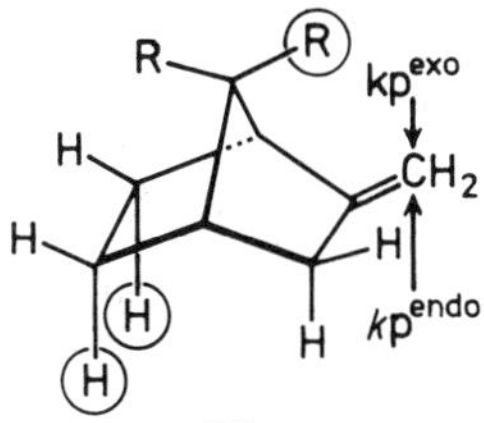 73c

2-Methylenenorbornanes (**73a** and **73b**), deuterated in either the 3-exo or the 3-endo position, provided the exo/endo ratios given in Table VIII, after the intermolecular isotope effects of 1.14 and 1.02, respectively, were taken into consideration [114]. Again, if the molecule was unsubstituted at C-7, the exo face was preferentially attacked, whereas 7,7-dimethyl substitution enforced the endo attack. Let us compare the partial rate constants k_P^{endo} and k_P^{exo} of these compounds with the partial rate constants k_P determined for β-pinene (**75**) [233, 248], $\Delta^{4(10)}$-carene (**77**) [80, 83], and methylenecyclopentane [112].*

If k_P (methylenecyclopentane) = 5×10^4 liter/mole sec is taken as a reference, singlet oxygen attack on $\Delta^{4(10)}$-carene (**77**) as well as on the exo-face of 2-methylenenorbornane (**73**) appears to occur unhindered. In addition, since k_r-values for 1-methylcyclopentene and methylenecyclopentane (see Table II) differ by a factor of less than four, the enthalpies of activation, $\Delta H^\ddagger$, may even be comparable between the systems containing trisubstituted double bonds and those containing exocyclic double bonds. Provided that this is the case for the above mentioned substrates, we may conclude (1) that the singlet oxygen attack on $\Delta^{4(10)}$-carene (**77**) occurs preferentially

* The k_r value for methylenecyclopentane, 2×10^5 liter/mole sec, has to be divided by 4 in order to account for the statistics.

CH_2OOH
$+ {}^1O_2$
$kp = 0.37 \times 10^4$ l/mole sec
75 76

CH_2OOH CH_2OOH
$+ {}^1O_2$
+
75% 25%
kp: 5.3×10^4 1.8×10^4 l/mole sec
77 78 79

on its α-face, (2) that the endo hydrogens at C-5 and C-6 of 2-methylenenorbornane (**73a**) and 2-methylene-7,7-dimethylnorbornane (**73b**) hinder an endo reaction with ${}^1O_2$ to the same extent as they do in the case of the two 2-methylnorbornenes (**70a** and **70b**), and (3) that this hindrance is executed towards the approaching singlet oxygen rather than towards the developing C–O bond. The latter conclusion may be drawn since, although an interaction between the endo hydrogens and the developing C–O bond might be expected to occur with the 2-methylnorbornenes (**70a**, **70b**), it seems rather unlikely to take place with the 2-methylenenorbornanes (**73a**, **73b**).

As expected, a syn methyl group at C-7 of 2-methylenenorbornane (marked in the stereochemical formula **73c**) hinders an exo attack, although to a lesser extent than in 2,7,7-trimethylnorbornene (**70b**). The k_P value of β-pinene is, however, smaller than expected by a factor of about five. In our opinion, this result indicates that our assumption of equal activation enthalpies for the various systems is indeed only approximately true since a steric hindrance on the α-face of β-pinene appears to be rather unlikely in view of the results obtained with α-pinene.

Many naturally occurring olefins contain the 10-methyl-$\Delta^{1(9)}$-octalin system as a rather rigid structural element. As 10-methyl-$\Delta^{1(9)}$-octalin itself (**80**) [83, 242], steroidal olefins such as cholesterol (**83**) react with singlet oxygen practically stereospecifically on the α-face of the ring system [83,

CH_3
8 9 1
80
≡
H CH_3 H H H
$+ {}^1O_2$
CH_3
OOH
81
but no
CH_3 OOH H
82
CH_3 CH_3
HO 5
83
≡
HO H H H CH_3 CH_3 C_8H_{17} H H H
$+ {}^1O_2$
CH_3 CH_3 C_8H_{17}
HO OOH
84
$+ R\cdot/{}^3O_2$
CH_3 CH_3 C_8H_{17}
HO OOH
85
+
CH_3 CH_3 C_8H_{17}
HO OOH
86
+
CH_3
HO O
87
+
CH_3
O
88

182, 183, 236].* The stereospecific (or at least stereoselective) oxygenation of a great many of the substrates reported in Section III is certainly due to steric inhibition exercised by an axial methyl group towards a singlet oxygen approach on the β-face of these molecules, as well as towards the development of an axial C–O bond that experiences a 1,3-diaxial interaction with this axial methyl group.

Steric effects may alter the reaction pathway. When 1-methyl-4a,5,6,7,-8,8a-*trans*-hexahydronaphthalene (**89**) was reacted with singlet oxygen, a 1,4-cycloaddition reaction took place to over 70%. If, however, the 4*a*-methyl derivative (**91**) was applied, the Diels–Alder reaction was entirely replaced by an ene-reaction which occurred exclusively on the α-face of the molecule [224]. Whereas the absence of any reaction with $^1O_2$ on the β-face is certainly due to the steric effect of the axial 4*a*-methyl group, the evasion of the singlet oxygen reaction from the 1,4-cycloaddition reaction to the ene-reaction

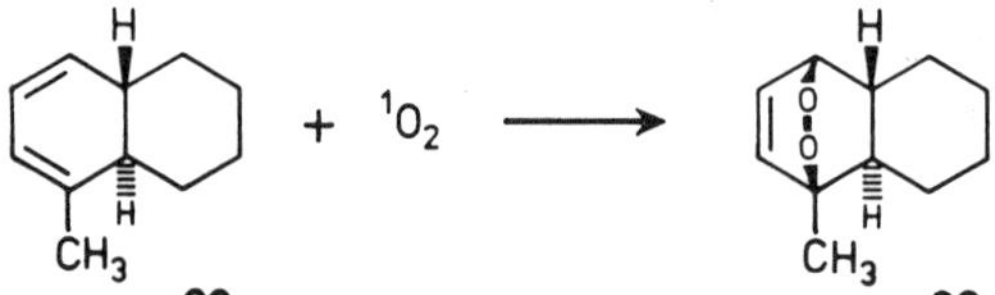

H H CH3 89 + $^1O_2$ → H O O H CH3 90

CH3 H CH3 91 + $^1O_2$ → CH3 HOO CH3 92 (85-90%) + CH3 HOO H CH2 93 (10-15%)

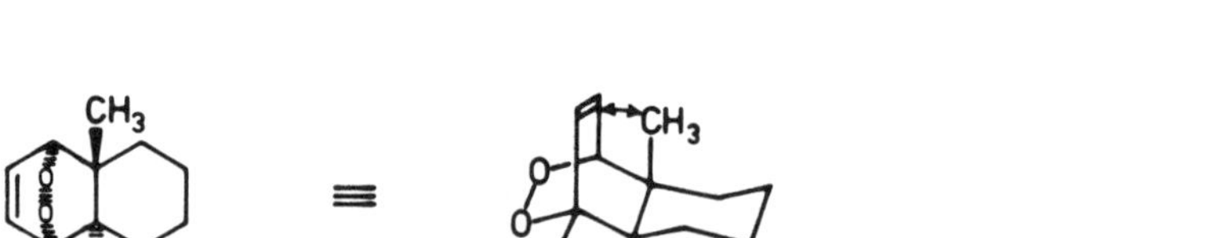

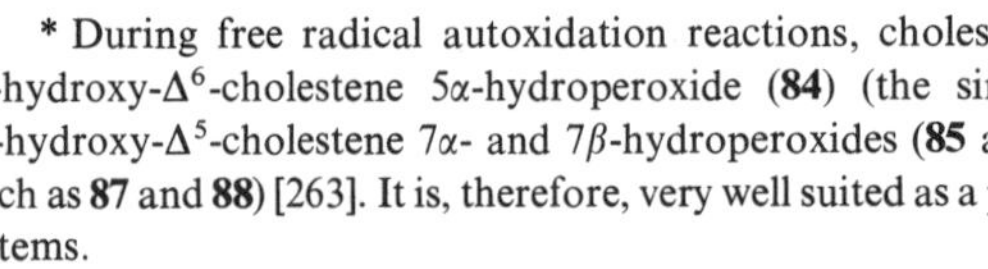

CH3 H CH3 94 ≡ CH3 O O CH3 H

* During free radical autoxidation reactions, cholesterol does not produce a trace of 3β-hydroxy-Δ^6-cholestene 5α-hydroperoxide (**84**) (the singlet oxygen product) but rather 3β-hydroxy-Δ^5-cholestene 7α- and 7β-hydroperoxides (**85** and **86**) and some ensuing products (such as **87** and **88**) [263]. It is, therefore, very well suited as a probe for singlet oxygen in oxidizing systems.

on the α-face is attributed to a severe steric 1,3-interaction in the Diels–Alder adduct (**94**) [224], although such adducts can be enforced with stronger electrophiles than singlet oxygen.

An interesting competition between steric and electronic effects in determining the stereospecificity of the singlet oxygen reaction with olefins **95** and **97** and with the conjugated diene (**96**) was recently observed by Paquette *et al.* [203, 204].

From the product distribution with olefin **95**, the less congested face appears to be the one which is trans to the cyclopropane ring, i.e., $k^{\alpha} \gg k^{\beta}$. In diene **96**, where the ethylene bridge is replaced by a hydrazide moiety, the same stereospecificity is observed, and thus $k^{\alpha}_{DA} \gg k^{\beta}_{DA}$. With olefin **97**, however, only the product which occurs via a β-attack of singlet oxygen is obtained, although the environment above and below the double bond

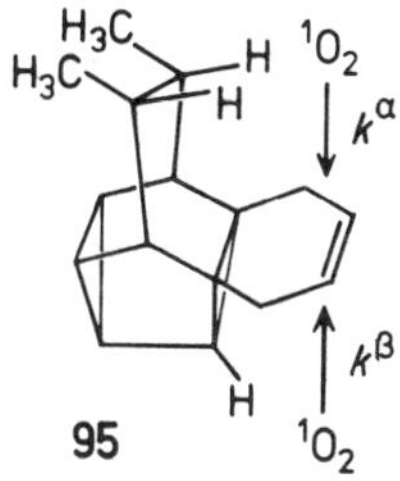

α-Product / β-Product

95 100 : 0

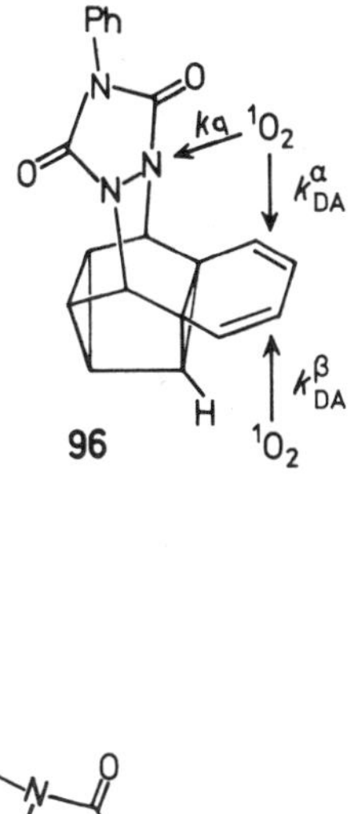

96 100 : 0

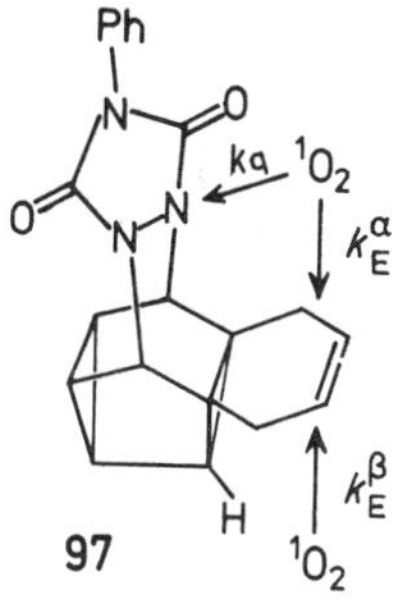

97 0 : 100

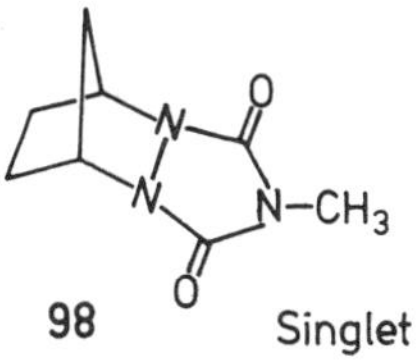

98 Singlet oxygen quencher

is the same as in diene **96**. Because of the latter statement, one would predict that $k^{\alpha}_{E} \gg k^{\beta}_{E}$. Amines are known to quench $^{1}O_2$ to $^{3}O_2$ rather efficiently [90, 192, 201, 241, 286] and so does the hydrazide (**98**) [203]. Thus, when $^{1}O_2$ approaches the α-face of **96** and **97**, it may experience quenching to its triplet ground state in competition with reaction to a Diels–Alder adduct or an ene product, respectively. One may therefore conclude, that the order of decreasing reactivity for diene **96** is given by $k_{DA}^{\alpha} > k_q > k_{DA}^{\beta}$, whereas the order of decreasing reactivity for olefin **97** should be $k_q > k_E^{\alpha} > k_E^{\beta}$, so that with the latter compound only approaches of $^1O_2$ towards the sterically more hindered β-face will remain successful in producing ene products. Paquette *et al.* [204] have related the observed product ratios to the lowest ionization potentials of the diene, hydrazide, and ene moieties (which increase in this order), and since the interaction between $^1O_2$ and a substrate decreases with increasing ionization potentials [121], the following order of reactivities is indicated: $k_{DA}^{\alpha} > k_q > k_E^{\alpha}$, which is indeed the order of reactivities observed with 1,3-cyclohexadienes, amines, and cyclohexenes (for a compilation of such data, see Gollnick [90]).

E. Conformational Effects

The conformation of an olefin may alter the reaction pathway as was discussed for the *s-cis-* and *s-trans*-1,3-diene system. Conformations may also determine whether, for example, a methyl group exerts steric hindrance (as an axial group in the neighborhood of a double bond) or not (as an equatorial group).

The conformational effects we shall discuss now, however, will deal with the conformation of the allylic hydrogen that is involved in the ene-reaction. As was already mentioned, cholesterol (**83**) reacts with $^1O_2$ to 3β-hydroxy-Δ^6-cholestene 5α-hydroperoxide (**84**) as practically the sole product [236]. Using cholesterol which was either deuterated in the 7α- or 7β-position, Nickon and Bagli [182, 183] showed that the formation of the 5α-hydroperoxide involved the 7α-hydrogen, i.e., that allylic hydrogen which is located on the same face of the molecule where the C–O bond formation takes place. From this and other experiments with steroids, the authors postulated a cis cyclic, but not necessarily concerted, mechanism in which $^1O_2$ approaches the π system perpendicular to the double bond

HO Rβ Rα **83** $\xrightarrow[\text{Pyridine}]{^1O_2}$ HO OOH R **84**

	R_α	R_β	**84**: R (%)
83a	D	H	91.5 H, 8.5 D
83b	H	D	95 D, 5 H

plane and, in addition, preferentially uses that allylic hydrogen which is oriented approximately orthogonal to the olefinic plane [42].

Such an orientation does not only provide for a closest approach between the allylic hydrogen and the oxygen atom which is not involved in C–O bonding, but also for a maximum overlap of the p orbitals of the developing new $\pi_{C=C}$ bond.

An approximately 1:1 mixture of secondary and tertiary allylic hydroperoxides is obtained from 1,1-dimethyl-2-alkylethylenes of the general formula **103** (see Table IX, [83] and Section III). In all cases, a trans-substituted double bond is created in the tertiary allylic hydroperoxide **104** which is easily explained by applying the postulate of the preferred participation of "axial" allylic hydrogens in ene-reactions with singlet oxygen.

$$+\ {}^1O_2 \longrightarrow$$

(**99**) $R^1 = R^2 = R^3 = H$
(**100**) $R^1 = CH_3, R^2 = R^3 = H$
(**101**) $R^1 = R^2 = CH_3, R^3 = H$
(**102**) $R^1 = R^2 = R^3 = CH_3$;

Table IX Effect of Allyl–H Substitution by Methyl Groups on Product Distributions and Partial Rate Constants of the Singlet Oxygen ($^1\Delta_g$) Reactions with 1,1-Dimethyl-2-alkylethylenes in Methanol at Room Temperature

Compound	Tertiary hydroperoxide		Secondary hydroperoxide	
	%[a]	k_p[b] (liter/mol sec)	%	k_p (liter/mol sec)
2-Methyl-2-butene (**99**)	54	140×10^4	46	110×10^4
2-Methyl-2-pentene (**100**)	55	42×10^4	45	35×10^4
2,4-Dimethyl-2-pentene (**101**)			>95	10.5×10^4
2,4,4-Trimethyl-2-pentene (**102**)			100	3.3×10^4

[a] Product distribution from Murray [176] and Schenck and Schulte-Elte [245].
[b] k_p = partial rate constant calc. from k_r values of Table II.

Less than 5% of the tertiary allylic hydroperoxide **107** is obtained when 2,4-dimethyl-2-pentene (**101**) is reacted with singlet oxygen. The olefin appears to assume conformation **101a** rather than **101b** at room temperature so that $^1O_2$ attack on C-2 to give **107** cannot compete with $^1O_2$ attack on C-3 which affords the secondary allylic hvdroperoxide **108** [76, 83]. The partial rate constants for the formation of the secondary allylic hydroperoxides of Table IX decrease by a factor of about three for each allylic hydrogen at C-4 in 2-methyl-2-butene (**99**) substituted by a methyl group. This indicates that steric hindrance is exerted by these additional methyl groups on the reaction of singlet oxygen at C-3 of olefins **100**, **101**, and **102**.

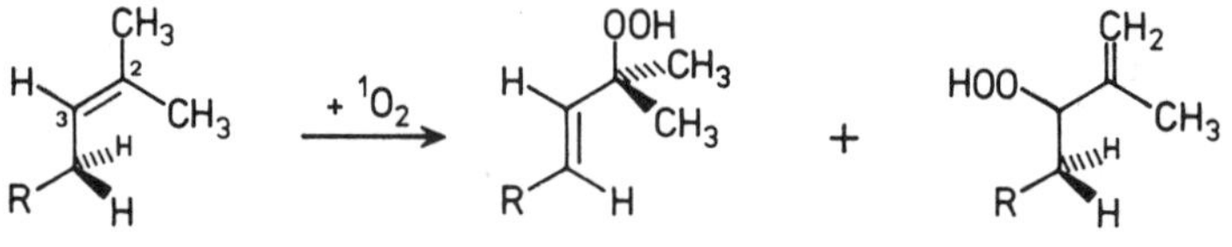

103 + $^1O_2$ → 104 + 105

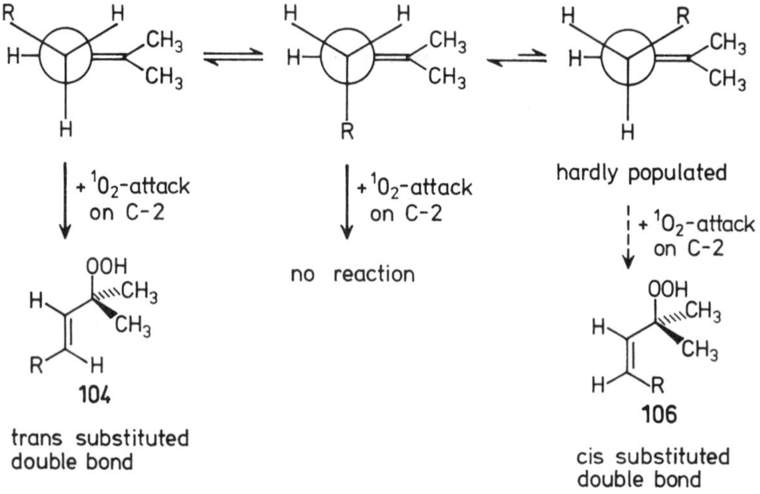

+ $^1O_2$-attack on C-2 → 104, trans substituted double bond

+ $^1O_2$-attack on C-2 → no reaction

hardly populated; + $^1O_2$-attack on C-2 → 106, cis substituted double bond

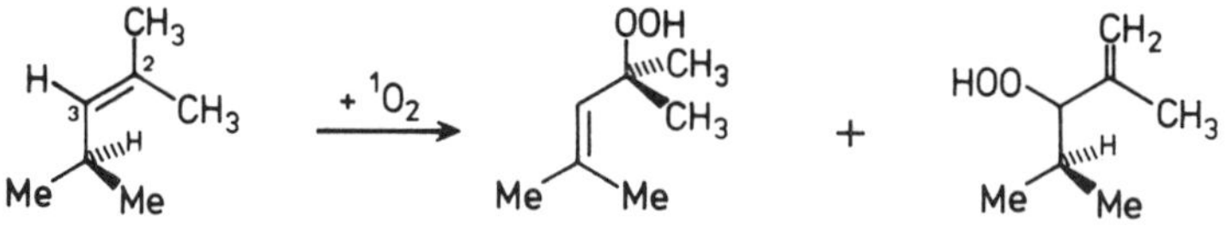

101 + $^1O_2$ → 107 <5% + 108 >95%

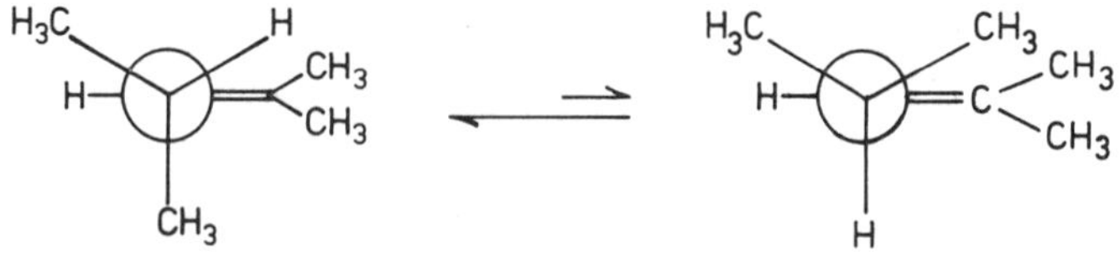

101a ⇌ 101b

In agreement with the observations made with 2,4-dimethyl-2-pentene (**101**), the singlet oxygen reaction with (*R*)(−)-*cis*-2-deuterio-5-methyl-3-hexene (**109**) affords mainly the allylic hydroperoxide **111** (94%) rather than **110** (6%) because, for the formation of **110**, the unfavorable allylic hydrogen at C-5 has to be used [267]. Less than 1% of the allylic hydroperoxide **112** with a cis-substituted double bond is formed, and furthermore, if the reaction is carried out in acetone, the main hydroperoxide **111** consists of a 1:1 mixture of **111a** and **111b**. These results are in accord with the assumption that the equilibrium conformation of **109** at room temperature is represented by **109a** ⇄ **109b** rather than by any other conformation (such as **109c** and **109d**), and that those allylic hydrogens are preferentially abstracted during the ene-reaction which approach a position orthogonal to the olefinic plane.*

109 —($^1O_2$, Acetone)→ 110 6% + 111 94% + 112 <1%

$^1O_2$-Attack on C-4:

109a ⇄ 109b ⇄ 109c ⇄ 109d

111a 111b 112a 112b

< 1%

* Note that this assertion does not violate the Curtin–Hammett principle: assume that the equilibrium is maintained during the $^1O_2$ reaction with **109,** and that, furthermore, the reaction constants for **109a → 111a, 109b → 111b,** etc., are practically the same (i.e., we neglect the small isotope effect). In this case, the ratio of products corresponds to the ratio of conformations **109a** through **109d.**

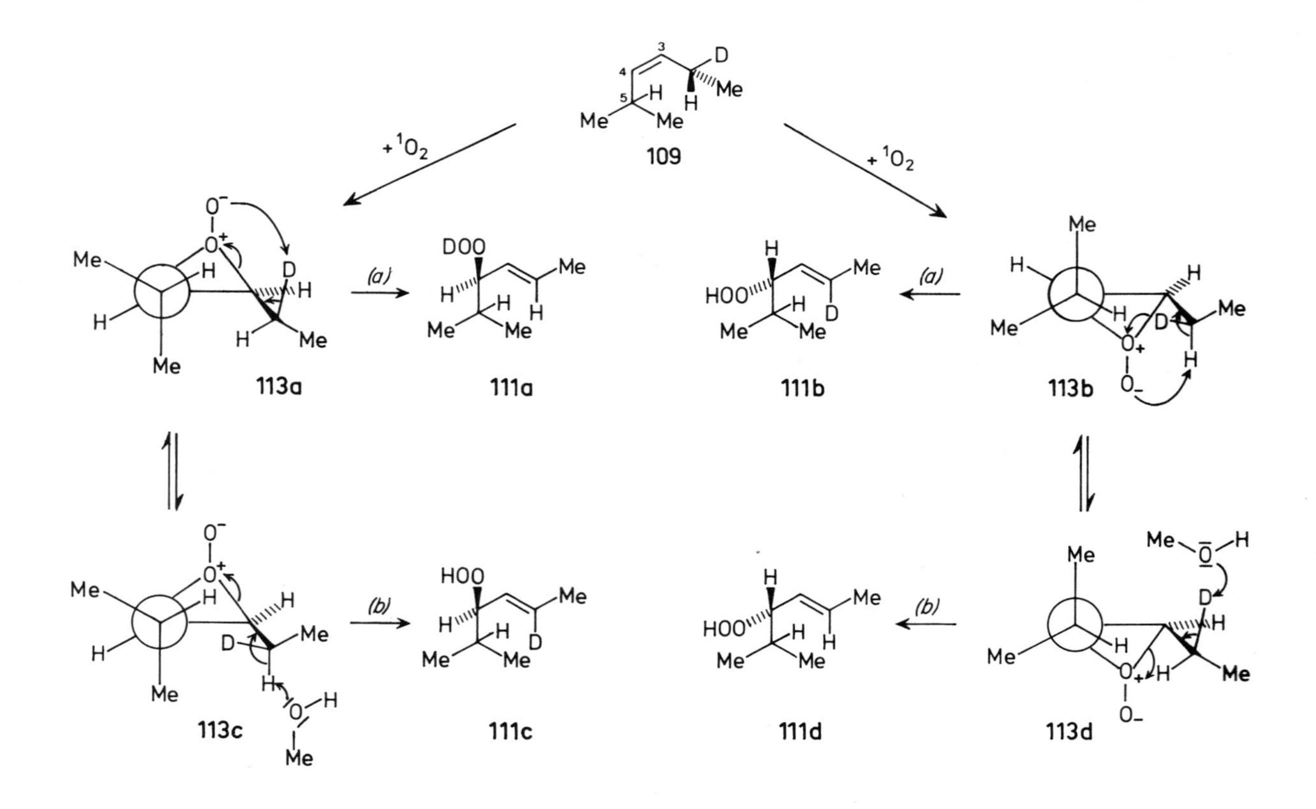

109
+ $^1O_2$
+ $^1O_2$
113a
111a
111b
113b
(a)
(a)
113c
111c
111d
113d
(b)
(b)
DOO
HOO
HOO
HOO

In methanol, the allylic hydroperoxide **111** consists of a 2:3:1 mixture of **111a**, **111b**, and **111c** [267]. The authors explain this result by suggesting the intermediacy of perepoxides **113a–d**, of which **113c** may rearrange to **111c** by a simultaneous opening of the perepoxide ring and an allylic hydrogen abstraction by the solvent methanol. The absence of hydroperoxide **111d** is explained by assuming that a rather remarkable deuterium isotope effect is operative in hydrogen abstraction by methanol [reaction pathways (*b*)], in contrast to the competing intramolecular allylic hydrogen abstraction [reaction pathways (*a*)], which thus leads to an enhanced formation of **111b** as compared to the formation of **111a** in methanolic solution.

However, we may easily explain the results obtained in methanolic solution by invoking a radical reaction to take place in addition to the singlet oxygen reaction, especially since **109** is expected to be relatively unreactive toward an attack of $^1O_2$ (see Table II, *cis*-4-methyl-2-pentene). Due to the deuterium isotope effect, *H*-abstraction rather than *D*-abstraction from C-2 of **109** should occur, generating radicals **114a** and **114b**. The equilibrium composition should consist mainly of **114a** which gives rise to a 1:1 mixture of **111c** and **111b** (**114b** would lead to isomers of **112a** and **112b**). Thus, the singlet oxygen reaction should contribute 68% of the product **111** (34% **111a** + 34% **111b**), whereas the radical reaction should contribute 32% (16% **111b** and 16% **111c**).*

109 —(+ R•, (−RH))→ 114a ⇌ 114b

114a —(1) + $^3O_2$; 2) + [H])→ 111c + 111b

1:1

* For a further argument against the perepoxide intermediacy see Section II,G.

Let us now consider simple cyclic olefins such as 1-methylcycloalkenes (**115**), which yield allylic hydroperoxides with endocyclic (**116**, **117**) and exocyclic double bonds (**118**), the ratio of which depends rather strongly on the ring size as do the rate constants for the overall reactions, k_r, and the partial rate constants, k_p (see Table X).

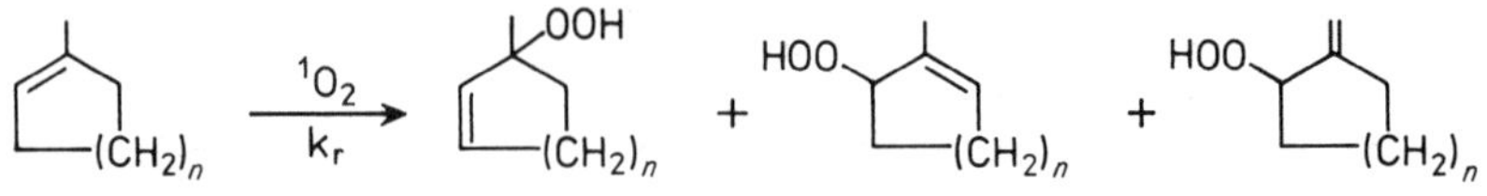

115 116 117 118

Table X Effect of Ring Size on the Product Distribution, Rate Constants, and Partial Rate Constants of the Singlet Oxygen ($^1\Delta_g$ O_2) Reaction with 1-Methylcycloalkenes in Methanol at Room Temperature

				116		117		118	
Compound	n	I.P.[a]	k_r (liter/mole sec)	Yield (%)[b]	k_p (liter/mole sec)	Yield (%)[b]	k_p (liter/mole sec)	Yield (%)[b]	k_p (liter/mole sec)
115a	1	8.65	7.3×10^5	53	39×10^4	43	31×10^4	4	3×10^4
115b	2	8.59	1.2×10^5	36	4×10^4	20	2×10^4	44	5×10^4
115c	3	8.57	1.1×10^6	48	53×10^4	48	53×10^4	4	4×10^4
115d	4	8.56	1.8×10^5	31	6×10^4	42	8×10^4	27	5×10^4

[a] From Potzinger and Kuhn [207].
[b] From Foote [59].

Since no steric hindrance may be expected to be exerted by these olefins on the attacking singlet oxygen, and furthermore since their ionization potentials are rather similar, the ring size effect appears to be mainly due to conformational effects. From partial rate constants it is apparent that the formation of allylic hydroperoxides, which involves the allylic hydrogens of the methyl group, is practically independent of the ring size. With 1-methylcyclohexene (**115b**), the ratios of the partial rate constants approximately agree with those evaluated by assuming that cyclohexenes acquire a twisted conformation (**119**) and that quasiaxial (a′) allylic hydro-

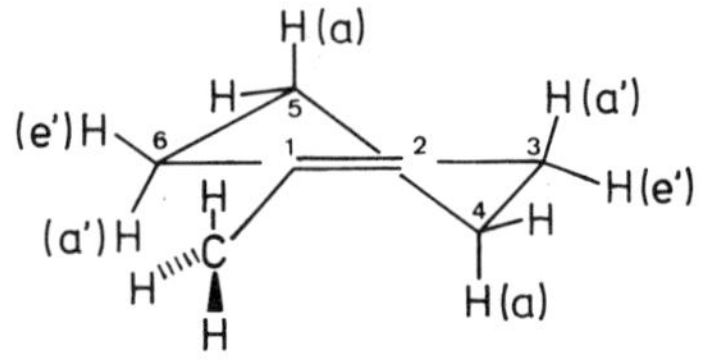

119

gens are better suited than quasiequatorial (e′) allylic hydrogens for the ene-reaction with singlet oxygen.

A comparison of 1-methylcyclopentene (**115a**) with 1-methylcycloheptene (**115c**) shows that the latter reacts a little faster with $^1O_2$ which may be attributed at least partly to its slightly lower ionization potential. Furthermore, the cycloheptene ring is rather flexible as compared with cyclohexene and the rather rigid cyclopentene ring. Therefore, the availability of the endocyclic allylic hydrogens of **115c** and **115a** as compared with those of **115b** (see stereochemical formula **119**) may contribute to the enhanced partial rate constants for the production of **116c** and **117c** as well as of **116a** and **117a**, respectively. In addition, a steric effect may lower the partial rate constants for the formation of **116b** and **117b** from 1-methylcyclohexene: if we consider the $^1O_2$ attack on C-1 (C-2) of **119** so that the quasiaxial (a′) hydrogen at C-3 (C-6) is used, the axial (a) hydrogen at C-5 (C-4) may produce a certain steric effect on the approaching $^1O_2$ but probably an even larger steric effect on the developing C—O bond at C-1 (C-2). Thus, a 1,3-diaxial interaction in the transition state and a conformational effect of the allylic hydrogens, operative in singlet oxygen reactions with cyclohexenes, add up to result in a lowering of the overall and partial rate constants by a about one order of magnitude (compare also k_r values of cyclopentene and cyclohexene, Table II).

(+)-Limonene (**120**) is very well suited as a probe for singlet oxygen [54, 83, 85, 87, 249], (see Table XI), since the product compositions obtained from $^1O_2$ reactions and free radical autoxidation reactions [251] are rather distinct.

120 (+)

1) + $^1O_2$
2) Reduction

121 (+) + 122 (+) + 123 (−) + 124 (−) + 125 (+) + 126 (+)

Table XI $^1O_2$ Reaction with (+)-Limonene[a]

Sensitizer	E_T (kcal/mole)	**121** (%)	**122** (%)	**123** (%)	**124** (%)	**125** (%)	**126** (%)	$[\alpha]_D^{20°}$ of **124**
Benzophenone	68.5	27	27	7	18	8	13	−18
Triphenylene	66.5	38	10	3	8	20	21	−141
Quinoline	62.0	33	12	4	10	20	21	n.d.
Pyrene	48.7	36	9	4	9	20	22	−176
Rose Bengal	39.5–42.2	34	10	5	10	21	20	−178
Methylene blue	34.0	36	10	4	9	21	19	−178
Microwave discharge of								
$^3O_2$		41	9	3	7	22	18	
CO_2		30	12	9	13	28	8	
$NaOCl + H_2O_2$		34	9	7	9	24	18	−131

[a] Photosensitized oxygenation in methanol at 20°C: from Gollnick *et al.* [85]. Microwave discharges, methanolic solutions at −50°C; from Gollnick and Schade [87]. Sodium hypochlorite–H_2O_2, aqueous alcoholic solutions at 20°C; from Foote *et al.* [54].

$^1O_2$ attacks (+)-limonene locospecifically at the trisubstituted double bond since its ionization potential (8.60 eV) is lower than that of the disubstituted double bond (9.05 eV). There is some stereo- and regioselectivity occurring, mainly if not exclusively due to the availability of the allylic hydrogens. Thus, as with 1-methylcyclohexene (**115b**) itself, the formation of secondary hydroperoxides with exocyclic double bonds (see the reduction products **125** and **126**) proceeds with an overall yield of about 40%. Alcohols **125** and **126** are obtained in a 1:1 ratio indicating that the isopropylene group at C-4 does not exert a steric effect toward an $^1O_2$ attack. The same conclusion is drawn from the fact that the tertiary alcohol (**121**) is preferably produced as compared with the tertiary alcohol (**122**), although the $^1O_2$ attack occurs on the β-face of **120** (i.e., cis to the isopropylene group). The 3:1 ratio of alcohols **121**/**122** and the 2:1 ratio of alcohols **124**/**123** thus depend on the availability of the allylic hydrogens as has been extensively discussed in this paragraph. The isolation of alcohol **124** may furthermore serve to ascertain that a singlet oxygen reaction rather than a free radical reaction takes place; starting with (+)-limonene, the *trans*-carveol **124** has to have a highly negative specific rotation if generated by a singlet oxygen reaction (see Table XI), whereas **124** has to be racemic if produced in a free radical autoxidation reaction. The results obtained

CH_3

+R• at C-3   3   6   +R• at C-6

127   120 (+)   128

1) $+ {}^3O_2$
2) + [H]
3) Reduction

1) $+ {}^3O_2$
2) + [H]
3) Reduction

OH   +   OH   OH   +   OH

121 (+) 12%   122 (+) 11%   123 racemic   124 racemic

19%   22%

O   O

130 13%   129 racemic 13%

with benzophenone as a typical ${}^3(n, \pi^*)$-sensitizer thus show that, in addition to a singlet oxygen reaction, a free radical autoxidation is operative (see Table XI). Obviously, hydrogen abstraction by the ${}^3(n, \pi^*)$-sensitizer can compete with energy transfer from this sensitizer to triplet oxygen, ${}^3\Sigma_g^{-}O_2$ [85].

F. Secondary Products

Allylic isomerizations [22, 239], dehydrations [239], and fragmentations [237] may proceed subsequent to the formation of allylic hydroperoxides and may thus sometimes obscure the reaction pathway of the oxygenation reaction.

Allylic isomerization and dehydration provide products which cannot be differentiated from those produced via free radical pathways. The fragmentation, on the other hand, yields products which could also originate from 1,2-dioxetanes generated by a [2 + 2] cycloaddition reaction of singlet oxygen to the olefin.

Rearranged Ene-Product

($^1O_2$)

Ene-Product

Dehydration Product

Fragmentation Product

If allylic alcohols (R^4 = OH) are reacted with $^1O_2$, dehydration to α-epoxy ketones and H_2O_2 elimination to give α,β-unsaturated ketones may occur [184, 187].

In order to avoid or minimize secondary reactions, the ene-reaction may be carried out in the presence of radical scavengers as well as at low temperatures (room temperature to about −100°C), because the ene-reaction with singlet oxygen may occur even at −100°C with acceptable rates as long as the solvent does not become too viscous. Furthermore, before a product analysis by separation techniques is executed, reduction of the allylic hydroperoxides to the corresponding allylic alcohols (under retention of configuration) is generally highly recommended.

Ene-Product

Dehydration Product + H_2O

H_2O_2 Elimination Product + H_2O_2

G. Mechanism of the Ene-Reaction

Table XII summarizes the effects which are operative in ene-reactions with singlet oxygen. Provided that all other effects are favorable, electronic effects determine the locospecificity of the ene-reaction with polyolefins. Likewise, the conformations of the allylic hydrogens will determine the regiospecificity of the $^1O_2$ attack in open chain and cyclic olefins and may

Table XII Ene-Reactions with Singlet Oxygen ($^1\Delta_g\ O_2$)

Effect	Due to	Rate enhancement by	Specificity (selectivity) generated
Electronic	π electron density (π_{CC}-ionization potential) (π_{CC} highest filled orbital)	Increased π electron density (Decreased π_{CC}-ionization potential) (Enhanced π_{CC} HFO)	Loco
Conformational	Allylic hydrogens	Decreasing Φ,Φ'; thus axial > equatorial	Regio (and stereo)
Steric hindrance	Bulky groups	Decreasing bulkiness; enhancing distance from reaction center	Stereo (and regio)

contribute to a stereospecific (or stereoselective) attack on cyclohexenoid systems. Finally, bulky groups may protect one face of the C=C bond of cyclic olefins against a singlet oxygen attack and thus produce a stereospecific ene-reaction; they may also contribute to the regiospecific (or regioselective) ene-reactions with singlet oxygen.

The rather small activation energies as well as the comparatively large exothermicities favor an early, reactantlike transition state for $^1O_2$ reactions with olefins. Primary deuterium isotope effects have been found to be rather small, $k_{allyl-H}/k_{allyl-D} \sim 1.1$–$1.8$ [114, 141, 147, 189], indicating that hydrogen abstraction is rather little developed in the transition state.

With propene as a model olefin, Figs. 2 and 3 describe the interactions between propene as a (nucleophilic) ene and singlet oxygen as an (electrophilic) enophile [275].

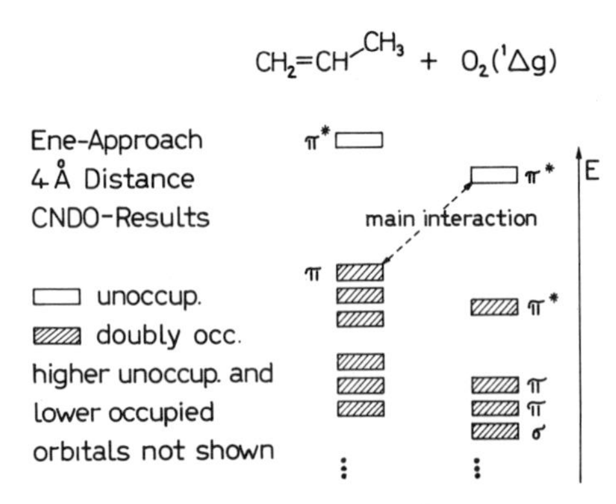

Fig. 2. MO energy levels for $CH_2{=}CHCH_3 + {}^1\Delta_gO_2$. The interaction between propene–π and O_2–π^* is favored by low energy difference and fitting symmetry. This interaction corresponds to a description of oxygen as an electrophile and the olefin as a nucleophile.

Comparison of the two ene-reactions of Fig. 3 reveals an interesting difference between the two reactions with respect to the transition states involved. The main interactions between the LUMO's of the enophiles ($^1O_2$ and ethylene) and the HOMO of propene result in the same bonding combinations, especially in a three-center bonding interaction between C-1, C-2, O-1 and C-1, C-2, C-1′, respectively. The other important interaction, that between the HOMO of the enophile and the LUMO of propene (corresponding to a back-donation interaction), however, depends on the nature of the enophile. Thus, with ethylene as the enophile, the interaction between the HOMO of the enophile and the LUMO of propene cancels the three-center bonding interaction mentioned above and therefore gives only bonding interactions between C-1 and C-1′ as well as between the allylic hydrogen of propene and C-2′. With singlet oxygen as the enophile, however, there is no interaction between the HOMO of $^1O_2$ and the LUMO of propene. Consequently, neither bonding nor antibonding effects appear so that a three-center bonding interaction prevails in the transition state of the ene-reaction with singlet oxygen.

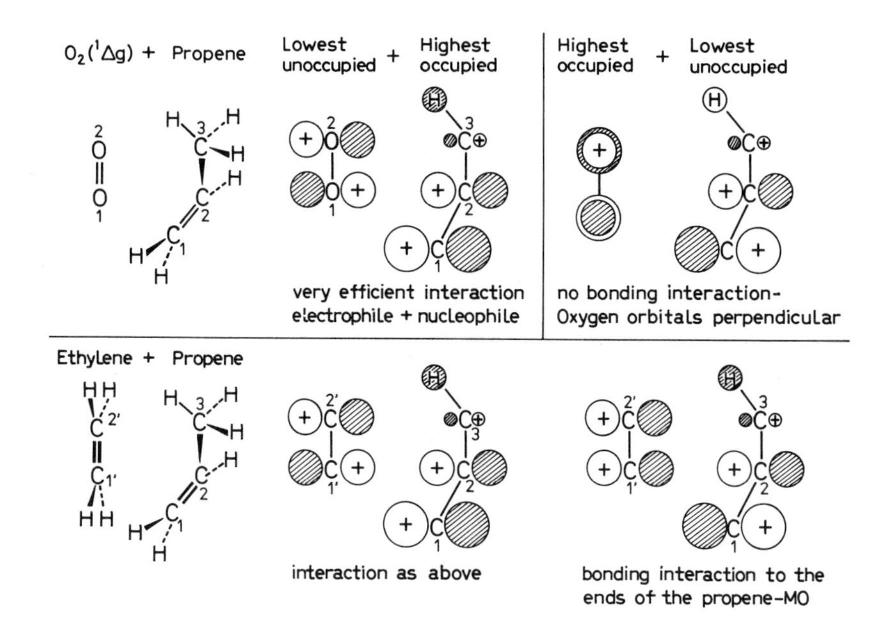

Fig. 3. Interaction of frontier orbitals.

In agreement with the results discussed in this and the preceding paragraphs, we may postulate an energy profile of the ene-reaction with singlet oxygen as shown in Fig. 4.

According to the MO considerations, a cyclic transition state is involved in which the interaction between C-1 of the olefin and O-1 of the singlet oxygen is rather strong whereas the interaction between C-2 and O-1 is relatively weak as is the interaction between the allylic hydrogen of the olefin and O-2 of singlet oxygen. After the reaction passed through the

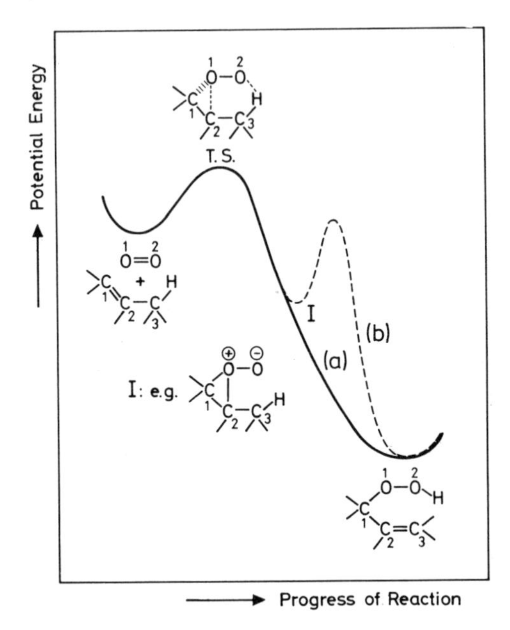

Fig. 4. Potential energy changes during the ene-reaction with singlet oxygen.

transition state, the ene-product may be formed either directly according to path (*a*) or, according to path (*b*), via an intermediate I, for example a perepoxide.*

The transition state appears to be somewhat polar, but since it is located closely to the starting reactants, only a slight effect of solvent polarity on the reaction rate should be expected, independent of whether a zwitterionic

* A perepoxide intermediate was suggested by Sharp [261] and Kopecky and Reich [138]. Kearns and collaborators [51, 119] suggested a 1,2-dioxetane as an intermediate; however, after 1,2-dioxetanes were shown to cleave to carbonyl fragments rather than to rearrange to allylic hydroperoxides [14, 15, 139, 140, 166, 226], they were abandoned as intermediates and perepoxides were favored instead [52, 120, 121]. Obviously, radicals, biradicals such as

O—O·
|
—C—Ċ—CH<
| |

and zwitterions as

OO⁻ |
| |
—C—C—CH<
| +

OO⁺
|
—C—C̤—CH<
| ⁻

are very unlikely intermediates [42, 55, 82, 83, 121, 147] since, among other arguments discussed in this chapter (a) radical scavengers are ineffective, (b) cis-trans isomerizations do not occur [76, 85, 147], and (c) Markovnikov-directing effects have not been observed.

perepoxide is subsequently formed or not. The (small) solvent effect observed therefore does not allow one to rule against the intermediacy of a perepoxide, in contrast to recent suggestions [55, 99].

The experimental results reported in the paragraphs on electronic, steric, and conformational effects appear to be in accord with a concerted reaction as well as with a two-step reaction via perepoxide intermediates. This is especially the case if we take into account that the geometry of the transition state and that of the perepoxide are very similar; arguments solely based on geometrical considerations should therefore not be suited to distinguish between reaction pathways (*a*) and (*b*) as long as we do not allow severe conformational changes to occur in a long-lived perepoxide intermediate.

In order to illustrate this point, let us consider the ene-reactions of olefins **131** and **134** with singlet oxygen which both produce mixtures of hydroperoxides **132** and **135** in a 67:33 and 40:60 ratio, respectively [123]. Provided that radical reactions are not responsible for the rather efficient formation of the "wrong" hydroperoxides, the occurrence of such hydroperoxides, **135** from **131** and **132** from **134**, can neither be explained by reaction path (*a*) nor by reaction path (*b*), since the availability of allylic quasiaxial and quasiequatorial hydrogens is practically the same in conformationally stable olefins (e.g. **134**) and perepoxides (e.g. **136** and **137**). In order to explain their results, the authors [123] suggested that either sufficiently long-lived perepoxides obtain energetically unfavorable conformations in which quasiaxial hydrogens are presented, or that there is a previously undetected antarafacial component to the ene-reaction.

H H H H 131 (a) (b) OOH 132 (b') $^{+}$O $^{-}$O H H H H 133

136   (b'_1)

(b_1)

134   (a) 135 (OOH)

(b_2)

137

Results which do not favor the intermediacy of a perepoxide include the absence of Markovnikov-directing effects in asymmetrically substituted olefins and the absence of Wagner–Meerwein rearranged products, e.g., from α- and β-pinene and norbornene derivatives.

Furthermore, if a perepoxide were involved as an intermediate, one would expect a preferential participation of the more acidic allylic hydrogen to take place in the final stage of the ene-reaction. This, however, appears not to be the case, since 2-methyl-4-aryl-2-butenes (**138**) afford approximately 1:1 mixtures of hydroperoxides **139** and **140**, independently of whether the meta- and para-substituents are electron-donating or electron-accepting groups [59].

138 $\xrightarrow{+\,^1O_2}$ 139 (47%) + 140 (53%)

Similarly, one would expect the predominant formation of hydroperoxide **142** from 1,1-dicyclopropylethylene **141** if the acidity of the allylic hydrogen would determine the course of rearrangement of the assumed perepoxide to the ene product. The ratios of hydroperoxides **142** and **143** obviously do not agree with this expectation. They agree, however, quite well with those expected from considerations along the lines of discussion executed in the paragraph on conformational effects (see Table XIII).

141 $\xrightarrow{+\ ^1O_2}$ 142 + 143

Table XIII Ene-Reaction of 1,1-Dicyclopropylethylenes with Singlet Oxygen in Acetone [218]

	Products	
141	**142** (%)	**143** (%)
R = R′ = H ′	60	40
R = H, R′ = CH_3	37	63
R,R′ = $-[CH_2]_2-$	100	0
= $-[CH_2]_3-$	27	73
= $-[CH_2]_4-$	75	25

Furthermore, if the acidity of the allylic hydrogen atom would determine the ene product formation from the assumed perepoxide intermediate, ketone **144** in the presence of pyridine should afford the α,β-unsaturated 3-keto-6α-hydroperoxide **146** with a hydrogen at C-4 rather than **147** with a deuterium at the C-4 position. The latter should be obtained if a concerted ene-reaction occurs. Experimentally, the product consists nearly exclusively of the deuterated α,β-unsaturated 3-keto-6α-hydroperoxide **147** [71], thus disfavoring the intermediacy of perepoxide **145**. This result also mitigates against the intermediacy of perepoxides **113a–d** (Section IIE) during the singlet oxygen reaction with (*R*)(−)-*cis*-2-deuterio-5-methyl-3-hexene (**109**), especially since the allylic hydrogen of **109** is less acidic than those of ketone **144**, and since methanol is a weaker base than pyridine.

144 + $^1O_2$ —(b)✕→ 145 → 146

144 + $^1O_2$ —(a)→ 147

However, the concerted ene mechanism appeared to be eliminated when Kearns and co-workers [52] reported on the trapping of the intermediate perepoxide by azide ions. They showed that, in the presence of sodium azide, 2,3-dimethyl-2-butene (**148**) and $^1O_2$ undergo the formation of the azido hydroperoxide **149**, and that 1,2-dimethylcyclohexene (**152**) stereospecifically affords the *trans*-azido hydroperoxide **153**.

$$(CH_3)_2C{=}C(CH_3)_2\ (148) + {}^1O_2 + N_3^- \longrightarrow CH_3{-}C(CH_3)(OOH){-}C(N_3)(CH_3){-}CH_3\ (149)$$

$$(CH_3)_2C{=}C(H)CH_2CH_3\ (100) + {}^1O_2 + N_3^- \longrightarrow CH_3{-}C(CH_3)(OOH){-}CH(N_3){-}CH_2CH_3\ (150)$$

$$+\ CH_3{-}C(CH_3)(N_3){-}CH(OOH){-}CH_2CH_3\ (151)$$

150:151 > 95:5

$$\mathbf{152} + {}^1O_2 + N_3^- \longrightarrow \mathbf{153}$$

152 153

Simultaneously, Gollnick [86], Foote [63], and Kearns [94] and their co-workers demonstrated that the formation of allylic hydroperoxides and azido hydroperoxides occur via different pathways: whereas the allylic hydroperoxides are formed via an interaction between the olefin and $^1O_2$, the azido hydroperoxides are produced via the primary interaction between $^1O_2$ and N_3^-. The intermediates formed by the latter interaction may subsequently react with the olefin to azido hydroperoxides.

If the interaction between $^1O_2$ and N_3^- results in an electron transfer from azide ions to singlet oxygen, superoxide ions and N_3 radicals are formed.

$$^1O_2 + N_3^- \rightarrow \cdot O_2^- + N_3\cdot$$

The nearly regiospecific production of azido hydroperoxide **150** from 2-methyl-2-pentene (**100**) [63] is then easily explained by a predominant addition of $N_3\cdot$ to the less substituted C atom of the double bond. The tertiary carbon radical so formed

$$(CH_3)_2\dot{C}-\underset{\displaystyle N_3}{\underset{|}{C}}HCH_3$$

adds $^3O_2$ to give a peroxy radical which subsequently dehydrogenates the solvent to give **150**. The stereospecific formation of the *trans*-azido hydroperoxide **153** from 1,2-dimethylcyclohexene (**152**) may, however, cast some doubt on this explanation.

Gollnick and co-workers [74, 86] investigated the reaction of $^1O_2$ with α-terpinene (**39**) in the presence of azide ions as well as the electrolysis of N_3^- (to produce N_3 radicals) in the presence of α-terpinene (see also refs. 89, 90). If α-terpinene is reacted with $^1O_2$, ascaridole (**40**) is the sole product, which slowly rearranges either thermally or photochemically to the *cis*-diepoxide **154**. If the $^1O_2$ reaction is carried out in the presence of N_3^-, the ascaridole production is predominant even at rather large concentrations of sodium azide. There is, in addition, the formation of further products which (after reduction) were unequivocally shown to have the structures given in Table XIV and of which the two 1-hydroxy-2-azido-Δ^3-menthenes are *cis–trans* isomers. The distribution of these products is independent of the N_3^- concentration but depends to some extent on the temperature

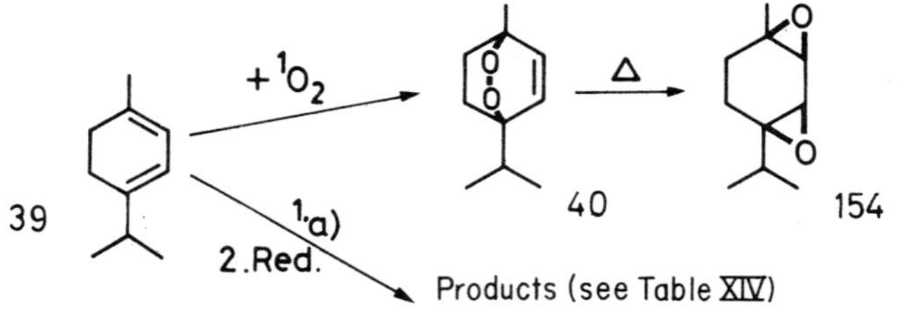

39 $\xrightarrow{+{}^1O_2}$ 40 $\xrightarrow{\Delta}$ 154

39 $\xrightarrow{1. (a)\ 2. Red.}$ Products (see Table XIV)

Table XIV Product Distributions Obtained with α-Terpinene in Photosensitization, Microwave Discharge, and Electrolysis Experiments in the Presence of $^3O_2$ and NaN_3

Reaction conditions, pathway (*a*)	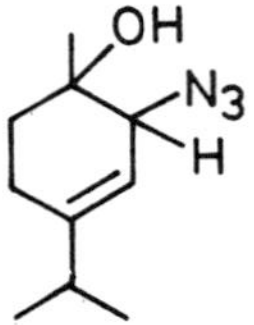 OH, N_3, H	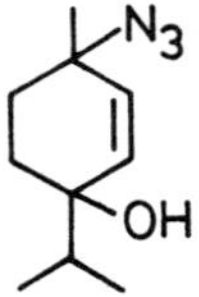 N_3, OH	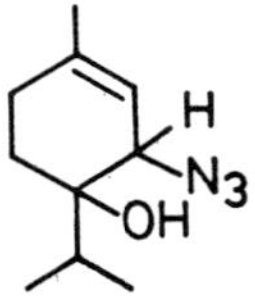 H, N_3, OH	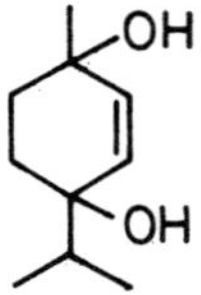 OH, OH	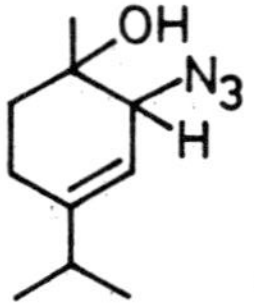 OH, N_3, H	**40**	**154**
Photosensitization in $CH_3OH/H_2O/{}^3O_2/N_3^-/\alpha$-T[a]; 20°C;	44	28		14	14	+	+
Microwave discharge of $^3O_2$; $^1O_2$-stream in $CH_3OH/H_2O/N_3^-/\alpha$-T.; −30°C	62	23		11	4	+	+
Electrolysis of N_3^-; in the presence of $CH_3OH/H_2O/{}^3O_2/\alpha$-T.; 20°C	47	26		14	13	−	−

[a] α-T, α-terpinene.

as may be seen by comparing the experiments in which $^1O_2$ was generated by photosensitization at 20°C and by microwave discharge of $^3O_2$ at -30°C.

If the electrolysis of N_3^- (anodic oxidation of $N_3^- \rightarrow N_3\cdot$) was executed in the presence of α-terpinene and N_2, none of the products of Table XIV was formed. If, however, the solution was saturated with $^3O_2$, the azido alcohols and the diol of Table XIV were obtained in exactly the same proportions at 20°C as were observed in the photosensitized reactions. But no traces of ascaridole (**40**) and its rearrangement product (**154**) were found in the electrolysis experiments. On the other hand, neither electrolysis of $^3O_2$ (cathodic reduction of $^3O_2 \rightarrow \cdot O_2^-$) in the presence of α-terpinene and N_3^- nor treatment of α-terpinene and sodium azide in acetonitrile with potassium superoxide, KO_2, result in the formation of ascaridole, the diepoxide, or the products of Table XIV [75, 92].

The formation of the azido alcohols of Table XIV may be explained by assuming that N_3 radicals preferentially add to the less substituted carbon atoms of the 1,3-diene system of α-terpinene. The diol production (Table XIV), however, is not easily understood, especially since superoxide ions do not appear to be involved in the formation of any of the products of Table XIV.

If the interaction between $^1O_2$ and N_3^- as well as between $^3O_2$ and $N_3\cdot$ result in bond formation with subsequent protonation and hydrogenation, respectively, the hitherto unknown, presumably rather short-lived hydroperoxy azide, N_3O_2H, may be responsible for the stereo- and regiospecific formation of azido hydroperoxides and for the production of glycols.

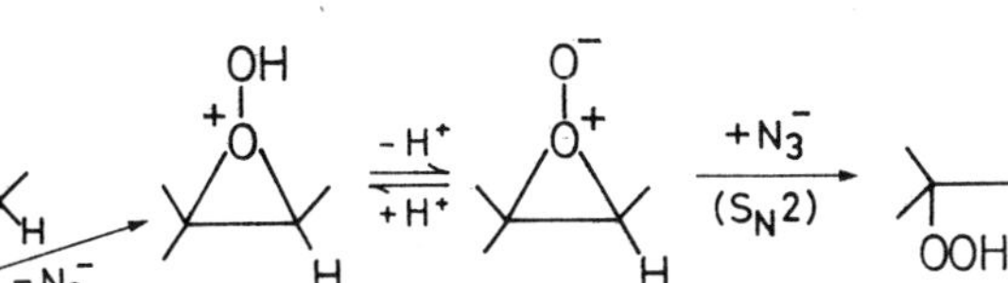

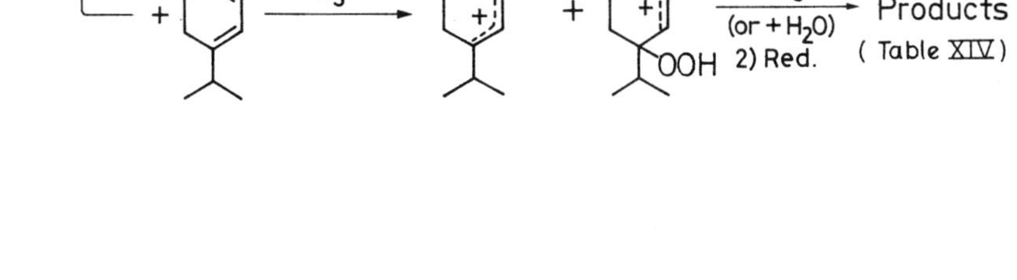

$$N_3^- + {}^1O_2 \longrightarrow N_3O_2^- \xrightarrow{[+H^+]} N_3{-}OOH$$

$$N_3\cdot + {}^3O_2 \longrightarrow N_3O_2\cdot \xrightarrow{[+H\text{-}Don.]} N_3{-}OOH$$

$N_3{-}OOH$; $-N_3^-$; $-H^+$ / $+H^+$; $+N_3^-$ (S_N2); OH; O^-; N_3; OOH; H

$-N_3^-$; OOH; 1) $+N_3^-$ (or $+H_2O$) 2) Red. → Products (Table XIV)

The kinetics of the photosensitized reactions in the presence of N_3^- are in agreement with the discussed mechanisms; they disagree strongly with the intermediacy of a perepoxide that is formed via an interaction between the olefin (or conjugated diene) and $^1O_2$, and which is subsequently trapped by the azide ion [75, 86].

Further evidence against the intermediacy of a perepoxide during the ene-reaction was provided by Kopecky and van de Sande [141] who showed that the product composition (**156** + **157** from **155** + $^1O_2$) differs substantially from that obtained from a 1.6:1 mixture of **158**/**159** + base. This result is taken as evidence against a common intermediate for both reactions.

$$(H_3C)_2C{=}C(CD_3)_2 \; (\mathbf{155}) + {}^1O_2 \longrightarrow H_2C{=}C(CH_3){-}C(OOH)(CD_3){-}CD_3 \; (\mathbf{156}) + H_3C{-}C(OOD)(CH_3){-}C(CD_3){=}CD_2 \; (\mathbf{157})$$

1.38 : 1

$$H_3C{-}C(OOH)(CH_3){-}C(Br)(CD_3){-}CD_3 \; (\mathbf{158}) + H_3C{-}C(Br)(CH_3){-}C(OOH)(CD_3){-}CD_3 \; (\mathbf{159}) \xrightarrow{+\ \text{base}} (\mathbf{156}) + (\mathbf{157})$$

158 : **159** = 1.6 : 1; **156** : **157** = 2.17 : 1

The authors assume the intermediacy of a perepoxide during the allylic hydroperoxide formation from the bromo hydroperoxide mixture, **158**/**159**, in order to explain the OOH migration that has obviously taken place. No azido hydroperoxides were observed when the rearrangement took place in the presence of rather large amounts of sodium azide. Perepoxides (if indeed intermediates) thus appear to be rather short-lived.

So far, there is no rigorous proof available that the ene-reaction with singlet oxygen occurs as a concerted reaction. However, there is also no need to postulate the intermediacy of a perepoxide, although theoretical considerations applying SCF CNDO/2 methods [103, 104, 120] and MINDO/3 calculations [43, 275] appear to prefer this mechanism.* On the contrary, the overwhelming bulk of experimental results can easily be explained by the assumption of a concerted ene-reaction, whereas the assumption of a perepoxide intermediate appears to be incompatible with a number of experimental results. The concerted ene-mechanism should therefore be preferred over the two-step mechanism via perepoxides, at least from a heuristic point of view.†

* For criticisms of these methods, see Hehre [96], and Pople [206]. Furthermore, recent *ab initio* studies by Harding and Goddard [93a] indicate that the intermediacy of perepoxides in reactions of singlet oxygen is highly unlikely.

† In the meantime, perepoxides are also discussed as intermediates in the [2 + 2] cycloaddition reactions of $^1O_2$ to olefins [95, 116].

III. TABULAR SURVEY† OF THE ENE-REACTION OF VARIOUS OLEFINS WITH SINGLET OXYGEN

Singlet oxygen acceptor trivial name	Reaction conditions‡ (1) Method of $^1O_2$ generation, solvent, temperature (°C), additives (2) Subsequent transformations	Isolated products§ Yield (%)* Product Ratio (numbers in italics) m.p. or b.p. $[\alpha]_D^{r.t.}$ (c)**	Ref.

A. Olefins with Acyclic Double Bonds

1. Ethylene Derivatives with Aliphatic Substituents

Acceptor: $R^1R^2C{=}CR^3R^4$. Conditions: (1) A, A(P),[a] B, C, F (2) Red.

Products: (I) $CH_2{=}C(R^1)$… allylic alcohol $R^1(R^2)C{=}C$–$C(OH)R^3R^4$ [b]; (II) HO–$C(R^1)(R^2)$–$C(R^3){=}R^4$

	R¹	R²	R³	R⁴	(I) A	(I) B	(I) C	(I) F	(II) A	(II) B	(II) C	(II) F	Ref.
(a)	H	Me	H	*i*-Pr	*96; 97.3*	*94*	*98.1*	—	*4; 2.7*	*6*	*1.9*	—	41, 54, 55, 59, 61, 62, 68, 73, 85, 99, 205, 245, 248, 262, 267, 283 cf. 258a, 200a
(b)	Me	H	H	*i*-Pr	*93.4*	—	*96.5*	—	*6.6*	—	*3.5*	—	
(c)	Me	Me	H	Me	*51; 45; 43.1; 55[a]; 47*	*48*	*29.1*	—	*49; 55; 56.9; 45[a]; 53*	*52*	*70.9*	—	
(d)	Me	H	Et	Me	*52.6*	—	*63.3*	—	*45[c]*	—	*36.7*	—	
(e)	Me	Me	H	Et	*49; 49.8; 45*	*51*	*42.6*	*46*	*51; 50.2; 55*	*49*	*57.4*	*53*	
(f)	H	Me	Et	Me	*70.0*	—	*80*	—	*20.4[d]*	—	*20*	—	
(g)	Me	Me	H	*i*-Pr	*95*	—	—	—	*5*	—	—	—	
(h)	Me	Me	H	*t*-Bu	*100*	—	—	—	None	—	—	—	
(i)	Me	Me	H	CH_2Ph	*54; 41*	—	—	—	*46; 59*	—	—	—	
(k)	Me	*i*-Pr	Me	Me	*39.5*	*38.7*	—	—	*60.5*	*61.3*	—	—	
(l)	Me	Me	Ph	Me	*28.7*	*28.8*	—	—	*71.3*	*71.2*	—	—	
(m)	Me	Me	H	$(CH_2)_2CHMeEt$	*40*	—	—	—	*60*	—	—	—	

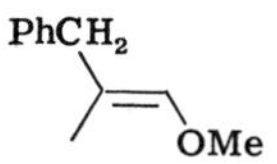	(1) A: TPP, Bz., +8° (2) $(Ph)_3P$	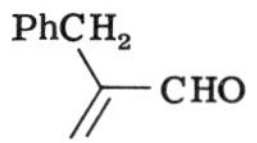 (90%) via hydroperoxide	218b 218c
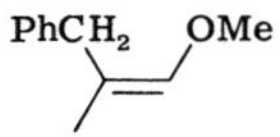	(1) A: TPP, Bz., +8° (2) $(Ph)_3P$	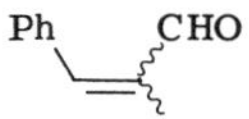 (60%) Z,E mixture 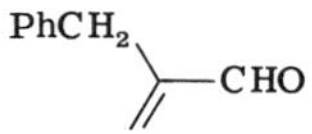 (30%) via hydroperoxides	218b 218c

Footnotes † Within the groups below, the parent compounds are generally arranged in the order of increasing number of carbon atoms, substitution and position (Steroids) of the reacting double bond, ring size, and number of rings.

‡ r.t., liquid phase, unless otherwise stated.

Abbreviations: allyl. rearr., allylic rearrangement; red., reduction; sens., sensitizer.

Sensitizers: CH, Chlorophyll; EO, Eosin; ER, Erythrosin; FL, Fluorescein; HD, Heterocoerdianthrone(dibenzo[a.j.]perylenedione-8,16-); HP, Hematoporphyrin; MB, Methylene Blue; RF, Riboflavin; RB, Rose Bengal; TPP, Tetraphenylporphin;–(P), polymer bound or adsorbed sensitizer.

Solvents: Ac., Acetone; AcOH, acetic acid; Ac_2O, acetic acid anhydride; An., acetonitrile; BuOH, butanol; Bz., benzene; Chlf., chloroform; Cy., cyclohexane; DMF, dimethylformamide; EtOH, ethanol; Mchl., methylene chloride; MeOH, methanol; Hx., hexane; PrOH, propanol; Py., pyridine.

Methods of Singlet Oxygen Generation:

A Photosensitization; A(P) heterogeneous sensitization by polymer bound sensitizer.
B Hydrogen peroxide, hypochlorite.
C Microwave discharge.
D Thermal ozonide decomposition.
E Thermal aceneperoxide decomposition.
F Potassium perchromate decomposition.
G Superoxide anion, diacyl peroxide.

§ Secondary (and/or unidentified) products omitted except when the corresponding primary products could not be isolated.

* Based on starting olefin or on conversion (oxygen uptake). Cf. original publications.

** In Chloroform (c = 1) or neat (α_D) unless otherwise stated; (c), concentration.

(*Continued*)

TABLE—*Continued*

Singlet oxygen acceptor trivial name	Reaction conditions‡	Isolated products§	Ref.
	(1) A: RB or other sens. (*vide infra*), NaN_3, 12% aq. MeOH (2) $NaBH_4$	OH, OH	64
	Ratio azido alcohols/allylic alcohols is highest with poor sensitizers (Rhodamin B, e.g.) and lowest with RB (8.44 and 0.074, resp.)	N_3 OH, OH N_3	
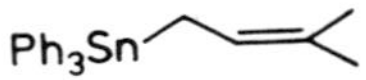 Ph_3Sn	A: RB, Bz. – MeOH	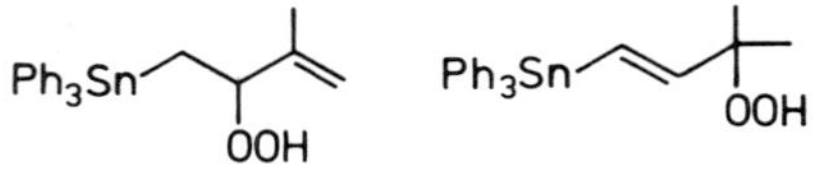 Ph_3Sn OOH, Ph_3Sn OOH total (95%)	247
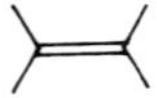	Solid, gas, or liquid phase, r.t. or low temp. A, B, C, D, E, F *vide infra*	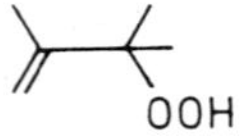 OOH b.p.$_{12}$ 54°–56°C	
	A: RB or MB, MeOH	*100* (90%)	20, 23, 60, 64, 73, 85, 237, 266, 280
	A(P): RB(P), Mchl.	(82%)	19,23a, 191, 227
	B: NaOCl, H_2O_2, MeOH	(64%)	53, 57, 280
	C: microwave discharge in O_2 or CO_2, Hg/HgO	*100* (45%)	24, 46, 72, 73, 87, 229, 284

Substrate	Conditions	Products (yield)	References
	D: $Ph_3P \cdot O_3$ or other ozonides, Mchl., $-78°C \rightarrow$ r.t.	(81%)	13, 174–177, 228
	E: 9,10-Diphenylanthracene peroxide or polymer analog, Bz., N_2, reflux	(25%)[e]	217, 276, 279
	F: K_3CrO_8, MeOH–H_2O 4:6	(35%)[e]	205
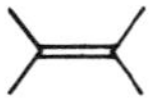	(1) A: RB (or MB), N_3^- MeOH, H_2O 12:8 (2) Na_2SO_3	HO ; HO, N_3 Azido hydroperoxide increases and oxidation rate decreases with increasing [N_3^-]	52, 86, 92
	A: MB (or Pyrene), NaN_3, MeOH	N_3 HOO ; N_3, HOO (70%)	94
	A: MB, MeOH	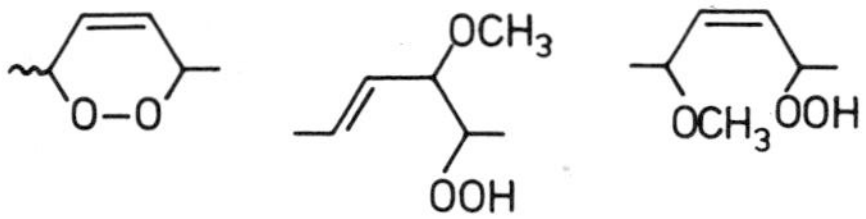 O–O ; OCH_3, OOH ; OCH_3 OOH	94
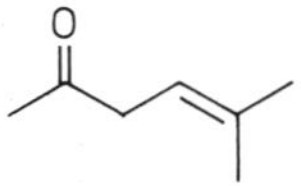 O	A: RB, Py.	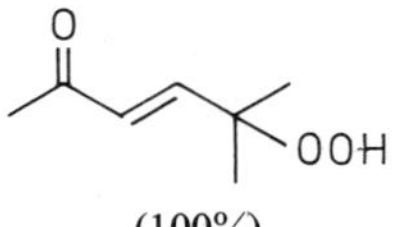 O, OOH (100%)	71

(Continued)

TABLE—*Continued*

Singlet oxygen acceptor trivial name	Reaction conditions‡	Isolated products§			Ref.
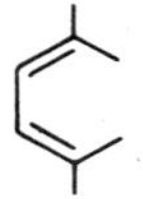 EtO	A: MB, *vide infra*	HOO	O	O	94, 95
	An.	*100*	*1*[f]		
	Mchl.	*10*	*1*		
	Ac.	*5*	*1*		
	MeOH	*10*	*26*		
	MeOH–H_2O, 7:3	*10*	*55*		
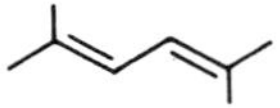	A: MB (or Pyrene), NaN_3, MeOH	N_3 OOH	N_3 OOH		94
		total (53%)			
EtO	A: RB, MeOH (or Ac)	O (19%; 23% in Ac.)	O (19%; 21%)		7
		O EtO (Trace; 13%)	O EtO (Trace; 13%)		
		Ene mode products			
		O (6%; 4%)	O HO (31%; none)		
		Dioxetane mode products			

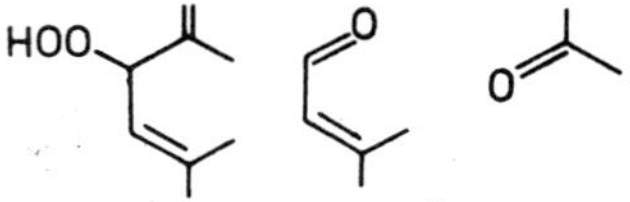

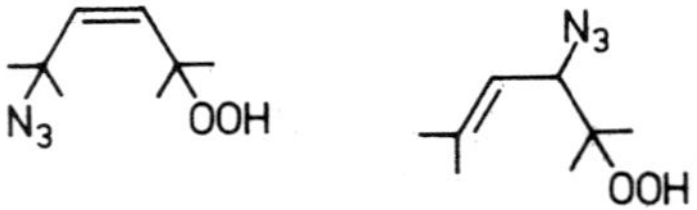

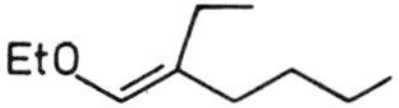

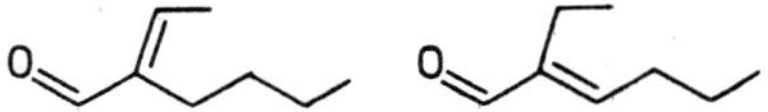

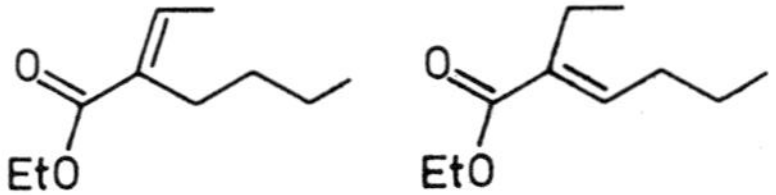

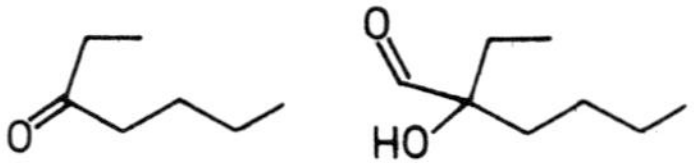

Substrate	Conditions	Products	Ref.
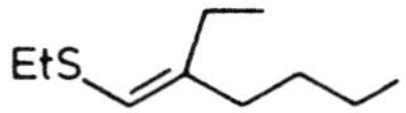 EtS	A: RB, solvent and temperature dependence *vide infra*	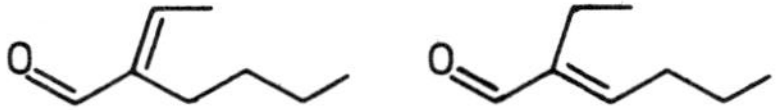 Ene mode products[g]; 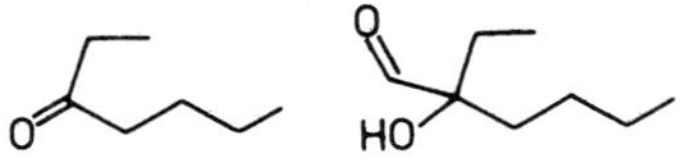 Dioxetane mode products[g] (HO). Exclusively dioxetane but no ene mode products in alcoholic solvents and in acetone at $-78°C$. Ratio, ene/dioxetane products at 20°C in acetone 22/78, benzene 18/22, CCl_4 36/44.	7
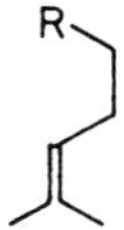 R; R = $HOCH_2CH_2CHCH_3$ (β-Citronellol), $CH_3COOCH_2CH_2CHCH_3$ (β-Citronellylacetate), $OHCCH_2CHCH_3$ (β-Citronellal), $HOOCCH_2CHCH_3$, $HOCH_2CH{=}CCH_3$ [*cis*(Nerol) or *trans*(Geraniol)], $CH_2{=}CHC(OH)CH_3$ (β-Linalool)	(1) A: RB, MeOH (2) Na_2SO_3	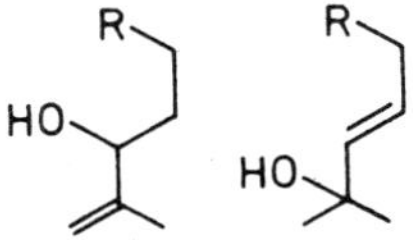 R, HO: *40*; R, HO: *60*; total (90%)	194, 195, 199, 244
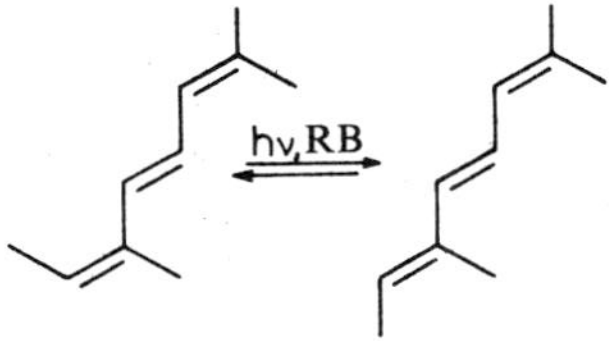 *1* Alloocimene $\underset{}{\overset{h\nu, RB}{\rightleftarrows}}$ *2*	A: RB, MeOH, hydroquinone	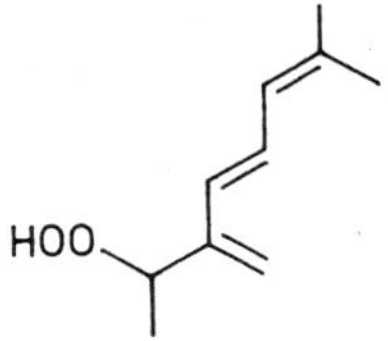 HOO (30%) besides numerous other products b.p.$_{0.00005}$ 40°C; dec. >75°C	136, 235

(Continued)

TABLE—*Continued*

Singlet oxygen acceptor trivial name	Reaction conditions‡	Isolated products§	Ref.
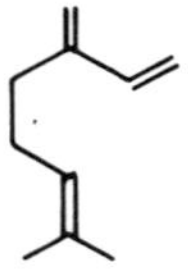 β-Myrcene	A: RB, Mchl., MeOH 95:5 or CH, neat	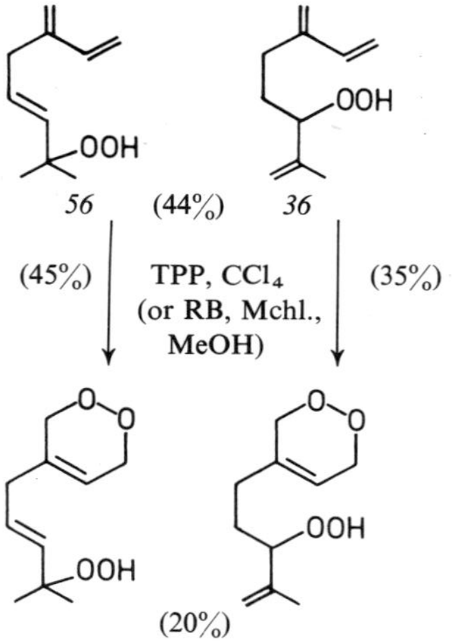 OOH 56 (44%) OOH 36; (45%) TPP, CCl_4 (or RB, Mchl., MeOH) (35%); O–O OOH; O–O OOH (20%)	125, 137, 155
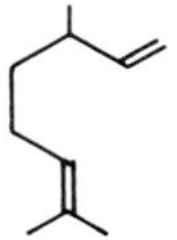 Dihydromyrcene	(1) A: RB, neat (2) $Na_2S_2O_5$, pH 10	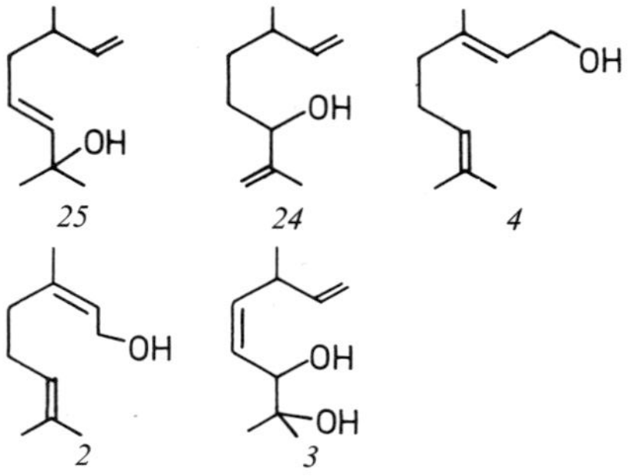 OH 25; OH 24; OH 4; OH 2; OH OH 3	10

Substrate	Conditions	Products	Ref.
 [+]-Dihydromyrcene	(1) A: RB, MeOH (2) Na_2SO_3	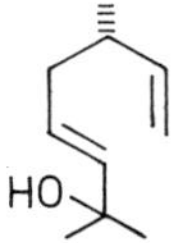 *51* b.p.$_{1.3}$ 58°C; $[\alpha]_D$ +7.4 (95%) (1, EtOH); 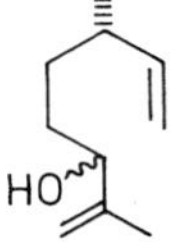 *49* b.p.$_{1.3}$ 70°C; $[\alpha]_D$ +5.8 (1, EtOH)	130
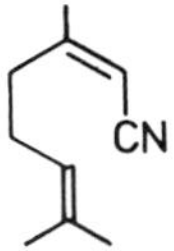 Geranonitrile	(1) A (2) Na_2SO_3	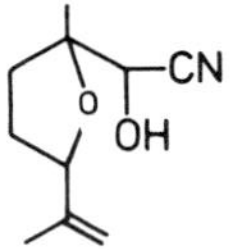 Via mixture of two hydroperoxides	29
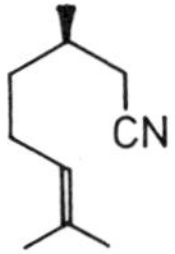 (−)-Citronellonitrile	(1) A: RB, MeOH (2) Na_2SO_3	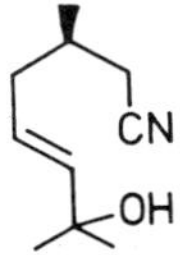 *55* (75%) 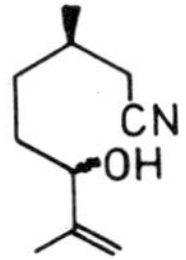 *45*	28
2.6- and 2.7-Dimethyl-2.6-octadiene		cf. Tanielian [269c]	
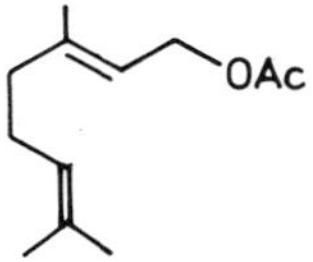 Geranyl acetate	(1) A: MB or CH, EtOH (2) $NaBH_4$	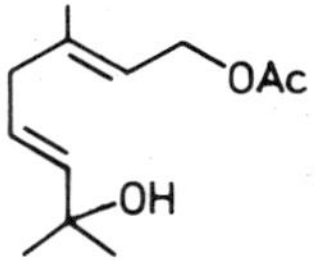 *1* 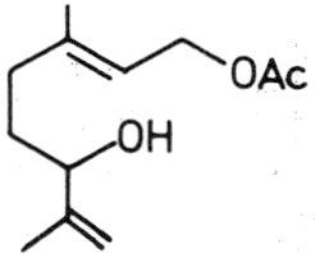 *1*	47

(*Continued*)

Singlet oxygen acceptor trivial name	Reaction conditions‡	Isolated products§	Ref.
 C_3H_7 C_3H_7 all-*cis*-$\Delta^{4,7}$-Undecadiene	(1) A: RB or Na–CH, MeOH, 30°C B: H_2O_2, NaOCl, MeOH, −5–0°C (2) $(MeO)_3P$, MeOH	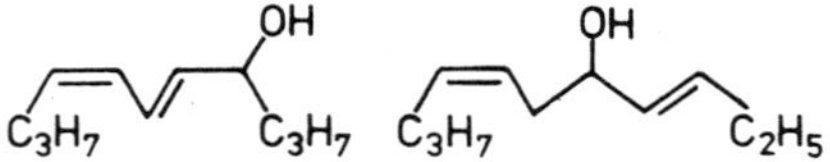 OH OH C_3H_7 C_3H_7 C_3H_7 C_2H_5 A: mixture (54%) b.p.$_{0.5}$ 68–75°C B: mixture *60* *40* (<1%)	35
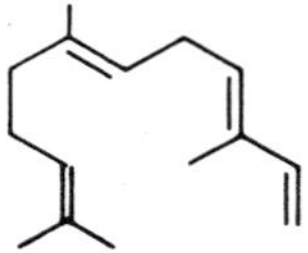 α-Farnesene	(1) A: EO, EtOH (2) $NaBH_4$	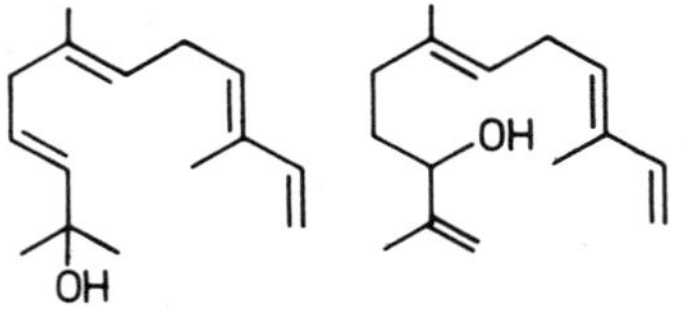 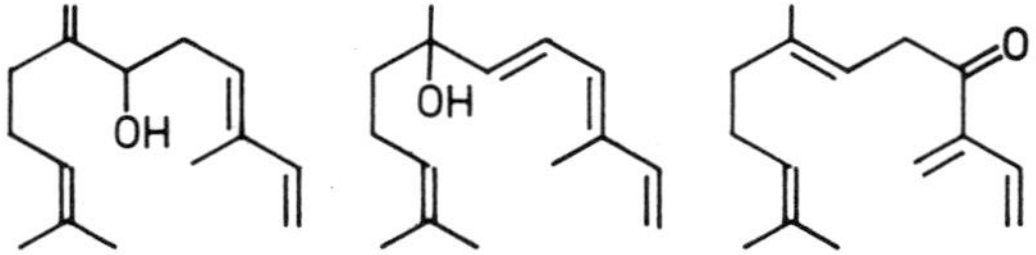 OH OH OH OH Not isolated	31
$CH_3(CH_2)_7CH=CH(CH_2)_7COOCH_3$ *cis*-Δ^9-Methyl oleate	A: CH, Hx.	$CH_3(CH_2)_7CH(OOH)CH=CH(CH_2)_6COOCH_3$ + 10-Δ^8-OOH $CH_3(CH_2)_6CH=CHCH(OOH)(CH_2)_7COOCH_3$ Together (80%) *1* *1* 9-Δ^{10}-OOH besides di- and higher hydroperoxides	32, 36, 50, 93, 127, 210, 284a, 269a, 269b
$CH_3(CH_2)_4(CH=CHCH_2)_2(CH_2)_6COOCH_3$ all-*cis*-$\Delta^{9,12}$-Methyl linoleate	A: CH, Hx.	Four isomeric monohydroperoxides (44%) (9-$\Delta^{10,12}$; 10-$\Delta^{8,12}$; 12-$\Delta^{9,13}$; 13-$\Delta^{9,11}$)	

Substrate	Conditions	Products (%)	Ref.
$CH_3CH_2(CH=CHCH_2)_3(CH_2)_6COOCH_3$ all-*cis*-$\Delta^{9,12,15}$-Methyl linolenate	A: CH, Hx.	Six isomeric monohydroperoxides (36%) (9-$\Delta^{10,12,15}$; 10-$\Delta^{8,12,15}$; 12-$\Delta^{9,13,15}$; 13-$\Delta^{9,11,15}$; 15-$\Delta^{9,12,16}$; 16-$\Delta^{9,12,14}$)	
$CH_3(CH_2)_4(CH=CHCH_2)_4(CH_2)_2COOC_2H_5$ all-*cis*-$\Delta^{5,8,11,14}$-Ethyl arachidonate	A: CH, Hx.	Eight isomeric monohydroperoxides (29%) (5-$\Delta^{6,8,11,14}$; 6-$\Delta^{4,8,11,14}$; 8-$\Delta^{5,9,11,14}$; 9-$\Delta^{5,7,11,14}$; 11-$\Delta^{5,8,12,14}$; 12-$\Delta^{5,8,10,14}$; 14-$\Delta^{5,8,11,15}$; 15-$\Delta^{5,8,11,13}$)	

2. Ethylene Derivatives with Cyclic Substituents

Substrate	Conditions	Products (%)		Ref.
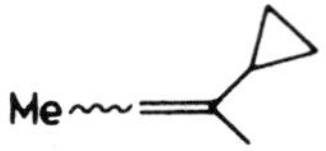 Me~~= (cyclopropyl, Me) cis and trans	A: Bz., 3°C	OOH 36	HOO 64	218a, 218c cf. 200a
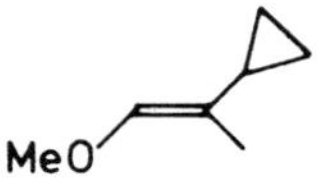 MeO (cyclopropyl)	A: Bz., 3°C	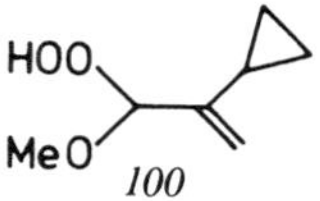 HOO MeO 100		218a, 218c 104, 93b cf. 200a
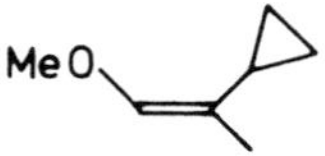 MeO (cyclopropyl)	A: Bz., 3°C	HOO MeO 28	MeO HOO 72	218a, 218c cf. 200a
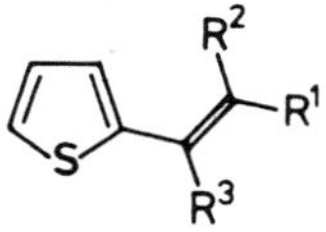 (thienyl) R^2 R^1 R^3	A: TPP, CCl_4, Na lamps	O–O R^2 R^1 S R^3	R^2 R^1 OOH S	157
$R^1 = R^2 = Me$; $R^3 = H$		(75%)	none	
$R^1 = R^2 = R^3 = H$		(10%) b.p.$_{0.3}$ 40°C	none	
$R^1 = Me$; $R^2 = R^3 = H$		(51%)	none	
$R^1 = Ph$; $R^2 = R^3 = H$		(72%) m.p. 83°C	none	
$R^1 = R^2 = H$; $R^3 = Me$		(27%)	(10%)	

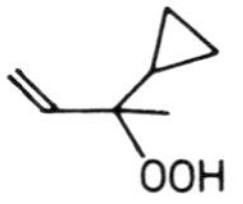

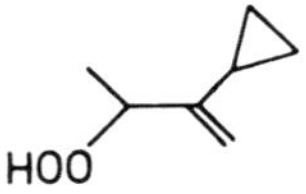

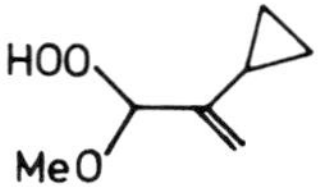

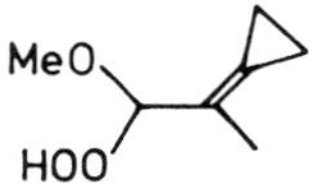

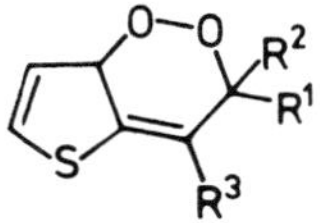

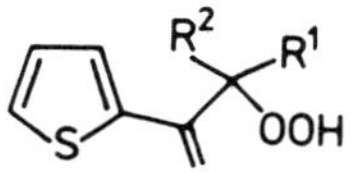

(Continued)

Singlet oxygen acceptor trivial name	Reaction conditions‡	Isolated products§		Ref.
$R^1 = R^2 = R^3 = Me$		(25%)	(28%)	
$R^1 = R^2 = H; R^3 = Ph$		(15%) m.p. 72°C	none	
$R^1 = \alpha$-naphthyl; $R^2 = R^3 = H$		(77%) m.p. 117–118°C	none	
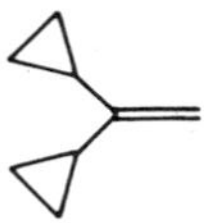	(1) A (2) Ph_3P	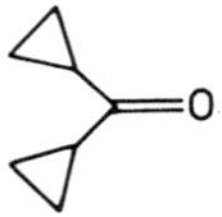 *30* besides polymer *70* Presumably via hydroperoxide and subsequent Hock-cleavage		67b cf. 218c
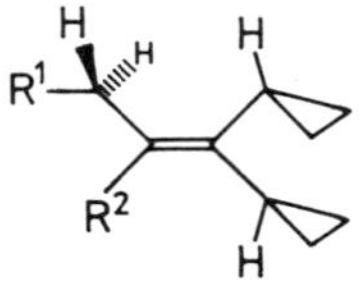	(1) A: EO, Ac., 3°–5°C (2) Ph_3P	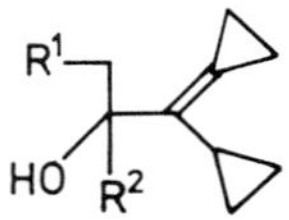 (80%)	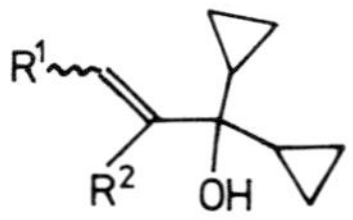	218 218c
$R^1 = R^2 = H$		*60*	*40*	
$R^1 = H; R^2 = Me$		*37*	*63*	
$R^1R^2 = -(CH_2)_2-$		*100*	*0*	
$R^1R^2 = -(CH_2)_3-$		*27*	*73*	
$R^1R^2 = -(CH_2)_4-$		*75*	*25*	
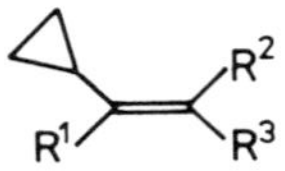	(1) A (2) Ph_3P	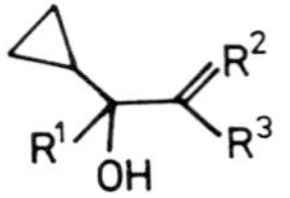	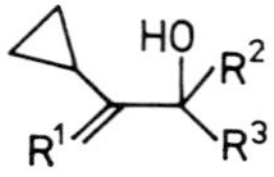	67b cf. 200a, 218c
(a) $R^1 = R^2 = R^3 = Me$		(a) *45*	*55*	
(b) R^1 = cyclopropyl; $R^2 = Me; R^3 = H$		(b) *25*	*75*	
(c) R^1 = cyclopropyl; $R^2 = R^3 = Me$		(c) *65*	*35*	

Substrate	Conditions	Product(s)	Comments	Ref.
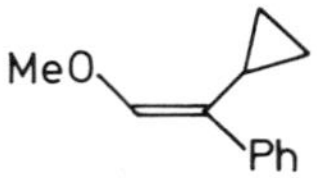 MeO, cyclopropyl, Ph (E) Z isomer gives epoxide, presumably via cyclic peroxide	A: Bz., 3°C	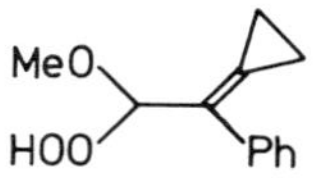 MeO, HOO, cyclopropylidene, Ph *100*		218a 218c
C_6H_{11}—=—OCH_3	(1) A: TPP, Bz., +8°C (2) $(Ph)_3P$	C_6H_{10}=CHCHO (~60%)	via hydroperoxide besides 1% dioxetane and 10% deconjugated aldehyde or ester	218b
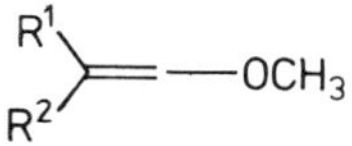 R^1, R^2 =—OCH_3 (a) $R^1 = R^2$ = cyclopropyl (b) R^1,R^2 = 2-norbornylidene (c) R^1,R^2 = cyclohexylidene (d) R^1,R^2 = cycloheptylidene (e) R^1,R^2 = 1-spiro[3.5]nonylidene	(1) A: TPP, Bz., +8°C (2) $(Ph)_3P$	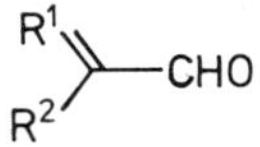 R^1, R^2 =—CHO (75–85%) via hydroperoxide (b) besides 1% dioxetane		218b
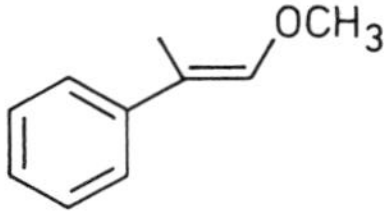 Ph, OCH_3 (E) Z-isomer gives cyclic peroxide	A: Dinaphthylenthiophene, Freon 11, −78°C	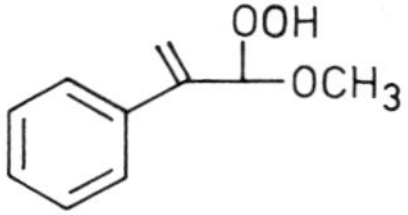 OOH, OCH_3 No. 1,4 product formed		144a 63a cf. 218a
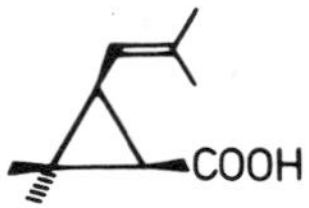 COOH *cis*-Chrysanthemic acid	(1) A: RB, MeOH (slow, 3 days) (2) Me_2S	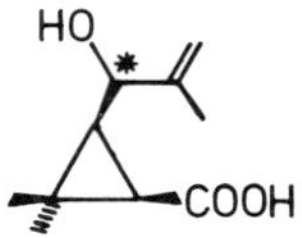 HO, *, COOH (15%) m.p. 162°–163°C 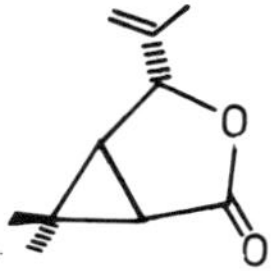 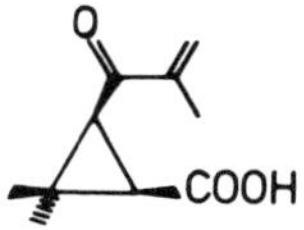 O, O (6%) O, COOH (10%) m.p. 65°–67°C		223

(*Continued*)

TABLE—*Continued*

Singlet oxygen acceptor trivial name	Reaction conditions‡	Isolated products§	Ref.
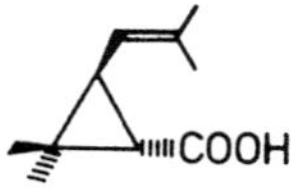 COOH *trans*-Chrysanthemic Acid	(1) A: RB, MeOH (2 days) (2) Me_2S	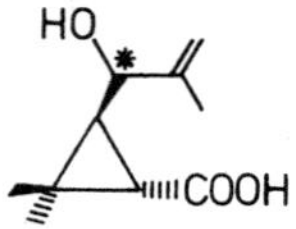 HO, *, COOH 2 epimers (25%) m.p. 114°–4.5°C (34%) m.p. 171–2°C	223
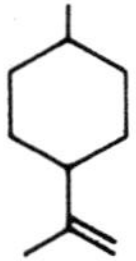 Δ^8-*p*-Menthene	A: RB, MeOH	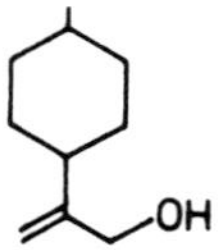 OH (13.6%)	129, 132
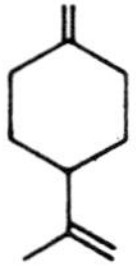 $\Delta^{1(7),8}$-*p*-Menthadiene	A: RB, MeOH	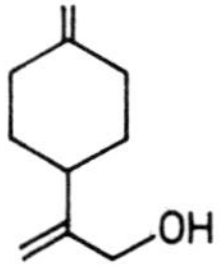 OH (12.5%)	129, 132
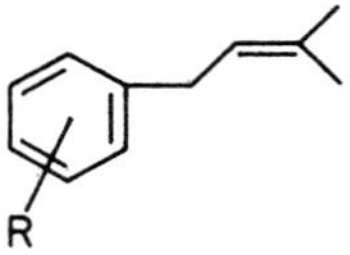 R R = *p*-OMe, *p*-Me, *m*-Me, H, *p*-Cl, *m*-Cl	(1) A: RB, MeOH, Py. (2) $NaBH_4$	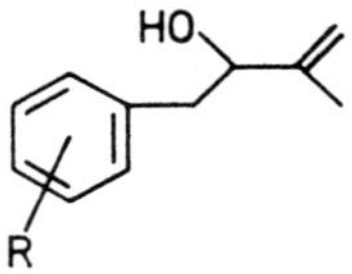 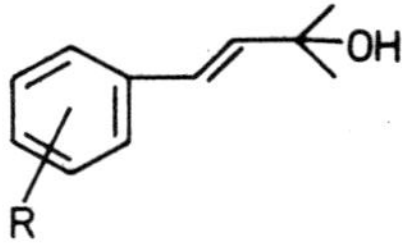 HO, R About *13* OH, R *10*[h]	68

Substrate	Conditions	Products (%)		References
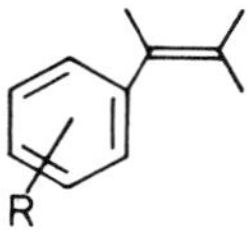 R^i = p-OMe, *p*-Me, *m*-Me, H, *m*-OMe, *p*-Cl, *m*-Cl, *m*-CN, *p*-CN	(1) A: RB, Py.-MeOH 2:98, 2°C B: NaOCl, H_2O_2, MeOH, *t*-BuOH, 2,6-di-*tert*-butyl phenol 10^{-2} *M*, 2°C (2) $NaBH_4$	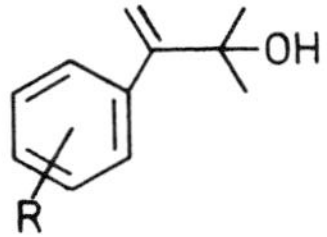 *73*[j]	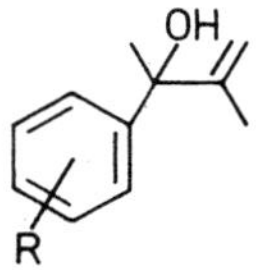 *27*	41, 55, 61 cf. 218a
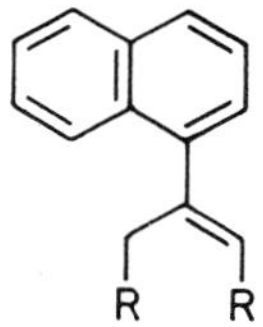 R = H R = $-(CH_2)_2-$ R = $-(CH_2)_3-$	A: TPP, CCl_4 Na lamps	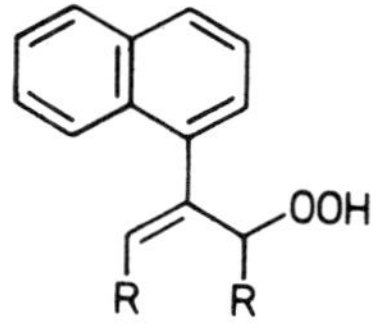 (90%) (>90%) (>90%)		156
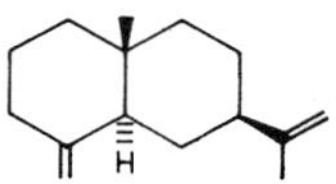 (+)-β-Selinene	(1) A: RB, MeOH (2) Na_2SO_3	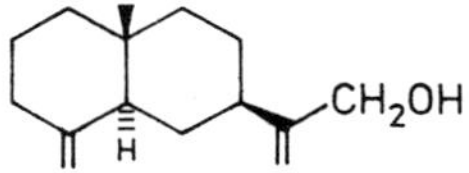 (69%) $[\alpha]_D$ +32.5		133
(+)-Davanone	(1) A: RB, Py. (2) Ph_3P	(19.1%)	(15.9%)	270
		Via cyclization of the primary α,β-unsaturated γ-hydroperoxy ketone		

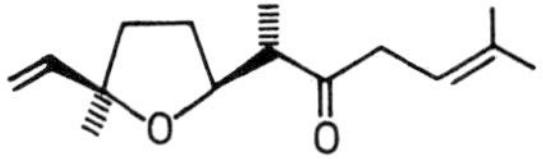

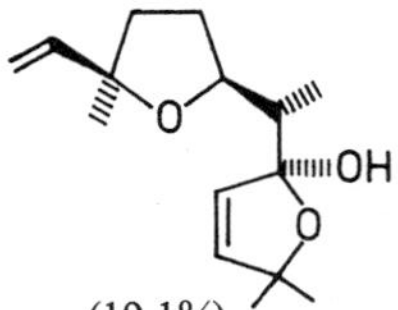

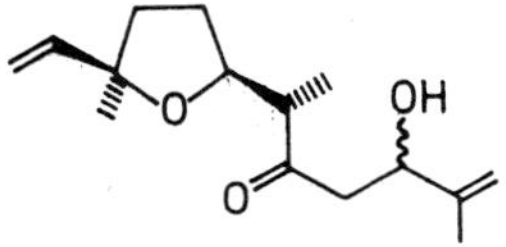

(*Continued*)

Singlet oxygen acceptor trivial name	Reaction conditions‡	Isolated products§		Ref.
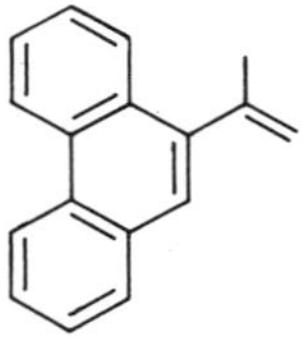	A: TPP, CCl_4, 5°C	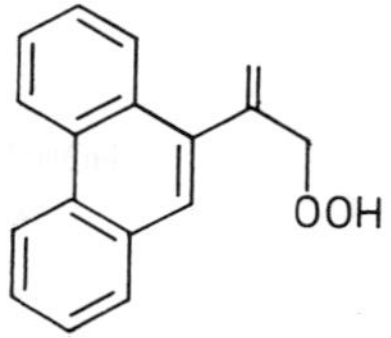 OOH No 1,4-peroxide and no formylphenanthrene formed		157a
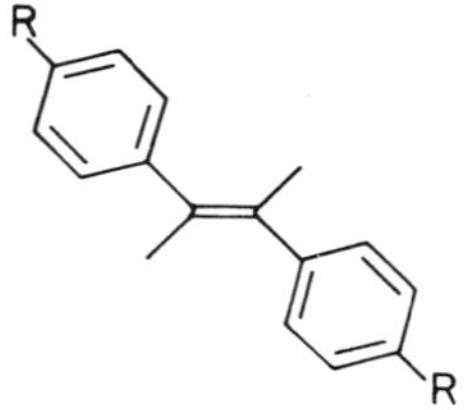 R, R	(a,b,c) A: *vide infra* (d) B: NaOCl, H_2O_2, EtOH, H_2O 10°C	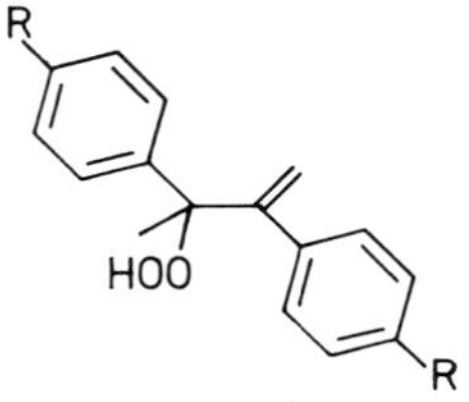 R, HOO, R	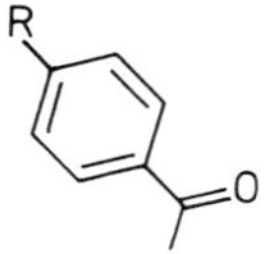 R, O	220 37
(a) R = H, cis or trans. RB, MeOH or MB, Mchl.		almost quantitative	none	
(b) R = OMe, trans, RB, MeOH		(80%)	(7%)	
(c) R = OMe, trans, RB, Ac.		(95%)	none	
(d) R = OMe, cis				
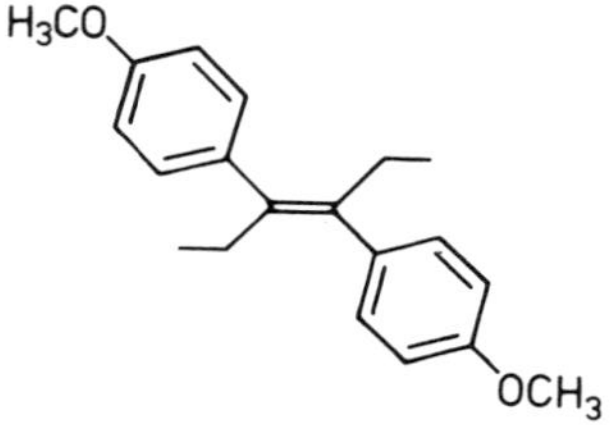 H_3CO, OCH_3	A: *vide infra*	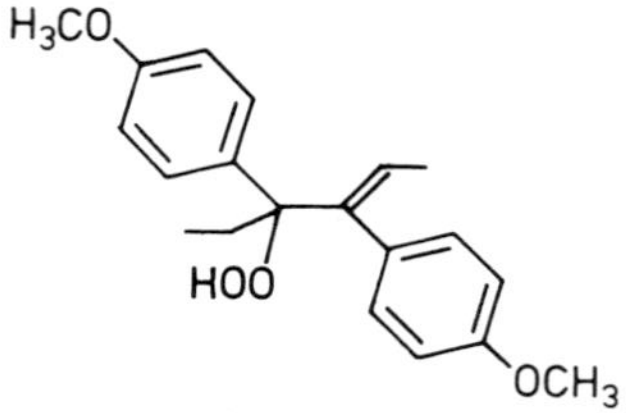 H_3CO, HOO, OCH_3	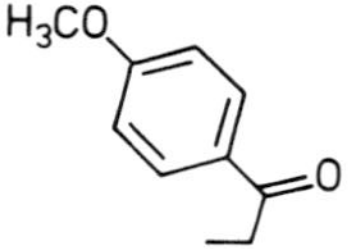 H_3CO, O	220
Stilbestroldimethyl ether	RB, MeOH	(10%)	(13%)[k]	

	RB, Ac. RB, MeOH–Py., 50:1 MB, Mchl.	(54%) (73%) (5%)	(2%) (5%) (60%)	
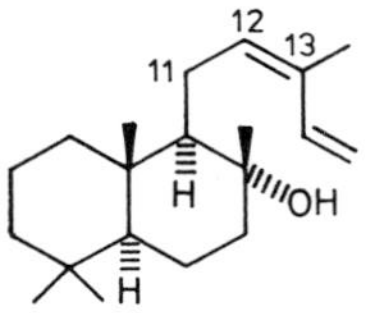 (12*Z*)-Abienol	(1) A: RB, MeOH (2) Ph_3P, toluene	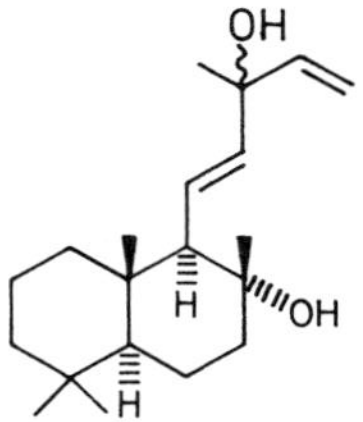 two epimers *46* (11*E*, 13*S*) m.p. 108°–12°C *22* (11*E*, 13*R*) m.p. 115°–17°C	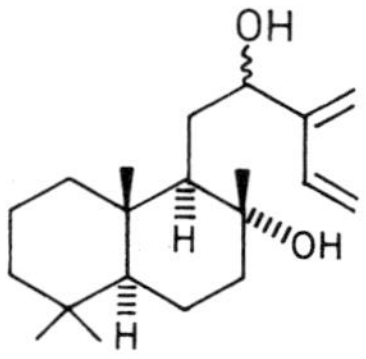 two epimers *2* (12*S*) m.p. 147°–49°C *1* (12*R*) m.p. 114°–15°C	275a
		Besides 1.4 adducts and other products		
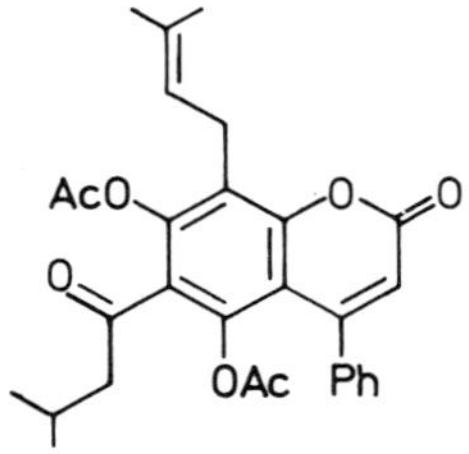 Mammeisine	A: HP, MeOH	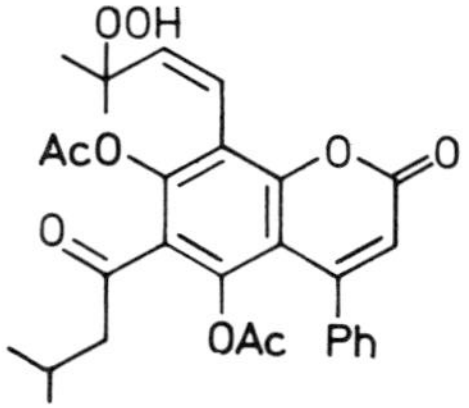		65

(*Continued*)

Singlet oxygen acceptor trivial name	Reaction conditions‡	Isolated products§	Ref.
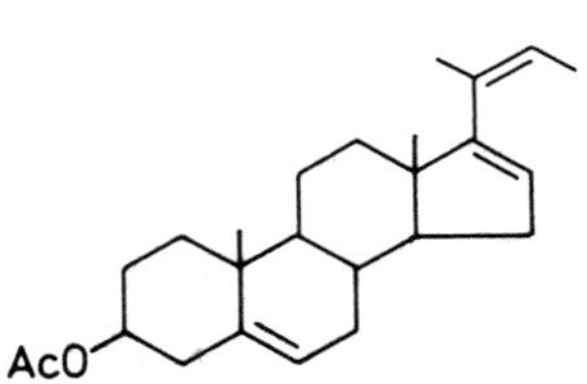 AcO	A: MB, MeOH-Hx., 1:1	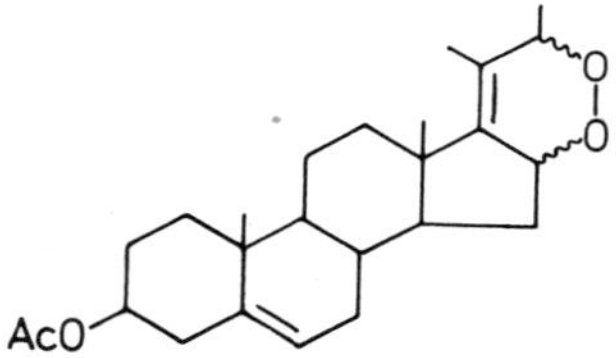 AcO O O 2 epimers (17.4%) Zn dust reduction gave 16β,22β-diol *1* m.p. 187°–191°C; $[\alpha]_D$ −75.9 and 16α,22α-diol *2* m.p. 184°–187°C; $[\alpha]_D$ −44.3 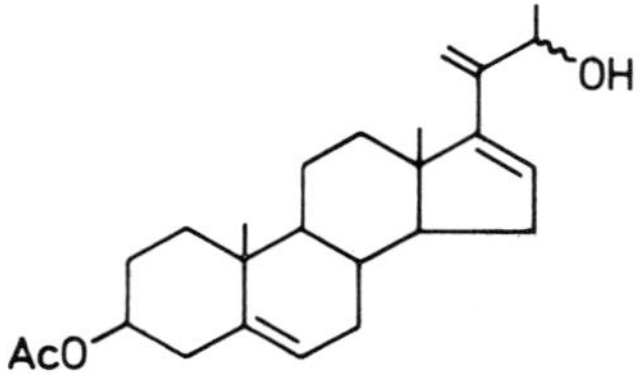 AcO OH 2 epimers (62%) Zn dust reduction and acetylation gave two diacetates m.p. 104°–107°C; $[\alpha]_D$ −84.7 and m.p. 119°–122°C	135
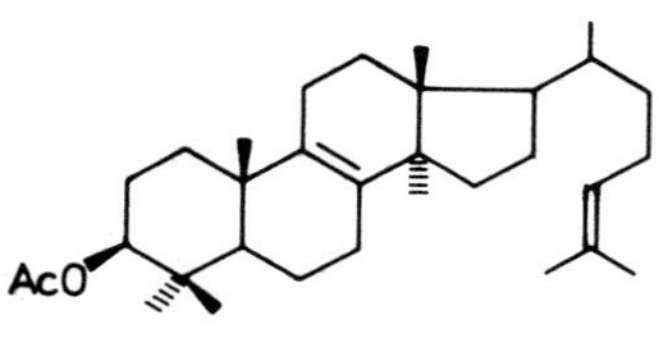 AcO	(1) A: HP, Py. (2) $NaBH_4$ (3) Ac_2O	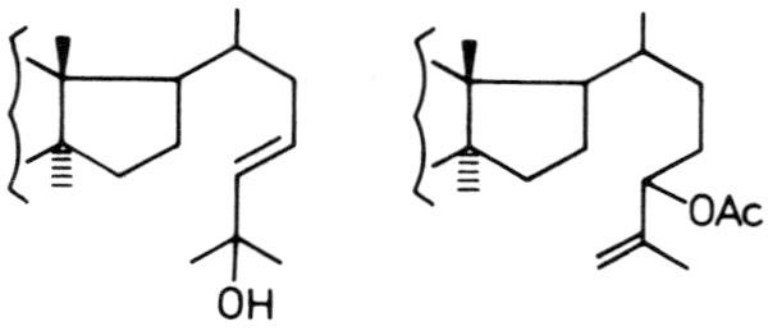 OH OAc	65

Substrate	Conditions	Product	Ref.
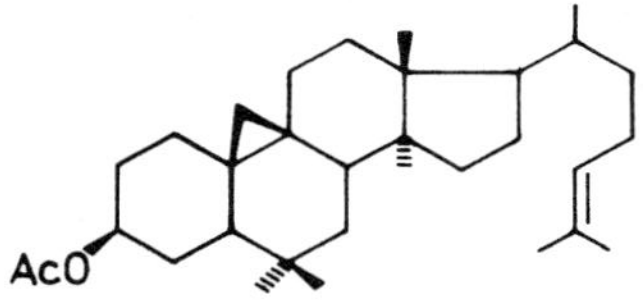 Cycloartenol acetate	(1) A: HP, Py. (2) $LiAlH_4$ (3) Ac_2O	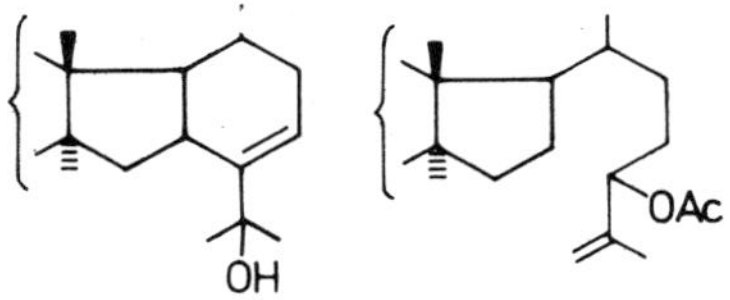 m.p. 139°–141°C $[\alpha]_D$ +45 (1.1)	65
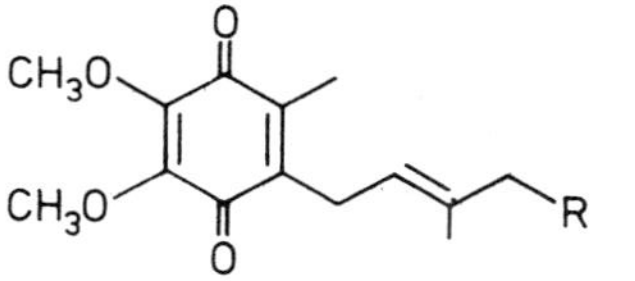 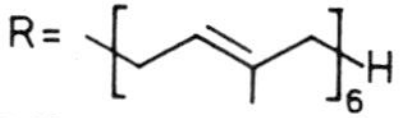 7-Ubiquinone, R = $[\ ]_6H$	solid, sun	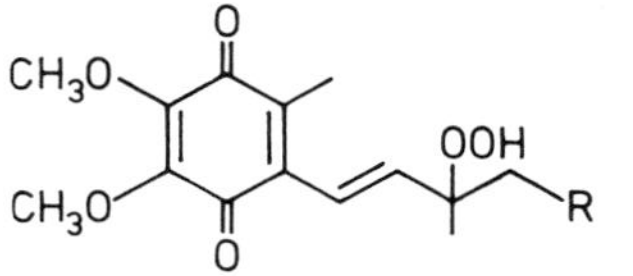 (6%) yellow-red oil; besides several other products	169
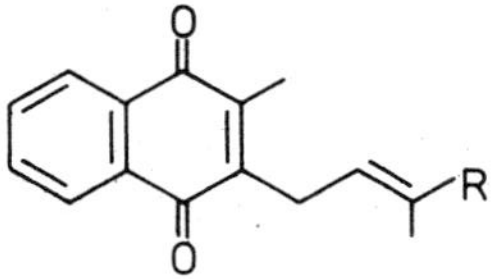 (a,b,c) R = $C_{16}H_{33}$ (Vitamin K_1, Phylloquinone)	A: (a) Bz., air, Hg-lamp (b) Cy., air (c) MB, EtOH, ~700 nm	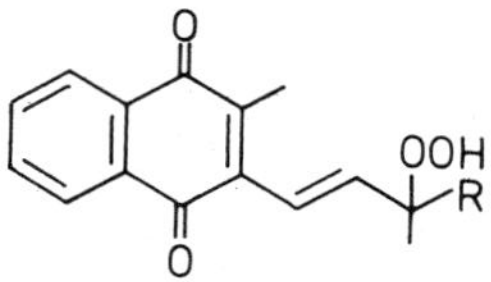 (a) (b) (10%), yellow oil, besides 6,10,14-trimethylpentadecan-2-one (50%) (c) (75%), yellow oil	200, 264
(b,c) R = Me		(b) (11%) m.p. 90°–94°C, yellow (c) (75%) m.p. 90°–94°C, yellow	
(b) R = $C_{41}H_{69}$		(b) (8%) besides C_{43}-ketone (7 F) (50%) (c) side chain oxygenation	
(b) R = $C_{41}H_{67}$		(b) (8%) besides C_{43}-ketone (8 F) (3%) Note: 7 F and 8 F denote number of double bonds	

(*Continued*)

TABLE—*Continued*

Singlet oxygen acceptor trivial name	Reaction conditions‡	Isolated products§	Ref.
B. Olefins with Semicyclic Double Bonds			
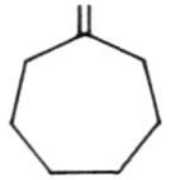	(1) A: RB, MeOH (2) Na_2SO_3	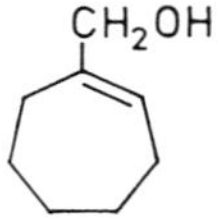 CH_2OH (78%)	132
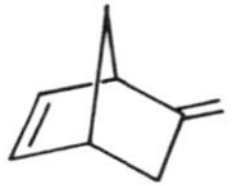	A: MB, An., 0°C	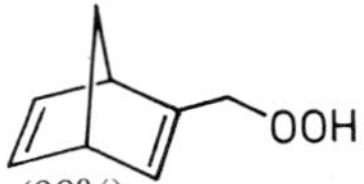 OOH (98%) 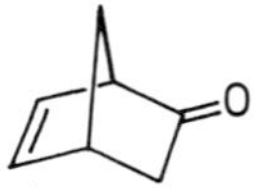 traces Besides isomeric epoxides (0.2% exo, 1.8% endo)	116b
 R R (a) R = H (b) R = Me	(1) A: MB, An., 5–10°C (2) Ph_3P or $NaBH_4$	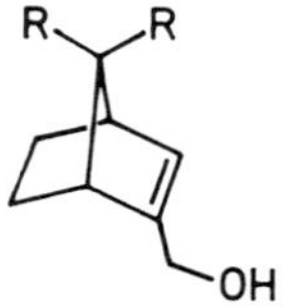  R R OH O *95* *4* solely none	112, 114
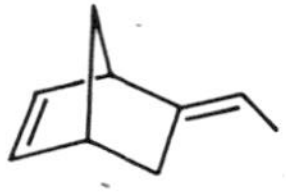	(1) A: RB, MeOH (2) Na_2SO_3	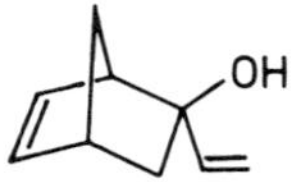 OH (35%) b.p.$_{40}$ 103–107°C 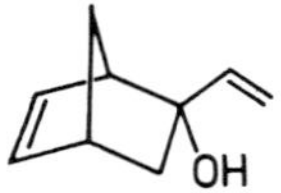 OH (12%) b.p.$_{40}$ 103–107°C 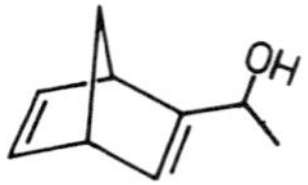 OH (23%) b.p.$_{29}$ 100–102°C	5, 278

Substrate	Conditions	Products	Ref.
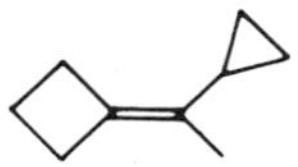	(1) A: EO, Ac. (2) Ph_3P	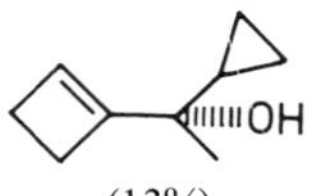 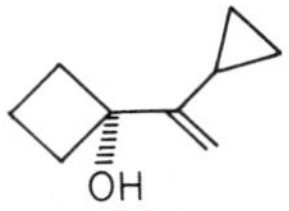 OH (13%); OH (87%)	218 cf. 218c
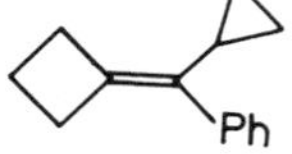 Ph	A: Bz., 3°C	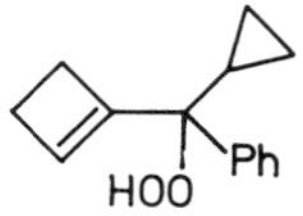 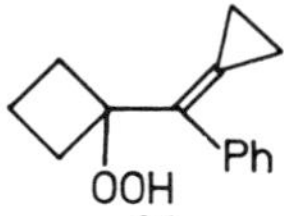 HOO Ph *14*; OOH Ph *86*	218a
 (+)-Sabinene	A: MB, *i*-PrOH	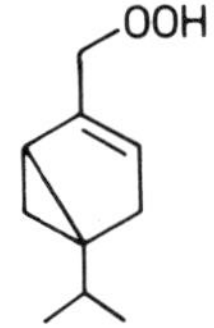 OOH; b.p.$_{0.01}$ 27.5–29°C; $[\alpha]_D$ +82.2	39
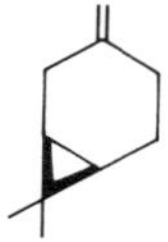 (+)-$\Delta^{4(10)}$-Carene	(1) A: RB, MeOH (2) Na_2SO_3	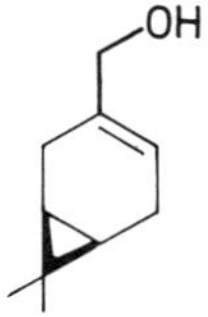 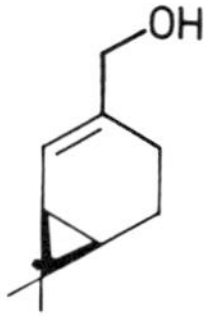 OH *25*; OH *75* $[\alpha]_D$ +66.5 (2.2, Bz.)	80
β-Pinene	(1) A: MB, An. (or *i*-PrOH) (2) Ph_3P (or Na_2SO_3 or $NaBH_4$)	  OH; O; *99.9* total (72.5%) *0.01*; Presumably via 2(10)-dioxetane	113, 141a, 233, 234

(*Continued*)

TABLE—*Continued*

Singlet oxygen acceptor trivial name	Reaction conditions‡	Isolated products§	Ref.
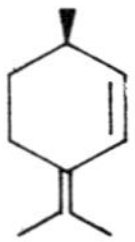 $\Delta^{2,4(8)}$-*p*-Menthadiene	(1) A: RB, MeOH (2) Na_2SO_3	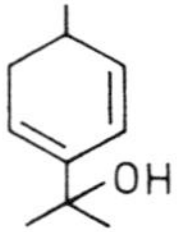 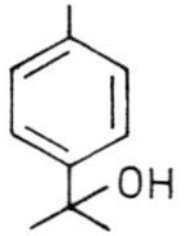 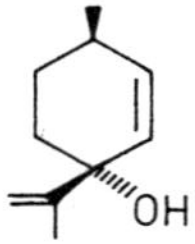 OH; OH; OH total (54.6%) *79*; *8*; *13*	132
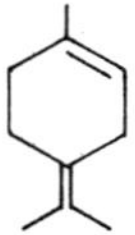 Terpinolene	(1) A: RB, MeOH (2) Na_2SO_3	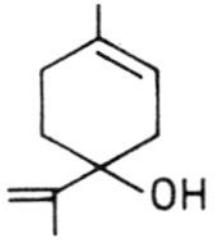 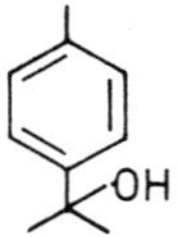 OH; OH total (79.5%) *2*; *1*	132, 233
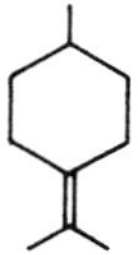 $\Delta^{4(8)}$-*p*-Menthene	(1) A: RB, MeOH (2) Na_2SO_3 or Ph_3P 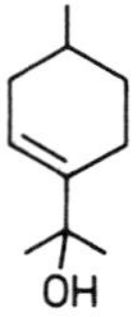 	 OH; OH; HO total (90%) *22*[l]; *32*; *46*	132, 257
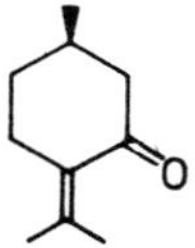 (+)-Pulegone	(1) A: RB, MeOH (or RB(P), Mchl., −10°C, or HP, Mchl.) (2) Ph_3P (or $SnCl_2$)	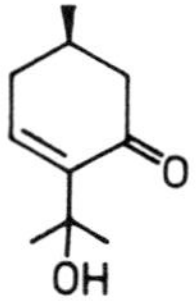 O, OH; O, OH *69*; *6* $[\alpha]_D$ −69.8; $[\alpha]_D$ +216.9	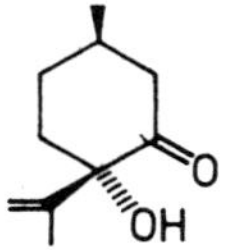 49, 257

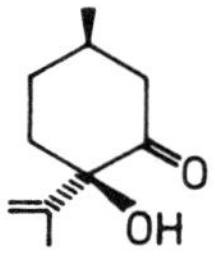

Substrate	Conditions	Products	Ref.
(−)-*cis*-Pulegol	(1) A: RB, MeOH, or DMF (or HP, Mchl., or EO, MeOH)	*12* total (82%) $[\alpha]_D$ −132; m.p. ~20°C; *6* $[\alpha]_D$ −62.1; *21* $[\alpha]_D$ −27.2; *51* m.p. 76°C; $[\alpha]_D$ +15.9; *12* m.p. 79°C; Presumably via 8-hydroperoxide 2 isomers *6* m.p. 54°C; $[\alpha]_D$ −32 *4* m.p. 45°C; $[\alpha]_D$ +45 total (86%)	257
 (+)-Homoverbenene	A: RB, MeOH	 m.p. 26°C Besides CH_3CHO, presumably via hydroperoxide	246

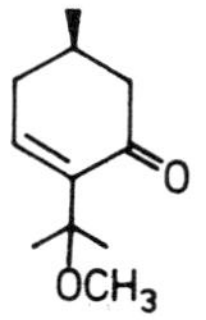

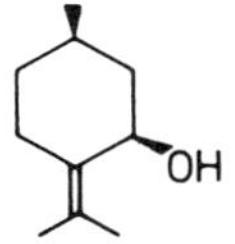

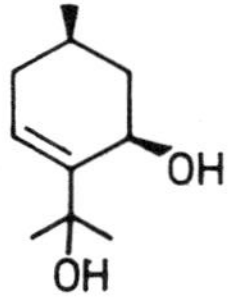

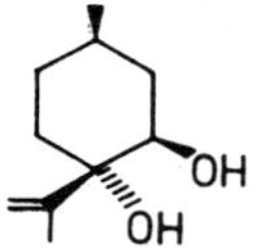

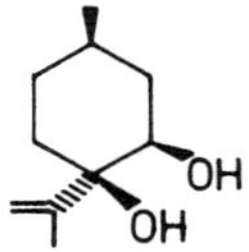

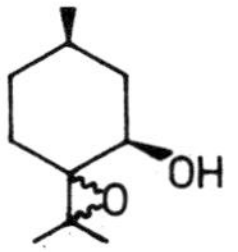

(*Continued*)

TABLE—*Continued*

Singlet oxygen acceptor trivial name	Reaction conditions‡	Isolated products§	Ref.
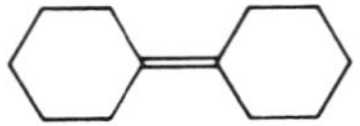	A: RB, *i*-PrOH	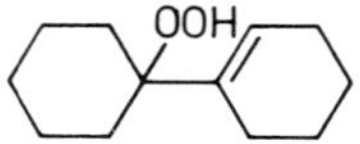 OOH (70%) m.p. 38–40°C	237 cf. 218c
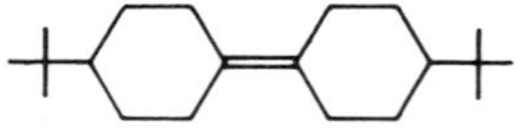 syn or anti	(1) A: *meso*-TPP, Bz. (2) $NaBH_4$	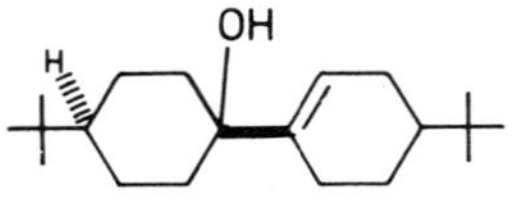 OH *60* from syn *33* from anti m.p. 154–55.5°C; 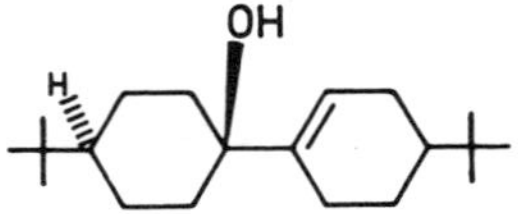 OH *40* from syn *67* from anti m.p. 180.5–83°C	122, 123
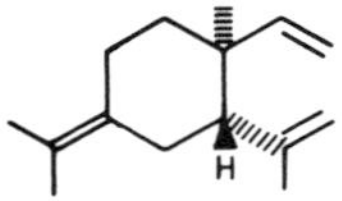 (+)-γ-Elemene	(1) A (2) Red.	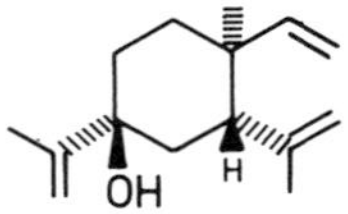 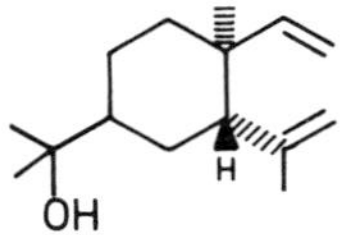 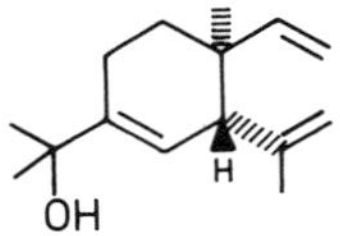 OH; OH; OH	193, 199
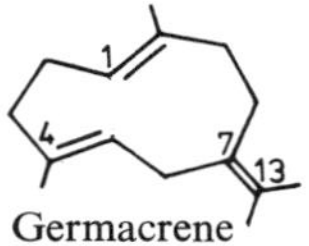 Germacrene	(1) A: MB (2) $NaBH_4$	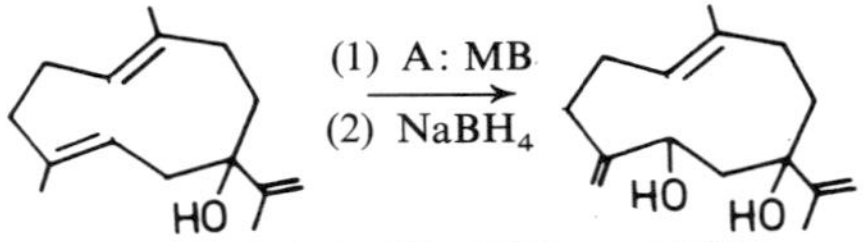 HO total (50–60%) —(1) A: MB (2) $NaBH_4$→ HO HO (40%)	222

Substrate	Conditions	Product(s)	Ref.
(+)-Isovalencene; 1:1-mixture	(1) A: RB, Bz., MeOH 1:1, Hydroquinone (2) BF_3–Etherate	(>90%)[m]; 1:1-mixture	165, 199
	(1) A (2) $NaBH_4$	2 C-7 epimers *2; 1*	222
(a) R = Me (b) R = H Allogibberic Acid	(1) A: HP, Py. (2) KI	(a) R = Me m.p. 161°–163°C	11
	A	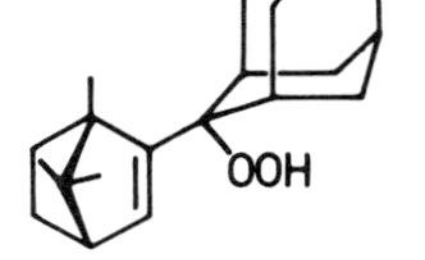 (tentative structure)	148a

(Continued)

Singlet oxygen acceptor trivial name	Reaction conditions‡	Isolated products§	Ref.
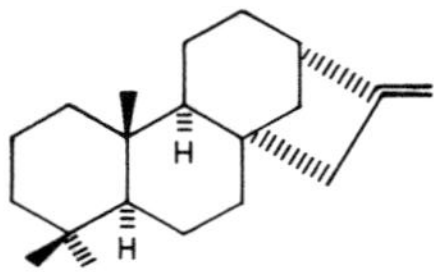 (±)-Kaurene	(1) A: HP, Py. (2) $LiAlH_4$	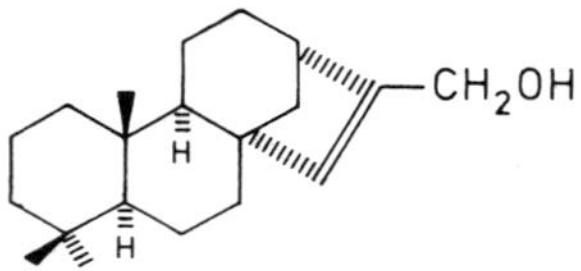 CH_2OH (63%) m.p. 122.5°–123.5°C	16
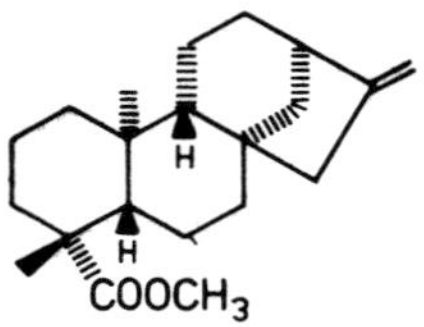 $COOCH_3$ Methyl(−)-Kaur-16-en-19-oate	(1) A: HP, Py. (2) NaI, AcOH, EtOH	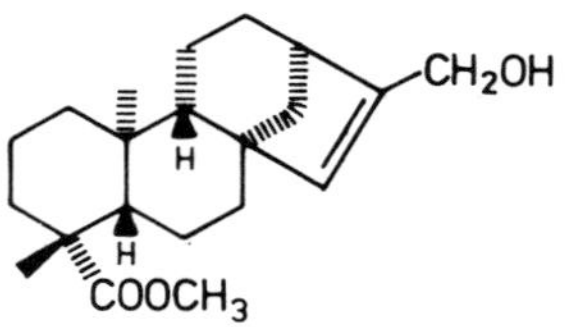 CH_2OH, $COOCH_3$ (30%) m.p. 125°–126°C; $[\alpha]_D$ −15(0.1)	9
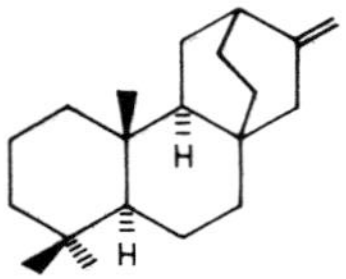 (±)-Atisirene	(1) A: HP, Py. (2) $LiAlH_4$	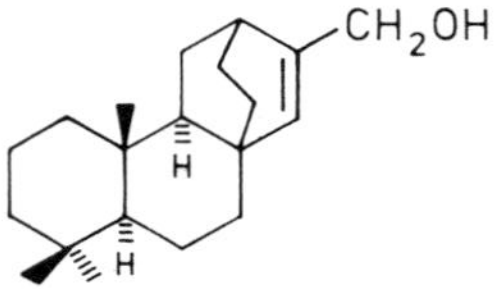 CH_2OH (75%) m.p. 114°–115°C	16
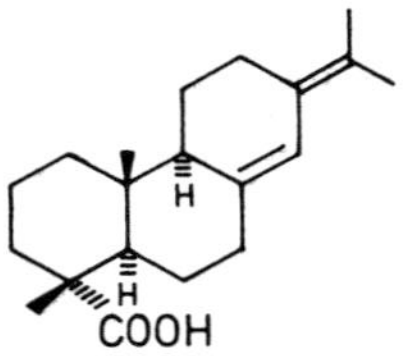 COOH Neoabietic Acid	(1) A: ER B, EtOH 95%	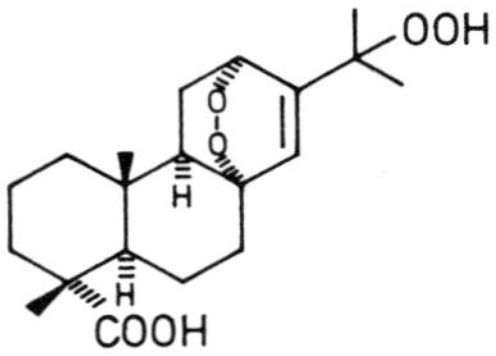 OOH, COOH (69%) m.p. 176°C dec.; $[\alpha]_D^{27}$ +94.4 (1.08, EtOH)	254

Substrate	Conditions	Product	Ref.
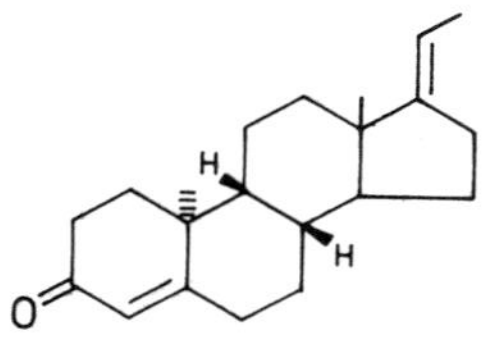	(1) A: HP, Py. (2) Ac_2O	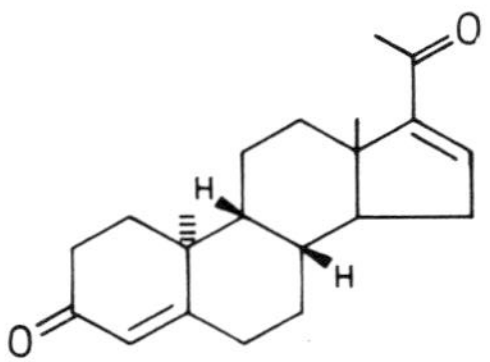 (88%) m.p. 165.5°–167°C presumably via Δ^{16}-20-hydroperoxide	142
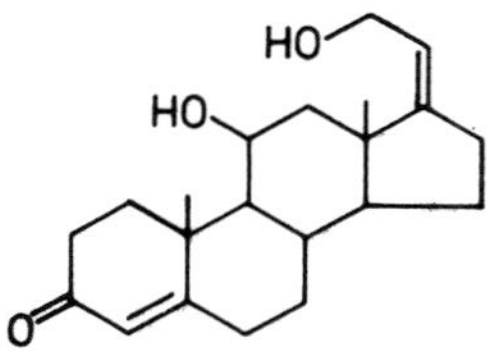	A: HP, Py.	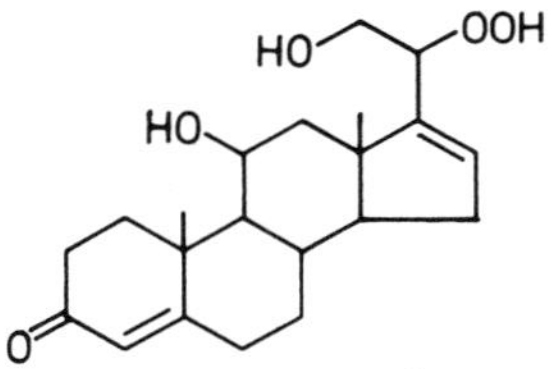	252a
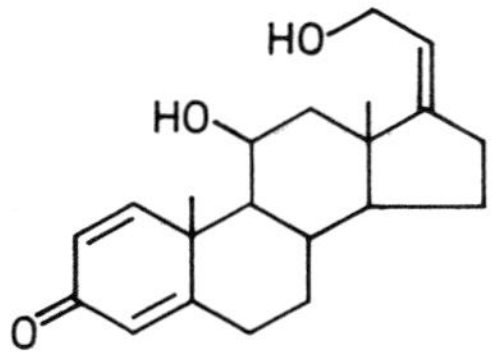	A: HP, Py.	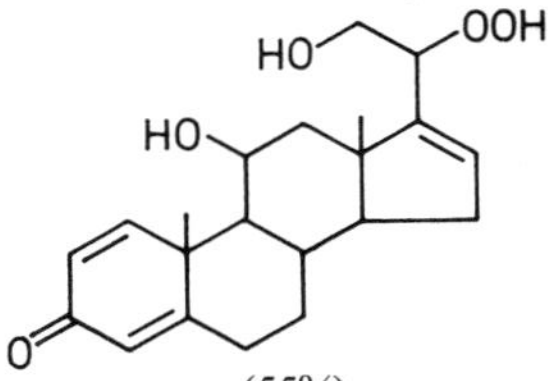 (55%)	252a
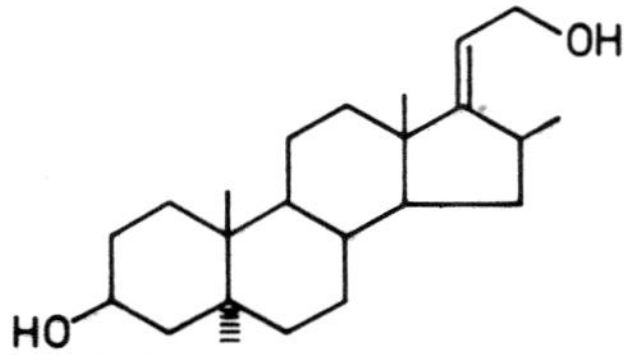 16α-epimer rather unreactive	(1) A: HP, Py. (2) Ac_2O	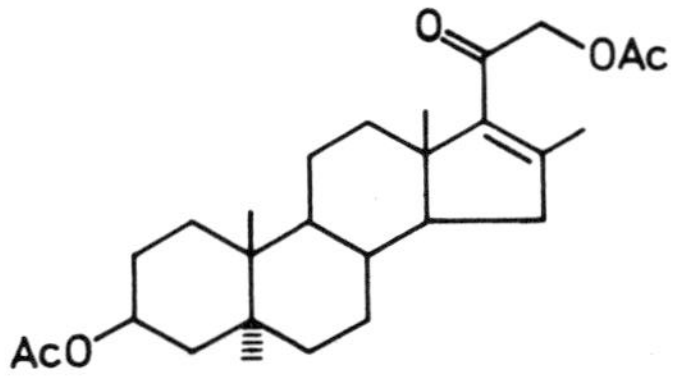	252a

(*Continued*)

TABLE—*Continued*

Singlet oxygen acceptor trivial name	Reaction conditions‡	Isolated products§	Ref.
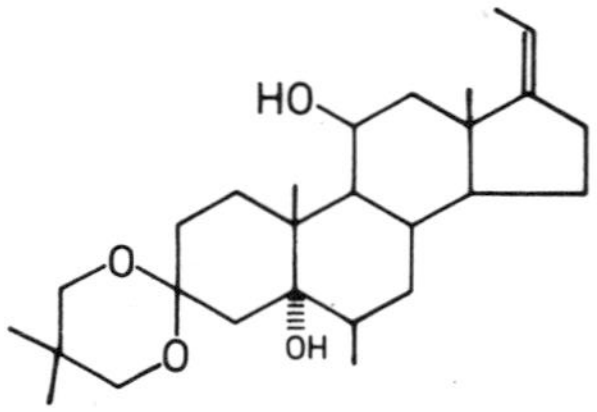	A: HP, Py.	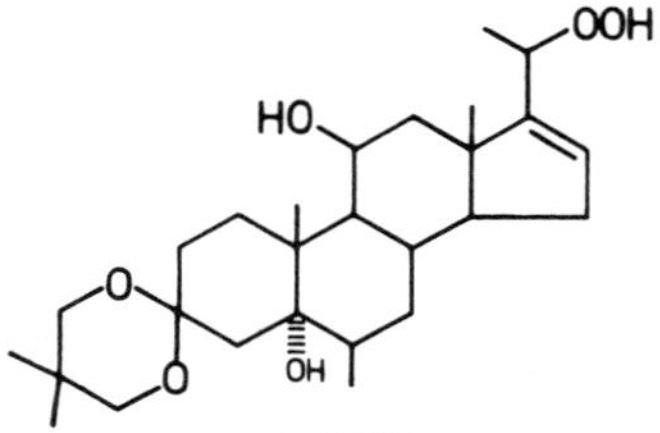 (> 70%)	252a
		(10%)	

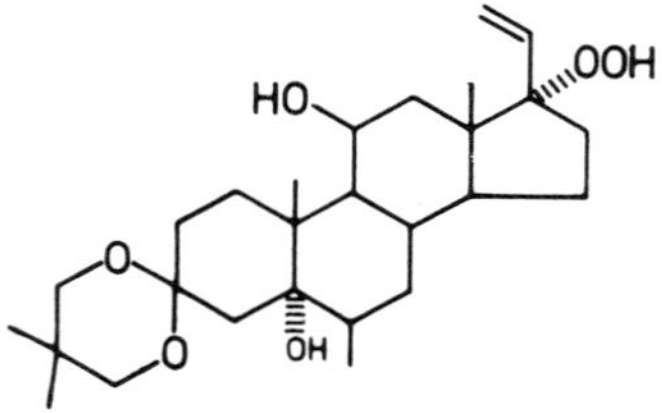

C. Olefins with Endocyclic Double Bonds

1. Monocyclic Olefins

Singlet oxygen acceptor trivial name	Reaction conditions‡	Isolated products§	Ref.
	A: sens. and solvent dependent, see r.h.s.	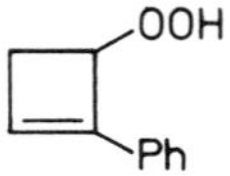 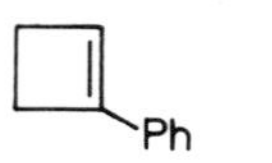 TPP, CCl_4: none; TPP, Chlf.: *20*; MB, An.: *28*; RB, MeOH: *28*	115

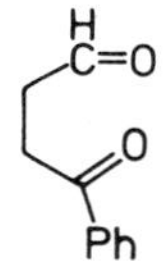	40	60	47	29		
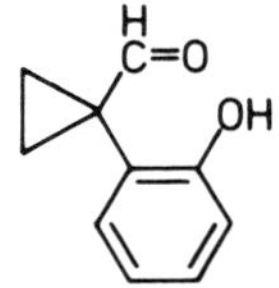	none	none	10	33		
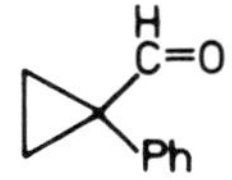	none	trace	trace	none		
		(besides polymer, ad 100)				
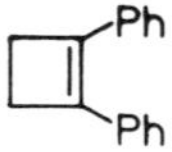 A: (a) MB, MeOH (or Ac.) 520 nm (b) MB, Mchl. 520 nm	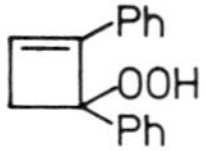 (a) (95%)	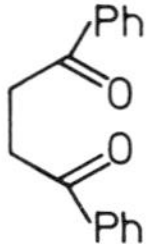 (a) (5%) (b) (40%)				259
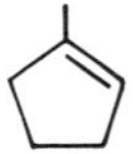 (1) A F (2) Red.	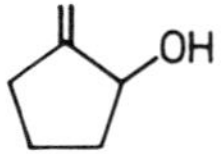 A: 4 F: 5	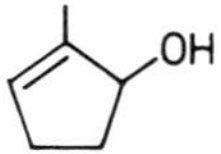 43 41	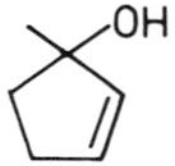 53 54			59, 205, 233, 234
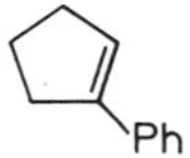 A: MB, An.	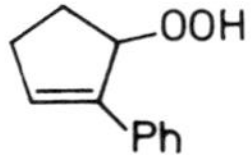					115

(Continued)

TABLE—*Continued*

Singlet oxygen acceptor trivial name	Reaction conditions‡	Isolated products§		Ref.
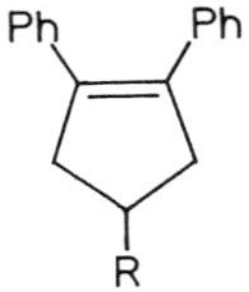 Ph, Ph	A: MB, Chlf. or Mchl.	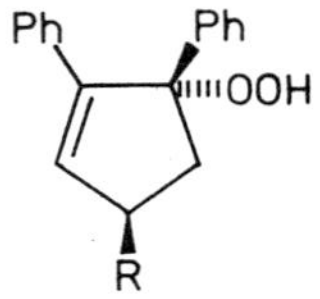 Ph, Ph, OOH, R		213, 214
R = H		(70%) m.p. 40°–41°C		
R = OH		(65%) m.p. 177°–178°C		
R = =O		(80%) m.p. 151°–153°C		
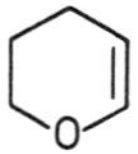 O	A: RB, Ac. or other solvents *vide infra*	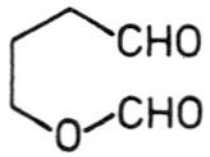 CHO, O, CHO	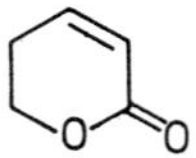 O, O	15, 19, 234
		Via dioxetane	Presumably via hydroperoxide	
	RB	Bz. *0.094*	*1*	
	RB	Ac. *0.80*	*1*	
	RB	Mchl. *2.67*	*1*	
	RB	An. *5.51*	*1*	
	RB(P)	Mchl. *2.70*	*1*	
	TPP	Mchl. *2.70*	*1*	
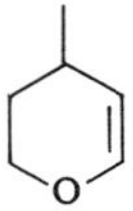 O	A: TPP, Bz., or other solvents and sens., *vide infra*	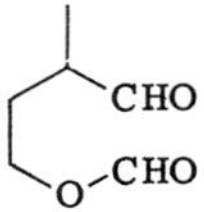 CHO, O, CHO	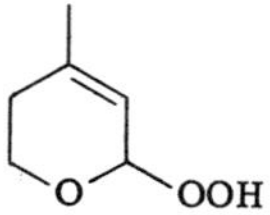 O, OOH	67c
		24	*76*	
	TPP, Cy.	*0.15*	*1*	
	TPP, Bz.	*0.20*	*1*	
	TPP, neat	*0.28*	*1*	
	TPP, Bz.-MeOH 5:1	*0.51*	*1*	
	TPP, CS_2	*0.67*	*1*	
	TPP, MeOH-Bz. 2:1	*0.89*	*1*	

Substrate	Conditions	Products (yield)			Ref.
	RB, MeOH	*1.50*	*1*		
	MB, Formamide	*1.78*	*1*		
	MB, An.	*5.25*	*1*		
	4.4-Dimethyl-2.3-dihydropyrane gives exclusively dioxetane mode products	(as dihydropyrone after vpc)			
	(1) A: MB C (2) $(PhO)_3P$	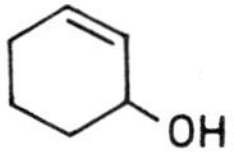 OH *100*			73, 233
OMe	A	O–O, OMe	OOH, OMe	OOH, OMe	67c, 67d
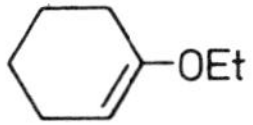 OEt	A: RB, (a) MeOH (b) Ac.	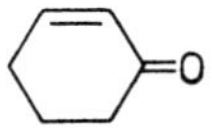 O (a) (14%) (b) (42%)	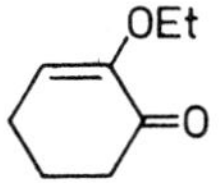 OEt, O (8%) (12%)	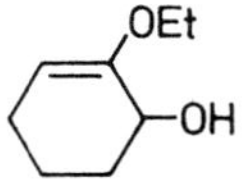 OEt, OH (11%) ene mode products (8%)	7
No ene mode products from the corresponding thioether		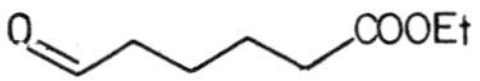 O, COOEt (a) (1%) (b) (9%)	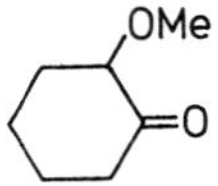 OMe, O (17%) (none)	dioxetane mode products	

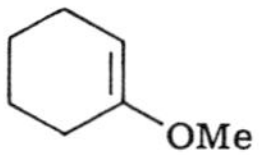

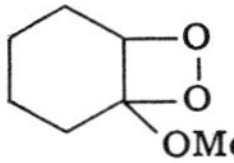

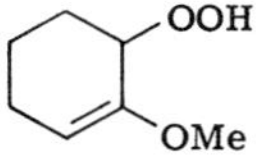

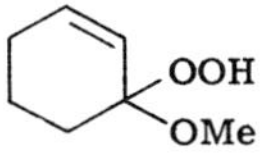

Substrate	Conditions	Product		A, A(P)	B	C	D	F[n]	G[o]	Ref.
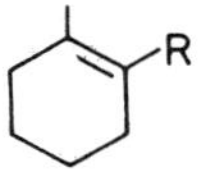 R (a) R = H (b) R = CH_3	(1) A, B, C, D, F, G (2) Red.	R, OH	(a)	*44*	*44*					A: 52, 55, 59, 73, 139, 225 233, 234, 245, 280
			(b)	*89; 87; 90: 92*	*91*	*95*	*95; 96*	*82*	*92*	

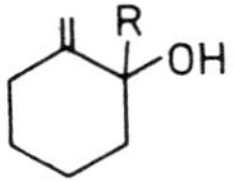

(*Continued*)

TABLE—*Continued*

Singlet oxygen acceptor trivial name	Reaction conditions‡	Isolated products§							Ref.
		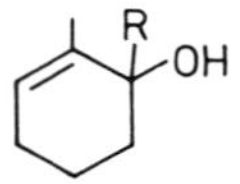 R, OH	(a) *20*	*20*					A(P): 19, 227 B: 55, 280 C: 73 D: 175, 228 E: 205 G: 38a
			(b) *11 ; 13* *10 ; 8*	*9*	*5*	*5 ; 4*	*18*	*5*	
		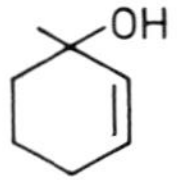 OH	(a) *36*	*36*					
		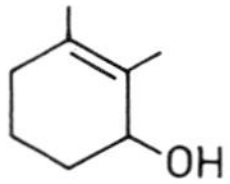 OH	(b)					*3*	
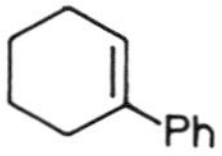 Ph	A: MB, An. C	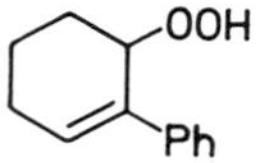 OOH, Ph C: (33%) *3*	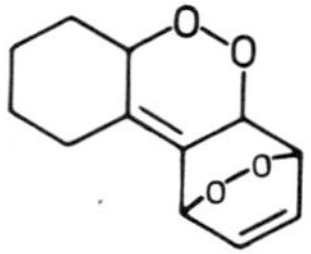 O–O, O–O *1*						115, 229
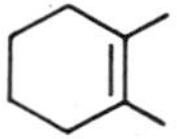	(1) A: FL, H_2O, Ac. 6:4, 1 *M* NaN_3 (2) Na_2SO_3	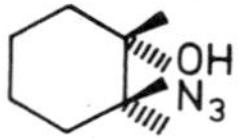 OH, N_3 (60%)							52

Substrate	Conditions	Products	References
(*1R*)-(+) *p*-Menth-3-ene	(1) A: RB, MeOH (2) Na_2SO_3	2 isomers trans $[\alpha]_D$ −22 (10) cis $[\alpha]_D$ +167.0 (10) 2 isomers trans $[\alpha]_D$ −176.0	196
(+) (+)-Isosylvestrene	(1) A: RB, MeOH (or other sens) (2) Na_2SO_3	*37* $[\alpha]_D$ −105.0 (2.1, Bz.) 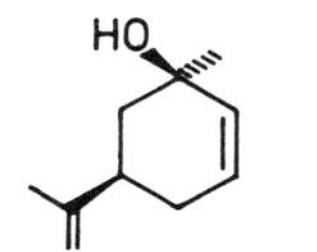 *4* $[\alpha]_D$ −74.0 (2.1, Bz.) *5* $[\alpha]_D$ +168.0 (4.0, Bz.) *10* $[\alpha]_D$ −20.7 (1.1, Bz.)	74, 83, 85

(*Continued*)

TABLE—*Continued*

Singlet oxygen acceptor trivial name	Reaction conditions‡	Isolated products§	Ref.
		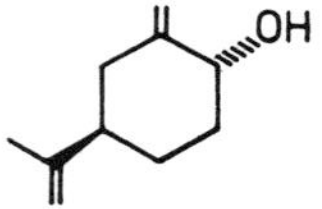 *26* $[\alpha]_D$ −40.0 (2.0, Bz.)	
		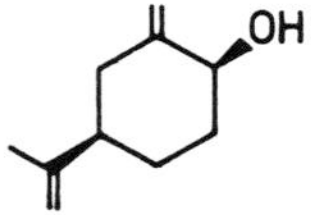 *17* $[\alpha]_D$ +31.2 (1.45, Bz.)	
 (+)-Carvomenthene	(1) A: (a) MB, *i*-PrOH (b) CH, EtOH (or other) (2) Na_2SO_3 (a) total (89%)	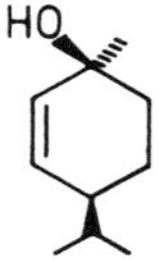 (a) *33*, (b) *27* m.p. 21°–23°C $[\alpha]_D$ −24.7	(a) 249 (b) 126 233, 234, 250, 251
		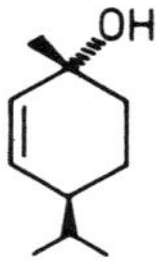 (a) *10*, (b) *10* b.p.$_{\cdot 4}$ 79.4°C $[\alpha]_D$ +80 (4, EtOH)	

Substrate	Conditions	Products and yields (%)	References
		18 15	
		b.p.$_{4}$ 81.6°C: $[\alpha]_D$ +94.9 (12, EtOH)	
		12 17	
		b.p.$_{4}$ 89.2°C; $[\alpha]_D$ −113.9 (4.1, EtOH)	
		none	
		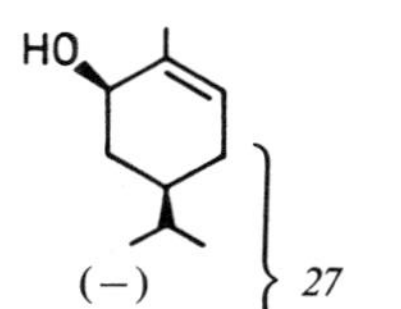 (−) } 27 19	
		m.p. 18°C; $[\alpha]_D$ +6.7	
		(b) *cis*-Piperitol [(+)-Δ^1-*cis*-3-OH] *2* and (+)-1,2-carvomenthene epoxide *9* were also found	
(+)-Limonene	(1) A: RB, MeOH (or *t*-BuOH, MeOH, 1:1 or *i*-PrOH) or other sens., MeOH B: H_2O_2, NaOCl (or $Ca(OCl)_2$ +2.6-di-	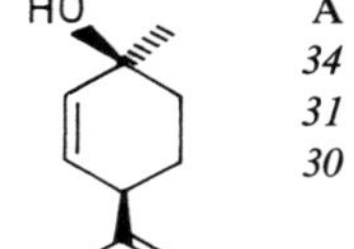 A B C *34* *34* *41* *31* *30*	54, 55, 85, 87, 225, 233, 234, 249–251, 280

(*Continued*)

TABLE—*Continued*

Singlet oxygen acceptor trivial name	Reaction conditions‡	Isolated products§	Ref.
	tert-butyl phenol), *t*-BuOH, MeOH 1:1) C: microwave discharge, MeOH, −50°C in O_2 (or in CO_2) (2) $LiAlH_4$ or Na_2SO_3 or $NaBH_4$ or $Na_2S_2O_3$ + KI total (76%)	m.p. 18°C; $[\alpha]_D$ +67.3	
		OH; *10* *10* *9* *11* *9* *7*	
		b.p.$_{6.5}$ 86.5°C; $[\alpha]_D$ +180.2	
		HO; *20* *18* *18* *19* *21* *23*	
		b.p.$_5$ 85°C; $[\alpha]_D$ +92.4	
		HO; *21* *21* *22* *23* *24* *25* *26*	
		m.p. 24°C; $[\alpha]_D$ +5.7	
		HO; *10* *10* *7* *9* *9* *8* *10*	
		$[\alpha]_D$ −178	

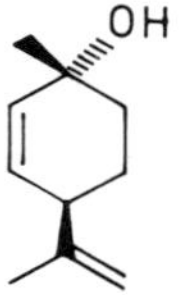

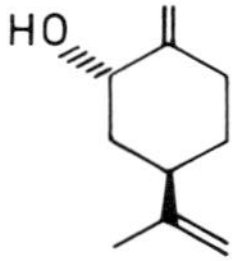

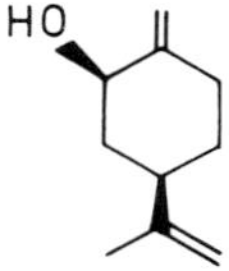

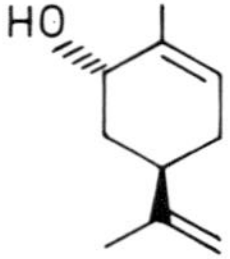

		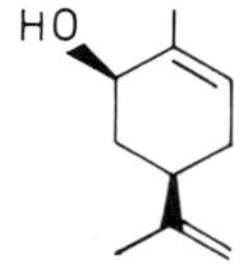 (−)	5 4 3 7 3	
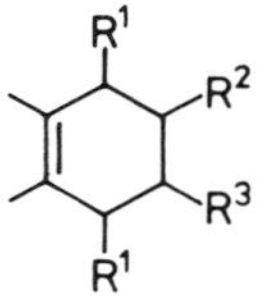 $R^1 = H$; R^2,R^3 = *cis*-COOMe $R^1 = H$; R^2,R^3 = *trans*-COOMe R^1 = *cis*-Me; R^2,R^3 = *cis*-COOMe R^1 = *cis*-Me; R^2 = *cis*-COOMe; R^3 = *trans*-COOMe	A: RB, MeOH	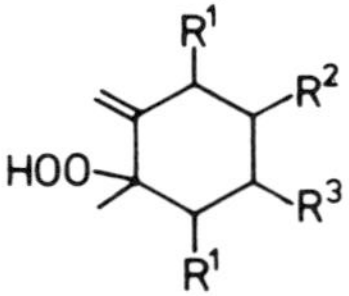 *80* *74* *71* *60*	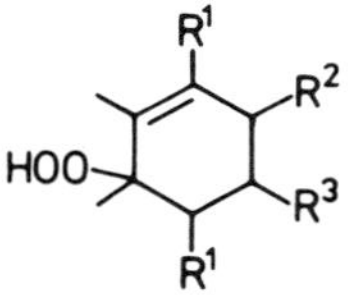 total (98%) *20* total (95%) *26* total (95%) *29* total (95%) *40*	245
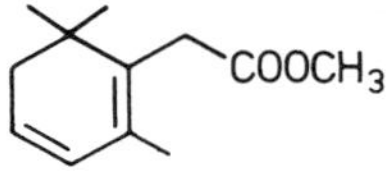 Methyl homosafranate	A: RB, MeOH	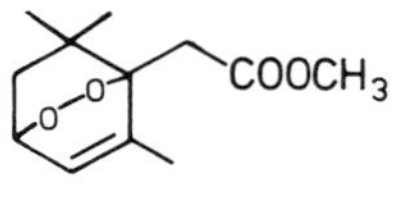 (53%)	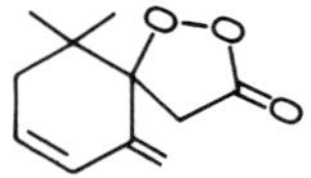 (9%) b.p.$_{0.001}$ 90°C	40
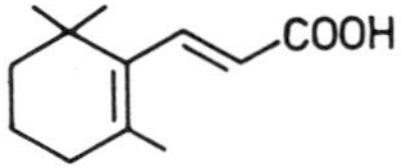	(1) A: RB, MeOH (2) $(MeO)_3P$ (3) CH_2N_2	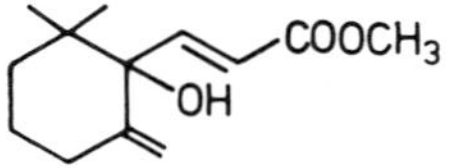 (33%)	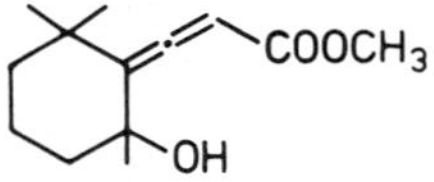 (7%)	21
		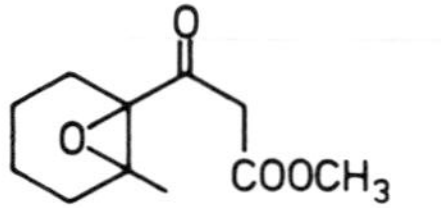 (5%)	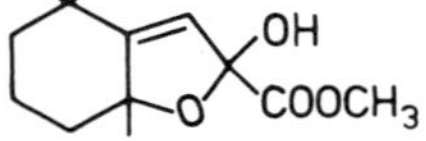 2 epimers (38%)	

(*Continued*)

TABLE—*Continued*

Singlet oxygen acceptor trivial name	Reaction conditions‡	Isolated products§	Ref.
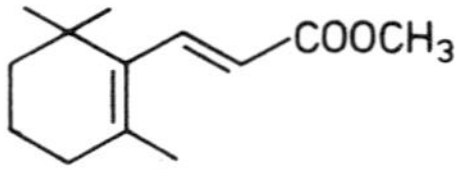	(1) A: RB, MeOH, 0.01 *N* NaOH (2) SiO_2–Chrom.	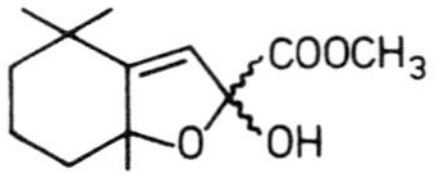 Mixture of isomers (24%) 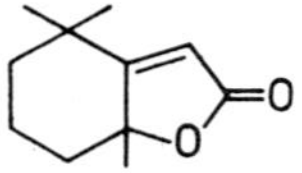 (5%) 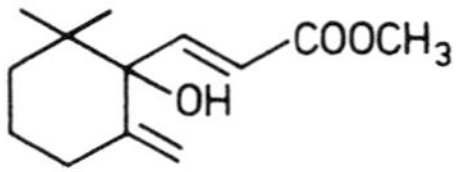 ~(15%) 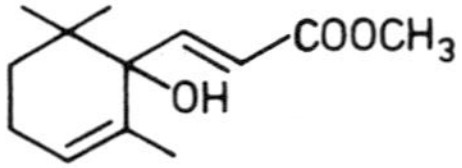 ~(2%)	170
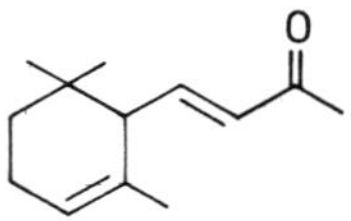 α-Ionone	(1) A: RB, MeOH, 0°C, Na lamp (2) Ph_3P	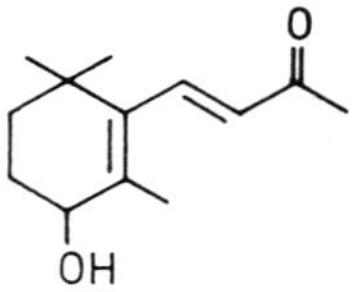	171
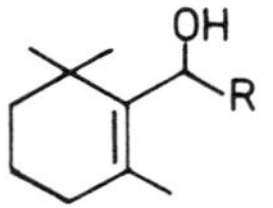 β-Damascoles (a) R =  (b) R =  (c) R = 	(1) A: RB, MeOH, NaOAc (2) Na_2SO_3, <15°C	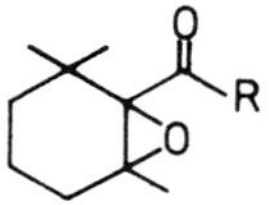 (a) (48%) b.p.$_{0.5}$ 80°C (b) (42%) (c) (50%) 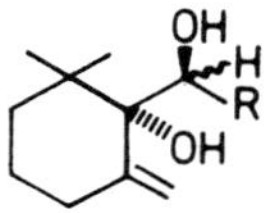 (30%) *threo* (25%) (27%) m.p. 64°C	258

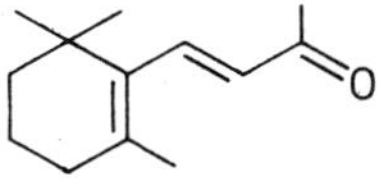 β-Ionone	(1) A: (a) RB, MeOH, 0.01 n NaOH (b) RB, MeOH (2) (a) SiO_2–Chrom. (b) $(MeO)_3P$	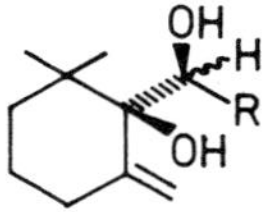 (a) (6%) m.p. 86°C *erythro* (b) (8%) (c) (4%) m.p. 56°C	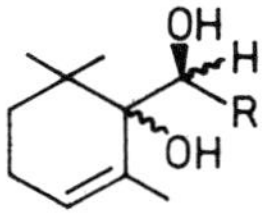 (4%)[p] (3%)[p] (5%)[p]	(a) 21 (b) 170
		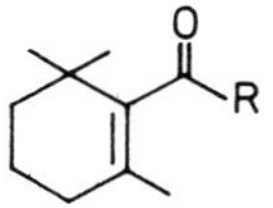 (a) (1%)		
		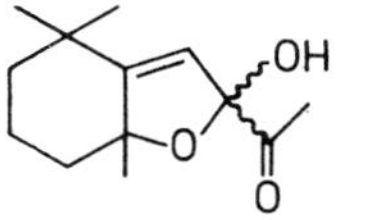 Mixture of isomers (a) *28* (b) *45*	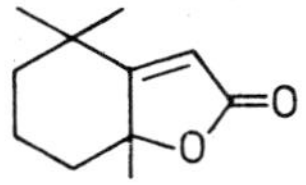 (a) *44* (b) *3*	
		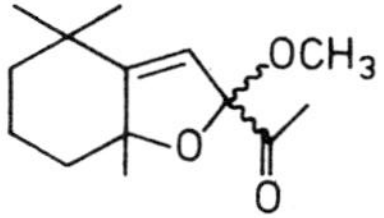 Mixture of isomers (a) None (b) *26*	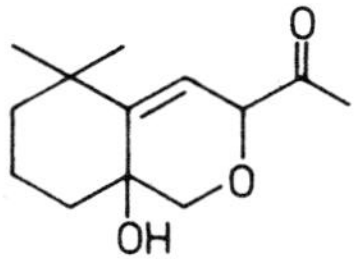 (a) *12* (b) None	
		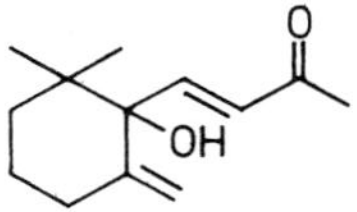 (a) *8* (b) *24*	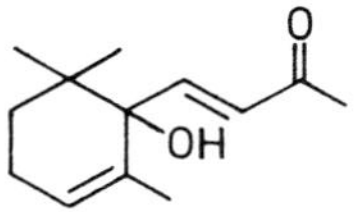 (a) *2* (b) None	

(*Continued*)

TABLE—*Continued*

Singlet oxygen acceptor trivial name	Reaction conditions‡	Isolated products§	Ref.
		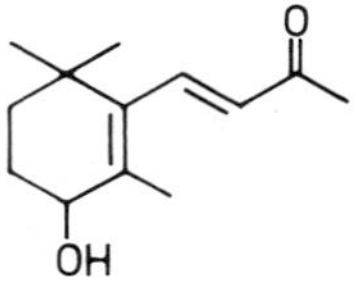 (a) *6* (b) None; 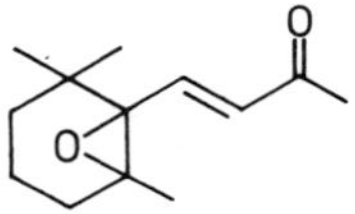 (a) None (b) *2*	
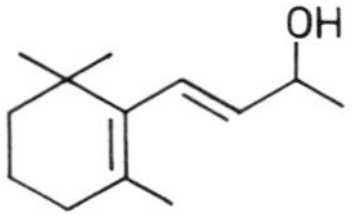 β-Ionol	A: RB, alkali traces	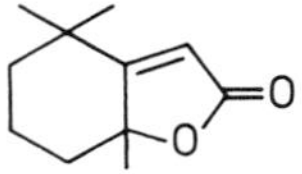 Major product 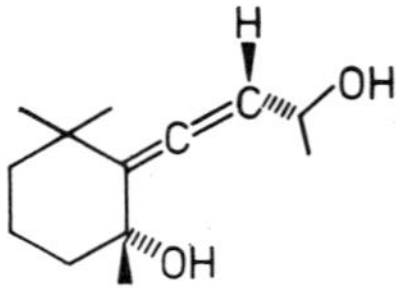	105, 106
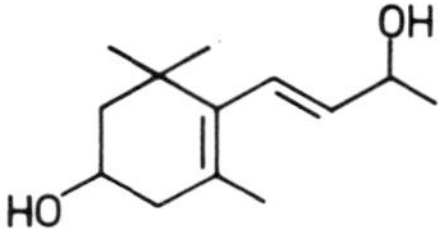 3-Hydroxy-β-ionol	A: RB, MeOH, alkali traces	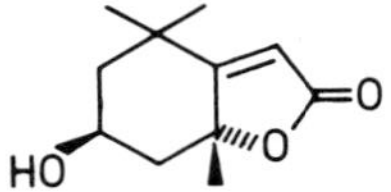 m.p. 143.5°–144°C 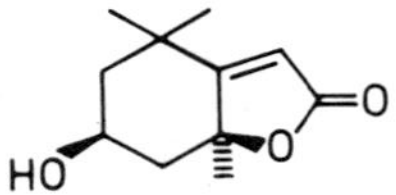 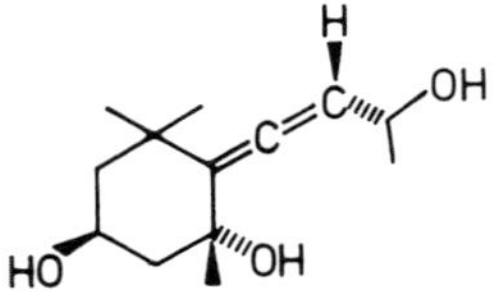	107, 274

Substrate	Conditions	Product(s) and yield(s) (%)			Refs.
	(1) A: RB, MeOH B: H_2O, NaOCl, Diglyme, −20°C (2) $NaBH_4$ total (volatile products) (85%)		A	B	21, 58, 274
			8	*10*	
		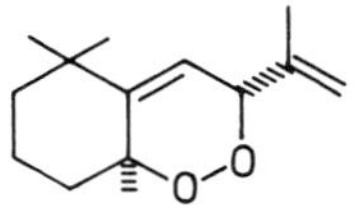	*61*	*51*	
		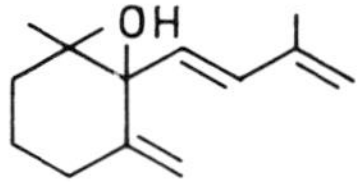	*27*	*33*	
		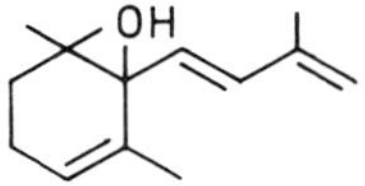	*4*	*5*	
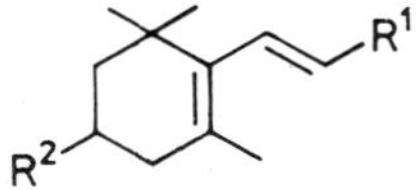 R^1 = cis or trans $C(CH_3)$=$CHCOOCH_3$ R^2 = H or OMe	(1) A: RB, MeOH, 0°C, Na lamp (2) Ph_3P	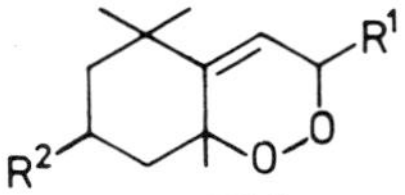 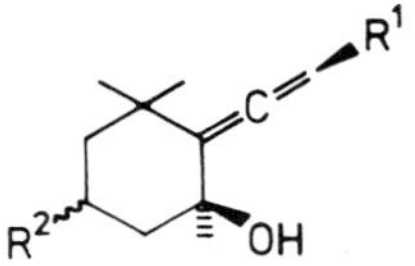 (40–50%) (20–30%) (5–10%)	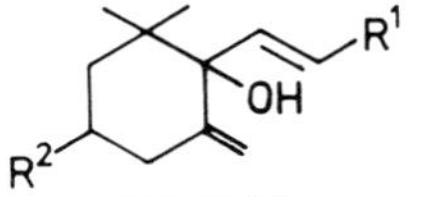		38, 170, 171, 173

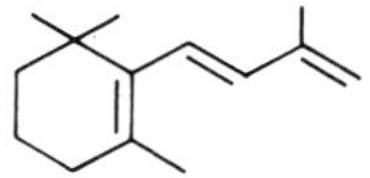

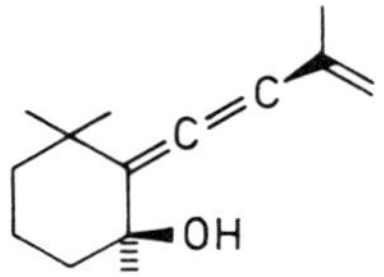

(Continued)

Singlet oxygen acceptor trivial name	Reaction conditions‡	Isolated products§	Ref.
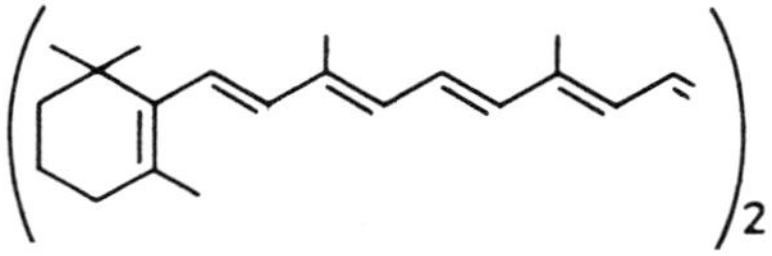 all-*trans*-β-Carotene	(1) A: CH (2) Al_2O_3 bas.	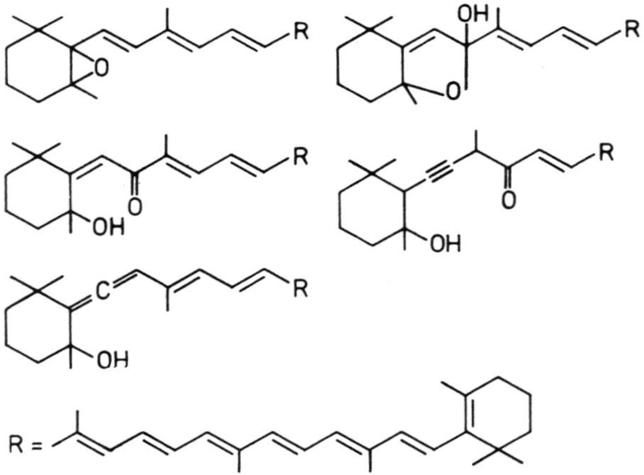 R =	106, 108, 260, 271
	(1) A: MB, MeOH (2) Pd–C, H_2, EtOAc	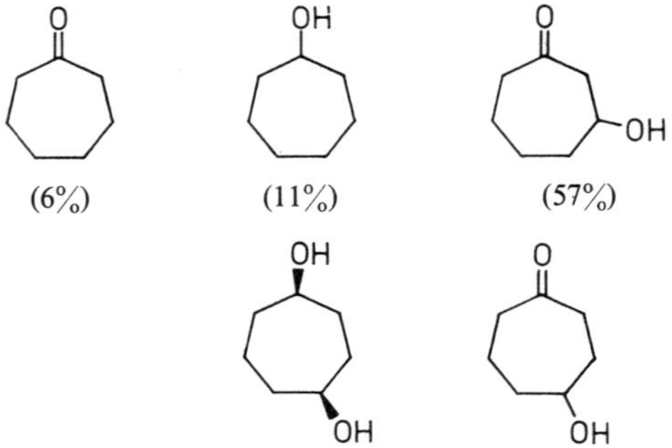 (6%) (11%) (57%) (6%) (20%)	124 cf. 168b 7a

Substrate	Conditions	Products	Ref.
	A	4 / 48 / 48	59
	(1) A: MB, EtOH (2) Na_2SO_3	(44%) b.p.$_2$ 80°C	160, 234
	(1) A: RB or MB, MeOH B: NaOCl, H_2O_2 (2) Na_2SO_3	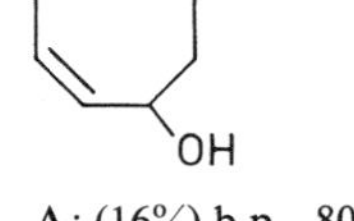 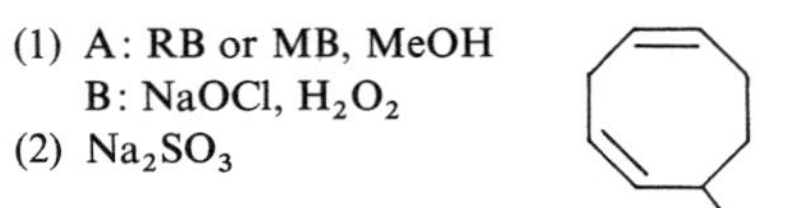 A: (16%) b,p.$_2$ 80°C	160
Reacts 7 times as fast as cyclooctene	(1) A: RB, MeOH (2) Na_2SO_3	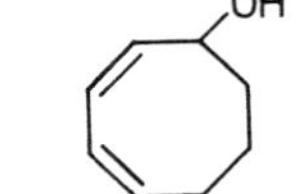 (39%) b.p.$_3$ 105°C	101
	A	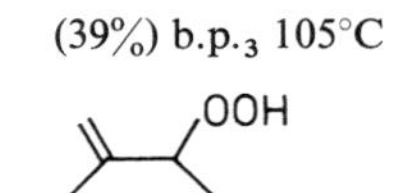 27 / 42 / 31	59

(*Continued*)

Singlet oxygen acceptor trivial name	Reaction conditions‡	Isolated products§	Ref.
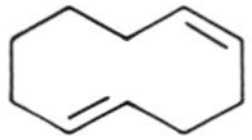 cis, trans (*cis,cis*-1,6-isomer: no reaction)	(1) A: RB, MeOH (2) Na_2SO_3	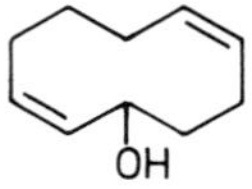 OH (44%) m.p. 96°–96.5°C	101
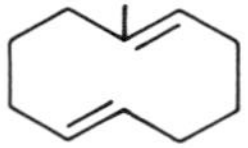	A: MB, MeOH, 10°C	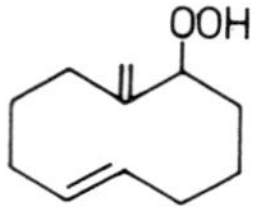 OOH (55%)	34
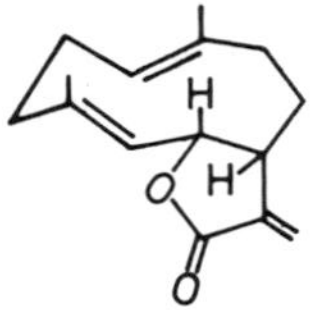 Costunolide	A: MB, EtOH	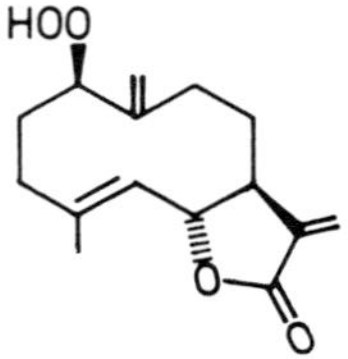 HOO Peroxycostunolide (excellent yield) m.p. 141°C dec.; $[\alpha]_D$ +171 (0.20, Ac.)	48a
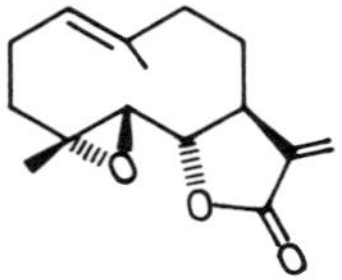 Parthenolide	A: MB	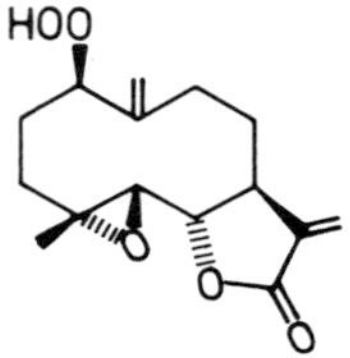 HOO Peroxypartenolide m.p. 190°C dec.; $[\alpha]_D$ +27 (0.21, Ac.)	48a cf. 44a

Substrate	Conditions	Products	Ref.
Germacrene	(1) D: $(PhO)_3P.O_3$, $-45°C$ (2) $LiAlH_4$	(50%) Neither of the products obtained by method A (cf. Section III,B) could be detected	222
Germacrone	(1) A: RB, MeOH (2) Na_2SO_3	(2.5%) m.p. 91.5°–92.5°C (besides complex mixture)	101
Agerol	(1) A: MB (2) $NaBH_4$	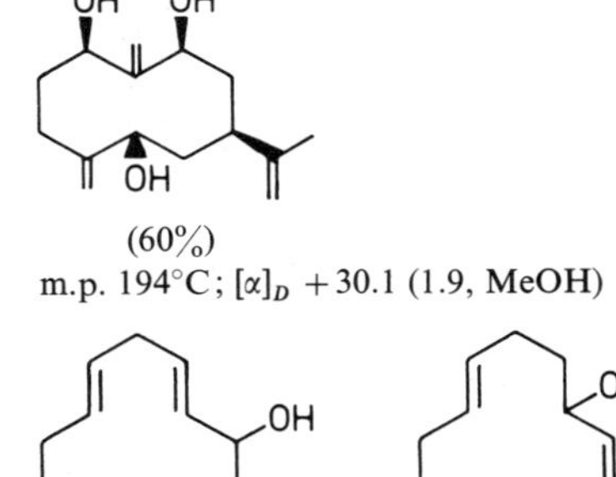 (60%) m.p. 194°C; $[\alpha]_D$ +30.1 (1.9, MeOH)	17
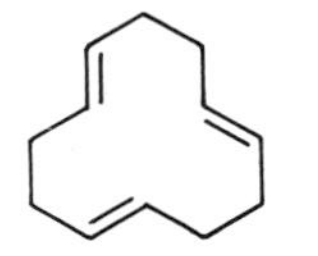 all-*trans*	(1) A: RB, MeOH (2) Na_2SO_3	(41%) *1* : *1.3* (besides other products)	101, 147

(*Continued*)

Singlet oxygen acceptor trivial name	Reaction conditions‡	Isolated products§	Ref.
2. Bicyclic Olefins			
	(1) A: RB or MB, Mchl., MeOH 9:1 (2) $NaBH_4$	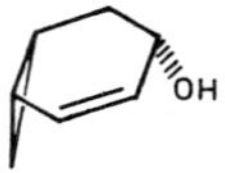 OH *100* (syn)	203
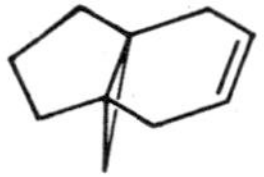	As above	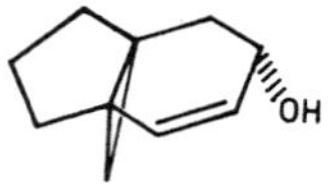 OH *100* (syn)	203
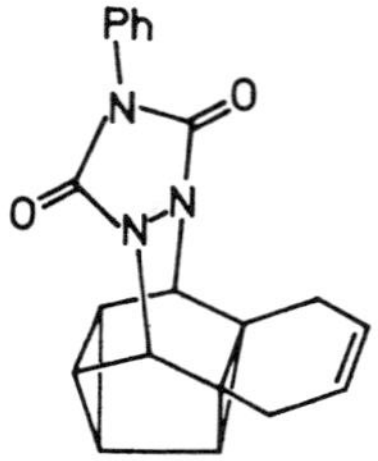 Ph N O O N N	As above	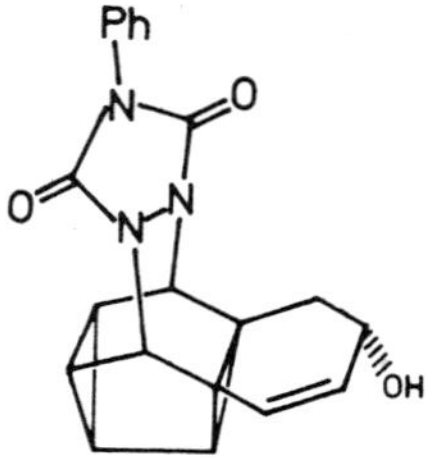 Ph N O O N N OH *91* (syn)	203
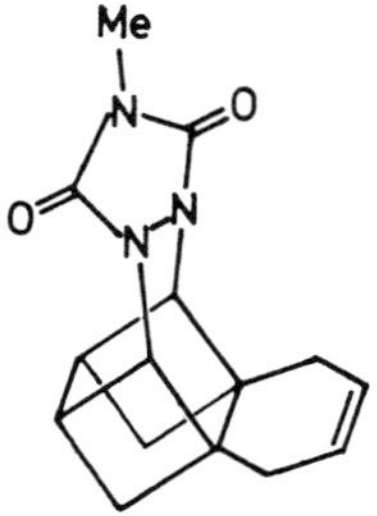 Me N O O N N	(1) A: RB or MB, Mchl, MeOH 9:1 (2) $NaBH_4$	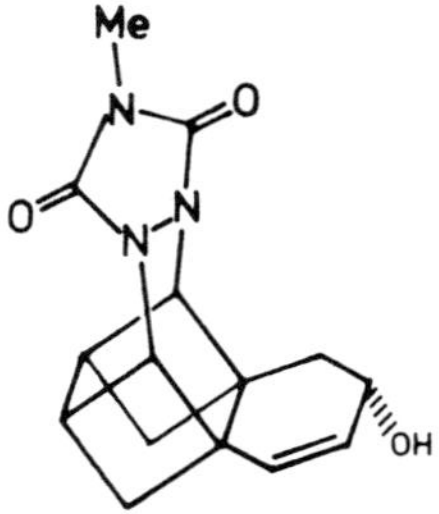 Me N O O N N OH *90* (syn)	203

Substrate	Conditions	Product(s) and Yield(s) (%)		Refs.
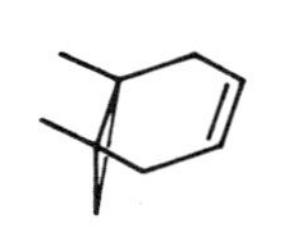	(1) A: RB or MB, Mchl., MeOH 9:1 (2) $NaBH_4$	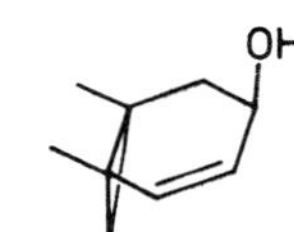 *100* (anti)		203
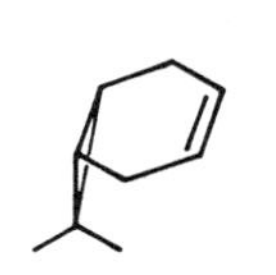	As above	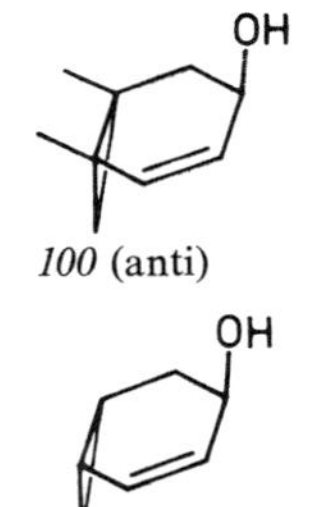 *100* (anti)		203
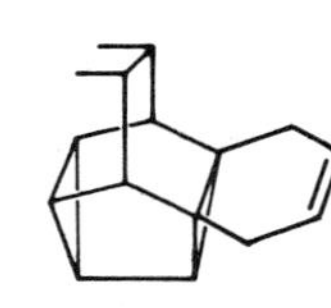	As above	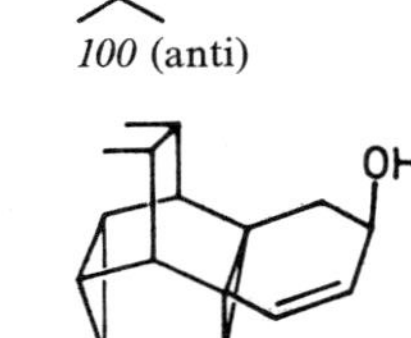 *100* (anti)		203
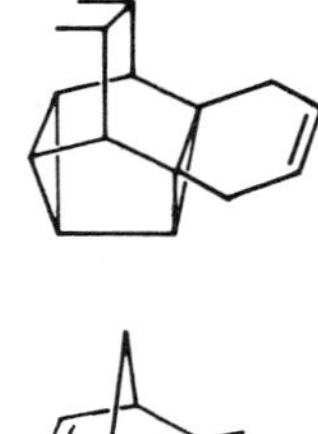	(1) A: MB, An., 0°C (2) Ph_3P or $NaBH_4$	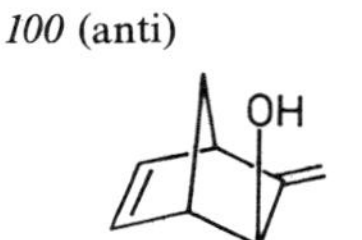 *7, 33*	(88%) 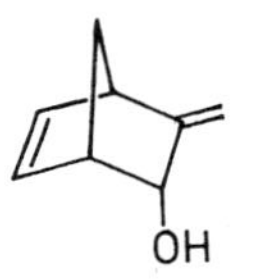 *1*	116b

(*Continued*)

TABLE—*Continued*

Singlet oxygen acceptor trivial name	Reaction conditions‡	Isolated products§	Ref.
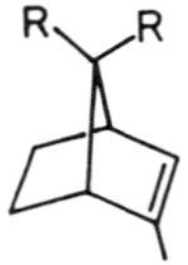 R, R (structure) (a) R = H (b) R = Me	(1) A: MB, An., 5°–10°C (2) Ph_3P or $NaBH_4$	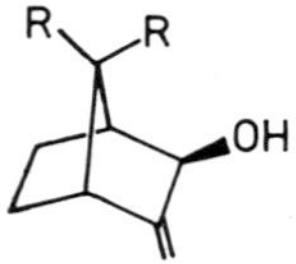 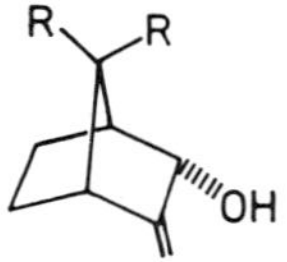 R, R, OH (structure); R, R, OH (structure) (a) 98.5 (85%) 1.5 (b) (~95%) exo-endo alcohol mixture not separated	5, 112, 114 cf. 116a
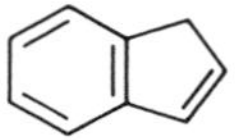 (structure)	(1) A: MB or RB, MeOH, 0°C (2) Na_2SO_3	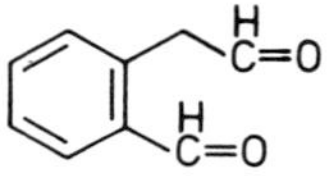 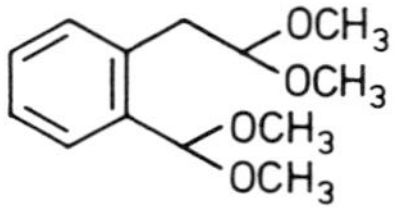 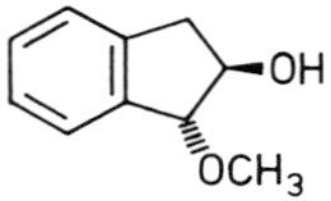 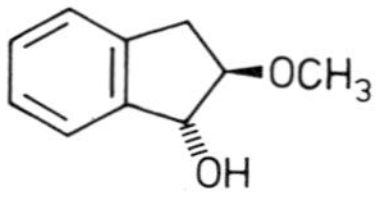 H, C=O, H, C=O (structure) (6%); OCH_3, OCH_3, OCH_3, OCH_3 (structure) (74%) OH, OCH_3 (structure) (16%); OCH_3, OH (structure) (4%)	51, 83
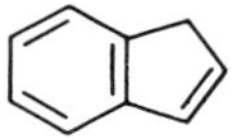 (structure)	(1) A: MB, H_2O, Ac. 6:4, 1 *M* NaN_3 (2) Na_2SO_3	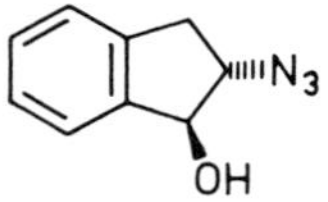 N_3, OH (structure) (besides other azides)	52

Substrate	Conditions	Product(s)		Ref.
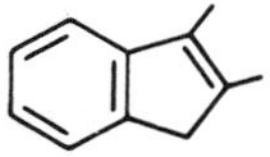 (a) $R^1 = R^2 = Me$ (b) $R^1 = H$; $R^2 = Me$ (c) $R^1 = Me$; $R^2 = H$	(1) A: MB, Mchl., 0°C (2) Na_2SO_3	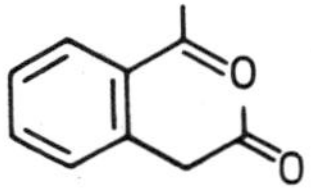 (a) (30%)	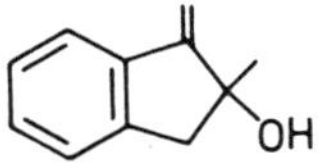 (65%)	51
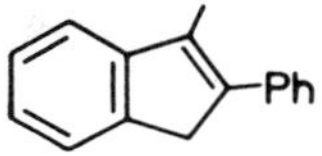	(1) A: RB, MeOH, Ac. 1:1 −78°C, ≥450 nm	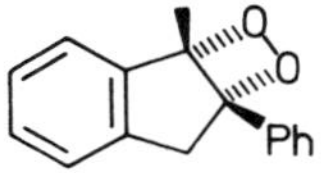 (54%) m.p. 69°–70°C dec.	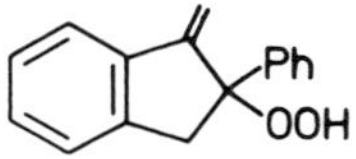 (3%)	25
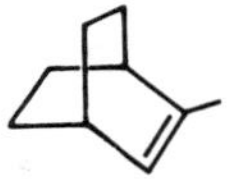	(1) A: RB, MeOH, Py. (0.2%), 2,6-di-*tert*-butyl-4-methyl phenol (2) $NaBH_4$	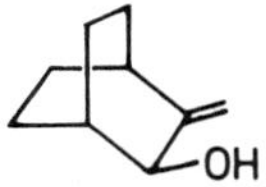 m.p. 28°–30°C		68
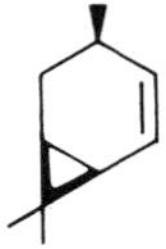 (−)-*cis*-Δ^2-Carene [No reaction with (−)-*trans*-Δ^2-Carene]	(1) A: RB, MeOH (2) Na_2SO_3	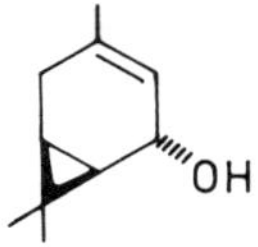 $[\alpha]_D$ +97.6 (2.1, Bz.)		80

(*Continued*)

TABLE—*Continued*

Singlet oxygen acceptor trivial name	Reaction conditions‡	Isolated products§			Ref.
 (+)-α-Thujene	(1) A: RB, MeOH (2) Na_2SO_3	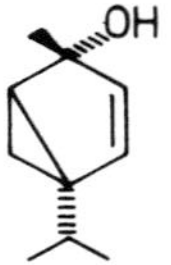 *78.5* $[\alpha]_D$ −58.5 total (71.5%)	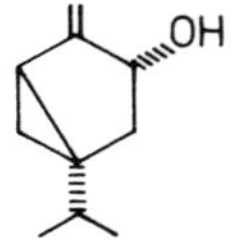 *19.5* m.p. 43.5°–45°C; $[\alpha]_D$ −59.5	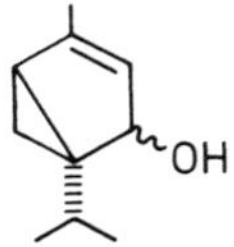 *2* $[\alpha]_D$ +49.8 Presumably secondary product (via allyl/rearr. of the tertiary alcohol)	131
 (+)-Δ^3-Carene	(1) A: RB, MeOH (2) Na_2SO_3	 *50* b.p.$_{1.5}$ 74°–76°C $[\alpha]_D$ −289 (3.6, Bz.)	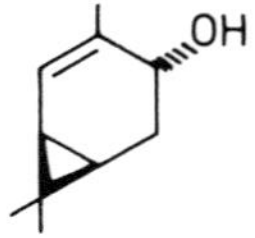 *28* b.p.$_3$ 93°–94°C $[\alpha]_D$ +203.8 (3.2, Bz.)	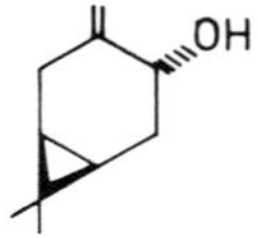 *22* m.p. 56°–57°C $[\alpha]_D$ −118.7 (3.4)	79, 250
 (+)-Δ^4-Carene	(1) A: RB, MeOH (2) Na_2SO_3	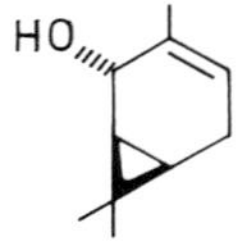 *18* $[\alpha]_D$ +132 (2.2, Bz.)	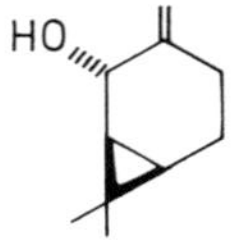 *82* $[\alpha]_D$ +33.0 (1.9, Bz.)		80
 α-Pinene	(1) A: MB, An. (or *i*-PrOH) B: H_2O_2, $Ca(OCl)_2$, 2.6-di-*tert*-butyl phenol, −20°C				54, 55, 76, 113, 233, 234 cf. 141a

Substrate	Conditions	Product(s)							Ref.
	F: no product (2) $NaBH_4$ (or Na_2SO_3 or Ph_3P)	A: (0.04%) b.p.$_{14}$ 83°C 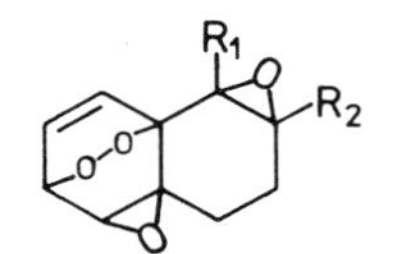 A: (99.3%) B: 85[q] b.p.$_{17}$ 95°–97°C	A: (0.01%) m.p. 41°–43°C 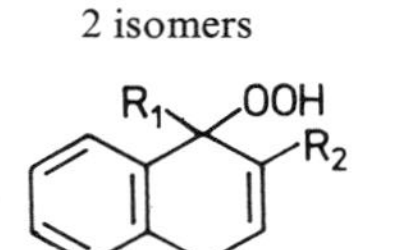 A: (0.8%)						
	A: MB, Mchl., $P(OEt)_3$	(55%) m.p. < 20°C							272a
R¹ R² (a) H Ph (b) Ph H (c) Ph Ph (d) Me Ph (e) Me H (f) H H	A: RB, Ac., −78°C		(a)	(b)	(c)	(d)	(e)	(f)	27
		2 isomers	none	*100*	*65*	*10*	*60*	*75*	
			100	none	*35*	*85*	*40*	*25*	
			none	none	none	*5*	none?	none	

(Continued)

TABLE—*Continued*

Singlet oxygen acceptor trivial name	Reaction conditions‡	Isolated products§	Ref.
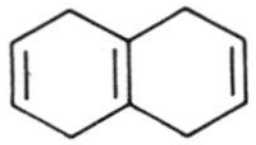	A	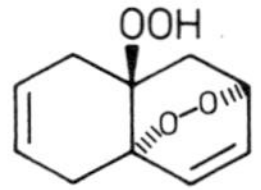 OOH	240
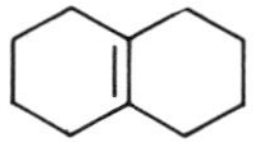	A: RB, *i*-PrOH B: NaOCl, H_2O_2, *i*-PrOH, 10°C	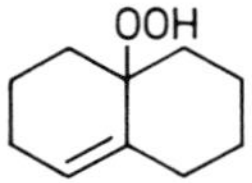 OOH A: (84%) m.p. 59°–60°C B: (12%)	57, 237, 281
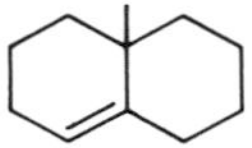	A	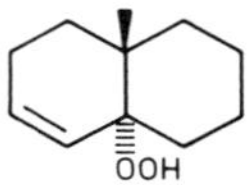 OOH	242
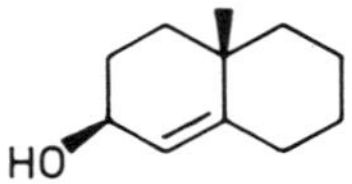 HO	(1) A: HP, Py. (2) $LiAlH_4$ (3) CrO_3	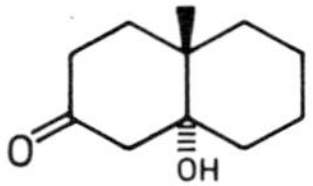 O OH (13%) m.p. 155°–156°C via 1(9)-oxido-2-ketone (?)	149
 (+)-Nopadiene	A: RB, MeOH	 O (23%) m.p. 26°C (besides CH_3CHO, presumably via hydroperoxide)  O–O (35%) $[\alpha]_D$ −68.9	246

Substrate	Conditions	Product(s)	Yield (%)	Ref.
 α-Ethylapopinene	(1) A: RB, MeOH (2) Na_2SO_3	 (39%) b.p.$_{0.015}$ 66°C; $[\alpha]_D$ +23.53		246
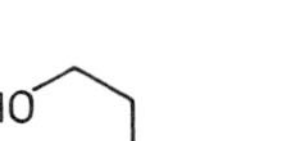 (−)-Nopol	(1) A: RB, MeOH (2) Na_2SO_3	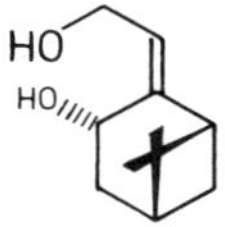 (50%) $[\alpha]_D$ +41.3		246
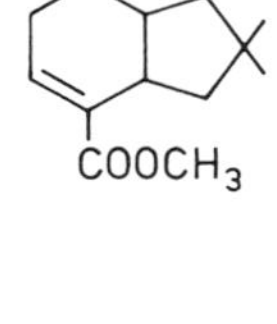	(1) A: RB (2) Ph_3P	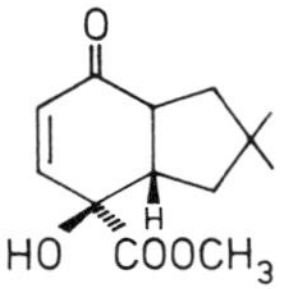 m.p. 111°C (presumably via dienol)		97
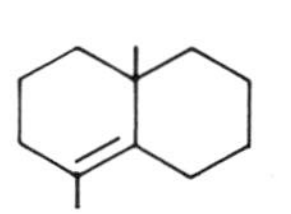	(1) A: HP, Py. B: NaOCl, H_2O_2, EtOH (2) $LiAlH_4$	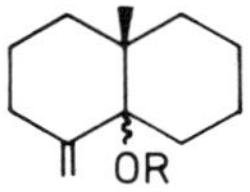 R = H 2 isomers	R = OH, trans: m.p. 76°–77°C A: total (82%) B: total (20%) trans *45* *36* cis *22* *18*	150
		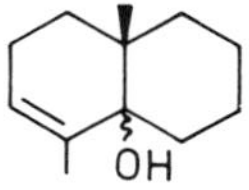 2 isomers	trans *19* cis *9* *34*	

(*Continued*)

TABLE—*Continued*

Singlet oxygen acceptor trivial name	Reaction conditions‡	Isolated products§		Ref.
		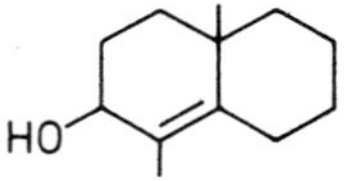 HO	4 11	
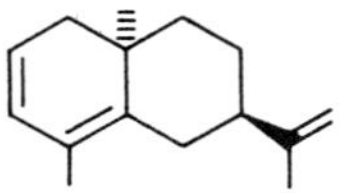	A: EO Y, Bz., EtOH 1:1	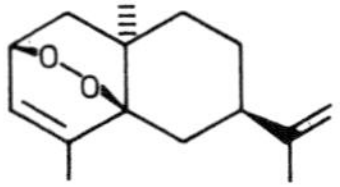 (62%) m.p. 62°–63°C $[\alpha]_D$ −29.0 (5.5)	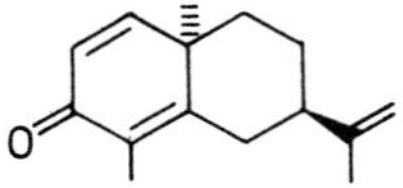 (22%) (presumably via hydroperoxide)	12
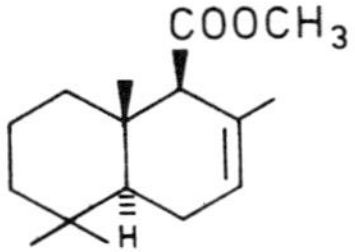 Drimenic acid methyl ester ($COOCH_3$, H)	(1) A: HP, Py. or RB, Xylene, *i*-PrOH 1:1 (2) KI	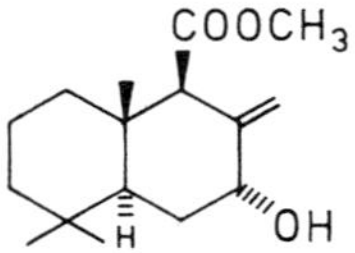 $COOCH_3$, OH, H (33%) m.p. 108°C	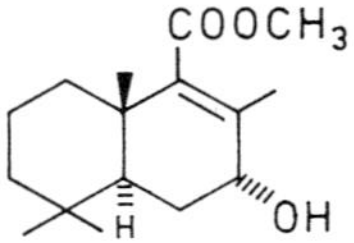 $COOCH_3$, OH, H (23%) m.p. 100°C [besides 7-keto ester (33%)]	128
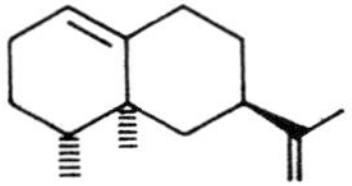 (+)-Valencene	A: sens, MeOH.-Bz. 1:1	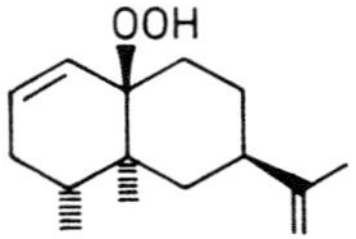 OOH (72%)	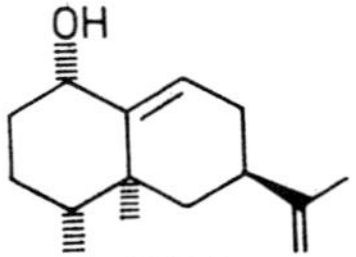 OH (22%)	199, 255

Substrate	Conditions	Products	Ref.
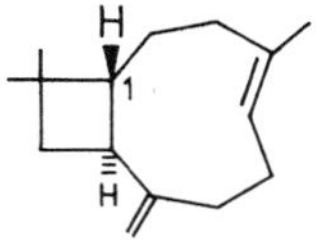 (−)-Caryophyllene	(1) A: RB or other sens., MeOH (2) Na_2SO_3	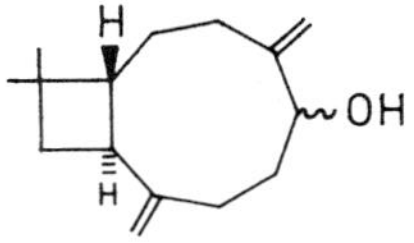 product 1: 2 isomers cis (16%) m.p. 49°–50°C $[\alpha]_D$ +59.6 (Bz.) trans (64%) $[\alpha]_D$ +34.65 (Bz.) 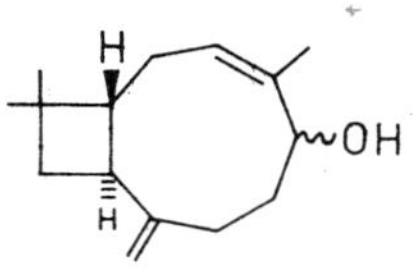 product 2: 2 isomers cis (11%) m.p. 88°–90°C $[\alpha]_D$ +137 (Bz.) trans (2%) m.p. 72°–73°C $[\alpha]_D$ −71.2 (Bz.)	84, 85, 256
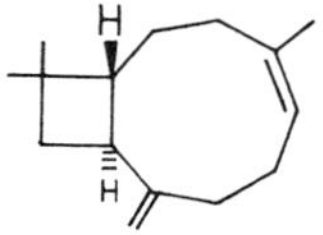 (−)-Isocaryophyllene	(1) A: RB, MeOH (2) Na_2SO_3	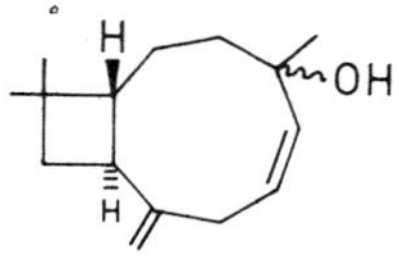 2 isomers cis (23%) m.p. 127°C $[\alpha]_D$ −123.8 (Bz.) trans (18%) m.p. 86°–87°C $[\alpha]_D$ +169 (Bz.) besides product 1 cis (20%); trans (22%) product 2 cis (3%); trans (14%)	
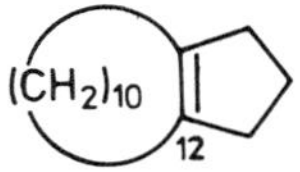 $(CH_2)_{10}$, 12	A: MB, ROH	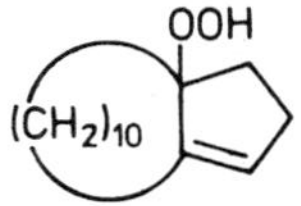 OOH, $(CH_2)_{10}$ (> 50%) m.p. 69°C [besides $\Delta^{11(12)}$ isomer (< 50%)	197

(*Continued*)

Singlet oxygen acceptor trivial name	Reaction conditions‡	Isolated products§	Ref.
3. Tri- and Polycyclic Olefins, Including Diterpenes			
 Homosemibullvalene	(1) A: RB, MeOH (2) $NaBH_4$	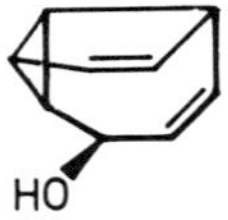 HO (35%) m.p. 88.5°–89.5°C (besides dialdehyde (via 1,2-dioxetane) and unidentified products)	221
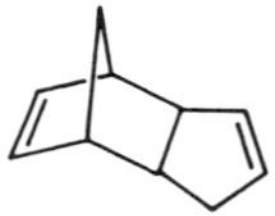	(1) A: RB, MeOH (2) Na_2SO_3	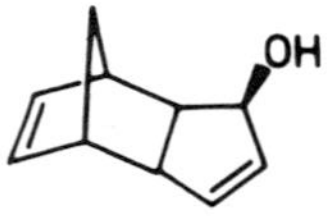 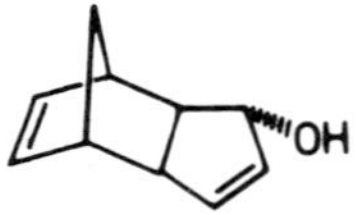 OH OH *9* (72%) *1*	101, 225
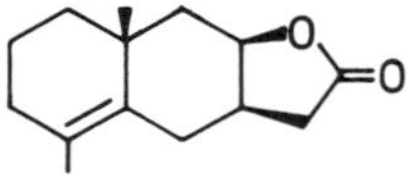	A: HP, Py.	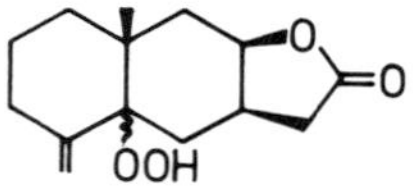 O O OOH 2 isomers trans (53%) m.p. 142°–143.5°C cis (22%) m.p. 150.5°–152°C (no hydroperoxides with endocyclic double bonds detected)	151–153
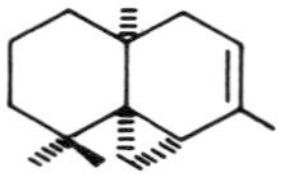 (−)-Thujopsene	(1) A: (a) RB, MeOH, Na_2CO_3 (b) MB or EO yellow (2) (a) Ph_3P (b) Na_2SO_3	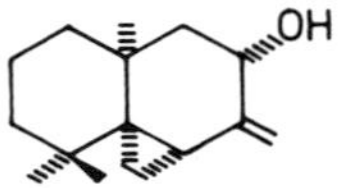 OH OH (a) *65* m.p. 84°–86°C 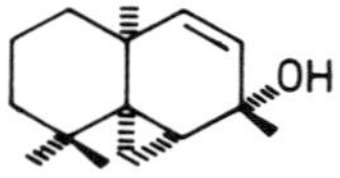 (a) *32* m.p. 110°–112°C	109, 198 269d

		$[\alpha]_D$ −20 (9.0, CCl_4) (b) (27%) m.p. 83.5°–85°C; $[\alpha]_D$ −45.1 (8.6, CCl_4) (b) (17%) m.p. 93°–95°C	
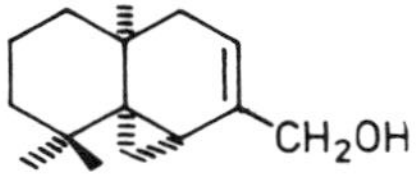	A: MB, MeOD	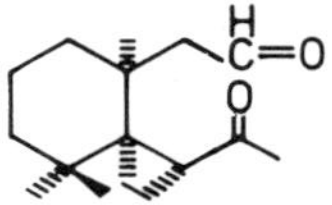	110
Thujopsenol	(1) A: MB, MeOH (2) Red.	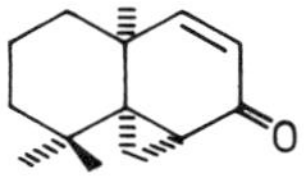 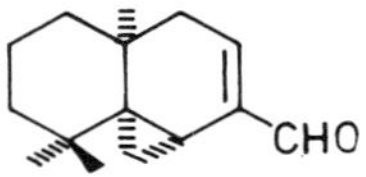 (15%) Mayurone; (13%) Thujopsenal; 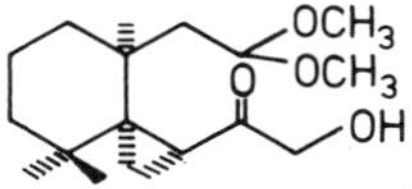 (28%)[n]	268
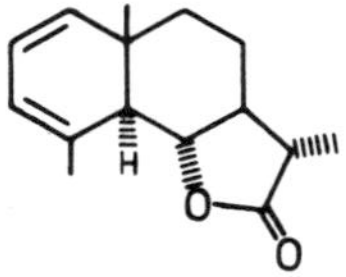 6,11-β-*H*-Eudesma-1,3-dien-6,13-olide	(1) A: EO, Py., EtOH, 0°C (2) Al_2O_3	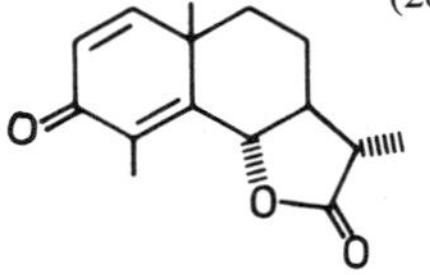 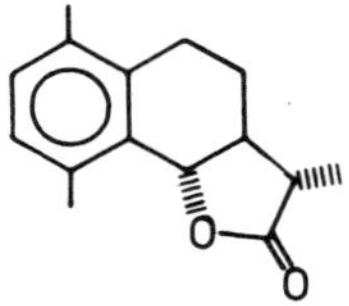 Santonine m.p. 169°–171°C; Hyposantonine (presumably via hydroperoxide)	102
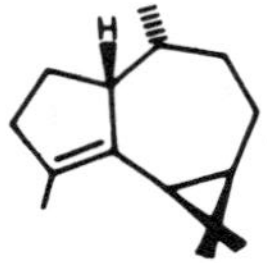 Gurjunene	(1) A: MB, MeOH, Bz. 2:1 (2) $NaBH_4$	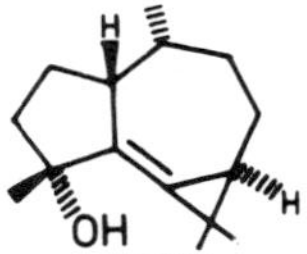 (47%) Essentially different products were obtained when the reaction mixture was worked up by chromatography without precedent reduction	111, 269

(*Continued*)

TABLE—*Continued*

Singlet oxygen acceptor trivial name	Reaction conditions‡	Isolated products§	Ref.
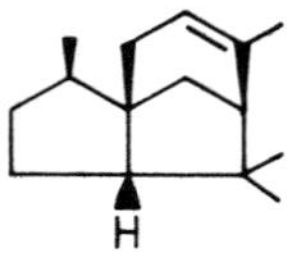 α-Cedrene	(1) A (2) Red.	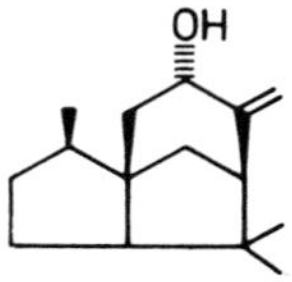	78
H_3COOC; R[1]; R[2] R^1 = OMe; R^2 = H R^1 = OAc; R^2 = H R^1 = OMe; R^2 = OAc R^1 = OAc; R^2 = OAc Methyl podocarpa-6,8,11,13-tetraen-19-oates	A: HP, Py.	H_3COOC; R[1] (93%) m.p. 174°–176°C (in MeOH mixture with corr. 7α-hydroperoxide) (80%) m.p. 141°–143°C $[\alpha]_D$ +146 (0.36) (85%) m.p. 174°–176°C (80%) m.p. 141°–143°C	30
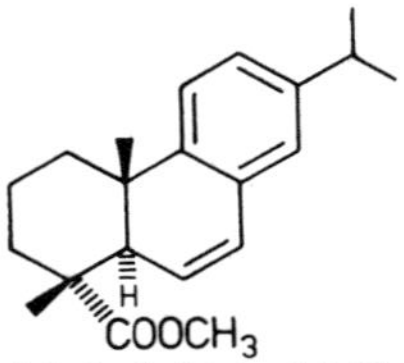 $COOCH_3$ Methyl abieta-6,8,11,13-tetraen-18-oate	A: HP, Py.	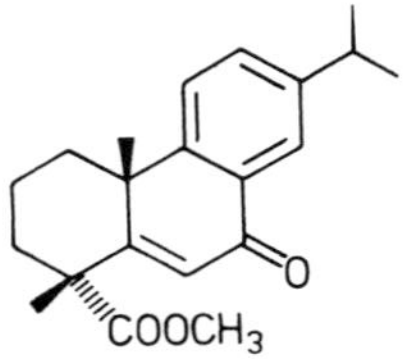 $COOCH_3$ (72%)	30

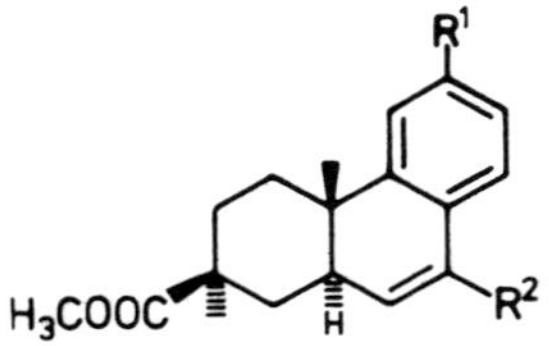

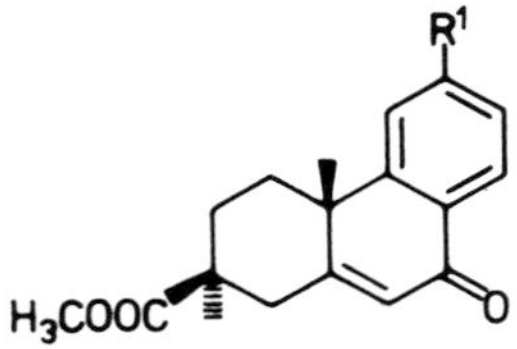

Substrate	Conditions	Product	Ref.
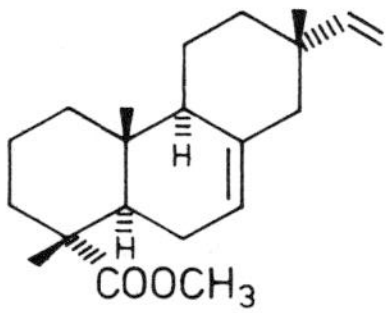 Methyl isopimarate	(1) A: HP, Py. (2) $NaBH_4$	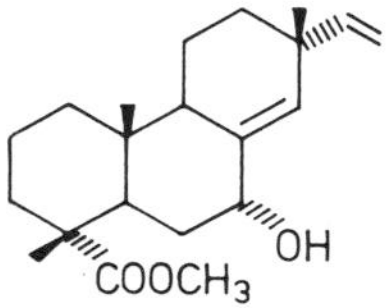 m.p. 75°–76°C $[\alpha]_D$ −77 (1.0)	65
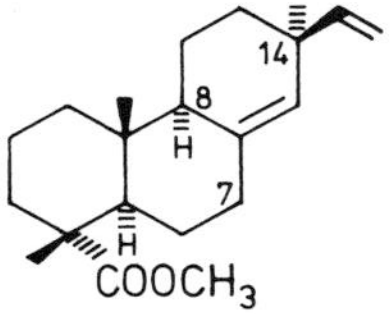 Methyl pimarate	(1) A: HP, Py. (2) $NaBH_4$	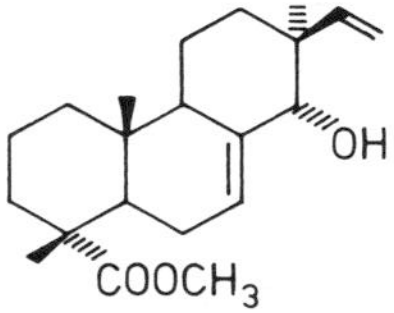 b.p.$_{0.01}$ 150°–160°C $[\alpha]_D$ −17 (0.6); 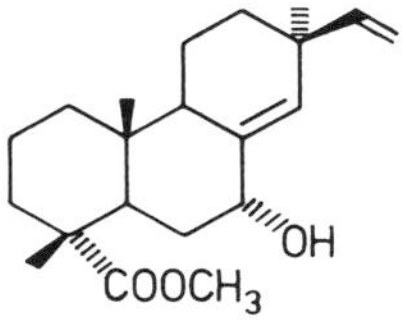 m.p. 120°–121°C $[\alpha]_D$ +24(0.75)	65
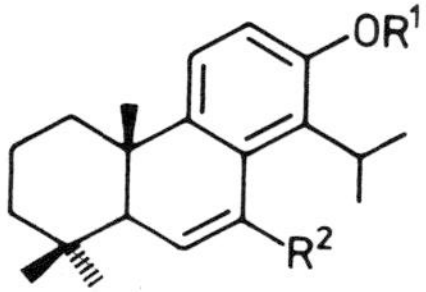 R^1 = Ac; R^2 = H R^1 = Me; R^2 = OAc R^1 = Ac; R^2 = OAc R^1 = Me; R^2 = H Totara-6,8,11,13-tetraenes	A: HP, Py.	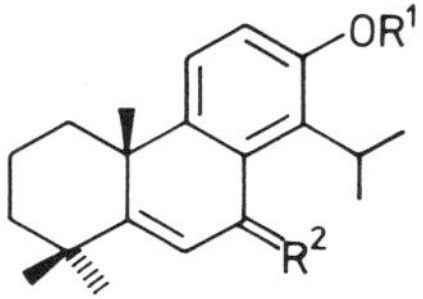 R^2 = α-OOH (62%) R^2 = =O (91%) m.p. 150°–152°C R^2 = =O (85%) m.p. 163°–164°C R^2 = α-OOH (74%) m.p. 140°–145°C dec.	30

(*Continued*)

TABLE—*Continued*

Singlet oxygen acceptor trivial name	Reaction conditions‡	Isolated products§	Ref.
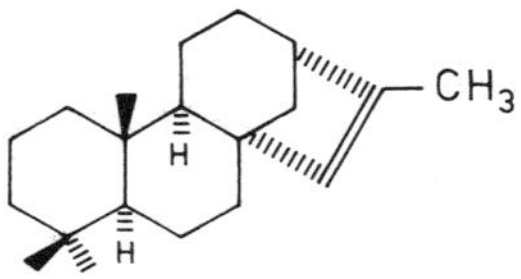 (±)-Isokaurene	(1) A: HP, Py. (2) $LiAlH_4$	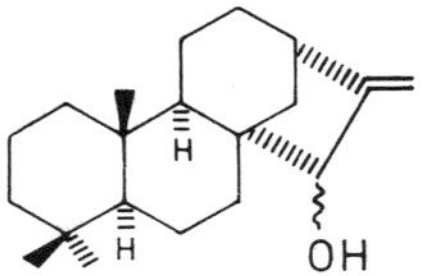 2 isomers (22%) m.p. 126°–127°C [second epimer (22%) not isolated in pure state]	16
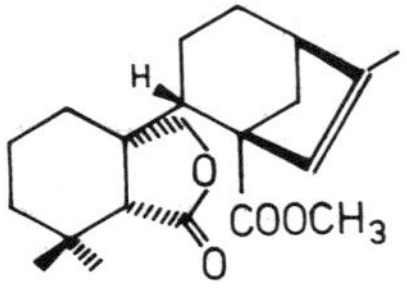	A: HP, Py.	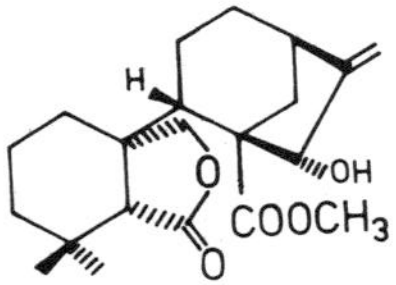 (32%) m.p. 193°–195°C (besides epoxide)	70
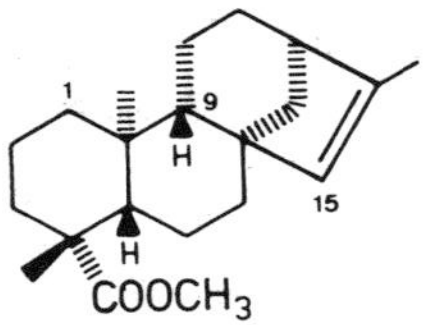 Methyl-(−)-kaur-15-en-19-oate	(1) A: HP, Py. (2) NaI, AcOH, EtOH	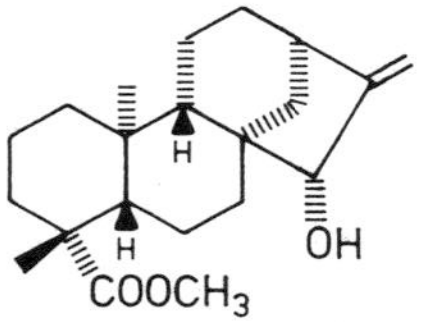 (42%) m.p. 114°–116°C $[\alpha]_D$ −95 (0.1)	9

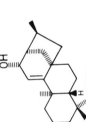

Starting material	Reaction conditions	Products	Ref.
(±)-Isoatisirene	(1) A: HP, Py. (2) $LiAlH_4$	2 isomers: (19%) m.p. 126°–127°C (29%) m.p. 153°C	16
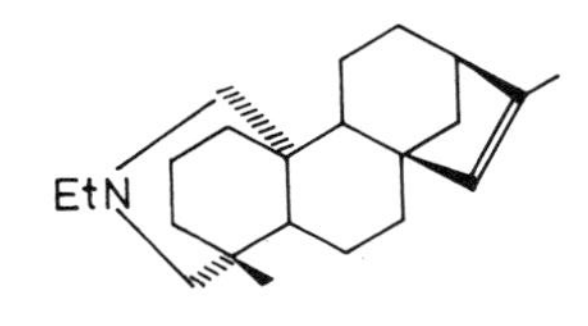	(1) A: HP (2) $LiAlH_4$	acetate m.p. 85°–86°C	154
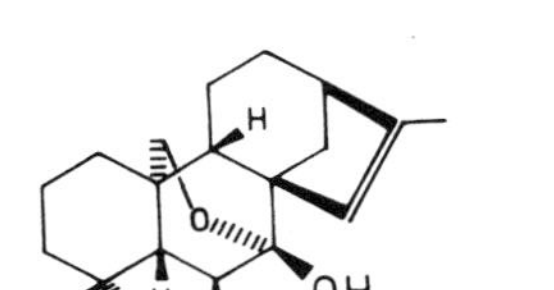	(1) A: HP, Py. (2) KI	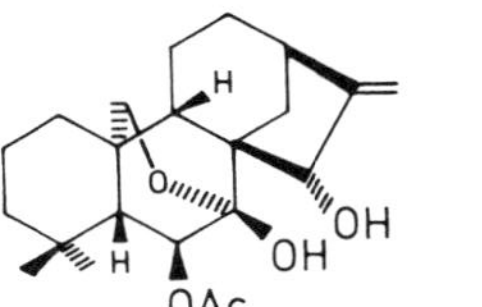 (71.5%) m.p. 187°–189°C 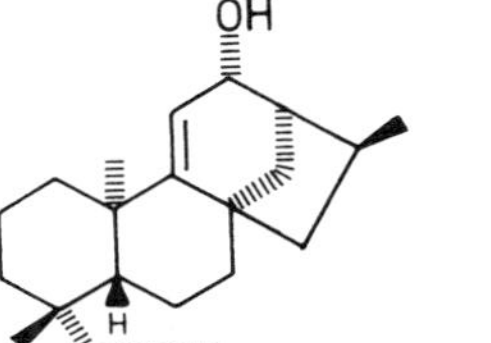 small amount m.p. 204°–204.5°C	69, 70
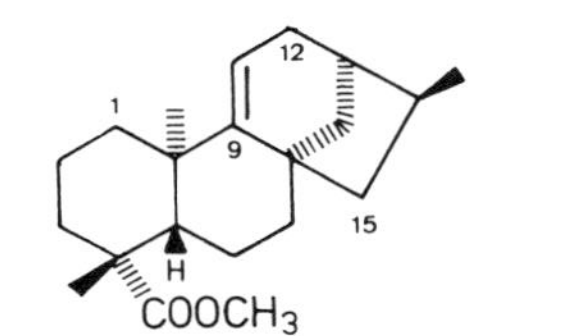 Methyl-(−)-kaur-9(11)-en-19-oate	(1) A: HP, Py. (2) KI, AcOH, EtOH	(10%) (presumably via allyl. rearr. of the 9α- to the 12α-hydroperoxide	9

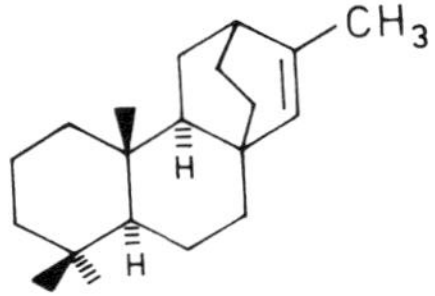

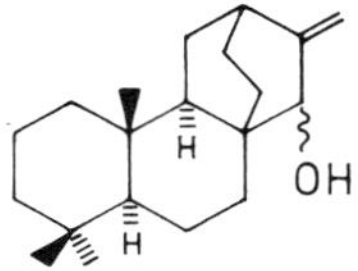

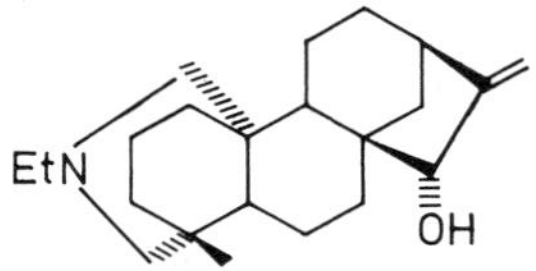

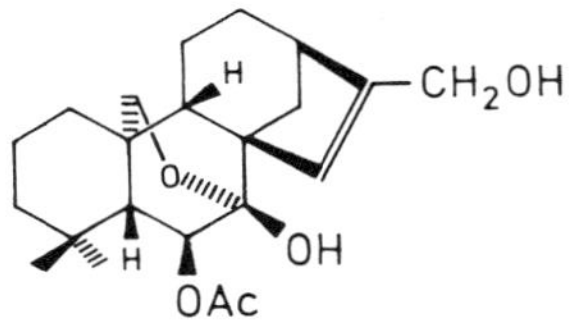

(Continued)

TABLE—*Continued*

Singlet oxygen acceptor trivial name	Reaction conditions‡	Isolated products§	Ref.
4. Steroids and Triterpenoids			
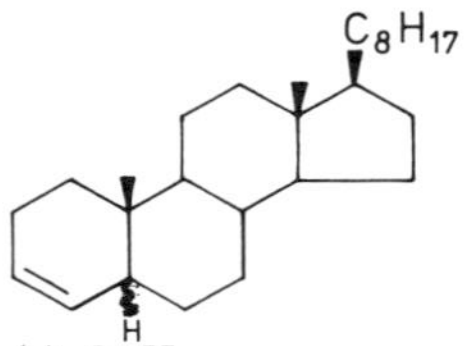 C_8H_{17}; H (a) 5α-H (b) 5β-H	(1) A: HP, Py. (2) NaI, AcOH	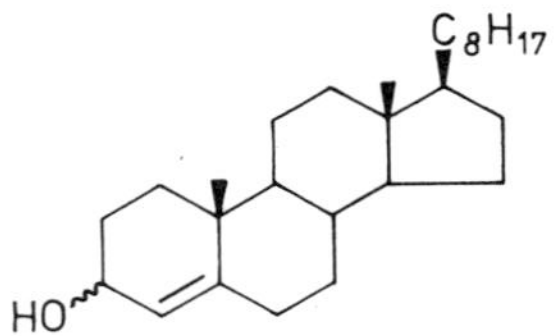 C_8H_{17}; HO (a) Two isomers: 3α-OH (73%) 3β-OH (11%) (besides Δ^4-3-ketone (8%) and other minor products) (b) Two isomers: 3α-OH (8%) 3β-OH (44%) (besides Δ^4-3-ketone (15%) and other minor products)	185
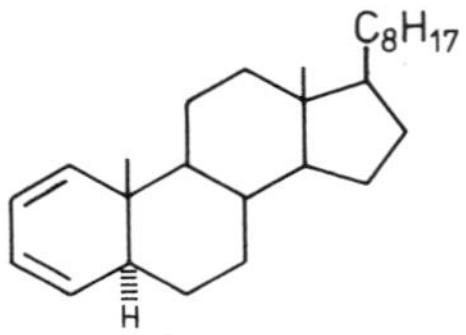 C_8H_{17}; H	(1) A: EO, Bz., EtOH, Py. 25:25:1 (2) Al_2O_3 bas.	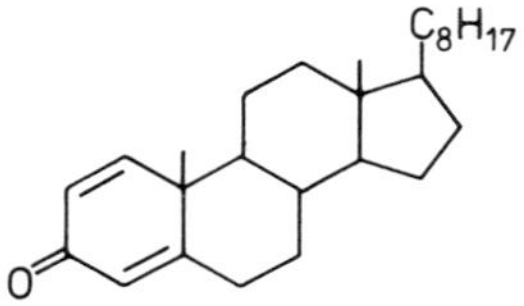 C_8H_{17}; O m.p. 108°–111°C	102
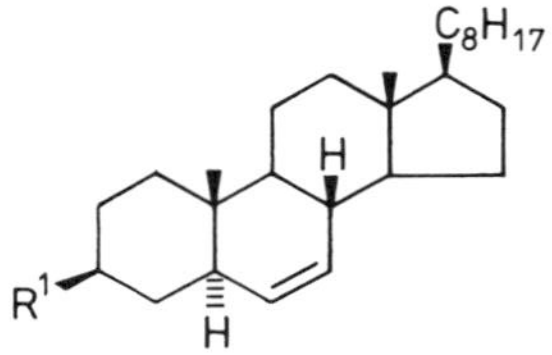 C_8H_{17}; R^1; H; H	(1) A: HP, Py. (2) (b) (c) NaI, AcOH (3) (c) PhCOCl, Py.	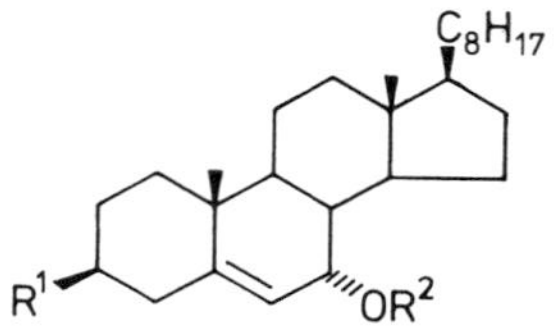 C_8H_{17}; R^1; OR^2	182, 183

(a) R^1 = OAc (b) R^1 = OH (c) R^1 = H 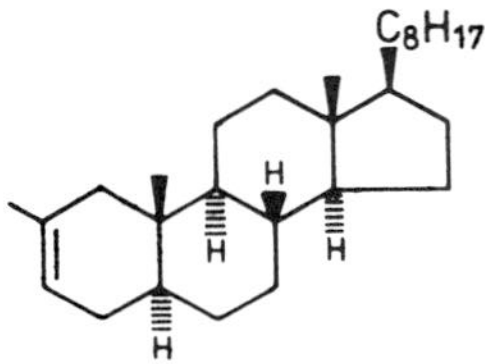	(1) A: HP, Py., 35°C (2) NaI	(a) R^2 = OH (30%)^r m.p. 142°–142.5°C $[\alpha]_D$ −137 (2.06) (b) R^2 = H (60%)^r m.p. 177°–179°C $[\alpha]_D$ −92 (c) R^2 = Bz (31%)^r m.p. 133°–133.5°C $[\alpha]_D$ −236 (1.46) 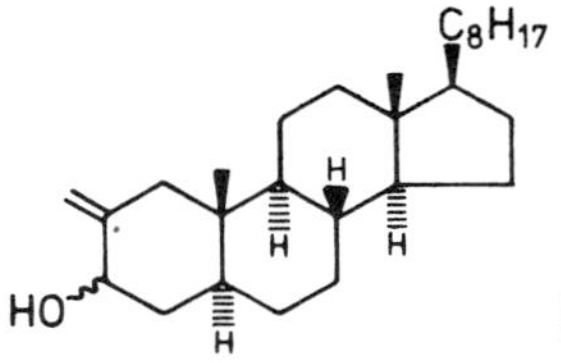 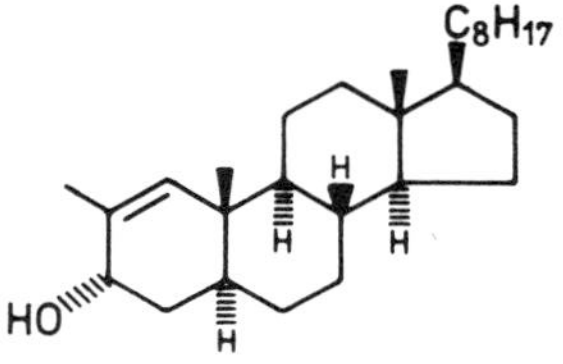 3α-OH *57* 30 3β-OH *13* Allylic alcohols (60%) besides ketones (7%) and unidentified products (6%)	189, 190
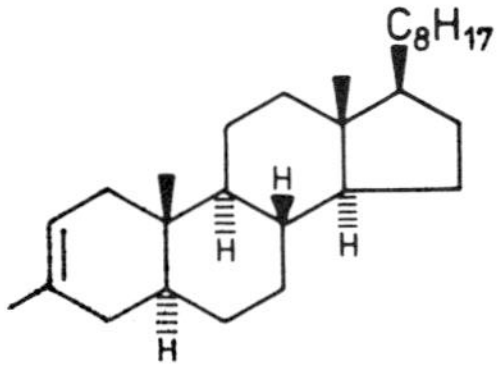	(1) A: HP, Py. (2) NaI	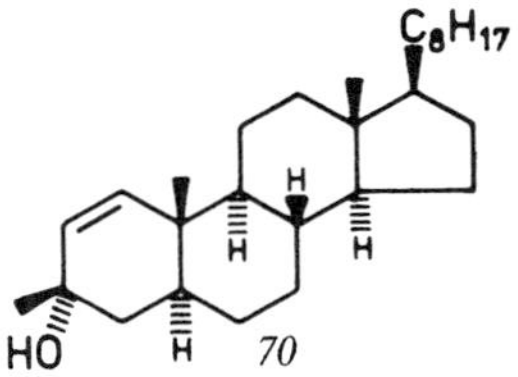 *70* m.p. 107°–108°C $[\alpha]_D$ +9 (3.34) 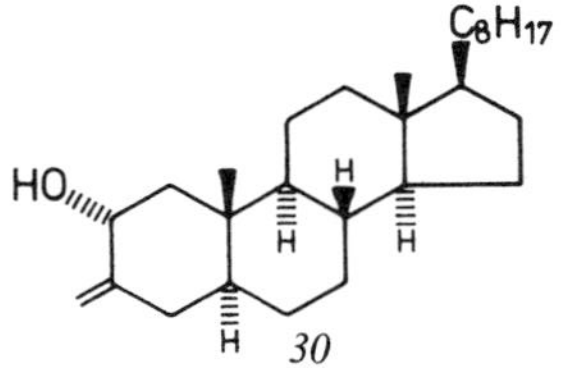 *30* allylic alcohols (77%) besides ketone (6%)	189, 190
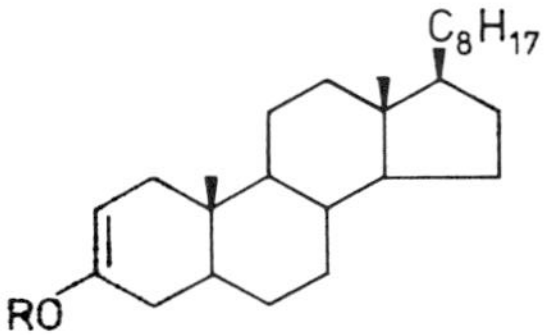 (a) R = Ac (b) R = Bz	A: HP, Py.	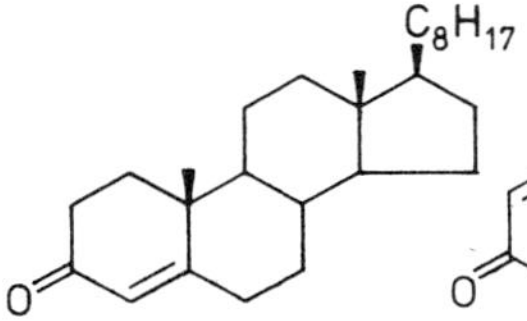 (a) (5%) m.p. 80°C $[\alpha]_D$ +85 (b) none 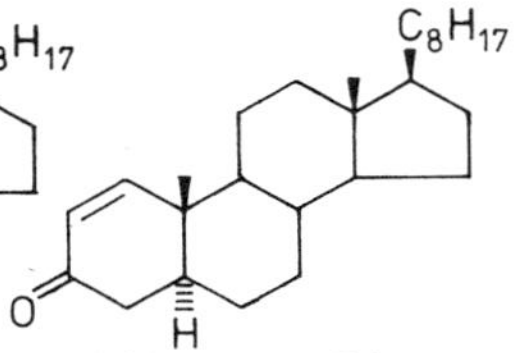 (a) (4%) m.p. 110°C $[\alpha]_D$ +59 (b) (10%) m.p. 110°C	208

(*Continued*)

Singlet oxygen acceptor trivial name	Reaction conditions‡	Isolated products§	Ref.
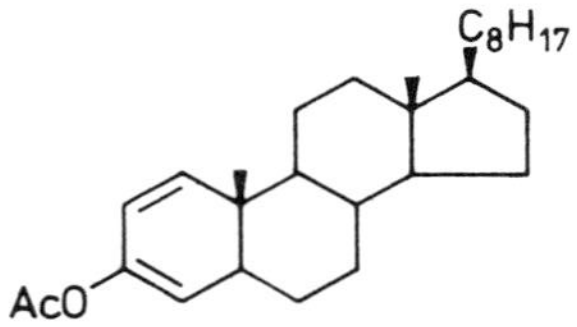 C_8H_{17}, AcO	A: HP, Py.	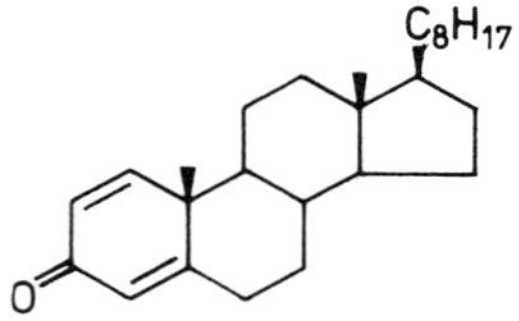 C_8H_{17}, O (35%) presumably via $\Delta^{1,4}$-3-acetoxy-3-hydroperoxide; 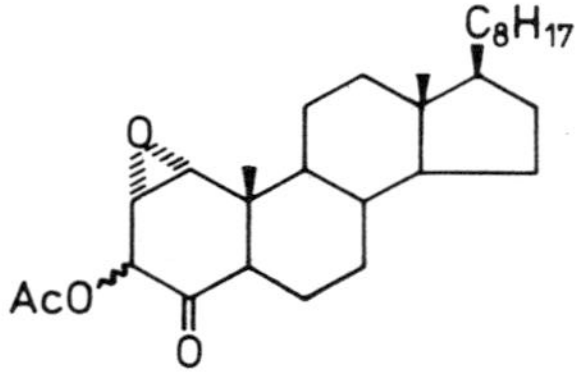 C_8H_{17}, O, AcO, O (15%) via endoperoxide 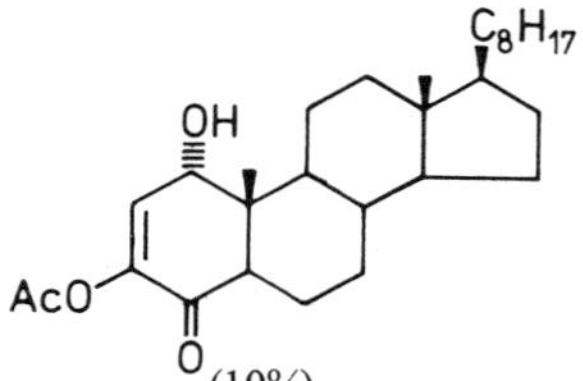 (?); C_8H_{17}, OH, AcO, O (10%) via endoperoxide 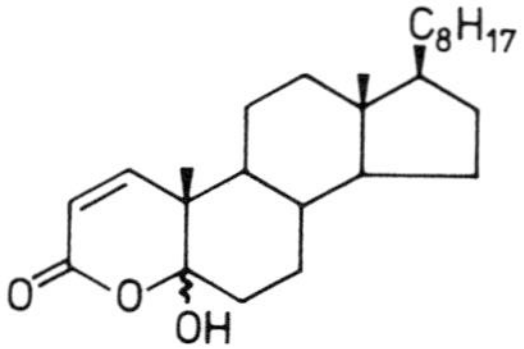 (?); C_8H_{17}, O, O, OH (10%) via dioxetane (?)	208
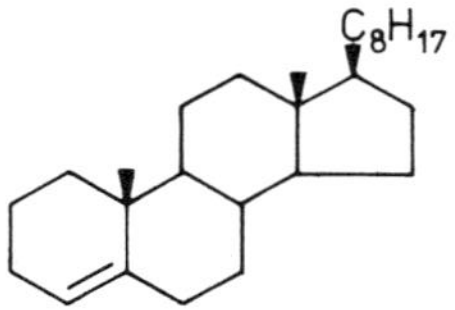 C_8H_{17}	(1) A: HP, Py. (2) KI	C_8H_{17}, OR $\xrightarrow{CHCl_3}$ C_8H_{17}, RO; R = OH: quantitative allyl. rearr. 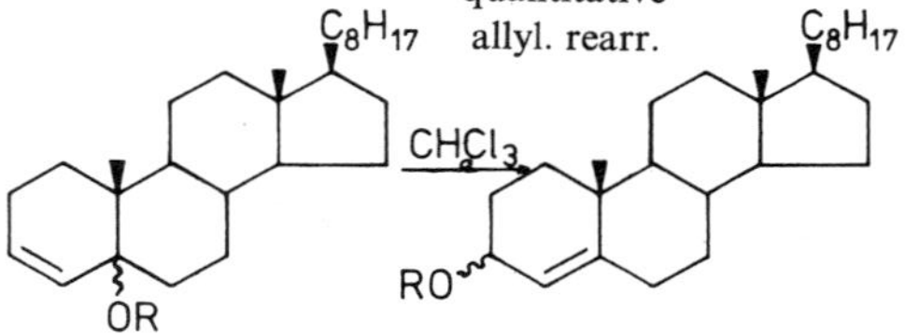	188

		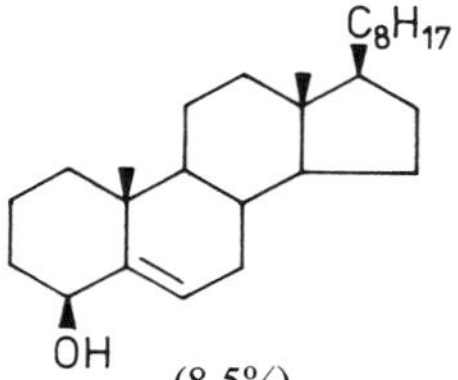 R = H: 2 isomers 5β-OH (2%) 5α-OH (29%) m.p. 72°–74°C $[\alpha]_D$ −15 C_8H_{17} OH (8.5%)	R = H: 2 isomers 3β-OH (7.5%) m.p. 113°–120°C 3α-OH (37%)	
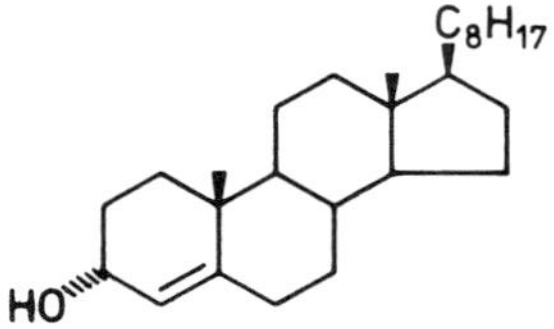 C_8H_{17} HO (3α-OAc: no reaction)	A: HP, Py. (slow)	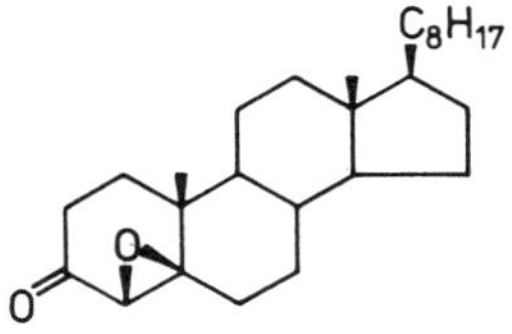 C_8H_{17} O O (75%) m.p. 116°–118°C	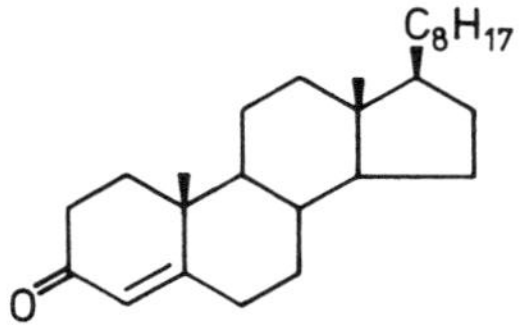 C_8H_{17} O (15%) m.p. 79–81°C	184, 187
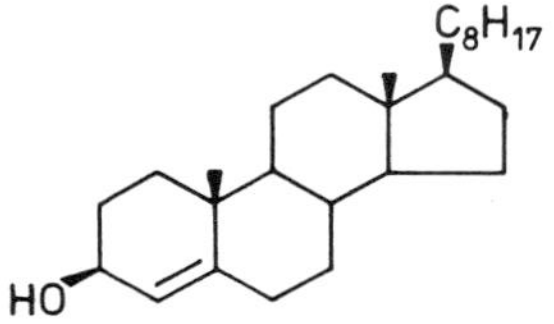 C_8H_{17} HO (3β-OAc, -OBz, -OMe: no reaction)	A: (a) HP, Py. (or other sens., *vide infra*) (rapid, cf. 3α-OH)	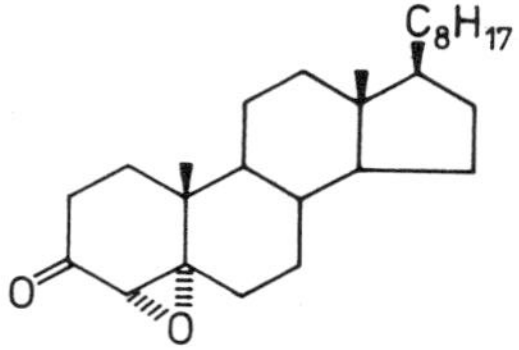 C_8H_{17} O O (a) (75%) m.p. 123°C $[\alpha]_D$ −39; favored by low-energy sens. and low conc.	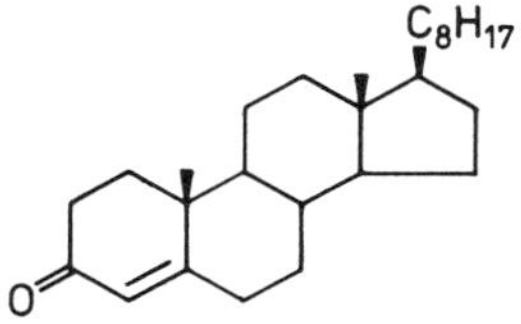 C_8H_{17} O (a) (16%) m.p. 78.5–79°C favored by high energy sens.	117, 118, 184, 187

(*Continued*)

TABLE—*Continued*

Singlet oxygen acceptor trivial name	Reaction conditions‡	Isolated products§	Ref.
		product ratio[s] (a) HP or Clorin e_6, Py. 4.5 1 (b) MB, Py. 3 1 (c) RB, Py. 1.2 1 (d) ER B, Py. 1 1.6 (e) EO Yellow, Py. 1 3.1 (f) RF, Py.–MeOH, 4:1 1 30	
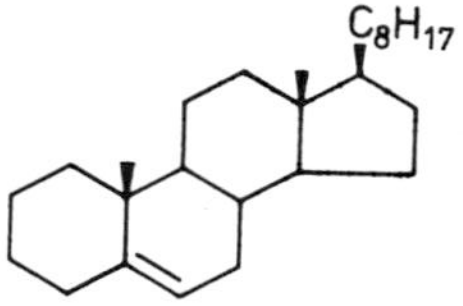 C_8H_{17}	(1) A: HP, Py.	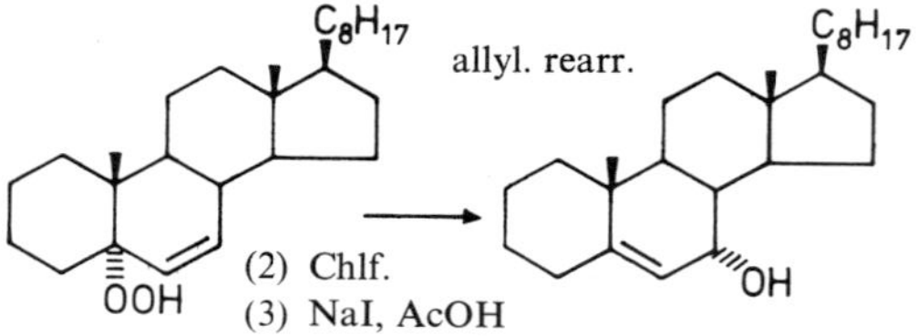 C_8H_{17}, OOH → (allyl. rearr.; (2) Chlf. (3) NaI, AcOH) C_8H_{17}, OH Not isolated in pure state; m.p. 62°–64°C; $[\alpha]_D$ −107	187
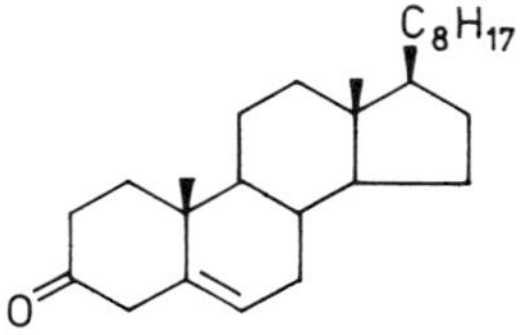 C_8H_{17}, O	(1) A: HP + MB, Py. (2) NaI	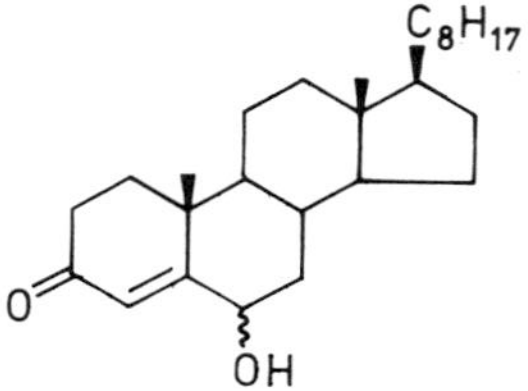 C_8H_{17}, O, OH (16%)[t] 6α-OH m.p. 156°C $[\alpha]_D$ +81 6β-OH m.p. 194°C $[\alpha]_D$ +27 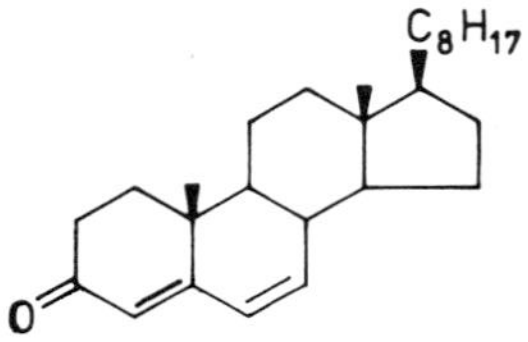 C_8H_{17}, O (35%)[u] m.p. 75°–76°C	186, 236

Substrate	Conditions	Product(s)	Ref.
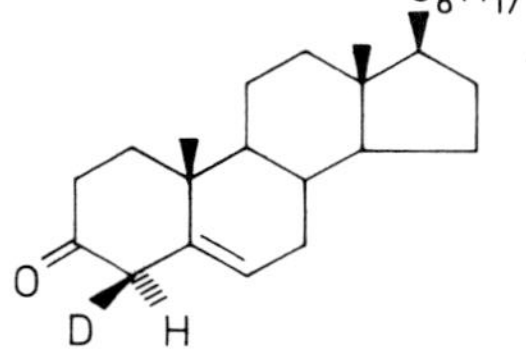 C_8H_{17} O D H 90% D	(1) A: RB, Py. (2) NaI, AcOH	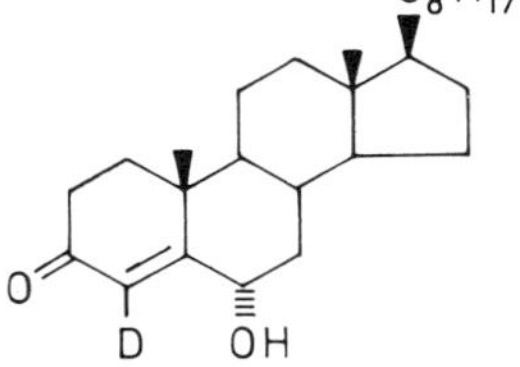 C_8H_{17} O D OH 85% D	71
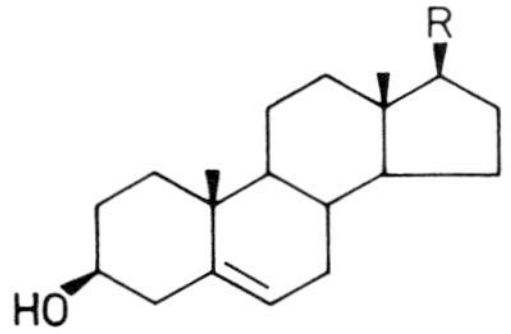 R HO (a) $R = C_8H_{17}$ (b) $R = COCH_3$	(1) A: HP, Py.	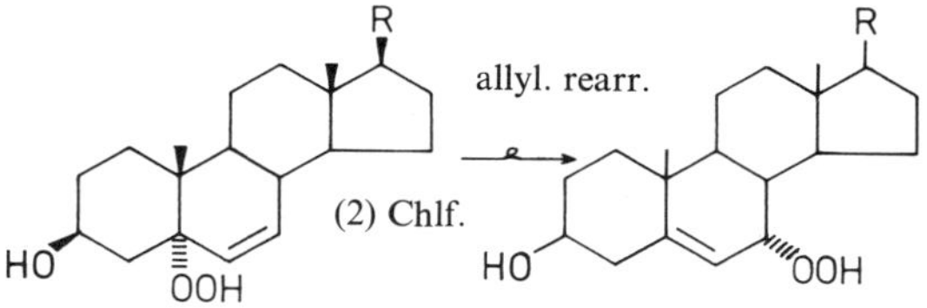 R HO OOH → allyl. rearr. (2) Chlf. → R HO OOH (a) (75%) m.p. 145°–148°C dec $[\alpha]_D$ +16 (1.0); (77%) m.p. 154°–156°C dec $[\alpha]_D$ −139 (2.4) (b) (66%) m.p. 157°C dec. $[\alpha]_D$ +91.5 (1.5); (100%) m.p. 157°C dec. $[\alpha]_D$ −108.7 (1.0) 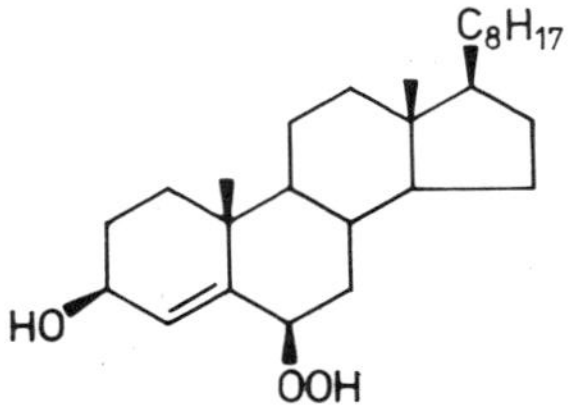 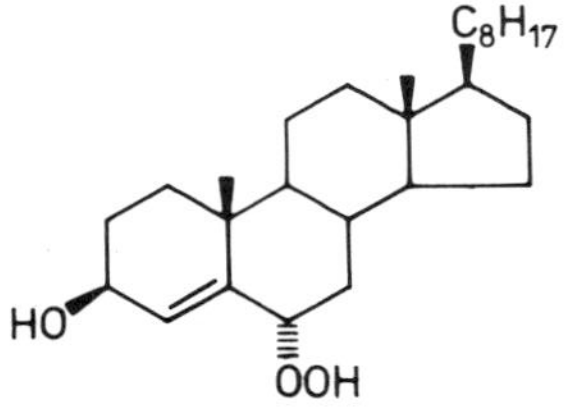 C_8H_{17} HO OOH; C_8H_{17} HO OOH (a) (2.2%) m.p. 162°–163°C; (1.3%) m.p. 152°–156°C	143–146, 236, 238, 239

(Continued)

TABLE—*Continued*

Singlet oxygen acceptor trivial name	Reaction conditions‡	Isolated products§	Ref.
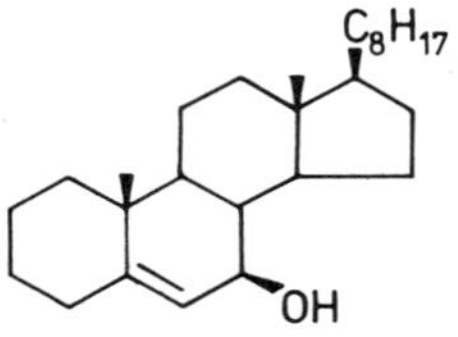 C_8H_{17}; OH	A: HP, Py.	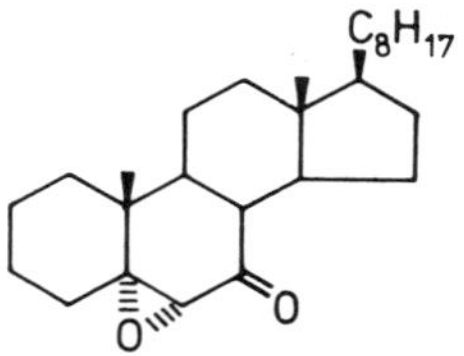 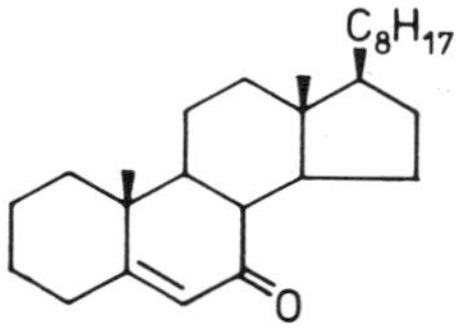 C_8H_{17}; O; O — Not isolated. C_8H_{17}; O — Not isolated (~20%)	184, 187
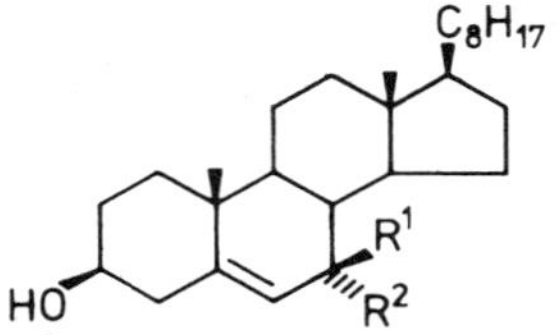 C_8H_{17}; HO; R^1; R^2 — $R^1 = H$; $R^2 = D$; $R^1 = D$; $R^2 = H$	A: HP, Py.	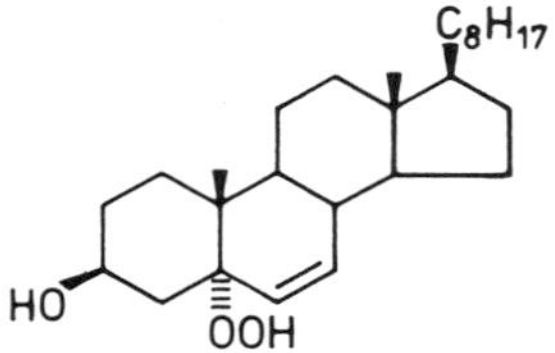 C_8H_{17}; HO; OOH — m.p. 150.5°–151.5°C; 8.5% D retention; m.p. 150.5°–151.5°C; 95% D retention	182, 183, cf. 236, 238
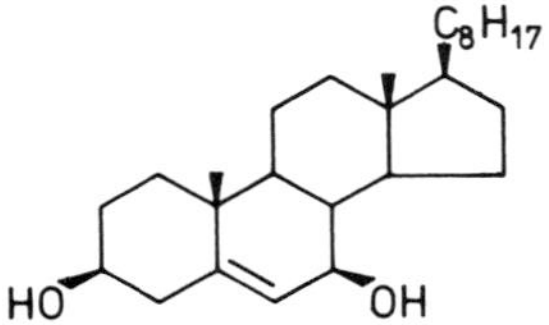 C_8H_{17}; HO; OH	A: HP, Py.	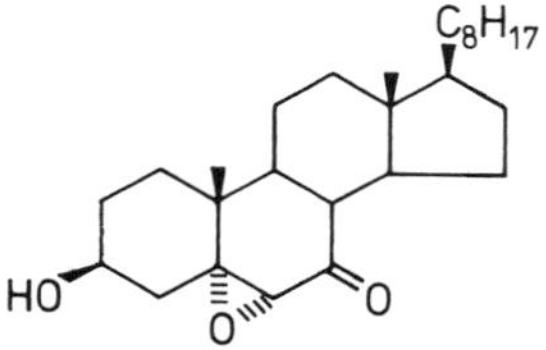 C_8H_{17}; HO; O; O — (56%) m.p. 165°–166.5°C; $[\alpha]_D$ +94. 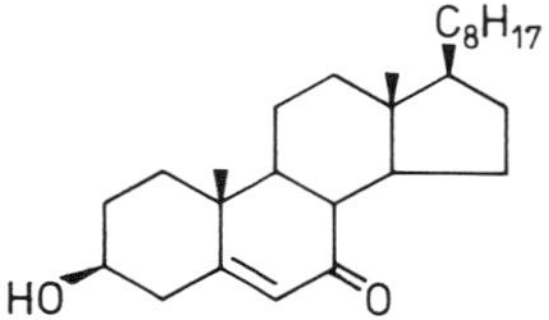 C_8H_{17}; HO; O — (17%)	184, 187

Reactant	Conditions	Product	Ref.
C_8H_{17}, AcO	(1) A:[v] BR or HP, Py. (2) Ra–Ni/H_2 or Ph_3P (3) Ac_2O	C_8H_{17}, OH, OAc, AcO m.p. 205°–210°C dec. $[\alpha]_D$ −2.5 C_8H_{17}, OH, OH m.p. 70°–90°C $[\alpha]_D$ −1.0	48, 252
CH_2OAc, H, AcO	(1) A: RB, Bz., MeOH, 4:1, Na Lamp (2) Al_2O_3 total (80%)	CH_2OAc, O, AcO *60* m.p. 206°–207°C (100%) ↓ A: RB, MeOH, Na lamp CH_2OAc, O, O, AcO *20* m.p. 235°–237°C	172

(*Continued*)

TABLE—*Continued*

Singlet oxygen acceptor trivial name	Reaction conditions‡	Isolated products§	Ref.
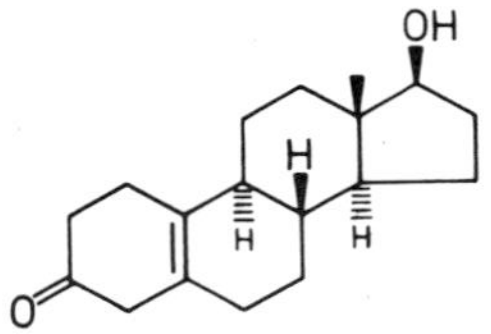	(1) A: EO, Ac., $-70°C$ (2) NH_4I	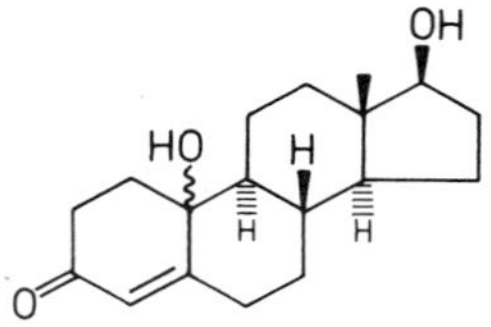 2 isomers 10β-OH: *80* m.p. 210°–215°C 10α-OH: *13* m.p. 214°–215°C	71, 164
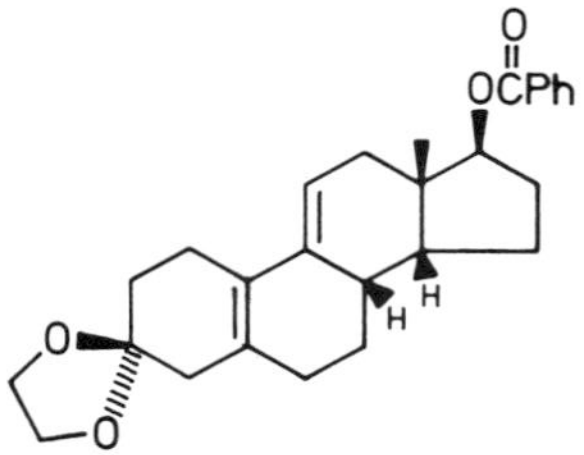	A: MB or RB, Ac.	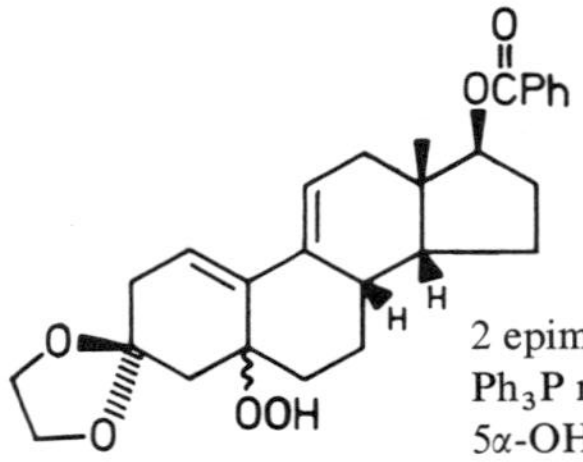 2 epimers Ph_3P reduction gave 5α-OH m.p. 147°–148°C 5β-OH m.p. 135°C 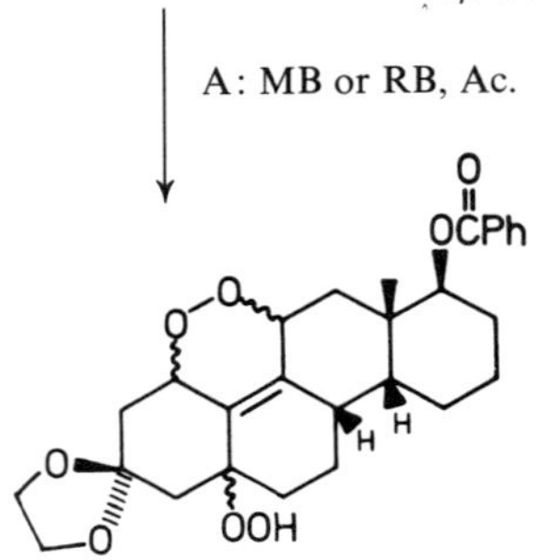 ↓ A: MB or RB, Ac.	163

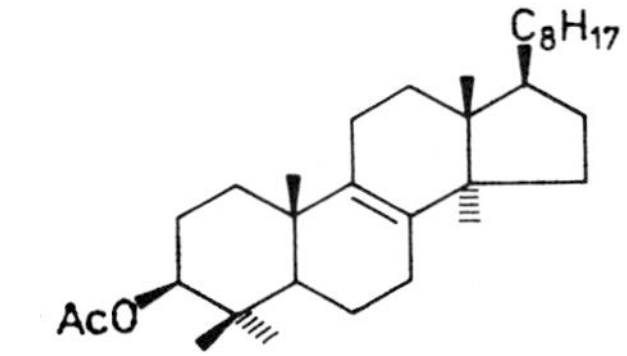 AcO, C_8H_{17}	A: HP, Py., p-$NO_2C_6H_4SO_2Cl$	(74%) 4 isomers 1α,11α-epidioxy-5α-OOH (14.5%) m.p. 227°C 1β,11β-epidioxy-5α-OOH (21.5%) m.p. 210°–212°C 1α,11α-epidioxy-5β-OOH (26.5%) m.p. 218°C 1β,11β-epidioxy-5β-OOH (11.5%) m.p. 220°C 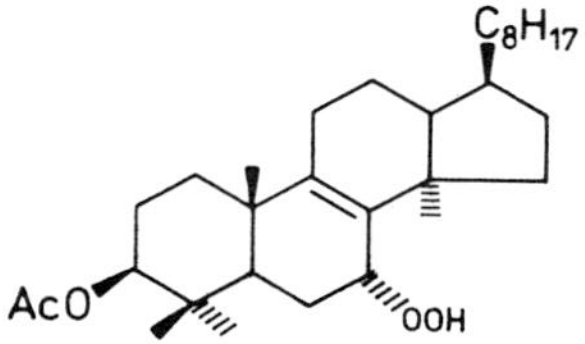 AcO, C_8H_{17}, OOH m.p. 175°–176°C; $[\alpha]_D$ +35.2 (0.83) presumably formed by allylic rearrangement of Δ^7-9α-hydroperoxide (besides 5 other products)	66, 67
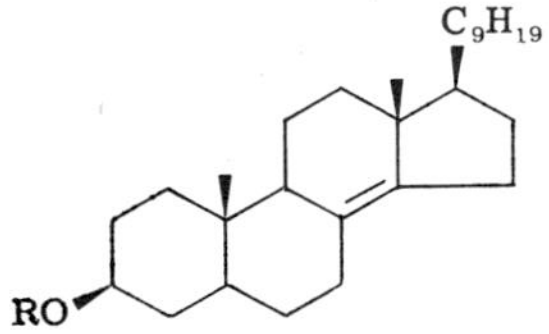 RO, C_9H_{19} (a) R = Ac (b) R = H	(1) A: HP, Py. (2) Raney-Ni, H_2 or $(Ph)_3P$	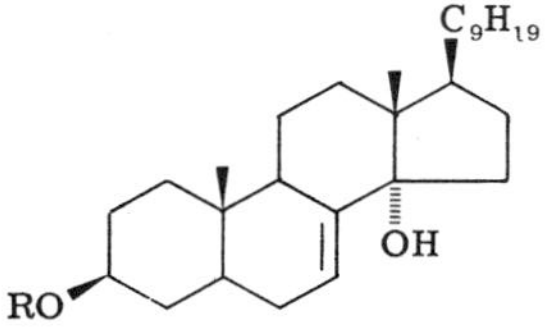 RO, C_9H_{19}, OH (a) (76%) m.p. 151°–156°C $[\alpha]_D$ +24 (b) m.p. 140°–150° $[\alpha]_D$ +26	48

(Continued)

TABLE—*Continued*

Singlet oxygen acceptor trivial name	Reaction conditions‡	Isolated products§	Ref.
		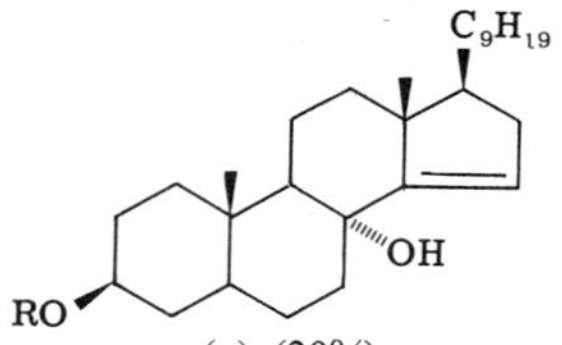 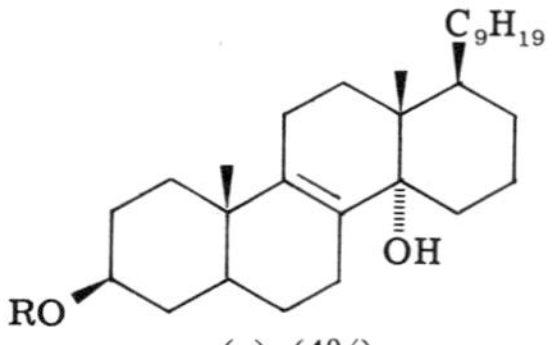 RO, C_9H_{19}, OH (a) (20%); RO, C_9H_{19}, OH (a) (4%)	
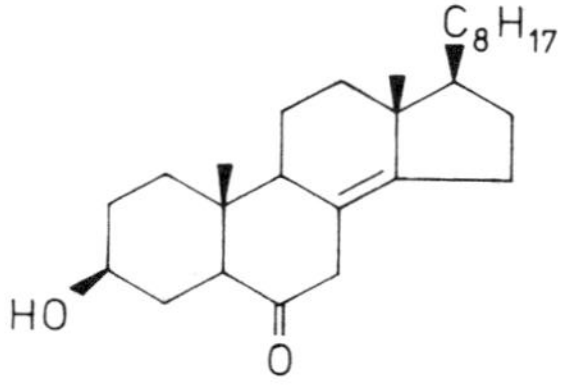 HO, O, C_8H_{17}	(1) A: RB, Py. (2) NaI, AcOH	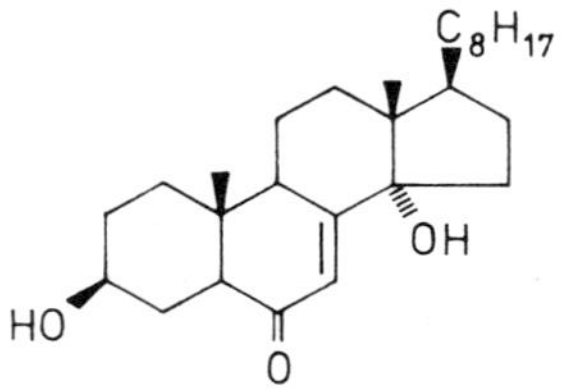 HO, O, OH, C_8H_{17} (> 70%)	71
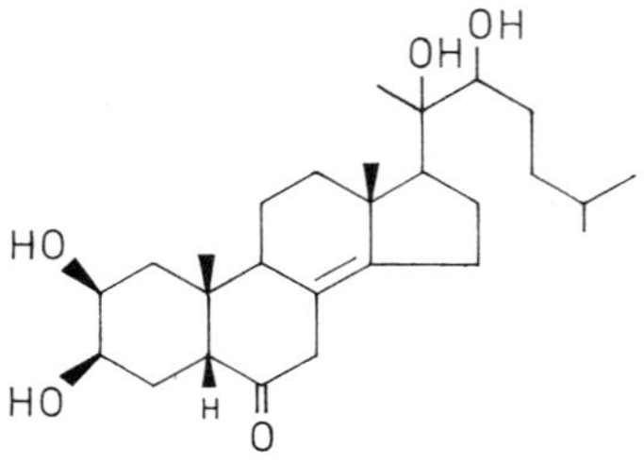 HO, HO, H, O, OH, OH	(1) A: RB, Py. (2) NaI, AcOH	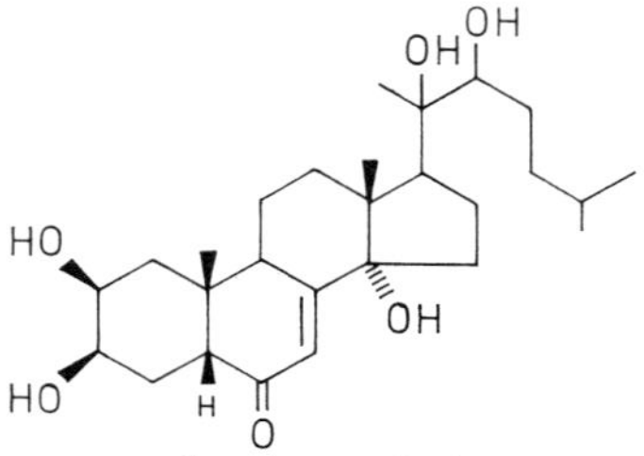 HO, HO, H, O, OH, OH, OH almost quantitative	71

Substrate	Conditions	Product(s)	Ref.
OH, H, H	A: MB or TPP, Ac. or Ac.–Bz. 75:25	OH, HOO, H, H, HO (96%) 2 epimers dec. 90°C	162

D. Miscellaneous Heterocompounds Which May Also Undergo Ene-Type Reactions

Substrate	Conditions	Product(s)	Ref.
R, R′, $OSiMe_3$, $OSiMe_3$ (a) R = H, R′ = *t*-Bu (b) R = R′ = Me	A	R, R′, O, Me_3SiOO, $OSiMe_3$ —MeOH, 5°–10°C→ R, R′, O, HOO, OH (a) b.p. 67°–69°C (a) m.p. 69°–70°C	2, 3, 3a cf. 3b
Me_3SiO, Ph	A: TPP, CCl_4	Me_3SiO, Ph, HOO (not isolated) —Ph_3P or heat or SiO_2→ O, Ph O, $OOSiMe_3$, Ph (not isolated) —SiO_2→ O, OOH, Ph	219

(*Continued*)

TABLE—*Continued*

Singlet oxygen acceptor trivial name	Reaction conditions‡	Isolated products§	Ref.
OSiMe3	A: TPP or MB, CCl_4, Chlf. or An., 0 . . −20°C	OOSiMe3 (95%) b.p.$_{0.025}$ 51°C	116a
	A: MB or RB, MeOH, 0 . . −78°C	OOSiMe3 Δ (→) OOH (15%) (presumably via polar intermediate)	116a
OSiMe3 •R R = H, 2-Me, 3-Me, 2-Et, 4-*i*-Pr, 2-*i*-Pr-5-Me	A: RB, THF, −78°C	HOO OSiMe3 •R not isolated (1) Ph_3P (2) MeOH → O •R (~20–80%)	67a
		OSiMe3 HOO •R (1) Ph_3P (2) MeOH → α-Hydroxyketone	

Substrate	Conditions	Products	Ref.
		not isolated (~1–40%) only with R = H, 3-Me, 4-*i*-Pr, 2-*i*-Pr-5-Me 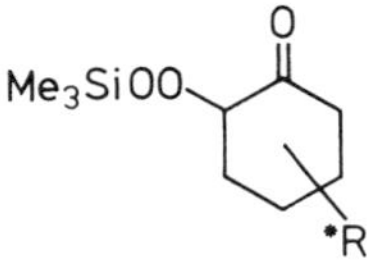 Me_3SiOO–(cyclohexanone, *R) (1) Ph_3P (2) MeOH → α-Hydroxyketone not isolated (~7–30%) only with R = H, 4-*i*-Pr, 2-*i*-Pr-5-Me	
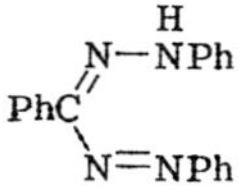 PhC(=N–NHPh)–N=NPh	A: MB, MeOH B: NaOCl, H_2O_2	PhCOOH PhH PhOH* N_2 * B: none	18
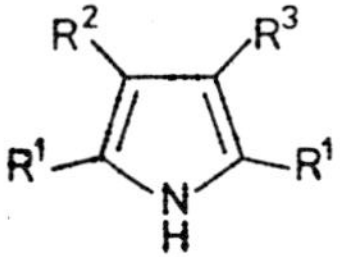 R^1, R^2, R^3-pyrrole (NH) (a) $R^1 = R^2 = R^3 = Ph$ (b) $R^1 = Ph; R^2 = R^3 = H$ (c) $R^1 = R^2 = t\text{-Bu}; R^3 = H$ (d) $R^1 = t\text{-Bu}; R^2 = R^3 = H$	A: (a,b) MB or HD, Chlf., CS_2 or MeOH (c,d) MB, Mchl.	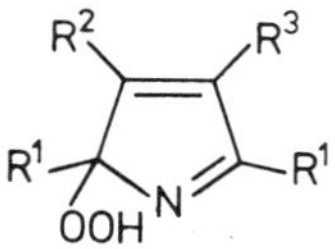 2-OOH-2H-pyrrole (R^1, R^2, R^3) (a) (80–90%) m.p. 170°–172°C dec. (b) (50%) m.p. 168°–173°C dec. (c) m.p. 142°–143°C (d) m.p. 138°–139°C	45, 209, 212
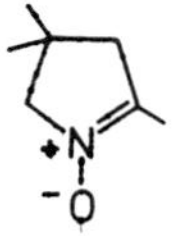 2,4,4-trimethyl-1-pyrroline N-oxide	A: MB, $CDCl_3$, −63°C $NaBH_4$, −30°C ←	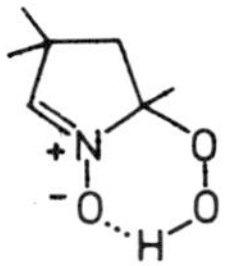 hydroperoxide N-oxide (intramolecular O···H–O) (100%) dec. > 0°C	33

(*Continued*)

TABLE—*Continued*

Singlet oxygen acceptor trivial name	Reaction conditions‡	Isolated products§	Ref.
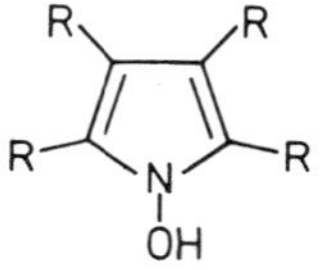 R = Ph	A: HD or MB, Chlf. $NH_2OH \cdot HCl$ ←	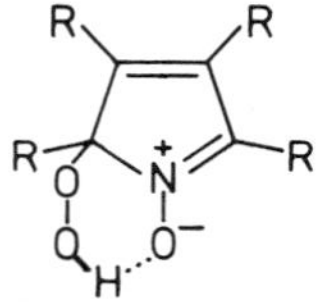 m.p. 208°–210°C dec.	211
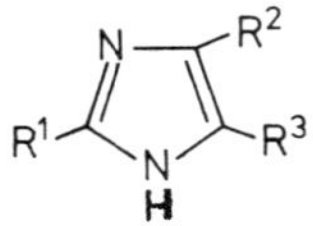 $R^1 = R^2 = R^3$ = Ph (Lophine) R^1 = p-$MeOC_6H_4$; $R^2 = R^3$ = Ph $R^1 = R^2 = R^3$ = p-$MeOC_6H_4$	A: Mchl., EtOH, 95:5, −78°C (or MB, Chlf.) B: NaOCl, H_2O_2, MeOH,	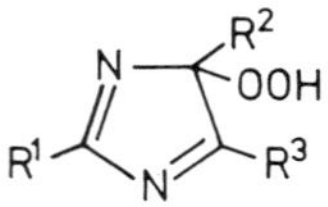 A: (47 or 68%) B: (41%) m.p. 110°C dec. A: (55%) m.p. 125°–126°C dec. A: (63%) m.p. 127°–128°C dec.	265, 277, 282
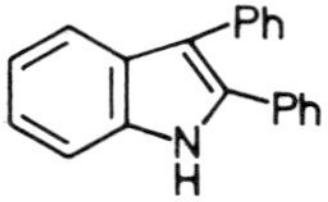	A: RB, MeOH, 0°C	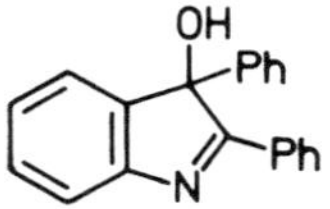 (58%) m.p. 190.5°–192.5°C 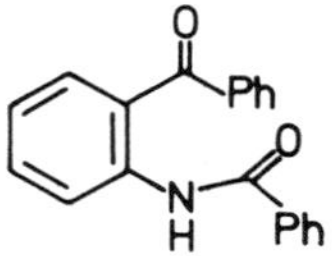 (37%)	181
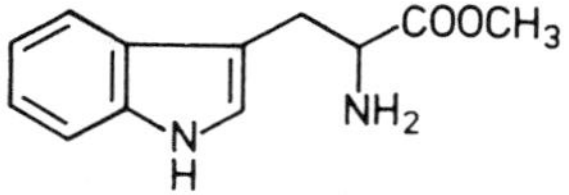	A: RB, MeOH, Bz. 1:50	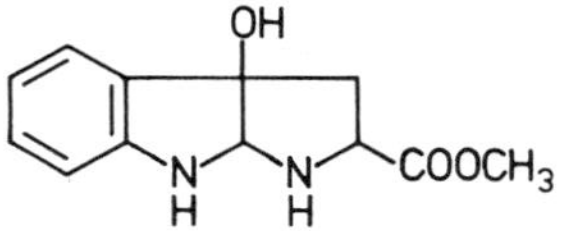 (3.7%) m.p. 166°–167°C	178, 181b, 181c, 220a

Substrate	Conditions	Products	References
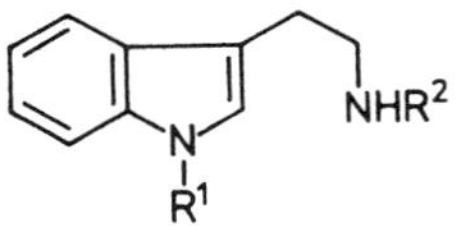 (a) $R^1 = H; R^2 = COOMe$ (b) $R^1 = H; R^2 = Me$ (c) $R^1 = Me; R^2 = COOMe$	(1) (a,b) A: RB, MeOH, Py. 95:5; 0°C (2) (a,b) Me_2S or $NaBH_4$ or Al_2O_3, SiO_2 (1) (c) A: −70°C (2) (c) Me_2S	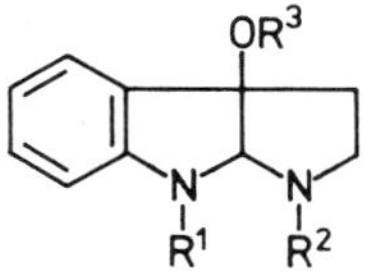 (a) (1) $R^3 = OH$ (41%); (2) $R^3 = H$ (71%) m.p. 126°–127°C (b) (1) $R^3 = H$ m.p. 151°C (c) (1) $R^3 = OH$; (2) $R^3 = H$ (91%)	179, 180, 181a, 181b, cf. 181d 181c, 220a
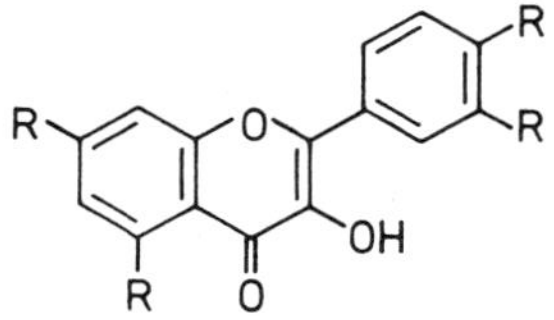 (a) R = OH Quercetin (b) R = OMe (c) R = H	(1) A: RB, Py. (2) CH_2N_2	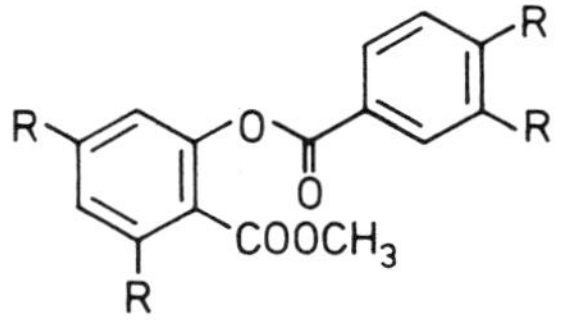 + CO + CO_2 (besides hydrolysis products) (b) (77%) m.p. 143°–144°C (c) (44%) m.p. 83°–84°C	158, 159

[a] Using ion exchange resin-bound sensitizers [283].
[b] With compounds (a) and (f) the double bond migrates in the direction of R^2 instead of R^1.
[c] Besides *2.4* 2-hydroxy-3-methyl-3-pentene [73].
[d] Besides *9.6* 2-hydroxy-3-methyl-3-pentene [73].
[e] After reduction.
[f] Product ratio decreases and photooxygenation rate increases with solvent polarity. At −78°C in MeOH the primary 1,2-dioxetane has been isolated.
[g] Besides S-containing cleavage products.
[h] No significant substituent effect on rate and product distribution.

(*Continued*)

[i] In order of decreasing oxygenation rate; the rates correlate with Hammett's σ.
[j] The product is invariably 2.7:1 and does not depend on substituents or $^1O_2$ source.
[k] The ketone was shown not to be a thermal secondary product of the hydroperoxide.
[l] Traces only according to Klein and Rojan [132].
[m] Via mixture of tertiary allylic hydroperoxides (not isolated).
[n] Besides other products.
[o] G: KO_2, 18-crown-6, benzoyl peroxide, Bz.
[p] Not isolated.
[q] Besides other products, *15*; see [54, 76].
[r] Besides 7-ketones.
[s] After Nickon [187]; cf. Kearns [117] for a dependence on sens. concentration.
[t] Both isomers not isolated in pure state; data refer to authentic material.
[u] Presumably formed via 5α-alcohol.
[v] Via 7α-monohydroperoxide as a primary product, isolated as $3\beta,7\alpha$-diacetoxy-$\Delta^{8(14)}$-cholestene (m.p. 136°–138°C. $[\alpha]_D$ +3.1°).

ACKNOWLEDGMENTS

The authors would like to thank Mrs. I. Schneider for preparing the drawings and Mrs. L. Heidemann for typing the manuscipt. One of the authors (K. G.) would also like to express his gratitude to the Deutsche Forschungsgemeinschaft, Bonn-Bad Godesberg, and the Fonds der chemischen Industrie, Frankfurt am Main, for the continuous support of his work.

REFERENCES

1. Acharya, S. P., *Tetrahedron Lett.* p. 4117 (1966).
2. Adam, W., and Liu, J.-C., *J. Am. Chem. Soc.* **94,** 2894 (1972).
3. Adam, W., Cueto, O., and Ehrig, V., *J. Org. Chem.* **41,** 370 (1976).
3a. Adam, W., Alzérreca, A., Liu, J.-Ch., and Yany, F., *J. Am. Chem. Soc.* **99,** 5768 (1977).
3b. Adam, W., and del Fiero, J., *J. Org. Chem.* **43,** 1159 (1978).
4. Adams, D. R., and Wilkinson, F., *J. Chem. Soc., Faraday Trans. 2* **68,** 586 (1972).
5. Adams, W. R., and Trecker, D. J., *Tetrahedron* **28,** 2361 (1972).
6. Adams, W. R., *4th Ed. Methoden Org. Chem.* (*Houben-Weyl*), **4,** Part 5b, II, 1465 (1975).
7a. Asao, T., Yagihara, M., and Kitahara, Y., *Bull. Chem. Soc. Jpn.* **51,** 2131 (1978).
7. Ando, W., Watanabe, K., Suzuki, J., and Migita, T., *J. Am. Chem. Soc.* **96,** 6766 (1974).
8. Ashford, R. D., and Ogryzlo, E. A., *J. Am. Chem. Soc.* **97,** 3604 (1975).
9. Banerjee, A. K., Martin, A., Nakano, T., and Usubillaga, A., *J. Org. Chem.* **38,** 3807 (1973).
10. Banthorpe, D. V., Young, M. R., and Fordham, W. D., *Chem. Ind.* (*London*) p. 901 (1973).
11. Barnes, M. F., Durley, R. C., and MacMillan, J., *J. Chem. Soc. C* p. 1341 (1970).
12. Barrett, H. C., and Büchi, G., *J. Am. Chem. Soc.* **89,** 5665 (1967).
13. Bartlett, P. D., and Mendenhall, G. D., *J. Am. Chem. Soc.* **92,** 210 (1970).
14. Bartlett, P. D., and Schaap, A. P., *J. Am. Chem. Soc.* **92,** 3223 (1970).
15. Bartlett, P. D., Mendenhall, G. D., and Schaap, A. P., *Ann. N.Y. Acad. Sci.* **171,** 79 (1970).
16. Bell, R. A., Ireland, R. E., and Mander, L. N., *J. Org. Chem.* **31,** 2536 (1966).
17. Bellesia, F., Pagnoni, U. M., and Trave, R., *Tetrahedron Lett.* p. 1245 (1974).
18. Bellin, J. S., and Mahoney, R. D., *J. Am. Chem. Soc.* **94,** 7991 (1972).
19. Blossey, E. C., Neckers, D. C., Thayer, A. L., and Schaap, A. P., *J. Am. Chem. Soc.* **95,** 5820 (1973).
20. Boden, R. M., *Synthesis* p. 783 (1975).
21. Brenner, M., Thesis, University of California, Los Angeles (1969); Univ. Microfilms 69-16,898.
22. Brill, W. F., *J. Am. Chem. Soc.* **87,** 3286 (1965).
23. Brkic', D., Forzatti, P., Pasquon, I., and Trifiro, F., *J. Photochem.* **5,** 23 (1976).
23a. Brkic', D., Forzatti, P., Pasquon, I., and Trifiro, F., *J. Molec. Catalysis* **3,** 173 (1977/78).
24. Broadbent, A. D., Gleason, W. S., Pitts, J. N., Jr., and Whittle, E., *J. Chem. Soc., Chem. Commun.* p. 1315 (1968).
25. Burns, P. A., and Foote, C. S., *J. Am. Chem. Soc.* **96,** 4339 (1974).
26. Burns, P. A., Foote, C. S., and Mazur, S., *J. Org. Chem.* **41,** 899 (1976).
27. Burns, P. A., and Foote, C. S., *J. Org. Chem.* **41,** 908 (1976).
28. Butsugan, Y., Yoshida, S., Muto, M., and Bito, T., *Tetrahedron Lett.* p. 1129 (1971).
29. Butsugan, Y., Yoshida, S., Bito, T., and Muto, M., *Nippon Kagaku Kaishi* p. 1617 (1973).
30. Cambie, R. C., and Hayward, R. C., *Aust. J. Chem.* **27,** 2001 (1974).
31. Cavill, G. W. K., and Coggiola, I. M., *Aust. J. Chem.* **24,** 135 (1971).
32. Chan, H. W.-S., *J. Am. Oil Chem. Soc.* **54,** 100 (1977).

33. Ching, T.-Y., and Foote, C. S., *Tetrahedron Lett.* p. 3771 (1975).
34. Chung, S.-K., and Scott, A. I., *J. Org. Chem.* **40,** 1652 (1975).
35. Clements, A. H., van den Engh, R. H., Frost, D. J., Hoggenhout, K., and Nooi, J. R., *J. Am. Oil Chem. Soc.* **50,** 325 (1973).
36. Cobern, D., Hobbs, J. S., Lucas, R. A., and Mackenzie, D. J., *J. Chem. Soc. C* p. 1897 (1966).
37. Collins, D. J., and Hobbs, J. J., *Aust. J. Chem.* **20,** 1905 (1967).
38. Dalle, J.-P., Mousseron-Canet, M., and Mani, J.-C., *Bull. Soc. Chim. Fr.* p. 232 (1969).
38a. Danen, W. C., and Arudi, R. L., *J. Am. Chem. Soc.* **100,** 3944 (1978).
39. Dässler, H. G., *Justus Liebigs Ann. Chem.* **622,** 194 (1959).
40. Demole, E., and Enggist, P., *Helv. Chim. Acta* **51,** 481 (1968).
41. Denny, R. W., Ph.D. Thesis, University of California, Los Angeles (1969), *Diss. Abstr. B* **30,** 2604 (1969).
42. Denny, R. W., and Nickon, A., *Org. React.* **20,** 133 (1973).
43. Dewar, M. J. S., and Thiel, W., *J. Am. Chem. Soc.* **97,** 3978 (1975).
44. Doleiden, F. H., Fahrenholtz, S. R., Lamola, A. A., and Trozzolo, A. M., *Photochem. Photobiol.* **20,** 519 (1974).
44a. Doskotch, R. W., El-Feraly, F. S., Fairchild, E. H., and Huang, C.-T., *J. Chem. Soc. Chem. Commun.* p. 402 (1976).
45. Dufraisse, C., Rio, G., Ranjon, A., and Pouchot, O., *C.R. Hebd. Seances Acad. Sci.* **261,** 3133 (1965).
46. Dumas, J. L., *Bull. Soc. Chim. Fr.* p. 658 (1976).
47. Dunphy, P. J., *Chem. Ind.* (*London*) p. 731 (1971).
48. Eisfeld, W., Dissertation, University of Göttingen (1958).
48a. El-Feraly, F. S., Chan, Y.-M., Fairchild, E. H., and Doskotch, R. W., *Tetrahedron Lett.* p. 1973 (1977).
49. Ensley, H. E., and Carr, R. V. C., *Tetrahedron Lett.* p. 513 (1977).
50. Fedeli, E., Camurati, F., and Jacini, G., *J. Am. Oil Chem. Soc.* **48,** 787 (1971).
51. Fenical, W., Kearns, D. R., and Radlick, P., *J. Am. Chem. Soc.* **91,** 3396 (1969).
52. Fenical, W., Kearns, D. R., and Radlick, P., *J. Am. Chem. Soc.* **91,** 7771 (1969).
53. Foote, C. S., and Wexler, S., *J. Am. Chem. Soc.* **86,** 3879 (1964).
54. Foote, C. S., Wexler, S., and Ando, W., *Tetrahedron Lett.* p. 4111 (1965).
55. Foote, C. S., *Acc. Chem. Res.* **1,** 104 (1968).
56. Foote, C. S., and Denny, R. W., *J. Am. Chem. Soc.* **90,** 6233 (1968).
57. Foote, C. S., Wexler, S., Ando, W., and Higgins, R., *J. Am. Chem. Soc.* **90,** 975 (1968).
58. Foote, C. S., and Brenner, M., *Tetrahedron Lett.* p. 6041 (1968).
59. Foote, C. S., *Pure Appl. Chem.* **27,** 635 (1971).
60. Foote, C. S., and Uhde, G., *Org. Photochem. Synth.* **1,** 60 (1971).
61. Foote, C. S., and Denny, R. W., *J. Am. Chem. Soc.* **93,** 5162 (1971).
62. Foote, C. S., and Denny, R. W., *J. Am. Chem. Soc.* **93,** 5168 (1971).
63. Foote, C. S., Fujimoto, T. T., and Chang, Y. C., *Tetrahedron Lett.* p. 45 (1972).
63a. Foote, C. S., private communication, cited in Inagaki *et al.* [104].
64. Forbes, E. J., and Griffiths, J., *J. Chem. Soc., Chem. Commun.* p. 427 (1967).
65. Fourrey, J. L., Rondest, J., and Polonsky, J., *Tetrahedron* **26,** 3839 (1970).
66. Fox, J. E., Scott, A. I., and Young, D. W., *J. Chem. Soc., Chem. Commun.* p. 1105 (1967).
67. Fox, J. E., Scott, A. I., and Young, D. W., *J. Chem. Soc., Perkin Trans. 1* p. 799 (1972).
67a. Friedrich, E., and Lutz, W., *Angew. Chem.* **89,** 426 (1977).
67b. Frimer, A. A., Rot, D., and Sprecher, M., *Tetrahedron Lett.*, p. 1927 (1977).
67c. Frimer, A. A., Bartlett, P. D., Boschung, A. F., and Jewett, J. C., *J. Am. Chem. Soc.* **99,** 7977 (1977).

67d. A. A. Frimer, *J. Org. Chem.* **42,** 3194 (1977).
68. Fujimoto, T. T., Dissertation, University of California, Los Angeles (1972); Univ. Microfilm, 72-18125; *Diss. Abstr. B.* **32,** 6910 (1972).
69. Fujita, E., Fujita, T., and Katayama, H., *Tetrahedron Lett.* p. 2577 (1969).
70. Fujita, E., Fujita, T., and Katayama, H., *Tetrahedron* **26,** 1009 (1970).
71. Furutachi, N., Nakadaira, Y., and Nakanishi, K., *J. Chem. Soc., Chem. Commun.* p. 1625 (1968).
72. Gleason, W. S., Broadbent, A. D., Whittle, E., and Pitts, J. N., Jr., *J. Am. Chem. Soc.* **92,** 2068 (1970).
73. Gleason, W. S., Rosenthal, I., and Pitts, J. N., Jr., *J. Am. Chem. Soc.* **92,** 7042 (1970).
74. Gollnick, K., unpublished results.
75. Gollnick, K., Schade, G., and Haisch, D., unpublished results.
76. Gollnick, K., and Schenck, G. O., *Pure Appl. Chem.* **9,** 507 (1964).
77. Gollnick, K., and Kotsonis, F., unpublished results; Kotsonis, F., M.S. Thesis, University of Arizona, Tucson (1969).
78. Gollnick, K., Schade, G., and Hartmann, H., unpublished results.
79. Gollnick, K., Schroeter, S., Ohloff, G., Schade, G., and Schenck, G. O., *Justus Liebigs Ann. Chem.* **687,** 14 (1965).
80. Gollnick, K., and Schade, G., *Tetrahedron Lett.* p. 2335 (1966).
81. Gollnick, K., and Schenck, G. O., *in* "1,4-Cycloaddition Reactions" (J. Hamer, ed.), p. 225. Academic Press, New York, 1967.
82. Gollnick, K., *Adv. Chem. Ser.* **77** III, 78 (1968).
83. Gollnick, K., *Adv. Photochem.* **6,** 1 (1968).
84. Gollnick, K., and Schade, G., *Tetrahedron Lett.* p. 689 (1968).
85. Gollnick, K., Franken, T., Schade, G., and Dörhöfer, G., *Ann. N.Y. Acad. Sci.* **171,** 89 (1970).
86. Gollnick, K., Haisch, D., and Schade, G., *J. Am. Chem. Soc.* **94,** 1747 (1972).
87. Gollnick, K., and Schade, G., *Tetrahedron Lett.* p. 857 (1973).
88. Gollnick, K., *Plenary Lect. on "Photo-Oxidations," EUCHEM Res. Conf. on Useful Preparative Aspects of Photochemistry, Sept. 1–5, 1975.* Ghent.
89. Gollnick, K., *in* "Radiation Research: Biochemical, Chemical and Physical Perspectives" (O. F. Nygaard, H. I. Adler, and W. K. Sinclair, eds.), p. 590. Academic Press, New York, 1975.
90. Gollnick, K., *in* "Singlet Oxygen Reactions with Polymers" (B. Ranby and J. F. Rabek, eds.), p. 111. Wiley, New York, 1978.
91. Gordon, A. J., and Ford, R. A., "The Chemist's Companion." Wiley (Interscience), New York, 1972.
91a. Gorman, A. A., and Rodgers, M. A. J., *Chem. Phys. Lett.* **55,** 52 (1978).
92. Haisch, D., Ph.D. Dissertation, University of München (1976).
93. Hall, G. E., and Roberts, D. G., *J. Chem. Soc. B* p. 1109 (1966).
93a. Harding, L. B., and Goddard, W. A., III, *J. Am. Chem. Soc.* **99,** 4520 (1977).
93b. Harding, L. B., and Goddard, W. A., III, *Tetrahedron Lett.* p. 747 (1978).
94. Hasty, N. M., Merkel, P. B., Radlick, P., and Kearns, D. R., *Tetrahedron Lett.* p. 49 (1972).
95. Hasty, N. M., and Kearns, D. R., *J. Am. Chem. Soc.* **95,** 3380 (1973).
96. Hehre, W. J., *J. Am. Chem. Soc.* **97,** 5308 (1975).
97. Helmlinger, D., de Mayo, P., Nye, M., Westfelt, L., and Yeats, R. B., *Tetrahedron Lett.* p. 349 (1970).
98. Herron, J. T., and Huie, R. E., *J. Chem. Phys.* **51,** 4164 (1969).
99. Higgins, R., Foote, C. S., and Cheng, H., *Adv. Chem. Soc.* **77,** 102 (1968).

100. Hollinden, G. A., and Timmons, R. B., *J. Am. Chem. Soc.* **92,** 4181 (1970).
101. Horinaka, A., Nakashima, R., Yoshikawa, M., and Matsuura, T., *Bull. Chem. Soc. Jpn.* **48,** 2095 (1975).
102. Huffman, J. W., *J. Org. Chem.* **41,** 3847 (1976).
103. Inagaki, S., and Fukui, K., *J. Am. Chem. Soc.* **97,** 7480 (1975).
104. Inagaki, S., Fujimoto, H., and Fukui, K., *Chem. Lett.* p. 749 (1976).
105. Isoe, S., Hyeon, S. B., Ichikawa, H., Katsumura, S., and Sakan, T., *Tetrahedron Lett.* p. 5561 (1968).
106. Isoe, S., Hyeon, S. B., and Sakan, T., *Tetrahedron Lett.* p. 279 (1969).
107. Isoe, S., Katsumura, S., Hyeon, S. B., and Sakan, T., *Tetrahedron Lett.* p. 1089 (1971).
108. Isoe, S., Hyeon, S. B., Katsumura, S., and Sakan, T., *Tetrahedron Lett.* p. 2517 (1972).
109. Itô, S., Takeshita, H., and Muroi, T., *Tetrahedron Lett.* p. 3091 (1969).
110. Itô, S., Takeshita, H., and Hirama, M., *Tetrahedron Lett.* p. 1181 (1971).
111. Itô, S., Takeshita, H., Hirama, M., and Fukazawa, Y., *Tetrahedron Lett.* p. 9 (1972).
112. Jefford, C. W., Laffer, M. H., and Boschung, A. F., *J. Am. Chem. Soc.* **94,** 8904 (1972).
113. Jefford, C. W., Boschung, A. F., Moriarty, R. M., Rimbault, C. G., and Laffer, M. H., *Helv. Chim. Acta* **56,** 2649 (1973).
114. Jefford, C. W., and Boschung, A. F., *Helv. Chim. Acta* **57,** 2242 (1974).
115. Jefford, C. W., and Rimbault, C. G., *Tetrahedron Lett.* p. 2479 (1976).
116. Jefford, C. W., Boschung, A. F., and Rimbault, C. G., *Helv. Chim. Acta* **59,** 2542 (1976).
116a. Jefford, C. W., and Rimbault, C. G., *J. Am. Chem. Soc.* **100,** 6437 (1978).
116b. Jefford, C. W., and Rimbault, C. G., *J. Org. Chem.* **43,** 1908 (1978).
117. Kearns, D. R., Hollins, R. A., Khan, A. U., and Radlick, P., *J. Am. Chem. Soc.* **89,** 5456 (1967).
118. Kearns, D. R., Hollins, R. A., Khan, A. U., Chambers, R. W., and Radlick, P., *J. Am. Chem. Soc.* **89,** 5455 (1967).
119. Kearns, D. R., *J. Am. Chem. Soc.* **91,** 6554 (1969).
120. Kearns, D. R., Fenical, W., and Radlick, P., *Ann. N.Y. Acad. Sci.* **171,** 34 (1970).
121. Kearns, D. R., *Chem. Rev.* **71,** 395 (1971).
122. Kellogg, R. M., and Kaiser, J. K., *Contrib. Pap., Useful Prep. Asp. Photochem., 1975* (1975).
123. Kellog, R. M., and Kaiser, J. K., *J. Org. Chem.* **40,** 2575 (1975).
124. Kende, A. S., and Chu, J. Y.-C., *Tetrahedron Lett.* p. 4837 (1970).
125. Kenney, R. L., and Fisher, G. S., *J. Am. Chem. Soc.* **81,** 4288 (1959).
126. Kenney, R. L., and Fisher, G. S., *J. Org. Chem.* **28,** 3509 (1963).
127. Khan, N. A., *Chem. Ind.* (*London*) p. 1000 (1973).
128. Kitahara, Y., Kato, T., Suzuki, T., Kanno, S., and Tanemura, M., *J. Chem. Soc., Chem. Commun.* p. 342 (1969).
129. Klein, E., and Rojahn, W., *Dragoco Rep.* (*German Ed.*) **11,** 123 (1964).
130. Klein, E., and Rojahn, W., *Chem. Ber.* **97,** 2700 (1964).
131. Klein, E., and Rojahn, W., *Chem. Ber.* **98,** 3045 (1965).
132. Klein, E., and Rojahn, W., *Tetrahedron* **21,** 2173 (1965).
133. Klein, E., and Rojahn, W., *Dragoco Rep.* (*German Ed.*) **15,** 159 (1968).
134. Koch, E., *Tetrahedron* **24,** 6295 (1968).
135. Kocór, M., and Wojciechowska, W., *Bull. Acad. Pol. Sci., Ser. Sci. Chim.* **21,** 809 (1973).
136. Koerner von Gustorf, E., Grevels, F.-W., and Schenck, G. O., *Justus Liebigs Ann. Chem.* **719,** 1 (1968).
137. Kondo, K., and Matsumoto, M., *Tetrahedron Lett.* p. 391 (1976).
138. Kopecky, K. R., and Reich, H. J., *Can. J. Chem.* **43,** 2265 (1965).
139. Kopecky, K. R., van de Sande, J. H., and Mumford, C., *Can. J. Chem.* **46,** 25 (1968).

140. Kopecky, K. R., and Mumford, C., *Can. J. Chem.* **47,** 709 (1969).
141. Kopecky, K. R., and van de Sande, J. H., *Can. J. Chem.* **50,** 4034 (1972).
141a. Kropf, H., and Kasper, B., *Justus Liebigs Ann. Chem.* p. 2232 (1975).
142. Krubiner, A. M., Saucy, G., and Oliveto, E. P., *J. Org. Chem.* **33,** 3548 (1968).
143. Kulig, M. J., and Smith, L. L., *J. Org. Chem.* **38,** 3639 (1973).
144. Kulig, M. J., and Smith, L. L., *J. Org. Chem.* **39,** 3398 (1974).
144a. Lerdal, D., and Foote, C. S., *Tetrahedron Lett.* p. 3227 (1978).
145. Lier, J. E. Van, and Smith, L. L., *J. Org. Chem.* **35,** 2627 (1970).
146. Litt, F. A., Ph.D. Thesis, John Hopkins University, Baltimore, Maryland (1967); *Diss. Abstr. B* **28,** 4501 (1968).
147. Litt, F. A., and Nickon, A., *Adv. Chem. Ser.* **77,** 118 (1968).
148. Long, C. A., and Kearns, D. R., *J. Am. Chem. Soc.* **97,** 2018 (1975).
148a. McCapra, F., and Beheshti, I., *J. Chem. Soc., Chem. Commun.* p. 517 (1977).
149. Marshall, J. A., and Fanta, W. I., *J. Org. Chem.* **29,** 2501 (1964).
150. Marshall, J. A., and Hochstetler, A. R., *J. Org. Chem.* **31,** 1020 (1966).
151. Marshall, J. A., and Cohen, N., *J. Am. Chem. Soc.* **87,** 2773 (1965).
152. Marshall, J. A., Cohen, N., and Hochstetler, A. R., *J. Am. Chem. Soc.* **88,** 3408 (1966).
153. Marshall, J. A., and Hochstetler, A. R., *Tetrahedron Lett.* p. 55 (1966).
154. Masamune, S., *J. Am. Chem. Soc.* **86,** 290 (1964).
154a. Matheson, I. B. C., Lee, J., and King, A. D., *Chem. Phys. Lett.* **55,** 49 (1978).
155. Matsumoto, M., and Kondo, K., *J. Org. Chem.* **40,** 2259 (1975).
156. Matsumoto, M., and Kondo, K., *Tetrahedron Lett.* p. 3935 (1975).
157. Matsumoto, M., Dobashi, S., and Kondo, K., *Tetrahedron Lett.* p. 4471 (1975).
157a. Matsumoto, M., Dobashi, S., and Kondo, K., *Bull. Chem. Soc. Jpn.* **51,** 185 (1978).
158. Matsuura, T., Matsushima, H., and Sakamoto, H., *J. Am. Chem. Soc.* **89,** 6370 (1967).
159. Matsuura, T., Matsushima, H., and Nakashima, R., *Tetrahedron* **26,** 435 (1970).
160. Matsuura, T., Horinaka, A., Yoshida, H., and Butsugan, Y., *Tetrahedron* **27,** 3095 (1971).
161. Matsuura, T., Horinaka, A., and Nakashima, R., *Chem. Lett.* p. 887 (1973).
162. Maumy, M., and Rigaudy, J., *Bull. Soc. Chim. Fr.* p. 1879 (1975).
163. Maumy, M., and Rigaudy, J., *Bull. Soc. Chim. Fr.* p. 1487 (1974).
164. Maumy, M., and Rigaudy, J., *Bull. Soc. Chim. Fr.* p. 2021 (1976).
165. Maurer, B., Fracheboud, M., Grieder, A., and Ohloff, G., *Helv. Chim. Acta* **55,** 2371 (1972).
166. Mazur, S., and Foote, C. S., *J. Am. Chem. Soc.* **92,** 3225 (1970).
167. Merkel, P. B., Nilsson, R., and Kearns, D. R., *J. Am. Chem. Soc.* **94,** 1030 (1972).
168. Merkel, P. B., and Kearns, D. R., *J. Am. Chem. Soc.* **94,** 7244 (1972).
168a. Monroe, B. M., *J. Phys. Chem.* **82,** 15 (1978).
168b. Mori, A., and Takeshita, H., *Chem. Lett.* p. 395 (1978).
169. Morimoto, H., Imada, I., and Goto, G., *Justus Liebigs Ann. Chem.* **735,** 65 (1970).
170. Mousseron-Canet, M., Mani, J.-C., and Dalle, J.-P., *Bull. Soc. Chim. Fr.* p. 608 (1967).
171. Mousseron-Canet, M., Dalle, J.-P., and Mani, J.-C., *Tetrahedron Lett.* p. 6037 (1968).
172. Mousseron-Canet, M., and Chabaud, J.-P., *Bull. Soc. Chim. Fr.* p. 245 (1969).
173. Mousseron-Canet, M., Dalle, J.-P., and Mani, J.-C., *Photochem. Photobiol.* **9,** 91 (1969).
174. Murray, R. W., and Kaplan, M. L., *J. Am. Chem. Soc.* **91,** 5358 (1969).
175. Murray, R. W., Lin, J. W.-P., and Kaplan, M. L., *Ann. N.Y. Acad. Sci.* **171,** 121 (1970).
176. Murray, R. W., *Chem. Eng. News* **48** (19), 34 (1970).
177. Murray, R. W., Lumma, W. C., Jr., and Lin, J. W.-P., *J. Am. Chem. Soc.* **92,** 3205 (1970).
178. Nakagawa, M., Kaneko, T., Yoshikawa, K., and Hino, T., *J. Am. Chem. Soc.* **96,** 624 (1974).
179. Nakagawa, M., Okajima, H., and Hino, K., *J. Am. Chem. Soc.* **98,** 635 (1976).
180. Nakagawa, M., Yoshikawa, K., and Hino, T., *J. Am. Chem. Soc.* **97,** 6496 (1975).

181. Nakagawa, M., Ohyoshi, N., and Hino, T., *Heterocycles* **4,** 1275 (1976).
181a. Nakagawa, M., Okajima, H., and Hino, T., *J. Am. Chem. Soc.* **99,** 4424 (1977).
181b. Nakagawa, M., and Hino, T., *Heterocycles* **6,** 1675 (1977).
181c. Nakagawa, M., Watanabe, H., Kodato, S., Okajima, H., Hino, T., Flippen, J. L., and Witkop. B., *Proc. Natl. Acad. Sci. U.S.A.* **74,** 4730 (1977).
181d. Nakagawa, M., Chiba, J., and Hino, T., *Heterocycles* **9,** 385 (1978).
182. Nickon, A., and Bagli, J. F., *J. Am. Chem. Soc.* **81,** 6330 (1959).
183. Nickon, A., and Bagli, J. F., *J. Am. Chem. Soc.* **83,** 1498 (1961).
184. Nickon, A., and Mendelson, W. L., *J. Am. Chem. Soc.* **85,** 1894 (1963).
185. Nickon, A., Schwartz, N., DiGiorgio, J. B., and Widdowson, D. A., *J. Org. Chem.* **30,** 1711 (1965).
186. Nickon, A., and Mendelson, W. L., *J. Org. Chem.* **30,** 2087 (1965).
187. Nickon, A., and Mendelson, W. L., *J. Am. Chem. Soc.* **87,** 3921 (1965).
188. Nickon, A., and Mendelson, W. L., *Can. J. Chem.* **43,** 1419 (1965).
189. Nickon, A., Chuang, V. T., Daniels, P. J. L., Denny, R. W., DiGiorgio, J. B., Tsunetsugu, J., Vilhuber, H. G., and Werstiuk, E., *J. Am. Chem. Soc.* **94,** 5517 (1972).
190. Nickon, A., DiGiorgio, J. B., and Daniels, P. J. L., *J. Org. Chem.* **38,** 533 (1973).
191. Nilsson, R., and Kearns, D. R., *Photochem. Photobiol.* **19,** 181 (1974).
192. Ogryzlo, E. A., and Tang, C. W., *J. Am. Chem. Soc.* **92,** 5034 (1970).
193. Ohloff, G., Vial, C., and Thomas, A. F., unpublished results.
194. Ohloff, G., Klein, E., and Schenck, G. O., *Angew. Chem.* **73,** 578 (1961).
195. Ohloff, G., Schulte-Elte, K.-H., and Willhalm, B., *Helv. Chim. Acta* **47,** 602 (1964).
196. Ohloff, G., and Uhde, G., *Helv. Chim. Acta* **48,** 10 (1965).
197. Ohloff, G., Becker, J., and Schulte-Elte, K.-H., *Helv. Chim. Acta* **50,** 705 (1967).
198. Ohloff, G., Strickler, H., Willhalm, B., Borer, C., and Hinder, M., *Helv. Chim. Acta* **53,** 623 (1970).
199. Ohloff, G., *Pure Appl. Chem.* **43,** 481 (1975).
200. Ohmae, M., and Katsui, G., *Vitamins* **35,** 116 (1967).
200a. Orfanopoulos, M., Bellarmine Grdina, Sr. M., and Stephenson, L. M., *J. Am. Chem. Soc.* **101,** 275 (1979).
201. Ouannès, C., and Wilson, T., *J. Am. Chem. Soc.* **90,** 6527 (1968).
202. Paquette, L. A., and Liotta, D. C., *Tetrahedron Lett.* p. 2681 (1976).
203. Paquette, L. A., Liao, C. C., Liotta, D. C., and Fristad, W. E., *J. Am. Chem. Soc.* **98,** 6412 (1976).
204. Paquette, L. A., Liotta, D. C., Liao, C. C., Wallis, T. G., Eickman, N., Clardy, J., and Gleiter, R., *J. Am. Chem. Soc.* **98,** 6413 (1976).
205. Peters, J. W., Bekowies, P. J., Winer, A. M., and Pitts, J. N., Jr., *J. Am. Chem. Soc.* **97,** 3299 (1975).
206. Pople, J. A., *J. Am. Chem. Soc.* **97,** 5306 (1975).
207. Potzinger, P., and Kuhn, H. J., unpublished results.
208. Pusset, J., Guénard, D., and Beugelmans, R., *Tetrahedron* **27,** 2939 (1971).
209. Ramasseul, R., and Rassat, A., *Tetrahedron Lett.* p. 1337 (1972).
210. Rawls, H. R., and van Santen, P. J., *J. Am. Oil Chem. Soc.* **47,** 121 (1970).
211. Rio, G., Ranjon, A., and Pouchot, O., *Bull. Soc. Chim. Fr.* p. 4679 (1968).
212. Rio, G., Ranjon, A., Pouchot, O., and Scholl, M. J., *Bull. Soc. Chim. Fr.* p. 1667 (1969).
213. Rio, G., and Charifi, M., *Bull. Soc. Chim. Fr.* p. 3598 (1970).
214. Rio, G., and Bricout, D., *Bull. Soc. Chim. Fr.* p. 3557 (1971).
215. Rio, G., Bricout, D., and Lecombe, L., *Tetrahedron Lett.* p. 3583 (1972).
216. Robin, M. B., "Higher Excited States of Polyatomic Molecules," Vol. 2. Academic Press, New York, 1975.

217. Rosenthal, I., and Acher, A. J., *Isr. J. Chem.* **12,** 897 (1974).
218. Rousseau, G., Le Perchec, P., and Conia, J. M., *Tetrahedron Lett.* p. 45 (1977).
218a. Rousseau, G., Le Perchec, P., and Conia, J. M., *Tetrahedron Lett.* p. 2517 (1977).
218b. Rousseau, G., Le Perchec, P., and Conia, J. M., *Synthesis* **1978,** 67
218c. Rousseau, G., Le Perchec, P., and Conia, J. M., *Tetrahedron* **34,** 3475, 3483 (1978).
219. Rubottom, G. M., and Lopez Nieves, M. I., *Tetrahedron Lett.* p. 2423 (1972).
220. Saito, I., and Matsuura, T., *Chem. Lett.* p. 1169 (1972).
220a. Saito, I., Matsuura, T., Nakagawa, M., and Hino, T., *Acc. Chem. Res.* **10,** 346 (1977).
221. Sakai, M., Harris, D. L., and Winstein, S., *J. Org. Chem.* **37,** 2631 (1972).
222. Sam, T. W., and Sutherland, J. K., *J. Chem. Soc., Chem. Commun.* p. 424 (1972).
223. Sasaki, T., Eguchi, S., and Ohno, M., *Synth. Commun.* **1,** 75 (1971).
224. Sasson, I., and Labovitz, J., *J. Org. Chem.* **40,** 3670 (1975).
225. Sato, T., and Murayama, E., *Bull. Chem. Soc. Jpn.* **47,** 715 (1974).
226. Schaap, A. P., and Bartlett, P. D., *J. Am. Chem. Soc.* **92,** 6055 (1970).
227. Schaap, A. P., Thayer, A. L., Blossey, E. C., and Neckers, D. C., *J. Am. Chem. Soc.* **97,** 3741 (1975).
228. Schaap, A. P., Kees, K., and Thayer, A. L., *J. Org. Chem.* **40,** 1185 (1975).
229. Scheffer, J. R., and Ouchi, M. D., *Tetrahedron Lett.* p. 223 (1970).
230. Schenck, G. O., *Naturwissenschaften* **35,** 28 (1948).
231. Schenck, G. O., German Patent 933,925 (1943).
232. Schenck, G. O., and Kinkel, K. G., *Naturwissenschaften* **38,** 355 (1951).
233. Schenck, G. O., Eggert, H., and Denk, W., *Justus Liebigs Ann. Chem.* **584,** 177 (1953).
234. Schenck, G. O., *Angew. Chem.* **64,** 12 (1952).
235. Schenck, G. O., Koerner von Gustorf, E., Meyer, K.-H., and Schänzer, W., *Angew Chem.* **68,** 304 (1956).
236. Schenck, G. O., Gollnick, K., and Neumüller, O. A., *Justus Liebigs Ann. Chem.* **603,** 46 (1957).
237. Schenck, G. O., and Schulte-Elte, K.-H., *Justus Liebigs Ann. Chem.* **618,** 185 (1958).
238. Schenck, G. O., and Neumüller, O.-A., *Justus Liebigs Ann. Chem.* **618,** 194 (1958).
239. Schenck, G. O., Neumüller, O.-A., and Eisfeld, W., *Justus Liebigs Ann. Chem.* **618,** 202 (1958).
240. Schenck, G. O., and Köller, H., unpublished; Köller, H., Diplomarbeit, University of Göttingen (1958).
241. Schenck, G. O., and Gollnick, K., *J. Chim. Phys.* **55,** 892 (1958).
242. Schenck, G. O., and Eisfeld, W., unpublished; Eisfeld, W., Diplomarbeit, University of Göttingen (1959).
243. Schenck, G. O., and Koch, E., *Z. Elektrochem.* **64,** 170 (1960).
244. Schenck, G. O., Ohloff, G., and Klein, E., DB Patent 1,137,730 (June 22, 1967); April 8, (1961).
245. Schenck, G. O., and Schulte-Elte, K.-H., unpublished; Schulte-Elte, K.-H., Dissertation, University of Göttingen (1961).
246. Schenck, G. O., and Helms, G., unpublished; Helms, G., Dissertation, University of Göttingen (1961).
247. Schenck, G. O., Koerner von Gustorf, E., and Köller, H., *Angew. Chem.* **73,** 707 (1961).
248. Schenck, G. O., and Gollnick, K., *Forschungsber, Landes Nordrhein-Westfalen* **1256,** (1963).
249. Schenck, G. O., Gollnick, K., Buchwald, G., Schroeter, S., and Ohloff, G., *Justus Liebigs Ann. Chem.* **674,** 93 (1964).
250. Schenck, G. O., Gollnick, K., Buchwald, G., Ohloff, G., Schade, G., and Schroeter, S., *Angew. Chem.* **76,** 582 (1964).

251. Schenck, G. O., Neumüller, O.-A., Ohloff, G., and Schroeter, S., *Justus Liebigs Ann. Chem.* **687,** 26 (1965).
252. Schenck, G. O., Eisfeld, W., and Neumüller, O.-A., *Justus Liebigs Ann. Chem.* p. 701 (1975).
252a. Schneider, W. P., and Ayer, D. E., *Proc. Int. Congr. Hormo. Steroids 2nd, 1966* p. 254 (1967).
253. Schönberg, A., "Preparative Organic Photochemistry," p. V. Springer-Verlag, Berlin and New York, 1958.
254. Schuller, W. H., and Lawrence, R. V., *J. Am. Chem. Soc.* **83,** 2563 (1961).
255. Schulte-Elte, K.-H., Fracheboud, M., and Ohloff, G., unpublished results.
256. Schulte-Elte, K.-H., and Ohloff, G., *Helv. Chim. Acta* **51,** 494 (1968).
257. Schulte-Elte, K.-H., Gadola, M., and Müller, B. L., *Helv. Chim. Acta* **54,** 1870 (1971).
258. Schulte-Elte, K.-H., Müller, B. L., and Ohloff, G., *Helv. Chim. Acta* **54,** 1899 (1971).
258a. Schulte-Elte, K.-H., Müller, B. L., and Rauteustrauch, V., *Helv. Chim. Acta* **61,** 2777 (1978).
259. Schultz, A. G., and Schlesinger, R. H., *Tetrahedron Lett.* p. 2731 (1970).
260. Seely, G. R., and Meyer, T. H., *Photochem. Photobiol.* **13,** 27 (1971).
261. Sharp, D. B., *Abstr., 138th Meet., Am. Chem. Soc., 1960* 79P, p. 146 (1960).
262. Sharp, D. B., and LeBlanc, J. R., *Prepr., Div. Petrol. Chem., Am. Chem. Soc.* **5,** No. 2, C57 (1960); *C.A.* **57,** 14923f (1962).
263. Smith, L. L., Teng, J. I., Kulig, M. J., and Hill, F. L., *J. Org. Chem.* **38,** 1763 (1973).
264. Snyder, C. D., and Rapoport, H., *J. Am. Chem. Soc.* **91,** 731 (1969).
265. Sonnenberg, J., and White, D. M., *J. Am. Chem. Soc.* **86,** 5685 (1964).
266. Steer, R. P., Sprung, J. L., and Pitts, J. N., Jr., *Environ. Sci. Technol.* **3,** 946 (1969).
267. Stephenson, L. M., McClure, D. E., and Sysak, P. K., *J. Am. Chem. Soc.* **95,** 7888 (1973).
268. Takeshita, H., Sato, T., Muroi, T., and Itô, S., *Tetrahedron Lett.* p. 3095 (1969).
269. Takeshita, H., Hirama, M., and Itô, S., *Tetrahedron Lett.* p. 1775 (1972).
269a. Terao, J., and Matsushita, S., *J. Am. Oil Chem.* **54,** 234 (1977).
269b. Terao, J., and Matsushita, S., *Agric. Biol. Chem.* **41,** 2467 (1977).
269c. Tanielian, C., and Chaineaux, J., *J. Photoch.* **9,** 19 (1978).
269d. Takeshita, H., Hatsui, T., and Shimooda, I., *Tetrahedron Lett.* p. 2889 (1978).
270. Thomas, A. F., and Dubini, R., *Helv. Chim. Acta* **57,** 2076 (1974).
271. Tsukida, K., Cho, S.-C., and Yokota, M., *Chem. Pharm. Bull. Jpn.* **17,** 1755 (1969).
272. Turner, D. W., *Adv. Phys. Org. Chem.* **4,** 31 (1966).
272a. Turner, J. A., and Herz, W., *J. Org. Chem.* **42,** 1657 (1977).
273. Vedeneyev, V. I., Gurvich, L. V., Kodrat'yev, V. N., Medvedev, V. A., and Frankevich, Ye. L., "Bond Energies, Ionisation Potentials and Electron Affinities." Arnold, London, 1966; VEB Deutscher Verlag für Grundstoffindustrie, Leipzig, 1971.
274. Ville, T. E. de, Hora, J., Hursthouse, M. B., Toube, T. B., and Weedon, B. C. L., *J. Chem. Soc., Chem. Commun.* p. 1231 (1970).
275. Wagner, H. U., and Gollnick, K., unpublished results.
275a. Wahlberg, I., Karlsson, K., Curvall, M., Nishida, T., and Enzell, C. R., *Acta Ch. Scand.* **B32,** 203 (1978).
276. Wasserman, H. H., and Scheffer, J. R., *J. Am. Chem. Soc.* **89,** 3073 (1967).
277. Wasserman, H. H., Stiller, K., and Floyd, M. B., *Tetrahedron Lett.* p. 3277 (1968).
278. Wasserman, H. H., *Ann. N.Y. Acad. Sci.* **171,** 108 (1970).
279. Wasserman, H. H., Scheffer, J. R., and Cooper, J. L., *J. Am. Chem. Soc.* **94,** 4991 (1972).
280. Wexler, S., Ph.D. Thesis, University of California, Los Angeles (1966); Univ. Microfilms, 66-11,921; *Diss. Abstr. B* **27,** 1837 (1966).

281. Wharton, P. S., Hiegel, G. A., and Coombs, R. V., *J. Org. Chem.* **28,** 3217 (1963).
282. White, E. H., and Harding, M. J. C., *Photochem. Photobiol.* **4,** 1129 (1965).
283. Williams, J. R., Orton, G., and Unger, L. R., *Tetrahedron Lett.* p. 4603 (1973).
284. Winer, A. M., and Bayes, K. D., *J. Phys. Chem.* **70,** 302 (1966).
284a. Yamauchi, R., and Matsushita, S., *Agric. Biol. Chem.* **41,** 1425 (1977).
285. Young, R. H., Brewer, D., and Keller, R. A., *J. Am. Chem. Soc.* **95,** 375 (1973).
286. Young, R. H., Martin, R. L., Feriozi, D., Brewer, D., and Kayser, R., *Photochem. Photobiol.* **17,** 233 (1973).

9

Reactions of Singlet Oxygen with Heterocyclic Systems

HARRY H. WASSERMAN and BRUCE H. LIPSHUTZ

I. Introduction 430
II. Furans 431
  A. Alkyl Furans 431
  B. Aryl Furans 434
  C. Benzofurans 441
  D. Polyfurans and Furanophanes 443
III. Pyrroles 447
  A. Alkyl Pyrroles 449
  B. Aryl Pyrroles; Epoxide Formation 455
  C. Pyrrolophanes 459
  D. Bilirubin 463
IV. Indoles 467
  A. Alkyl Derivatives 467
  B. Tryptophan and Related Systems 469
  C. Rearrangements to Benzoxazines 471
  D. Other Reactions of 3-Peroxy Derivatives 475
V. Imidazoles 481
  A. Alkyl- and Aryl-Substituted Derivatives 483
  B. Model Systems Related to Histidine Oxidation 488
VI. Purines 490
VII. Oxazoles 498
  A. Mechanism of Rearrangement to Triamides 499
  B. Solvent Effects 501
VIII. Thiazoles 502
IX. Thiophenes 503
  References 506

Singlet Oxygen

ISBN 0-12-736650-4

I. INTRODUCTION

In reviewing the reactions of singlet oxygen with selected heterocyclic systems, special attention has been given to the manner in which the initially formed intermediates, such as hydroperoxides, dioxetanes, or endoperoxides rearrange to the observed products. Except for reactions at very low temperatures, these peroxides have usually not been isolated in normal laboratory reactions of singlet oxygen, and it is the subsequent decomposition of these species which provides the basis for the rich source of chemical transformations which have been uncovered in the course of recent studies in singlet oxygen chemistry. As will be shown in this section, the great diversity among the products formed in these reactions results not so much from the variation in the mode of attack of singlet oxygen on the particular heterocyclic system, but from the varied processes involved in the breakdown of the peroxidic intermediates. In these subsequent reactions the effects of solvent, temperature, substituents, and geometry all play an important role in determining the nature of the products obtained.

The particular heterocyclic substrates reviewed here have been chosen to illustrate typical pathways of singlet oxygen reactions. No attempt has been made to survey in an exhaustive manner all known reactions of $^1O_2$ with heterocyclic compounds.

Addition of singlet oxygen to a heterocyclic system usually occurs by one of three methods: (a) 1,4-addition to the 1,3-diene system as frequently observed in furans, pyrroles, oxazoles, thiazoles, imidazoles, and purines; (b) dioxetane formation, often observed in benzofurans, certain imidazoles, and purines; (c) hydroperoxide formation by typical ene-type reactions. Some of these processes may be preceded by the initial formation of a zwitterionic species in which the heteroatom releases electrons to the electrophilic singlet oxygen through an adjacent double bond as shown in the enaminelike reaction pictured below.

$$\text{—N—C=C} \quad \text{O=O} \longrightarrow \text{—N}^{+}\text{=C—C—O—O}^{-}$$

Very recently (Saito *et al.*, 1977a), new evidence has been obtained showing how such zwitterions may be trapped in the case of *N*-substituted indoles. Other evidence has shown that perepoxides or related zwitterions may be intermediates in the reactions of certain enol ethers with singlet oxygen (Jefford and Rimbault, 1978).

II. FURANS

Some of the earliest information on the reactions of singlet oxygen with heterocyclic systems was derived from the studies of Schenck and co-workers on the dye-sensitized photooxidation of furans.

The uptake of oxygen generally involves 1,4-addition to the furan ring, with the formation of a transannular (2,5)-endoperoxide resembling the monoozonide of a cyclobutadiene. Unstable intermediates of this type may then undergo a variety of further reactions such as hydrolysis, rearrangement, reduction, or polymerization depending on solvent, substitution pattern, or other conditions.

Thus, the parent compound, furan (**1**) is oxidized to the peroxide (**2**) which decomposes explosively at $-10°C$ but may be isolated from photo-oxygenation reactions at $-100°C$ in a solvent system containing methanol/*n*-propanol/acetone (2:2:1) (Schenck and Koch, 1966). With triphenyl-phosphine, **2** is reduced to maleic dialdehyde (**4**), while methanolysis yields the *pseudo*-ester (**3**).

O_2, $h\nu$, Sens., solvent, $-100°C$; Et_2O → polymer; O_2, $h\nu$, Sens., CH_3OH; CH_3OH; Ph_3P, $-70°C$

(**1**) (**2**) (**3**) (**4**)

In ether solution, the same photooxygenation reaction yields only products of polymerization. Other endoperoxides isolated from aryl-substituted furans will be discussed later.

A. Alkyl Furans

Photooxidation of alkyl furans may, in some cases, yield stable hydro-peroxides derivable from such ozonidelike intermediates. For example, 2-methylfuran (**5**) reacts with singlet oxygen in methanol to yield the meth-oxyhydroperoxide (**6**) (Foote *et al.*, 1967). This reaction may be viewed as a solvolytic ring opening of intermediate **7**. The fact that none of the isomeric product (**8**) is formed is in accord with the expected preference for attack of methanol at the tertiary position of the endoperoxide. Treatment of the

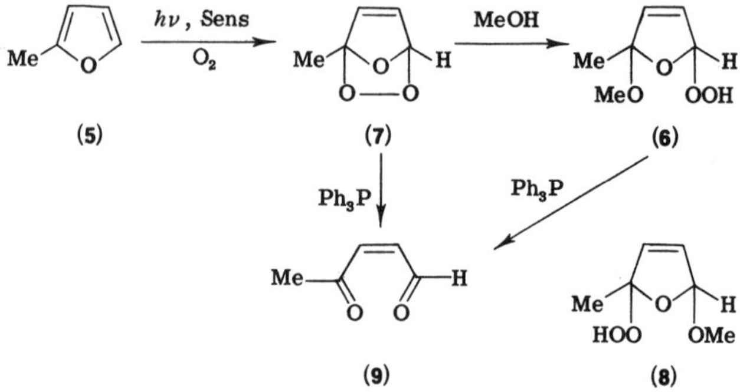

(5) —$h\nu$, Sens, O_2→ (7) —MeOH→ (6)

(7) —Ph_3P→ (9); (6) —Ph_3P→ (9)

(8)

crude photooxidation mixture with triphenylphosphine gives the enedione (**9**).

In the same manner, 2,5-dimethylfuran (**10**) and dicyclohexanofuran (**12**) yield methoxy hydroperoxides **11** and **13** on photooxidation in methanol. When the reactions are carried out in nonalcoholic solvents such as benzene/petroleum ether, unstable intermediates are formed which give the methoxy hydroperoxides on treatment with methanol.

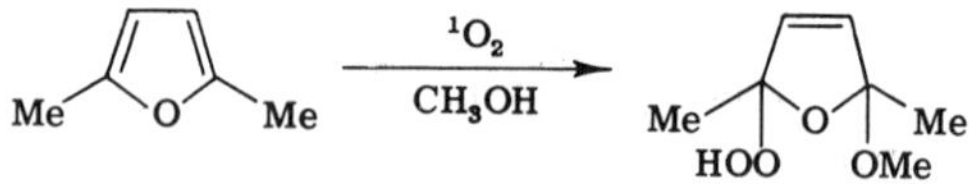

(10) —$^1O_2$, CH_3OH→ (11)

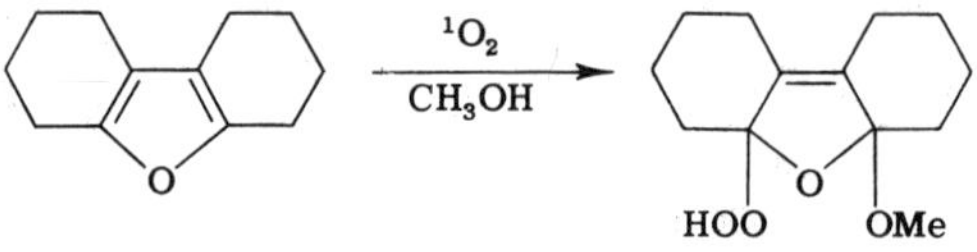

(12) —$^1O_2$, CH_3OH→ (13)

Similarly, photooxidation of menthofuran (**14**) in methanol yields a stable monomeric peroxidic species which was shown (Foote *et al.*, 1967) to be **15**. Again, the reaction outcome is consistent with the solvolytic opening of an intermediate peroxide at the more favored tertiary position.

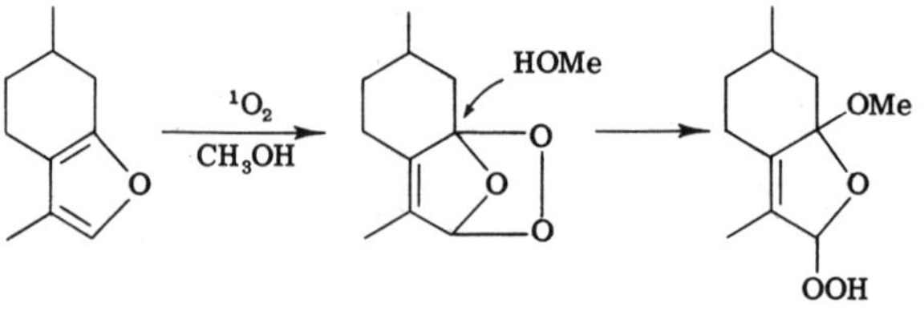

(14) —$^1O_2$, CH_3OH→ [HOMe] → (15)

Baeyer–Villiger-like rearrangements also appear to be generally involved in the breakdown of transannular peroxides formed from a variety of 2-substituted furans. The reaction of furfural (**16**) with singlet oxygen in ethanol at room temperature leading to the *pseudo*-ester (**18**) is typical

$^1O_2$, EtOH; (*a*); EtOH; (*b*); EtOH

(**16**) (**17**) (**19**) (**18**)

(Schenck, 1953). Here, the conversion may be pictured in terms of path *a*, involving the migration of a formyl group. Alternatively, the intermediate peroxide (**17**) may undergo solvolysis to the ethoxy hydroperoxide (path b) and then fragmentation to the product. The latter process may very probably involve a homolytic cleavage of the O–O bond with loss of a formyl radical. This type of breakdown is analogous to that suggested for the α-carbon cleavage reactions observed in pyrrole photooxidation. Table I lists a number of 2-substituted furans which undergo similar oxidative cleavage reactions.

Table I Pseudo Ester (**18**) Formation from Photooxidation of 2-Substituted Furans

Compound	ψ-Ester (%)	Ref.
2-furyl–COOH	98	Schenck (1948)
2-furyl–CH_2OCH_3	98	Schenck (1952)
2-furyl–$CH_2NHCOCH_3$	98	Schenck (1948, 1952)
2-furyl–CH_2OH	86	Schenck and Appel (1946)

Recent applications of alkyl furan photooxidations are found in the preparation of intermediates in the synthesis of camptothecin (Meyers *et al.*, 1973), portulal (Kanazawa *et al.*, 1975), and trigol (Heather *et al.*, 1974).

B. Aryl Furans

While it has generally not been possible to isolate the ozonidelike peroxides formed in the reactions of singlet oxygen with alkyl furans, there have been cases among aryl-substituted derivatives where these species have been prepared at low temperature and their chemical reactions studied. Among the relatively stable monomeric endoperoxides which have been reported are **20**, **21**, and **22** derived from 2,5-diphenyl-3,4-di(*p*-bromophenyl)furan, (Lutz *et al.*, 1962), 2,5-diphenylfuran (Dufraisse *et al.*, 1967; Trozzolo and Fahrenholtz, 1970), and 1,3-diphenylisobenzofuran (Dufraisse and Ecary, 1946), respectively.

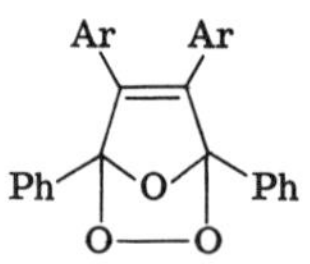

Ar = p-BrC_6H_4
(20)

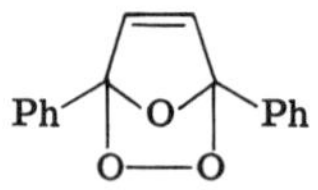

(21)

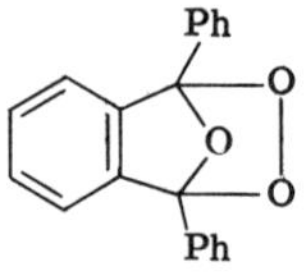

(22)

The endoperoxide (**20**) isolated by Lutz could be transformed to **23** by iodide, while, in the solid state it undergoes spontaneous conversion to **24**, **25**, and **26**.

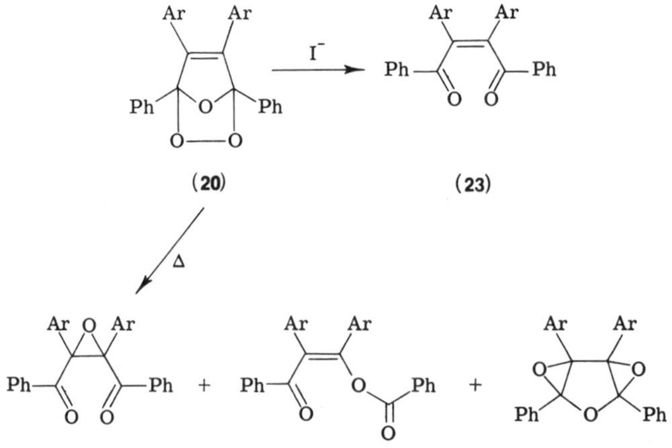

(20) —I^-→ (23)

(20) —Δ→ (24) + (25) + (26)

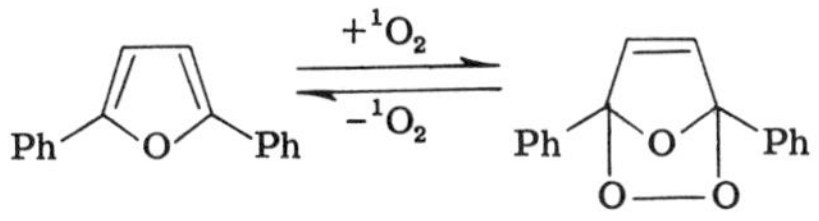

(27) (21)

Dye-sensitized photooxygenation of 2,5-diphenylfuran (**27**) in carbon disulfide (Schenck, 1947; Martel, 1957) leads to the formation of *cis*-dibenzoylethylene. At $-70°C$, however, the endoperoxide (**21**) could be isolated and the furan (**27**) regenerated on warming to room temperature. When (**21**) was allowed to warm up in the presence of cyclohexadiene, the endoperoxide (**28**) was obtained, showing that (**21**) is capable of acting as a singlet oxygen transfer agent (Trozzolo and Fahrenholtz, 1970). Oxidations of other typical singlet oxygen acceptors by **21** yield the expected products.

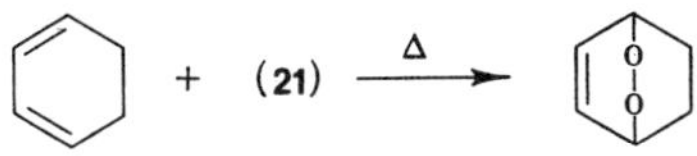

(28)

The monomeric endoperoxide (**22**), isolated in the photooxidation of 1,3-diphenylisobenzofuran (**29**) at temperatures below $-78°C$, explodes on warming to 18°C. If **22** is allowed to warm up in solution it yields *o*-dibenzoylbenzene (**30**).

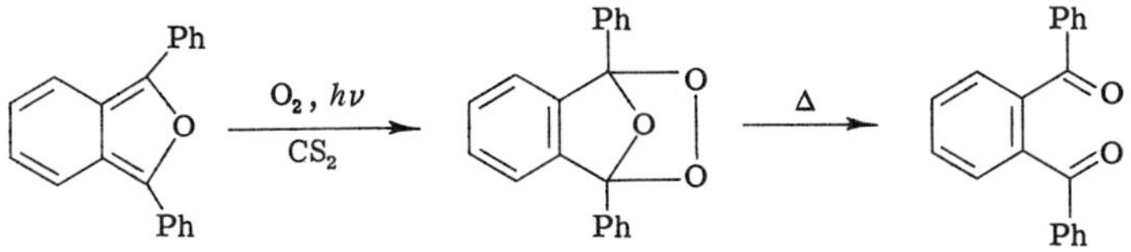

(29) (22) (30)

Despite extensive investigation of the reaction of singlet oxygen with **29**, the mode of decomposition of **22** to **30** involving the apparent loss or transfer of an atom of oxygen, has not been established. A possible pathway for this decomposition may involve an intermediate carbonyl oxide (**31**). Such a species could form a dimer (Murray and Higley, 1974) which may decompose to **30** and $^1O_2$. Alternatively, the carbonyl oxide could react directly with another molecule of **29** as shown to form two molecules of **30**.

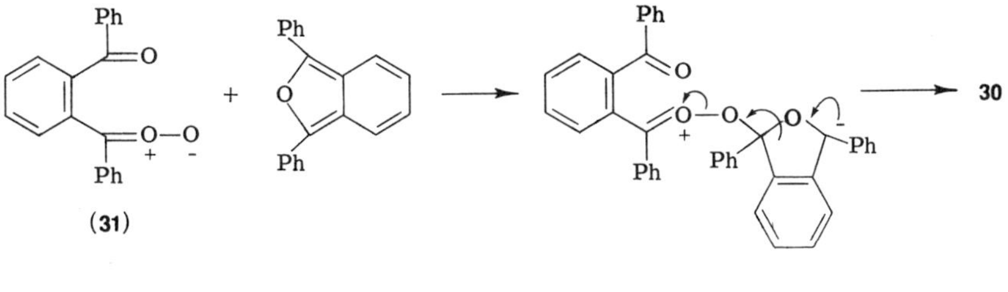

(31)

In more recent studies on the reaction of (**29**) with singlet oxygen (H. H. Wasserman and F. Vinick, unpublished results, 1973; Stevens *et al.*, 1973) the product (**32**) of a Baeyer–Villiger rearrangement has been isolated.

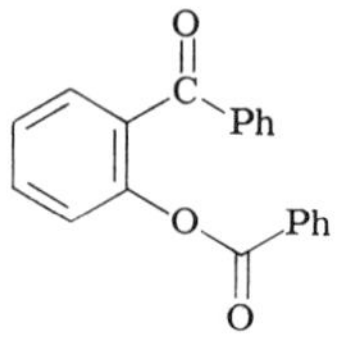

(**32**)

Because of the readiness with which 1,3-diphenylisobenzofuran (**29**) is converted to *o*-dibenzoylbenzene (**30**) by singlet oxygen, this reaction has frequently been used as a diagnostic test for the presence of singlet oxygen in both chemical and biological processes (Chan, 1970). This result may, however, lead to erroneous conclusions since it has been shown that **29** is very susceptible to autoxidation and may lead to **30** under conditions in which singlet oxygen is not an intermediate (Mendenhall and Howard, 1975). Thus, an earlier suggestion that catalysis by lipoxidases mimics the reaction of singlet oxygen, rested, in part, on the observation that soybean lipoxidases were effective in catalyzing the conversion of 1,3-diphenyliso-benzofuran (**29**) to **30** (Chan, 1971). While the above conversion may have been brought about by singlet oxygen, rapid autoxidation could also have been a competing process, and later evidence (Baldwin *et al.*, 1971) served to refute the earlier conclusions. For a fuller discussion of this work, see Chapter 12.

Aryl-substituted furans give products analogous to those observed in alkyl furan–singlet oxygen reactions for the most part, through initial endoperoxide formation. In mechanistic studies on the photooxidations of 2-aryl furans, Young *et al.* (1972) concluded that such endoperoxide formation is a concerted rather than a stepwise process. A Hammett plot provided rho (ρ) values which could be correlated with the polarity of the solvents employed. While Young's results argue against a completely dipolar intermediate of the type shown below, his evidence favors a "slightly polar" transition state.

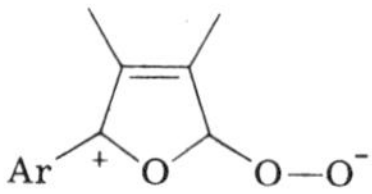

Photooxidation of tetraphenylfuran (**33**) may take varied courses depending on solvent, temperature, and other conditions. Wasserman and Liberles (1960) showed that in methanol, using methylene blue as sensitizer, *cis*-dibenzoylstilbene (**34**) and the dimethoxydihydrofuran (**35**) are the principal products. On the other hand, in aqueous acetone, the epoxide (**36**) is formed along with the enol benzoate (**37**). In concurrent work, Lutz *et al.* (1962) were able to isolate the bisepoxide (**38**) in addition to the above products.

O_2, $h\nu$, Sens., MeOH → (34) + (35)

(33)

O_2, $h\nu$, H_2O, Me_2CO → (36) + (37)

(38)

(39) —MeOH→ (40) —H_2O→ (41)

(40) —MeOH→ (35)

(41) → (34)

Formation of **35** most probably takes place by solvolysis of the endoperoxide (**39**) in methanol via the methoxyhydroperoxide (**40**). The conversion of intermediate (**40**) to *cis*-dibenzoylstilbene could take place by a solvolytic reaction with water present in the alcohol leading to the corresponding hemiacetal (**41**). This transformation has been studied in the alkyl furan series (Foote *et al.*, 1967) and appears to be a common type of reaction sequence in furan photooxidations carried out in alcoholic solvents.

The formation of 3,4-epoxides such as **36** in the reactions of singlet oxygen with aryl furans (and aryl pyrroles) is an unusual reaction, the mechanism of which remains to be clarified. A possible pathway for this 3,4-oxidation may involve the zwitterion (**42**) produced directly from **33** or formed by ring-opening of a 2,3-dioxetane derived from zwitterion (**43**). Intermediate **42** could be trapped by H_2O forming the hydroperoxide (**44**) and suffer fragmentation in a manner similar to the process observed by Nickon and Mendelson (1963) in their studies on the photooxygenation of allylic alcohols (e.g., **45** → **46**).

(**42**) ← ← (**43**) ← **33**

H_2O^*

(**44**) ⇌ → (**36**)

(**45**) $\xrightarrow[\text{"ene"}]{^1O_2}$ → (**46**)

* = O-18

Evidence against this mechanism was provided by H. H. Wasserman and F. Vinick (unpublished results, 1973) in studies of this reaction in the presence of $H_2{}^{18}O$. As shown by high resolution mass spectrometry, the diketoepoxide (**36**) contained none of the oxygen-18 enrichment, a result to be expected from the process outlined in the conversion: **42** → **36**.

An alternate route to the epoxide (**36**) might involve an intermediate carbonyl oxide (**47**). Such a peroxidic species could yield the epoxide by an inter- or intramolecular oxygen transfer as shown. Evidence for this type of intermediate was provided by Wasserman and Saito (1975) using diphenyl sulfide (DPS) as a trapping agent. In the presence of DPS no epoxide was formed and the product (**34**) was accompanied by diphenyl sulfoxide. An analogous deoxygenation reaction takes place in the reaction of singlet oxygen and DPS with diphenyldiazomethane where a carbonyl oxide (**48**) is considered to be the intermediate.

43 ⟶ (**47**) $\xrightarrow{(34)}$ ⟶ **36** + **34**

(**47**) $\xrightarrow{Ph_2S}$ (**34**) + $Ph_2\overset{+}{S}-O^-$

$$Ph_2CN_2 + {}^1O_2 \longrightarrow Ph_2C{=}\overset{+}{O}{-}O^- \ (\mathbf{48}) \xrightarrow{Ph_2S} Ph_2CO + Ph_2\overset{+}{S}{-}O^-$$

Other aryl furan-singlet oxygen reactions related to the work of Wasserman and Liberles are found in studies of Basselier *et al.* (1970a,b, 1971). The fused ring system (**49**) yields the epoxy diketone (**50**) and the enol benzoate (**51**) analogous to the products (**36** and **37**) formed in the tetraphenylfuran case.

With the 3,4-bridged system (**52**), a mixture of enol benzoates (**55** and **56**) is obtained. It was suggested that both endoperoxide (**53**) and dioxetane (**54**) are involved as intermediates (Scheme I). However, it is quite conceivable that product (**56**) is actually formed by an intramolecular rearrangement of **55**, along the lines of reactions extensively studied by Lutz and co-workers (1962). It may therefore not be necessary to postulate the intermediacy of the strained dioxetane (**54**).

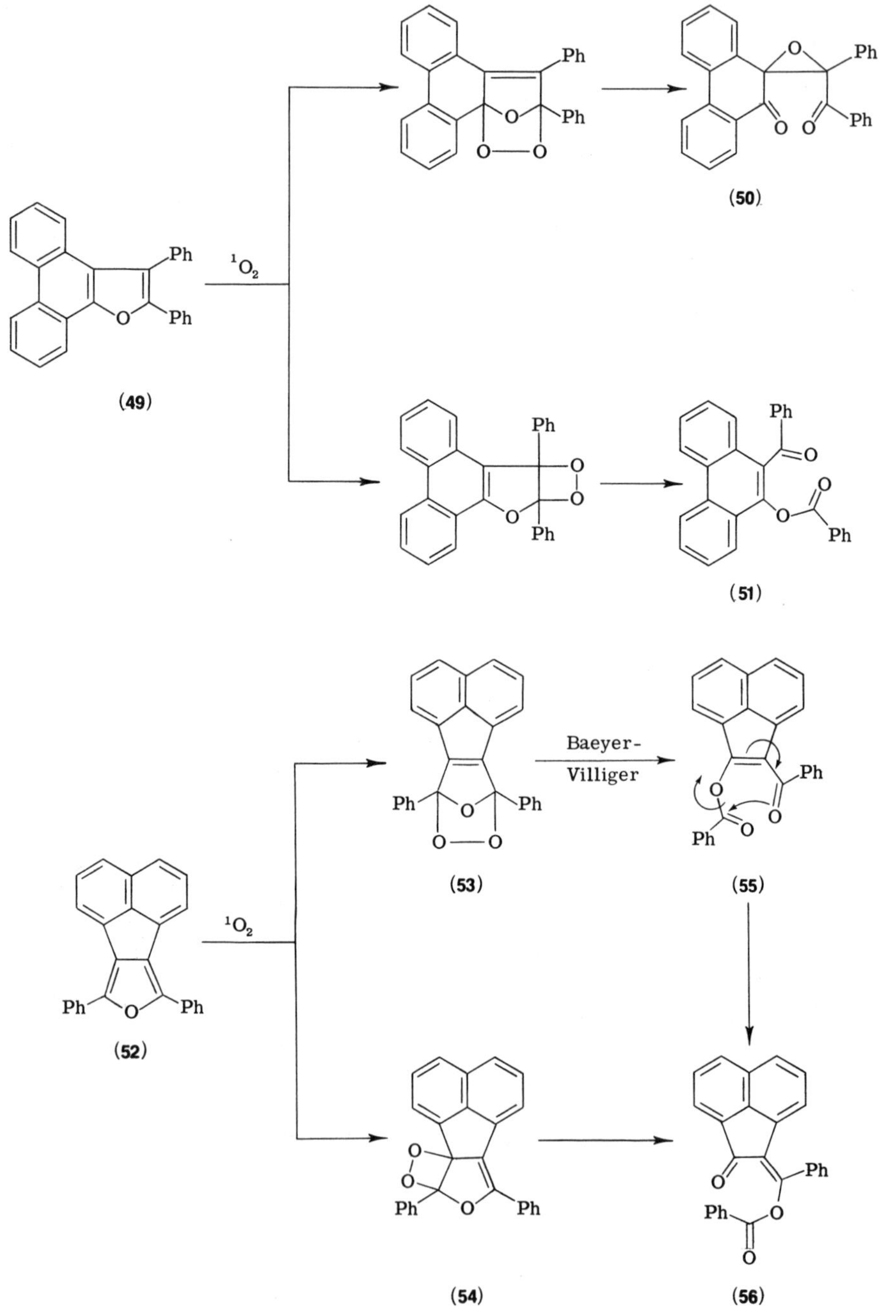

$^1O_2$

(49)

(50)

(51)

$^1O_2$

Baeyer-Villiger

(52)

(53)

(55)

(54)

(56)

Scheme 1

C. Benzofurans

Dioxetanes appear to be intermediates in the photooxidation of 3,4-benzofurans (Basselier *et al.*, 1971), as shown by the formation of the keto esters (**58**) from **57**. In the presence of thiourea, reduction of the intermediate dioxetanes to diols appears to take place.

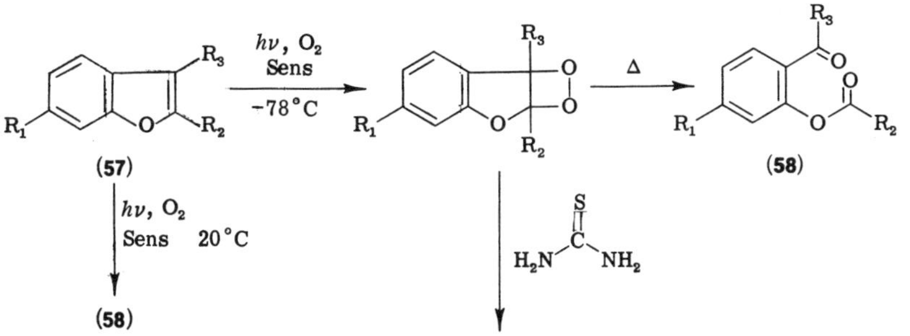

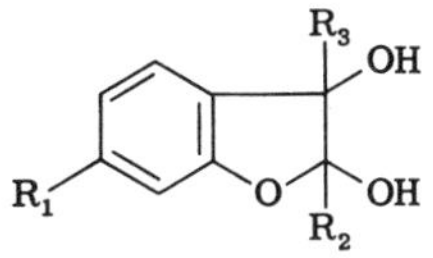

$h\nu$, O_2, Sens, −78°C; Δ; $h\nu$, O_2, Sens 20°C; H_2N–C(=S)–NH_2

(57) (58) (58)

R_1	R_2	R_3
H	Me	Me
OMe	Ph	H
OMe	Me	Ph

Related aryl furan photooxidations have been observed in the furocoumarin series (Musajo and Rodighiero, 1962). These reactions are of special interest because of the photosensitization which takes place when human skin comes into contact with plants containing furocoumarins such as xanthotoxin (**59**), bergapten (**60**), and the psoralens followed by exposure of the skin to sunlight.

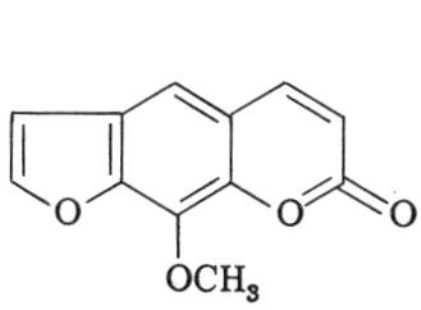

(59) 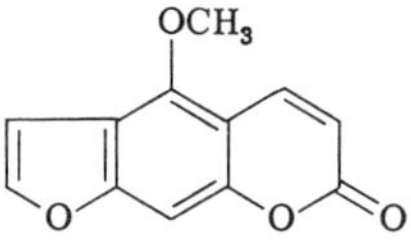 (60)

Evidence has been presented (Krauch *et al.*, 1965) to show that oxygen is not involved in the photodynamic action of the natural furocoumarins. It is nevertheless interesting to note that the three products reported to form in the photolysis of methylpsoralen (**61**) (Caporale *et al.*, 1965) appear to result from oxidative processes related to $^1O_2$ reactions. If oxygen were taken up during irradiation in aqueous methanolic solution in the presence of the sensitizer, flavin mononucleotide, peroxidic intermediates could result, followed by familiar reactions of cleavage and hydrolysis. A rationalization for the production of compounds **64**, **67**, and **71** is shown in Scheme 2.

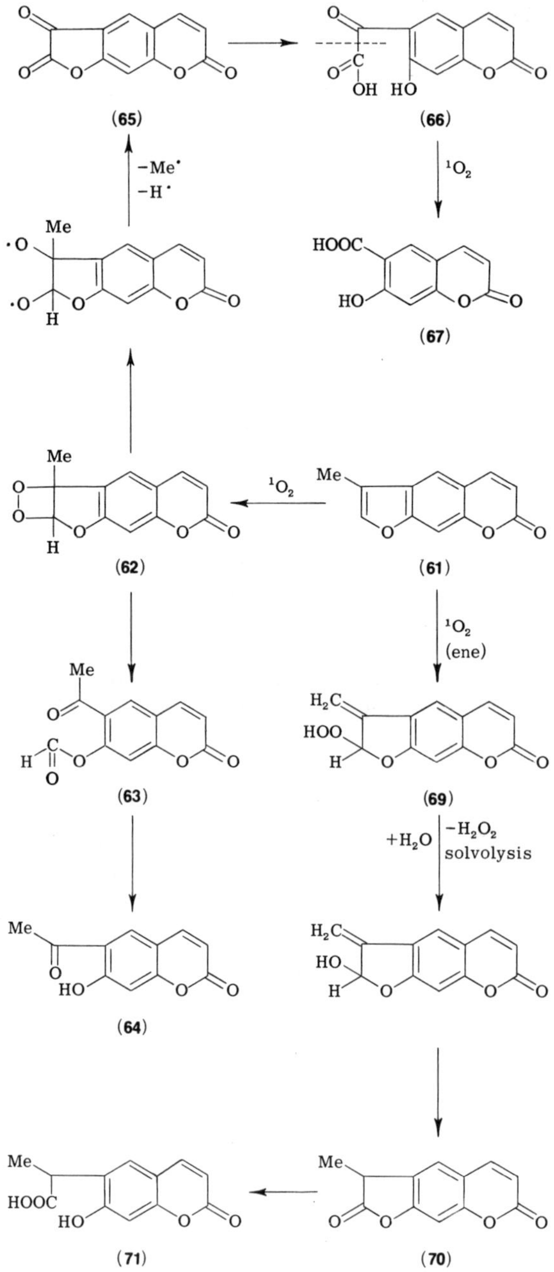

(65)

(66)

(67)

(62)

(61)

(63)

(69)

(64)

(71)

(70)

Scheme 2

The methyl ketone (**64**) appears to be derived from the formation of a dioxetane (**62**) at the 2,3-position followed by the usual type of ring cleavage and hydrolysis of the resulting keto formate (**63**).

The carboxylic acid (**67**) may be assumed to form by homolysis of the oxygen–oxygen bond in **62** with accompanying loss of a methyl radical to form the intermediate (**65**) along the lines of similar cleavage reactions observed in the pyrrole series (Quistad and Lightner, 1971b). Ando *et al.* (1975) has provided other examples of related dioxetane cleavage reactions accompanied by carbon–carbon bond rupture. Conversion of **65** to **67** may take place through the α-keto carboxylic acid (**66**) by the action of a second equivalent of singlet oxygen. Recent studies by Jefford *et al.* (1976) have shown that α-keto acids (**68**) undergo oxidative decarboxylation on exposure to singlet oxygen.

$$2\,R{-}\overset{O}{\overset{\|}{C}}{-}\overset{O}{\overset{\|}{C}}{-}OH \xrightarrow{^1O_2} 2\,R{-}\overset{O}{\overset{\|}{C}}{-}OH + 2\,CO_2$$

(**68**)

To explain the formation of the carboxylic acid (**71**) in the above process, one may visualize a conventional "ene" singlet oxygen oxidation in which the β-methyl protons and the double bond of the furan ring participate. Solvolysis of the intermediate allylic hydroperoxide (**69**) by water followed by a tautomeric shift would give lactone (**70**), the precursor of the observed phenolic acid (**71**). Such ene reactions on alkylbenzofurans have previously been observed in reactions with benzyne (H. H. Wasserman and J. Fernandez, unpublished results, 1968). In this case, the formation of **71** represents an unusual example in the furan series where side chain oxidation by singlet oxygen competes favorably with attack on the heterocyclic nucleus.

Recently, Matsumoto *et al.* (1977) have found that vinyl benzofurans are readily photooxygenated to give endoperoxides as shown.

R₁ R₂ O $\xrightarrow{^1O_2}$ O O R₁ R₂ O

D. Polyfurans and Furanophanes

Photooxidations of furans have been extended to systems containing more than one furan ring. Here, the possibility exists of forming polyunsaturated ketones by the hydrolytic decomposition of intermediate endoperoxides. Thus, Wasserman and Doumaux (1962) have shown that the reaction of the bisfuran (**72**) with singlet oxygen in methanol followed by hydrogenation yields the tetraketone (**73**).

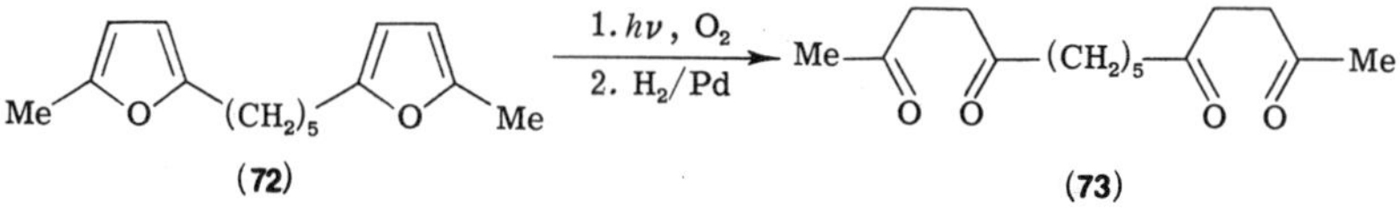

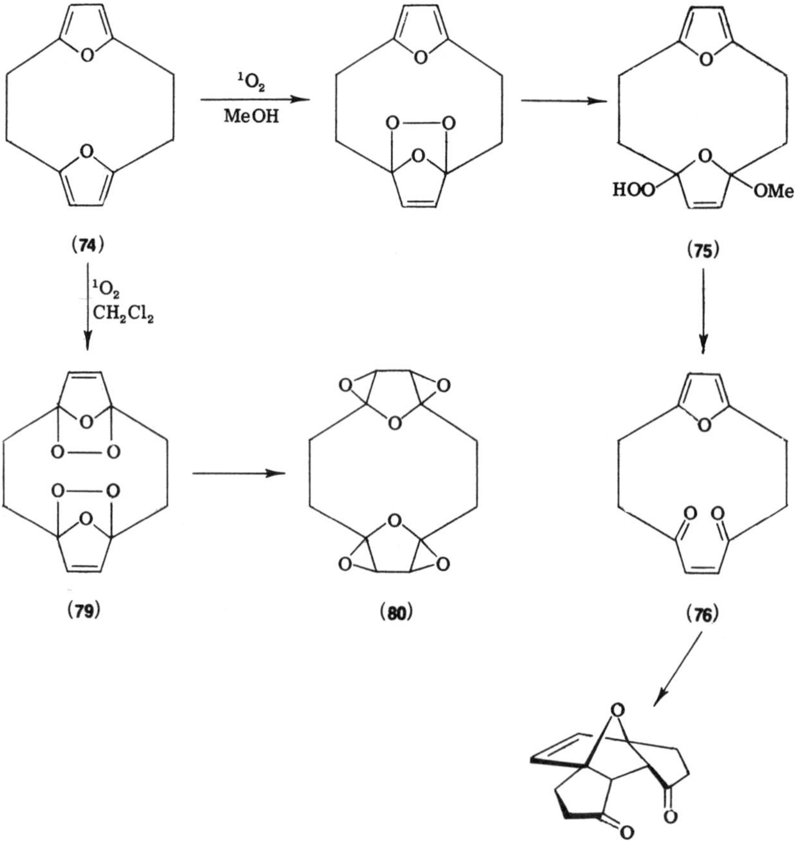

(**77**)

In the photooxidation of the bisfuranophane (**74**) in methanol, a tetraketone was not obtained, but, instead, polycyclic system (**77**) resulted. This reaction takes place most probably by endoperoxide formation, followed by solvolysis to the methoxy hydroperoxide (**75**) and then enedione formation (**76**). Molecular models show that the arrangement of the dienophile (enedione) and the diene (furan) components in **76** is very favorable for an internal Diels–Alder reaction, thereby accounting for the ready formation of **77**. The tendency for intramolecular cyclization in this system must be quite strong since the Diels–Alder addition takes place even at the methoxy-hydroperoxide stage in **75** forming the polycyclic system (**78**) (Katz *et al.*, (1968).

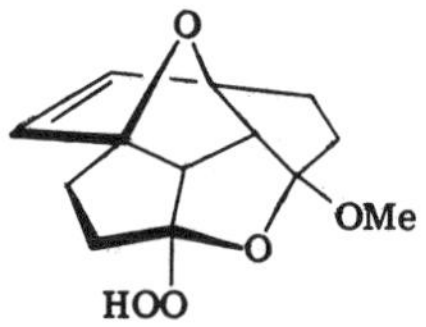

O
OMe
O
HOO
(**78**)

A dramatic solvent effect has been observed in the photooxidation of **74** in dichloromethane. In this nonpolar solvent, the exclusive product isolated is the tetraepoxide (**80**), presumably formed by rearrangement of the intermediate bisendoperoxide (**79**) (Wasserman and Kitzing, 1969).

The furanonaphthalenophane (**81**) undergoes very rapid reaction with $^1O_2$ in methanol forming polycyclic system **83**. Here, oxidative conversion of the furan to the enedione (**82**) is followed by direct attack on the naphthalene ring which is relatively susceptible to intramolecular Diels–Alder addition.

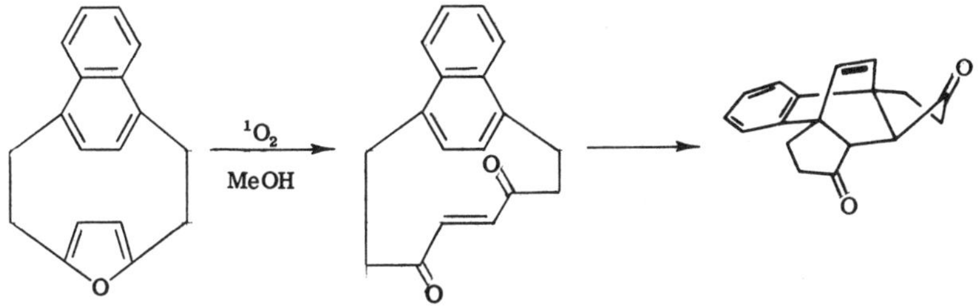

$\xrightarrow[\text{MeOH}]{^1O_2}$

(**81**) (**82**) (**83**)

Photooxidation of the furanocyclophane (**84**) in methanol leads to a mixture of products, which, after hydrogenation, can be separated into three compounds: **87**, **89**, and **92**. The formation of **87** appears to involve a typical Baeyer–Villiger rearrangement of endoperoxide (**85**) to **86**. Similarly, the (solvolytic) conversion of **85** to **88**, the enedione precursor of **89**, is not an unusual pathway for furan photooxidation in alcoholic solvents.

On the other hand, the more complex product (**92**), characterized by X-ray crystallographic analysis seems to have been formed by an uncommon process. It was suggested that the dehydro-derivative (**91**) forms by internal oxygen transfer to the neighboring benzene ring by the ozonidelike intermediate product **85**, as shown in Scheme 3. Formation of epoxide **90**, fol-

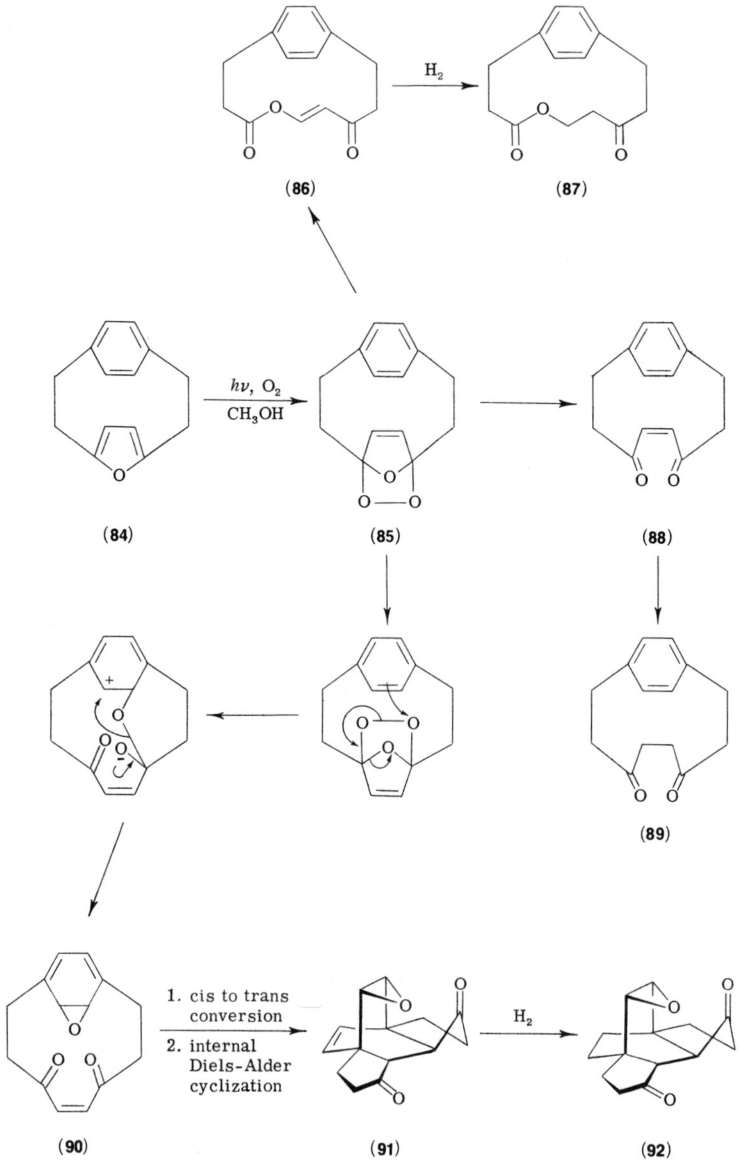

Scheme 3

lowed by intramolecular Diels–Alder reaction (after isomerization to the trans form), would lead to the unsaturated polycyclic intermediate **(91)** which would then be reduced by hydrogenation to the observed product (**92**). Alternatively, in the conversion of **84** to **90**, a carbonyl oxide intermediate analogous to **47** may be involved.

As was observed in the bisfuran case, change of solvent from methanol to dichloromethane modified the reaction course, yielding the epoxy ketone (**94**), presumably by a two-stage rearrangement of the endoperoxide (**85**) through the bisepoxide (**93**) (Wasserman and Kitzing, 1969).

84 $\xrightarrow[CH_2Cl_2]{^1O_2}$ (85) ⟶ (93) ⟶ (94)

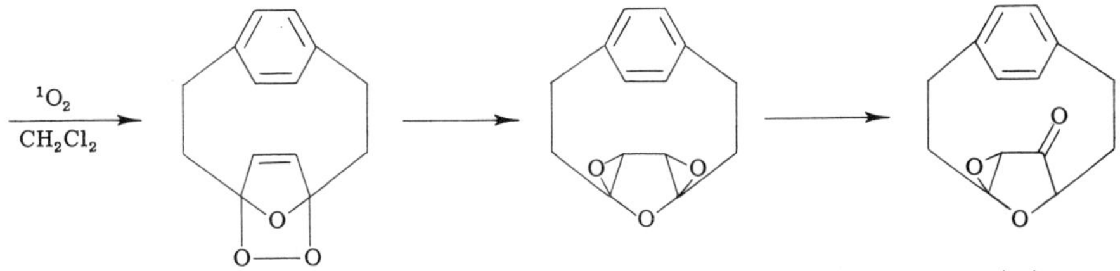

III. PYRROLES

It has long been known that pyrroles, particularly alkyl pyrroles, are sensitive to air and light, and require special handling to minimize oxidative decomposition.

In early work (Ciamician and Silber, 1912) the reaction products of pyrrole with oxygen were reported as undefined polymeric peroxidic substances. Later, a monomeric product from the autoxidation of 2,4-dimethylpyrrole was formulated as **95** by Metzger and Fischer (1936) but subsequently shown by Hoft *et al.* (1968) to be **96**.

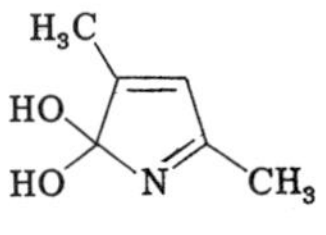

(95) 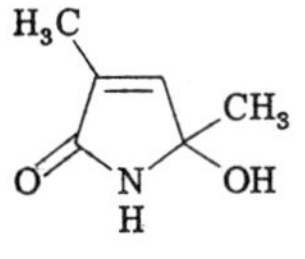 (96)

Autoxidation of trialkylpyrroles has also yielded dimeric products such as the lactam (**97**) from 2,3,4-trimethylpyrrole (Hoft *et al.*, 1967).

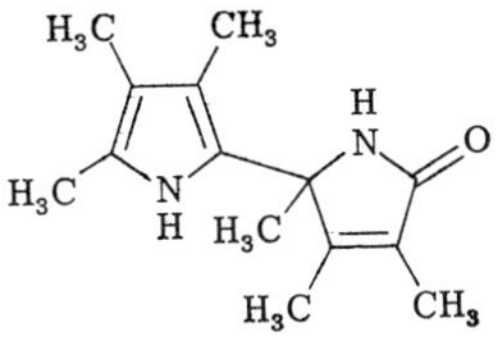

(97)

While there are numerous references such as the above to the autoxidation of pyrroles, the earliest report on the isolation of a product in the photosensitized oxygenation of pyrroles was that of Bernheim and Morgan (1939). Methylene blue photooxidation of pyrrole (**98a**) and *N*-methylpyrrole (**98b**) in aqueous solution afforded crystalline products, which were initially identified incorrectly, but later shown by de Mayo and Reid (1962) to be the 5-hydroxy-Δ^3-pyrrolin-2-ones (**99**).

$O_2/h\nu$, methylene blue, H_2O

(**98a**) R = H
(**98b**) R = CH_3

(**99a**) 32%
(**99b**) 48%

Reaction conditions (solvent, dilution, temperature, etc.) play a key role in pyrrole oxidations, and significantly affect the nature of the products obtained. Thus, the photooxidation of pyrrole (**98a**) in methanol produces none of the hydroxy derivative (**99a**) but instead only maleimide (**100**) and 5-methoxy-Δ^3-pyrrolin-2-one (**101**) (Quistad and Lightner, 1971a).

$O_2/h\nu$, methylene blue, MeOH

(**98a**) (**100**) (**101**)

In general, pyrrole photooxidations are not useful for preparative purposes since they lead to mixtures of products and, unless carefully controlled, result in considerable decomposition (Martel, 1957). Dilution appears to be an important factor. For example, while earlier workers (Martel, 1957) reported only tar formation in the photooxidation of 2,3,4,5-tetraphenylpyrrole (**102**), Wasserman and Liberles (1960) found that this reaction could be made to proceed smoothly in dilute methanol solution to give **103** and **104** in excellent yield. The important role played by dilution was also observed by de Mayo and Reid (1962) in repeating the early photooxidation studies on pyrrole. Here again, unless the reaction is run in a large amount of solvent, extensive decomposition takes place.

$^1O_2$, MeOH, 85%

(**102**) (**103**) (**104**)

A. Alkyl Pyrroles

Lightner and co-workers have examined the oxidation by singlet oxygen of a series of alkyl-substituted pyrroles (**105**) in methanol. In general, three types of products (**106**, **107**, and **108**) are formed in varying amounts as summarized in Table II.

(105) (106) (107) (108)

Table II Oxidation of Alkyl-Substituted Pyrroles by Singlet Oxygen

Substituents on **105**	Isolated yields (%) **106**	**107**	**108**	Ref.
R_2 = Me; R_3 = R_4 = Et	40	11	3	Quistad and Lightner (1971b)
R_2 = H; R_3 = R_4 = Et	0	33	34	Quistad and Lightner (1971a)
R_2 = R_4 = Me; R_3 = H	48	16	0	Lightner and Low (1972b)
R_2 = Me; R_3 = R_4 = H	15	20	0	Lightner and Low (1972b)
R_2 = R_4 = Me; R_3 = Et	16	14	3	Lightner and Crandall (1973c)
R_2 = R_3 = H; R_4 = Me	10	22	13	Lightner and Low (1972a)
R_2 = *t*-Bu; R_3 = R_4 = H	12	23	0	Lightner and Pak (1975)
R_2 = R_3 = H; R_4 = *t*-Bu	11	34	31	Lightner and Pak (1975)

One may account for the formation of these products on the basis of an initial Diels–Alder type of addition of singlet oxygen to form endoperoxides corresponding to **109**. Solvolysis of **109** to the methoxy hydroperoxide (**110**) in methanol is analogous to the process observed in the furan series by Foote *et al.* (1967) and Katz *et al.* (1968), where species similar to **110** have been isolated. Loss of H_2O from **110** would yield the lactam (**111**).

(109) (110) (111) (112)

An alternative route for the formation of **111** may involve ring opening of the 1,4-peroxide to form a 2-hydroperoxy intermediate (**112**) analogous to the hydroperoxides isolated by Rio and Masure (1972) (aryl pyrroles), Ramasseul and Rassat (1972) (di-*tert*-butyl pyrroles), and by White and Harding (1964) (imidazoles). Methanol addition to the reactive C=N group of **112** followed by β-elimination of H_2O from the peroxide affords the product.

A different mode of decomposition of peroxide (**109a**) has been suggested by Quistad and Lightner (1971b) to explain the formation of the 5-hydroxy lactams (**115**) and the maleimides (**116**). In this view, homolysis of the O–O endoperoxide bond yields the diradical (**113**) which may react further by either of two pathways: (*a*) rearrangement to **115**, possibly through **114**; (*b*) loss of a hydrogen atom and an alkyl radical to form the maleimide (**116**).

Et Et H N H Me O—O (**109a**) → Et Et H N H Me O· O· (**113**) (*a*) → [Et Et O= H₂N O —Me] (**114**) → Et Et O N H Me OH (**115**); (*b*) → Et Et O N H O (**116**)

In more recent studies on the mechanism of dye-sensitized photooxygenation of pyrroles, Lightner and co-workers (1976) have analyzed the formation of products (**120**), (**121**), and (**122**) from *N*-substituted pyrroles (**117**) in various solvents. Their low temperature NMR and $H_2{}^{18}O$ studies are consistent with the intermediacy of endoperoxides (**118**) which undergo methanolysis or hydrolysis to form hydroperoxides (**119**). The latter may then decompose to yield **120** or **121**. The authors note that formation of 5-hydroxylactams (**121**) may also result from pathways in which H_2O does not participate. One possibility would involve a β-elimination of the proton on the α-carbon atom.

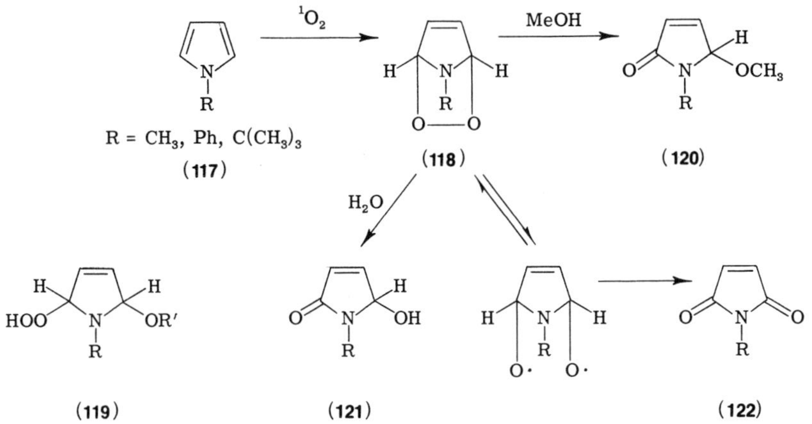

The formation of hydroxy lactams of type **124** obtained from the photooxidation of *N*-substituted pyrroles (**123**) has been applied by Franck and Auerbach (1971) to the synthesis of a model mitomycin. In particular, a

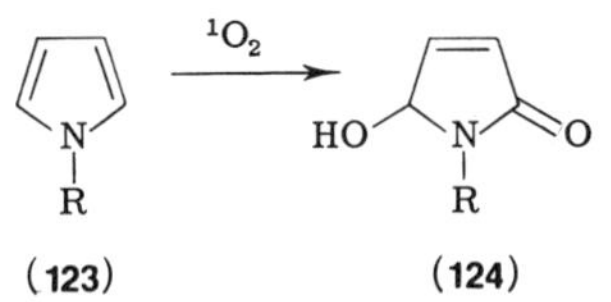

phenyl substituted system (e.g. **123**, R = Ph) was of interest as it was envisioned that the product of oxidation (**125**) would contain certain features related to the mitomycin antibiotics (**126**). Thus, photooxidation of heterocycle **127** in aqueous THF containing 10% pyridine proceeded smoothly to

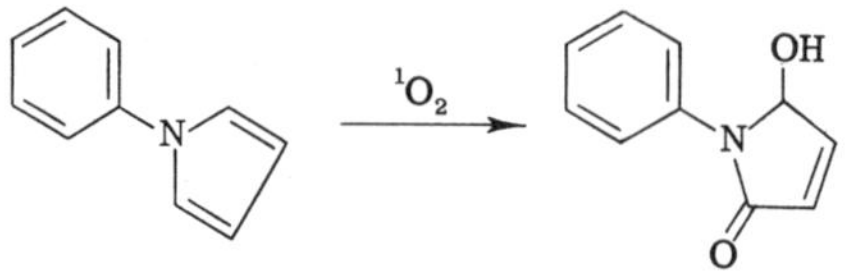

(123) (125) 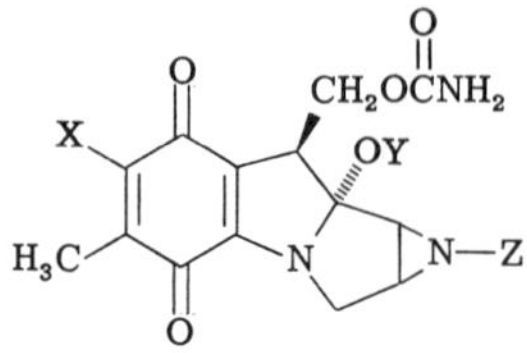 (126)

	X	Y	Z
Mitomycin A	MeO	Me	H
Mitomycin B	MeO	H	Me
Mitomycin C	NH_2	Me	H
Perfiromycin	NH_2	Me	Me

afford the indole lactam (**129**) in ca. 70% yield. Similar reactions performed without added pyridine gave low yields of the desired product. Presumably, the presence of base in the reaction medium serves to facilitate β-elimination of the initially formed endoperoxide (**128**).

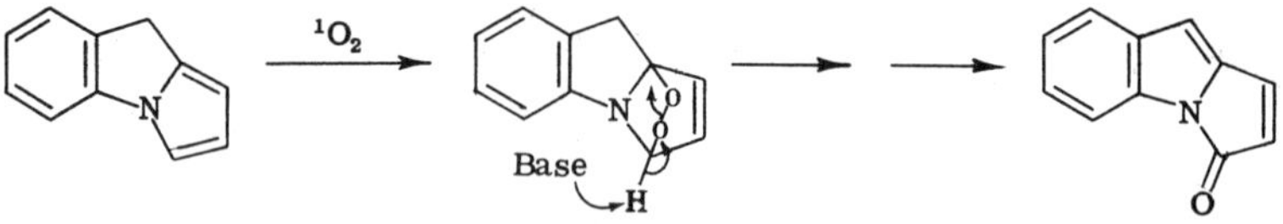

(127) (128) (129)

More recently, Kametani *et al.* (1978) photooxygenated the keto-pyrrole (**127a**) in methanol to form the hydroperoxide (**128a**), a product of special interest in connection with synthesis in the mitomycin field.

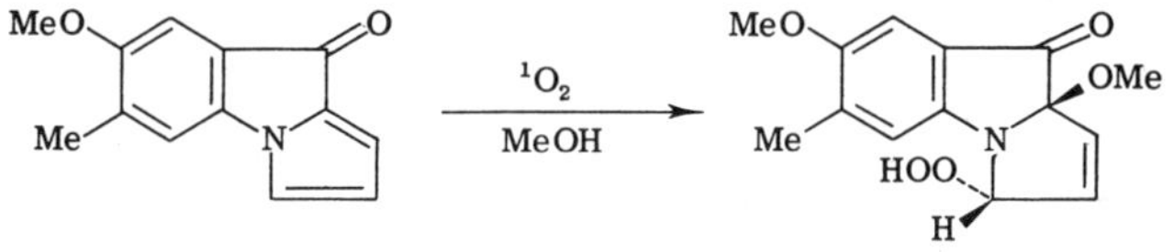

(127a) (128a)

Pyrroles substituted with electron-attracting (carbonyl) groups are normally not very reactive toward singlet oxygen. However, Lightner and Quistad (1973) have shown in studies with pyrrole aldehydes (Table III) that the success of the reaction depends on the degree of alkylation of the pyrrole nucleus. Similar effects have been observed in the case of pyrrole α-carboxylic esters (Lightner and Crandall, 1973a).

Recent studies on the photooxidation of 3,4-diethylpyrrole (**129a**), 2,5-dimethylpyrrole (**133**), and 3,4-diethyl-2,5-dimethylpyrrole (**138**) have provided some unusual results. The reaction of 3,4-diethylpyrrole with singlet oxygen yields diethylmaleimide (**131**) as one of the major products along with the expected 5-methoxypyrrolinone (**132**) (Quistad and Lightner,

Table III Reaction of Pyrrole Aldehydes with Singlet Oxygen

O_2, $h\nu$, CH_3OH, Rose Bengal

Substituents	Isolated yields (%)
$R_1 = R_2 = Me; R_3 = Et; R_4 = H$	53
$R_1 = R_2 = Et; R_3 = Me; R_4 = H$	53
$R_1 = R_3 = Me; R_2 = R_4 = H$	41
$R_1 = Me; R_2 = R_3 = R_4 = H$	10
$R_1 = R_2 = R_4 = H; R_4 = Me$	10
$R_1 = R_2 = R_4 = H; R_3 = Me$	7
$R_1 = R_2 = R_3 = R_4 = H$	0

1971a). Formation of **131** presumably results from the homolytic cleavage of the O–O bond in **130** followed by loss of two hydrogen atoms.

(129a) $\xrightarrow{^1O_2}$ (130) $\longrightarrow$ $\xrightarrow{-2H\cdot}$ (131)

(130) $\xrightarrow{CH_3OH}$ (132)

By contrast, the singlet oxygen reaction of 2,5-dimethylpyrrole (**133**) in methanol yields a mixture of products including **134** and **135**, the previously unreported products of α-methyl oxidation. The lactam (**136**), resulting from alkyl group cleavage of intermediate (**137**) was also found (Low and Lightner, 1972).

(133) $\xrightarrow[MeOH]{^1O_2}$ (136) + (134) + (135)

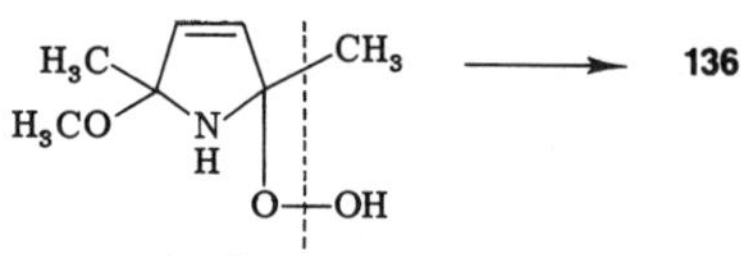

(137)

Similarly, photooxidation of the tetrasubstituted derivative (**138**) affords several products as shown in Scheme 4, including the dealkylated derivatives

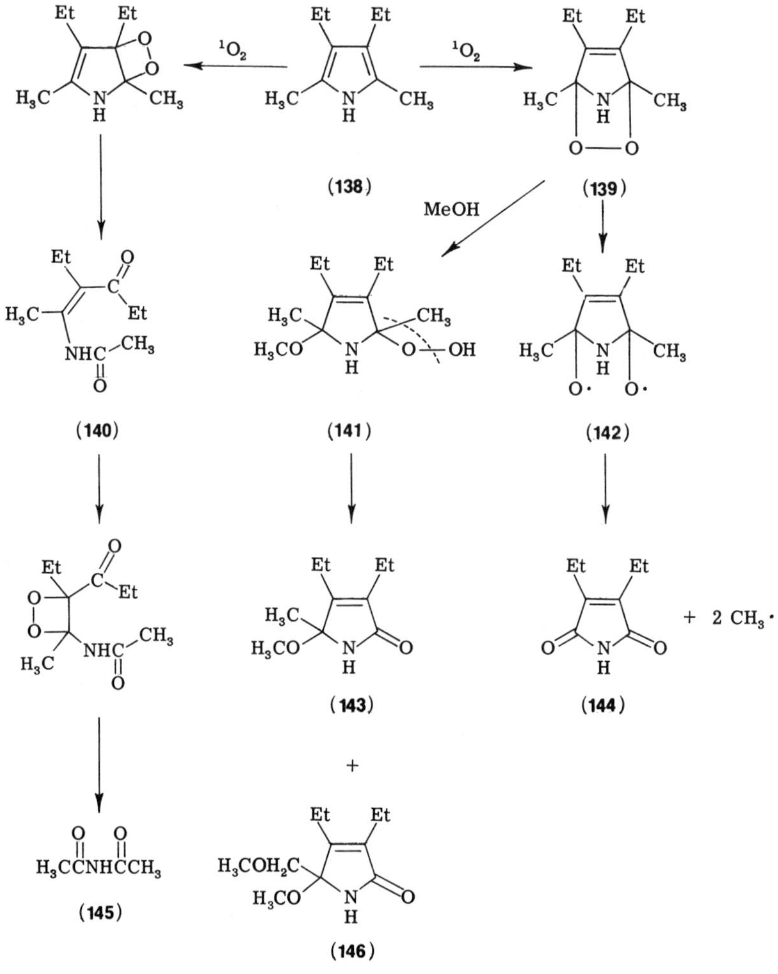

Scheme 4

(**143**) and (**144**) along with the more highly oxidized lactam (**146**) (Lightner and Quistad, 1972b).

Formation of **141** is analogous to the production of methoxy hydroperoxides in the furan series, while **140** and **145** appear to result from conventional types of dioxetane cleavage. Homolytic cleavage of the endoperoxide bridge in **139** would yield a diradical (**142**) which could further fragment to carbonyl groups (**144**) and methyl radicals, a route which, as the authors note, is similar to the known decomposition of the *tert*-butoxy radical, forming acetone and methyl radical (Hiatt, 1971). Formation of **143** appears to involve loss of a methyl radical from **141** along with homolytic cleavage of the O–O bond in the hydroperoxide. Dealkylation is also observed in the photooxidation of 3,4-diethyl-2-methylpyrrole (Quistad and Lightner, 1971b).

Though there is little precedent for the formation of **134**, **135**, and **146** in pyrrole photooxidations, it is possible that these products may arise from endoperoxide (**139**), by an elimination–addition sequence as suggested in Scheme 5. α-Hydroperoxides of type **147** have been isolated as intermediates

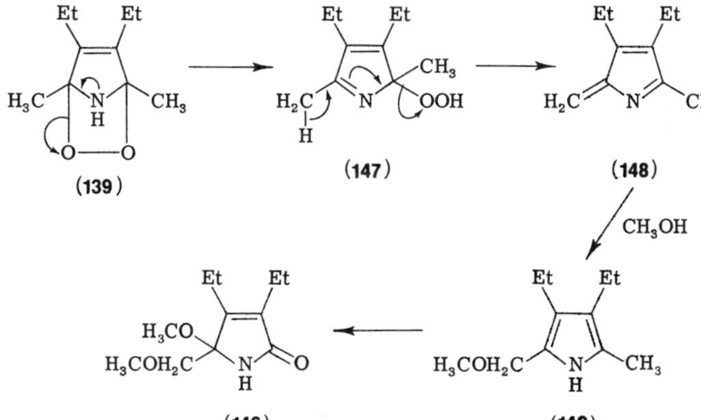

(**139**) → (**147**) → (**148**) $\xrightarrow{CH_3OH}$ (**149**) → (**146**)

Scheme 5

in aryl pyrrole photooxidation (Dufraisse *et al.*, 1965; Ramasseul and Rassat, 1972). Elimination of hydrogen peroxide yielding **148**, followed by addition of methanol would provide the intermediate 2-methyl-3,4-diethyl-5-methoxymethylpyrrole (**149**). Conversion of **149** to the dimethoxy derivative (**146**) could take place by an oxidative dealkylation similar to that proposed for the formation of **143** from **141**.

B. Aryl Pyrroles; Epoxide Formation

Formation of 3,4-epoxides has been observed in a number of instances in photooxidation of aryl-substituted pyrroles. Thus, in reactions reminiscent of the aryl furan oxidations, Wasserman and Liberles (1960) found that on photooxidation in dilute methanol, 2,3,4,5-tetraphenylpyrrole (**150**) yields the epoxide (**152**) along with **153**, the product of cleavage at the 2,3-position.

Somewhat later, Dufraisse *et al.* (1965) and Rio *et al.* (1966, 1969) reported that the initial product in this reaction is the unstable α-hydroperoxide (**151**), which can be reconverted to **150** on heating. In boiling

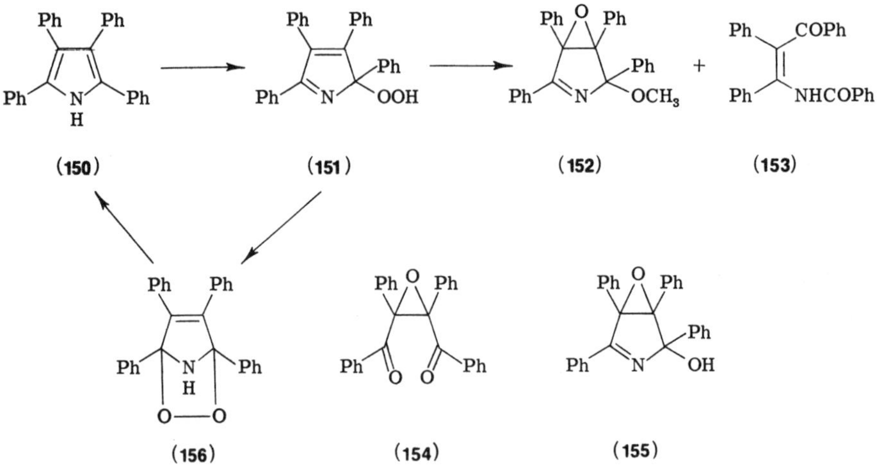

(**150**) (**151**) (**152**) (**153**)

(**156**) (**154**) (**155**)

methanol the hydroperoxide (**151**) yields the epoxide (**152**) and the 2,3-cleavage product (**153**) along with *cis*-dibenzoylstilbene oxide (**154**) and the epoxide (**155**).

One of the mechanisms for rationalizing the formation of products (**152**), (**153**), (**154**), and (**155**) involves the cyclization of α-hydroperoxide (**151**) to the dioxetane (**157**). Rearrangement of **157**, possibly through **158**, would lead to the observed products.

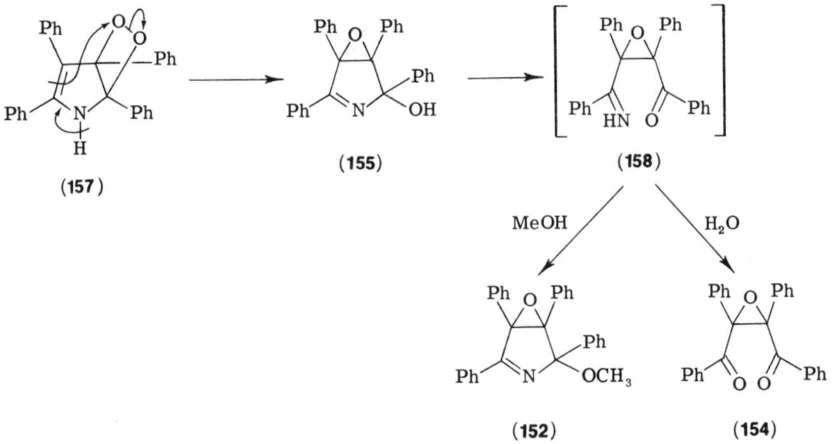

(**157**) (**155**) (**158**)

MeOH H_2O

(**152**) (**154**)

In connection with the conversion of **157** to **155**, it is interesting to note the recent results of Nishinaga and Rieker (1976) from the oxidation of the phenol (**159**) by molecular oxygen in KO-*tert*-Bu/*tert*-BuOH. The epoxy ketone (**161**), isolated as an intermediate in the reaction, appears to form by a related process via the hydroperoxide anion (**160**).

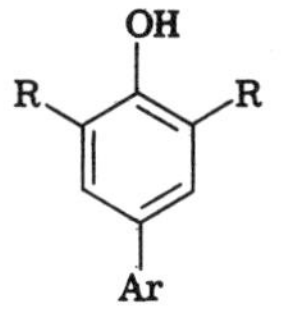

(159)

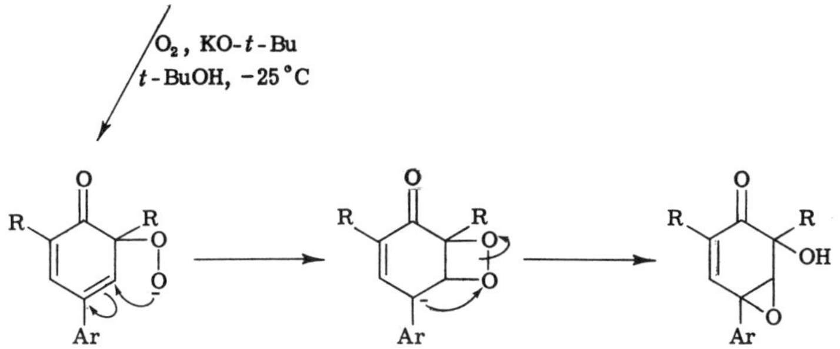

O_2, KO-*t*-Bu
t-BuOH, −25 °C

(160)

(161)

Thermal regeneration of the parent tetraphenylpyrrole (**150**) from α-hydroperoxide (**151**) takes place most probably through the 2,5-endoperoxide (**156**) which, like the furan analog (Trozzolo and Fahrenholtz, 1970) may break down to yield singlet oxygen and the pyrrole.

Photooxidations of 2,5-di-*tert*-butylpyrrole and 2,3,5-tri-*tert*-butylpyrrole (**162**) (Ramasseul and Rassat, 1972) yield stable hydroperoxides (**163**) and (**163a**) which, like the polyphenyl counterparts, form the *cis*-epoxydione (**164**). Other products formed from the α-hydroperoxide are the alcohol (**165**) and the unsaturated lactam (**166**).

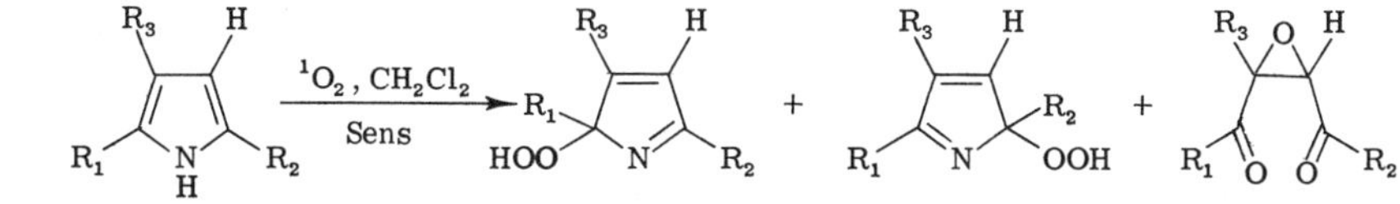

$^1O_2$, CH_2Cl_2
Sens

(162a) $R_1 = R_2 = t$-Bu, $R_3 = H$
(162b) $R_1 = R_2 = R_3 = t$-Bu

(163) (163a) (164)

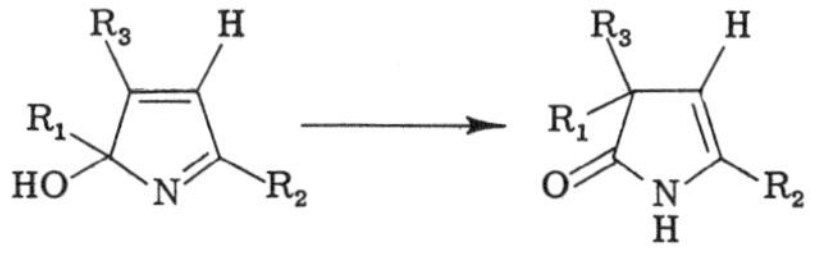

(165) (166)

In other work leading to 3,4-epoxide formation (Wasserman and Miller, 1969), 1-methyl-2,3,5-triphenylpyrrole (**167**, R = CH_3) yields the dibenzoyl oxide (**168**) on photooxidation, while the *N*-phenyl derivative (**167**, R = Ph) forms a mixture of Schiff bases (**169a,b**). Modes of formation of the Schiff bases (**169a,b**) are outlined in Scheme 6. The initial product of oxygen uptake

$^1O_2$, CH_2Cl_2
R = CH_3

(**168**)

(**167**)

R = Ph

(**169a**) R_1 = H,, R_2 = Ph
(**169b**) R_1 = Ph, R_2 = H

by **167** (R = Ph) could be either the zwitterionic peroxide (**170**), the dioxetane (**171**), or the endoperoxide (**172**). Since **170, 170a**, **171**, and **172** appear to be readily interconvertible, it would be difficult to distinguish among these labile intermediates experimentally. The carbonyl oxide (**170a**) could serve as an epoxidizing agent, transferring oxygen intramolecularly, or to another molecule, as shown in part-structure **170b**.

There are other reactions where 3,4-epoxides appear to be intermediates in the oxidations of pyrroles with singlet oxygen. Inhoffen (1968) has reported that the irradiation of chlorin–phlorins with visible light in the presence of oxygen leads to products in the bacteriochlorin series. The oxidation (part structures shown) was formulated as an ene-type reaction involving the methylene bridge hydrogens and one of the double bonds of the pyrrole ring (**173**). Formation of an epoxide, followed by reaction with solvent

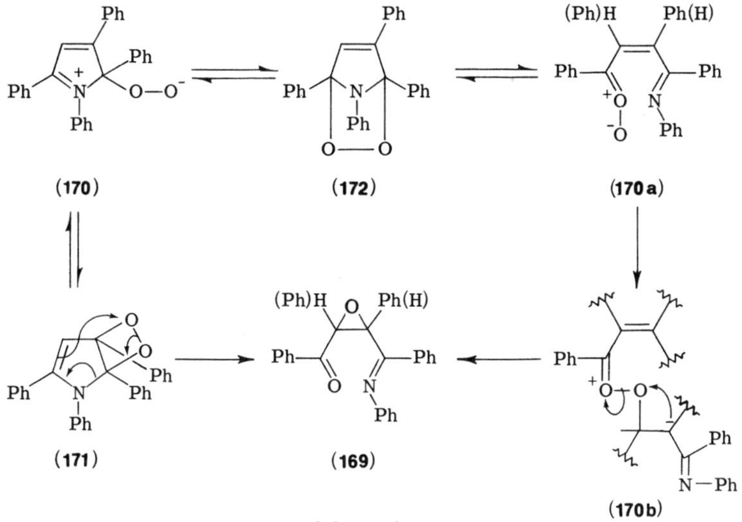

(170) (172) (170a)

(171) (169) (170b)

Scheme 6

(either water or methanol), would yield the observed hydroxy ether (**174**) (stereochemistry not shown).

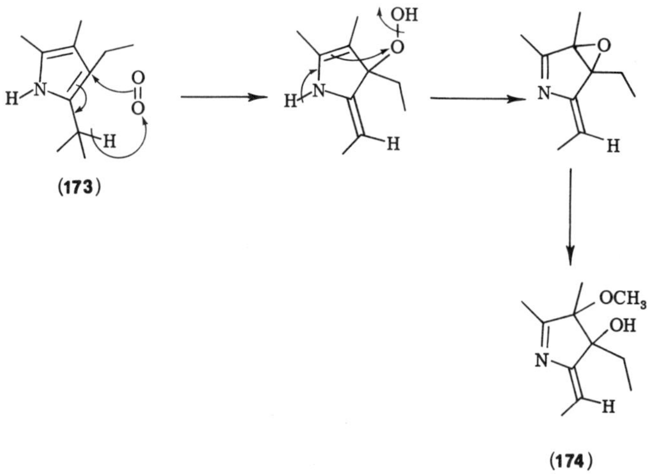

(173)

(174)

C. Pyrrolophanes

More recently, it has been shown (H. H. Wasserman and L. P. Fenocketti, unpublished results, 1975) that the reaction of the heterophane (**175**) with singlet oxygen gives a mixture of products which, on hydrogenation, yields the cyclic ether (**177**) along with a small amount of the dione (**178**). Here, the most reasonable pathway for the formation of product **177** appears to involve an intermediate 3,4-epoxide (**176**) as observed in the oxidation of polyphenyl- and *tert*-butylpyrroles.

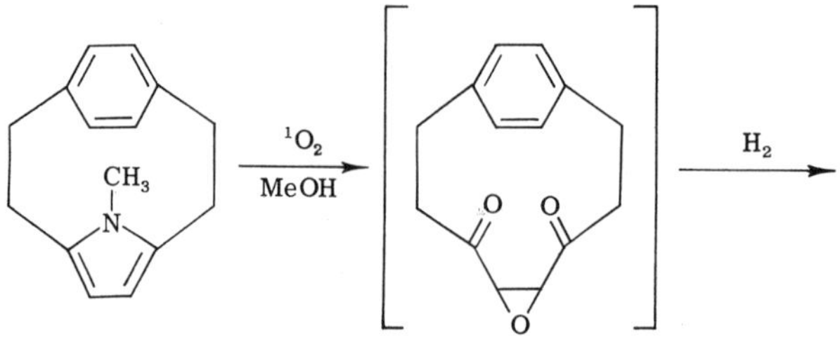

CH_3 N $^1O_2$ MeOH O O O H_2

(**175**) (**176**)

O O O + O O

(**177**) (**178**)

In other studies on the photooxygenation of [2.2](2,5)pyrrolophanes it was found that the results were strongly governed by solvent effects. [2.2]-Paracyclo(2,5)pyrrolophane (**179**) afforded [8]paracyclophane-3,6-dione (**180**) (40%) upon reaction with singlet oxygen in chloroform followed by hydrogenation. The yield of the dione was only 4% in methanol.

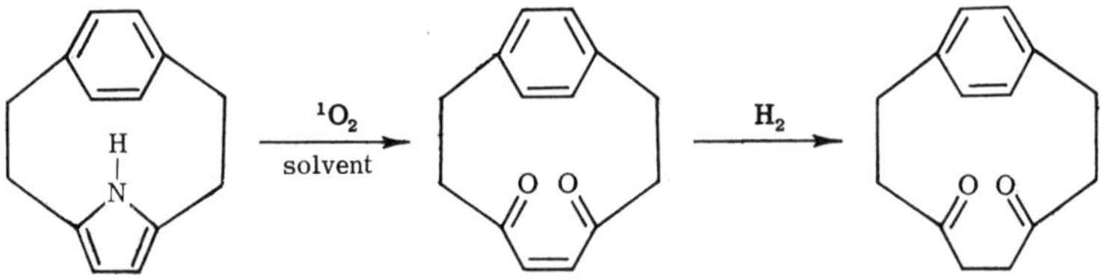

H N $^1O_2$ solvent O O H_2 O O

(**179**) (**180**)

Under the above conditions of oxidation followed by hydrogenation of naphthalenopyrrolophanes (**181**), a mixture of [8](1,4)-naphthalenophane-3,6-dione (**182**), and the indacendione (**183**) was obtained. These products appear to form by processes analogous to those observed in the furanophane series.

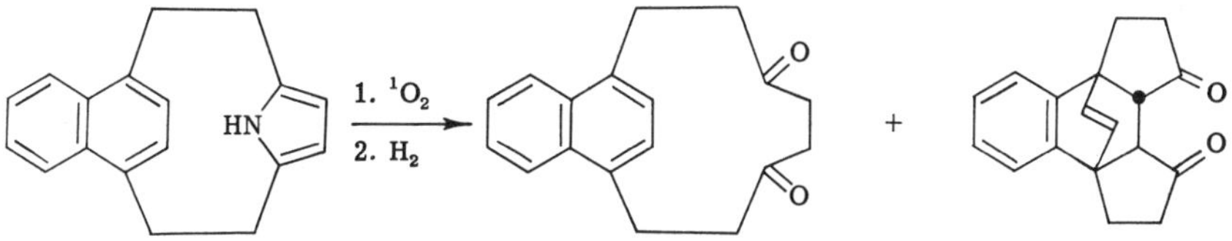

(181) (182) (183)

Products similar to those obtained in the photooxidation of polyphenylpyrroles have also been observed in some of the less highly substituted aryl pyrroles. Thus, photosensitized oxidation of 1-phenylpyrrole (**184**) in methanol yields both the hydroxy and methoxy unsaturated lactams (**185**) and (**186**) (Lightner *et al.*, 1974).

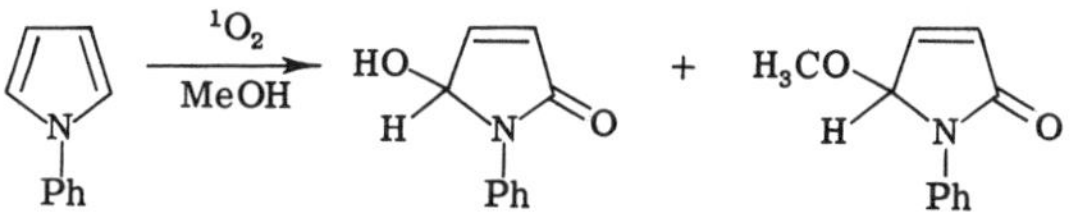

(184) (185) (186)

Rio and Masure (1972) have similarly reported formation of 5-hydroxy or 5-methoxy-Δ^3-pyrrolin-2-ones (**188**) from the photooxidations of 2,4-diphenyl- and 2,3,5-triphenylpyrrole (**187a**) and **187b** in methanol.

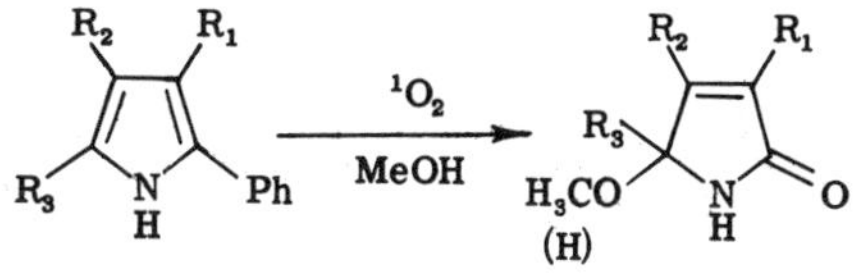

(**187a**) $R_1 = R_3 = H$, $R_2 = Ph$
(**187b**) $R_1 = R_3 = Ph$, $R_2 = H$

(188)

Photooxidation of 2,3,4,5-tetraphenylpyrrole (**189**) in methanolic KOH yields product **192** (Wasserman and Liberles, 1960). A likely pathway for this rearrangement is shown in Scheme 7. The α-hydroperoxy anion (**190**), first formed, is converted in KOH to the alkoxide (**191**) followed by a benzylic acidlike rearrangement.

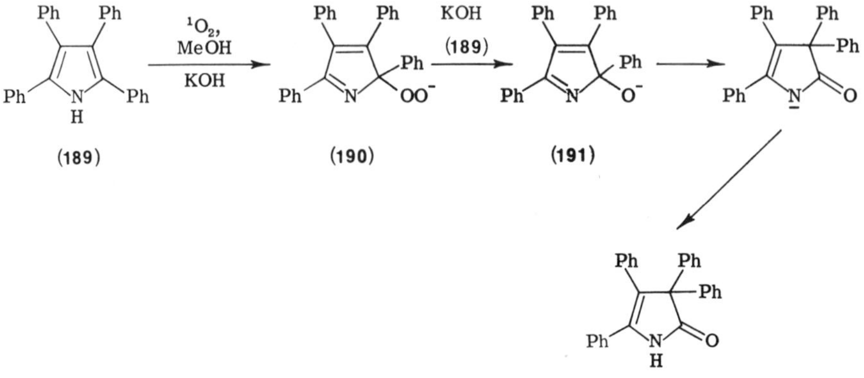

(189) (190) (191) (192)

Scheme 7

A related reaction has been observed in the case of triphenylimidazole with singlet oxygen in the presence of diphenyl sulfide (Wasserman and Saito, 1975). 1-Benzoyl-2,3,4,5-tetraphenylpyrrole (**193**) similarly gives the epoxide (**195**) (Ranjon, 1971), while the 1-amino derivative (**194**) forms the pyrazine (**197**) via **196** (Rio and Lecas-Nawrocka, 1971).

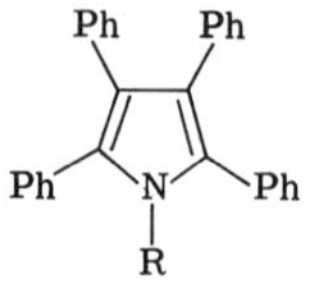

(**193**) R = COPh
(**194**) R = NH_2

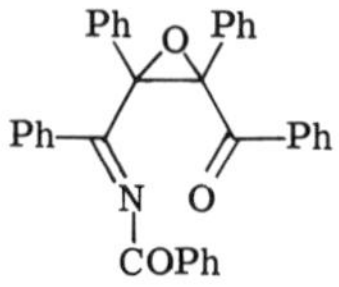

(**195**)

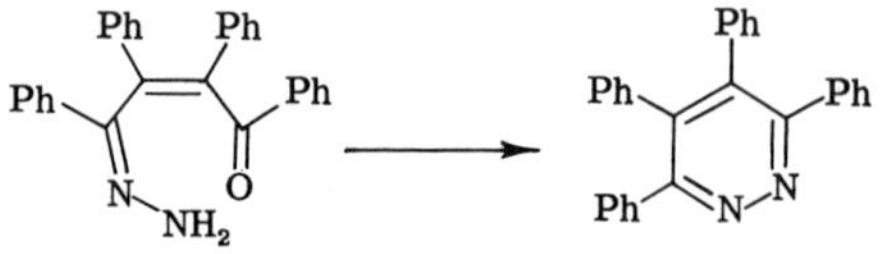

(**196**) (**197**)

The reaction of singlet oxygen with **194** also leads to a mixture of pyrazoles (**198**) and (the hydrolysis product) **199**. Compound **198** forms by cleavage of the 2,3-bond of **194**, most probably via a dioxetane.

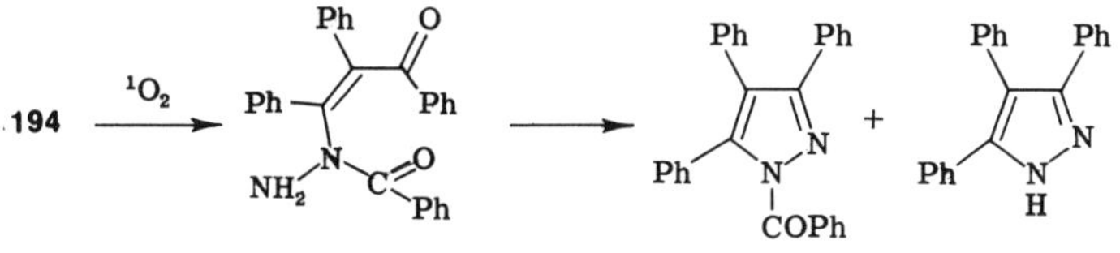

194 $\xrightarrow{^1O_2}$ [Ph(NH$_2$N(COPh))C=C(Ph)COPh] ⟶ (198) + (199)

(198) (199)

D. Bilirubin

Much of the interest in pyrrole photooxidations stems from the observation by Cremer *et al.* (1958) that hyperbilirubinemia (neonatal jaundice), somewhat commonly found in the prematurely born, may be treated by irradiation of the infant with visible light (λ at ca 450 nm). Such phototherapy effectively decreases the level of bilirubin (BR)(200) in the serum by converting it to water-soluble derivatives which can be excreted (Schmid, 1971). This prevents retarded motor development, cerebral palsy, or, in some cases, death. Questions have, however, been raised by researchers concerning adverse side effects resulting from the indiscriminate photodynamic action by singlet oxygen during phototherapy (Odell *et al.*, 1972). In addition, the toxicity of the photoproducts formed is not known.

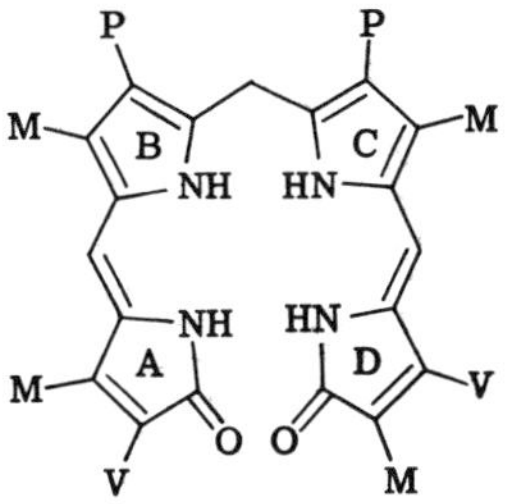

(200)

M = CH_3
V = $CH{=}CH_2$
P = CH_2CH_2COOH

Recently, competitive studies on the photooxidation of BR in the presence of cholesterol (Lightner and Norris, 1974) have revealed that complete destruction of BR occurs while the cholesterol remains unchanged. These findings are in complete agreement with the kinetic data reported by Foote and Ching (1975), who found BR to be the most active known acceptor of singlet oxygen with a rate approaching that of diffusion control. These results, considered along with the fact that BR is an inefficient sensitizer for singlet oxygen (McDonagh, 1971), suggest that the *in vivo* photodecomposition of BR is controlled by BR itself in a very rapid reaction with whatever singlet oxygen it has made available.

In tracing the path involved in the reaction of BR with singlet oxygen it is interesting to note the results of oxidation of a model system, the pyrrolic unsaturated lactam (**201**) containing many of the structural features of one half of the bilirubin molecule. Rose Bengal sensitized photooxidation of **201** in methanol yielded **202** and **205** as the main products (Lightner and Quistad, 1972a)

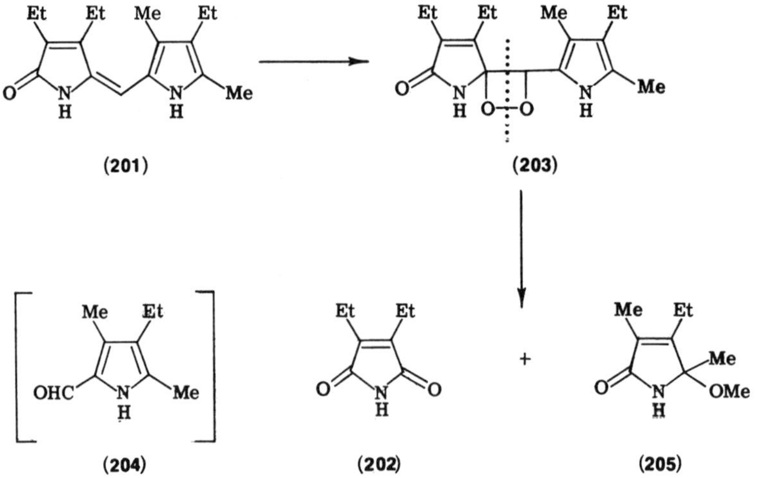

(201) (203) (204) (202) (205)

The imide (**202**) could have resulted from cleavage of the dioxetane (**203**) formed by an enamine-type of reaction with oxygen. The other product of dioxetane cleavage (**204**) was not observed. However, it was shown (Lightner and Quistad, 1973) that under the conditions of the reaction this pyrrole aldehyde is readily converted to the methoxy lactam (**205**), perhaps by the route shown below.

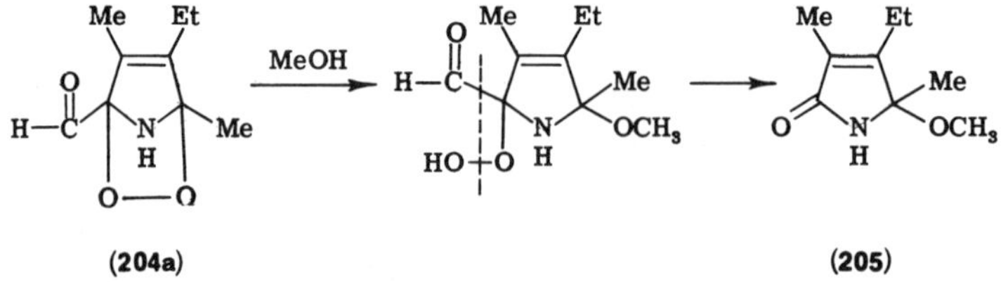

MeOH

(204a) (205)

An alternate mode of formation of **205** would involve the endoperoxide (**206**) which could undergo methanolysis and then dealkylation by homolytic cleavage as shown.

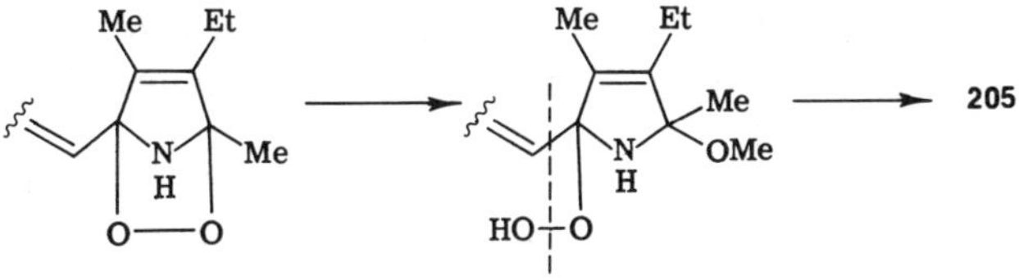

→ 205

(206)

In vitro studies on the photooxidation of BR have been carried out by three groups of workers. McDonagh (1971) predicted the formation of the dipyrrylmethane dialdehyde (**207**) which was subsequently isolated by Bonnett and Stewart (1975). The other fragment expected by the cleavage of the bisdioxetane (**208**) would be methylvinylmaleimide (**209**), and this was isolated both by Lightner and Quistad (1972a), and by Bonnett and Stewart (1972a, 1975). In other work, Lightner and Quistad (1972c) isolated hematinic acid (**212**) and the propentdyopents (**210**) and (**211**). Biliverdin (BV) (**213**) was also shown to be a product of BR photooxidation (Lightner *et al.*, 1973).

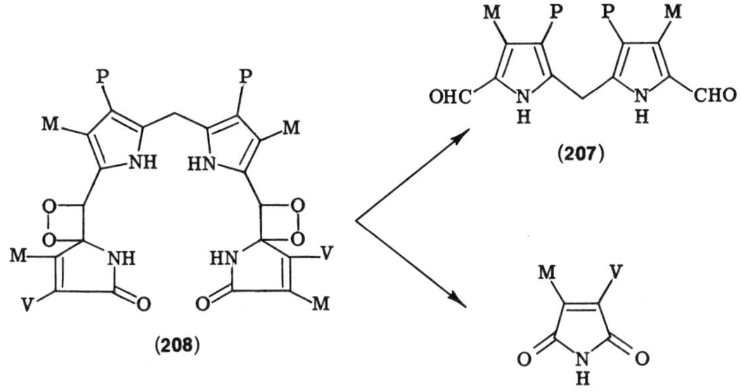

(207)

(208)

(209)

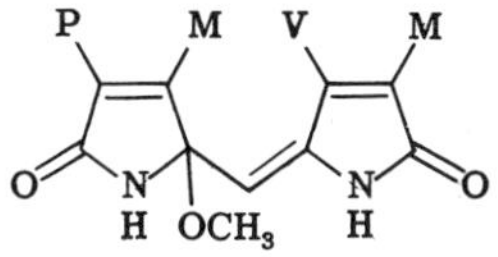

(210)

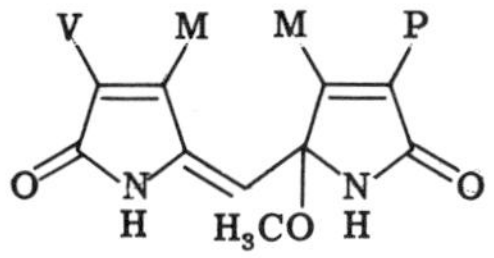

(211)

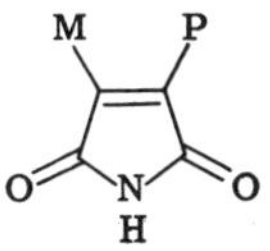

(212)

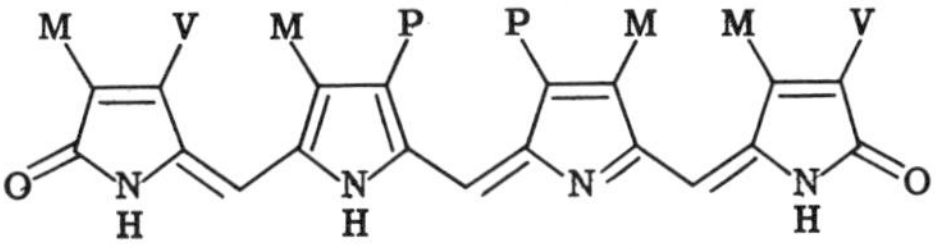

(213)

Methoxy lactams (**210**) and (**211**) could have formed either by a mechanism analogous to the deformylation process outlined above (**204**→**205**) or through the endoperoxides of the central two pyrrole rings, B and C (**200**). Thus, endoperoxide formation involving ring B would yield **211** by methanolysis and cleavage (Bonnett and Stewart, 1972a) as shown below.

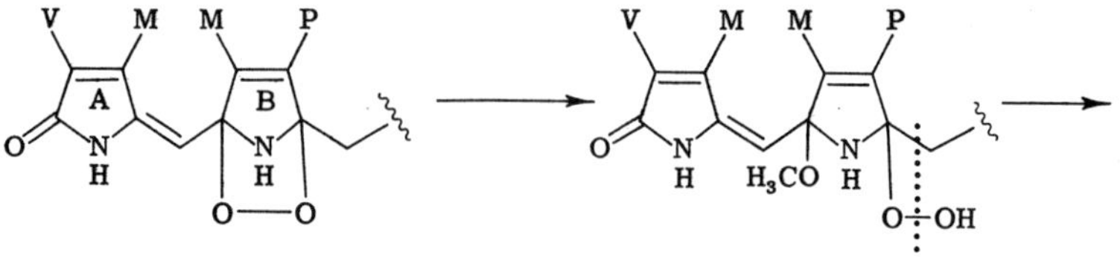

While the photooxidation of BR in hydroxylic solvents affords a minor amount of BV, a considerably improved yield (38%) may be isolated from the reaction when it is carried out in chloroform (Lightner *et al.*, 1973). That BV may be the primary photoproduct derived from BR, as well as the precursor to all other BR derivatives, has been suggested by Ostrow and Branham (1970). It has been shown, however, that the photodegradation of BR is rapid in comparison with that of BV (Bonnett and Stewart, 1972b).

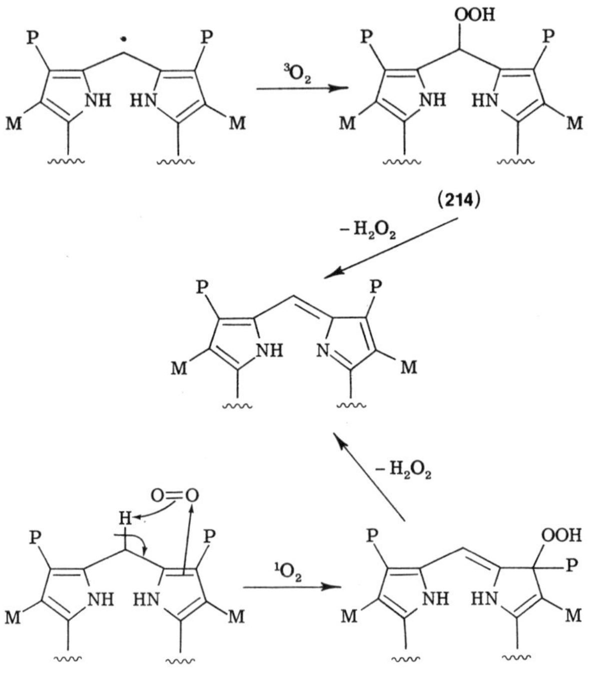

(**214**)

(**215**)

Based on these results, Lightner *et al.* (1973) have proposed that BV is not a key intermediate in the photooxidation of BR.

These conclusions have subsequently been confirmed and extended by Lightner and Crandall (1973b) who reported that, while BV is considerably more stable to photooxidation than is BR, it nonetheless undergoes photooxygenation to afford similar products of reaction. These include methylvinylmalemide (**209**), the ester of hematinic acid (**212**), and propentdyopents (**210**) and (**211**).

The manner in which BV is produced in the photooxidation of BR is still unclear. This may involve a radical process whereby hydrogen abstraction by a triplet sensitizer at the methylene bridge is followed by reaction with oxygen to form hydroperoxide (**214**). Alternatively, singlet oxygen may take part in an ene-reaction with BR affording hydroperoxide (**215**). Elimination of hydrogen peroxide from either **214** or **215** would then give BV (Lightner, 1974).

IV. INDOLES

The reactions of indole derivatives with singlet oxygen are of particular interest in connection with the photodynamic degradation of tryptophan residues in proteins, and the possibility that these oxidations may represent model reactions for the biological oxidation of tryptophan catalyzed by monooxygenases and dioxygenases. Beyond their significance in the biological aspects of singlet oxygen chemistry, indole systems have provided an important group of substrates for studying the scope and variability of singlet oxygen reactions with organic molecules.

A. Alkyl Derivatives

Alkyl-substituted indoles have long been known to be susceptible to autoxidation, a reaction in which uptake of oxygen leads to 3-hydroperoxyindolenines (Sundberg, 1970). While these hydroperoxides have been isolated in certain cases, they usually undergo further reactions, such as cleavage of the 2,3-bond, or addition of an indole residue to the reactive C=N bond of the indolenine. There is evidence (*vide infra*) that the formation of the 3-hydroperoxides takes place by an enamine-type reaction through zwitterionic intermediates.

The reactions of 2-methylindole (**216**), 1,2-dimethylindole (**220**), and 1,2,3-trimethylindole (**223**) show some of the typical transformations encountered in indole–singlet oxygen chemistry (Saito *et al.*, 1972a). Thus, photooxidation of **216** affords the oxidation product (**219**) most probably by addition of an indole molecule to intermediate (**218**). In the formation of **222** from the *N*-methyl derivative (**220**), the zwitterion (**221**) is the most likely intermediate.

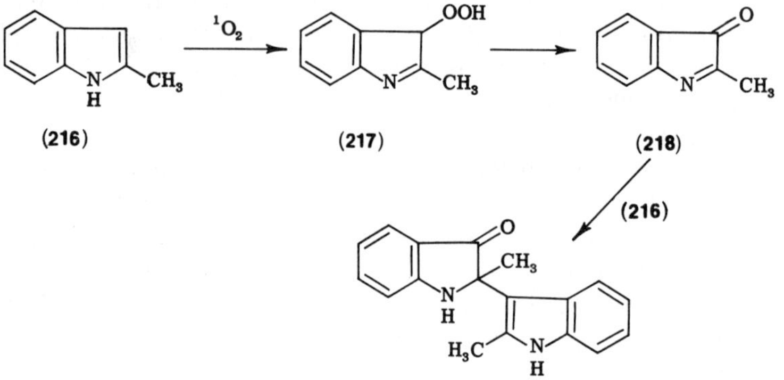

(216) (217) (218) (219)

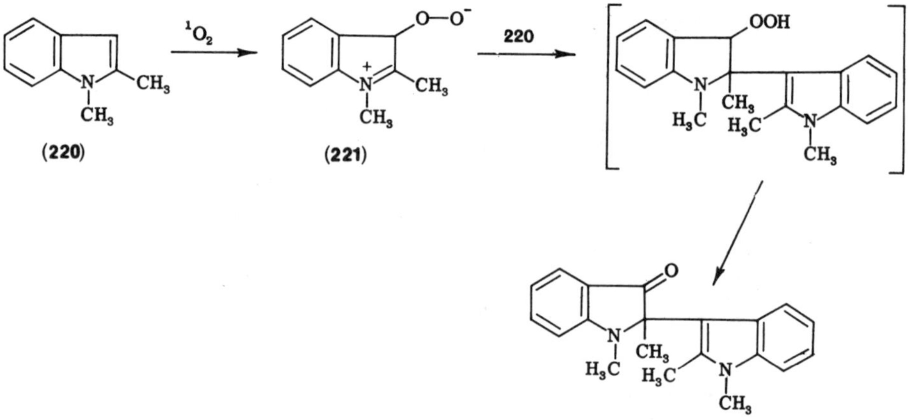

(220) (221) (222)

In the case of 1,2,3-trimethylindole (**223**), only oxidative cleavage of the 2,3-double bond occurs to give **224**. This result parallels the findings of Evans (1971) in the photooxidation of 3-methylindole (**225**), where only *o*-formamidoacetophenone (**226**) and *o*-aminoacetophenone (**227**) are observed.

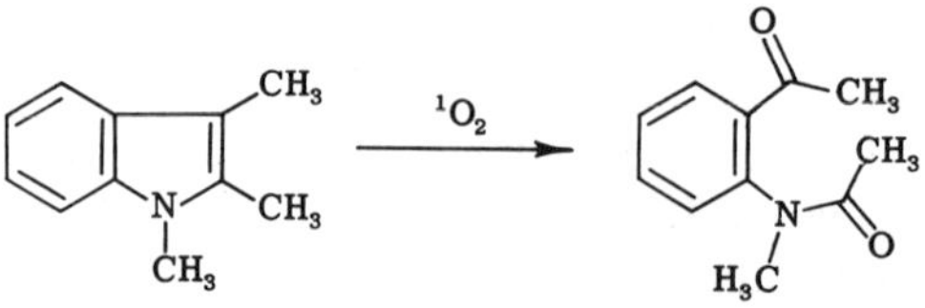

(223) (224)

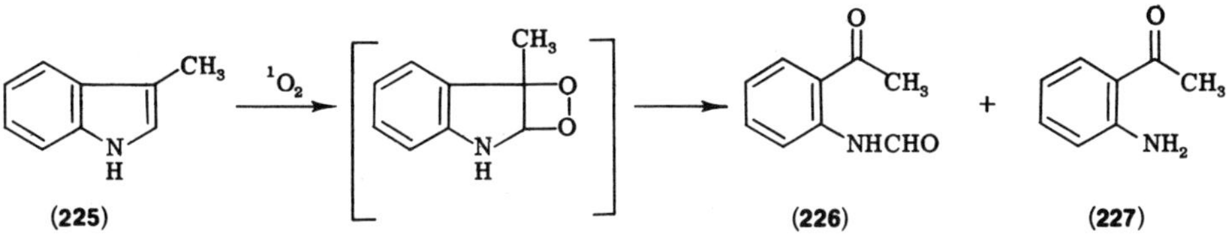

(225) (226) (227)

The dioxetane cleavage which accounts for the formation of product (**226**) from **225** is typical of many indole–singlet oxygen reactions. Recent examples of this oxidative cleavage are found in the conversion of indoles (**228**) and (**230**) to products (**229**) and (**231**). The rearrangement product (**232**) is formed along with **231** (see Section IV,C).

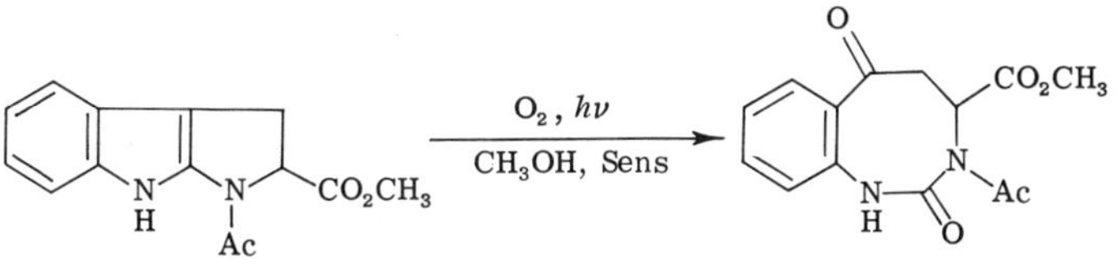

(228) (229)

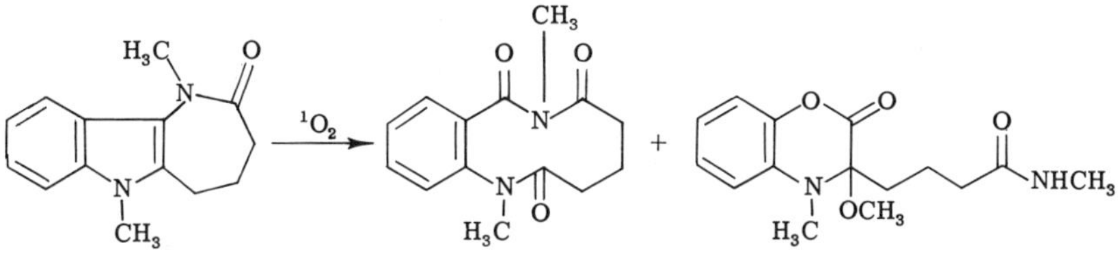

(230) (231) (232)

B. Tryptophan and Related Systems

A number of photooxidation studies have been carried out on tryptophan (**233**), and this work has been recently reviewed (Fontana and Toniolo, 1976; Saito *et al.*, 1977b). Savige (1971) reported the formation of *N*-formylkynurenine (**234**) and its hydrolysis product (**235**), presumably through the 2,3-dioxetane. In the presence of ammonia (pH 8–9), photooxidation gave the quinazoline (**236**) as the main product. *N*-Formylkynurenine was shown not to be an intermediate in the formation of **236**. Analogous reactions have been observed in the photooxidation of 2,3-disubstituted indoles in alcohol containing ammonium acetate (Maeda *et al.*, 1974).

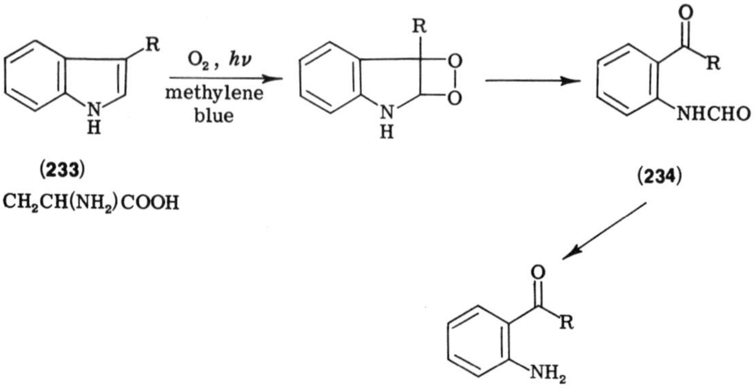

(233)
R = $CH_2CH(NH_2)COOH$

O_2, $h\nu$
methylene blue

(234)

(235)

A possible pathway for the formation of **236** may involve hydroperoxide **237**, which, upon addition of ammonia to the reactive 2-position followed by fragmentation in a conventional manner, would yield **238**, the precursor of the observed quinazoline.

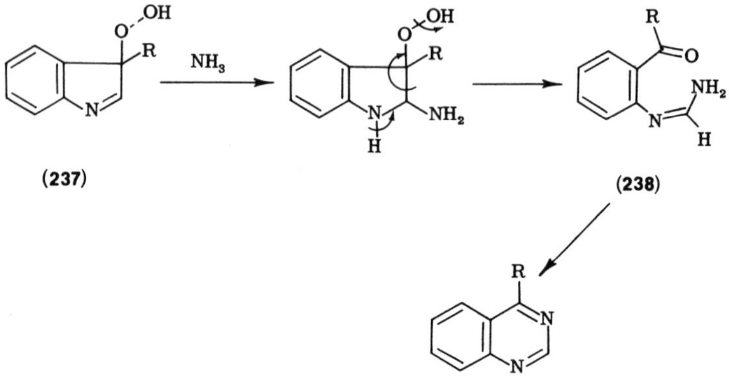

(237)

NH_3

(238)

(236)

In the reaction of tryptophan with singlet oxygen in the presence of acetic acid, the β-carboline (**242**) is formed (Cauzzo and Jori, 1972). Here, the role of singlet oxygen may only involve the oxidation of **240** to **241**, since formation of **240** from the tryptophan derivative (**239**) is a nonoxidative process.

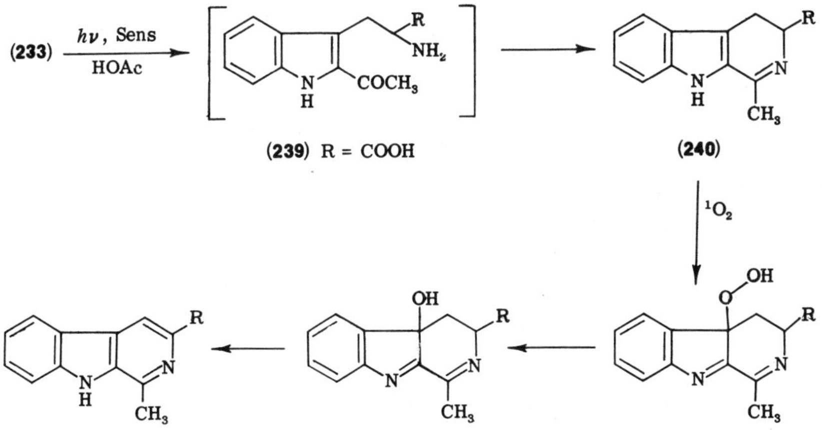

(233) $\xrightarrow[\text{HOAc}]{h\nu,\ \text{Sens}}$ [(239) R = COOH] → (240) $\xrightarrow{^1O_2}$ → (241) → (242)

C. Rearrangements to Benzoxazines

Eskins (1972) recently reported a novel rearrangement of **243** in the proflavin-sensitized photooxidation in methanol to yield the benzoxazine (**245**). It was suggested that the rearrangement may take place through a 3,4-epoxide analogous to that observed in tetraphenylpyrrole photooxidations. However, it is more likely that an intermediate methanol-addition product (**244**) rearranges to (**245**) by an acid-catalyzed ring expansion.

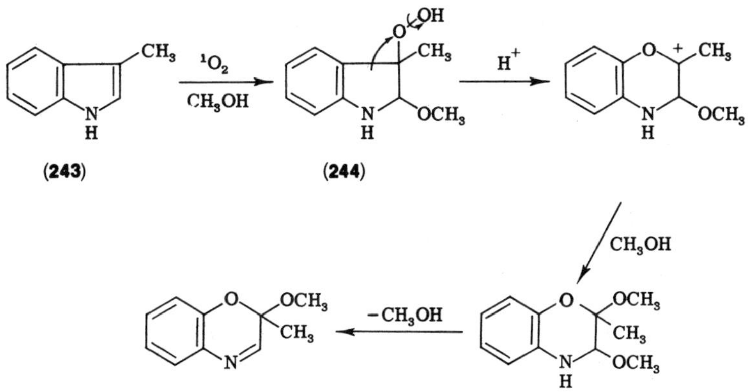

(243) $\xrightarrow[CH_3OH]{^1O_2}$ (244) $\xrightarrow{H^+}$ $\xrightarrow{CH_3OH}$ $\xrightarrow{-CH_3OH}$ (245)

In accord with this view, recent studies on the reaction of *N*-methyltryptophol (Saito *et al.*, 1975, 1977a) (**246**) with singlet oxygen yielded a 3-hydroperoxyindoline (**247**) which underwent rearrangement in HCl to the 2,3-dihydro-1,4-benzoxazine (**248**). Similarly, *N*-methylindole-3- propionic acid (**249**) and tryptamine hydrochloride (**252**) readily underwent photooxidation in methanol directly to the rearranged dihydrobenzoxazines (**251**) and (**253**), respectively.

CH_2CH_2OH $\xrightarrow{^1O_2}$ OOH $\xrightarrow[CH_3OH]{H^+}$ OCH_3

(**246**) (**247**) (**248**)

$CH_2CH_2CO_2H$ $\xrightarrow[CH_3OH]{^1O_2}$ OCH_3 $\longrightarrow$ OCH_3

(**249**) (**250**) (**251**)

$CH_2CH_2NH_2\cdot HCl$ $\xrightarrow[CH_3OH]{^1O_2}$ OCH_3

(**252**) (**253**)

A related rearrangement to a benzoxazine derivative (**254**→**256**) via hydroperoxide **255** has been observed by Nakagawa (1976b) on treatment of **255** with acid in methanol.

$CH_2CH(CO_2CH_3)NHCO_2CH_3$ $\xrightarrow{^1O_2}$ OOH, CO_2CH_3, CO_2CH_3

(**254**) (**255**)

H^+ / MeOH

OCH_3, $COOCH_3$, CO_2CH_3

(**256**)

Studies on the oxidations of **246** and **249**, have provided evidence (Saito *et al.*, 1975) for the intermediacy of zwitterionic intermediates in sensitized indole photooxidations. In recent work, Saito *et al.* (1977a) have shown that a zwitterionic species such as **257** is efficiently intercepted intermolecularly and intramolecularly by alcohols or amines at low temperatures. Inter-

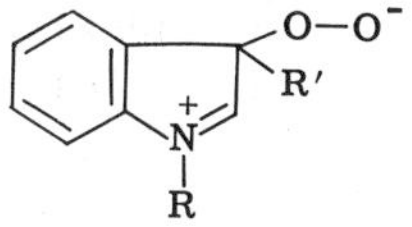

(257)

mediate (**257**) would normally be expected to undergo cyclization to a dioxetane followed by cleavage of the C-2–C-3 bond. However, in cases where the side chain contains a nucleophilic group such as OH or NH_2, intramolecular addition to the C=N bond appears to be favored leading to the isolation of hydroperoxides (**247**) and (**255**).

Considerable evidence exists to suggest that the 3-hydroperoxyindolenines (e.g., **217**) initially formed in the reactions of indoles with both triplet and singlet oxygen may act as oxidizing agents capable of transferring what is equivalent to positive hydroxyl to donor systems. Such acceptors include tertiary amines, enamines, and other indole systems. The recent work of Nakagawa *et al.* (1974) provides interesting examples of this type of reaction. Thus, photosensitized oxygenation of tryptophan methyl ester (**258**) in benzene yields a small amount of **259** along with other products, presumably

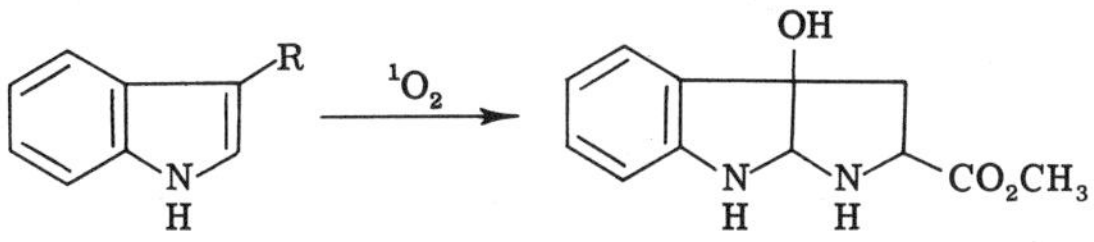

$\xrightarrow{^1O_2}$

(258) **(259)**

R = $CH_2CH(NH_2)CO_2CH_3$

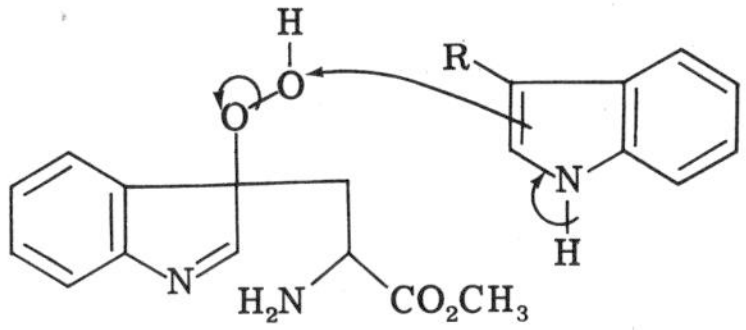

(260)

through the hydroperoxyindolenine derivative (**260**). Isolation of **259** from the reaction may be a consequence of the tendency of hydroperoxides such as **260** to undergo O–O bond cleavage with transfer of OH^+ to donor molecules.

Photooxygenation of N_b-methyltryptamine (Nakagawa *et al.*, 1975b) (**261**) under similar conditions (benzene) yields the hydroxyoxazinoindole (**264**). In methanol at 0°C, however, the 3-hydroperoxy derivative (**262**) can be isolated. This product undergoes rearrangement to the *N*-oxide (**263**), the precursor of (**264**). Formation of **263** most probably occurs by intermolecular transfer of oxygen to the tertiary amino function of another molecule. The postulated rearrangement of **263** to **264** pictured above is supported by independent studies on the peracid oxidation of **265** which yields only **264** and none of the expected *N*-oxide product (**263**). Other significant evidence

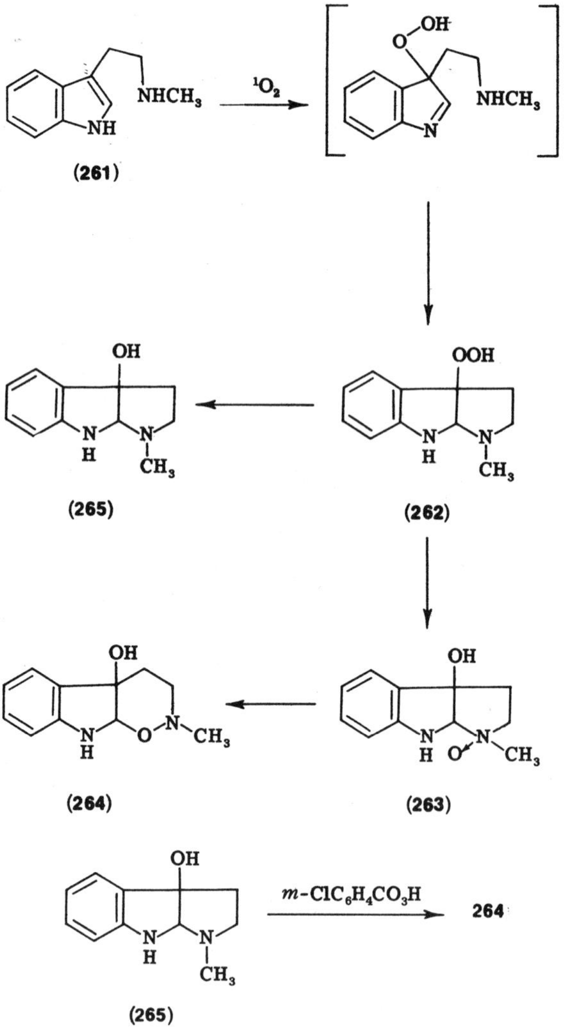

NHCH$_3$ $\xrightarrow{^1O_2}$ [OOH ... NHCH$_3$]

(**261**)

(**265**) (**262**)

(**264**) (**263**)

(**265**) $\xrightarrow{m\text{-}ClC_6H_4CO_3H}$ **264**

favoring the oxygen transfer mechanism from hydroperoxide to tertiary amine was obtained from the reaction of a model 3-hydroperoxide (**266**) with **265** whereby the oxazinoindole (**264**) was obtained in high yield (Nakagawa *et al.*, 1975b).

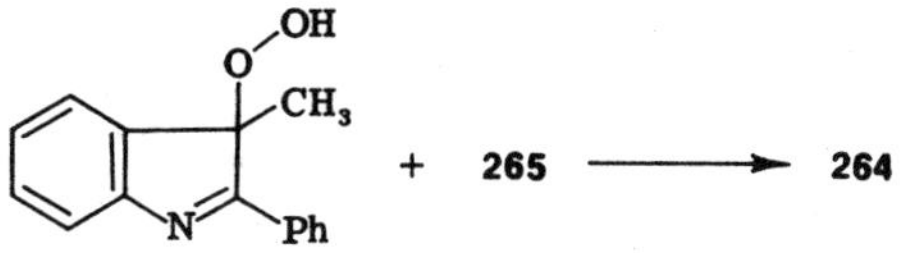

(**266**) + **265** → **264**

D. Other Reactions of 3-Peroxy Derivatives

The ready conversion by photosensitized oxygenation of N_b-methyltryptamine (Nakagawa *et al.*, 1975b) (**261**) to the 3a-hydroxypyrroloindole (**265**) and the 4a-hydroxyoxazinoindole (**264**) via the 3a-hydroperoxyindole (**262**) are of special interest as possible model reactions in connection with postulated biosynthetic pathways for formation of alkaloids such as sporidesmin A (Safe and Taylor, 1972) (**267**) brevianamide E (Birch and Wright, 1970) (**268**), and hunteracine chloride (Burnell *et al.*, 1974) (**269**).

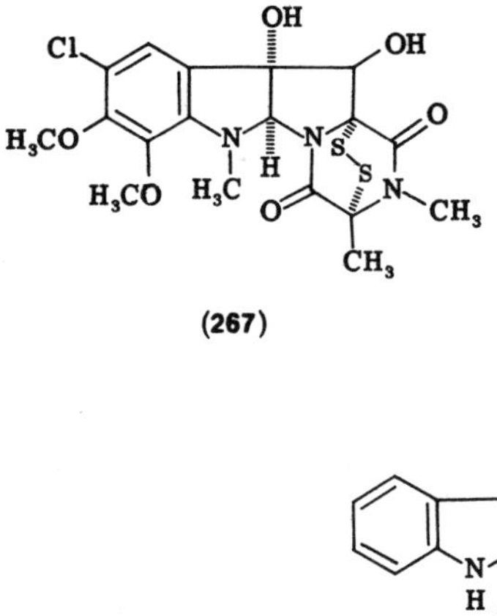

(**267**)

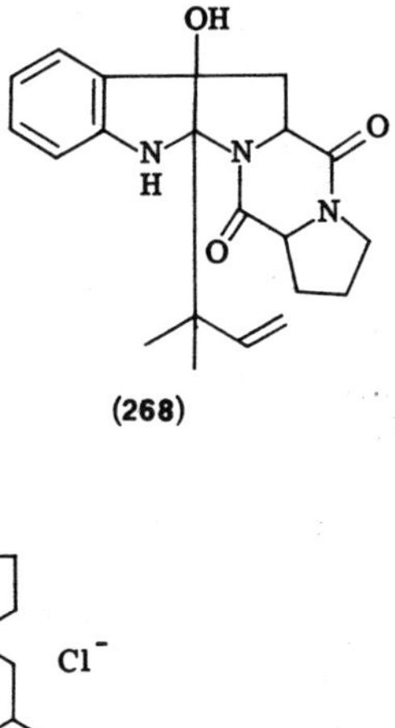

(**268**)

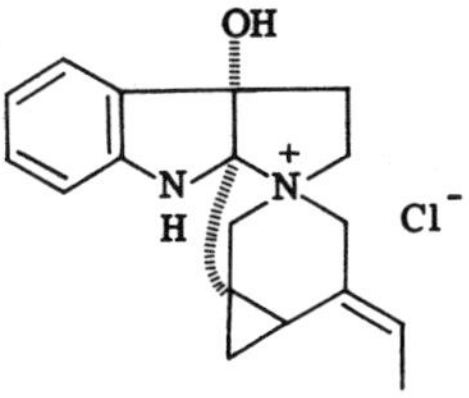

(**269**)

Recent studies (Nakagawa *et al.*, 1975a) on the Rose Bengal photosensitized oxygenation of the quinolizidinium chloride (**270a**) in water yielded **271**. Formation of **271** is reminiscent of the ready autoxidation of 2,3-dialkylindoles to 2-acylindoles observed in a variety of indoles and carbazoles (Leete, 1961; Wasserman and Floyd, 1963; Leete and Chen., 1963; Taylor, 1962; McLean and Dmitrienko, 1971).

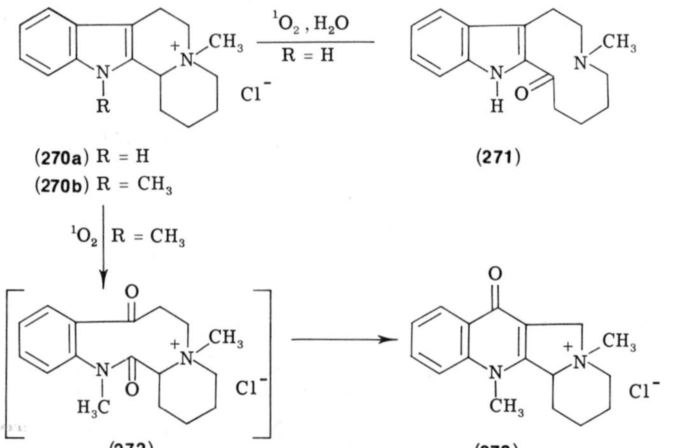

(**270a**) R = H
(**270b**) R = CH_3

(**271**)

(**272**) (**273**)

The reaction may proceed through the 3-hydroperoxyindolenine (**274**) followed by tautomerization to the enaminium salt (**275**), and then an acid-catalyzed solvolytic displacement of hydrogen peroxide by H_2O. Under

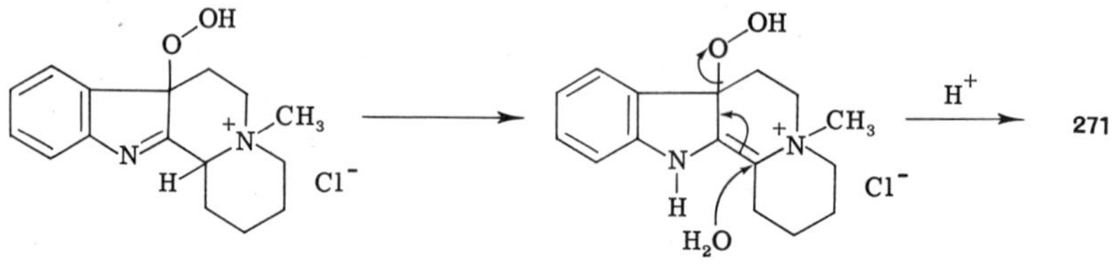

(**274**) (**275**)

similar conditions, the *N*-methyl derivative (**270b**) was converted to product (**273**), presumably by cyclization of the 2,3-cleavage product (**272**).

Photosensitized oxidation of the free base (**276**) at 0°C in methanol containing thiourea (M. Nakagawa, unpublished results, 1977) yields the

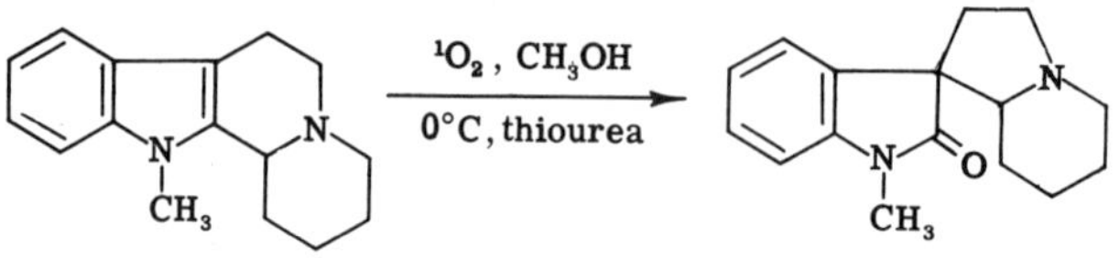

(**276**) (**277**)

oxindole (**277**). Here the reaction appears to involve a pinacol-type of rearrangement of the diol formed after oxygen uptake and reduction of the intermediate peroxidic species by thiourea.

276 $\xrightarrow{^1O_2}$ [O–O⁻; N⁺; CH$_3$; N] ⟶ [O–H; N; H$_3$C; O; H; N] $\xrightarrow{H^+}$ (277)

Another possible pathway for the conversion of tryptophan to formylkynurenine by photosensitized oxidation is revealed in the reactions of N_b-acyl derivatives of tryptophan with singlet oxygen (Nakawawa *et al.*, 1976b). Thus, hydroperoxide (**255**), isolated from the Rose Bengal-sensitized photooxygenation of **254** in methanol/pyridine was transformed during silica gel chromatography with CH_2Cl_2 to a mixture of **278**, **279**, and **280**.

254 $\xrightarrow{^1O_2}$ [O–OH; N H; N; R; CO_2CH_3]

(255)
R = CO_2CH_3

OH; N H; N; R; CO_2CH_3
(18%)
(278)

O; R; $NHCO_2CH_3$; NHCHO
(17%)
(279)

O; R; NCO_2CH_3; CHO; NH_2
(44%)
(280)

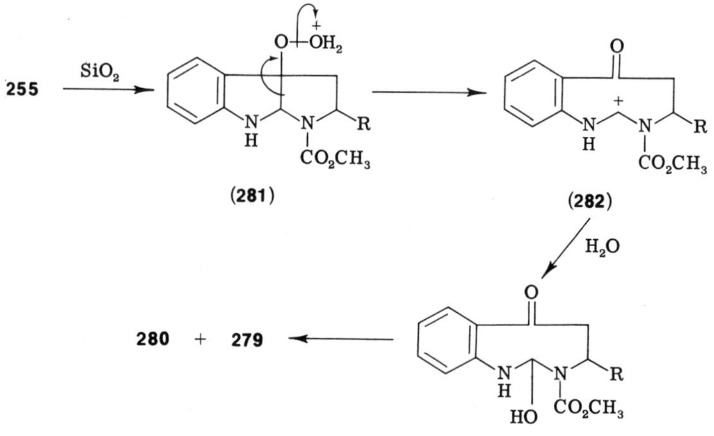

(**281**) (**282**)

(**283**)

Formation of **279** and **280** from **255** seems to be acid-catalyzed; e.g. **255** → **281** → **282** → **283** → **279** and **280**. The preferred formation of the N_b-formyl derivative (**280**) is, however, unclear.

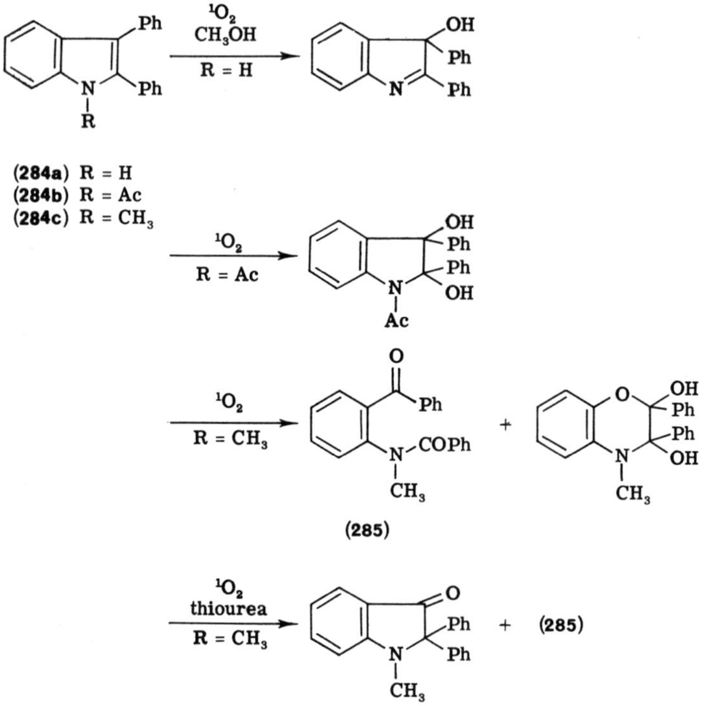

(**284a**) R = H
(**284b**) R = Ac
(**284c**) R = CH_3

(**285**)

(**286**)

Scheme 8

Studies on 2,3-diphenylindole (**284a**), the *N*-acetyl derivative (**284b**), and the *N*-methyl derivative (**284c**) outlined in Scheme 8 illustrate the variations observed in product formation depending on the substitution pattern on nitrogen (Nakagawa *et al.*, 1976a).

It is likely that formation of the *N*-methyl-2,2-diphenyl-3-oxindole (**286**) results from a benzilic acid-like rearrangement of the oxyanion (**288**) through an initially produced zwitterion (**287**) as shown.

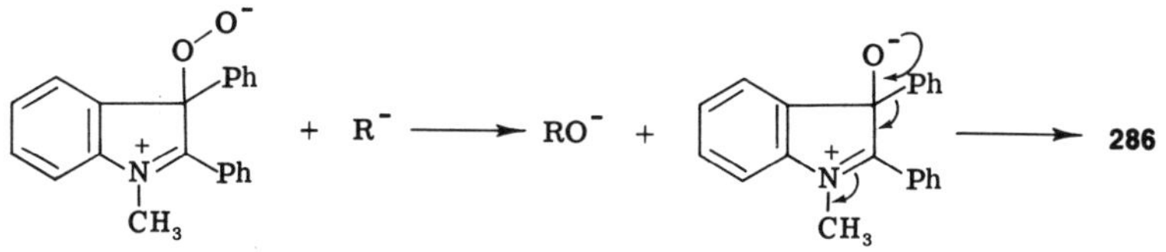

(287) + R^- ⟶ RO^- + (288) ⟶ 286

Photooxygenation of *N*-methyl-1,2,3,4-tetrahydrocarbazole (**289a**) in a variety of solvents sensitized by Rose Bengal gave **291a** (90%), a reaction characteristic of the oxidative cleavage of the enamine double bond (Saito *et al.*, 1972b). On the other hand, the *N*-acetyl derivative (**289b**), afforded substantial amounts (54%) of the diol (**294**) in addition to the cleavage product (**291b**). The authors suggest that formation of **294** provides support for the existence of zwitterionic intermediates such as **292**. Dioxetanes (e.g., **290**) are, perhaps, less likely intermediates since they do not usually undergo nucleophilic cleavage of the O–O bond, except by the action of strong nucleophiles (Richardson and Hodge, 1970, 1971a,b).

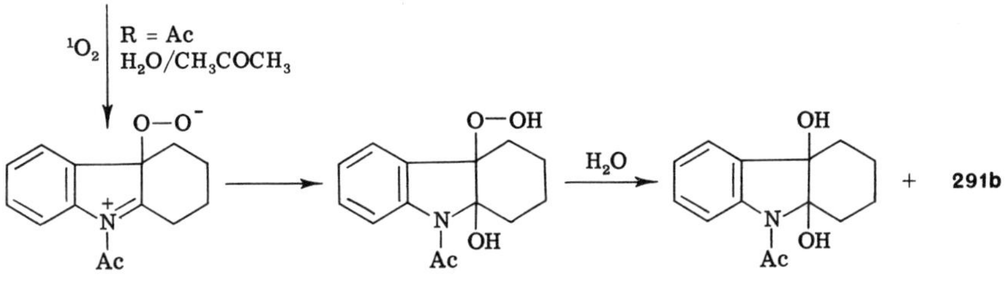

(289a) $R = CH_3$
(289b) $R = Ac$

(289) —$^1O_2$, $R = CH_3$⟶ [(290)] ⟶ (291a)

(289) —$^1O_2$, $R = Ac$, H_2O/CH_3COCH_3⟶ (292) ⟶ (293) —H_2O⟶ (294) + 291b

An interesting example of unsensitized photooxygenation is that of pyrano[3,4-b]indol-3-(9*H*) (**295**), which yields the 2,3-diacylindole (**297**) (Rokach *et al.*, 1971). An endoperoxide seems to be the intermediate as shown by ^{18}O-labeling experiments but it is not certain whether singlet oxygen is involved.

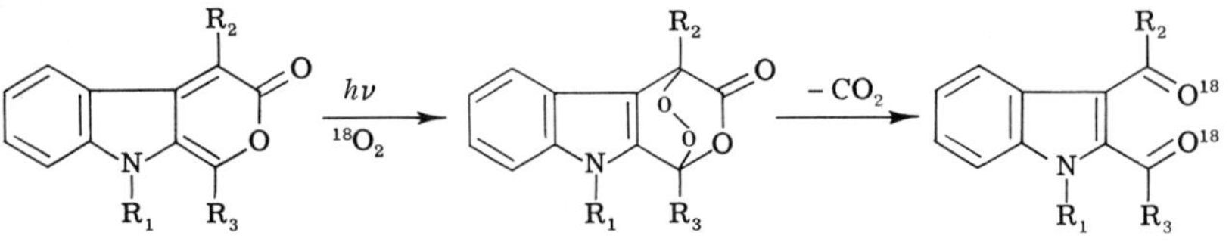

(295) (296) (297)

In very recent work, Matsumoto and Kondo (1977) have shown that, in aprotic solvents, 1-methyl-3-vinylindoles (**298**) react with singlet oxygen to yield 1,4-endoperoxides of type **299** with complete retention of stereochemistry. In protic solvents, however, 1-methyl-3-formylindole (**300**) is formed. The authors provide evidence suggesting that **300** as well as products **301** and **302** are produced by decomposition of the dioxacarbazole (**299**).

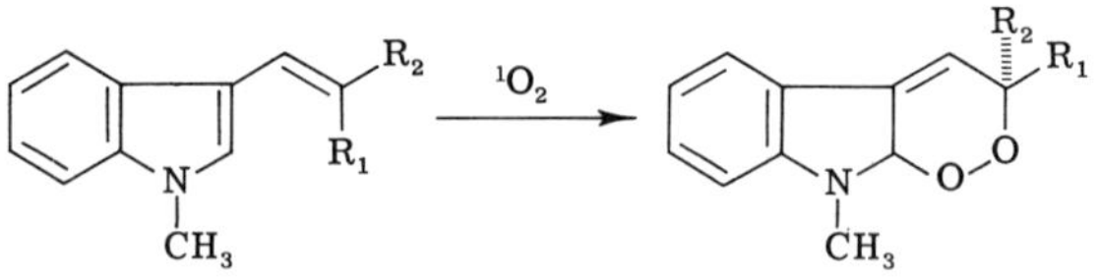

(298) (299)

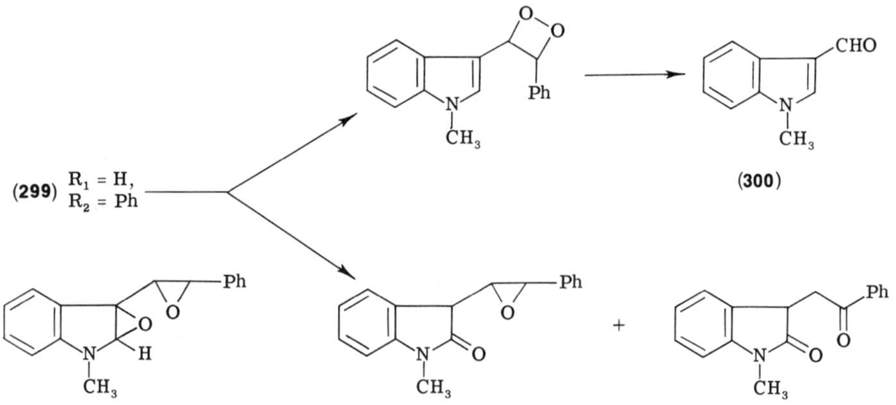

(299) R_1 = H, R_2 = Ph

(300)

(303) (301) (302)

It is reasonable to assume here that **301** and **302** are formed from **303** which in turn may be derived from **299** by a well-precedented rearrangement, including the thermal conversion of ascaridole (**304**) to the bisepoxide (**305**) (Runquist and Boche, 1968). The dioxetane precursor of **300** most probably

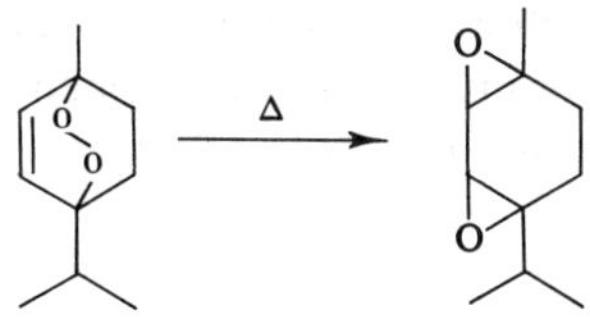

(304) (305)

arises from the zwitterion (**306**). Intermediate (**306**) may be formed as the initial product of the reaction or may be derived from **299** by electron release from nitrogen to oxygen as shown.

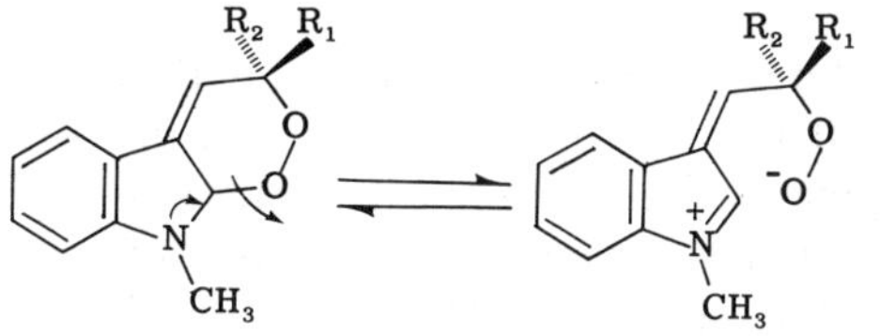

(299) (306)

V. IMIDAZOLES

An early observation of chemiluminescence in the reaction of 2,4,5-triphenylimidazole (lophine) (**307**), with air in the presence of base was made by Radziszewski in 1877. Dufraisse *et al.* (1957) later isolated a peroxide from this reaction and formulated this product incorrectly as the endoperoxide (**308**). Shortly thereafter, two groups of investigators (Sonnenberg and White, 1964; White and Harding, 1964) showed that the peroxidic product is actually the hydroperoxide (**309**) which undergoes decomposition with chemiluminescence through the dioxetane (**310**).

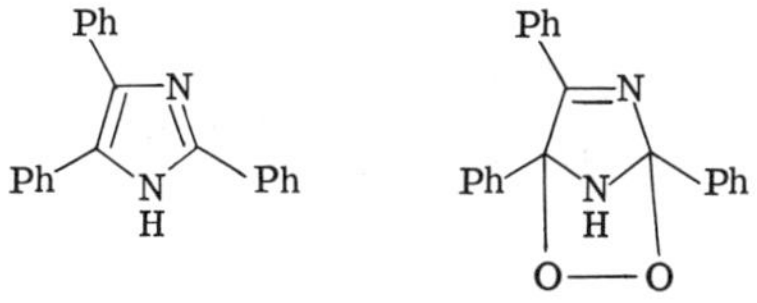

(307) (308)

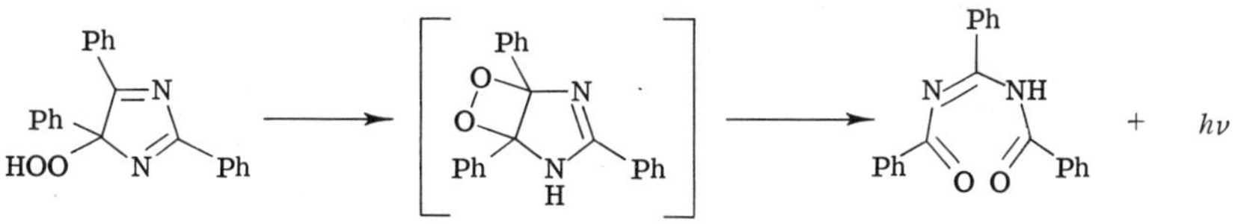

(309) (310)

Other phenyl-substituted imidazoles have given similar products on photooxidation, as in the conversion of tetraphenylimidazole (**311**) to *N*,*N*′-dibenzoyl-*N*-phenylbenzamidine (**312**) (Ranjon, 1971; Wasserman *et al.*, 1968).

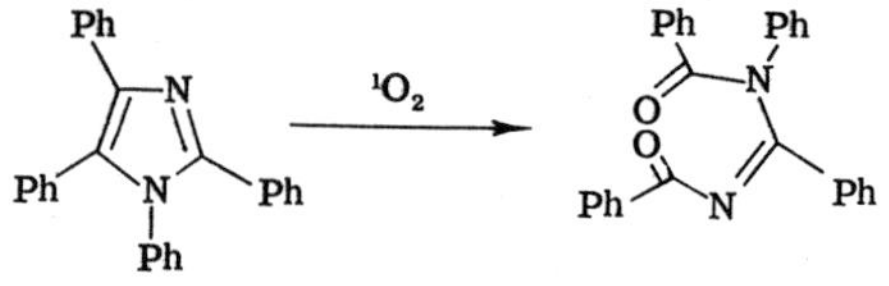

Ph, Ph, N, N, Ph, Ph — $^1O_2$ → Ph, Ph, N, O, O, Ph, N, Ph

(**311**) (**312**)

The major interest in imidazole photooxidation is based on the reasonable assumption that this system is involved in the destruction by oxygen and light of enzymes such as lysozyme, ribonuclease, chymotrypsin, insulin, and phosphoglucomutase. A considerable body of work has been carried out on this so-called "photodynamic effect" by Weil and Seibles (1955), Ray and Koshland (1962), and Nakatani (1960a,b) showing that the inactivation of these enzymes can be correlated with the destruction of histidine residues and, more specifically, with oxidative breakdown of the imidazole ring.

Up until recently, however, few details were known about the mode of breakdown of alkyl imidazoles on reaction with singlet oxygen. Matsuura and Ikari (1969) showed that on photooxidation, 2-methylimidazole under-

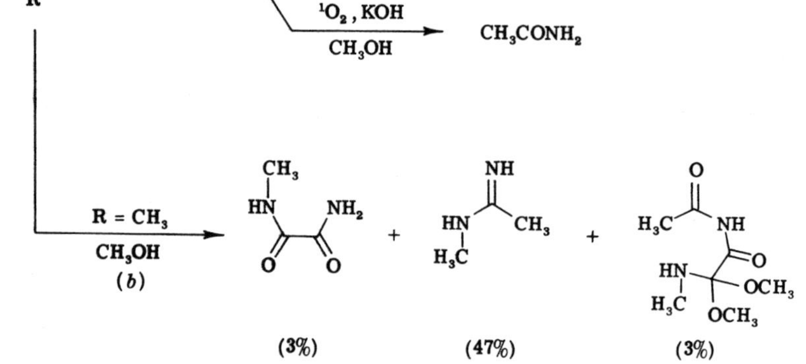

R = H (*a*): $^1O_2$, CH_3OH → H_2N, NH_2, O, O (2%) (**313**) + NH, H_2N, CH_3 (49%) (**314**)

$^1O_2$, KOH, CH_3OH → CH_3CONH_2

N, N, CH_3, R

R = CH_3, CH_3OH (*b*) → CH_3, HN, NH_2, O, O (3%) + NH, HN, CH_3, H_3C (47%) + O, H_3C, NH, O, HN, OCH_3, H_3C, OCH_3 (3%)

Scheme 9

went extensive breakdown (path *a*) yielding acetamidine (**314**) and a small amount of oxamide (**313**). Similar results were obtained from 1,2-dimethylimidazole (path *b*) as shown in Scheme 9. Table IV summarizes the results of dye-sensitized photooxidation of a number of phenyl-substituted imidazole derivatives.

A. Alkyl- and Aryl-Substituted Derivatives

More extensive studies by H. H. Wasserman and M. S. Wolff (unpublished data, 1970) on a variety of alkyl- and aryl-substituted imidazoles have revealed the main pathways of reaction with singlet oxygen. Table V summarizes the products formed in a series of methylene blue-sensitized photooxidations carried out mostly in methanol solution. From these results, it appears that two main pathways of oxidation may be observed, depending on the substitution pattern of the imidazole. Thus, in methanol, when the 2- or 5-position bears a hydrogen atom (e.g., **315**), a carbonyl group is formed at the site of the hydrogen-bearing carbon atom along with the uptake of two equivalents of solvent to form **319**. The reaction appears to take place through the endoperoxide (**316**), possibly after initial formation of the zwitterion (**321**), followed by solvolysis to **317**, loss of water to form **318** and then addition of methanol to the imino group. None of the alternative, less-favored product of solvolysis (**320**) is observed.

R_2, N, R_1, N, H, H — $^1O_2$ → R_2, N, R_1, N, H, H, O—O — CH_3OH → R_2, N, R_1, H, H_3CO, N, H, OOH

(**315**) (**316**) (**317**)

R_2, N, R_1, H, HOO, N, H, OCH_3 — R_2, N, H, R_1, N+, H, O—O$^-$ — R_2, N, R_1, H_3CO, N, H, O

(**320**) (**321**) (**318**)

R_2, H_3CO, NH, R_1, H_3CO, N, H, O

(**319**)

Table IV Photooxidation of Imidazoles

Imidazole	Solvent	Product(s)		Ref.
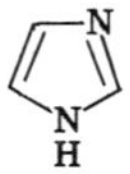 N, N H	CH_3OH	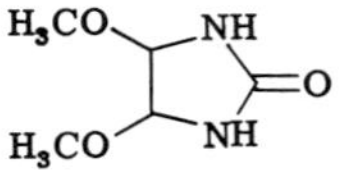 H_3CO, NH, O, H_3CO, NH (9%)		Wasserman *et al.* (1968)
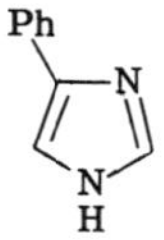 Ph, N, N H	CH_3OH	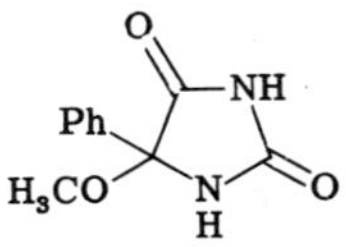 O, NH, Ph, H_3CO, N H, O (15%)	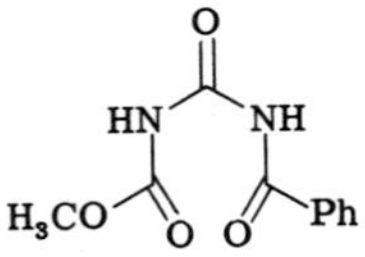 O, HN, NH, H_3CO, O, O, Ph (2%)	Wasserman *et al.* (1968)
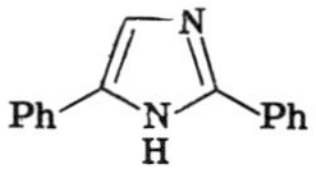 N, Ph, N H, Ph	CH_3OH	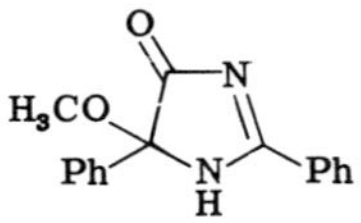 O, N, H_3CO, Ph, N H, Ph		Stiller (1968)

Substrate	Solvent	Products			Reference
	CH_3OH	(24%)	(6%)	(5%)	Wasserman *et al.* (1968)
	CH_3OH (**A** + **B**) CS_2 (**C** + **D**)	$PhCONH_2$ + $PhCO_2CH_3$ (**A**) (64%) (**B**) (58%)	(**C**)	(**D**)	Wasserman *et al.* (1968) Sonnenberg and White (1964) White and Harding (1964)
	CH_3OH	(97%)			Wasserman *et al.* (1968) Ranjon (1968)

Table V[a] Photooxidation of Imidazoles

Imidazole	Solvent	Product(s)	
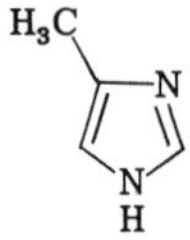	CH_3OH	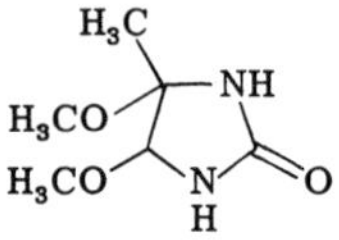 (13%)	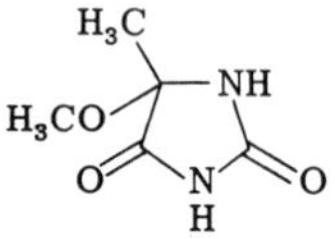 (9%)
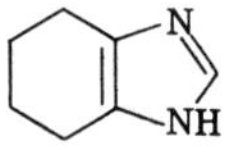	CH_3OH	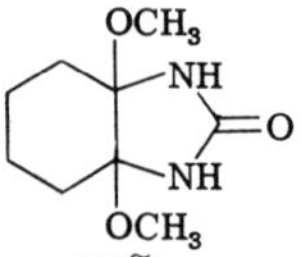 (42%)	
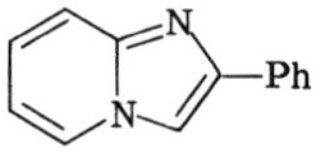	CH_3OH	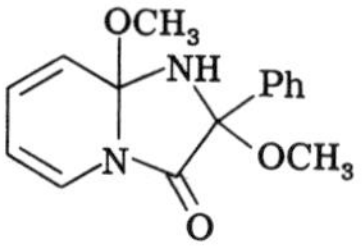 (12%)	
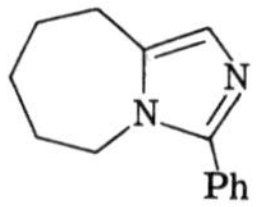	CH_3OH	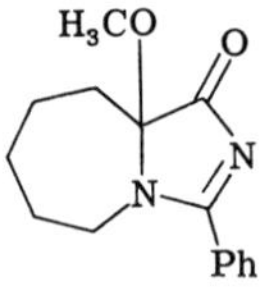 (60%)	
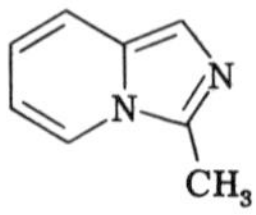	CH_3OH	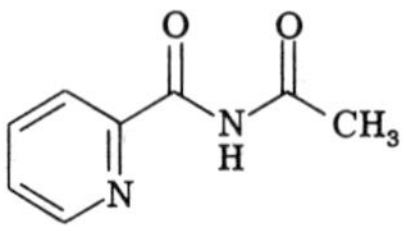 (42%)	
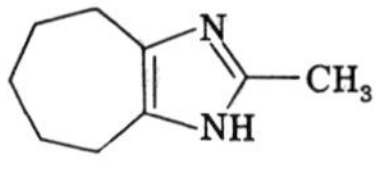	(**A**) CH_3OH (**B**) CH_2Cl_2	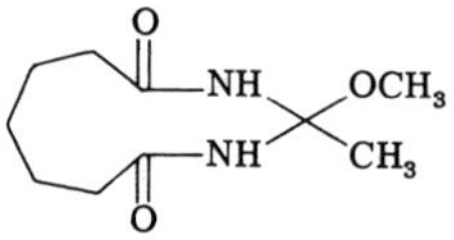 (**A**) (15%)	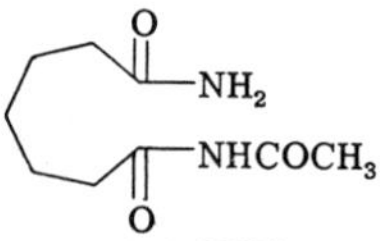 (**B**) (33%)
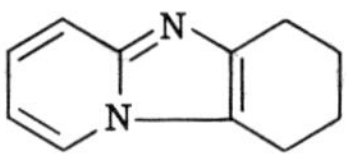	(**A**) CH_3OH (**B**) CH_2Cl_2	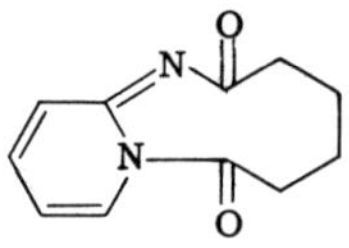 (**A**) (46%) (**B**) (52%)	
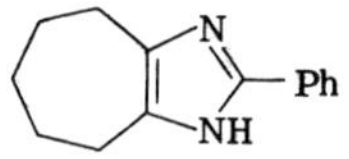	(**A**) CH_3OH (**B**) CH_2Cl_2	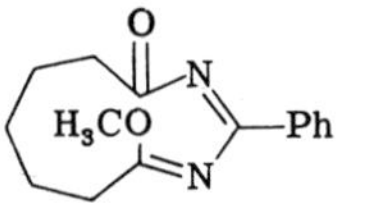 (**A**) (75%)	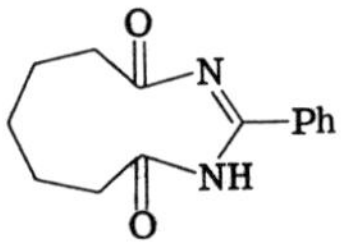 (**B**) (61%)

[a] From H. H. Wasserman and M. S. Wolff (unpublished data, 1970).

In cases where the imidazole ring has no proton at either the 2- or 5-position (**322**, $R_1 = R_2 = R_3 =$ alkyl), the reaction course changes leading to the formation of dicarbonyl compounds (**324**), presumably through dioxetanes (**323**).

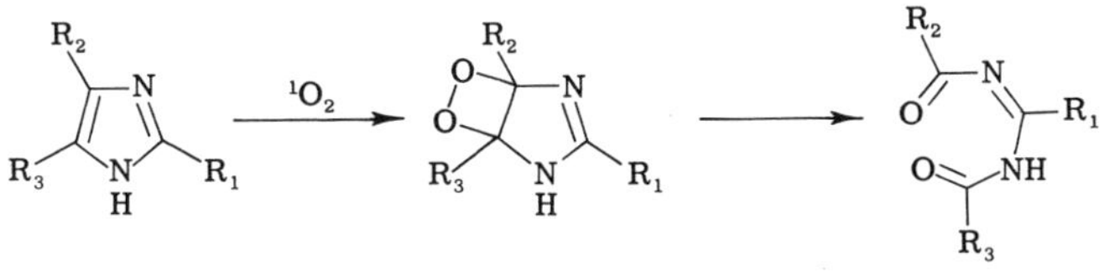

(**322**) (**323**) (**324**)

Another possible mode of oxygenation is observed in imidazoles having a proton at the 4-position where carbon–hydrogen cleavage may compete with carbon–carbon cleavage in the decomposition of intermediate dioxetanes. For example, the photooxidation of **325** in methanol may lead to products of type **326** or **327**, possibly by a β-elimination.

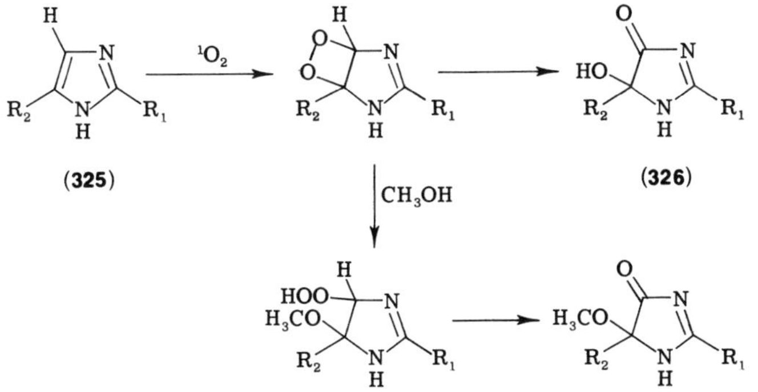

(**325**) (**326**) (**327**)

It is conceivable that all of the above results may be explained by postulating a common intermediate zwitterion (**328**) in the reactions of imidazoles with singlet oxygen. This type of dipolar species (**328**) has often been suggested as an initially formed intermediate in the reactions of singlet oxygen with many types of substrates containing the enamine grouping (Foote *et al.*, 1975; Dewar and Thiel, 1975; Wasserman and Terao, 1975; Saito *et al.*, 1977a). Cyclization of **328** at the 2-position would give the 2,5-endoperoxide (**329**) or, at the 4-position, the 4,5-dioxetane (**330**). The choice of reaction path thus depends on the nature of the substituents on the imidazole ring. Solvent effects commonly responsible for divergent pathways in many singlet oxygen reactions have thus far not appeared to be of major importance in imidazole photooxidations.

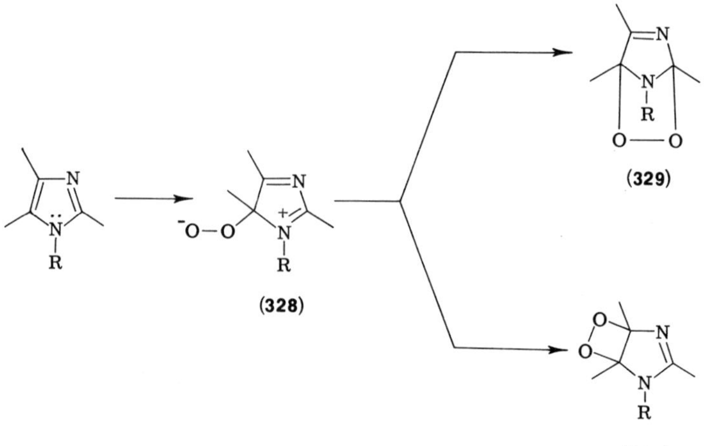

(328)

(329)

(330)

B. Model Systems Related to Histidine Oxidation

Studies on model systems, as discussed above, help to clarify the probable modes of imidazole breakdown in enzyme deactivation by photooxidation. In such cases it has been shown that histidine destruction by two equivalents of oxygen can be correlated with loss of biological activity. More specific studies have been directed toward elucidating the nature of such histidine photooxidation. Thus, Irie and co-workers (1968) were able to identify seventeen compounds from the photooxidation of *N*-benzoylhistidine (**331**) at pH 11. These products presumably result from second-stage

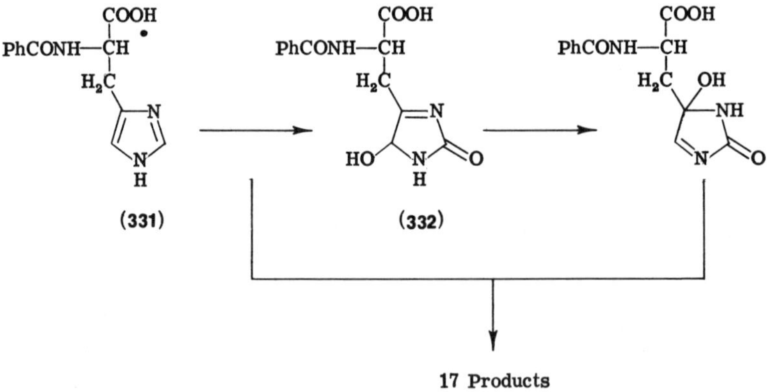

(331)

(332)

17 Products

intramolecular reactions of a common precursor (**332**) which was not isolated.

Wasserman *et al.* (1968) found that *N*-benzoylhistidine methyl ester undergoes photooxidation in methanol to give a mixture of products which, on hydrolysis, yields aspartic acid in 65% yield.

$$\text{N-benzoylhistidine methyl ester } (CH_2CHNHCOPh,\ COOMe) \xrightarrow[2.\ H_3O^+]{1.\ {}^1O_2,\ CH_3OH} HOOCCH_2CH(NH_2)COOH \quad 65\%$$

Further evidence regarding the process by which two equivalents of singlet oxygen are involved in the oxidative cleavage of 5-substituted imidazoles is derived from the photooxidation of 5-phenylimidazole (**333**) leading to *N,N'*-dibenzoyl-*N*-phenylbenzamidine (**334**). Isolation of **334**, containing the intact carbon skeleton of the starting imidazole, suggests a rationale for the formation of aspartic acid from *N*-benzoylhistidine methyl ester, since a key feature of the latter oxidation involves the formation of a carbonyl group at the 2-position of (**333**).

As pictured in Scheme 10, uptake of the first equivalent of singlet oxygen gives the endoperoxide (**335**) which undergoes solvolysis, loss of water, and addition–elimination of methanol to yield **336**. One would then expect **336** to be oxidized by a second equivalent of singlet oxygen giving **334**, presumably through the 4,5-dioxetane.

The susceptibility of imidazolinones of type **336** (Scheme 10) to oxidative cleavage both in methanol and in dichloromethane has been shown separately in related systems, as pictured below (H. H. Wasserman and M. S. Wolff, unpublished data, 1970).

$$\text{4,5-diphenyl-1,3-R}_2\text{-imidazolin-2-one} \xrightarrow{{}^1O_2} (Ph{-}CO{-}NR)_2C{=}O$$

R = H, CH_3, Ph

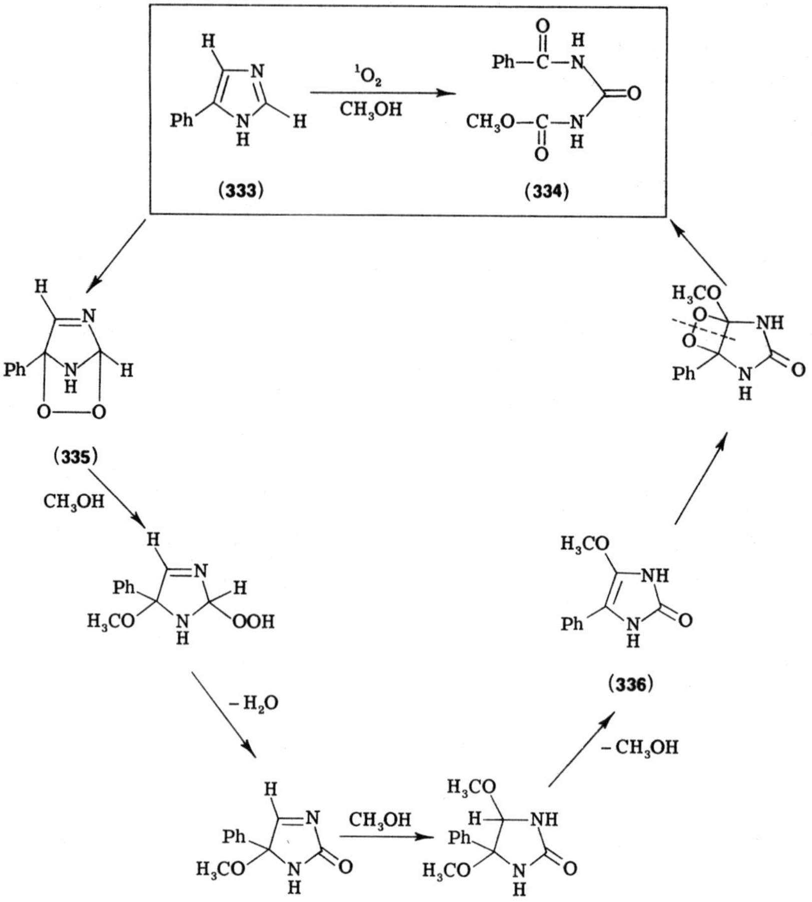

Scheme 10

VI. PURINES

The inactivation of nucleic acids by sensitized photooxygenation, correlated with the destruction of the guanine portion of the molecule, has led to numerous studies of the reaction of guanine and various purine derivatives with singlet oxygen (Simon and Van Vunakis, 1962; Wacker *et al.*, 1963; Sussenbach and Berends, 1963, 1964, 1965). From these investigations, a variety of oxidation products have been isolated and identified, but, as yet, the mechanism(s) of the primary processes in the photooxidation of nucleic acids has not been established.

Early work by Hallet *et al.* (1970) and Clagett *et al.* (1973; Clagett and Galen, 1971) showed the relationship between the reactivity of a large number of purine and pyrimidine bases in photodynamic reactions involving chemically produced singlet oxygen (hypochlorite–peroxide). Similar results were obtained by Rosenthal and Pitts (1971) who found that the percent destruction of purines and pyrimidines by singlet oxygen generated externally via microwave discharge of an oxygen stream (Table VI) were in agreement with Hallet's earlier findings.

The most extensive studies on the photooxidation of purine derivatives have been carried out by Matsuura *et al.* (1972). Their investigations, including the reactions of singlet oxygen with xanthine, theophylline, uric acid, and various other purines are summarized in Table VII.

The results in Table VII show that the reactions of singlet oxygen with purines appear to take place by attack on the imidazole portion of the molecule. As will be shown below, the products from these reactions are formed, most probably, by the decomposition of intermediate peroxidic species analogous to those postulated for the imidazole series.

Table VI Effect of Singlet Oxygen on Purine and Pyrimidine Bases[a]

	% destruction		
Base	Phosphate buffer pH = 6.8	Tris buffer pH = 8.5	Carbonate buffer pH = 10.5
Uric acid	65[b]	77[b]	83[b]
Guanine	21	56[c]	69[b]
Guanosine	23	53[c]	69[b]
Thymine	5	35	60[c]
Thymidine	3	16	50[c]
Uracil	4	10	28
Uridine	3	3	5
Theophylline	18	75[c]	57[b]
Theobromine	7	23	58[c]
Xanthosine	37	48[b]	58[b]
Hypoxanthine	5	12	17
Inosine	3	3	3
Caffeine	5	16	35
Cytosine	3	3	5
Cytidine	3	3	3
Adenine	3	3	3
Adenosine	3	3	4

[a] Time of exposure 3 hr unless otherwise specified.
[b] Time of exposure 45 min.
[c] Time of exposure 90 min.

Table VII Photooxidation of Purines

Purine	Solv./Sens.	Product(s)	Ref.
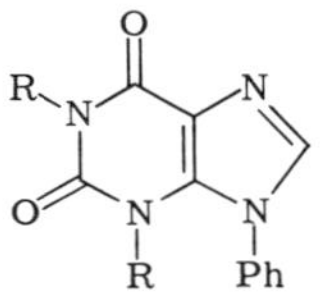 O, R, N, N, O, N, N, R, Ph R = H, CH_3	CH_3OH Rose Bengal	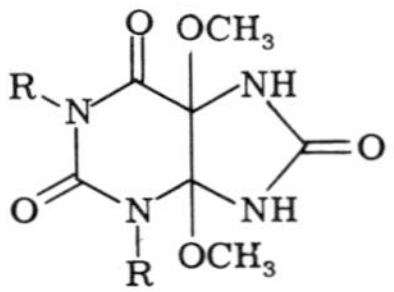 O, OCH_3, R, N, NH, O, O, N, NH, R, OCH_3	Matsuura and Saito (1968b, 1969a)
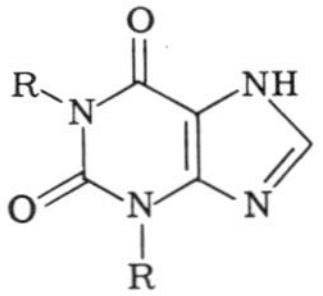 O, R, N, NH, O, N, N, R R = H (Xanthine) R = CH_3 (Theophylline)	NaOH, H_2O Rose Bengal	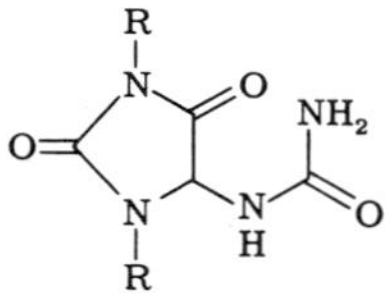 R, N, O, NH_2, O, N, N, H, O, R	Matsuura and Saito (1967, 1968a)
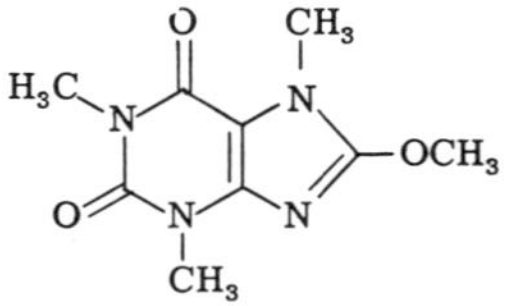 O, CH_3, H_3C, N, N, OCH_3, O, N, N, CH_3 8-Methoxycaffeine	CH_3OH, $CHCl_3$ (20:1) Rose Bengal	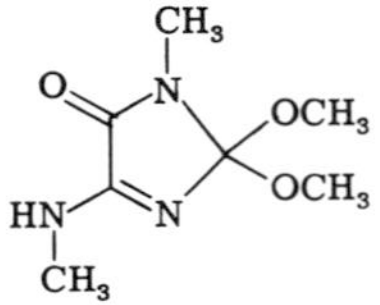 CH_3, O, N, OCH_3, OCH_3, HN, N, CH_3 + $CH_3NHCO_2CH_3$ + CO_2	Matsuura and Saito (1969c)
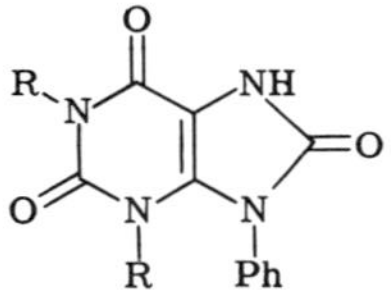 O, R, N, NH, O, O, N, N, R, Ph R = H, CH_3	CH_3OH Rose Bengal	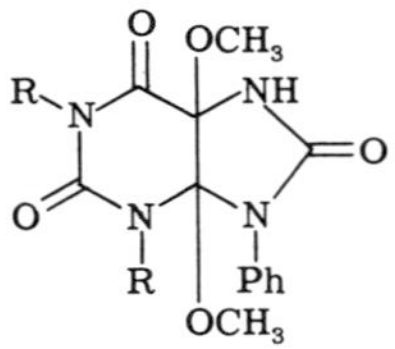 O, OCH_3, R, N, NH, O, O, N, N, R, Ph, OCH_3 R = H, CH_3 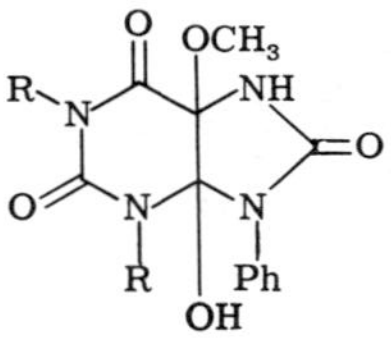 O, OCH_3, R, N, NH, O, O, N, N, R, Ph, OH R = CH_3	Matsuura and Saito (1969a)

Substrate	Conditions	Products			Reference
$R = CH_3, C_2H_5$	$CHCl_3$ Methylene blue	$R = CH_3, C_2H_5$	$R = C_2H_5$	$R = CH_3$	Matsuura and Saito (1968b, 1969b)
$R = CH_3, C_2H_5$	CH_3OH Rose Bengal	$R = CH_3, C_2H_5$	$R = CH_3$		Matsuura and Saito (1968b, 1969b)
Uric acid	NaOH, H_2O Rose Bengal				Matsuura and Saito (1967, 1968a)

Xanthine (**337a**) and theophylline (**337b**) undergo photooxidation in aqueous base to yield the corresponding allantoins (**341**), presumably by the process given in Scheme 11 in which the initial 2,5-endoperoxide (**338**) breaks down to an intermediate polycarbonyl derivative (**339**). The authors

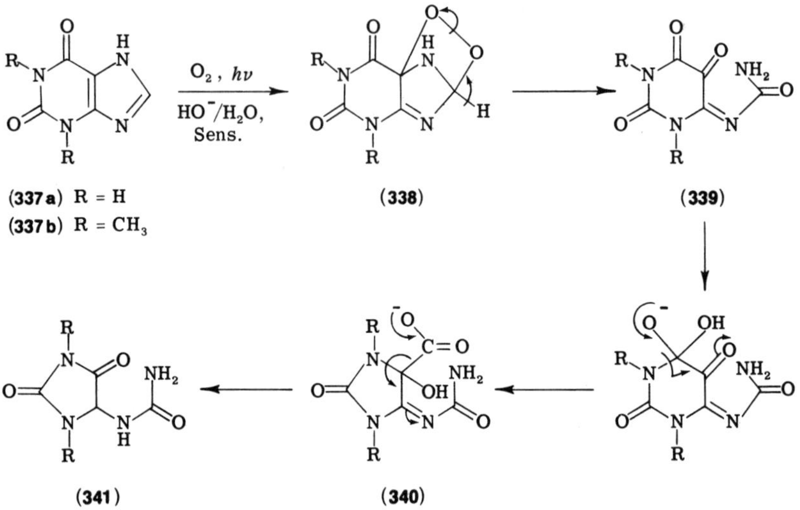

(**337a**) R = H
(**337b**) R = CH_3

O_2, $h\nu$
HO^-/H_2O,
Sens.

(**338**)

(**339**)

(**340**)

(**341**)

Scheme 11

O_2, $h\nu$
CH_3OH, Sens

(**342**) R = H, Me

CH_3OH

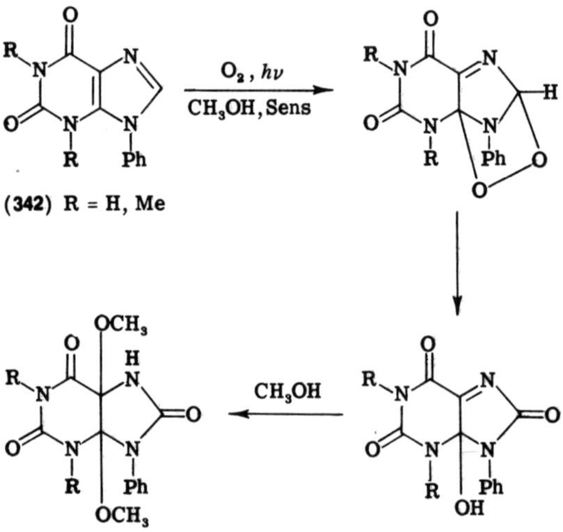

(**343**)

suggest that a benzilic acidlike rearrangement of **339** to **340** takes place, followed by decarboxylation to yield **341** (Matsuura and Saito, 1967; Simon and Van Vunakis, 1964).

In the case of 9-phenylxanthine (**342**), photooxidation in methanol yields the dimethoxy derivative (**343**) (Matsuura and Saito, 1969a) by a sequence similar to that discussed earlier in the imidazole series (H. H. Wasserman and M. S. Wolff, unpublished data, 1970).

A process by which 8-alkoxycaffeine and its alkyl derivatives (**344**) may yield the imidazolone (**350**) is shown in Scheme 12. Solvolysis of the initially formed endoperoxide (**345**) gives the dialkoxy hydroperoxide (**346**) which may undergo intramolecular transannular addition forming **347**. Fragmentation of **347** would then yield an isocyanate and **348**, the precursors of observed products **349** and **350** (Matsuura and Saito, 1969c).

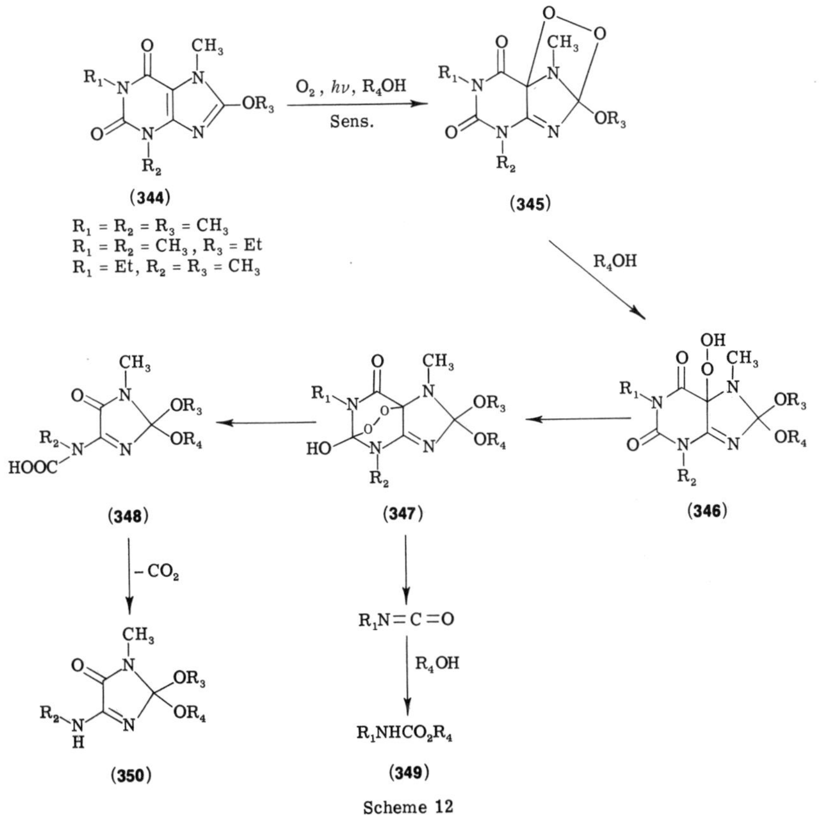

(**344**)

$R_1 = R_2 = R_3 = CH_3$
$R_1 = R_2 = CH_3, R_3 = Et$
$R_1 = Et, R_2 = R_3 = CH_3$

O_2, $h\nu$, R_4OH / Sens. → (**345**) → R_4OH → (**346**) → (**347**) → (**348**) → $-CO_2$ → (**350**)

(**347**) → $R_1N{=}C{=}O$ → R_4OH → $R_1NHCO_2R_4$ (**349**)

Scheme 12

To account for the products formed in methanol from tetramethyluric acid (**351**), Matsuura and Saito (1969b) suggest an initial enaminelike uptake of singlet oxygen to form the zwitterion (**352**) (Scheme 13). This species may react along two possible paths, yielding **353** (path *a*) or **354** (path *b*). In chloroform, a third mode of decomposition of the zwitterion (**352**) could lead to the spiro product (**355**) (path *c*).

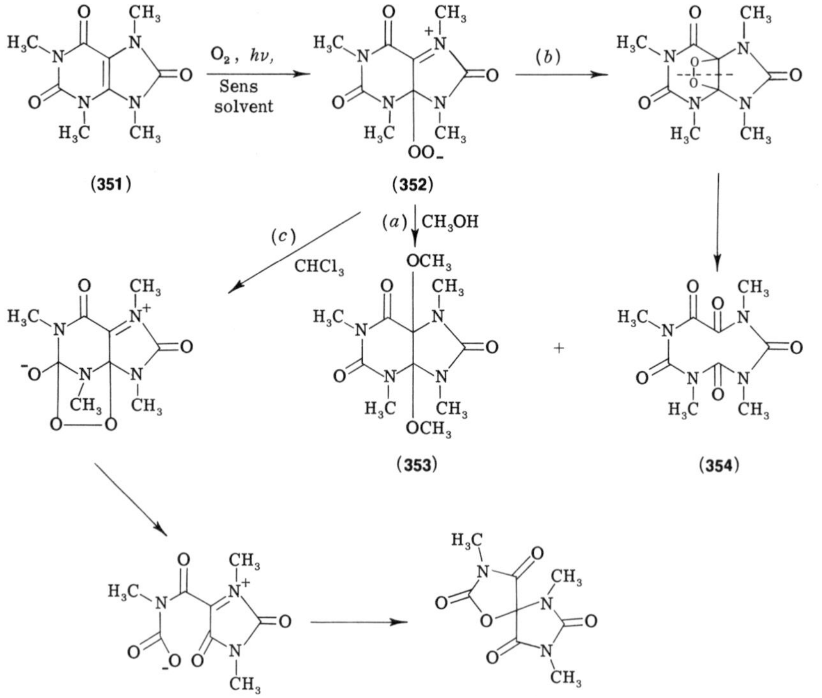

(351) —O_2, $h\nu$, Sens solvent→ (352) —(*b*)→ → (354)

(352) —(*a*) CH_3OH→ (353) + (354)

(352) —(*c*) $CHCl_3$→ → → (355)

Scheme 13

In light of the above work, it is interesting to examine the results of Sussenbach and Berends (1963, 1964, 1965) who showed that photooxidation of guanine (**356a**) in aqueous alkaline solution yields a complex mixture of products including guanidine (**358**), parabanic acid (**357**), and carbon dioxide, as well as hydrogen peroxide. Scheme 14 presents their results, and in addition, those of Sastry and Gordon (1966) who found that the photooxidation of guanosine (**356b**) afforded guanidine (**358**), urea, ribose, and ribosylurea.

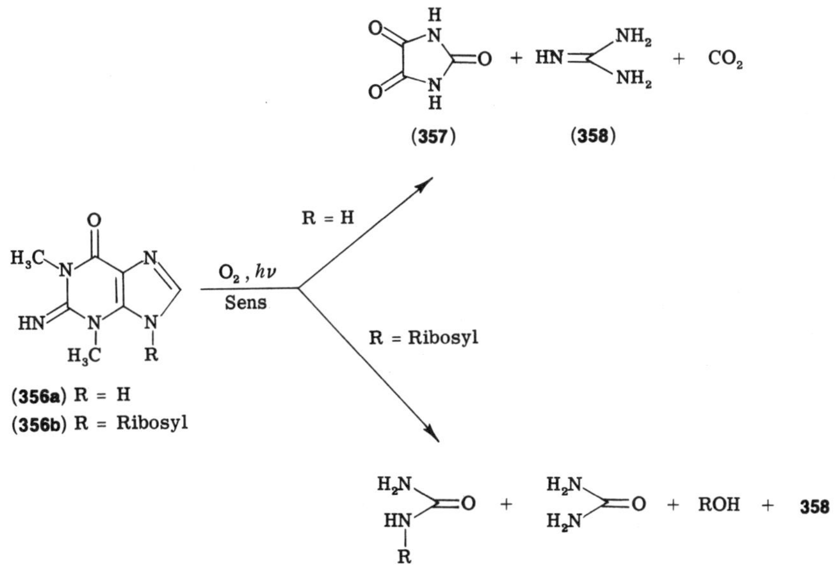

(357) (358)

O_2, $h\nu$
Sens

R = H

R = Ribosyl

(356a) R = H
(356b) R = Ribosyl

ROH + 358

Scheme 14

The formation of parabanic acid (**357**) is unusual in this series in that it appears to involve consumption of more than one equivalent of singlet oxygen per guanine molecule. One can account for the formation of all other products listed in Table VII on the basis of a reaction with only one equivalent of singlet oxygen. As pictured below in Scheme 15, the first equivalent of oxygen most probably reacts in the expected fashion with the imidazole part of guanine giving intermediate (**359**) which, in aqueous base. could lead to **360** and **361**. The mechanism of the conversion of **361** to parabanic acid is not immediately apparent. In basic medium **361** could undergo cleavage to form **362** which then reacts with a second molecule of $^1O_2$ to form **362a**. Decomposition of dioxetane (**362a**) as shown would yield parabanic acid (**357**) (Sussenbach and Berends, 1963). Alternatively, **361** could take up oxygen in base to form a trioxide (Wasserman and Van Verth, 1974) which could break down to form **357**.

Other evidence has been presented (Matsuura *et al.*, 1975; Kearns *et al.*, 1972; Knowles, 1971a,b; Knowles and Mautner, 1972) suggesting that photooxidation of guanosine and its derivatives may, indeed, be occurring by both a radical mechanism (Type I) as well as by the expected Type II singlet oxygen process.

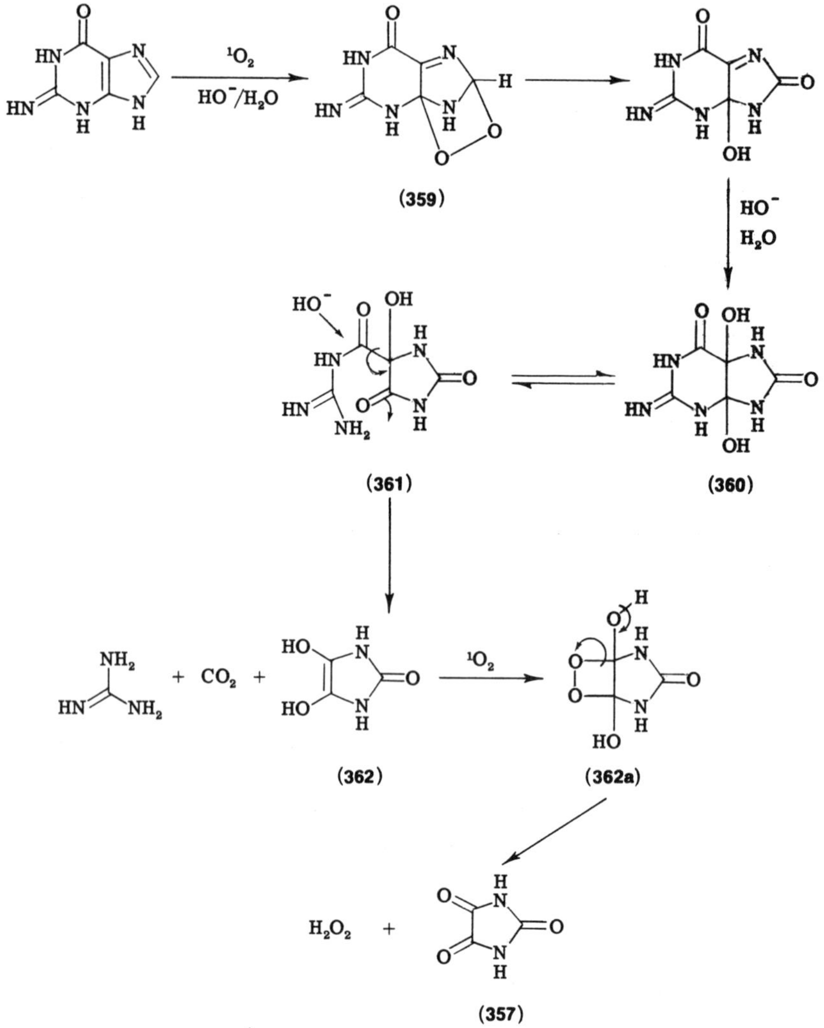

Scheme 15

VII. OXAZOLES

The relatively high reactivity of oxazoles toward singlet oxygen was first observed in studies with the natural product, pimprinine (**363**) (Wasserman and Floyd, 1966). Under oxygenation conditions in which the indole ring remained intact (methylene blue, methanol, 150 watt floodlamp) the oxazole residue in **363** suffered complete destruction, forming indole-3-

carboxylic acid (**364**). Similarly, 2-methyl-5-phenyloxazole (**365**) and 1,4-bis(5-phenyloxazole-2-yl)benzene (**367**) yielded benzoic (**366**) and terephthalic acids (**368**) as shown.

O_2, $h\nu$ / CH_3OH, methylene blue

(**363**) → (**364**)

$^1O_2$

(**365**) → (**366**)

$^1O_2$

(**367**) → (**368**)

A. Mechanism of Rearrangement to Triamides

More information on the nature of the oxazole–singlet oxygen reaction was obtained from the oxidations of the 2,4,5-trisubstituted derivatives (**369**) which gave triamides (**370**) in high yields (Wasserman and Floyd, 1966). This unusual reaction is pictured in terms of a three-step transformation shown below involving (*a*) 1,3-transannular peroxide formation, (*b*) Baeyer–

$^1O_2$

(**369a**) $R_1 = R_2 = R_3 = Ph$
(**369b**) $R_1 = R_2 = Ph$, $R_3 = CH_3$
(**369c**) $R_1 = R_3 = Ph$, $R_2 = CH_3$

(**370**)

Villiger-like rearrangement to the imino anhydride (**371**), and (*c*) an *O*-acyl to *N*-acyl transposition affording triamides (**370**).

369 —(*a*)→ [structure] —(*b*)→ (**371**) —(*c*)→ 370

Evidence supporting this sequence was obtained from photooxidation of oxazoles enriched with oxygen-18, and by the use of diphenylanthracene peroxide doubly labeled with oxygen-18 as the singlet oxygen source (Wasserman *et al*., 1972). Mass spectroscopic analysis of the triamide products gave results which were in accord with the sequence pictured above and which clearly ruled out the intermediacy of peroxidic intermediates such as the dioxetane (**372**). By the usual dioxetane cleavage, the latter could yield the isomeric imino anhydrides (**373**).

(**372**) → (**373**)

While the imino anhydride intermediates (**373**) were not isolated in the case of monocyclic oxazoles, these derivatives were formed as the main products when fused ring oxazoles (**374**) were oxidized by singlet oxygen (Wasserman and Lenz, 1976). Thus, reaction of $^1O_2$ with 2-phenylcyclohexeno-4,5-oxazole (**374**, $n = 1$) yielded **375** ($n = 1$) as the sole product. It appears that under the restrictions imposed by the geometry of the system (**375**, $n = 1$) the rearrangement to triamide (**375**→**376**) is inhibited. However, when the 4,5-polymethylene bridge is large enough ($n = 10$) the *O*-acyl to *N*-acyl type of migration may take place, and triamide (**376**) is formed (Wasserman and Druckrey, 1968).

(**374**) —$^1O_2$→ [structure] → (**375**)

(**376**)

B. Solvent Effects

As observed in the furan series, solvent effects may play an important role in governing the outcome of singlet oxygen reactions with oxazoles. Thus, during photooxygenation of 4,5-condensed oxazoles such as **377** in non-polar solvents (CH_2Cl_2) the intermediate endoperoxides (**378**) undergo fragmentation, affording cyano anhydrides (**379**) (Wasserman and Druckrey, 1968).

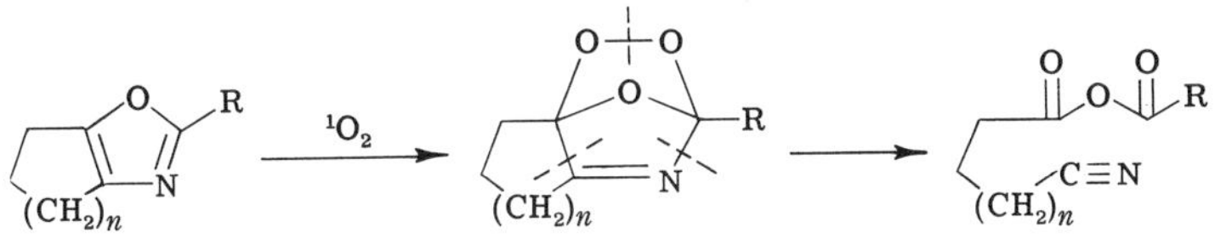

(**377**) $\xrightarrow{^1O_2}$ (**378**) ⟶ (**379**)

Formation of cyano anhydrides from intermediates (**378**) represents a novel type of peroxide breakdown not commonly found in singlet oxygen reactions. There does, however, exist an interesting analogy in a transformation recently observed during studies on the biogenesis of prostaglandins. When 8,11,14-eicosatrienoic acid is incubated with homogenates of the vesicular glands of sheep and the products, prostaglandins E_1 and F_1, removed from the reaction mixture, a monohydroxy acid fraction is recovered by silicic acid chromatography and identified as 12-hydroxy-8,10-heptadecadienoic acid (**381**). Both the double bonds were shown to have the trans configuration. It was suggested (Hamberg and Samuelsson, 1966) that this trans acid (**381** is formed from the same cyclic peroxide (**380**) which is postulated to be an intermediate in the biogenesis of the prostaglandins. The conversion of **380** to **381** has obvious features in common with the peroxide breakdown pictured in the generation of cyano anhydrides from oxazoles, i.e., **377** → **378** → **379**.

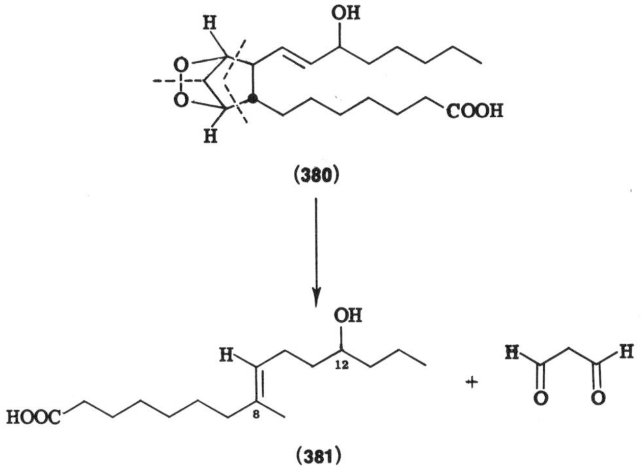

(**380**)

↓

(**381**) + OHC–CH₂–CHO

VIII. THIAZOLES

Thiazoles react with singlet oxygen in a manner similar to the oxidation of oxazoles. The photooxidation of 2,4,5-triphenylthiazole (**382**) in methanol yields benzil and benzamide, (path *a*) while in chloroform (path *b*), the thiotriamide (**384**) can be isolated (Matsuura and Saito, 1969d). The latter reaction appears to proceed via an endoperoxide which rearranges by a Baeyer–Villiger process to a thioimino anhydride (**383**), and then undergoes an *O*-acyl to *N*-acyl transposition. When the thiazole is fused to a six-membered ring as in **385**, methylene blue sensitized photooxidation takes a similar course. In this case, acyl migration does not take place because of steric constraints. Instead, the intermediate thioimino anhydride (**386**) undergoes tautomerization to the enamide (**387**) (Wasserman and Lenz, 1974).

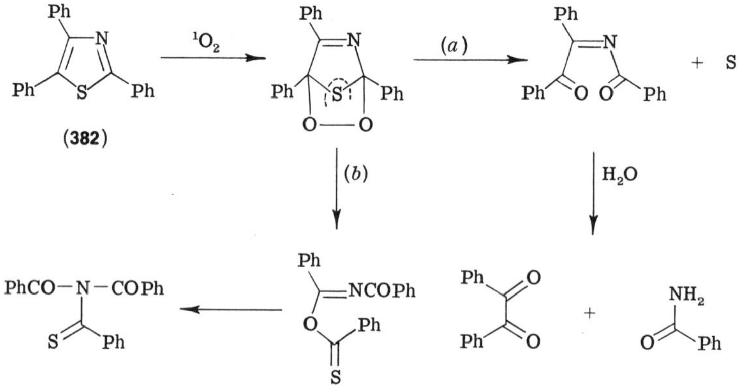

$^1O_2$ (*a*) + S (*b*) H_2O

PhCO—N—COPh

(382) **(383)** **(384)** **(385)** **(386)** **(387)**

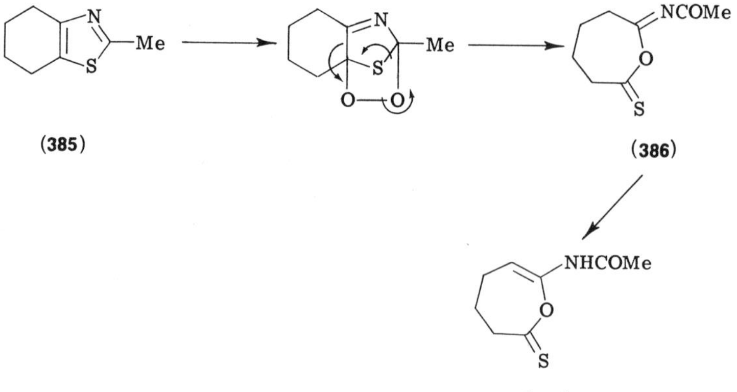

IX. THIOPHENES

Unlike the furans and pyrroles, thiophene and its tetraphenyl derivative are unreactive in dye-sensitized photooxygenations (Schenck and Krauch, 1962; Martel, 1957). When, however, thiophenes are substituted by alkyl groups, they do undergo attack by singlet oxygen (Skold and Schlessinger, 1970; Wasserman and Strehlow, 1970). Thus, it has been found that 2,5-dimethylthiophene (**388**) is readily photooxidized in methanol to the sulfine (**389**). In chloroform, the *trans*-diketone (**391**) is formed along with **389**. Skold and Schlessinger (1970) suggest that **391** results from an intermediate endoperoxide (**390**) which loses sulfur and undergoes cis to trans isomerization under the reaction conditions.

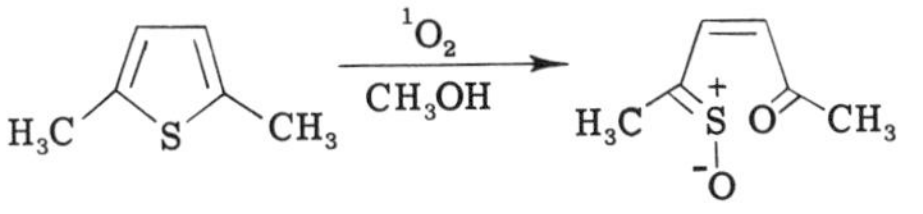

$^1O_2$ / CH_3OH

(**388**) (**389**)

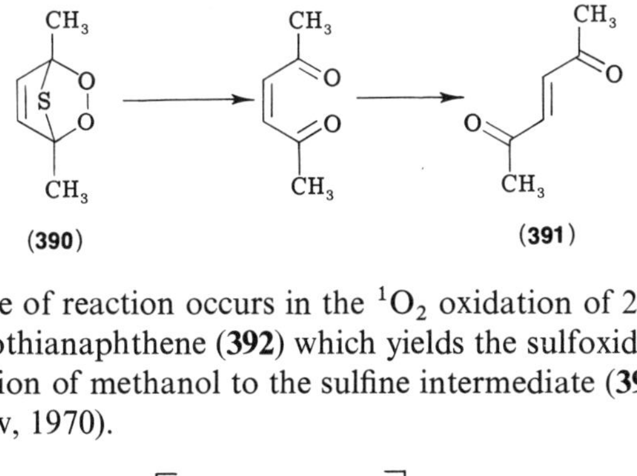

(**390**) (**391**)

A similar type of reaction occurs in the $^1O_2$ oxidation of 2,3-dimethyl-4,5,6,7-tetrahydrothianaphthene (**392**) which yields the sulfoxide (**394**), presumably by addition of methanol to the sulfine intermediate (**393**) (Wasserman and Strehlow, 1970).

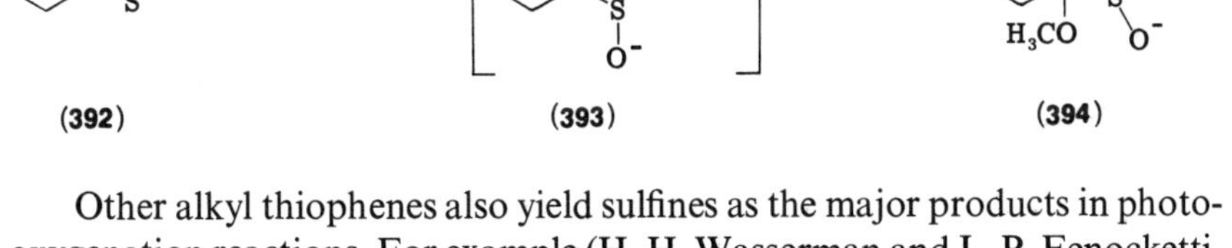

(**392**) (**393**) (**394**)

Other alkyl thiophenes also yield sulfines as the major products in photooxygenation reactions. For example (H. H. Wasserman and L. P. Fenocketti, unpublished results, 1977), 2,3,5-trimethylthiophene undergoes very rapid reactions with $^1O_2$ to afford a mixture of products corresponding to **395**.

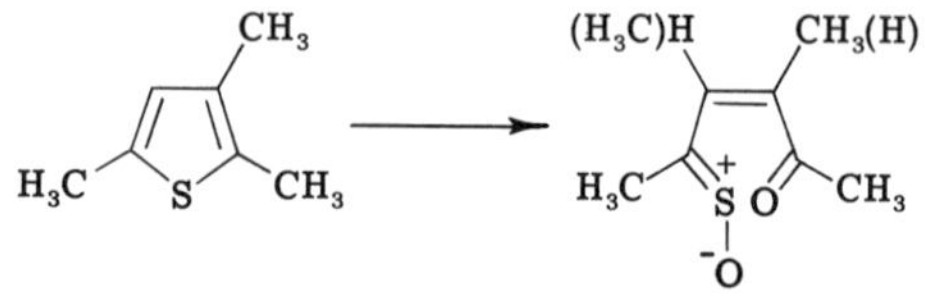

(395)

Various types of peroxide intermediates may be considered to be precursors of the sulfine (**389**), including the endoperoxide (**396**), the persulfoxide (**397**), the dioxetane (**398**), or zwitterionic peroxide (**399**). The evidence to date does not distinguish among these possibilities.

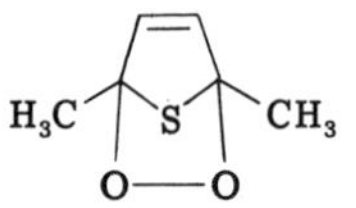

(396) 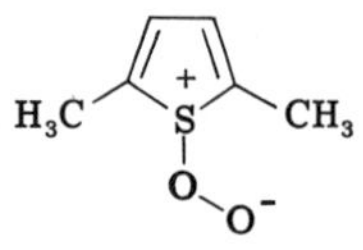 (397) 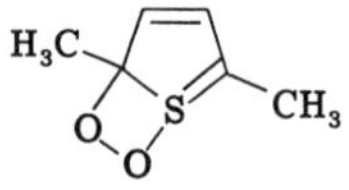 (398) 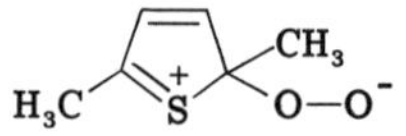 (399)

It is interesting to compare the possible formation of **389** from **400** with the related formation of the epoxy aldehyde (**402**) in the reaction of cyclopentadiene (**401**) with singlet oxygen (von Schulte-Elte *et al.*, 1969).

(396) → (400) → (397)

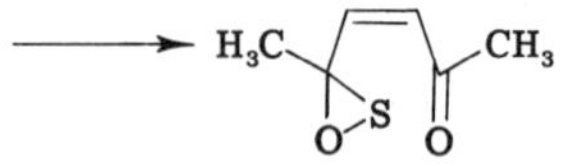

(401) (402)

Photooxidation of [2.2](2,5)thiophenophane (**403**) in chloroform–methanol consumes two molecules of oxygen per thiophenophane molecule, but no products have thus far been identified. If the oxidation is stopped after only one equivalent of oxygen is consumed and the reaction is worked up at ambient temperatures, the starting material is recovered quantitatively

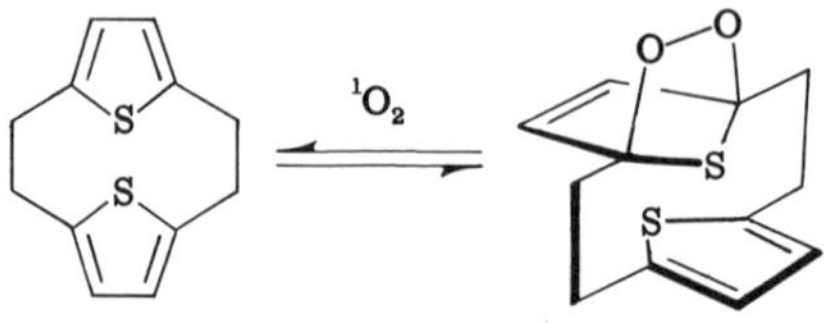

(403) (404)

(H. H. Wasserman and L. P. Fenocketti, unpublished results, 1975). These observations are in accord with the formation of a thermally unstable 2,5-endoperoxide (**404**) which reversibly releases oxygen.

Like 1,3-diphenylisobenzofuran, the corresponding thiophene (**405**) yields *o*-benzoylbenzophenone on photooxidation, presumably by loss of sulfur from the intermediate endoperoxide (**406**) (Theilacker and Schmidt, 1957).

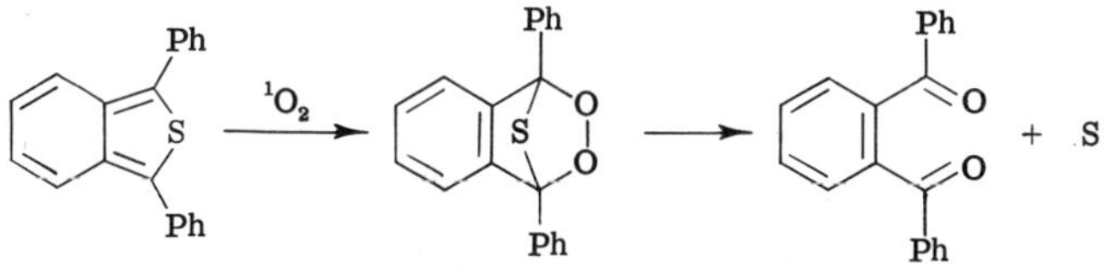

Ph Ph $^1O_2$ S O O + S

(**405**) (**406**)

In an interesting reaction related to the $^1O_2$–thiophene oxidation Hoffman and Schlessinger (1970) isolated the thioozonide (**408**) from the dye-sensitized photooxidation of 1,3,6,7-tetraphenylacenaphtho[5,6-*cd*]-thiopyran (**407**). On thermolysis, the ozonide (**408**) yielded the diketone (**409**) and the thioester (**410**).

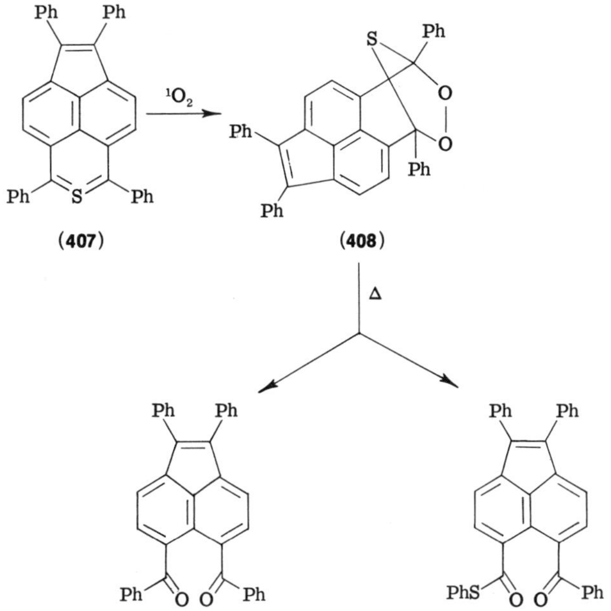

$^1O_2$ Δ Ph PhS

(**407**) (**408**)

(**409**) (**410**)

REFERENCES

Ando, W., Saiki, T., and Migita, T. (1975). *J. Am. Chem. Soc.* **97,** 5028–5029.

Baldwin, J. E., Chan, H. W.-S., and Swallow, J. C. (1971). *Chem. Commun.* pp. 1407–1408.

Basselier, J.-J., and LeRoux, J.-P. (1970a). *C.R. Hebd. Seances Acad. Sci., Ser. C* **270,** 1366–1369.

Basselier, J.-J., and LeRoux, J.-P. (1970b). *C.R. Hebd. Seances Acad. Sci., Ser C* **271,** 461–464.

Basselier, J.-J., Cherton, J.-C., and Caille, J. (1971). *C.R. Hebd. Seances Acad. Sci.* **273,** 514–517.

Bernheim, F., and Morgan, J. E. (1939). *Nature (London)* **144,** 290.

Birch, A. J., and Wright, J. J. (1970). *Tetrahedron* **26,** 2329–2344.

Bonnett, R., and Stewart, J. C. M. (1972a). *Chem. Commun.* pp. 596–597.

Bonnett, R., and Stewart, J. C. M. (1972b). *Biochem. J.* **130,** 895–897.

Bonnett, R., and Stewart, J. C. M. (1975). *J. Chem. Soc., Perkin Trans. 1* pp. 224–231.

Burnell, R. H., Chapelle, A., and Khalil, M. F. (1974). *Can. J. Chem.* **52,** 2327–2330.

Caporale, G., Rodrighiero, G., Giacomelli, C., and Ballotta, C. (1965). *Gazz. Chim. Ital.* **95,** 513–532.

Cauzzo, G., and Jori, G. (1972). *J. Org. Chem.* **37,** 1429–1433.

Chan, H. W.-S. (1970). *Chem. Commun.* pp. 1550–1551.

Chan, H. W.-S. (1971). *J. Am. Chem. Soc.* **93,** 2357–2358.

Ciamician, G., and Silber, P. (1912). *Ber. Dtsch. Chem. Ges.* **45,** 1842–1845.

Clagett, D. C., and Galen, T. J. (1971). *Arch. Biochem. Biophys.* **146,** 196–201.

Clagett, D. C., Kornhauser, A., Krinsky, N. I., and Huang, P.-K. C. (1973). *Photochem. Photobiol.* **18,** 63–69.

Cremer, R. J., Perryman, P. W., and Richards, D. H. (1958). *Lancet* **1,** 1094–1097.

de Mayo, P., and Reid, S. T. (1962). *Chem. Ind. (London)* pp. 1576–1577.

Dewar, M. J. S., and Thiel, W. (1975). *J. Am. Chem. Soc.* **97,** 3978–3986.

Dufraisse, C., and Ecary, S. (1946). *C.R. Hebd. Seances Acad. Sci.* **223,** 735–73 .

Dufraisse, C., Etienne, A., and Martel, J. (1957). *C.R. Hebd. Seances Acad. Sci.* **244,** 970–974.

Dufraisse, C., Rio, G., Ranjon, A., and Pouchot, O. (1965). *C.R. Hebd. Seances Acad. Sci.* **261,** 3133–3136.

Dufraisse, C., Rio, G., and Ranjon, A. (1967). *C.R. Hebd. Seances Acad. Sci. Ser. C* **264,** 516–519.

Eskins, K. (1972). *Photochem. Photobiol.* **15,** 247–252.

Evans, N. A. (1971). *Aust. J. Chem.* **24,** 1971–1973.

Fontana, A., and Toniolo, C. (1976). *In* "Progress in the Chemistry of Natural Products" (W. Herz, H. Grisebach, and G. W. Kirby, eds.), Vol. 33, p. 309. Springer-Verlag, Berlin and New York.

Foote, C. S., and Ching, T.-Y. (1975). *J. Am. Chem. Soc.* **97,** 6209–6214.

Foote, C. S., Wuesthoff, M. T., Wexler, S., Burstain, I. G., Denny, R., Schenck, G. O., and Schulte-Elte, K.-J. (1967). *Tetrahedron* **23,** 2583–2599.

Foote, C. S., Dzakpasu, A. A., and Lin, J. W.-P. (1975). *Tetrahedron Lett.* pp. 1247–1250.

Franck, R. W., and Auerbach, J. (1971). *J. Org. Chem.* **36,** 31–36.

Hallet, F. R., Hallet, B. P., and Snipes, W. (1970). *Biophys. J.* **10,** 305–315.

Hamberg, M., and Samuelsson, B. (1966). *J. Am. Chem. Soc.* **88,** 2349–2350.

Heather, J. B., Mittal, R. S. D., and Sih, C. J. (1974). *J. Am. Chem. Soc.* **96,** 1976–1977.

Hiatt, R. (1971). *Intra-Sci. Chem. Rep.* **5,** 163–182.

Hoffman, J. M., Jr., and Schlessinger, R. H. (1970). *Tetrahedron Lett.* pp. 797–800.

Hoft, E., Katritzky, A. R., and Nesbit, M. R. (1967). *Tetrahedron Lett.* pp. 3041–3044.

Hoft, E., Katritzky, A. R., and Nesbit, M. R. (1968). *Tetrahedron Lett.* p. 2028.

Inhoffen, H. H. (1968). *Pure Appl. Chem.* **17,** 443–460.

Irie, M., Tomita, M., and Ukita, T. (1968). *Tetrahedron Lett.* pp. 4933–4936.
Jefford, C. W., Boschung, A. F., Bolsman, T. A. B. M., Moriarty, R. M., and Melnick, B. (1976). *J. Am. Chem. Soc.* **98,** 1017–1018.
Jefford, C. W., and Rimbault, C. G. (1978). *J. Am. Chem. Soc.* **97,** 295–296.
Kametani, T., Ohsawa, T., Ihara, M., and Fukumoto, K. *J.C.S. Perkin 1*, in press.
Kanazawa, R., Kotsuki, H., and Tokoroyama, T. (1975). *Tetrahedron Lett.* pp. 3651–3654.
Katz, T. J., Balogh, V., and Schulman, J. (1968). *J. Am. Chem. Soc.* **90,** 734–739.
Kearns, D. A., Nilsson, R., and Merkel, P. B. (1972). *Photochem. Photobiol.* **16,** 117–124.
Knowles, A. (1971a). *Photochem. Photobiol.* **13,** 225–236.
Knowles, A. (1971b). *Photochem. Photobiol.* **13,** 473–487.
Knowles, A., and Mautner, G. N. (1972). *Photochem. Photobiol.* **15,** 199–207.
Krauch, C. H., Farid, S., Draft, S., and Wacker, A. (1965). *Biophysik* **2,** 301–302.
Leete, E. (1961). *J. Am. Chem. Soc.* **83,** 3645–3647.
Leete, E., and Chen, F. Y.-H. (1963). *J. Am. Chem. Soc.* **83,** 2013–
Lightner, D. A. (1974). *In* "Phototherapy in the Newborn: An Overview" (G. B. Odell, R. Schaffer, and A. P. Simopoulos, eds.), pp. 34–55. Natl. Acad. Sci., Washington, D.C.
Lightner, D. A., and Crandall, D. C. (1973a). *Chem. Ind.* (*London*) pp. 638–639.
Lightner, D. A., and Crandall, D. C. (1973b). *Tetrahedron Lett.* pp. 953–956.
Lightner, D. A., and Crandall, D. C. (1973c). *Experientia* **29,** 262–264.
Lightner, D. A., and Low, L. K. (1972a). *Chem. Commun.* pp. 625–626.
Lightner, D. A., and Low, L. K. (1972b). *J. Heterocycl. Chem.* **9,** 167–168.
Lightner, D. A., and Norris, R. D. (1974). *N. Engl. J. Med.* **290,** 1260.
Lightner, D. A., and Pak, C.-S. (1975). *J. Org. Chem.* **40,** 2726–2728.
Lightner, D. A., and Quistad, G. B. (1972a). *Nature* (*London*) *New Biol.* **236,** 203–205.
Lightner, D. A., and Quistad, G. B. (1972b). *Angew. Chem., Int. Ed. Engl.* **11,** 215–216.
Lightner, D. A., and Quistad, G. B. (1972c). *FEBS Lett.* **25,** 94–96.
Lightner, D. A., and Quistad, G. B. (1973). *J. Heterocycl. Chem.* **10,** 273–274.
Lightner, D. A., Crandall, D. C., Gertler, S., and Quistad, G. B. (1973). *FEBS Lett.* **30,** 309–312.
Lightner, D. A., Kirk, D. I., and Norris, R. D. (1974). *J. Heterocycl. Chem.* **11,** 1097–1098.
Lightner, D. A., Bisacchi, G. S., and Norris, R. D. (1976). *J. Am. Chem. Soc.* **98,** 802–807.
Low, L. K., and Lightner, D. A. (1972). *Chem. Commun.* pp. 116–117.
Lutz, R. E., Welstead, W. J., Jr., Bass, R. G., and Dak, J. I. (1962). *J. Org. Chem.* **27,** 1111–1112.
McDonagh, A. F. (1971). *Biochem. Biophys. Res. Commun.* **44,** 1306–1311.
McLean, S., and Dmitrienko, G. I. (1971). *Can. J. Chem.* **49,** 3642–3647.
Maeda, K., Mishima, T., and Hayashi, T. (1974). *Bull. Chem. Soc. Jpn.* **47,** 334–338.
Martel, J. (1957). *C.R. Hebd. Seances Acad. Sci.* **244,** 626–629.
Matsumoto, M., and Kondo, K. (1977). *J. Am. Chem. Soc.* **99,** 2393–2394.
Matsumoto, M., Dobashi, S., and Kondo, K. (1977). *Bull. Chem. Soc. Jpn.* **50,** 3026–3028.
Matsuura, T., and Ikari, M. (1969). *Kogyo Kagaku Zasshi* **72,** 179–183.
Matsuura, T., and Saito, I. (1967). *Chem. Commun.* pp. 693–694.
Matsuura, T., and Saito, I. (1968a). *Tetrahedron* **24,** 6609–6614.
Matsuura, T., and Saito, I. (1968b). *Tetrahedron Lett.* pp. 3273–3276.
Matsuura, T., and Saito, I. (1969a). *Tetrahedron* **25,** 541–547.
Matsuura, T., and Saito, I. (1969b). *Tetrahedron* **25,** 549–556.
Matsuura, T., and Saito, I. (1969c). *Tetrahedron* **25,** 557–564.
Matsuura, T., and Saito, I. (1969d). *Bull. Chem. Soc. Jpn.* **42,** 2973–2975.
Matsuura, T., Saito, I., and Kato, S. (1972). *Jerusalem Symp. Quantum Chem. Biochem.* **4,** 418–430.
Matsuura, T., Saito, I., and Inoue, K. (1975). *Photochem. Photobiol.* **21,** 27–30.
Mendenhall, G. D., and Howard, J. A. (1975). *Can. J. Chem.* **53,** 2199–2201.

Metzger, W., and Fischer, H. (1936). *Justus Liebigs Ann. Chem.* **527,** 1–37.
Meyers, A. I., Nolen, R. L., Collington, E. W., Narwid, T. A., and Strickland, R. C. (1973). *J. Org. Chem.* **38,** 1974–1982.
Murray, R. W., and Higley, D. P. (1974). *J. Am. Chem. Soc.* **96,** 3330–3332.
Musajo, L., and Rodighiero, G. (1962). *Experientia* **18,** 153–161.
Nakagawa, M., Kaneko, T., Yoshikawa, K., and Hino, T. (1974). *J. Am. Chem. Soc.* **96,** 624–625.
Nakagawa, M., Okajima, Y., Kobayashi, K., Asaka, T., and Hino, T. (1975a). *Heterocycles* **3,** 799–803.
Nakagawa, M., Yoshikawa, K., and Hino, T. (1975b). *J. Am. Chem. Soc.* **97,** 6496–6501.
Nakagawa, M., Ohyoshi, N., and Hino, T. (1976a). *Heterocycles* **4,** 1275–1280.
Nakagawa, M., Okajima, H., and Hino, T. (1976b). *J. Am. Chem. Soc.* **98,** 635–637.
Nakatani, M. (1960a). *J. Biochem.* (*Tokyo*) **48,** 633–639.
Nakatani, M. (1960b). *J. Biochem.* (*Tokyo*) **48,** 640–644.
Nickon, A., and Mendelson, W. L. (1963). *J. Am. Chem. Soc.* **85,** 1894–1895.
Nishinaga, A., and Rieker, A. (1976). *J. Am. Chem. Soc.* **98,** 4667–4668.
Odell, G. B., Brown, R. S., and Kopelman, A. E. (1972). *J. Pediatr.* **81,** 472–482.
Ostrow, J. D., and Branham, R. V. (1970). *Gastroenterology* **58,** 15–25.
Quistad, G. B., and Lightner, D. A. (1971a). *Chem. Commun.* pp. 1099–1100.
Quistad, G. B., and Lightner, D. A. (1971b). *Tetrahedron Lett.* pp. 4417–4420.
Radziszewski, B. (1877). *Ber. Dtsch. Chem. Ges.* **10,** 70–75.
Ramasseul, R., and Rassat, A., (1972). *Tetrahedron Lett.* pp. 1337–1340.
Ranjon, A. (1971). *Bull. Soc. Chim. Fr.* pp. 2068–2072.
Ray, W. J., Jr., and Koshland, D. E., Jr. (1962). *J. Biol. Chem.* **237,** 2493–2505.
Richardson, W. H., and Hodge, V. F. (1970). *Tetrahedron Lett.* pp. 2271–2274.
Richardson, W. H., and Hodge, V. F. (1971a). *Tetrahedron Lett.* pp. 749–751.
Richardson, W. H., and Hodge, V. F. (1971b). *J. Am. Chem. Soc.* **93,** 3996–4004.
Rio, G., and Lecas-Nawrocka, A. (1971). *Bull. Soc. Chim. Fr.* pp. 1723–1727.
Rio, G., and Masure, M. (1972). *Bull. Soc. Chim. Fr.* pp. 4610–4619.
Rio, G., Ranjon, A., and Pouchot, O. (1966). *C.R. Hebd. Seances Acad. Sci., Ser. C* **263,** 634–636.
Rio, G., Ranjon, A., Pouchot, O., and Scholl, M. J. (1969). *Bull. Soc. Chim. Fr.* pp. 1667–1673.
Rokach, J., McNeill, D., and Rooney, C. S. (1971). *Chem. Commun.* pp. 1085–1086.
Rosenthal, I., and Pitts, J. N., Jr. (1971). *Biophys. J.* **11,** 963–966.
Runquist, O., and Boche, J. (1968). *J. Org. Chem.* **33,** 4285–4286.
Safe, S., and Taylor, A. (1972). *J. Chem. Soc., Perkin Trans. 1* pp. 472–479.
Saito, I., Imuta, M., and Matsuura, T. (1972a). *Chem. Lett.* pp. 1173–1176.
Saito, I., Imuta, M., and Matsuura, T. (1972b). *Chem. Lett.* pp. 1197–1200.
Saito, I., Imuta, M., Matsugo, S., and Matsuura, T. (1975). *J. Am. Chem. Soc.* **97,** 7191–7193.
Saito, I., Takahashi, Y., Imuta, M., Matsugo, S., Kaguchi, H., and Matsuura, T. (1976). *Heterocycles* **5,** 53–58.
Saito, I., Imuta, M., Takahashi, Y., Matsugo, S., and Matsuura, T. (1977a). *J. Am. Chem. Soc.* **99,** 2005–2006.
Saito, I., Matsuura, T., Nakagawa, M., and Hino, T. (1977b). *Acc. Chem. Res.* **10,** 346–352.
Sastry, K. S., and Gordon, M. P. (1966). *Biochim. Biophys. Acta* **129,** 42–48.
Savige, E. W. (1971). *Aust. J. Chem.* **24,** 1285–1293.
Schenck, G. O. (1947). *Chem. Ber.* **80,** 289–297.
Schenck, G. O. (1948). *Angew. Chem.* **60,** 244–245.
Schenck, G. O. (1952). *Angew. Chem.* **64,** 12–23.
Schenck, G. O. (1953). *Justus Liebigs Ann. Chem.* **584,** 156–176.

Schenck, G. O., and Appel, R. (1946). *Naturwissenschaften* **33,** 122–123.
Schenck, G. O., and Koch, E. (1966). *Chem. Ber.* **99,** 1984–1990.
Schenck, G. O., and Krauch, C. H. (1962). *Angew. Chem.* **74,** 510.
Schmid, R. (1971). *N. Engl. J. Med.* **285,** 520–522.
Simon, M. I., and Van Vunakis, H. (1962). *J. Mol. Biol.* **4,** 488–499.
Simon, M. I., and Van Vunakis, H. (1964). *Arch. Biochem. Biophys.* **105,** 197–206.
Skold, C. N., and Schlessinger, R. H. (1970). *Tetrahedron Lett.* pp. 791–794.
Sonnenberg, J., and White, D. M. (1964). *J. Am. Chem. Soc.* **86,** 5685–5686.
Stevens, M. P., Nahavandi, F., and Razmara, F. (1973). *Tetrahedron Lett.* pp. 301–304.
Stiller, K. (1968). Ph.D. Dissertation, Yale University, New Haven, Connecticut.
Sundberg, R. J. (1970). *Org. Chem., Ser. Monogr.* **18,** 282.
Sussenbach, J. S., and Berends, W. (1963). *Biochim. Biophys. Acta* **76,** 154–156.
Sussenbach, J. S., and Berends, W. (1964). *Biochem. Biophys. Res. Commun.* **16,** 263–266.
Sussenbach, J. S., and Berends, W. (1965). *Biochim. Biophys. Acta* **95,** 184–185.
Taylor, W. I. (1962). *Proc. Chem. Soc., London* p. 247.
Theilacker, W., and Schmidt, W. (1957). *Justus Liebigs Ann. Chem.* **605,** 43–49.
Trozzolo, A. M., and Fahrenholz, S. R. (1970). *Ann. N.Y. Acad. Sci.* **171,** 61–66.
von Schulte-Elte, K. H., Willhalm, B., and Ohloff, G. (1969). *Angew. Chem.* **8,** 985–986.
Wacker, A., Turck, G., and Gerstenberger, A. (1963). *Naturwissenschaften* **50,** 377.
Wasserman, H. H., and Doumaux, A. R. (1962). *J. Am. Chem. Soc.* **84,** 4611–4612.
Wasserman, H. H., and Druckrey, E. (1968). *J. Am. Chem. Soc.* **90,** 2440–2441.
Wasserman, H. H., and Floyd, M. B. (1963). *Tetrahedron Lett.* pp. 2009–2012.
Wasserman, H. H., and Floyd, M. B. (1966). *Tetrahedron, Suppl.* **7,** 441–448.
Wasserman, H. H., and Kitzing, R. (1969). *Tetrahedron Lett.* pp. 5315–5318.
Wasserman, H. H., and Lenz, G. R. (1974). *Tetrahedron Lett.* pp. 3947–3950.
Wasserman, H. H., and Lenz, G. (1976). *Heterocycles* **5,** 409–412.
Wasserman, H. H., and Liberles, A. (1960). *J. Am. Chem. Soc.* **82,** 2086.
Wasserman, H. H., and Miller, A. H. (1969). *Chem. Commun.* pp. 199–200.
Wasserman, H. H., and Saito, I. (1975). *J. Am. Chem. Soc.* **97,** 905–906.
Wasserman, H. H., and Strehlow, W. (1970). *Tetrahedron Lett.* pp. 795–796.
Wasserman, H. H., and Terao, S. (1975). *Tetrahedron Lett.* pp. 1735–1738.
Wasserman, H. H., and Van Verth, J. E. (1974). *J. Am. Chem. Soc.* **96,** 585–586.
Wasserman, H. H., Stiller, K., and Floyd, M. B. (1968). *Tetrahedron Lett.* pp. 3277–3280.
Wasserman, H. H., Vinick, F. J., and Chang, Y. C. (1972). *J. Am. Chem. Soc.* **94,** 7180–7182.
Weil, L., and Seibles, T. S. (1955). *Arch. Biochem. Biophys.* **54,** 368–377.
White, E. H., and Harding, M. J. C. (1964). *J. Am. Chem. Soc.* **86,** 5686–5687.
Young, R. H., Martin, R. L., Chinh, N., Mallon, C., and Kayser, R. H. (1972). *Can. J. Chem.* **50,** 932–938.

10

The Oxidations of Electron-Rich Aromatic Compounds

ISAO SAITO and TERUO MATSUURA

I. Oxidation of Polycyclic Aromatic Hydrocarbons 511
A. Kinetics 512
B. Reactivity and Regiospecificity 514
C. Oxidation of Naphthalenes, Phenanthrenes, and Other Acenes . . 519
D. Chemical Behavior of Acene Endoperoxides 524
II. Oxidations of Monocyclic Aromatic Compounds 531
A. Electron Donor-Substituted Benzenes 531
B. 1,4-Cycloaddition to Styrene-Type Olefins 539
C. 1,4-Cycloaddition to Carbocyclic Nonbenzenoid Aromatic Compounds 541
III. Oxidations of Phenolic and Enolic Systems 551
A. Mechanisms for Photosensitized Oxygenation of Phenols 552
B. Oxidations of Phenols 557
C. Oxidations of Enolic Systems 563
References 569

I. OXIDATION OF POLYCYCLIC AROMATIC HYDROCARBONS

Singlet oxygen ($^1\Delta_g$) adds to conjugated diene systems such as cisoid 1,3-dienes or aromatic hydrocarbons yielding 1,4-endoperoxides. The reaction is thought to involve a concerted 1,4-cycloaddition analogous to the Diels–Alder reaction. A concerted combination of singlet oxygen and 1,3-dienes is symmetry allowed and is presumed to occur by way of a six-membered ring transition state (Kearns, 1971; Khan and Kearns, 1968). In general, the 1,4-cycloaddition proceeds with low activation energies. Electron-donating groups increase the reactivity of 1,3-dienes and aromatic

Singlet Oxygen

ISBN 0-12-736650-4

hydrocarbons, and, conversely, electron-withdrawing groups decrease their reactivity.

1,4-Cycloaddition of oxygen to polycyclic aromatic systems are largely achieved by a self-sensitized (direct) photooxygenation reaction. Earlier work in this area has been carried out chiefly in the laboratory of Dufraisse. Since the discovery of rubrene photobleaching (Moureu *et al.*, 1926), a large number of polycyclic aromatic systems such as naphthalenes, anthracenes, pentacenes, hexacenes, and azaanthracenes have been investigated. Since extensive reviews have been published (Gollnick and Schenck, 1967; Gollnick, 1968b; Rigaudy, 1968; Denny and Nickon, 1973), this section will be limited to the general features of the reaction, such as kinetics, reactivities of acenes, regiospecificity of singlet oxygen addition, and particular chemical properties of endoperoxides.

A. Kinetics

Most of the condensed aromatic hydrocarbons so far investigated are susceptible to photooxygenation without sensitizer and the oxidation proceeds via self-sensitized mechanisms involving singlet oxygen to form 1,4-endoperoxides. In fact, Gollnick demonstrated that a number of aromatic hydrocarbons can function as sensitizers comparable to dye sensitizers in the photooxygenation of certain singlet oxygen acceptors such as 2,5-dimethylfuran and tetramethylethylene (Gollnick, 1968a,b; Gollnick *et al.*, 1970). There is, however, a characteristic difference between dye and aromatic hydrocarbon sensitizers. As suggested by Bowen (1963) on the basis of the fluorescence quenching of anthracenes by oxygen, both singlet and triplet excited states of the aromatic hydrocarbons can take part in singlet oxygen formation, although the appreciable quenching is not always observable.

The kinetics of the photoperoxidation sensitized by polycyclic aromatic hydrocarbons developed by Stevens (1973), consider four possible interactions between the excited sensitizers ($^1S^*$ and $^3S^*$) and the ground state oxygen ($^3O_2$) shown in Eqs. (1)–(4), where the sensitizer S–T splitting $\Delta E_{ST} > 8000\ cm^{-1}$ for process (1) and the triplet energy $E_T > 8000\ cm^{-1}$ for process (2). The overall quantum yield of singlet-oxygen formation (Φ_Δ, experimentally available)

$$^1S^* + {}^3O_2 \xrightarrow{k_1} {}^3S^* + O_2(^1\Delta_g) \quad (1)$$

$$^3S^* + {}^3O_2 \xrightarrow{k_2} S + O_2(^1\Delta_g) \quad (2)$$

$$^1S^* + {}^3O_2 \xrightarrow{k_3} {}^3S^* + {}^3O_2 \quad (3)$$

$$^1S^* + {}^3O_2 \xrightarrow{k_4} S + O_2(^1\Delta_g) \quad (4)$$

is given by Eq. (5): Φ_{ST} is the quantum yield of the intersystem crossing; $P(O_2)$, the probability that a sensitizer singlet state is quenched by oxygen; α, the probability that oxygen quenching of the singlet state generates O_2 ($^1\Delta_g$); δ, the probability that oxygen quenching of the singlet state produces the sensitizer triplet state; and ε, the probability that oxygen quenching of the triplet state generates O_2 ($^1\Delta_g$).

$$\Phi_\Delta = \varepsilon\Phi_{ST} + P(O_2)[\alpha + \varepsilon(\delta - \Phi_{ST})] \tag{5}$$

Stevens and Algar (1968, 1969) demonstrated that Φ_Δ is linearly dependent on $P(O_2)$, independently available from oxygen quenching of sensitizer fluorescence (Fig. 1). The data lines in Fig. 1 extrapolate to a

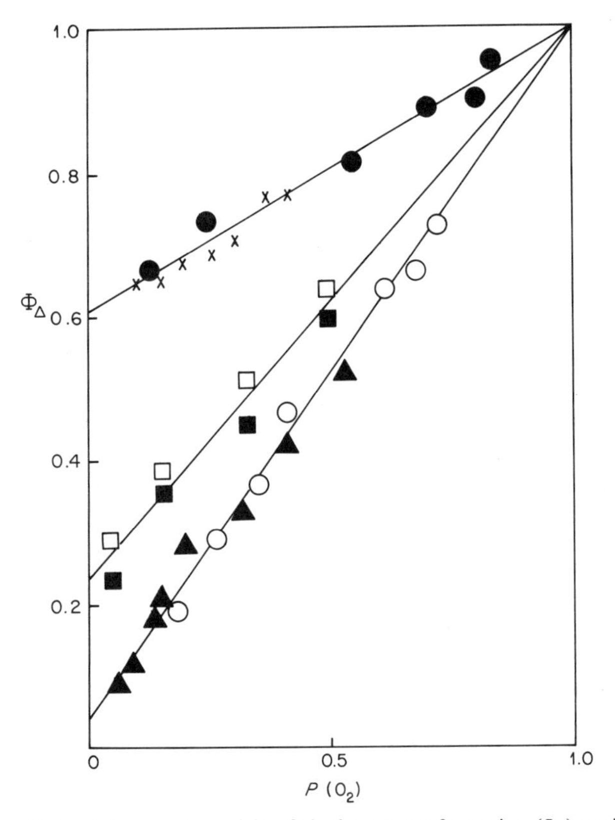

Fig. 1. Plots of the quantum yield of singlet oxygen formation (Φ_Δ) against the probability [$P(O_2)$] that a sensitizer singlet state is quenched by oxygen. (●) 9,10-Dimethyl-1,2-benzanthracene; (×) naphthacene; (□) anthanthrene with 9,10-dimethylanthracene acceptor; (■) anthanthrene with 9,10-dimethyl-1,2-benzanthrecene acceptor; (▲) rubrene; (○) 9,10-dimethylanthracene. Solvent, benzene at 25°C (Stevens, 1973).

limiting Φ_Δ of unity when $P(O_2) = 1$. It means $\alpha + \delta = 1$ with $\varepsilon = 1$, eliminating process (1). Quenching of the unsensitized photooxygenation of 9,10-dimethylanthracene by azulene, which quenches both singlet and triplet states, gives $\alpha \ll \delta > 0$, $k_3 \gg k_4$, and the value $k = 3.3 \pm 0.5 \times 10^9\ M^{-1}\ \text{sec}^{-1}$, which is lower than the diffusion-controlled rate constant $k_3 = 3.15 \pm 0.20 \times 10^{10}\ M^{-1}\ \text{sec}^{-1}$ (Algar and Stevens, 1970). The value of δ is independently found as 1.0 ± 0.2 for several aromatic hydrocarbons (Potashnik *et al.*, 1971). Thus, both excited singlet and triplet states of the sensitizer participate, as shown in Eqs. (2) and (3), in singlet oxygen formation by aromatic hydrocarbon sensitization.

The kinetics of the photooxygenation of anthracenes in the solid state have also been investigated as models for the environment in photobiological systems. Irradiation of anthracene in polymer films in the presence of oxygen leads to the formation of the photoperoxide predominantly over that of the anthracene photodimer (Cowell and Pitts, 1968). Sensitized photooxygenation of 9,10-diphenylanthracene, which was separated from the sensitizer by metal stearate films, has shown that singlet-oxygen reaction can be observed for a thickness up to 500 Å, and that half of singlet oxygen molecules are deactivated after a diffusion path of 115 Å (Schnuriger and Bourdon, 1968). Khan and Kasha (1970) have proposed for photocarcinogenesis with polycyclic aromatic compounds that the determining factor is whether the chromophoric residue remaining after bonding of such a molecule to a cell site [for example, Eq. (6)] can absorb light and sensitize the formation of singlet oxygen with reasonable efficiency.

binding site

chromophoric residue

(6)

B. Reactivity and Regiospecificity

Reactivity indices ($\beta = k_r/k_d$) (Foote, 1968a), rate constants (k_r), and activation energies ($E_F^\ddagger$) relative to rubrene [Eq. (7)], have been measured by Stevens *et al.* (1974b) for singlet oxygen addition to various aromatic hydrocarbons. They have estimated β values by a direct method for reactive substrates (Stevens and Algar, 1968) and by competitive methods or a "static" competitive technique for less reactive substrates (Wilson, 1966), and calculated k_r and $E_F^\ddagger$ values according to Eqs. (8) and (9), respectively, assuming the reported values $\tau = 4.2 \times 10^7\ M^{-1}\ \text{sec}^{-1}$ (Merkel and Kearns,

1972) and $\Delta S = -15.6 \pm 1.0$ eu (Koch, 1968). The results are shown in Table I. The β values are influenced by solvent effect;

$$\begin{aligned} M + {}^1O_2 &\xrightarrow{k_r} MO_2 \\ {}^1O_2 &\xrightarrow{k_d} {}^3O_2 \end{aligned} \tag{7}$$

$$k_r = 1/\beta\tau \tag{8}$$

$$k_r = (ekT/h)\exp(\Delta S/R)\exp(-E_F^{\ddagger}/RT) \tag{9}$$

for example, the β values for anthracene are 0.002 (CS_2), 0.013 ($CHCl_3$), and 0.11 (PhBr) (Denny and Nickon, 1973). The efficiency in carbon disulfide is attributed largely to its ability to prolong the lifetime of singlet oxygen (Merkel and Kearns, 1972). The results in Table I indicate that the reactivity increases with the substitution of electron-donating groups at the sites of oxygen addition in the order $H < \text{Ph} < \text{Me} \approx \text{MeO}$, reflecting the electrophilic nature of singlet oxygen.

Regarding the regiospecificity of singlet-oxygen addition, the earlier results mainly contributed from Dufraisse's school, established that anthracene, its homologs, and aza analogs give endoperoxides as depicted in Fig. 2, where the positions of oxygen attack are indicated by asterisks. The addition occurs generally at meso positions of an anthracene skeleton in these parent acenes. Various types of hydrocarbons shown in Fig. 2 are reported

Table I Reactivity Indices (β), Rate Constants (k_r), and Relative Activation Energies ($E_F^{\ddagger}$) for Singlet Oxygen Addition to Aromatic Hydrocarbons[a]

Acceptor	$10^3\beta$ (M)	$10^{-6}\,k_r$ ($M^{-1}\,\text{sec}^{-1}$)	$E_F^{\ddagger}$ (kcal/mole)
1,4-Dimethylnaphthalene	3400 + 1400	0.012	7.6
Anthracene	270 + 50	0.15	6.1
9-Phenylanthracene	100 + 20	0.42	5.5
9,10-Diphenylanthracene	35 + 10	1.2	4.9
9-Methylanthracene	13 + 3	3.2	4.3
9,10-Dimethylanthracene	2.0 + 0.5	21	3.1
9-Methoxyanthracene	17 + 5	2.5	4.5
9,10-Dimethoxyanthracene	3.0 + 0.8	14	3.4
1,2-Benzanthracene	870 + 250	0.048	6.7
9,10-Dimethyl-1,2-benzanthracene	3.0 + 1.0	14	3.4
Naphthacene	3.4 + 0.9	12	3.5
Rubrene	(1.0 + 0.1)	42	2.8
1,2,5,6-Dibenzanthracene	4400 + 1600	0.0095	7.8
Pentacene	0.010 + 0.002	4200	0.0
1,3-Diphenylisobenzofuran	0.060 + 0.005	700	1.1

[a] In benzene at 25°C (Stevens *et al.*, 1974b).

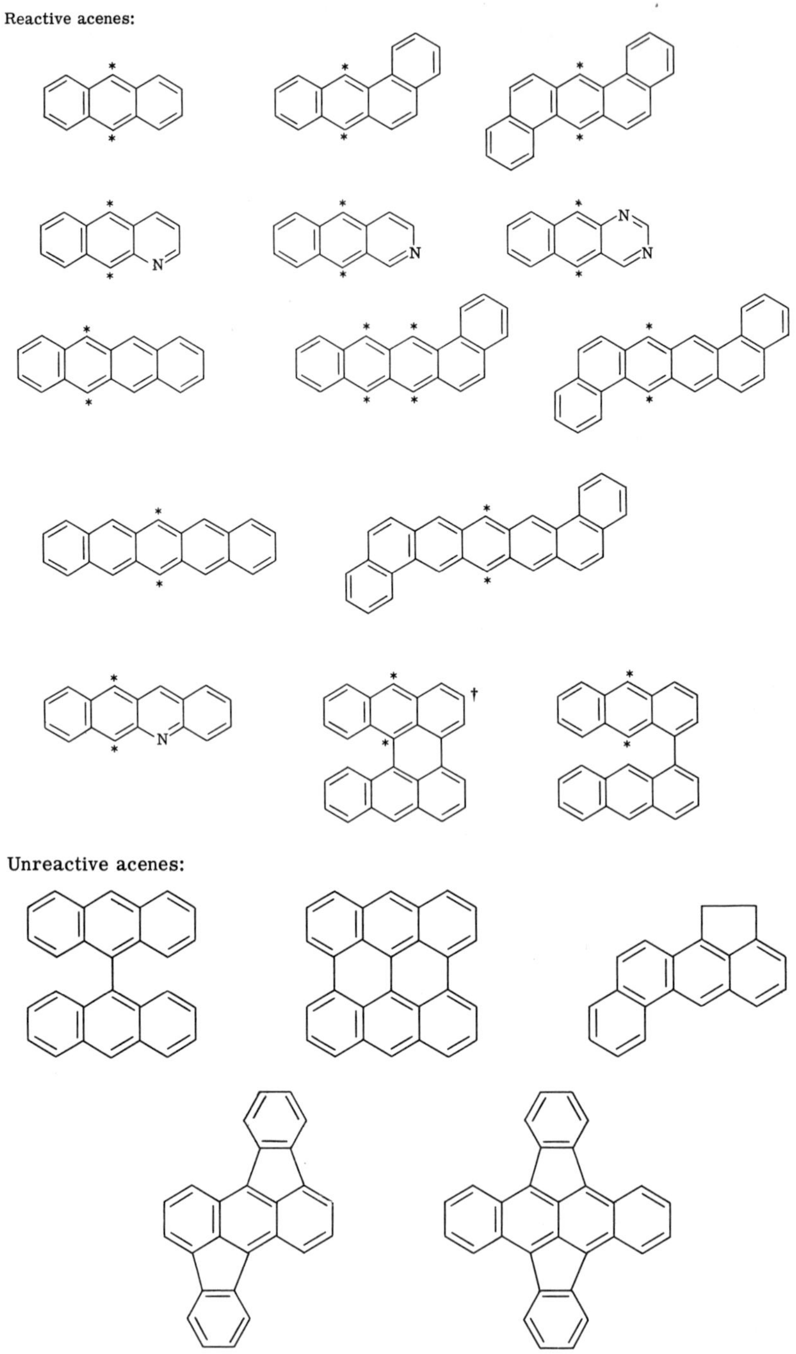

Reactive acenes:

Unreactive acenes:

Fig. 2. Regiospecificity for singlet-oxygen addition to acenes. The structure of the endoperoxide (†) has been recently established (Brockmann and Dicke, 1970).

to be unreactive, probably due to steric effects or to the rigidity of the molecules (Gollnick and Schenck, 1967).

One can alter regiospecificity by introducing electron-donating groups such as methyl, methoxy, and dimethylamino groups *at appropriate positions* of an aromatic nucleus (Rigaudy, 1968). As seen from typical examples of anthracene derivatives (**1**) given in Table II, the predominant formation of a 1,4-endoperoxide (**3**) over 9,10-endoperoxide (**2**) requires conditions that the electron-donating substituents are located at 1- and 4-positions but not at 2- and 3-positions.

Table II Regiospecificity in Singlet Oxygen Addition to Substituted Anthracenes.

(**1**) $\xrightarrow{^1O_2}$ (**2**) + (**3**)

Structure No.	Substituents X	Y	Z	Approximate ratio (**2**):(**3**)	Ref.
(**1a**)	H	Ph	H	100:0	Dufraisse and Velluz, 1940
(**1b**)	H	Me	H	35:65	Rigaudy *et al.*, 1971a
(**1c**)	H	OR (R = Me, Et, or CH_2Ph)	H	0:100	Rigaudy, 1968; Rigaudy *et al.*, 1967, 1969b; Dufraisse *et al.*, 1965a
(**1d**)	H	NMe_2, H	H	0:100	Chalvet *et al.*, 1970
(**1e**)	Ph	H	H	100:0	Rigaudy, 1968; Schönberg, 1968; Bowen, 1954
(**1f**)	Ph	Cl	H	100:0	Dufraisse and Velluz, 1940
(**1g**)	Ph	OMe, H	H	50:50	Rigaudy, 1968
(**1h**)	Ph	Me	H	30:70	Rigaudy *et al.*, 1971a
(**1i**)	Ph	OR (R = Me or Et)	H	0:100	Dufraisse *et al.*, 1965a; Shönberg, 1968
(**1j**)	Ph	NMe_2, Me	H	0:100	Rigaudy *et al.*, 1970
(**1k**)	Cl	H	H	100:0	Bowen, 1954
(**1l**)	H	H	OMe, H	100:0	Ranico, 1955
(**1m**)	Ph	H	Me	100:0	Mellier, 1955
(**1n**)	Ph	H	SMe, H	100:0	Ranico, 1955
(**1o**)	Ph	H	OMe	100:0	Chalvet *et al.*, 1970

Theoretical explanations for the regiospecific addition have been attempted. Calculation of the paralocalization energies of 1,4-dimethoxyanthracene (**1c**; R = Me) by Hückel MO method predicts a regiospecificity which is not in agreement with the experimental result (Chalvet *et al.*, 1968). However, Chalvet *et al.* (1970) have shown that a delocalized transition state model can satisfactorily explain the regiospecificity. The difference between the electron energies of two isomeric transition states [Eq. (10); $\Delta E = E_{9,10} - E_{1,4}$] has been calculated for anthracenes, (**1c**; R = Me), (**1d**), (**1f**), (**1i**; R = Me), and (**1o**). A positive ΔE value predicts 9,10-addition and a negative one predicts 1,4-addition, which are in good agreement with the experimental results.

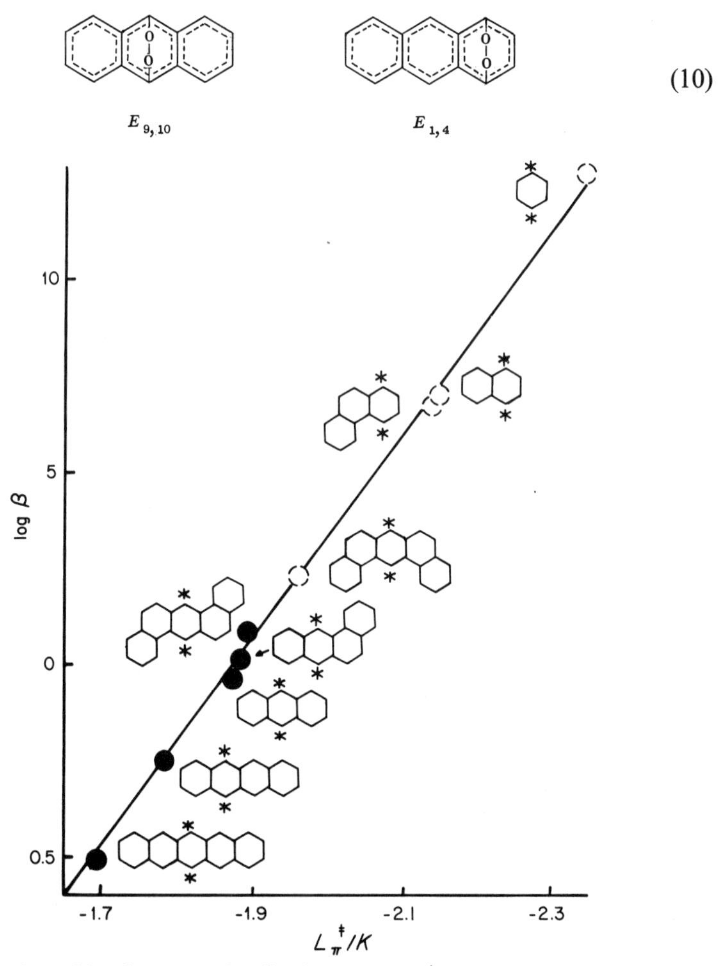

Fig. 3. Plots of log β against relocalization energy $L_\pi^\ddagger$ for singlet oxygen addition at sites indicated by asterisks. ●, Experimental; ◌, predicted values of β (Stevens *et al.*, 1974b).

In order to rationalize the reactivities of the parent acenes listed in Table I, Stevens *et al.* (1974) have tried to introduce a new parameter. They found that their log β values are linearly correlated to $L_{\pi}^{\ddagger}/K$ value (Fig. 3), where $L_{\pi}^{\ddagger}$ is the energy required to localize two electrons in the HOMO at the site of singlet-oxygen attack and K is the resonance integral, being estimated as -36 kcal. They also predicted β values for several aromatic hydrocarbons such as benzene, naphthalene, phenanthrene, and 1,2,7,8-dibenzoanthracene, which are known to be virtually unreactive to singlet oxygen.

If the orbital interaction concept (Fukui, 1971) is applicable to the interaction between the HOMO of an aromatic hydrocarbon and the LUMO of singlet oxygen, the regiospecificity may be qualitatively rationalized by a simple rule that the predominant 1,4-addition should occur at the positions with the largest atomic orbital coefficients of HOMO, which is easily calculated by the simple Hückel MO method for the aromatic molecules (Matsuura *et al.*, 1971).

C. Oxidation of Naphthalenes, Phenanthrenes, and Other Acenes

Benzene, naphthalene, and phenanthrene show practically no reactivity toward singlet oxygen. However, the introduction of electron-donating substituents into suitable positions (even for benzene; see Section II), causes singlet oxygen addition. The reaction of substituted naphthalenes has been extensively studied by Wasserman and Larsen (1972; also see Larsen, 1973) and by other workers (Rigaudy, 1968; Song and Moore, 1968; Rigaudy *et al.*, 1969a, 1971b, 1973b; Hart and Oku, 1972). They found that the dye-sensitized photooxygenation of the naphthalenes gave the corresponding 1,4-endoperoxides, either regiospecifically [see Fig. 4 for compounds (**4**)–(**13**)] or nonregiospecifically [see Fig. 5 for compounds (**14**)–(**17**)]. With a series of naphthalenes substituted by methyl groups only at one side ring, the reactivity was found to be in the order: **8** > **7** > **4** > 1-methylnaphthalene (Wasserman and Larsen, 1972; Larsen, 1973). Direct photooxygenation of certain methylnaphthalenes gives the corresponding naphthaldehydes, probably via a process not involving singlet oxygen (Rigaudy and Maumy, 1972). As will be described in Section I,D, many of the endoperoxides formed undergo reverse reaction to the parent naphthalenes and oxygen molecule even at room temperature. In Figs. 4 and 5, the half-life, $t_{1/2}$ (hr) at 25°C, is given for each endoperoxide. In most cases, the sensitized photooxygenation must be carried out at a lower temperature in order to obtain the endoperoxide.

Substituted naphthalenes listed in Fig. 6 have been reported to show no reactivity toward singlet oxygen (Wasserman and Larsen, 1972; Larsen, 1973). The unreactivity may be ascribed to the insufficient electron-richness

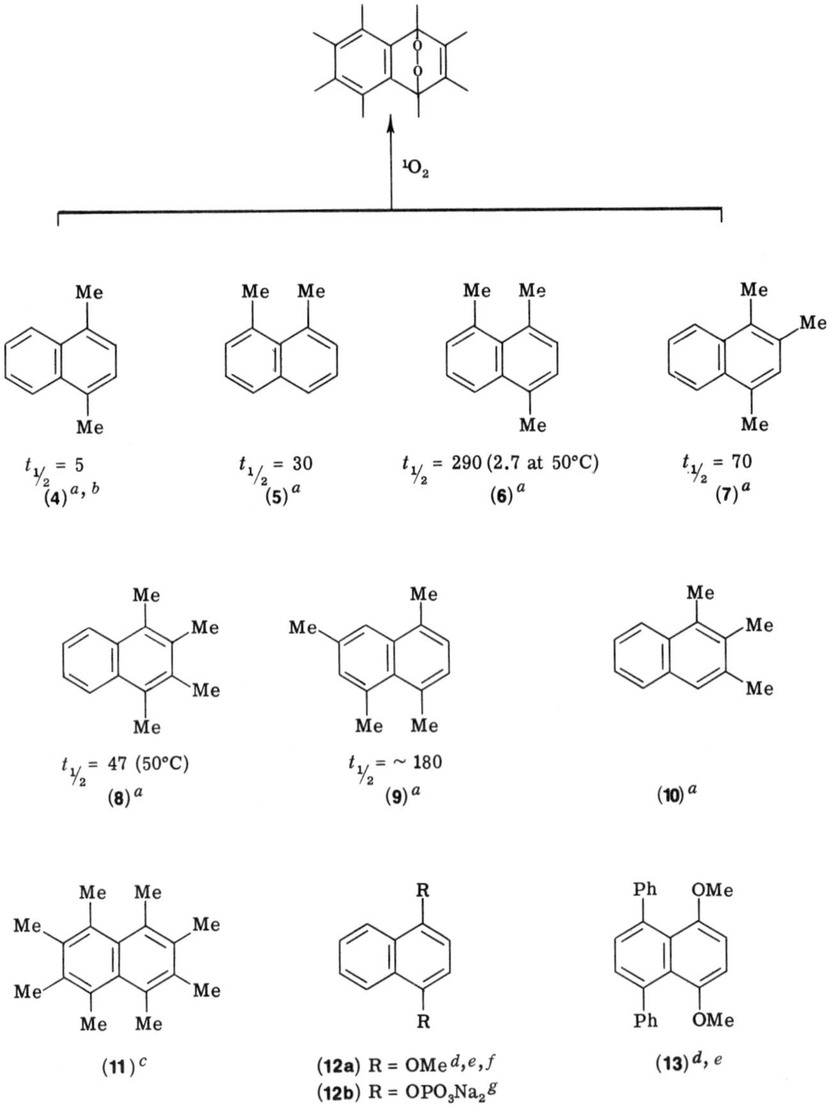

$^1O_2$

Me Me

$t_{1/2}$ = 5
(4)[a, b]

Me Me

$t_{1/2}$ = 30
(5)[a]

Me Me Me

$t_{1/2}$ = 290 (2.7 at 50°C)
(6)[a]

Me Me Me

$t_{1/2}$ = 70
(7)[a]

Me Me Me Me

$t_{1/2}$ = 47 (50°C)
(8)[a]

Me Me Me Me

$t_{1/2}$ = ~ 180
(9)[a]

Me Me Me

(10)[a]

Me Me Me Me Me Me Me Me

(11)[c]

R R

(12a) R = OMe[d,e,f]
(12b) R = OPO_3Na_2[g]

Ph OMe Ph OMe

(13)[d, e]

Fig. 4. Naphthalene derivatives giving an endoperoxide regiospecifically. $t_{1/2}$, Half-life (hr) of endoperoxides at 25°C. Key to references: (*a*) Wasserman and Larsen (1972), Larsen (1973); (*b*) Rigaudy *et al.* (1971b); (*c*) Hart and Oku (1972); (*d*) Rigaudy (1968); (*e*) Rigaudy *et al.* (1969a); (*f*) Rigaudy *et al.* (1973a); (*g*) Song and Moore (1968).

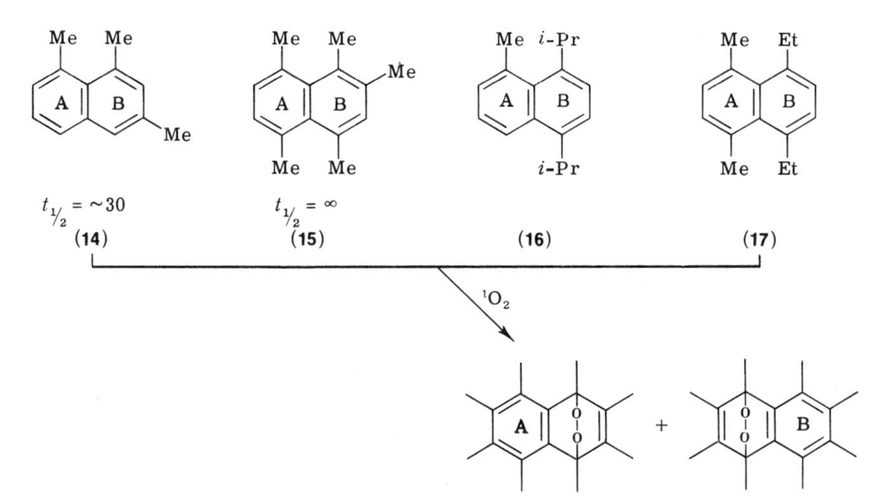

Fig. 5. Naphthalene derivatives giving an endoperoxide nonregiospecifically. $t_{1/2}$, Half-life (hr) of endoperoxide at 25°C. From Wasserman and Larsen (1972); Larsen (1973).

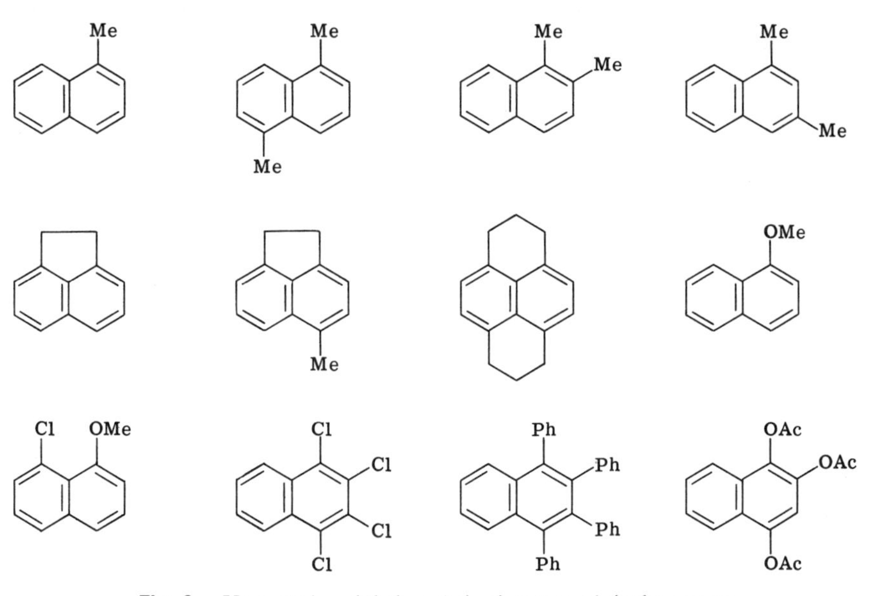

Fig. 6. Unreacted naphthalene derivatives toward singlet oxygen.

of the π systems or in certain cases to steric effects. Naphthalene analogs **18** and **19** of [2.2]-paracyclophane were reported to react with singlet oxygen in methanol to yield **20** and **21**, respectively [Eqs. (11) and (12)] (Wasserman and Keehn, 1966, 1967, 1972; Wasserman, 1970). The formation of these products is ascribed to the initial formation of an endoperoxide intermediate which undergoes solvolysis followed by rearrangement or internal Diels–Alder reaction followed by solvolysis. In the oxygenation of [2.2](2.5)furanonaphthane, singlet oxygen attacks exclusively on the more electron-rich furan ring (Wasserman, 1970).

Substituted phenanthrenes give two types of peroxides depending upon the position of substituents. For example, photooxygenation of **22** at 220°K has been suggested to give a dioxetane (**23**) which undergoes acid-catalyzed cleavage of 9,10-bond (Rio and Berthelot, 1972), whereas **24** yields a 1,4-endoperoxide (**25**) [Eqs. (13) and (14)] (Wasserman and Larsen, 1972; Larsen, 1973). Peroxide **25** is thermally unstable and reverts back to the starting material and oxygen, probably singlet oxygen.

Singlet oxygen reactions of particular types of compounds related to polycyclic aromatic hydrocarbons have been reported. Both 9,9′-bifluorenylidene (**26**) (Richardson and Hodge, 1970) and 10,10′-dimethyl-9,9′-biacridylidene (**27**) (McCapra and Hann, 1969; Janzen *et al.*, 1970) undergo cleavage of the double bond possibly via a dioxetane intermediate, being accompanied by chemiluminescence [Eq. (15)]. Photooxygenation of 10,10′-disubstituted phenafulvenes (**28**) yields crystalline epiperoxides (**29**) which undergo base-catalyzed rearrangement, as shown in [Eq. (16)], to phenalenones (**30**) (Morita and Asao, 1975).

MeOH

MeO

MeO

$^1O_2$

(11)

MeO

(**18**)

(**20**)

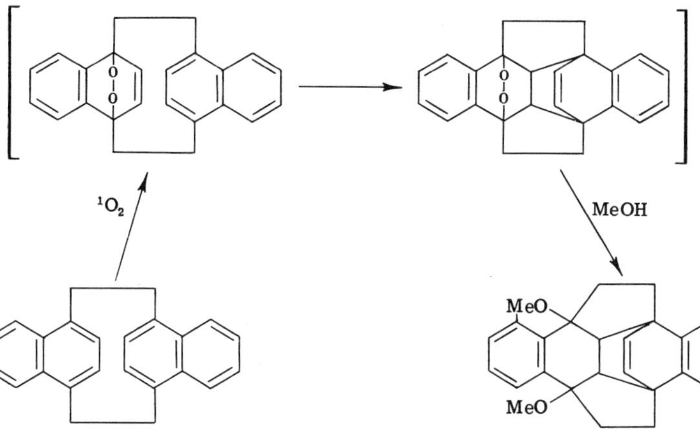

$^1O_2$

MeOH

MeO

MeO

(19) (21)

(12)

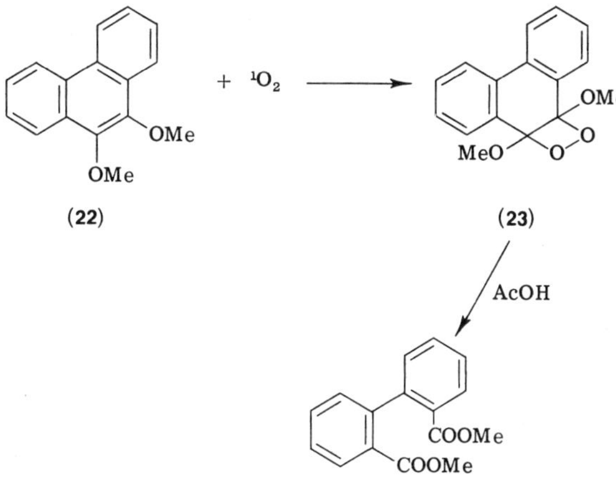

$+ \ ^1O_2$

OMe

OMe

OMe

MeO

(22) (23)

AcOH

COOMe

COOMe

(13)

Me

Me

$+ \ ^1O_2$

Me

Me

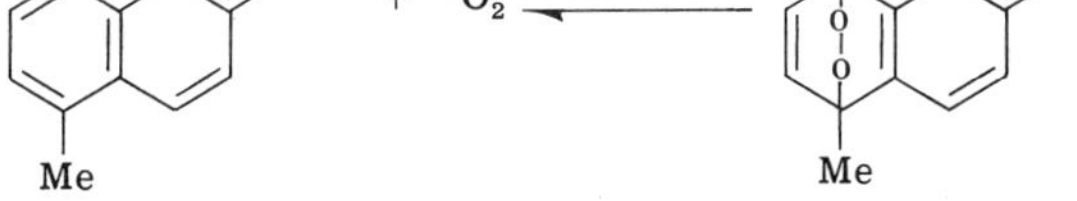

(24) (25)

(14)

(15)

(26) X = none
(27) X = NMe

(16)

R = Me or Ph
(28)

(29)

(30)

D. Chemical Behavior of Acene Endoperoxides

Acene endoperoxides are known to undergo various types of transformation reactions, either spontaneously or under selected reaction conditions. The reactions may be classified into the following three types of reactions: (1) reversion into the parent acene and singlet oxygen; (2) rearrangement or decomposition via a fission of the O–O bond; and (3) transformation into other types of peroxides. These reactions provide useful synthetic tools for introducing oxygen functions into acenes.

1. Regeneration of Singlet Oxygen

The decomposition of many acene endoperoxides being accompanied by the liberation of molecular oxygen has long been known. The ease of oxygen release depends upon the skeletal structure of the acene and the nature of the substituents in the meso positions (Bergmann and McLean, 1941). In some cases, for example, the endoperoxides of 9,10-diphenylanthracene (Dufraisse and LeBras, 1937), 1,4-dimethoxy-9,10-diphenylanthracene (Dufraisse and Velluz, 1942), or rubrene (5,6,11,12-tetraphenylnaphthalene) (Audubert, 1939) produce a chemiluminescence, the spectrum of which is in accord with the fluorescence spectrum of the parent acenes. Among the various possibilities [earlier proposals have been summarized by Gollnick (1968b)], the emitting species, i.e., the lowest excited singlet state of the parent acene, is most probably generated by a two-step energy transfer from singlet oxygen monomers ($^1\Delta_g$ and $^1\Sigma_g{}^+$) or their collisional pairs as shown in Eq. (17), judging from the following results: (i) Rubrene

$$AO_2 \xrightarrow{\Delta} A + {}^1O_2 \tag{17}$$

$$A + {}^1O_2 \longrightarrow AO_2$$

$$A + {}^1O_2\,[\text{or } ({}^1O_2)_2] \longrightarrow {}^3A^* + {}^3O_2\,[\text{or } 2\,{}^3O_2]$$

$${}^3A^* + {}^1O_2\,[\text{or } ({}^1O_2)_2] \longrightarrow {}^1A^* + {}^3O_2\,[\text{or } 2\,{}^3O_2]$$

$${}^1A^* \longrightarrow A + h\nu$$

reacts with singlet oxygen generated by a microwave discharge, emitting fluorescence with the concomitant formation of the rubrene endoperoxide (Wilson, 1969b) and (ii) chemiluminescence from the thermal decomposition of dibenzal diperoxide in the presence of an acene fluorescer such as anthracene and dibenzanthrone, occurs from the acene fluorescent state, which is produced by energy transfer from both triplet benzaldehyde and singlet oxygen (Abbott *et al.*, 1970).

Wasserman *et al.* (1972) have utilized the thermal decomposition of 9,10-diphenylanthracene 9,10-peroxide as a chemical singlet oxygen source. When a mixture of this peroxide and various singlet oxygen acceptors (usually 1:1 to 2:1 ratio) is heated in benzene, the oxidation products are obtained in good to excellent yield, indicating that the efficiency of liberation of singlet oxygen is very high. They have also demonstrated by kinetic studies that the reaction involves singlet oxygen but not direct bimolecular oxygen transfer from the peroxide to the acceptor.

Wasserman and Larsen (1972; Larsen, 1973) have examined the thermal stability of a series of methylnaphthalene endoperoxides, many of which also result in liberation of the parent naphthalene and singlet oxygen. They have suggested from the halflife values given in Figs. 4 and 5 that the more the relief of the steric ortho and peri strains in the parent naphthalenes by endoperoxide formation, the more stable are the endoperoxides.

The activation energies for the thermal decomposition of 9,10-diphenylanthracene 9,10-peroxide (Wasserman *et al.*, 1972) and 1,4,5-trimethylnaphthalene 1,4-peroxide (Wasserman and Larsen, 1972; Larsen, 1973) have been estimated as 27.8 ± 0.2 and 29 ± 0.5 kcal/mole, respectively. It is interesting to compare these values with the activation energy (29.8 kcal/mole) (Breitenback and Kastell, 1954) for the decomposition of anthracene 9,10-peroxide, which is known to decompose by a process involving a homolysis of the O–O bond.

2. *Reaction via O–O Bond Fission*

There are several types of reactions involving a homolytic or heterolytic fission of the O–O bond of acene peroxides. The reactions are principally the same as those of the endoperoxides of cyclic dienes. Rigaudy *et al.* have shown that the transformation of acene peroxides into bisepoxides can be achieved by photolysis or by thermolysis. For example, photolysis of 1,4-dibenzylanthracene 1,4-peroxide gives the corresponding bisepoxide in addition to fragmentation into the diester (**31**) and acetylene as shown in Eq. (18) (Rigaudy *et al.*, 1969b). A similar result is obtained with naphthalene 1,4-peroxides [Eq. (19)] (Rigaudy *et al.*, 1969a, 1971b, 1973a).

Acene peroxides unsubstituted at peroxide bond sites such as anthracene 9,10-peroxide have long been known to undergo thermal decomposition to the corresponding quinones by homolytic cleavage of the O–O bond (Gollnick and Schenck, 1967; Breitenback and Kastell, 1954). Recently, Rigaudy *et al.* (1972, 1973b, 1975) have shown that thermolysis of anthracene 9,10-peroxide (**32**) gives the products which are believed to derive from the anthracene bisepoxide (**33**). Heating of a highly diluted solution of **32** in *o*-dichlorobenzene under reflux gives benzocyclobutene (**34**) together with 9,10-anthraquinone and anthracene (Rigaudy *et al.*, 1973b, 1975). The

R = $PhCH_2O$-

$h\nu$

CH≡CH + (**31**) (COR, COR)

(18)

R = Me or OMe

(19)

formation of **34** is explained by a mechanism involving bisepoxide (**33**) and its valence tautomer (**35**). Evidence for the intermediacy of **33** and **35** is obtained by trapping reactions. Thus, prolonged refluxing of **32** in benzene in the presence of *N*-methylmaleimide produces two adducts (**36** and **37**) [Eq. (20)]. Similarly, thermolysis of 9,10-dimethylanthracene 9,10-peroxide in various solvents gives benzocyclobutene (**38**) and dibenzo-[*b*,*e*]oxepin (**39**) together with the parent anthracene, 9,10-anthraquinone, and **40** [Eq. (21)] (Rigaudy *et al.*, 1972).

(34) dimer

(32) Δ C_6H_6 or o-$C_6H_4Cl_2$ (33) (35)

(20)

(36) (37)

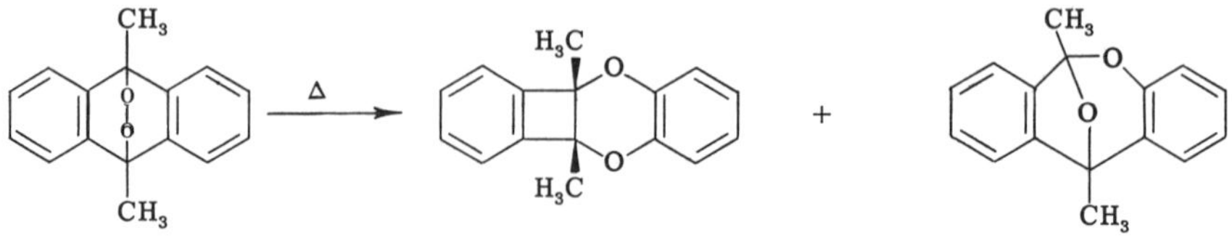

(38) (39)

(21)

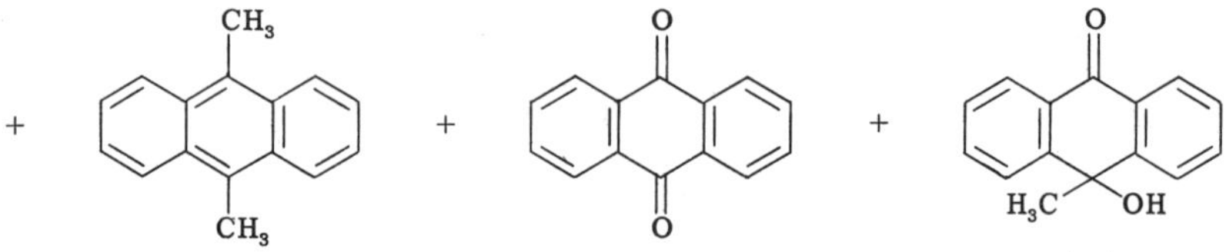

(40)

Base-catalyzed β-elimination is observable with meso-monosubstituted anthracene peroxides (**41**) which give rise to the corresponding quinols (**42**) [Eq. (22)] (Dufraisse *et al.*, 1957a,b). An unusual thermal rearrangement of an arene peroxide (**43**) giving **44** has been reported (Etienne and Beauvois, 1954). The reaction is suggested to involve an oxygen transfer to the adjacent ring [Eq. (23)].

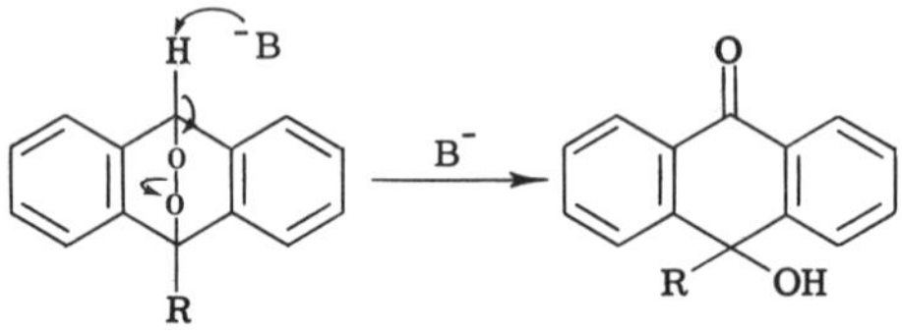

(22)

(41) (42)

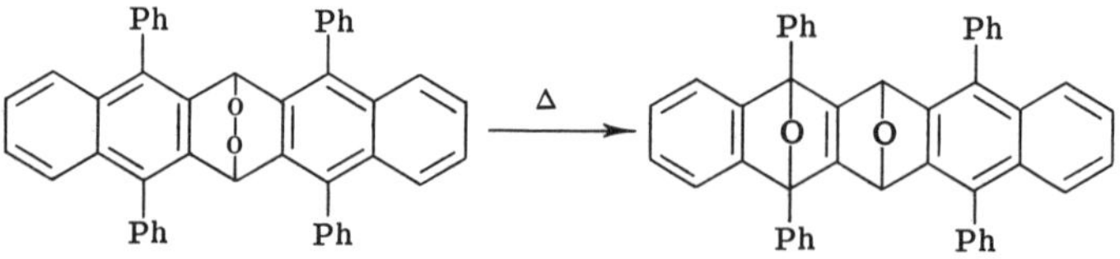

(23)

(43) (44)

Acid-catalyzed transformations of acene *meso*-peroxides have been studied extensively. Treatment of 9,10-diphenylanthracene 9,10-peroxide and its derivatives with strong acid in aqueous media gives **45** and **46**, whereas acid treatment under anhydrous conditions produces dibenzo-[*b*,*e*]-oxepin derivatives (**47**) [Eq. (24)] (Rigaudy and Brelière, 1972). Treatment of rubrene endoperoxide with perchloric acid has been reported to yield various products shown in Eq. (25) (Rigaudy and Kim Cuong, 1972).

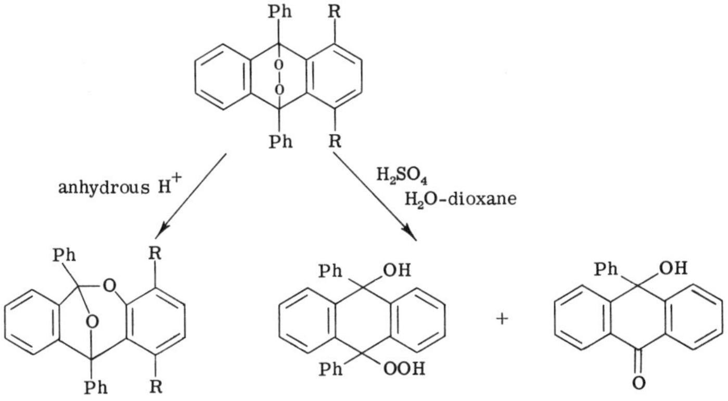

(24)

(47) (45) (46)

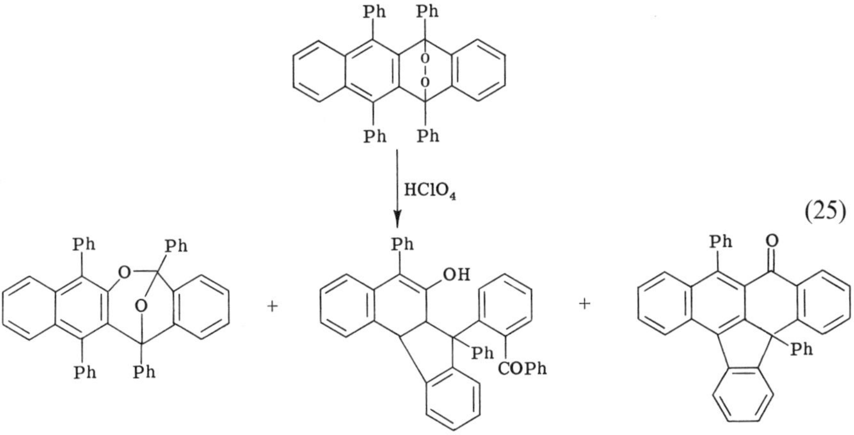

(25)

3. *Transformation into Other Types of Peroxides*

As exemplified in Eq. (24), acene peroxides readily suffer acid hydrolysis in aqueous media to give the corresponding hydroperoxides as primary products. A number of such examples have been known (Rigaudy, 1968; Rigaudy *et al.*, 1969a, 1973a). A typical reaction is shown in Eq. (26) (Rigaudy and Sparfel, 1972).

dil. acid, MeOH; 0.2 *N* HCl; 2 *N* HCl (26)

Acene 1,4-peroxides are known to undergo acid-catalyzed 1,2-bond cleavage reactions (Rigaudy *et al.*, 1973a; Baldwin *et al.*, 1968). For example, treatment of 1,4-dimethoxy-9,10-diphenylanthracene 1,4-peroxide (**48a**) in methanol with *p*-nitrobenzoic acid gives a ring cleavage product (**49a**) (Rigaudy *et al.*, 1973a). A mechanism involving rearrangement to a dioxetane (**50a**) has been proposed. The reaction is shown to be accompanied by a strong chemiluminescence (Eq. (27)). The mechanism for the

(**48a**) R = Ph
(**49b**) R = H
(**50**)
(**49**)
(27)

chemiluminescence has been studied by Wilson (1969a). The emission spectrum is well matched with the fluorescence spectrum of the parent anthracene. The luminescence quantum yield depends critically on the acidity of the medium. The kinetic data obtained are consistent with a mechanism shown in Eq. (28). A similar cleavage is observed with 1-(*N*,*N*-

$$AO_2 \xrightarrow{\Delta} A + O_2 \text{ (to a small extent)}$$
$$AO_2 \xrightarrow{H^+} AO_2'$$
$$AO_2' \longrightarrow \alpha P^* + (1-\alpha)P$$
$$P^* + A \longrightarrow P + {}^1A^*$$
$${}^1A^* \longrightarrow A + h\nu$$

AO_2 = (**48a**)
AO_2' = (**50a**)
P = (**49a**)
α = fraction of the excited state (P*) in P formed

(28)

dimethylamino)-9,10-diphenylanthracene 1,4-peroxide, which undergoes spontaneous decomposition to give an unstable cleavage product and its cyclized products [Eq. (29)] (Rigaudy *et al.*, 1970).

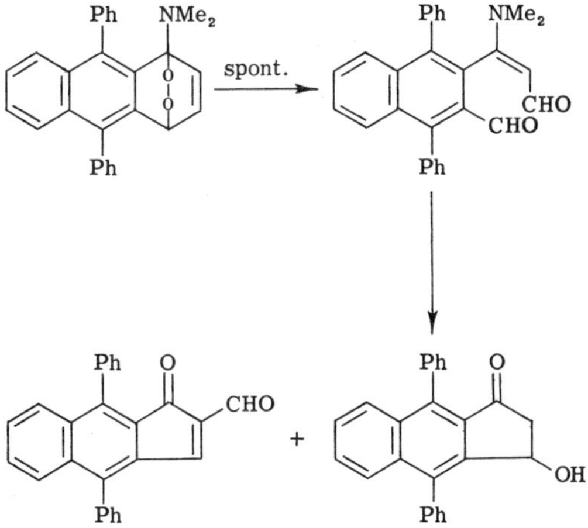

(29)

II. OXIDATIONS OF MONOCYCLIC AROMATIC COMPOUNDS

While 1,4-cycloaddition of singlet oxygen to polynuclear aromatic hydrocarbons or heterocycles is well established, addition to a benzene ring was not reported until recently. Highly electron donor-substituted benzenes, however, have been reported to undergo 1,4-cycloaddition with singlet oxygen to give an unstable arene peroxide intermediate as a primary product. Another important discovery on the additions of singlet oxygen to benzene rings is a 1,4-cycloaddition to styrene-type olefins. Such reactions, in which molecular oxygen becomes incorporated into aromatic nuclei, have also received much attention in connection with the biological oxidations of aromatic substrates catalyzed by oxygenases (Forster, 1962; Gibson *et al.*, 1970; Ziffer *et al.*, 1973; Daly *et al.*, 1972; Hayaishi, 1974). 1,4-Cycloaddition to carbocyclic nonbenzenoid aromatic compounds such as fulvenes, tropolones, arene oxides, and annulenes is also described in this section.

A. Electron Donor-Substituted Benzenes

Benzene and alkylbenzenes are usually resistant to the action of singlet oxygen. Suitably substituted polymethoxybenzenes and *N*,*N*-dimethylanilines, however, have been shown to react with singlet oxygen to give products which are believed to derive from 1,4- or 1,2-endoperoxides, although such endoperoxides of benzene derivatives have not been isolated so far (Saito and Matsuura, 1970; Saito *et al.*, 1970, 1972a,b,c, 1974). For example, photooxygenation of 1,2,4,5-tetramethoxybenzene (**51**) in methanol gives 2,5-dimethoxy-*p*-benzoquinone (**52**) and enone (**53**) [Eq. (30)] (Saito *et al.*, 1972c). The participation of singlet oxygen in the reaction has been shown by the photooxygenations using sensitizers of different triplet energies and by quenching experiments. The products are proposed as deriving from a common intermediate, arene 1,4-peroxide (**54**). The product (**53**) may very probably be formed by rearrangement of **54** to the diepoxide followed by addition of methanol to the epoxy rings.

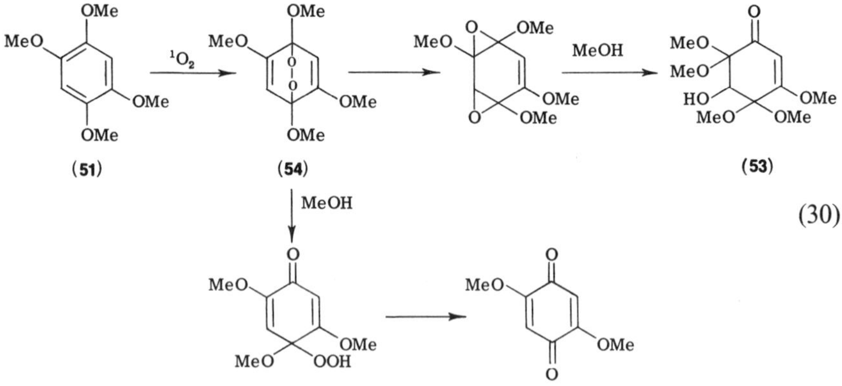

(30)

1,2,3,5-Tetramethoxybenzene (**55**), on sensitized photooxygenation in alcohol (R_2OH), yields enedione **56** which may be formed through an analogous endoperoxide (**57**) and diepoxide (**58**) [Eq. (31)] (Saito *et al.*, 1972c). Similar photooxygenation of hexamethoxybenzene (**59**) in methanol-d_4 produces **60**, which has been characterized as *o*-phenylenediamine adduct (**61**). Attempts to purify **60** by silica gel chromatography give rise to the formation of triketone hydrate **62** [Eq. (32)].

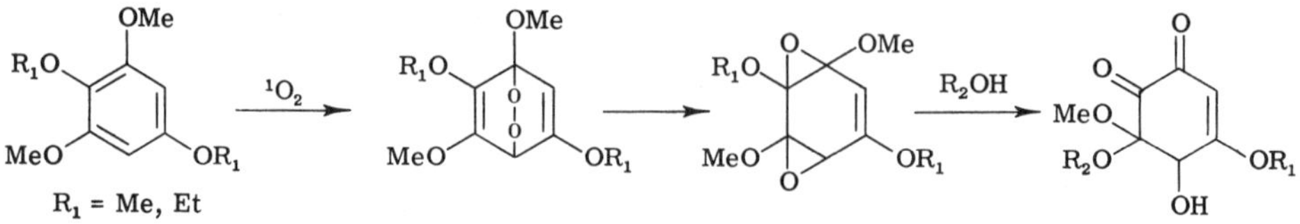

(31)

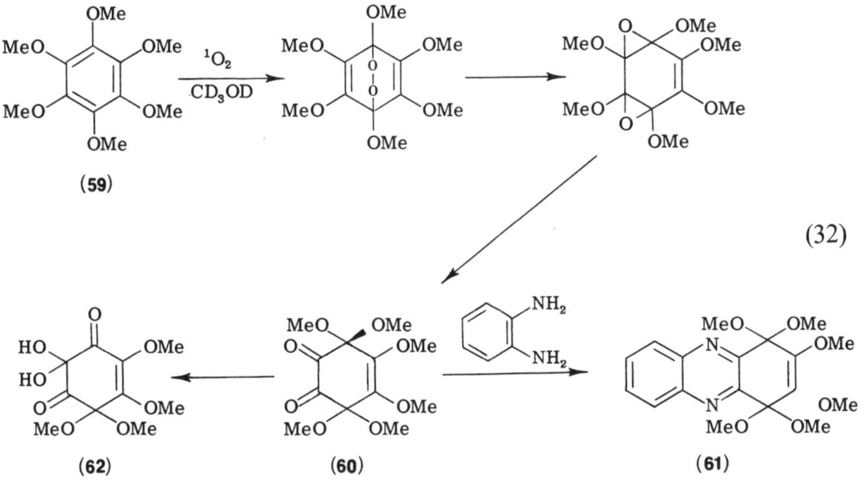

$$(32)$$

(59) (60) (61) (62)

By analogy with **55** or **59**, photooxygenation of pentamethoxybenzene (**63**) in methanol gives enedione **64**. In this case, a different type of epoxide intermediate (**65**) has been suggested to be involved as the precursor for **64** (see [Eq. (38)]) [Eq. (33)] (Saito *et al.*, 1972c).

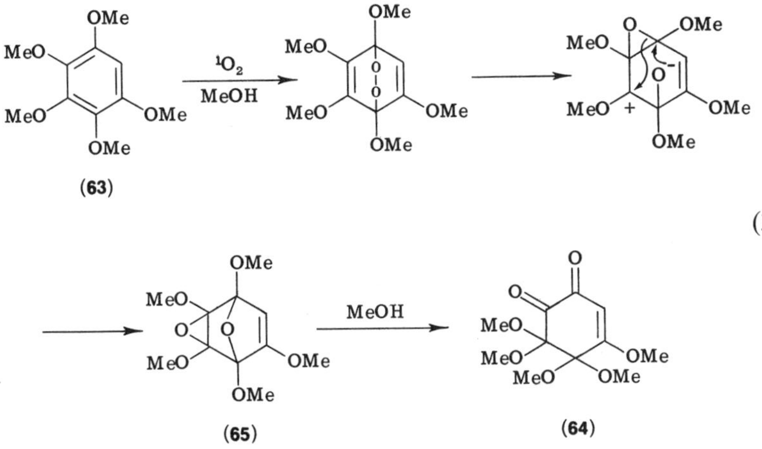

$$(33)$$

(63) (65) (64)

The relative rates of photooxygenation of a series of methoxybenzenes have been measured in methanol containing Rose Bengal or methylene blue as sensitizer (Saito *et al.*, 1972b). Highly substituted methoxybenzenes such as **51**, **55**, **59**, **63**, and 1,2,4-trimethoxybenzene are susceptible to photooxygenation, whereas less substituted ones such as anisole, dimethoxybenzenes, other trimethoxybenzenes, and 1,2,3,4-tetramethoxybenzene are shown to be unreactive under these conditions. The relative rates are compared with the half-wave oxidation potentials ($E_{1/2}$) (Zweig *et al.*, 1964) and the charge transfer band ($\nu_{c.t.}$) (Zweig, 1963) with tetracyanoethylene, indicating that methoxybenzenes having less than +1.24 V vs. saturated calomel electrode are susceptible to photooxygenation (Table III) (Saito *et al.*, 1972b). The rate constant for the most reactive methoxybenzene (**51**) is estimated to be $1.6 \pm 0.6 \times 10^6\ M^{-1}\ \text{sec}^{-1}$ (0.026 ± 0.004, β,M) by a competitive technique with 9,10-diphenylanthracene in benzene. For the reactive ones, there is a good correlation between the relative rate and $E_{1/2}$, while with the exception of 1,2,4-trimethoxybenzene, $\nu_{c.t.}$ is also linearly related to the relative rate (Fig. 7). The results indicate that reactivities of methoxybenzenes toward singlet oxygen are linearly related to their π-ionization potentials, and it seems likely that addition of singlet oxygen to the electron-rich aromatic π system may proceed by way of a charge-transferlike transition state.

Reaction of the dimethyl ether of 4,6-di-*tert*-butylresorcinol (**66**) with chemically or photochemically generated singlet oxygen gives the epoxy enone (**67**) [Eq. (34)], in contrast to the finding that the dimethyl ethers of 3,5-di-*tert*-butylcatechol and 2,5-di-*tert*-butylhydroquinone resist oxidation with singlet oxygen (Saito *et al.*, 1970). The relative reactivity of **66** is determined to be ca. $0.8\beta,M$ by competitive reaction (Saito *et al.*, 1972a).

Table III Relative Rate of Photooxygenation of Methoxybenzenes (Saito *et al.*, 1972b)

Compounds	Relative rate		$E_{1/2}$ (oxid), (V) *vs.* s.c.e.[a] in MeCN[b]	$\nu_{c.t.}$[c] (cm^{-1}) $\times 10^3$
	RB	MB		
Anisole	0	0	+1.76	19.7
1,2-Dimethoxybenzene	0	0	1.45	16.9
1,3-Dimethoxybenzene	0	0	—	18.2
1,4-Dimethoxybenzene	0	0	1.34	16.1
1,2,3-Trimethoxybenzene	0	0	1.42	19.4
1,3,5-Trimethoxybenzene	0[d]	0	1.49	18.2
1,2,4-Trimethoxybenzene	0.086	—	1.12	14.6
1,2,3,4-Tetramethoxybenzene	0	0	1.25	17.9
1,2,3,5-Tetramethoxybenzene	0.150	0.146	1.09	16.5
1,2,4,5-Tetramethoxybenzene	1.0	1.0	0.81	12.5
Pentamethoxybenzene	0.185	0.202	1.07	16.4
Hexamethoxybenzene	0.030	0.032	1.24	19.5

[a] Saturated calomel electrode.
[b] Zweig *et al.* (1964).
[c] With tetracyanoethylene in CH_2Cl_2 (Zweig, 1963).
[d] The compound is photooxygenated at an extremely slow rate.

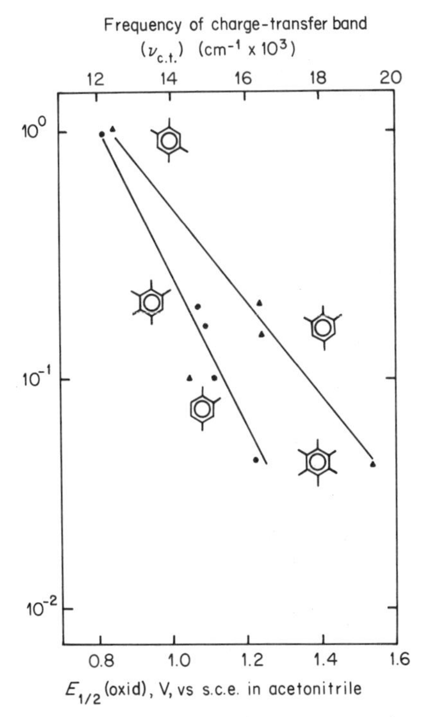

Fig. 7. Semilog plot of the relative rate of photooxygenation of methoxybenzenes vs half-wave oxidation potentials (●) and frequency of charge-transfer maxima with tetracyano-ethylene (▲) (Saito *et al.*, 1972b).

(34)

(66) **(67)**

Although tertiary amines are well known to be capable of quenching singlet oxygen (Ouannès and Wilson, 1968; Ogryzlo and Tang, 1970; Smith, 1972), it has also been shown that certain amines do react under photooxygenation conditions by way of Type I and singlet oxygen mechanisms (Lindner *et al.*, 1972; Gollnick and Lindner, 1973; Fisch *et al.*, 1970). Young *et al.* (1973) have observed that a Hammett rho plot gives a rho value of -1.39 for the quenching of singlet oxygen by a series of para- and meta-substituted *N,N*-dimethylanilines, and they have suggested that singlet oxygen undergoes physical quenching by the amines via partial charge-transfer intermediates. More recently, Saito *et al.* (1974) have shown that highly electron-donor substituted *N,N*-dimethylanilines are capable of undergoing 1,4- or 1,2-cycloaddition with singlet oxygen. Photooxygenation of 4,5-bis(*N,N*-dimethylamino)-*o*-xylene (**68**) gives **69** and **70**. The product ratio is highly sensitive to sensitizer type and solvent, and the addition of singlet oxygen quenchers to the reaction system inhibits the formation of **70** but has no significant effect on the yield of **69**. Based on these results, it has been suggested that a large portion of **69** results from a Type I process and that singlet oxygen is responsible for the formation of **70** as shown in Eq. (35).

N(Me)$_2$ N(Me)$_2$ (**68**) $\xrightarrow[CH_3OD]{^1O_2}$ N(Me)$_2$ N(Me)$_2$ ⟶ O N(Me)$_2$ N(Me)$_2$

(**68**) $\xrightarrow{h\nu/Sens/O_2}$ Me NCHO N(Me)$_2$ (**69**)

OH N(Me)$_2$ N(Me)$_2$ ⇌ O H N(Me)$_2$ N(Me)$_2$ (35)

$\xrightarrow{CH_3OD}$ O N(Me)$_2$ N(Me)$_2$ H D (**70**)

Unlike methoxybenzenes and **68** which give the corresponding 1,4-endoperoxide, 2,4-dimethoxy-*N,N*-dimethylaniline (**71**) produces a 1,2-cleavage product (**72**) upon photooxygenation in methanol [Eq. (36)] (Saito *et al.*, 1974). The product (**72**) is apparently derived from dioxetane **73** which may be formed by 1,2-cycloaddition to the benzene ring or by a methanol-assisted rearrangement of a 1,4-endoperoxide intermediate [see Eq. (50)].

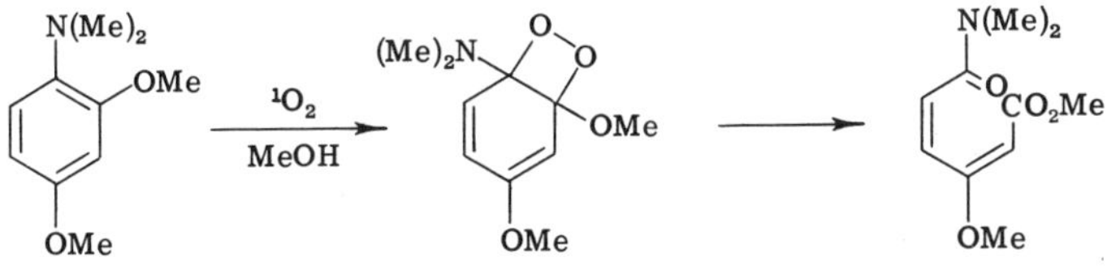

N(Me)$_2$ OMe OMe $\xrightarrow[\text{MeOH}]{^1O_2}$ (Me)$_2$N O O OMe OMe $\longrightarrow$ N(Me)$_2$ O CO$_2$Me OMe (36)

(71) (73) (72)

1,4-Cycloadditions to six-membered heteroaromatic rings have also been observed with 2,4,6-tri-*tert*-butylphosphorine derivatives **74** and **75**. Photooxygenation of **74** and **75** yields various types of products, which are believed to be formed through intermediate endoperoxides **76** and **77**, respectively [Eqs. (37–38)] (Dimroth *et al.*, 1972).

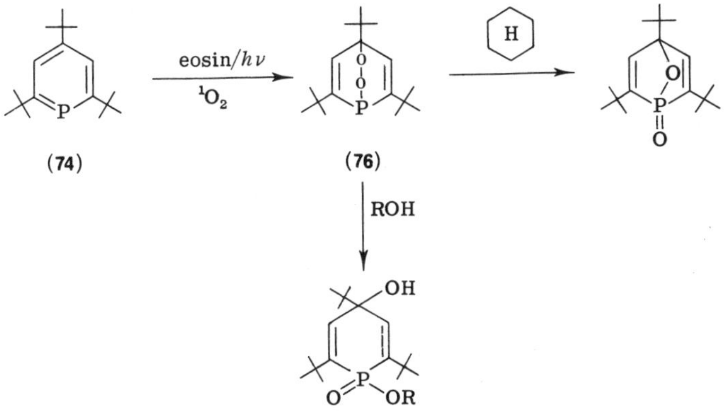

P $\xrightarrow[^1O_2]{\text{eosin}/h\nu}$ O O P $\xrightarrow{\text{H}}$ O P O

(74) (76)

ROH

OH O=P OR (37)

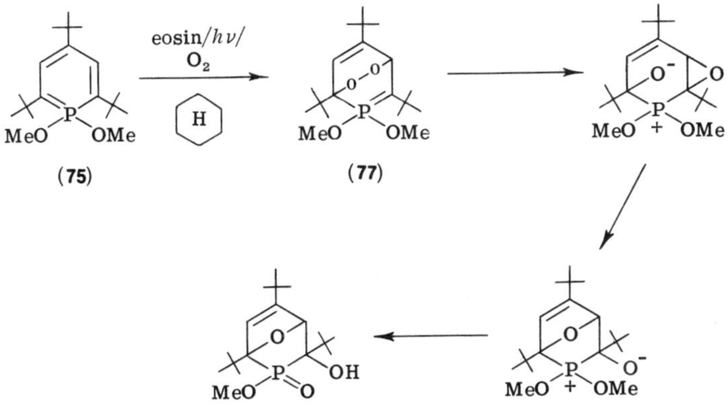

MeO P OMe $\xrightarrow[\text{H}]{\text{eosin}/h\nu/O_2}$ O O MeO P OMe $\longrightarrow$ O$^-$ O MeO P + OMe

(75) (77) (38)

O MeO P O OH $\longleftarrow$ O MeO P + OMe O$^-$

A number of reports have appeared on the photosensitized oxygenation of aromatic hydrocarbons, in which benzylic C–H groups have been oxidized (Häflinger *et al.*, 1956; Reiche and Gross, 1962; Mustafa *et al.*,

1966; Wei and Adelman, 1969; Bick *et al.*, 1971; Wasserman *et al.*, 1971). In general, the involvement of singlet oxygen has not been demonstrated and a Type I mechanism has been suggested for the reactions. A typical example is shown in Eq. (39) (Wasserman *et al.*, 1971).*

Me Me Me Me Me Me

MB/$h\nu$/O_2
MeOH

Me Me Me CH_2OMe Me Me + Me Me CH_2OMe CH_2OMe Me Me + Me Me Me O O Me

(39)

Photooxygenation of triphenylcarbonium ion has been reported to result in an oxidative destruction of the benzene ring [Eq. (40)] (Van Tamelen and Cole, 1970). The reaction of the excited triplet state of the carbonium ion with ground state triplet oxygen has been suggested. Similar type of oxidative cleavage of the benzene ring has been reported on unsensitized photooxygenation of benzene leading to mucondialdehyde (Wei *et al.*, 1967; Irina and Kurien, 1973).

Ph_3C^+ $h\nu$/O_2 → Ph Ph—C + O O H → AcOH → OAc Ph Ph—C O O H

Ph Ph—C^+ O O → Ph_2CO

AcOH ← OAc Ph—C Ph H O CHO

Ph_2CO

(40)

* Key to abbreviations: RB, Rose Bengal; MB, methylene blue; DNT, dinaphthalenethiophene; TPP, tetraphenylporphine; HP, hematoporphyrin.

B. 1,4-Cycloaddition to Styrene-Type Olefins

Recently, Foote *et al.* (1973) have shown that benzoannelated diene systems undergo 1,4-cycloaddition with singlet oxygen to give an unusual type of endoperoxide. Photooxygenation of 1,1-diphenyl-2-methoxyethylene (**78**) in benzene at room temperature produces four products, benzophenone, **79**, **80**, and **81**, an unusual peroxide directly bonded to the benzene ring. The photooxygenation at −78°C in ether gives the peroxide (**82**) [Eq. (41)].

Similar type of addition has been observed with indene derivatives. When photooxygenation of indene **83a** is carried out at −78°C in acetone solution, two equivalent moles of oxygen are taken up and the rearranged peroxide (**84a**) is obtained in good yield. Very similar results are obtained with substituted indenes (**83b–d**) [Eq. (42)] (Foote *et al.*, 1973; Burns *et al.*, 1976). The peroxides (**84**) undergo rapid rearrangement on heating to tetraepoxide (**85**), a benzene trioxide. The formation of these products suggests a 1,4-cycloaddition giving endoperoxide **86** in the first step. Rearrangement of this highly strained peroxide to benzene oxide (**87**) followed by addition of a second singlet oxygen molecule would give **84**. The greater rate of photooxygenation of indenes at low temperature suggests that the formation of **86** may be reversible at higher temperature. Analogous results have been obtained in the low temperature photooxygenation of 1,2-dihydronaphthalenes (Burns and Foote, 1976). These reactions provide a useful method for the incorporation of molecular oxygen into the benzene rings.

$(C_6H_4X)_2C{=}CHOMe$ (X = H, *p*-OMe) (**78**) $\xrightarrow[C_6H_6]{DNT/h\nu/O_2}$ (**79**) [C_6H_4X, CO_2Me, OH, X] + (**80**) [C_6H_4X, =O, O, X]

(**78**) $\xrightarrow[\text{ether, } -78°]{DNT/h\nu/O_2}$ (**82**) [C_6H_4X, H, OMe, O–O, X] ⟶ (**81**) [H, C_6H_4X, OMe, H, O–O, X] + $(C_6H_4X)_2CO$; (**81**) ⟶ (**80**) (41)

(**83a**) X = Y = H
(**83b**) X = Y = Ph
(**83c**) X = i-C_3H_7, Y = Ph
(**83d**) X = Me, Y = Ph

RB/$h\nu$/O_2/−75°C, acetone; Δ; RB/$h\nu$/O_2/−75°C, MeOH; $^1O_2$; Δ

(**86**) (**87**) (**88**) (**85**) (**84**) (42)

On the other hand, the photooxygenation of indenes in protic solvents such as methanol is known to give different types of products apparently derived from dioxetanes (Fenical *et al.*, 1969a,b). Recently, Burns and Foote (1974) have isolated the dioxetanes (**88b–d**) by low temperature (−78°C) photooxygenation of indenes (**83c–d**) in methanol.

Photooxygenation of benzhydrylidenecyclobutane (**89**) in chloroform produces a thermally stable bisperoxide (**90**) in good yield [Eq. (43)] (Rio *et al.*, 1972, 1973). An expected allylic hydroperoxide formation has not been observed.

MB/$h\nu$/O_2, $CHCl_3$; $^1O_2$

(**89**) (**90**) (43)

A similar type of 1,4-cycloadditions to vinyl-substituted heterocyclic (Matsumoto *et al.*, 1975) and polycyclic aromatic (Matsumoto and Kondo, 1975) systems has been reported. Thus, the photooxygenation of 1-vinylnaphthalene (**91**) and 9-vinylphenanthrene (**92**) in carbon tetrachloride at room temperature results in the stereospecific formation of the thermally stable endoperoxides **93** and **94**, respectively [Eqs. (44)–(45)] (Matsumoto and Kondo, 1975; Matsumoto *et al.*, 1977).

TPP/$h\nu$/O_2, CCl_4 (44)

(91)

(93a) $R_1 = R_2 = H$
(93b) $R_1 =$ Ph, $R_2 = H$
(93c) $R_1 =$ Me, $R_2 = H$
(93d) $R_1 = H$, $R_2 =$ Me

TPP/$h\nu$/O_2, CCl_4 (45)

(92)

(94a) $R_1 = R_2 = H$
(94b) $R_1 = R_2 =$ Me
(94c) $R_1 = H$, $R_2 =$ Me

C. 1,4-Cycloaddition to Carbocyclic Nonbenzenoid Aromatic Compounds

1. Fulvenes

Fulvene systems behave in many respects like five-membered heteroaromatic systems such as furans or pyrroles on photooxygenation to give 1,4-endoperoxides as primary products (Hasselmann, 1952; Koch, 1970). Photooxygenation of 1,2,3,4-tetraphenylfulvene (**95**) gives 1,4-endoperoxide (**96**) and the diepoxy derivative (**97**) resulting from photorearrangement of **96** [Eq. (46)] (Dufraisse *et al.*, 1957a; Basselier, 1964).

eosin/$h\nu$/O_2, EtOH (46)

(95) **(96)** **(97)**

Dye-sensitized photooxygenation of 6,6-dimethylfulvene (**98**) at room temperature gives three products: **99**, **100**, and 2(3*H*)-oxepinone (**101**) [Eq. (47)] (Storianetz *et al.*, 1971; Harada *et al.*, 1972b). The product ratio is highly sensitive to solvent, and **99** is obtained as a major product in methanol and in pyridine (Harada *et al.*, 1972b). On photooxygenation at $-70°C$, the peroxidic intermediate (**102**) has been isolated, and thermal isomerization and rearrangement of **102** by slow warming to room temperature gives **99** and **101** (Harada *et al.*, 1973). The formation of the products has been explained by a mechanism involving allene oxide (**103**) and its valence tautomer, cyclopropanone (**104**). The allene oxide (**103**) is proposed to be formed by rearrangement of 1,4-peroxide (**102**) analogous to the formation of 4,5-epoxy-*cis*-2-pentenal from cyclopentadiene (Schulte-Elte *et al.*, 1969; Adams and Trecker, 1971), or by way of 1,6-peroxide (**105**). The peroxide (**105**) can be formed by $[6\pi + 2\pi]$ cycloaddition or by rearrangement of **102**. Rearrangement of **104** accompanied by opening of the cyclopropane ring would give **101** [Eq. (47)].

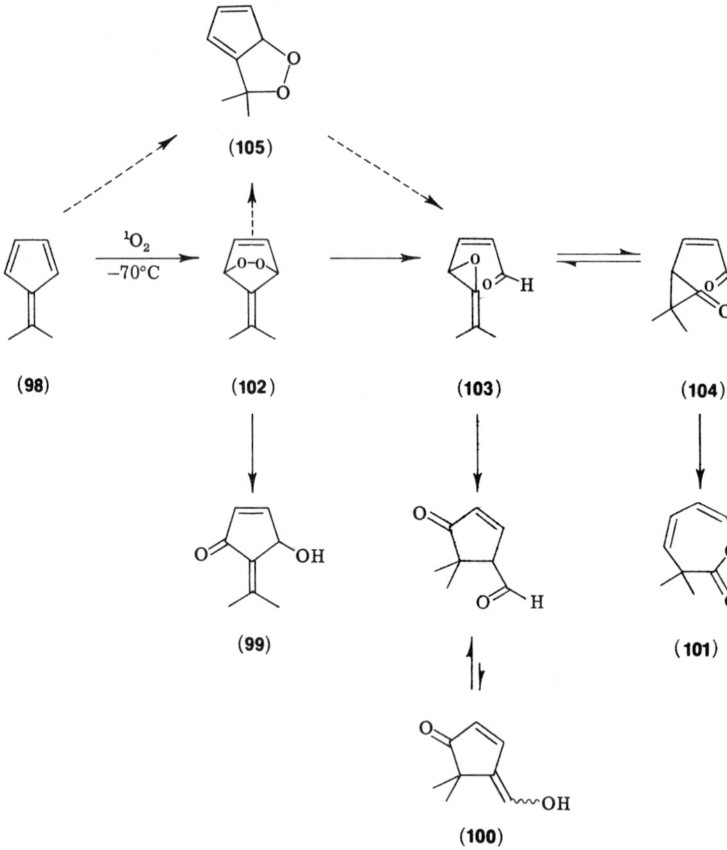

(105)

$^1O_2$, $-70°C$

(98) (102) (103) (104)

(47)

(99) (101)

(100)

Photooxygenation of **98** in alkaline methanolic solution gives α-furyl-isopropylketone (**106**) together with the methanol adducts (**107** and **108**) [Eq. (48)] (Harada *et al.*, 1972a). The formation of **106** is again explained in terms of the allene oxide (**103**), although direct proof for the intermediacy of the allene oxide and cyclopropanone in these reactions seems to be lacking,

102 → **103** $\xrightarrow{MeOH/OH^-}$ → (**106**)

(48)

99 $\xrightarrow{MeOH/OH^-}$ (**107**) + (**108**)

Photooxygenation of 6-methyl-6-ethyl-(**109a**), 6-methyl-6-phenyl-(**109b**), and 6,6-pentamethylenefulvene (**109c**) in benzene produces the corresponding 3,3-disubstituted 2(3*H*)-oxepinone (**110a–c**), respectively (Harada *et al.*, 1972b). 6-Phenyl-(**109d**) and 6-isopropylfulvene (**109e**) give 3-phenyl-2(7*H*)-oxepinone (**111d**) and an inseparable mixture of **110e** and **111e**, respectively [Eq. (49)] (Kawamoto *et al.*, 1972). Thus, the reaction provides a useful method for the preparation of oxepinones.

$\xrightarrow{^1O_2}$ (**110**) + (**111**)

(49)

(**109a**) R_1 = Et, R_2 = Me
(**109b**) R_1 = Me, R_2 = Ph
(**109c**) R_1 = R_2 = $-(CH_2)_5-$
(**109d**) R_1 = Ph, R_2 = H
(**109e**) R_1 = i-C_3H_7, R_2 = H

1,4-Endoperoxides have been isolated by low temperature photooxygenation of polyarylfulvenes (Le Roux and Basselier, 1970; Le Roux and Goasdoue, 1975). For example, when 6-methyl-1,2,3,4-tetraphenylfulvene (**112b**) in acetone containing methylene blue is irradiated at −70°C under oxygen with light through chromate filter, 1,4-endoperoxide (**113b**) is obtained in good yield (Le Roux and Goasdoue, 1975). On dissolving into a mixture of benzene and methanol, the peroxide (**113b**) rearranges rapidly to 1,2-dioxetane (**114**), which on heating decomposes to **115b** with the concomitant emission from the excited state of **115b**. A polar mechanism for the rearrangement has been proposed on the basis of the influence of solvent acidities and the substituent effect on the reaction [Eq. (50)].

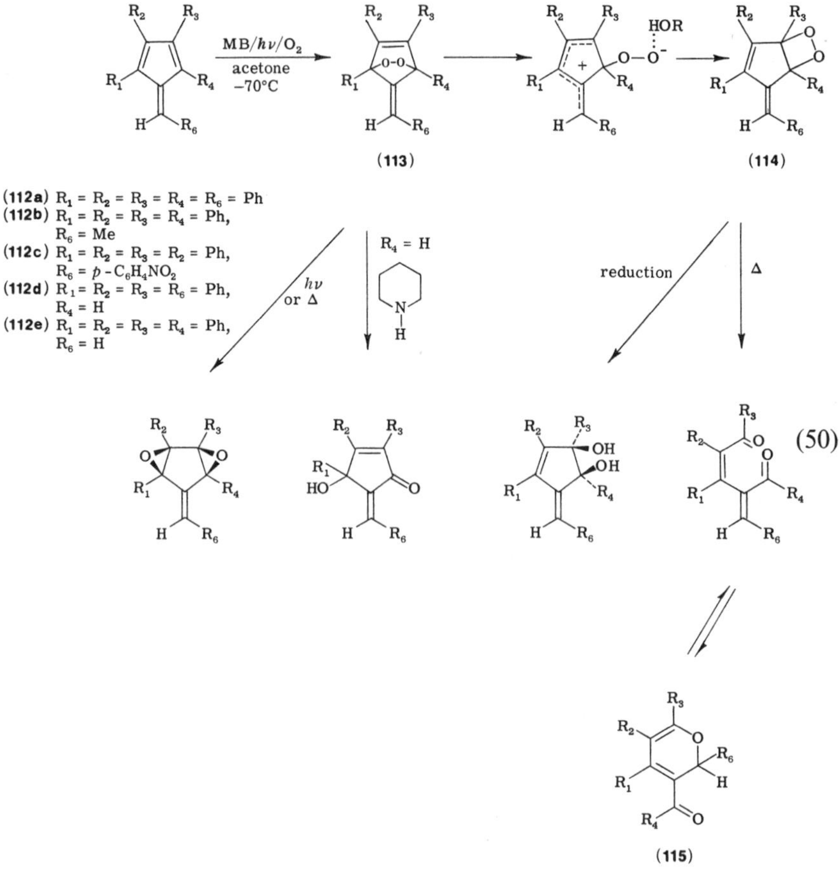

R_2 R_3 R_1 R_4 H R_6 — $MB/h\nu/O_2$, acetone, −70°C → (**113**) → HOR, $O{-}O^-$ → (**114**)

(**112a**) $R_1 = R_2 = R_3 = R_4 = R_6 = Ph$
(**112b**) $R_1 = R_2 = R_3 = R_4 = Ph$, $R_6 = Me$
(**112c**) $R_1 = R_2 = R_3 = R_2 = Ph$, $R_6 = p\text{-}C_6H_4NO_2$
(**112d**) $R_1 = R_2 = R_3 = R_6 = Ph$, $R_4 = H$
(**112e**) $R_1 = R_2 = R_3 = R_4 = Ph$, $R_6 = H$

$h\nu$ or Δ; $R_4 = H$, piperidine (N–H); reduction; Δ

(OH, OH products; R₁ HO ketone; diepoxide) (50)

(**115**)

The related compounds, spiro[2,4]hepta-4,6-diene (**116**), gives the diepoxide (**117**) and epoxy ketone (**118**) presumably by way of the unisolated endoperoxide (**119**) [Eq. (51)] (Takeshita *et al.*, 1973). Cycloaddition to the heptafulvene (**120**) in acetone gives a 1:1 mixture of endoperoxides (**121**) and (**122**), which on treatment with triethylamine produce the oxaazulene (**123**) [Eq. (52)] (Oda and Kitahara, 1969a).

MB/$h\nu$/O_2, $CHCl_3$

(116) (119) (117) (118)

(51)

RB/$h\nu$/O_2, acetone

(120) (121) + (122)

Et_3N

(123)

(52)

2. *Tropolones and Tropones*

Dye-sensitized photooxygenation of tropolone and benztropolone methyl ether has been investigated by Forbes and Griffiths (1966, 1967, 1968). Rose Bengal-sensitized photooxygenation of tropolone methyl ether (**124**) in carbon disulfide–methanol–ether (14:1:1.5) produces the diester (**125**). When the reaction is carried out in carbon disulfide, the sole monomeric product is lactone **126** which isomerizes to **127** in the work-up process.

The peroxidic intermediate (**128**) has been isolated by low temperature ($-20°C$) photooxygenation. An initial reaction of the peroxide (**128**) leading to **125** and **126** is considered to be the cleavage of the weak peroxide bond followed by rearrangement to **129**. The diradical (**129**) containing an effective allylic moiety would undergo rapid rearrangement to epoxy ketene **130**, a common intermediate for **125** and **126**. In the presence of alcohols, the ketene (**130**) would undergo rapid nucleophilic attack (**131**) to give the observed diester (**125**), whereas in protic solvents **130** can decompose by an intramolecular reaction giving lactone **126** [Eq. (53)]. Evidence for the intermediacy of the ketene (**130**) during decomposition of the peroxide (**128**) is obtained by low temperature infrared spectroscopy ($-30°C$), in which a new intense absorption band at 2110 cm^{-1}, a characteristic band of ketenes, has been detected (Forbes and Griffiths, 1968). However, an alternative pathway involving a direct attack of alcohols to **128**, not via **130**, cannot be ruled out.

(53)

Benztropolone methyl ethers (**132a–b**) are efficiently photooxygenated in the presence of sensitizer to give high yields of lactone **133a–b**. In this case, an analogous type of endoperoxide (**134**) and diradical (**135**) are proposed as the intermediates [Eq. (54)] (Forbes and Griffiths, 1966).

Photooxygenation of **132c** in methanol also produces **133b** (Forbes and Griffiths, 1967).

(**132a**) R_1 = H, R_2 = OMe
(**132b**) R_1 = OH, R_2 = OMe
(**132c**) R_1 = OH, R_2 = H

HP/$h\nu$/O_2

(**134**) (**135**) (**133**)

(54)

Tropone (**136**), on sensitized photooxygenation in acetone, gives a stable 1,4-endoperoxide (**137**). Some chemical reactions with **137** are shown in Eq. (55) (Oda and Kitahara, 1969b). Under similar conditions, 2,3-homotropone (**138**) gives endoperoxide **139** which can be transformed into 1,2-dioxocycloocta-3,5-diene (**140**) by reduction [Eq. (56)] (Ito *et al.*, 1975).

HP/$h\nu$/O_2, acetone; Et_3N; thiourea; Δ

(**136**) (**137**)

(55)

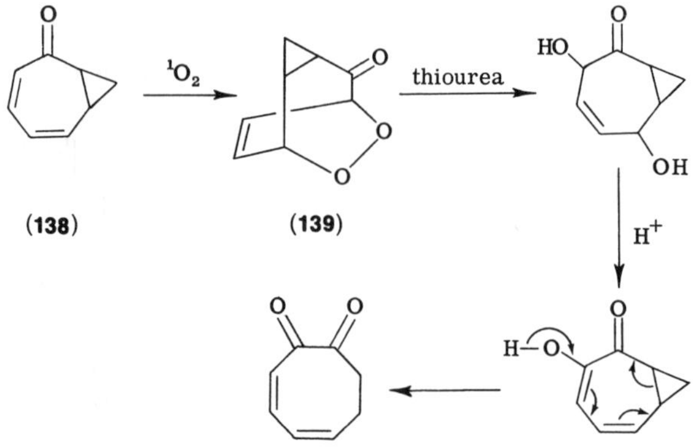

(138) (139) (140) (56)

3. *Arene Oxides and Annulenes*

In Section II,A, the intermediate benzene oxide (**87**) has been shown to undergo 1,4-cycloaddition with singlet oxygen. The reaction of the tautometic benzene oxide–oxepin system [(**141**) ⇌ (**142**)] with singlet oxygen generated from hypochlorite–hydrogen peroxide (Foote *et al.*, 1968) has been shown to yield 1,4-endoperoxide (**143**) which undergoes quantitative rearrangement to *trans*-benzene trioxide (**144**) (Vogel *et al.*, 1972) in chloroform at 45°C with a half-life of ca. 14 hr (Forster and Berchtold, 1972). The peroxide (**143**) can also be prepared by the reaction with singlet oxygen generated from the adduct of ozone and triphenyl phosphite (Murray and Kaplan, 1968, 1969), but the photoreaction of **142** in the presence of methylene blue and oxygen gives mainly phenol. Endoperoxide **143** reacts with triethylamine to give ketol **145**, and with triphenyl phosphite to give *trans*-

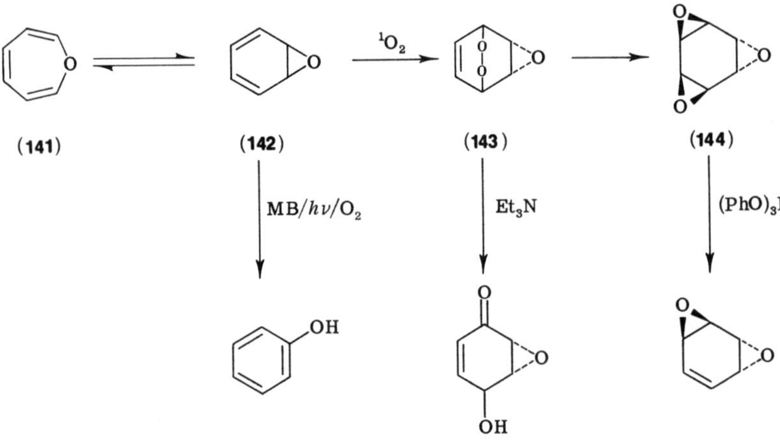

(141) (142) (143) (144) (145) (146) (57)

benzene dioxide (**146**) [Eq. (57)] (Forster and Berchtold, 1975). Addition of singlet oxygen to indan 8,9-oxide (**147**) is more facile than that to **142** giving a stable 1,4-endoperoxide (**148**) [Eq. (58)] (Forster and Berchtold, 1975).

(147) $\xrightarrow{^1O_2}$ (148) $\xrightarrow{(PhO)_3P}$ (58)

Methylene blue-sensitized photooxygenation of the 1-benzoxepin (**149**) system, where naphthalene oxide (**150**) does not contribute to any major extent (< 5%), provides the endoperoxide (**151**) in 60% yield. The peroxide (**151**) undergoes an unusual rearrangement to aldehyde **152** probably via the quinone methide (**153**) on deoxygenation with trimethyl phosphite [Eq. (59)] (Baldwin and Lever, 1973).

(150) ⟵// (149) $\xrightarrow{^1O_2}$ (151) $\xrightarrow{P(OMe)_3}$ → (153) CHO → (152) CHO (59)

Reaction of singlet oxygen with [$4n + 2$] annulenes has been studied extensively by Vogel *et al.* (1974). Methylene blue-sensitized photooxygenation of 1,6-methano[10]annulene (**154**) in dichloromethane gives 1,4-endoperoxide **155**, which on warming rapidly rearranges to diepoxide **156**. Extensive photooxygenation of **154** under the conditions produces tetraepoxide **157**, which may be formed via **156** by addition of a second singlet oxygen molecule [Eq. (60)] (Vogel *et al.*, 1974). Photooxygenation of

[14]annulene (**158**) under similar conditions gives **159**, **160**, and **161** [Eq. (61)] (Vogel *et al.*, 1974).

MB/$h\nu$/O_2
CH_2Cl_2

(**154**) (**155**)

(60)

$^1O_2$

(**157**) (**156**)

(**158**)

MB/$h\nu$/O_2
CH_2Cl_2

(61)

CHO
CHO

(**159**) (**160**) (**161**)

Methylene blue-sensitized photooxygenation of 1,6-imino[10]annulene (**162**) in dichloromethane at 10°–15°C gives 1,4-endoperoxide (**163**) (Vogel *et al.*, 1974; Schäfer-Ridder *et al.*, 1976). Treatment of the endoperoxide (**163**) with nitrosylchloride and triethylamine at −78°C provides the naphthalene-1,4-endoperoxide (**164**), which was not produced by singlet oxygen reaction with naphthalene (see Section I,C). The naphthalene endoperoxide (**164**) readily decomposes to naphthalene and singlet oxygen with half-life of 303 min at 20°C. Thermolysis of **163** gives diepoxide **165** which on similar treatment with nitrosyl chloride produces the naphthalene *cis*-diepoxide **166** [Eq. (62)] (Schäfer-Ridder *et al.*, 1976; Vogel *et al.*, 1976).

(162) $\xrightarrow{^1O_2}$ (163) $\xrightarrow[-78°C]{NOCl/Et_3N}$ (164) $\xrightarrow{\Delta}$ naphthalene + $^1O_2$

(163) $\xrightarrow{\Delta}$ (165) $\xrightarrow{NOCl/Et_3N}$ (166); (164) $\xrightarrow{P(Ph)_3}$

(62)

Photooxygenation of 1,6-oxido[10]annulene (**167**) under similar conditions gives trioxide **168**, whereas extensive photooxygenation of **167** gives naphthalene pentaoxide (**169**) [Eq. (63)] (Vogel *et al.*, 1974, 1976), although their stereochemistry is not yet clarified.

(167) $\xrightarrow[CH_2Cl_2]{MB/h\nu/O_2}$ (168) $\xrightarrow{^1O_2}$ (169) (63)

An analogous reaction is reported with methyl cyclohepta-2,4,6-triene-1-carboxylate (**170**). In this case, the intermediate endoperoxide (**171**) is considered to undergo further photochemical rearrangement to the diepoxide (**172**) [Eq. (64)] (Ritter *et al.*, 1974). Related photooxygenations of cycloheptatriene are well known (Schenck, 1957; Kende and Chu, 1970).

(170) CO_2Me $\xrightarrow{^1O_2}$ (171) CO_2Me $\xrightarrow{h\nu}$ (172) CO_2Me (64)

III. OXIDATIONS OF PHENOLIC AND ENOLIC SYSTEMS

In view of their significance as synthetic tools and in biosynthetic schemes for many natural products, oxidations of phenols and the related enolic systems have been studied extensively using various kinds of oxidizing reagents (Musso, 1963; Scott, 1964; Müller *et al.*, 1966; Taylor and Battersby, 1967). Photochemical oxidations of phenols have also been investigated as one of such oxidation methods. Unsensitized photooxidation of phenols by ultraviolet (2537 Å) irradiation in aqueous media is known to give a mixture of dimeric and hydroxylated products, where singlet oxygen is not involved (Johnson and Tamm, 1964; Joschek and Miller, 1966; Matsuura and Omura, 1974). The dye-sensitized photooxygenation, on the other hand, produces hydroperoxides as primary products, which often undergo secondary reactions to give more complex products. The products of the dye-sensitized photooxygenations are in many cases quite analogous to those of usual radical oxidations (Musso, 1963; Sosnovsky, 1971) and base-catalyzed autoxidations (Sosnovsky and Zaret, 1970).

Recent studies have demonstrated that both of Type I and singlet oxygen processes are involved in the dye-sensitized photooxygenation of phenols. Chemically generated singlet oxygen can also react with certain reactive phenols, although their reactivities toward singlet oxygen are in many cases very low. In addition, phenols have been known to serve as singlet oxygen quenchers.

In this section, a brief survey on the mechanisms for the sensitized photooxygenation of phenols as well as the chemistry of singlet oxygen reactions with phenols are described. Examples of the oxidative cleavage of aromatic enolic systems by singlet oxygen are also described.

A. Mechanisms for Photosensitized Oxygenation of Phenols

There are two major classes of reactions, Type I and Type II processes (Gollnick, 1968b; Foote, 1968a,b), open to the sensitizer triplet in the dye-sensitized photooxygenation of readily oxidizable substrates such as phenols or tertiary amines (see Scheme 1). In Type I process, the sensitizer triplet interacts with phenols resulting in hydrogen or electron transfer (k_1), whereas Type II process involves activation of molecular oxygen to give singlet oxygen (k_2) which can react with phenols. Depending on the sensitizer, the solvent (neutral or basic), and the concentrations of the substrate and oxygen, the mechanism of the photooxygenation may change, e.g.,

Type I Type II

$$\mathrm{PO\cdot\,(POH^{+})} + \mathrm{\cdot Sens{-}H\,(Sens^{\overline{\cdot}})} \underset{k_1}{\overset{\mathrm{POH}}{\longleftarrow}} {}^{3}\mathrm{Sens}^{*} \underset{k_2}{\overset{^{3}\mathrm{O}_2}{\longrightarrow}} {}^{1}\mathrm{O}_2 \underset{k_r}{\overset{\mathrm{POH}}{\longrightarrow}} \text{products}$$

$$\mathrm{PO\cdot\,(POH^{+})} \xrightarrow{^{3}\mathrm{O}_2} \text{products}$$

$$ {}^{1}\mathrm{O}_2 \xrightarrow[k_q]{\mathrm{POH}} {}^{3}\mathrm{O}_2 + \mathrm{POH} \qquad {}^{1}\mathrm{O}_2 \xrightarrow{k_d} {}^{3}\mathrm{O}_2$$

from Type I to Type II (Foote, 1976). There is a second competition to be considered for the interactions of singlet oxygen with phenols: phenols (POH) are known both to react (k_r) with singlet oxygen and quench it (k_q) without reaction.

Earlier reports on flash photolysis studies (Grossweiner and Zwicker, 1959; Zwicker and Grossweiner, 1963; Chrysochoos and Grossweiner, 1968) of the eosin-sensitized photooxidation of phenol and its derivatives have shown that the primary step involves electron or hydrogen transfer from phenol to the triplet state of the dye, resulting in the formation of phenoxy radical and semireduced dye radical. The situation, however, is more complex when sufficient oxygen is present in the reaction system. For example, it has been demonstrated that the rate constant for the formation of singlet oxygen is tenfold greater than that for the hydrogen abstraction of the triplet state of eosin from phenol in neutral aqueous solution under aerobic conditions (Zwicker and Grossweiner, 1963; Grossweiner and Zwicker, 1959).

Recent studies have demonstrated the intermediacy of singlet oxygen in some phenol photooxygenation reactions by competitive inhibition (Saito *et al.*, 1970; Matsuura *et al.*, 1969, 1972a), quenching techniques (Matsuura *et al.*, 1972a), and solvent isotope effect (Saito *et al.*, 1975a), and by showing that the same products and relative rates are obtained with chemically generated singlet oxygen (Saito *et al.*, 1972a; Matsuura *et al.*, 1972b; Pfoertner and Böse, 1970; Grams, 1971). These results should not be generalized to imply that singlet oxygen mechanism is always obtained from phenol oxidations. Indeed, Matsuura *et al.* (1972a) have presented evidence that both Type I and singlet oxygen mechanisms can occur in the photooxygenation of 2,6-di-*tert*-butyl phenol under one set of conditions (*vide infra*).

In general, the products of singlet oxygen reaction with phenols are almost the same as those of usual radical oxidations. 2,6-Di-*tert*-butyl-4-methyl phenol (**173a**) gives the hydroperoxide (**174a**) (Foote, 1976; Matsuura *et al.*, 1965). Products of phenoxy radical coupling are often observed, as well as products of secondary reactions of the initial hydroperoxides which are isolable only under mild conditions. For example, dye-sensitized photooxygenation of 2,6-di-*tert*-butyl phenol (**173b**) produces benzoquinone (**175**) and diphenoquinone (**176**) [Eq. (65)] (Matsuura *et al.*, 1965). Reaction of **173b** with chemically generated singlet oxygen, namely by thermolysis of 9,10-diphenylanthracene peroxide and of the triphenyl phosphite–ozone adduct, also produces the same products **175** and **176** under oxygen bubbling, whereas under nitrogen atmosphere, the reaction gives only the diphenoquinone (**176**) (Matsuura *et al.*, 1972a). Oxidation of **173b** in the solid phase using microwave discharge-generated singlet oxygen gives **175** as a sole product (Saito, Takahashi, Matsuura, unpublished). The results indicate that the hydrogen abstraction from the phenol by singlet oxygen does indeed occur to give a phenoxy radical **177** as an initial product, and that the benzoquinone (**175**) can be formed only at a sufficient ground-state oxygen concentration. A competitive inhibition and quenching experiments have shown that both singlet oxygen and the triplet sensitizer are responsible for the hydrogen abstraction in the photooxygenation. The phenoxy radical (**177**) thus formed can react with ground-state triplet oxygen to give a peroxy radical which is then converted to **175** probably via **174**. A semiquinone radical intermediate is detected by ESR techniques in the eosin-sensitized photooxygenation of phenols (Learer, 1971). On the other hand, two molecules of the radical couple to yield a biphenol which is dehydrogenated to the observed product (**176**) [Eq. (65)] (Matsuura *et al.*, 1972a).

(**173a**) R = Me
(**173b**) R = H
(**174**)
(**175**)
^{3}Sens* and $^1O_2$, R = H
SH
$^3O_2$
(**177**)
^{3}Sens*
(**176**)

(65)

Very similar results are obtained with methyl-substituted phenols. Dye-sensitized photooxygenation of 2,3,6-trimethylphenol (**178**) gives benzoquinone (**179**) and biphenol (**180**), and the product ratio is shown to be highly dependent on solvent (Pfoertner and Böse, 1970). Reaction with chemically generated singlet oxygen produces only **179**. A mechanism involving an electrophilic attack of singlet oxygen on the para position of the phenol followed by an intermolecular hydrogen transfer has been proposed [Eq. (66)].

Sens $/h\nu/O_2$ (66)

(**178**) (**179**) (**180**)

Mechanistic studies on singlet oxygen reaction with trialkylated phenols (**181**) leading to the hydroperoxides (**182**) have been made by Foote *et al.* (Foote, 1976; Thomas and Foote, 1973, 1978). They have observed a strong correlation between half-wave oxidation potentials of the phenols and the reaction rates (Table IV). A similar relationship is observed with methoxybenzenes (Saito *et al.*, 1972b). Direct spectroscopic evidence for the postulated phenoxy radical (**183**) has been obtained in the reaction (Thomas and Foote, 1973). The primary step of the reaction is suggested to be an electron transfer process from phenols to singlet oxygen, followed by rapid proton transfer. The phenoxy radicals thus formed may combine with hydroperoxy radical or react with singlet and/or triplet oxygen to give **182** [Eq. (67)] (Foote, 1976).

Table IV Reactivities of Phenols toward Singlet Oxygen[a]

Phenol	k_A (mole^{-1} sec^{-1})[b]	β (M)	$E_{1/2}$ (V)[c]
4-X-2,6-Di-*tert*-butyl			
X = C_6H_5	7.9×10^6	0.015	1.19
= CH_3	5.6×10^6	0.021	1.20
= $C_6H_5CH_2$	3.8×10^6	0.031	1.29
= $(CH_3)_3C$	3.4×10^6	0.034	1.33
= $HOCH_2$	1.5×10^6	0.079	1.35
= H	1.0×10^6	0.12	1.48
= Br	8.4×10^5	0.14	1.39
= $C_2H_5O_2C$	2.2×10^5	0.53	1.68
2,4,6-Triphenyl[d]	2.5×10^8	0.0005	1.02

[a] In methanol, methylene blue sensitized (Foote, 1976).
[b] Sum of reaction and quenching.
[c] In acetonitrile, 0.1 M tetraethylammonium perchlorate.
[d] This substrate quenches singlet oxygen but does not react to give isolable products.

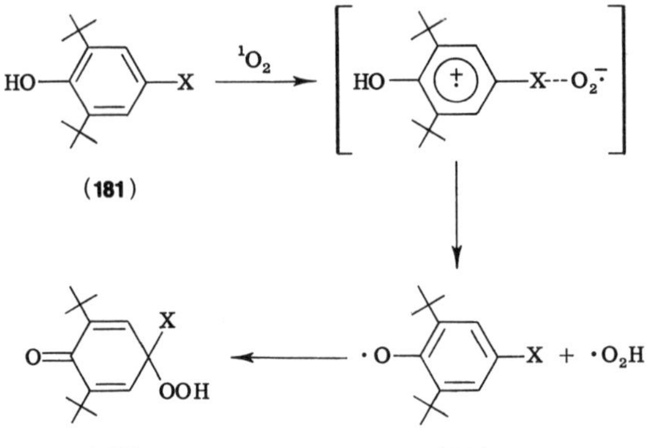

(181) (182) (183)

$X = C_6H_5$, Me, $C_6H_5CH_2$, CH_2OH, H, *t*-butyl, Br

(67)

A possibility that 1,4-cycloaddition of singlet oxygen to the aromatic system of electron-rich phenols has been pointed out by Saito *et al.* (1970, 1972a). 4,6-Di-*tert*-butyl-3-methoxyphenol (**184**) undergoes reaction with singlet oxygen generated by chemical and photochemical means in the same rate (0.27, β, *M*) to give the hydroperoxide (**185**). It has been suggested that the reaction proceeds by way of 1,4-endoperoxide **186** rather than phenoxy radical (**187**), since the reaction of the stable radical (**187**) generated

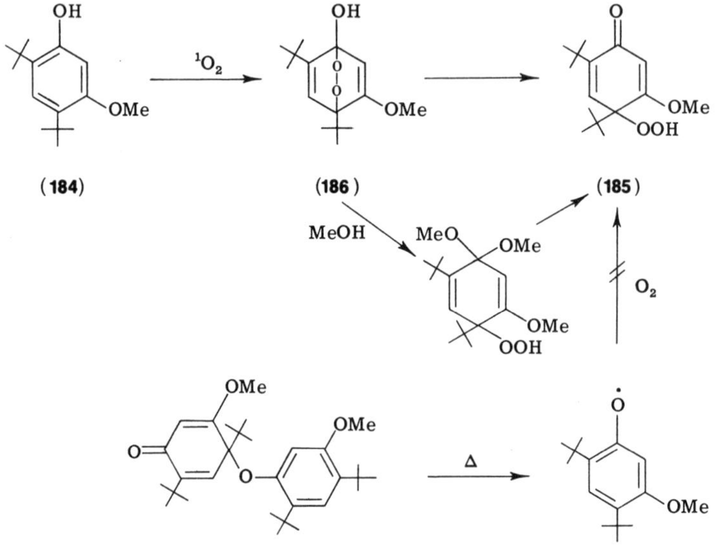

(184) (186) (185) (188) (187)

(68)

by thermolysis of dimer **188** with ground-state oxygen does not yield the observed product (**185**) but gives only **184** under the conditions [Eq. (68)] (Saito *et al.*, 1972a).

In summary, it seems likely that most of the reactive phenols, e.g., trialkylated phenols, undergo reaction with singlet oxygen by way of partial electron transfer intermediates, followed by rapid proton transfer to give phenoxy radicals. Highly electron-donor substituted phenols, on the other hand, may undergo 1,4-cycloaddition with singlet oxygen as has been observed with methoxybenzenes (see Section II,A). A Type I mechanism may also play an important role as well as singlet oxygen process in the dye-sensitized photooxygenation, especially of less reactive phenols at low oxygen concentration, although the mechanistic details for the reaction appears to remain to be clarified.

B. Oxidations of Phenols

Photooxygenation of para substituted phenols generally gives the corresponding 4-hydroperoxy-2,5-cyclohexadiene as a primary product. The hydroperoxides undergo secondary reactions such as elimination or reduction, depending upon solvent and para substituents, to give more complex products. For example, eosin-sensitized photooxygenation of 2,6-di-*tert*-butyl-4-methyl phenol (**173a**) and 2,4,6-tri-*tert*-butyl phenol (**173c**) in methanol gives **189** and **190** from the hydroperoxides **174a** and **174c**, respectively [Eq. (69)] (Matsuura *et al.*, 1965).

eosin/ $h\nu/O_2$ MeOH

OOH R

(**173a**) R = Me
(**173b**) R = H
(**173c**) R = *t*-butyl

(**174**)

MeOH

(69)

CH_2OMe

(**190**) (**189**)

Photooxygenation of 2,5-di-*tert*-butyl-4-methoxyphenol (**191**) in methanol produces benzoquinone (**192**) and formaldehyde by way of **193**. Analogously, 2,5-di-*tert*-butylhydroquinone (**191b**) gives **192**, which on further irradiation undergoes photorearrangement (Orland *et al.*, 1968) to give **194** and **195** [Eq. (70)] (Saito *et al.*, 1970; Matsuura *et al.*, 1970).

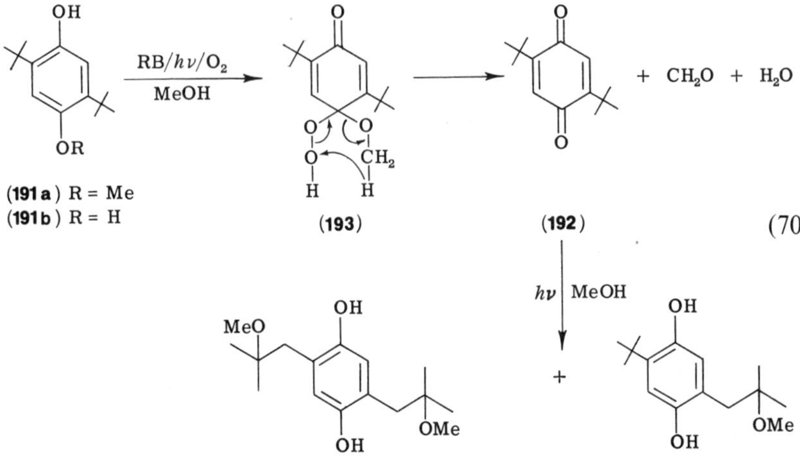

(**191a**) R = Me
(**191b**) R = H
(**193**) (**192**) (70)

(**194**) (**195**)

Dye-sensitized photooxygenation of para substituted phenols in alkaline aqueous media produces *p*-quinols in synthetically useful yields. The quinols are formed by displacement reaction of the initially formed hydroperoxides by hydroxy anion, liberating hydrogen peroxide. Thus, photooxygenation of *p*-hydroxyphenylacetic acid (**196a**) and phloretic acid (**196b**) in alkaline phosphate buffer give the corresponding *p*-quinol (**197a**) and **197b**, respectively (Saito *et al.*, 1975a). In the latter case, spirolactone (**198**) has also been obtained [Eq. (71)] (Matsuura *et al.*, 1967a; Saito *et al.*, 1975a). *p*-Phenethyl alcohol (**199a**) and *N*-acetyltyramine (**199b**) give the fused bicyclic product (**200a** and **200b**) in a stereospecific manner, respectively [Eq. (72)]. The participation of singlet oxygen in these reactions has been

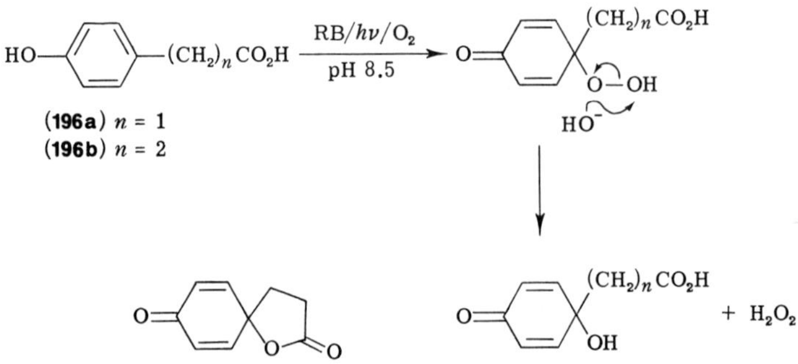

(**196a**) n = 1
(**196b**) n = 2

(71)

(**198**)

(**197a**) n = 1
(**197b**) n = 2

HO–C₆H₄–CH$_2$CH$_2$XH $\xrightarrow[\text{pH 9.0}]{RB/h\nu/O_2}$ (quinol with CH$_2$CH$_2$XH, OH) ⟶ (200) (72)

(199a) X = O
(199b) X = NCOCH$_3$

shown by quenching techniques and by solvent isotope effects (Saito *et al.*, 1975a).

Dye-sensitized photooxygenation of *p*-hydroxyphenylpyruvic acid (**201**) in its keto form in phosphate buffer gives quinol **197a**, which is converted into homogentisic acid (**202**) on treatment with alkali (pH > 12) [Eq. (73)] (Saito *et al.*, 1975a). The formation of **197a** is interpreted by a mechanism involving hydroperoxide **203** followed by intramolecular reaction of the hydroperoxy group with the keto group of the side chain giving a cyclic peroxide (**204**). The sequence of the reaction is discussed in connection with the mechanism of the action of *p*-hydroxyphenylpyruvate dioxygenase, where atmospheric oxygen is incoporated into both the hydroxy moiety and the carbonyl group of **202** [Eq. (73)] (Lindblad *et al.*, 1970; Nakai *et al.*, 1975).

HO–C₆H₄–CH$_2$COCO$_2$H (201) $\xrightarrow{{}^*O_2}$ (203) [CH$_2$COCO$_2$H, OOH] ⟶ (204) $\xrightarrow{-CO_2}$ (197a) [CH$_2$CO$_2$H, OH]; (197a) $\xrightarrow{\text{pH} < 2}$ lactone; (197a) $\xrightarrow{\text{pH} > 12}$ HO–C₆H₃(OH)–CH$_2$CO$_2$H (202) (73)

Photooxygenation of α-tocopherol, a biological antioxidant, has been studied to gain information on the mechanism by which tissues and lipids are protected from oxidative damage (Grams *et al.*, 1972). α-Tocopherol (**205**), on proflavin-sensitized photooxygenation in methanol, gives at least four products (**206**, **207**, **208**, and **209**). The reaction with chemically generated singlet oxygen also gives the same products but in a different ratio (Grams, 1971). The products are derived from the subsequent reactions of the hydroperoxide (**210**), although the intermediacy of a 1,4-endoperoxide has been proposed [Eq. (74)] (Grams *et al.*, 1972). α-Tocopherol is also known as a singlet oxygen quencher (Grams and Eskins, 1972; Foote and Ching, 1974; Fahrenholtz *et al.*, 1974; Stevens *et al.*, 1974a).

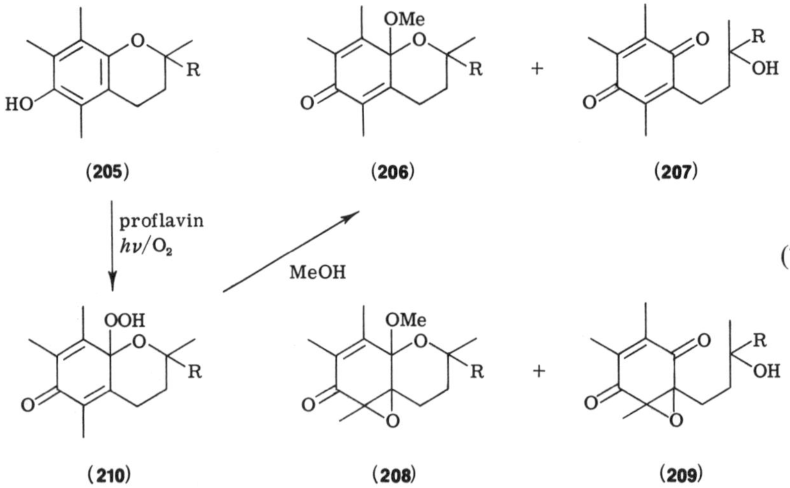

(205) (206) (207)

proflavin $h\nu/O_2$ MeOH (74)

(210) (208) (209)

Flash photolysis studies of the dye-sensitized photooxygenation of tyrosine have provided evidence that it reacts by a Type I mechanism in unbuffered aqueous solution (Chrysochoos and Grossweiner, 1968; Kepka and Grossweiner, 1972; Kasche and Lindqüist, 1965; Nemoto *et al.*, 1967). It has been suggested that singlet oxygen may react extremely slowly with neutral tyrosine and that the excited dye reacts rapidly with it (Weil, 1965; Foote, 1976). The products are not well characterized, but 3-(3,4-dihydroxyphenyl)alanine (DOPA) has been reported to be one of the products (Hara, 1960).

tertiary-Butylcatechol or its monomethyl ether undergoes oxidative cleavage of the aromatic ring. 3,5-Di-*t*-butylpyrocatechol (**211a**) and 4,6-di-*tert*-butyl guaiacol (**211b**) in methanol produce the lactonic acid (**212a**) and its methyl ester (**212b**), respectively (Matsuura *et al.*, 1972b). Phenoxy radical **213** formed by both Type I and singlet oxygen processes can account for the formation of the products. Reaction of **213** with ground-state oxygen followed by hydrogen abstraction would yield α-hydroperoxy ketone **214**, which then undergoes cleavage to give **212** probably via the dioxetane (**215**). The intermediacy of the phenoxy radical (**213**) is supported by the fact that oxygenation of **213** (R = Me) generated by thermolysis of dimer **216** also produces **212b** under the conditions [Eq. (75)]. Analogously, photooxygenation of 3,6-di-*tert*-butyl guaiacol (**217**) in methanol gives α,α′-di-*tert*-butyl muconate (**218**) together with **219** and **220** [Eq. (76)] (Matsuura *et al.*, 1972b).

(75)

(**211a**) R = H
(**211b**) R = Me

RB/$h\nu$/O_2

(**213**)

$^3O_2$

(**212a**) R = H
(**212b**) R = Me

(**216**)

RH

(**215**)

(**214**)

(76)

(**217**)

Sens /$h\nu$/O_2 MeOH

$^3O_2$ RH

(**220**) + (**219**) + (**218**)

Although photooxygenation of 4,6-di-*tert*-butyl resorcinol (**221**) and its methyl ether (**184**) by visible light gives the corresponding hydroperoxide [Eq. (68)], irradiation of **221** in methanol with a high pressure mercury lamp (Pyrex filter) produces the ring cleavage products (**222** and **223**) [Eq. (77)] (Saito *et al.*, 1972a). The keto ester (**222**) can be produced by reaction with chemically generated singlet oxygen under oxygen bubbling, indicating that not only singlet oxygen but also sufficient amounts of ground-state oxygen are necessary for the formation of **222**. A mechanism involving phenoxy radical **224** has been proposed for the complicated multistep reaction [Eq. (77)] (Saito *et al.*, 1972a).

In connection with lignin synthesis in the plant, dye-sensitized photooxygenation of isoeugenol (**225**) has been carried out. Dimeric products **226** and **227** have been isolated [Eq. (78)] (Eskins *et al.*, 1972). The participation of singlet oxygen in the reaction has not been demonstrated.

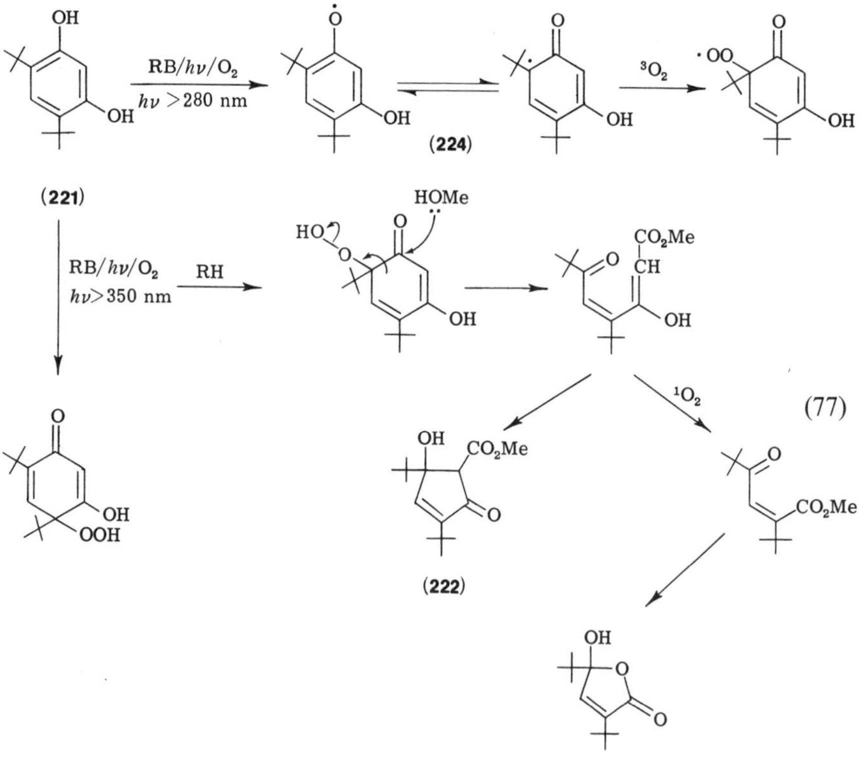

(221) (224) (222) (223)

(77)

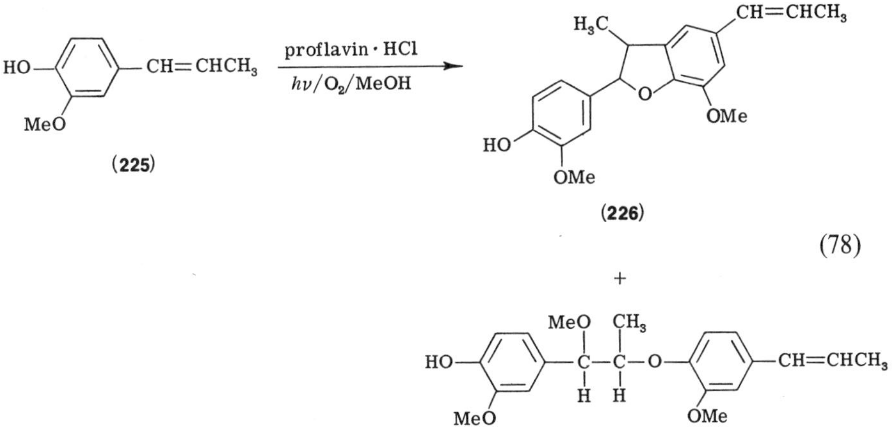

HO—C$_6$H$_3$(OMe)—CH=CHCH$_3$ (225) $\xrightarrow[h\nu/O_2/MeOH]{\text{proflavin} \cdot \text{HCl}}$ (226) + (227) (78)

Related photooxygenations of 9-hydroxy-10-alkylanthracenes (**228**) leading to the corresponding hydroperoxides (**229**) have been demonstrated [Eq. (79)] (Denny and Nickon, 1973).

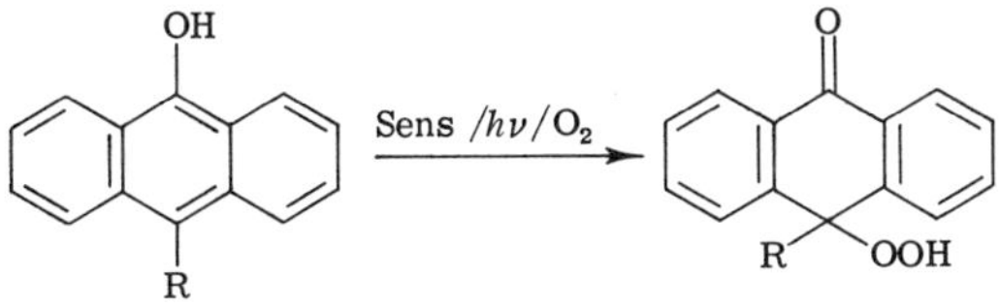

(228) $\xrightarrow{\text{Sens}/h\nu/O_2}$ (229) (79)

R = Me, Et, n-C_3H_7, CH=CHMe, C_6H_6, etc.

C. Oxidations of Enolic Systems

Heterocyclic compounds bearing —C=C—XH (X = O, NR) moieties have been suggested to undergo reactions with singlet oxygen formally analogous to the "ene" reaction. Such examples include pyrroles, indoles, imidazoles, and purines. The hydroperoxidic products are isolable in several systems (White and Harding, 1965; Dufraisse *et al.*, 1965b; Ramasseul and Rassat, 1972; Nakagawa *et al.*, 1975; Saito *et al.*, 1975b). In analogous fashions, enols and enolates are expected to undergo reaction with singlet oxygen to give α-hydroperoxy ketones. α-Hydroperoxy ketones, if formed, are known to undergo base-catalyzed (Richardson *et al.*, 1974b; Sawaki and Ogata, 1975) or photochemical (Richardson *et al.*, 1974a) α-cleavage reactions to give two carbonyl fragments via the dioxetane (path *a*) and/or the carbonyl addition intermediate (path *b*). Indeed, there are many examples in which activated double bonds of enols or enolates suffer oxidative cleavage reaction with singlet oxygen. For these reactions, α-hydroperoxy ketone or its equivalent has been generally accepted as the intermediate, although such intermediates have not been isolated. An alternative mechanism involving concerted 1,2-cycloaddition to the electron-rich double bond seems also probable.

$$\text{—C=C(OY)—} \xrightarrow{^1O_2} \text{—C(OOY)—C(=O)—} \xrightarrow[\text{base or } h\nu]{(a)} \text{—C—C(OY)— (O—O dioxetane)} \longrightarrow \text{—C(=O)—} + \text{YO—C(=O)—}$$

Y = H or $^-$

$$(b)\ \text{RO}^-: \quad \text{—C(OOY)—C(O}^-\text{)(OR)—} \longrightarrow \text{—C(=O)—} + \text{RO—C(=O)—}$$

For example, sensitized or unsensitized photooxygenation of diacetylfilicinic acid (**230**) in methanol containing excess of methoxide gives **231** (Young and Hart, 1967). Chemically generated singlet oxygen also produces **231**. The initial product has been suggested to be **232**, which, on further reactions including a benzylic acid-type rearrangement with solvent and base, can produce the observed product [Eq. (80)].

$$(\mathbf{230}) \xrightarrow{\text{MeOH/NaOMe}} \text{enolate} \xrightarrow{^1O_2} (\mathbf{232}) \xrightarrow{\text{MeOH},\ -CO_2} (\mathbf{231}) \qquad (80)$$

5,6-Diphenyluracil (**233**), on sensitized photooxygenation in liquid ammonia at $-70°C$, gives an unstable peroxide (**234**), which on warming breaks down to the cleavage product (**235**) [Eq. (81)] (Vickers and Foote, 1970). A similar type of reaction is known to occur with *N*-methylpyridinium-3-oxide which results in cleavage of the 2,3-bond to give aminolactone (**236**) probably via the dioxetane intermediate [Eq. (82)] (Mori *et al.*, 1975).

Sens/$h\nu$/O_2, liq. NH_3

(**233**) (**234**) (**235**) (81)

MB/$h\nu$/O_2, $CHCl_3$

MeNCHO

(**236**) (82)

In connection with the metabolism of quercetin (**237a**) catalyzed by quercetinase, a dioxygenase, leading to a depside (**238a**) and carbon monoxide, sensitized photooxygenation of 3-hydroxyflavones **237b** and **237c** has been investigated. Both compounds yield the depside (**238**) together with carbon monoxide and carbon dioxide. These products have been proposed to form through a common hydroperoxide intermediate (**239**). While **239** decomposes to **238** and carbon monoxide via an intermediate (**240**) (path *a*), it goes to **238** and carbon dioxide through rearrangement to a dioxetane (**241**) followed by further oxidation (path *b*) [Eq. (83)] (Matsuura *et al.*, 1967b, 1970).

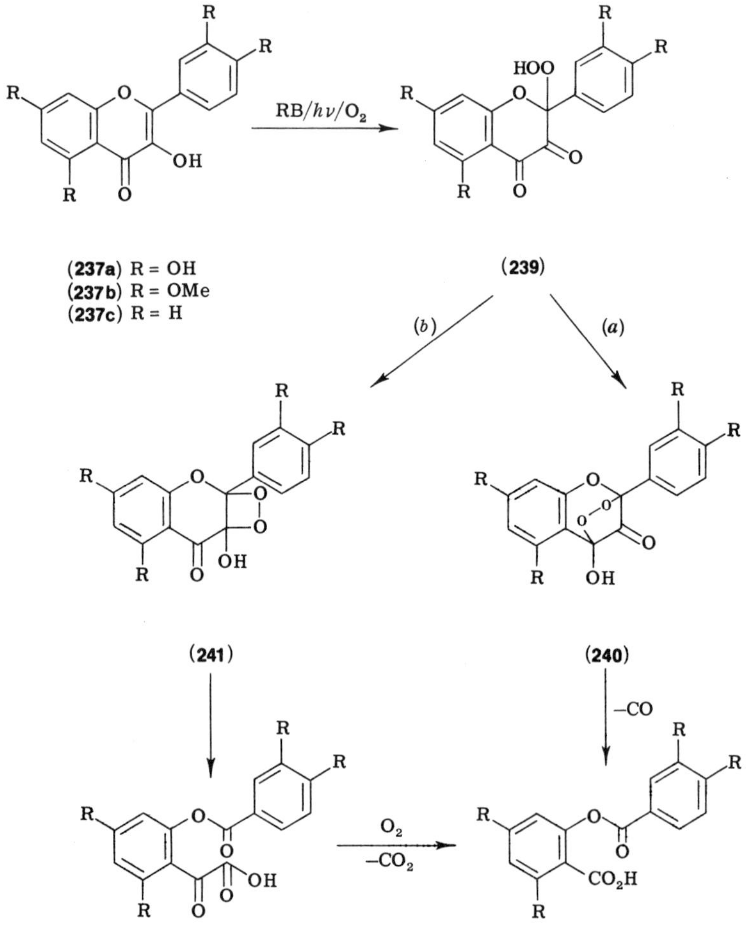

RB/$h\nu$/O_2

(**237a**) R = OH
(**237b**) R = OMe
(**237c**) R = H

(**239**)

(*b*) (*a*)

(83)

(**241**) (**240**)

$-CO$

O_2
$-CO_2$

(**238**)

There are several known examples in which the enol tautomer of the substrate undergoes reaction with singlet oxygen. Methylene blue-sensitized photooxygenation of 3-phenylisocoumaranone (**242**) in methanol gives **243** and **244**, which result from the reaction of the enol tautomer (**245**) [Eq. (84)] (Padwa *et al.*, 1975). Reaction of singlet oxygen with *p*-hydroxyphenyl pyruvic acid in methanol, which exists in its enol form, produces *p*-hydroxybenzaldehyde and oxalic acid (Saito *et al.*, 1975a).

4-Arylazo-1-naphthol (**246**) undergoes dye-sensitized photooxygenation in methanol to give **247** and **248** (Griffith and Hawkins, 1972). In this

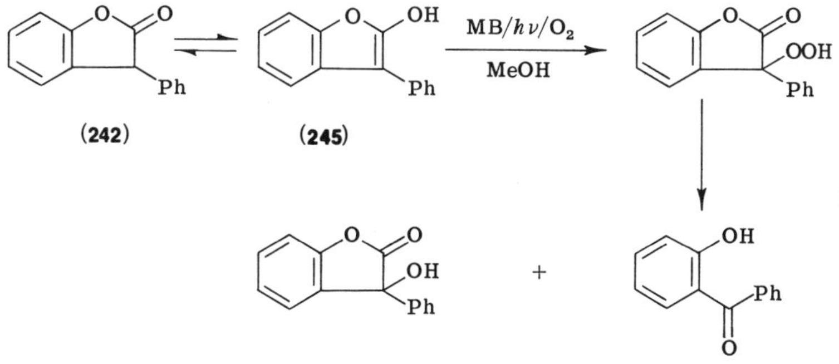

(242) (245) (244) (243)

(84)

case, the attack of singlet oxygen on the less-favored tautomer (**249**) (**246**: **249** = 4:1 in methanol) has been proposed [Eq. (85)].

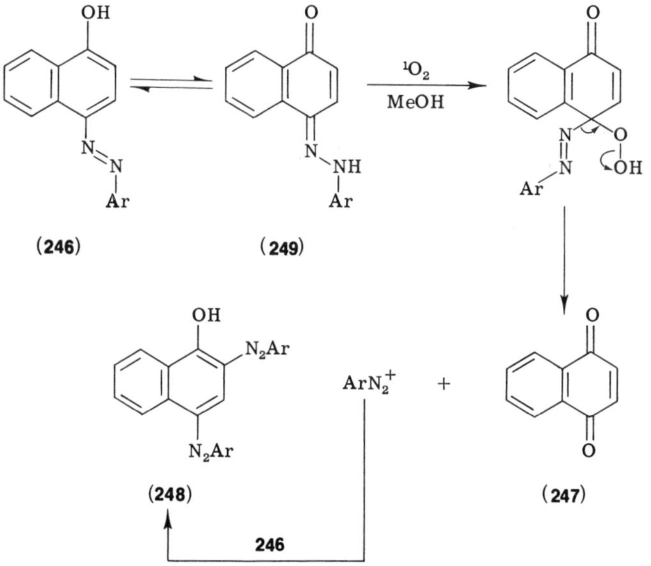

(246) (249) (248) (247) 246

(85)

Unsensitized photooxygenation of photochromic compound **250**, which produces photoenol **251** on normal photolysis, is reported to give peroxide **252** and its secondary decomposition product (**253**), probably with the attack of singlet oxygen on the photoenol [Eq. (86)] (Henderson and Ullman, 1955).

(250) R = H, Ph (251) (252)

(86)

(253)

There are a few exceptional examples that are formally recognized as 1,3-cycloadditions of singlet oxygen to aromatic zwitterion systems. The intermediate for the reaction is suggested to be a 1,3-endoperoxide which can be formed by the ene reaction involving the enolate anion system followed by addition of the peroxy anion, although a concerted 1,3-cycloaddition can not be ruled out. 2,4,6-Triphenylpyrylium oxide is sensitive not only to photooxygenation but also to autoxidation, yielding lactone **254** probably via 1,3-endoperoxide (**255**) [Eq. (87)] (Wasserman and Pavia, 1970). Unsensitized photooxygenation of diphenylbenzopyrylium oxide (**256**), which is known to be in photoequilibrium with zwitterion **257**, yields an anhydride (**258**), probably via an ozonide type of 1,3-endoperoxide (**259**) [Eq. (88)] (Suld and Price, 1962; Ullman and Henderson, 1967).

Sens $/h\nu/O_2$

(255)

(87)

(254)

(256) $\underset{h\nu'}{\overset{h\nu}{\rightleftharpoons}}$ (257) $\xrightarrow{^1O_2}$ (259) → (258) (88)

REFERENCES

Abbott, R. S., Ness, S., and Hercules, D. M. (1970). *J. Am. Chem. Soc.* **92,** 1128.
Adams, W. R., and Trecker, D. J. (1971). *Tetrahedron* **27,** 2631.
Algar, B. E., and Stevens, B. (1970). *J. Phys. Chem.* **74,** 3029.
Audubert, R. (1939). *Trans. Faraday Soc.* **35,** 197.
Baldwin, J. E., and Lever, O. W., Jr. (1973). *Chem. Commun.* p. 344.
Baldwin, J. E., Basson, H. H., and Krauss, H., Jr. (1968). *Chem. Commun.* p. 984.
Basselier, J.-J. (1964). *C. R. Hebd. Seances Acad. Sci.* **258,** 2851.
Bergmann, W., and McLean, M. J. (1941). *Chem. Rev.* **28,** 367.
Bick, I. R. C., Bremner, J. B., and Wiriyachitra, P. (1971). *Tetrahedron Lett.* p. 4795.
Bowen, E. J. (1954). *Trans. Faraday Soc.* **50,** 97.
Bowen, E. J. (1963). *Adv. Photochem.* **1,** 23.
Breitenback, J. W., and Kastell, A. (1954). *Monatsh. Chem.* **85,** 676.
Brockmann, H., and Dicke, F. (1970). *Chem. Ber.* **103,** 7.
Burns, P. A., and Foote, C. S. (1974). *J. Am. Chem. Soc.* **96,** 4339.
Burns, P. A., and Foote, C. S. (1976). *J. Org. Chem.* **41,** 908.
Burns, P. A., Foote, C. S., and Mazur, S. (1976). *J. Org. Chem.* **41,** 899.
Chalvet, O., Daudel, R., Ponce, C., and Rigaudy, J. (1968). *Int. J. Quantum Chem.* **2,** 521.
Chalvet, O., Daudel, R., Schmid, G. M., and Rigaudy, J. (1970). *Tetrahedron* **26,** 365.
Chrysochoos, J., and Grossweiner, L. I. (1968). *Photochem. Photobiol.* **8,** 193.
Cowell, G. W., and Pitts, J. N., Jr. (1968). *J. Am. Chem. Soc.* **90,** 1106.
Daly, J. W., Jerina, D. M., and Witkop, B. (1972). *Experientia* **28,** 1129.
Denny, R. W., and Nickon, A. (1973). *Org. React.* **20,** 133.
Dimroth, V. K., Chatzidakis, A., and Schaffer, O. (1972). *Angew. Chem.* **84,** 526.
Dufraisse, C., and LeBras, J. (1937). *Bull. Soc. Chim. Fr.* [5] **4,** 349.
Dufraisse, C., and Velluz, L. (1940). *C. R. Hebd. Seances Acad. Sci.* **211,** 790.
Dufraisse, C., and Velluz, L. (1942). *Bull. Soc. Chim. Fr.* [5] **9,** 171.

Dufraisse, C., Etienne, A., and Basselier, J.-J. (1957a). *C. R. Hebd. Seances Acad. Sci.* **244,** 2209.
Dufraisse, C., Rio, G., and Burris, W. A. (1957b). *C. R. Hebd. Seances Acad. Sci.* **244,** 2674.
Dufraisse, C., Rigaudy, J., Basselier, J.-J., and Cuong, N. K. (1965a). *C. R. Hebd. Seances Acad. Sci.* **260,** 5031.
Dufraisse, C., Rio, G., Ranjon, A., and Pouchol, O. (1965b). *C. R. Hebd. Seances Acad. Sci.* **261,** 3133.
Eskins, K., Glass, C., Rohwedder, W., Kleiman, R., and Sloneker, J. (1972). *Tetrahedron Lett.* p. 861.
Etienne, A., and Beauvois, C. (1954). *C. R. Hebd. Seances Acad. Sci.* **239,** 64.
Fahrenholtz, S. R., Doleiden, F. H., Trozzolo, A. M., and Lamola, A. A. (1974). *Photochem. Photobiol.* **20,** 519.
Fenical, W., Kearns, D. R., and Radlick, P. (1969a). *J. Am. Chem. Soc.* **91,** 3396.
Fenical, W., Kearns, D. R., and Radlick, P. (1969b). *J. Am. Chem. Soc.* **91,** 7771.
Fisch, M. H., Gramain, J. C., and Oleson, J. A. (1970). *Chem. Commun.* p. 13.
Foote, C. S. (1968a). *Acc. Chem. Res.* **1,** 104.
Foote, C. S. (1968b). *Science* **162,** 963.
Foote, C. S. (1976). *In* "Free Radicals in Biology" (W. A. Pryor, eds.), Vol. 2, p. 85. Academic Press, New York.
Foote, C. S., and Ching, T.-Y. (1974). *Photochem. Photobiol.* **20,** 511.
Foote, C. S., Wexler, S., Ando, W., and Higgins, R. (1968). *J. Am. Chem. Soc.* **90,** 975.
Foote, C. S., Mazur, S., Burns, P. A., and Lerdal, D. (1973). *J. Am. Chem. Soc.* **95,** 586.
Forbes, E. J., and Griffiths, J. (1966). *Chem. Commun.* p. 896.
Forbes, E. J., and Griffiths, J. (1967). *J. Chem. Soc. C* p. 601.
Forbes, E. J., and Griffiths, J. (1968). *J. Chem. Soc. C* p. 575.
Forster, C. H., and Berchtold, G. A. (1972). *J. Am. Chem. Soc.* **94,** 7939.
Forster, C. H., and Berchtold, G. A. (1975). *J. Org. Chem.* **40,** 3743.
Forster, J. W. (1962). *In* "Oxygenases" (O. Hayaishi, ed.), p. 241. Academic Press, New York.
Fukui, K. (1971). *Acc. Chem. Res.* **4,** 57.
Gibson, D. T., Gardini, G. E., Maseles, F. C., and Kallia, R. E. (1970). *Biochemistry* **9,** 631.
Gollnick, K. (1968a). *Adv. Chem. Ser.* **77,** 78.
Gollnick, K. (1968b). *Adv. Photochem.* **6,** 1.
Gollnick, K., and Lindner, J. M. E. (1973). *Tetrahedron Lett.* p. 1903.
Gollnick, K., and Schenck, G. O. (1967). *In* "1,4-Cycloaddition Reactions" (J. Hamer, ed.), p. 255. Academic Press, New York.
Gollnick, K., Franken, T., Schade, G., and Dörhöfer, G. (1970). *Ann. N. Y. Acad. Sci.* **171,** 89.
Grams, G. W. (1971). *Tetrahedron Lett.* p. 4823.
Grams, G. W., and Eskins, K. (1972). *Biochemistry* **11,** 606.
Grams, G. W., Eskins, K., and Inglett, G. E. (1972). *J. Am. Chem. Soc.* **94,** 866.
Griffith, J., and Hawkins, C. (1972). *Chem. Commun.* p. 463.
Grossweiner, L. I., and Zwicker, E. F. (1959). *J. Chem. Phys.* **31,** 1141.
Häflinger, O., Brossi, A., Chopard-Dit-Jean, L. H., Walter, M., and Schneider, O. (1956). *Helv. Chim. Acta* **39,** 2053.
Hara, H. (1960). *Nogei Kagaku Zasshi* **34,** 493.
Harada, N., Suzuki, S., Uda, H., and Ueno, H. (1972a). *Chem. Lett.* p. 803.
Harada, N., Suzuki, S., Uda, H., and Ueno, H. (1972b). *J. Am. Chem. Soc.* **94,** 1777.
Harada, N., Uda, H., Ueno, H., and Utsumi, S. (1973). *Chem. Lett.* p. 1173.
Hart, H., and Oku, A. (1972). *Chem. Commun.* p. 254.
Hasselmann, J. (1952). Ph.D. Dissertation, Göttingen University.

Hayaishi, O., ed. (1974). "Molecular Mechanisms of Oxygen Activation." Academic Press, New York.
Henderson, W. A., Jr., and Ullman, E. F. (1955). *J. Am. Chem. Soc.* **87,** 5424.
Irina, J., and Kurien, K. C. (1973). *Chem. Commun.* p. 738.
Ito, Y., Oda, M., and Kitahara, Y. (1975). *Tetrahedron Lett.* p. 239.
Janzen, G., Lopp, I. G., and Happ, J. W. (1970). *Chem. Commun.* p. 1140.
Johnson, A. W., and Tamm, S. W. (1964). *Chem. Ind. (London)* p. 1425.
Joschek, H.-I., and Miller, S. I. (1966). *J. Am. Chem. Soc.* **88,** 3269 and 3273.
Kasche, V., and Lindqüist, L. (1965). *Photochem. Photobiol.* **4,** 923.
Kawamoto, A., Kosugi, H., and Uda, H. (1972). *Chem. Lett.* p. 807.
Kearns, D. R. (1971). *Chem. Rev.* **71,** 395.
Kende, A. S., and Chu, J. Y.-C. (1970). *Tetrahedron Lett.* p. 4837.
Kepka, A., and Grossweiner, L. I. (1972). *Photochem. Photobiol.* **14,** 621.
Khan, A. U., and Kasha, M. (1970). *Ann. N. Y. Acad. Sci.* **171,** 24.
Khan, A. U., and Kearns, D. R. (1968). *Adv. Chem. Ser.* **77,** 143.
Koch, E. (1968). *Tetrahedron* **24,** 6295.
Koch, E. (1970). *Angew. Chem.* **82,** 313.
Larsen, D. L. (1973). Dissertation, Yale University, New Haven, Connecticut.
Learer, I. H. (1971). *Aust. J. Chem.* **24,** 891.
Le Roux, J. P., and Basselier, J.-J. (1970). *C. R. Hebd. Seances Acad. Sci., Ser. C* **271,** 461.
Le Roux, J. P., and Goasdoue, C. (1975). *Tetrahedron* **31,** 2761.
Lindblad, B., Lindstedt, G., and Lindstedt, S. (1970). *J. Am. Chem. Soc.* **92,** 7466.
Lindner, J. M. E., Kuhn, H. J., and Gollnick, K. (1972). *Tetrahedron Lett.* p. 1705.
McCapra, F., and Hann, R. A. (1969). *Chem. Commun.* p. 442.
Matsumoto, M., and Kondo, K. (1975). *Tetrahedron Lett.* p. 3935.
Matsumoto, M., Dobashi, S., and Kondo, K. (1975). *Tetrahedron Lett.* p. 4471.
Matsumoto, M., Dobashi, S., and Kondo, K. (1977). *Tetrahedron Lett.* p. 2329.
Matsuura, T., and Omura, K. (1974). *Synthesis* p. 173.
Matsuura, T., Omura, K., and Nakashima, R, (1965). *Bull. Chem. Soc. Jpn.* **38,** 1358.
Matsuura, T., Nishinaga, A., Matsuo, K., Omura, K., and Oishi, Y. (1967a). *J. Org. Chem.* **32,** 3457.
Matsuura, T., Matsushima, H., and Sakamoto, H. (1967b). *J. Am. Chem. Soc.* **89,** 3880.
Matsuura, T., Yoshimura, N., Nishinaga, A., and Saito, I. (1969). *Tetrahedron Lett.* p. 1669.
Matsuura, T., Matsushima, H., and Nakashima, R. (1970). *Tetrahedron* **26,** 435.
Matsuura, T., Saito, I., Imuta, M., and Kato, S. (1971). *Symp. Oxid. React., 5th, 1971* (Osaka) p. 63.
Matsuura, T., Yoshimura, N., Nishinaga, A., and Saito, I. (1972a). *Tetrahedron* **28,** 4933.
Matsuura, T., Matsushima, H., Kato, S., and Saito, I. (1972b). *Tetrahedron* **28,** 5119.
Mellier, M.-T. (1955). *Ann. Chim. (Paris)* [12] **10,** 666.
Merkel, P. B., and Kearns, D. R. (1972). *J. Am. Chem. Soc.* **94,** 7244.
Mori, A., Ohta, S., and Takeshita, H. (1975). *Heterocycles* **2,** 243.
Morita, N., and Asao, T. (1975). *Chem. Lett.* p. 71.
Moureu, C., Dufraisse, C., and Dean, P. M. (1926). *C. R. Hebd. Seances Acad. Sci.* **182,** 1440 and 1584.
Müller, E., Riecker, A., Scheffler, S., and Moosmayer, A. (1966). *Angew. Chem.* **78,** 98.
Murray, R. W., and Kaplan, M. L. (1968). *J. Am. Chem. Soc.* **90,** 537 and 4161.
Murray, R. W., and Kaplan, M. L. (1969). *J. Am. Chem. Soc.* **91,** 5358.
Musso, H. (1963). *Angew. Chem.* **75,** 965.
Mustafa, A., Asker, W., and Sobby, M. E. D. (1966). *J. Org. Chem.* **25,** 1519.

Nakagawa, M., Yoshikawa, K., and Hino, T. (1975). *J. Am. Chem. Soc.* **97,** 6496.
Nakai, C., Nozaki, O., Hayaishi, O., Saito, I., and Matsuura, T. (1975). *Biochem. Biophys. Res. Commun.* **67,** 590.
Nemoto, M., Usui, Y., and Koizumi, M. (1967). *Bull. Chem. Soc. Jpn.* **40,** 1035.
Oda, M., and Kitahara, Y. (1969a). *Angew. Chem.* **81,** 702.
Oda, M., and Kitahara, Y. (1969b). *Tetrahedron Lett.* p. 3275.
Ogryzlo, E. A., and Tang, C. W. (1970). *J. Am. Chem. Soc.* **92,** 5034.
Orland, C. M., Jr., Mark, H., Bose, A. K., and Manchas, M. S. (1968). *J. Org. Chem.* **33,** 2512.
Ouannés, C., and Wilson, T. (1968). *J. Am. Chem. Soc.* **90,** 6527.
Padwa, A., Dehm, D., Oine, T., and Lee, G. A. (1975). *J. Am. Chem. Soc.* **97,** 1837.
Pfoertner, K., and Böse, D. (1970). *Helv. Chim. Acta* **53,** 1553.
Potashnik, R., Goldschmidt, C. R., and Ottolenghi, M. (1971). *Chem. Phys. Lett.* **9,** 424.
Ramasseul, R., and Rassat, A. (1972). *Tetrahedron Lett.* p. 1337.
Ranico, R. (1955). *Ann. Chim. (Paris)* [12] **10,** 695.
Reiche, A., and Gross, H. (1962). *Chem. Ber.* **90,** 91.
Richardson, W. H., and Hodge, V. (1970). *J. Org. Chem.* **38,** 1216.
Richardson, W. H., Hodge, V. H., and Montgomery, F. C. (1974a). *J. Am. Chem. Soc.* **96,** 4688.
Richardson, W. H., Hodge, V. H., Stiggall, D. L., Yelvington, M. B., and Montgomery, F. C. (1974b). *J. Am. Chem. Soc.* **96,** 6625.
Rigaudy, J. (1968). *Pure Appl. Chem.* **16,** 169.
Rigaudy, J., and Brelière, C. (1972). *Bull. Soc. Chim. Fr.* pp. 1390 and 1399.
Rigaudy, J., and Kim Cuong, N. (1972). *Bull. Soc. Chim. Fr.* p. 1407.
Rigaudy, J., and Maumy, M. (1972). *Bull. Soc. Chim. Fr.* p. 3936.
Rigaudy, J., and Sparfel, D. (1972). *Bull. Soc. Chim. Fr.* p. 3441.
Rigaudy, J., Cohen, N. C., and Cuong, N. K. (1967). *C. R. Hebd. Seances Acad. Sci., Ser. C* **264,** 1851.
Rigaudy, J., Deletang, C., and Basselier, J.-J. (1969a). *C. R. Hebd. Seances Acad. Sci., Ser. C* **268,** 344.
Rigaudy, J., Dupont, R., and Nguyen, K. C. (1969b). *C. R. Hebd. Seances Acad. Sci., Ser. C* **269,** 416.
Rigaudy, J., Defoin, A., and Cuong, N. K. (1970). *C. R. Hebd. Seances Acad. Sci., Ser. C* **271,** 1258.
Rigaudy, J., Guillaume, J., and Maurette, D. (1971a). *Bull. Soc. Chim. Fr.* p. 144.
Rigaudy, J., Maurette, D., and Cuong, N. K. (1971b). *C. R. Hebd. Seances Acad. Sci., Ser. C* **273,** 1553.
Rigaudy, J., Moreau, M., and Kim Cuong, N. (1972). *C. R. Hebd. Seances Acad. Sci., Ser. C* **274,** 1589.
Rigaudy, J., Deletang, C., Sparfel, D., and Kim Cuong, N. (1973a). *C. R. Hebd. Seances Acad. Sci., Ser. C* **276,** 1215.
Rigaudy, J., Baranne-Lafont, J., Moreau, M., and Kim Cuong, N. (1973b). *C. R. Hebd. Seances Acad. Sci., Ser. C* **276,** 1607.
Rigaudy, J., Baranne-Lafont, J., Defoin, A., and Kim Cuong, N. (1975). *C. R. Hebd. Seances Acad. Sci., Ser. C* **280,** 527.
Rio, G., and Berthelot, J. (1972). *Bull. Soc. Chim. Fr.* p. 822.
Rio, G., Bricout, D., and Lacombe, M. (1972). *Tetrahedron Lett.* p. 3583.
Rio, G., Bricout, D., and Lacombe, M. (1973). *Tetrahedron* **29,** 3553.
Ritter, A., Bayer, P., Leitich, J., and Schomburg, G. (1974). *Justus Liebigs Ann. Chem.* p. 835.
Saito, I., and Matsuura, T. (1970). *Tetrahedron Lett.* p. 4987.
Saito, I., Kato, S., and Matsuura, T. (1970). *Tetrahedron Lett.* p. 239.

Saito, I., Yoshimura, N., Arai, T., Nishinaga, A., and Matsuura, T. (1972a). *Tetrahedron* **28,** 5131.
Saito, I., Imuta, M., and Matsuura, T. (1972b). *Tetrahedron* **28,** 5307.
Saito, I., Imuta, M., and Matsuura, T. (1972c). *Tetrahedron* **28,** 5313.
Saito, I., Abe, S., Takahashi, Y., and Matsuura, T. (1974). *Tetrahedron Lett.* p. 4001.
Saito, I., Chujo, Y., Shimazu, H., Yamane, M., Matsuura, T., and Cahnmann, H. J. (1975a). *J. Am. Chem. Soc.* **97,** 5272.
Saito, I., Imuta, M., Matsugo, S., and Matsuura, T. (1975b). *J. Am. Chem. Soc.* **97,** 7191.
Sawaki, Y., and Ogata, Y. (1975). *J. Am. Chem. Soc.* **97,** 6983.
Schäfer-Ridder, M., Brocker, U., and Vogel, E. (1976). *Angew. Chem.* **88,** 262.
Schenck, G. O. (1957). *Angew. Chem.* **69,** 579.
Schnuriger, B., and Bourdon, J. (1968). *Photochem. Photobiol.* **8,** 361.
Schönberg, A. (1968). "Preparative Organic Photochemistry." Springer-Verlag, Berlin and New York.
Schulte-Elte, K. H., Willhalm, B., and Ohloff, G. (1969). *Angew. Chem.* **81,** 1045.
Scott, A. I. (1964). *Q. Rev., Chem. Soc.* **18,** 347.
Smith, W. F., Jr. (1972). *J. Am. Chem. Soc.* **94,** 186.
Song, P.-S., and Moore, T. A. (1968). *J. Am. Chem. Soc.* **90,** 6507.
Sosnovsky, G. (1971). *Org. Peroxides* **2,** 280.
Sosnovsky, G., and Zaret, E. H. (1970). *Org. Peroxides* **1,** 517.
Stevens, B. (1973). *Acc. Chem. Res.* **6,** 90.
Stevens, B., and Algar, B. E. (1968). *J. Phys. Chem.* **72,** 3468.
Stevens, B., and Algar, B. E. (1969). *J. Phys. Chem.* **73,** 1711.
Stevens, B., Small, R. D., Jr., and Perez, S. R. (1974a). *Photochem. Photobiol.* **20,** 515.
Stevens, B., Perez, S. R., and Ors, J. A. (1974b). *J. Am. Chem. Soc.* **90,** 6846.
Storianetz, W., Shulte-Elte, K. H., and Ohloff, G. (1971). *Helv. Chim. Acta* **54,** 1913.
Suld, G., and Price, C. C. (1962). *J. Am. Chem. Soc.* **84,** 2090.
Takeshita, H., Kanamori, H., and Hatsui, T. (1973). *Tetrahedron Lett.* p. 3139.
Taylor, W. I., and Battersby, A. R., eds. (1967). "Oxidative Coupling of Phenols." Dekker, New York.
Thomas, M., and Foote, C. S. (1973). Ph.D. Thesis, University of California at Los Angeles.
Thomas, M., and Foote, C. S. (1978). *Photochem. Photobiol.* **27,** 683.
Ullman, E. F., and Henderson, W. A., Jr. (1967). *J. Am. Chem. Soc.* **89,** 4390.
Van Tamelen, E. E., and Cole, T. M., Jr. (1970). *J. Am. Chem. Soc.* **92,** 4123.
Vickers, R. S., and Foote, C. S. (1970). *Boll. Chim. Farm.* **109,** 599.
Vogel, E., Altenbach, H.-J., and Sommerfeld, C.-D. (1972). *Angew. Chem., Int. Ed. Engl.* **11,** 939.
Vogel, E., Alscher, A., and Wilms, K. (1974). *Angew. Chem.* **86,** 407.
Vogel, E., Klug, H.-H., and Schäfer-Ridder, M. (1976). *Angew. Chem.* **88,** 268.
Wasserman, H. H. (1970). *Ann. N. Y. Acad. Sci.* **171,** 108.
Wasserman, H. H., and Keehn, P. M. (1966). *J. Am. Chem. Soc.* **88,** 4522.
Wasserman, H. H., and Keehn, P. M. (1967). *J. Am. Chem. Soc.* **89,** 2770.
Wasserman, H. H., and Keehn, P. M. (1972). *J. Am. Chem. Soc.* **94,** 298.
Wasserman, H. H., and Larsen, D. L. (1972). *Chem. Commun.* p. 253.
Wasserman, H. H., and Pavia, D. C. (1970). *Chem. Commun.* p. 1459.
Wasserman, H. H., Mariano, P. S., and Keehn, P. M. (1971). *J. Org. Chem.* **36,** 1765.
Wasserman, H. H., Scheffer, J. R., and Cooper, J. L. (1972). *J. Am. Chem. Soc.* **94,** 4991.
Wei, K. S., and Adelman, A. H. (1969). *Tetrahedron Lett.* p. 3297.
Wei, K. S., Mani, J., and Pitts, J. N., Jr. (1967). *J. Am. Chem. Soc.* **89,** 4225.

Weil, L. (1965). *Arch. Biochem. Biophys.* **110,** 57.
White, E. H., and Harding, M. J. C. (1965). *Photochem. Photobiol.* **4,** 1129.
Wilson, T. (1966). *J. Am. Chem. Soc.* **88,** 2898.
Wilson, T. (1969a). *Photochem. Photobiol.* **10,** 441.
Wilson, T. (1969b). *J. Am. Chem. Soc.* **91,** 2387.
Young, R. H., and Hart, H. (1967). *Chem. Commun.* p. 827.
Young, R. H., Martin, R. L., Feriozi, D., Brewer, D., and Kayser, R. (1973). *Photochem. Photobiol.* **17,** 233.
Ziffer, H., Jerina, D. M., Gibson, D. T., and Kobal, V. M. (1973). *J. Am. Chem. Soc.* **95,** 4048.
Zweig, A. (1963). *J. Phys. Chem.* **67,** 506.
Zweig, A., Hodson, W. G., and Jura, W. H. (1964). *J. Am. Chem. Soc.* **86,** 4124.
Zwicker, F., and Grossweiner, L. I. (1963). *J. Phys. Chem.* **67,** 549.

11

Role of Singlet Oxygen in the Degradation of Polymers

MARTIN L. KAPLAN and ANTHONY M. TROZZOLO

I. Introduction 575
II. Degradation of Polymers 577
A. Polyethylene 577
B. Polypropylene 580
C. Polystyrene 583
D. Polydienes 585
E. Miscellaneous Polymers 588
III. Biopolymers and Related Species 589
A. Amino Acid Derivatives and Peptides 589
B. Erythropoietic Protoporphyria 590
C. Neonatal Jaundice 591
References 592

I. INTRODUCTION

In addition to water, fresh air and sunlight have been generally regarded as the most bountiful gifts of Nature by mankind since the dawn of recorded history. Yet, more than a hundred years ago it was recognized that natural rubbers were degraded by these very agents that mankind holds so dear. The problem then was not too different from those which are encountered today in the deterioration of synthetic plastics. For example, Indian telegraph wire coatings of natural gums were found to gradually lose their advantageous properties and thereby caused substantial financial loss. Investigation of this phenomenon led to the conclusion that oxidation (from exposure to air and accelerated by light) was responsible for the deterioration of the rubber (Hofmann, 1861). Subsequently, a report ap-

Singlet Oxygen

ISBN 0-12-736650-4

peared (Miller, 1865) on the decay of gutta-percha and caoutchouc used in submarine cables. The author said, ". . . alternate exposure to moisture and dryness, particularly if at the same time the sun's light has access, is rapidly destructive of the gutta-percha, rendering it brittle, friable, and resinous in aspect, and in chemical properties."

Science, in some ways, has made rapid strides in this area. Additives are known which can extend the lifetime of both natural and synthetic materials to several decades. Polymers can be "tailor-made" with chemical and physical properties for almost any use. The specter of oxidative degradation, however, remains. Slowly, though, progress is being made in the understanding of the mechanism by which high molecular weight molecules are oxidized. One mechanistic aspect of the overall degradative scheme where light is involved, namely, the role of singlet oxygen, is the subject of this chapter.

Within the past two decades evidence has been accumulating at an ever increasing rate indicating the intermediacy of singlet molecular oxygen in the facile oxidations of a variety of organic materials. These materials range from simple olefins to the most complicated biological polymers. Reviews of the subject are now proliferating in the chemical and biological literature. One significant mode of action of molecular oxygen is as a dienophile, and this reaction was the subject of a thorough review (Arbuzov, 1965). Subsequently, Foote (1968a) gave historical perspective to the field and demonstrated that chemically generated excited oxygen and photosensitized oxygenations had a common intermediate ($^1\Delta_g\ O_2$). In the same year, Foote (1968b) tied the new knowledge of the mechanism of photosensitized oxidations to reactions of biological importance (e.g., "photodynamic action"). Further comprehensive treatments examined the role of dioxetane intermediates (Kearns and Khan, 1969) in photodynamic effects and the possible biological implications of singlet oxygen in general (Wilson and Hastings, 1970). The First International Conference on Singlet Molecular Oxygen and its Role in Environmental Sciences was held in 1969 and its proceedings demonstrated the world-wide interest in the subject that had arisen in a few short years (Trozzolo, 1970). The possible role of singlet oxygen has also been discussed in terms of its importance in the chemistry of urban atmospheres (Pitts, 1969, 1971). General reviews of the physical and chemical aspects of singlet oxygen have also appeared (Kaplan, 1971; Wayne, 1969; Kearns, 1971).

Other chapters in this volume give detailed accounts of methods of generating singlet oxygen, descriptions of the basic reactions of singlet oxygen, and discussions of quenching, isotope effects, and other phenomena related to the basic chemistry involved.

The intention of the authors of this chapter was to gather together the experimental data which implicate singlet molecular oxygen in the oxidative degradation of plastics, rubbers, and biopolymers and to assess its importance in the overall schemes which eventually lead to dramatic deleterious effects in their useful properties. Some attention has already been given to this subject, particularly in the case of rubber and plastics (Cicchetti, 1970; Trozzolo, 1971; Rabek and Ranby, 1975).

II. DEGRADATION OF POLYMERS

Electronically excited molecular oxygen in its $^1\Delta_g$ state is a metastable species of high reactivity. Nonetheless, it has been shown (Pitts *et al.*, 1969) to have a sufficient lifetime at atmospheric pressure (half-life 5×10^{-2} sec) and a large enough mean diffusion path in thin films (115 ± 20 Å) (Schnuriger and Bourdon, 1968) to warrant consideration by polymer degradation specialists as a causative agent in various oxidation processes.

A. Polyethylene

Polyethylene (PE) is a polymer nominally consisting of linear chains of methylene units. On the basis of simple analogy with the stability of *n*-alkane molecules one would suppose that PE would exhibit relatively good stability toward photodegradation. However, it has long been known that various structural anomalies are found in processed PE. Not only are small amounts of three types of olefinic impurities found, i.e., vinyl, vinylene, and vinylidene (Rugg *et al.*, 1954), but the ultraviolet absorption spectrum indicates the presence of carbonyl functions as well (Pross and Black, 1950; Burgess, 1952; Achhammer *et al.*, 1959). The formation of a triene structure has also been reported when radical polymerized PE was subjected to ultraviolet radiation (Heacock *et al.*, 1968). The olefinic structures would be particularly vulnerable to attack by singlet oxygen.

From the known photochemistry of carbonyl functions, a mechanism was proposed which explained, to a large extent, the experimental findings during the initial stages of the oxidative photodegradation of PE (Trozzolo and Winslow, 1968). This mechanism initially involved the absorption of light by the ketone groups present followed by a so-called Norrish Type II [Eq. (1)] cleavage (Calvert and Pitts, 1966; Turro, 1965).

$$\mathrm{RC(=O)CH_2CH_2CH_2R'} \underset{}{\overset{h\nu}{\rightleftharpoons}} \mathrm{RC^{\bullet}(OH)CH_2CH_2\dot{C}HR'} \longrightarrow \mathrm{RC(OH){=}CH_2}\ (\rightarrow \mathrm{RC(=O)CH_3}) + \mathrm{CHR'{=}CH_2} \tag{1}$$

This reaction was suggested to be the major source of chain scission during the oxidative photodegradation of PE. Heskins and Guillet (1968) showed that the Norrish Type II cleavage was a major source of chain scission in the photolysis of the carbonyl groups of the ethylene–carbon monoxide copolymer. The effect of this cleavage in a polymer network is to generate a methyl ketone and terminal olefin in close proximity to each other.

The second step in the mechanism required the resulting methyl ketone to be photoexcited to an n–π^* triplet state which was physically quenched by ground state O_2 molecules with the formation of singlet excited oxygen molecules. The feasibility of the n–π^* ketone quenching step in solution photochemistry is well documented (Trozzolo and Fahrenholtz, 1970; Gollnick *et al.*, 1970). That electronic energy transfer is possible within a bulk polymer has also been shown (Fox, 1973). Once the singlet oxygen molecule is formed in the vicinity of the terminal olefin, reaction should be rapid [Eq. (2)]. In fact, singlet oxygen molecules have been shown to diffuse through 10 nm films with loss of only half the excited species (Schnuriger and Bourdon, 1968), which would imply that singlet molecular oxygen formation at the site of chain scission is not necessarily a requirement for reaction with the olefinic groups. In solution, the photooxidation of terminal olefins was shown to be very slow (Kopecky and Reich, 1965). Model compound studies with 1-olefins dispersed on a solid substrate have shown them to be quite susceptible to attack by microwave-generated gas phase singlet oxygen (Kaplan and Kelleher, 1971), thus providing evidence for the third step in the proposed mechanism. Hydroperoxides may be formed in this manner.

$$\text{Type II cleavage of } n\text{-}\pi^* \text{ excited state} \downarrow$$

$$\diagup CH_2 \diagdown CH_2 \diagup CH_2 \diagdown C(=O)CH_3 \quad (\text{Ketone}) \qquad H_2C{=}CH{-}CH_2 \diagup CH_2 \diagdown \quad (\text{Olefin}) \tag{2}$$

$$(\text{ketone})^3 n\text{-}\pi^* + O_2 \longrightarrow \text{ketone} + {}^1O_2$$

$${}^1O_2 + RCH_2CH{=}CH_2 \longrightarrow RCH{=}CHCH_2OOH \longrightarrow \text{further reactions}$$

One group has reported that saturated carbons also demonstrate some slight reactivity toward excited molecular oxygen (Kaplan and Kelleher, 1971). Although this work has been criticized (Carlsson and Wiles, 1973, 1975), unambiguous conclusions cannot yet be drawn.

The primary question remaining to be answered about this mechanism concerns the lack of correlation of oxidative photodegradation with carbonyl group concentration. Studies have shown that the rates of photooxygenation of both polyethylene containing long-chain aliphatic ketones, and of the ethylene–carbon monoxide copolymer containing varying amounts of carbonyl groups are independent of the bulk concentrations of those carbonyl functions (Winslow *et al.*, 1969). Further, the data has also shown that as films of branched polyethylene became thicker, oxygen uptake during photooxidation (3000 Å light) became slower (Fig. 1). These results are consistent with the suggestion that oxidation is more rapid at the surface than in the bulk, perhaps due to ozone-initiated degradation and that it is likely that photodegradation begins at the surface, thus making internal carbonyl groups relatively ineffective as photosensitizers.

The mechanism proposed for the oxidative photodegradation of PE specifically requires ketones to act as the photosensitizing groups. However, there may be other photoinitiators. Singlet oxygen has been presumed to arise by a number of other photosensitization reactions also. For example, dyes (Egerton, 1964), white pigments (Egerton, 1964; Morand, 1968), and polynuclear aromatics (Carlsson and Wiles, 1973) promote the photodegradation of various polymers. Further, ozone, which is found in fairly high concentrations in some urban areas, can undergo decomposition by ultraviolet light (Jones and Wayne, 1971) or by various air pollutants

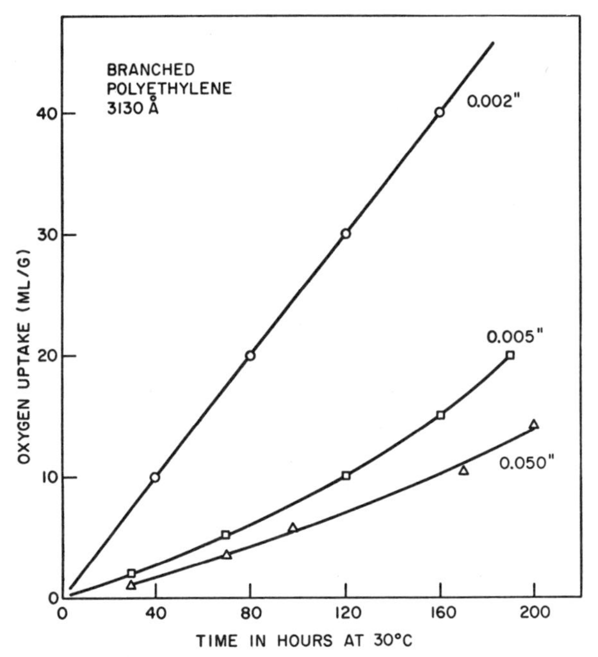

Fig. 1. Effect of film thickness on the photooxidation rate of branched polyethylene under RPR 3000-A lamps at 30°C (Winslow *et al.*, 1969).

(Murray *et al.*, 1970), and result in singlet oxygen. Within a partially oxidized polymer, singlet oxygen may arise by a mechanism involving the decomposition of hydroperoxides through the self-termination reaction of the resulting *sec*-peroxy radicals, as proposed by Howard and Ingold (1968). The question of whether externally generated gas phase singlet oxygen can react heterogeneously with solid PE is not yet answered conclusively. Two laboratories using microwave generated $^1\Delta_g\,O_2$ (Furakawa *et al.*, 1970; Cook and Miller, 1974) have found opposing results on this question (Breck *et al.*, 1974; S. R. Fahrenholtz, unpublished results).

In sum, the available evidence strongly suggests that singlet molecular oxygen is involved in the early stages of the degradation of PE. It is quite likely that hydroperoxide formation on the surface of the polymer is a primary result of reaction with singlet oxygen. The products which are observed later, e.g., acetone, acetaldehyde, water, and oxides of carbon (Rugg *et al.*, 1954; Burgess, 1953) along with evidence for both scission and crosslinking are undoubtedly the consequence of free radical chain reactions initiated by the decomposition of photogenerated hydroperoxides.

B. Polypropylene

Like PE, pure polypropylene (PP) also should not absorb light in the ultraviolet or visible spectral regions. However, just as in the case of PE, the processing of PP results in the formation of measurable amounts of various functional groups. One infrared spectroscopic study of process degradation, by Adams and Goodrich (1970) resulted in the identification of carboxylic acid, vinyl olefins, aldehyde, methyl ketone, and as a minor product, γ-lactone. Carlsson and Wiles (1969) have simulated the extrusion and molding conditions which PP might encounter and determined the nature of the polymeric carbonyl products formed during such thermal oxidation. Their results are in contrast to the previous work since they found no aldehydes, and only traces of carboxylic acids. They also did not assign the γ-lactone structure, although their data exhibited the absorption. By comparison of their polymer spectra with those of model compounds three types of ketonic products (**A**, **B**, **C**) were postulated.

$$-\mathrm{C(CH_3)(H)}-\mathrm{CH_2}-\mathrm{C(=O)}-\mathrm{CH_2}-\mathrm{C(CH_3)(H)}- \quad \textbf{(A)}$$

$$-\mathrm{C(CH_3)(H)}-\mathrm{CH_2}-\mathrm{C(=O)}-\mathrm{CH_3} \quad \textbf{(B)}$$

$$-\mathrm{CH_2}-\mathrm{C(H_3C)(H)}-\mathrm{C(=O)}-\mathrm{C(CH_3)(H)}-\mathrm{CH_2}- \quad \textbf{(C)}$$

Clearly then, polyolefins such as PP are subject, by thermal processing, to the development of functional groups which could be potential absorbers of sunlight and consequently major influences on the photostability of the polymer.

When this preoxidized PP was photodegraded, in vacuum, by ultraviolet irradiation, and the volatile products were analyzed, ketones **A** and **B** were shown to be the main components observed in the carbonyl IR absorption bands. In addition, product analysis provided further evidence for the mechanism proposed by Trozzolo and Winslow (1968). Ketone **A** was found to dissociate predominantly by a Norrish Type I mechanism [Eq. (3)].

$$-CH_2C(CH_3)(H)CH_2C(=O)CH_2C(CH_3)(H)- \quad (\mathbf{A}) \xrightarrow{h\nu}$$

$$\xrightarrow{\geq 90\%} -C(CH_3)(H)CH_2\cdot + O{=}\dot{C}CH_2C(CH_3)(H)- \longrightarrow CO + \cdot CH_2C(CH_3)(H)-$$

$$\xrightarrow{\leq 10\%} \left\{ \begin{array}{l} -CH{=}C(CH_3)H \\ -CH_2C(=CH_2)H \end{array} \right\} + CH_2{=}C(OH)(H)CH_2C(CH_3)(H)- \longrightarrow CH_3C(=O)-CH_2C(CH_3)(H)- \qquad (3)$$

Alternatively, ketone **B** decomposed primarily by a Norrish Type II process [Eq. (4)].

$$CH_2C(CH_3)(H)CH_2C(=O)CH_3 \quad (\mathbf{B}) \xrightarrow{h\nu}$$

$$\xrightarrow{\sim 15\%} -CH_2C(CH_3)(H)CH_2\cdot + CH_3\dot{C}O \longrightarrow CH_3CHO; \quad CH_3\dot{C}O \longrightarrow CH_3\cdot + CO$$

$$\xrightarrow{\sim 85\%} \left\{ \begin{array}{l} -CH{=}C(CH_3)H \\ -CH_2C(=CH_2)H \end{array} \right\} + CH_2{=}C(OH)CH_3 \longrightarrow CH_3COCH_3 \qquad (4)$$

In the cases of Norrish Type II cleavage, particularly with ketone B, the olefin formed conceivably would be subject to singlet oxygen attack. This excited oxygen could be formed by energy transfer from ketone triplet to ground state oxygen molecules. There is evidence from the attempted Rose Bengal solution photosensitized oxidation of "pure" atactic PP that singlet oxygen is ineffective in oxidizing the pure polymer (Mill *et al.*, 1973). Although this result would argue against the saturated polymer backbone as being susceptible to singlet oxygen attack, it does not minimize the possibility that singlet oxygen generated very near to polymeric olefins (as might occur following a Norrish Type II cleavage) could attack the double bond and form hydroperoxides.

Although direct evidence for reactions of $^1\Delta_g\,O_2$ with PP is sparse, a number of studies when examined together implicate these reactions in the oxidative photodegradation scheme. For example, $^1\Delta_g\,O_2$ has long been known to react with olefins, and that olefins are certainly present as impurities in processed PP. Also, many of the commercial Ni(II) chelates used to stabilize PP have been shown to be extremely efficient $^1\Delta_g\,O_2$ quenchers (Flood *et al.*, 1973; Carlsson *et al.*, 1972, 1973). This quenching efficiency, however, has been demonstrated to be significantly reduced on going from liquid phase model reactions to solid phase polymer reactions (Bystritskaya and Karpukhin, 1975). Nevertheless, such quenchers exhibit remarkable effectiveness in protecting polypropylene against the ravages of light-induced oxidation (Briggs and McKellar, 1968). The origins of singlet oxygen within the polymer are a related matter which are still subject to speculation. As mentioned previously, ketones are known photosensitizers for $^1\Delta_g\,O_2$ production and, in fact, PP was shown to photooxidize readily with benzophenone as sensitizer (Karpukhin and Slobodetskaya, 1974). Recently, attention has focussed on polynuclear aromatic (PNA) impurities and their role in singlet oxygen production. Large concentrations of PNA are released into the atmosphere from automobile exhausts (Begeman, 1964) and can concentrate very readily in polyolefin materials exposed to this air (Partridge, 1966; Bonstead and Charlesby, 1967; Pivovarov *et al.*, 1971). Also PNA compounds are known to photosensitize the production of $^1\Delta_g\,O_2$ in the gas phase (Wasserman *et al.*, 1969; Kearns *et al.*, 1969). When anthracene was deliberately introduced into PP film, it markedly accelerated the photodegradation of the polymer (Carlsson and Wiles, 1975). Furthermore, the simultaneous inclusion of a nickel chelate (which could act as a $^1O_2$ quencher) prevented the anthracene sensitization and even stabilized the film beyond the lifetime of the additive-free sample. Additional support for the importance of PNA in photochemical degradation of polyolefins comes from the report that efficient UV stabilizers for PP (such as 2-hydroxybenzophenones and substituted triazoles) quench PNA excited states in polyolefins thereby suggesting that energy transfer to ground state oxygen is not effective in their presence (Pivovarov and Lukovnikov, 1968).

Strong evidence has been presented which argues for the consideration of singlet oxygen in any overall oxidative photodegradation scheme for PP. However, it should be pointed out that much of the effectiveness of the Ni(II) compounds may not be in their ability to quench excited oxygen or PNA triplet states, but may be due simply to their demonstrated capacity to decompose PP hydroperoxides (Chien and Boss, 1972). Not even considered in the above discussion, because of the paucity of information on the subject, is the possible role of charge-transfer complexes of polyolefins with oxygen (Tsuji and Seiki, 1970) in the generation of singlet oxygen.

C. Polystyrene

The light stability of polystyrene (PS) has been a subject of considerable interest for many years (Savides, 1968). When PS is exposed to ultraviolet irradiation for long periods in the presence of oxygen, yellowing sets in and precedes the deterioration of mechanical properties (Matheson and Boyer, 1952). In the absence of oxygen, coloration is negligible except when 254 nm irradiation is used (Grassie and Weir, 1965). Characteristically, the discoloration is concentrated on the surface facing the light source and seems to act as a protective coating toward further degradation. The origin of the yellow color has been the subject of much speculation. Since the development of color is accompanied by carbonyl group formation, various mechanisms have been proposed on that basis. One such suggestion involves the free radical formation of a hydroperoxide with subsequent photolytic hydroperoxide decomposition leading to a ketone and to chain scission [Eq. (5)].

$$\sim\!\overset{\displaystyle H}{\underset{\displaystyle C_6H_5}{C}}\!-\!CH_2\!-\!\overset{\displaystyle OOH}{\underset{\displaystyle C_6H_5}{C}}\!-\!CH_2\!-\!\overset{\displaystyle H}{\underset{\displaystyle C_6H_5}{C}}\!\sim \;\longrightarrow\; \sim\!\underset{\displaystyle C_6H_5}{C}\!=\!CH\!-\!\underset{\displaystyle C_6H_5}{C}\!=\!O \;+\; H_3C\!-\!\overset{\displaystyle H}{\underset{\displaystyle C_6H_5}{C}}\!\sim \;+\; H_2O \qquad (5)$$

Evidence for the viability of the hydrogen abstraction step was provided by Tryon and Wall (1961). When PS which was deuterated in the α-position was photooxidized, the rate of oxidation was halved. Deuteration in the β-position left the rate unaffected. Unfortunately, these facts do not, a priori, distinguish free radical chain process from the controversial reaction proposed by Rabek and Ranby (1974). Their mechanism involves the abstraction of the tertiary hydrogen atom by singlet oxygen resulting in the same hydroperoxide as in Eq. (5). The only evidence concerning direct attack of singlet oxygen on PS is, however, negative. When microwave-generated $^1O_2$ was passed over PS for several hours, no reaction was observed (MacCallam and Rankin, 1974).

Alternative structures have been put forward to explain the yellowing of PS when it is exposed to light. One of these structures is a quinomethane (**1**) (Achhammer *et al.*, 1953) and another a benzalacetophenone (**2**) (Tryon and Wall, 1961):

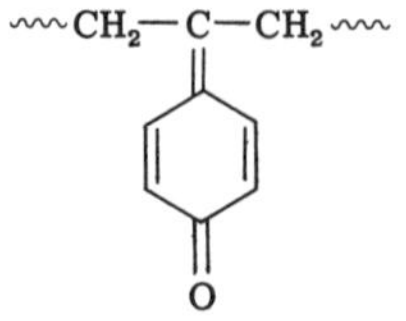

(1) 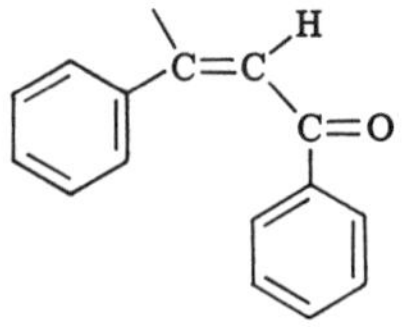 (2)

Grassie and Weir (1965), on the other hand, suggested that the yellow color is not the result of oxidation but is caused by conjugated carbon–carbon unsaturation. Also, they proposed a mechanism for hydroperoxide decomposition for which the end result would be an aryl ketone next to an olefin [Eq. (6)].

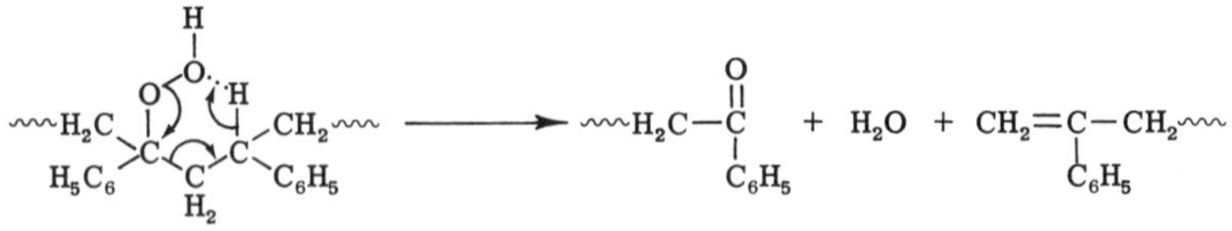

$$\sim H_2C-C(=O)-C_6H_5 + H_2O + CH_2=C(C_6H_5)-CH_2\sim \quad (6)$$

Here, again, as in the cases of PE and PP, is the possibility of $^1\Delta_g\ O_2$ formation by quenching of triplet ketone, and subsequent reaction with the olefin.

Another possible mode of singlet oxygen attack and color development was proposed by Rabek and Ranby (1974). This involved reaction with the benzene rings in PS with formation of dioxetane intermediates followed by ring opening with final production of conjugated dialdehydes [Eq. (7)]. This assignment was made on the basis that the 1740 cm^{-1} infrared band which developes in photooxidized PS is similar to positions of bands in photooxidized benzene. The presence of such a dialdehyde structure in photodegraded PS does not seem tenable since reported spectra for all the mucondialdehyde isomers have carbonyl absorption maxima below 1700 cm^{-1} (Nakajima *et al.*, 1959). Much more likely candidates for the 1740 cm^{-1} band would be an unsaturated anhydride or lactone (Avram and Mateescu, 1972) possibly the result of the oxidation of the benzene rings proceeding past the carbonyl stage.

The present evidence in the case of PS photooxidation is not at all convincing in implicating singlet oxygen involvement.

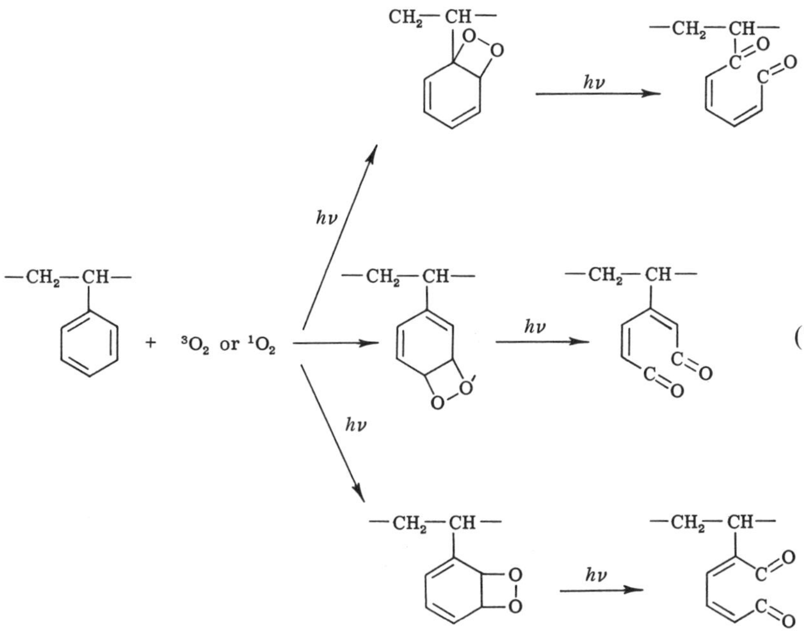

(7)

D. Polydienes

Natural rubber (Saunders, 1973) is a hydrocarbon polymer collected from the South American tree *Hevea brasiliensis*. The latex which flows from the tree is a dispersion of rubber in water, with various proteins acting as the dispersing agents. The bulk latex is diluted and then coagulated with acetic or formic acid. Depending on the method of drying of the coagulum the resultant product is called "smoked sheet" or "pale crepe." Gutta-percha is another rubbery material obtained from trees grown primarily in Malaysia and Indonesia.

Chemically, *Hevea* latex and gutta-percha latex have been shown to be 1,4-polyisoprenes. However, the former is a high cis-polymer while the latter is high-trans (Cunneen and Higgins, 1963). Both polymers can now be successfully synthesized stereospecifically.

Commercial use of rubber requires that natural rubber be vulcanized so that plasticity be reduced and elasticity increased. The principal chemical change occurring during vulcanization is a cross-linking in which discreet rubber chains are converted into a three-dimensional network. The procedures and reagents leading to the final vulcanization are many and varied and they have been shown to affect the quantum yield for chain scissions, as a function of wavelength (Morand, 1966). For example, mercaptobenzothiazole and sulfur considerably decrease the rate of degradation by solar ultraviolet light while zinc oxide protects only against radiation below 370 nm but exhibits a sensitizing effect in the near visible region. This effect is believed to be due to singlet oxygen formation occurring via a ZnO–oxygen complex [Eq. (8)] (Egerton, 1949; Trozzolo, 1971).

$$
\begin{aligned}
ZnO + O_2 &\xrightarrow{h\nu} ZnO^{+}\cdots O_2^{-} \\
O_2^{-} &\longrightarrow {}^1O_2 + e \\
ZnO^{+} + e &\longrightarrow ZnO
\end{aligned}
\tag{8}
$$

The photooxidation of nonconjugated olefins has been exhaustively studied by several groups over the years. Various quantum yield data have been collected and used as the basis for making generalizations about the overall degradation mechanism (Bateman, 1946; Bateman and Gee, 1948, 1951; Hart and Matheson, 1948; Morand, 1968; Mill *et al.*, 1968). This work appears to point to an autocatalytic radical chain process when 254 nm or 313 nm light was used, especially after photooxidation has proceeded past the early stages. Much of this research must be criticized, however, because of the failure of the workers to take into account the role of sensitizing impurities and their contributions to the quantum yield determination.

As the understanding of the interactions of light, sensitizers and oxygen grew, and techniques became available for preparing singlet excited oxygen without the use of light, researchers were then able to isolate the role of ${}^1\Delta_g\ O_2$ from the overall degradation scheme.

Perhaps the first case reported in the literature of an intentional singlet oxygen reaction with a polymer was a study of the oxidation of polyisoprene in chlorobenzene sensitized by methylene blue (Mill *et al.*, 1968) which has been shown to be an efficient ${}^1\Delta_g\ O_2$ producer (Foote, 1968a,b). This reaction was demonstrated to be nonradical in nature by the fact that the initial rate of oxygen uptake was unimpeded by the addition of an oxidation inhibitor.

Kaplan and Kelleher (1970a, 1972) also have studied the reactions of polydienes with singlet oxygen. By generating ${}^1O_2$ *in situ* from the triphenyl phosphite ozonide (Murray and Kaplan, 1968) in solutions of high *cis-*, *trans-*, and vinylpolybutadienes (PBD) they showed that only the cis and trans polymers were susceptible to oxidation. An acrylonitrile–butadiene–styrene (ABS) polyblend treated in a similar manner provided evidence for the suggestion that the initial photooxidation of ABS occurs through attack at the PBD portion of the polymer. Also discussed was a series of events which might lead to a cyclic peroxide–hydroperoxide structure by the cycloaddition of ${}^1\Delta_g\ O_2$ across a conjugated double bond system [Eq. (9)].

```
—CH2  H2C—CH2  H2C—CH2    CH2—
   \  /       \  /       \  /
   HC═CH      HC═CH      HC═CH

            │ 1O2
            ↓

—CH2   HC—CH   H2C—CH2   CH2—
   \   //  \\   /      \  /
   HC—CH    HC—CH      HC═CH                (9)
   |          |
  HOO        OOH

            │ 1O2
            ↓

— CH2  HC═CH      CH2—CH2     CH2—
   |   /     \     |      \   /
   HC—CH      CH—CH      HC═CH
   |    \    /     |
  HOO    O—O      OOH
```

Similar structures had previously been put forward to account for some of the volatile products derived from polyisoprene autoxidation (Bolland and Hughes, 1949; Bevilacqua, 1961).

cis-PBD films have also been subjected to the chemical effects of gas phase singlet oxygen (Kaplan and Kelleher, 1970b). When the $^1\Delta_g\ O_2$ produced in a microwave discharge of ground state oxygen via a flow system was passed over *cis*-PBD films, hydroperoxides were formed at the surface. Subsequent heat treatment of the hydroperoxidized films resulted in the rapid degradation of the bulk film. This behavior strongly suggests that the initial surface photooxidation of polymer films tends to overcome the well-known induction periods observed in thermal autoxidations.

For a number of years it has been recognized that certain transition metal complexes, e.g., alkyl xanthates and dithiocarbamates are useful in the protection of rubber against light aging (Karmitz, 1958; Dunn, 1960). Recently, Zweig and Henderson (1975) in a comprehensive study, examined the singlet oxygen quenching ability of numerous chelate compounds in unsaturated polymer. They concluded that almost every type of dye molecule, with the notable exception of azo compounds, was an effective sensitizer for singlet oxygen production in polymer films. In addition, they were able to show that the efficiency of quenching of singlet oxygen by some of the metal chelates was comparable to the effectiveness of β-carotene (which is a near diffusion control quencher) (Foote *et al.*, 1970).

The effect of other singlet oxygen quenchers also has been demonstrated. Although rate studies of the gas phase quenching of $^1\Delta_g\ O_2$ by amines had been performed earlier (Ogryzlo and Tang, 1970), confirmation of the effectiveness of amines in solution awaited the work on the sensitized rubber photooxidation performed by Dalle *et al.* (1972). Shortly thereafter, these findings were extended to other amines and the quenching mechanism which was proposed involved the intermediacy of a charge-transfer complex (Young *et al.*, 1973).

The terpolymer of ethylene, propylene and ethylidene norbornene also was shown to react with singlet oxygen generated by dye sensitization, with the formation of at least two different hydroperoxides (Duynstee and Mevis, 1972). Heating of the photooxidized polymers above 100°C caused rapid crosslinking and deterioration of the rubber.

The clearest, least ambiguous inferences of the intermediacy of $^1\Delta_g$ O_2 in the oxidative photodegradation of polymers exist in the case of nonconjugated olefinic substances. Singlet oxygen, prepared in the gas phase, chemically in solution, or by photosensitization, reacts readily with the unsaturated polymers. In addition, materials used to prevent the photodegradation of polydienes are now recognized to be very efficient singlet oxygen quenchers.

E. Miscellaneous Polymers

The question of the reactivity of other polymers with singlet oxygen and its possible role in their oxidative photodegradation is wide open. There is very little experimental evidence in this regard although in a few cases, suggestions implicating singlet oxygen have been made.

Poly(vinyl chloride) films (PVC) have been subjected to gas phase singlet oxygen treatment (MacCallam and Rankin, 1974) and to photosensitized oxidation (Zweig and Henderson, 1975). In neither case was any reaction observed. In one study, however, $^1\Delta_g$ O_2 was invoked as a possible reactive intermediate during the photooxidation of PVC (Kwei, 1969).

Silicone containing polymers have been hydroxylated by treating surfaces with oxidizing plasmas (Hollahan and Carlson, 1970). Some hydroxylation of the cured and uncured polymethylsiloxanes was presumed due to the reaction of $^1\Delta_g$ O_2 with methyl groups in the polymer.

Cellulose is readily decomposed by ultraviolet irradiation, particularly if light absorbing dyes are present (Egerton, 1949). Some direct evidence for the reaction of cellulose with gas phase singlet oxygen has been presented (Kaplan and Kelleher, 1971) while its possible role in cellulose photooxidation has been discussed (Egerton, 1949; Bourdon and Schnuriger, 1967).

Whether, in fact, the involvement of singlet oxygen in the light-induced aging of polymers is of some importance and is an effect that must be inhibited in order to protect the polymers remains to be seen. In the meantime, many man-made polymers are rapidly degraded by oxygen and light (e.g., polyacetals and polyamides) and no adequate understanding of their susceptibility has been proposed. Until more data are collected and sifted, those instances where a clear connection between singlet oxygen attack and degradation has been demonstrated, must stand as models from which our knowledge of oxidative photodegradation will grow, and in the end push forward the development of better polymers and stabilization systems.

III. BIOPOLYMERS AND RELATED SPECIES

When living organisms absorb light in the presence of oxygen, oxidation processes can occur which often lead to detrimental chemical and biological effects. This type of "photodynamic effect" has been observed in virtually all classes of organisms. Comprehensive reviews have described the multitude of studies in this area of research (Gallo and Santamaria, 1972; Spikes and Livingston, 1969; Spikes and MacKnight, 1970; Foote, 1975; Grossweiner, 1976).

A wide variety of short-lived reactive species are possible in these systems; singlet oxygen is only one candidate, and the reactions are of an extremely complex nature. Since there are other reasonable alternative mechanisms via free radical formation (by electron transfer involving sensitizer and substrate or molecular oxygen, or by hydrogen abstraction from substrate or solvent), one of the most difficult problems is the unequivocal assignment of a role for singlet oxygen in these photosensitized oxidations. The types of mechanistic tests that can be used have been reviewed by Foote (1975) and Grossweiner (1976). However, there are several instances where a good case can be made for the intermediacy of singlet oxygen and these will be discussed briefly. The larger question of the possible role of singlet oxygen in a variety of biological systems is discussed by Krinsky in Chapter 12 in this volume.

A. Amino Acid Derivatives and Peptides

Photodynamic oxidation of free amino acids leads to a variety of products depending on the sensitizer and irradiation conditions. On the basis of flash photolysis studies, Grossweiner and co-workers (Grossweiner and Zwicker, 1963; Kepka and Grossweiner, 1971; Grossweiner and Kepka, 1972) concluded that tryptophan oxidation by $^1O_2$ appeared to be favorable at pH 7, particularly for high substrate concentrations.

The kinetic spectroscopic identification of the initial amino acid products under aerobic conditions was complicated by the short lifetime of $^1O_2$ and its relatively low reactivity which required using high substrate concentrations when the sensitizer triplet state is being quenched. Foote (1975) analyzed the product distributions reported in the literature for a number of photodynamic studies and suggested that a role involving initial reactions of $^1O_2$ seemed likely in (1) the oxidation of methionine to the sulfoxide; (2) the slow oxidation of cysteine to cysteic acid; (3) the cleavage of the imidazole ring of histidine; and (4) attack (particularly under alkaline conditions) on the phenolic ring of tyrosine. While limited information is available about the products in the photodynamic processes of enzymes, direct evidence for methionine sulfoxide has been found in a number of studies such as on phosphoglucomutase (Ray and Koshland, 1962), elastase (Jori *et al.*, 1974a), and lysozyme (Jori *et al.*, 1974b). Additional evidence for $^1O_2$ in the photooxidation of tryptophan residues to *N*-formylkynurenine has been obtained by Jori and co-workers in elastase (Jori *et al.*, 1973), in lysozyme (Jori *et al.*, 1974b), and in papain (Jori and Galiazzo, 1971). Walrant and Santus (1974) also studied the transformation in carbonic anhydrase.

Nilsson and Kearns (1973) have shown that photosensitized oxygenation of yeast alcohol dehydrogenase and trypsin with methylene blue led to deactivation rates which were about ten times greater in D_2O than in H_2O. The rate was unaffected by an increase in oxygen pressure and NaN_3 quenched the oxidation. These facts were taken to suggest that $^1O_2$ is the reactive intermediate. On the basis of histidine's reactivity with singlet oxygen and its lack of reactivity toward many triplet dye molecules, Foote (1975) concluded that it is probable that deactivation of most enzymes, such as trypsin, in which attack at histidine is involved, proceeds via a singlet oxygen mechanism under favorable conditions. However, the flavins appear to differ from other sensitizers in the deactivation of trypsin as shown by Ghiron and Spikes (1965). A study of the eosin-sensitized photoinactivation of lysozyme (Kepka and Grossweiner, 1973) showed that tryptophan is probably the major oxidation site and it was concluded that both singlet oxygen and Type I (radical producing) reactions contribute to the oxidation under various conditions. Earlier, singlet oxygen had been implicated in the photoinactivation of lysozyme (using acridine orange as sensitizer) by the D_2O technique (Schmidt and Rosenkranz, 1972). Churakova *et al.* (1973) also showed that the products of reaction of lysozyme with microwave generated $^1O_2$ were identical to those from the methylene blue sensitized photooxygenation.

B. Erythropoietic Protoporphyria

Among the several known porphyrias, erythropoietic protoporphyria (EPP) is associated with photosensitivity since patients with this disorder are subject to edema, erythrema, and ultimately lesions on exposure to light. This photosensitivity is related to the fact that the red blood cells of patients with erythropoietic protoporphyria (EPP) are hemolyzed upon irradiation with visible light due to photooxidation of membrane components sensitized by the large amount of free protoporphyrin present in the cells (Harber *et al.*, 1964; Tschurdy *et al.*, 1971; Hsu *et al.*, 1971; Schothurst *et al.*, 1972; Goldstein and Harber, 1972). Ludwig *et al.* (1967) have reported that preincubation of EPP red cells with α-tocopherol (vitamin E) inhibits the oxidative damage and homolysis which result from irradiation of the cells. Goldstein and Harber (1972) reported that incubation of EPP red cells with α-tocopheryl acetate affords similar protection. It was demonstrated (Lamola *et al.*, 1973) that 3β-hydroxy-5α-hydroperoxy-Δ^6-cholestene, the product of $^1O_2$ attack on cholesterol, is produced in good yield on irradiating aerated aqueous suspensions of normal red blood cell ghosts into which protoporphyrin had been incorporated. Addition of this hydroperoxide to normal red blood cells led to increased osmostic fragility of the cells. The tentative conclusion was that hemolysis in EPP is caused by photosensitized oxidation of cholesterol and other lipids in erythrocyte membranes. It has been reported that the photosensitivity of EPP patients can be substantially relieved by orally administered β-carotene (Mathews-Roth *et al.*, 1970, 1974; Harber, 1972; De La Faille *et al.*, 1972). While the mechanism of the remedial action is not known, the effect is too specific and spectacular to be caused by carotene absorption, and singlet oxygen quenching seems an attractive hypothesis.

C. Neonatal Jaundice

A common problem among newborn infants, particularly those who are born prematurely, is jaundice, which, if unchecked, may lead to brain damage (Bergsma *et al.*, 1970). The cause of the jaundice has been traced to the lack of glucuronyl transferase which influences the conversion of the lipid-soluble yellow pigment, bilirubin, to the water-soluble conjugate with glucuronic acid. As a result, excess bilirubin begins to deposit in the skin and brain. The common treatment for neonatal jaundice has been to irradiate the infant with light of wavelengths which can be absorbed by the bilirubin. The irradiation bleaches the bilirubin in the skin and apparently prevents brain damage (Bergsma *et al.*, 1970; Cremer *et al.*, 1957). Singlet oxygen, formed by bilirubin as sensitizer, has been shown to be the reactive intermediate in the destruction (McDonagh, 1971; Bonnett and Stewart, 1972). One might wonder whether the phototherapy produces any additional photodynamic damage. It is a fortunate circumstance that bilirubin reacts with singlet oxygen at such a large rate that it is consumed before other cellular components are affected. Lightner and Norris (1974) showed that cholesterol in liposomes containing bilirubin was not photooxidized, which is in accord with the fact that the rate of reaction of bilirubin with singlet oxygen is 5×10^3 greater than that of cholesterol (Fahrenholtz *et al.*, 1974; Doleiden *et al.*, 1974).

ACKNOWLEDGMENTS

The authors wish to acknowledge many stimulating discussions with F. H. Winslow, W. L. Hawkins, A. A. Lamola, and S. R. Fahrenholtz at Bell Labs and with R. D. Small at Notre Dame.

REFERENCES

Achhammer, B. G., Reiney, M. J., Wall, L. A., and Reinhart, F. W. (1953). *Natl. Bur. Stand. (U.S.), Circ.* **525,** 205.

Achhammer, B. G., Tryon, M., and Kline, G. M. (1959). *Kunststoffe* **49,** 600.

Adams, J. H., and Goodrich, J. E. (1970). *J. Polym. Sci., Part A-1* **8,** 1269.

Arbuzov, Yu. A. (1965). *Russ. Chem. Rev.* **34,** 558.

Avram, M., and Mateescu, Gh. (1972). "Infrared Spectroscopy," chapter 8. Wiley (Interscience), New York.

Bateman, L. (1946). *Trans. Faraday Soc.* **42,** 267.

Bateman, L., and Gee, G. (1948). *Proc. R. Soc. London, Ser. A* **195,** 376 and 391.

Bateman, L., and Gee, G. (1951). *Trans. Faraday Soc.* **47,** 155.

Begeman, C. R. (1964). *SAE, Tech. Prog. Ser.* **6,** 163.

Bergsma, D., Hsia, D. Y. Y., and Jackson, C., eds. (1970). "Bilirubin Metabolism in the Newborn." Williams & Wilkins, Baltimore, Maryland.

Bevilacqua, E. M. (1961). *In* "Autoxidation and Antioxidants" (W. O. Lundberg, ed.), Vol. II, p. 857. Wiley (Interscience), New York.

Bolland, J, L., and Hughes, H. (1949). *J. Chem. Soc.* p. 492.

Bonnett, R., and Stewart, J. C. M. (1972). *Biochem. J.* **130,** 895.

Bonstead, I., and Charlesby, A. (1967). *Eur. Polym. J.* **3,** 459.

Bourdon, J., and Schnuriger, B. (1967). *In* "Physics and Chemistry of the Organic Solid State" (M. M. Labes and A. Weissberger, eds.), Chapter 2. Wiley (Interscience), New York.

Breck, A. K., Taylor, C. L., Russell, K. E., and Wan, J. K. S. (1974). *J. Polym. Sci., Polym. Chem. Ed.* **12,** 1505.

Briggs, P. J., and McKellar, J. F. (1968). *J. Appl. Polym. Sci.* **12,** 1825.

Burgess, A. R. (1952). *Chem. Ind. (London)* p. 78.

Burgess, A. R. (1953). *Natl. Bur. Stand. (U.S.), Circ.* **525,** 149.

Bystritskaya, E. V., and Karpukhin, O. N. (1975). *Dokl. Akad. Nauk SSSR* **221,** 1100.

Calvert, J., and Pitts, J. N., Jr. (1966). "Photochemistry," p. 382. Wiley, New York.

Carlsson, D. J., and Wiles, D. M. (1969). *Macromolecules* **2,** 587.

Carlsson, D. J., and Wiles, D. M. (1973). *J. Polym. Sci., Polym. Lett. Ed.* **11,** 759.

Carlsson, D. J., and Wiles, D. M. (1975). *Rubber Chem. Technol.* **47,** 991.

Carlsson, D. J., Mendenhall, G. D., Suprunchuk, T., and Wiles, D. M. (1972). *J. Am. Chem. Soc.* **94,** 8960.

Carlsson, D. J., Suprunchuk, T., and Wiles, D. M. (1973). *J. Polym. Sci., Part B* **11,** 61.
Chien, J. C. W., and Boss, C. R. (1972). *J. Polym. Sci., Part A-1* **10,** 1599.
Churakova, N. I., Kravchenko, N. A., Serebryakov, F. P., Lavrov, I. A., and Kaversneva, E. D. (1973). *Photochem. Photobiol.* **18,** 201.
Cicchetti, O. (1970). *Adv. Polym. Sci.* **7,** 70.
Cook, T. J., and Miller, J. A. (1974). *Chem. Phys. Lett.* **25,** 396.
Cremer, R. J., Perryman, P. W., and Richards, D. H. (1957). *Lancet* **1,** 1094.
Cunneen, J. I., and Higgins, G. M. C. (1963). *In* "The Chemistry and Physics of Rubber-Like Substances" (L. Bateman, ed.), Chapter 2. Wiley, New York.
Dalle, J. P., Magous, R., and Mousseron-Canet, J. (1972). *Photochem. Photobiol.* **15,** 411.
De La Faille, H. B., Suurmond, D., Went, L. N., van Stevenick, J., and Schothurst, A. A. (1972). *Dermatologica* **145,** 389.
Doleiden, F. H., Fahrenholtz, S. R., Lamola, A. A., and Trozzolo, A. M. (1974). *Photochem. Photobiol.* **20,** 519.
Dunn, J. R. (1960). *J. Appl. Polym. Sci.* **4,** 151.
Duynstee, E. F. J., and Mevis, M. E. A. H. (1972). *Eur. Polym. J.* **8,** 1375.
Egerton, G. S. (1949). *J. Soc. Dyers Colour.* **65,** 764.
Egerton, G. S. (1964). *Nature (London)* **204,** 1153.
Fahrenholtz, S. R., Doleiden, F. H., Trozzolo, A. M., and Lamola, A. A. (1974). *Photochem. Photobiol.* **20,** 505.
Flood, J., Russell, K. E., and Wan, J. K. S. (1973). *Macromolecules* **6,** 669.
Foote, C. S. (1968a). *Acc. Chem. Res.* **1,** 104.
Foote, C. S. (1968b). *Science* **162,** 963.
Foote, C. S. (1975). *In* "Free Radicals in Biology" (W. A. Pryor, ed.), Vol. 1. Academic Press, New York.
Foote, C. S., Chang, Y. C., and Denny, R. W. (1970). *J. Am. Chem. Soc.* **92,** 5216.
Fox, R. B. (1973). *Pure Appl. Chem.* **34,** 235.
Furakawa, K., Gray, W. W., and Ogryzlo, E. A. (1970). *Ann. N.Y. Acad. Sci.* **171,** 175.
Gallo, U., and Santamaria, L. (1972). "Research Progress in Organic Biological and Medicinal Chemistry," Vol. III. Am. Elsevier, New York.
Ghiron, C. S., and Spikes, J. D. (1965). *Photochem. Photobiol.* **4,** 901.
Goldstein, B. D., and Harber, L. C. (1972). *J. Clin. Invest.* **51,** 892.
Gollnick, K., Franken, T., Schade, G., and Dorhofer, G. (1970). *Ann. N.Y. Acad. Sci.* **171,** 89.
Grassie, N., and Weir, N. A. (1965). *J. Appl. Polym. Sci.* **9,** 999.
Grossweiner, L. I. (1976). *Curr. Top. Radiat. Res. Q.* **11,** 141.
Grossweiner, L. I., and Kepka, A. G. (1972). *Photochem. Photobiol.* **16,** 305.
Grossweiner, L. I., and Zwicker, E. F. (1963). *J. Chem. Phys.* **39,** 2774.
Harber, L. C. (1972). *Arch. Dermatol.* **106,** 414.
Harber, L. C., Fleischer, A. S., and Baer, R. L. (1964). *J. Am. Med. Assoc.* **189,** 191.
Hart, E. J., and Matheson, M. S. (1948). *J. Am. Chem. Soc.* **70,** 784.
Heacock, J. F., Mallory, F. B., and Gay, F. P. (1968). *J. Poly. Sci. A1* **6,** 2921.
Heskins, M., and Guillet, J. E. (1968). *Macromolecules* **1,** 97.
Hofmann, A. W. (1861). *J. Chem. Soc.* **13,** 87.
Hollahan, J. R., and Carlson, G. L. (1970). *J. Appl. Polym. Sci.* **14,** 2499.
Howard, J. A., and Ingold, K. U. (1968). *J. Am. Chem. Soc.* **90,** 1056.
Hsu, J., Goldstein, B. D., and Harber, L. C. (1971). *Photochem. Photobiol.* **13,** 67.
Jones, L. T. N., and Wayne, R. P. (1971). *Proc. R. Soc. London, Ser. A* **321,** 409.
Jori, G., and Galiazzo, G. (1971). *Photochem. Photobiol.* **14,** 607.
Jori, G., Galiazzo, G., and Buso, O. (1973). *Arch. Biochem. Biophys.* **158,** 116.
Jori, G., Gennari, G., and Folin, M. (1974a). *Photochem. Photobiol.* **19,** 79.

Jori, G., Folin, M., Gennari, G., Galiazzo, G., and Buso, O. (1974b). *Photochem. Photobiol.* **19,** 419.
Kaplan, M. L. (1971). *Chem. Tech.* **1,** 621.
Kaplan, M. L., and Kelleher, P. G. (1970a). *J. Polym. Sci., Part A-1* **8,** 3163.
Kaplan, M. L., and Kelleher, P. G. (1970b). *Science* **169,** 1206.
Kaplan, M. L., and Kelleher, P. G. (1971). *J. Polymer Sci., Part B* **9,** 565.
Kaplan, M. L., and Kelleher, P. G. (1972). *Rubber Chem. Technol.* **45,** 423.
Karmitz, P. (1958). *Rev. Gen. Caoutch.* **35,** 913.
Karpukhin, O. N., and Slobodetskaya, E. M. (1974). *Polym. Sci. USSR* (*Engl. Transl.*) **16,** 1882.
Kearns, D. R. (1971). *Chem. Rev.* **71,** 395.
Kearns, D. R., and Khan, A. U. (1969). *Photochem. Photobiol.* **10,** 193.
Kearns, D. R., Khan, A. U., Duncan, C. K., and Kaki, A. H. (1969). *J. Am. Chem. Soc.* **91,** 1039.
Kepka, A. G., and Grossweiner, L. I. (1971). *Photochem. Photobiol.* **14,** 621.
Kepka, A. G., and Grossweiner, L. I. (1973). *Photochem. Photobiol.* **18,** 49.
Kopecky, K. R., and Reich, H. J. (1965). *Can. J. Chem.* **43,** 2265.
Kwei, K. P. S. (1969). *J. Polym. Sci.* **7,** 1075.
Lamola, A. A., Yamane, T., and Trozzolo, A. M. (1973). *Science* **179,** 1131.
Lightner, D. A., and Norris, R. D. (1974). *N. Engl. J. Med.* **290,** 1260.
Ludwig, G. D., Bilheimer, D., and Iverson, L. (1967). *Clin. Res.* **15,** 284 (abstr.).
MacCallam, J. R., and Rankin, C. T. (1974). *Makromol. Chem.* **175,** 2477.
McDonagh, A. F. (1971). *Biochem. Biophys. Res. Commun.* **44,** 1306.
Matheson, L. A., and Boyer, R. F. (1952). *Ind. Eng. Chem.* **44,** 867.
Mathews-Roth, M. M., Pathak, M. A., Fitzpatrick, T. B., Harber, L. C., and Kass, E. H. (1970). *N. Engl. J. Med.* **282,** 1231.
Mathews-Roth, M. M., Pathak, M. A., Fitzpatrick, T. B., Harber, L. C., and Kass, E. H. (1974). *J. Am. Med. Assoc.* **228,** 1004.
Mill, T., Irwin, K. C., and Mayo, F. R. (1968). *Rubber Chem. Technol.* **41,** 296.
Mill, T., Richardson, H., and Mayo, F. R. (1973). *J. Polym. Sci., Polym. Chem. Ed.* **11,** 2899.
Miller, W. A. (1865). *J. Chem. Soc.* **18,** 273.
Morand, J. (1966). *Rubber Chem. Technol.* **39,** 537.
Morand, J. (1968). *Rev. Gen. Caoutch. Plast.* **45,** 615 and 999.
Murray, R. W., and Kaplan, M. L. (1968). *J. Am. Chem. Soc.* **90,** 4161.
Murray, R. W., Lumma, W. C., Jr., and Lin, J. W. P. (1970). *J. Am. Chem. Soc.* **92,** 3205.
Nakajima, M., Tomida, I., and Takei, S. (1959). *Chem. Ber.* **92,** 163.
Nilsson, R., and Kearns, D. R. (1973). *Photochem. Photobiol.* **17,** 65.
Ogryzlo, E. A., and Tang, C. W. (1970). *J. Am. Chem. Soc.* **92,** 5034.
Partridge, R. H. (1966). *J. Chem. Phys.* **45,** 1679.
Pitts, J. N., Jr. (1969). *Adv. Environ. Sci.* **1,** 289.
Pitts, J. N., Jr. (1971). *In* "Chemical Reactions in Urban Atmospheres" (C. S. Tuesday, ed.), p. 3. Am. Elsevier, New York.
Pitts, J. N., Jr., Kahn, A. U., Smith, E. B., and Wayne, R. P. (1969). *Environ. Sci. Technol.* **3,** 241.
Pivovarov, A. P., and Lukovnikov, A. F. (1968). *Khim. Vys. Energ.* **2,** 220.
Pivovarov, A. P., Gak, Y. V., and Lukovnikov, A. F. (1971). *Vysokomol. Soedin., Ser. A* **13,** 2110.
Pross, A. W., and Black, R. M. (1950). *J. Soc. Chem. Ind., London* **69,** 113.
Rabek, J. F., and Ranby, B. J. (1974). *J. Polym. Sci., Part A-1* **12,** 273.
Rabek, J. F., and Ranby, B. J. (1975). *Polym. Eng. Sci.* **15,** 40.

Ray, W. J., Jr., and Koshland, D. E., Jr. (1962). *J. Biol. Chem.* **237,** 2493.

Rugg, F. H., Smith, J. J., and Bacon, R. C. (1954). *J. Polym. Sci.* **13,** 535.

Saunders, K. J. (1973). "Organic Polymer Chemistry," Chapter 18. Chapman & Hall, London.

Savides, C. (1968). *In* "Stabilization of Polymers and Stabilizer Processes" (R. F. Gould, ed.), Chapter 20. Am. Chem. Soc., Washington, D.C.

Schmidt, H., and Rosenkranz, P. (1972). *Z. Naturforsch., Teil B* **27,** 1436.

Schnuriger, B., and Bourdon, J. (1968). *Photochem. Photobiol.* **8,** 361.

Schothurst, A. A., van Steveninck, J., Went, L. D., and Suurmond, D. (1972). *Clin. Chim. Acta* **39,** 161.

Spikes, J. D., and Livingston, R. (1969). *Adv. Radiat. Biol.* **3,** 29.

Spikes, J. D., and MacKnight, M. L. (1970). *Ann. N.Y. Acad. Sci.* **171,** 149.

Trozzolo, A. M., ed. (1970). "International Conference on Singlet Molecular Oxygen and Its Role in Environmental Sciences," Ann. N.Y. Acad. Sci. No. 171. N.Y. Acad. Sci., New York.

Trozzolo, A. M. (1971). *In* "Polymer Stabilization" (W. L. Hawkins, ed.), p. 159. Wiley, New York.

Trozzolo, A. M., and Fahrenholtz, S. R. (1970). *Ann. N.Y. Acad. Sci.* **171,** 61.

Trozzolo, A. M., and Winslow, F. H. (1968). *Macromolecules* **1,** 98.

Tryon, M., and Wall, L. A. (1961). *In* "Autoxidation and Antioxidants" (W. O. Lundberg, ed.), Vol. II, p. 919. Wiley (Interscience), New York.

Tschurdy, D. P., Magnus, I. A., and Dalvas, J. (1971). *In* "Dermatology in General Medicine" (T. B. Fitzpatrick *et al.*, eds.), p. 1143. McGraw-Hill, New York.

Tsuji, K., and Seiki, T. (1970). *J. Polym. Sci., Part B* **8,** 817.

Turro, N. J. (1965). "Molecular Photochemistry." Benjamin, New York.

Walrant, P., and Santus, R. (1974). *Photochem. Photobiol.* **20,** 455.

Wasserman, E., Kuck, V. J., Delevan, W. M., and Yager, W. A. (1969). *J. Am. Chem. Soc.* **91,** 1040.

Wayne, R. P. (1969). *Adv. Photochem.* **7,** 311.

Wilson, T., and Hastings, J. W. (1970). *Photophysiology* **5,** 49.

Winslow, F. H., Matreyek, W., and Trozzolo, A. M. (1969). *Polym. Prepr., Am. Chem. Soc., Div. Polym. Chem. 10* (2), 1271.

Young, R. H., Martin, R. L., Geriozi, D., Brewer, D., and Kayser, R. (1973). *Photochem. Photobiol.* **17,** 233.

Zweig, A., and Henderson, W. A., Jr. (1975). *J. Polym. Sci.* **13,** 717 and 993.

12

Biological Roles of Singlet Oxygen

NORMAN I. KRINSKY

I. Introduction 597
Sources of $^1O_2$ 598
II. Identification of $^1O_2$ in Biological Systems 599
A. Inhibition by Specific Quenchers 599
B. Product Analysis 600
C. Chemiluminescence (CL) 604
D. Deuterium Enhancement 606
III. Biological Systems 606
A. Photodynamic Effects 606
B. Polymorphonuclear Leukocytes 606
C. Enzyme Systems 619
D. Miscellaneous Oxidant Effects 634
IV. Summary 635
References 636

I. INTRODUCTION

The science of photobiology introduced biologists to the presence of electronically excited molecules interacting with biological systems. This interaction was expressed in such processes as photosynthesis, vision, phototropism, photomorphogenesis, diurnal rhythms, and many others. In the field of bioluminescence, biologists became aware of the fact that certain specialized organisms can actually generate electronically excited species which are deactivated yielding a photon of visible light. Although these important areas have flourished and developed, the concept of electronically excited molecules has not had much impact in the mainstream of biochemical research.

However, the situation changed with the realization that a relatively long-lived electronically excited species of oxygen, the $^1\Delta_g$ state, was pro-

Singlet Oxygen

ISBN 0-12-736650-4

duced in both photosensitized and chemical reactions (Foote, 1968a,b). This electronically excited species of oxygen, referred to hereafter as singlet oxygen ($^1O_2$), was responsible for the photodynamic action—the destruction of cells and tissues in the presence of light, O_2, and a suitable sensitizing dye (Blum, 1941). With the realization that singlet oxygen survived long enough to react chemically, this species has caught the attention of biologists and biochemists and has been increasingly invoked as being responsible for a wide variety of oxidative effects in nature. It is the purpose of this review to organize the various observations into several related categories and then to evaluate whether the available evidence supports the contention for $^1O_2$ involvement in these systems. For prior discussions relating to $^1O_2$ and its involvement in biological systems, the reader is referred to the reviews of Foote (1968a, 1976), Wilson and Hastings (1970), Kearns (1971), Politzer *et al.* (1971), Bland (1976), and Krinsky (1977). A number of specific systems in which a good case can be made for the involvement of singlet oxygen in biological systems are discussed by Kaplan and Trozzollo in Chapter 11 of this volume.

Sources of $^1O_2$

A comprehensive review of the chemical sources of singlet oxygen is presented by Murray in Chapter 3 in this volume. This is an important

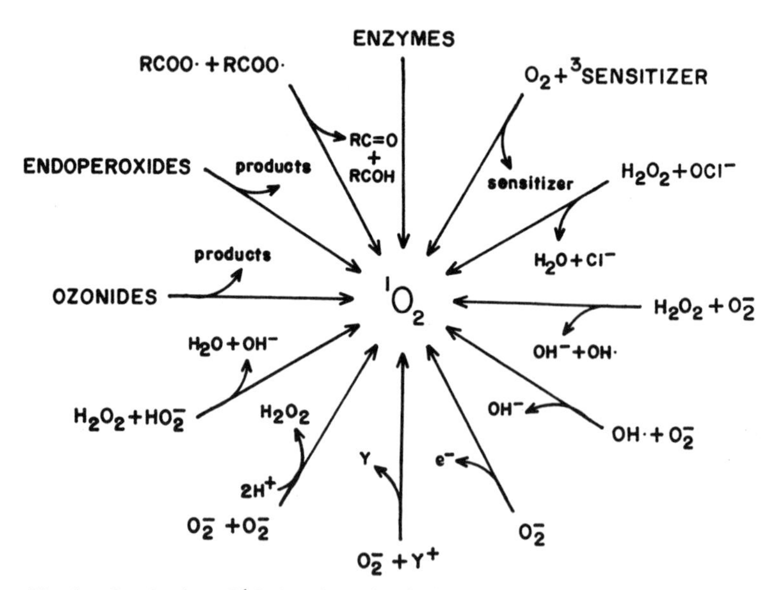

Fig. 1. Production of $^1O_2$ by photochemical, chemical, and biological systems [Reproduced by permission of Krinsky (1977) and Elsevier Scientific Publishing Co.].

area to take into consideration when one is concerned about the biological roles of $^1O_2$, for *in vitro* chemical reactions frequently serve as models for reactions occurring in biological systems. In addition to the chemical sources, the photochemical source of $^1O_2$ has now been well documented (Foote, 1968a,b). Krinsky (1977) has graphically presented some of the chemical and photochemical reactions involved in producing $^1O_2$ and these are shown in Fig. 1. Although, as will be discussed in Section III,B and Section III,C, there is much indirect evidence for $^1O_2$ production in biological systems, there has not as yet been a demonstration of $^1O_2$ being produced directly from an enzymatic reaction.

II. IDENTIFICATION OF $^1O_2$ IN BIOLOGICAL SYSTEMS

The key to understanding whether $^1O_2$ is involved in any biological system is to carefully analyze the techniques that have been used by various investigators to demonstrate the presence or absence of $^1O_2$. Much of this material is covered in other chapters in this volume. However, a few techniques that have been used will be discussed in some detail with respect to their application to biological systems.

A. Inhibition by Specific Quenchers

The details of the interaction between $^1O_2$ and its quenchers are documented in the contribution by Foote in Chapter 5 of this volume. From the point of view of using specific quenchers to characterize $^1O_2$ reactions in biological systems several factors must be taken into consideration. In the first place, it is important to understand whether the quencher has accessibility to $^1O_2$. For example, lipophilic quenchers such as carotenoids and tocopherols would not be expected to interact strongly with the $^1O_2$ that is produced in an aqueous medium. On the other hand, these compounds would have accessibility to biomembranes and would be presumed to be active there. This has been demonstrated for carotenoids in both *in vivo* systems (Mathews-Roth *et al.*, 1974) and *in vitro* systems (Anderson and Krinsky, 1973; Anderson *et al.*, 1974). There are additional cautions that must be observed in utilizing quenching data to identify $^1O_2$ in biological systems. For example, Davidson and Trethewey (1976) have demonstrated that certain $^1O_2$ quenchers can interact with the excited singlet state of dyes and therefore block any possibility of generating $^1O_2$ in a photosensitized system. This phenomenon has been known for carotenoid pigments for some time (Fujimori and Livingston, 1968) and has been used as one of the bases for explaining carotenoid protection *in vivo* (Krinsky, 1976).

B. Product Analysis

The process of analyzing products of chemical reactions has been used for many years to show identity or dissimilarity in the mechanism of the reactions under study. This technique has proven extremely useful and was used very successfully by Foote and Wexler (1964a,b) to identify $^1O_2$ as the oxidant in photosensitized oxidations. Detailed descriptions of the products that can be formed by the interaction of $^1O_2$ with various substrates can be found in the contributions by Bartlett and Landis, Gollnick and Kuhn, Saito and Matsuura, Schaap and Zaklika, and Wasserman and Lipshutz in Chapters 7, 8, 10, 6, and 9 of this volume, respectively. Because of the great emphasis that investigators have placed on the appearance of presumed specific $^1O_2$ products in biological systems, it is worth spending some time in analyzing the specificity of some of these reactions.

Extensive work has been done on the products of the photosensitized oxidation of both substituted furans and benzofurans, starting with the observations of Guyot and Catel (1906) that 1,3-diphenylisobenzofuran (DPBF) produced *o*-dibenzoylbenzene (DBB) when illuminated in a benzene solution. Since then there have been many similar reports that $^1O_2$ generated either photochemically (Dufraisse and Ecary, 1946; Weser *et al.*, 1975; Gorman *et al.*, 1976), chemically by the decomposition of diphenylanthracene endoperoxide (Wasserman *et al.*, 1972), by microwave irradiation (Corey and Taylor, 1964), or radical cation abstraction (Mayeda and Bard, 1973) results in the formation of DBB from DPBF. Other workers have relied solely on the decrease in the absorption of DPBF in the visible region of the spectrum as a basis for concluding that they were observing a $^1O_2$ catalyzed reaction. There are a number of reasons, apart from failing to identify the presumed specific $^1O_2$ product, which makes this type of analysis unsatisfactory as far as evidence for $^1O_2$ involvement. In the first place, Le Berre and Ratsimbazafy (1963) report that DPBF is converted to a "peroxide" intermediate in the presence of oxygen *in the dark*. This peroxide could be reduced to DBB and presumably represents the endoperoxide intermediate which has been identified as the $^1O_2$ addition product, as shown in Scheme 1.

DPBF $\xrightarrow[\text{dark}]{O_2}$ endoperoxide $\xrightarrow{\text{reduction}}$ *o*-DBB

Scheme 1

Similarly, King *et al.* (1975) report that DBB was formed from DPBF when the latter was incubated with several enzymes in the *absence* of substrate under conditions where $^1O_2$ could not have been formed.

In addition, we have observed (S. Kwong and N. I. Krinsky, unpublished observations, 1974) that DPBF is oxidized, as determined by a decrease in the absorbancy at 412 nm, not only by H_2O_2/OCl^-, but the rate of oxidation is even more rapid when the DPBF is reacted with OCl^- *alone*. Boyer *et al.* (1975) have also presented evidence that DBB is formed when DPBF is treated with *m*-chloroperoxybenzoic acid and argue quite convincingly that this reaction *does not* occur via a $^1O_2$ mechanism. These results would therefore place the conclusions reached by several authors regarding the generation of $^1O_2$ by various enzyme systems in doubt, even when that conclusion is based on the appearance of material having similar chromatographic properties to DBB. These include the observations on lipooxygenase (Chan, 1971a; Finazzi Agrò *et al.*, 1973), horseradish peroxidase (Chan, 1971b; Finazzi Agrò *et al.*, 1973), xanthine oxidase (Pederson and Aust, 1973; King *et al.*, 1975), microsomal enzymes (King *et al.*, 1975; Marnett *et al.*, 1975), lactoperoxidase (Piatt *et al.*, 1977), and the photodynamic destruction of melanoma tissue (Weishaupt *et al.*, 1976).

This is not to suggest that DPBF is useless insofar as $^1O_2$ mechanisms are concerned. It is known that DPBF is an excellent $^1O_2$ acceptor and based on DPBF inhibition, several groups have inferred that they have been studying $^1O_2$ reactions. Some reactions which have been attributed to $^1O_2$, such as the oxidation of DPBF by diperoxychromium(VI) oxide etherate (Chan, 1970) have now been refuted (Baldwin *et al.*, 1971), but the majority of reports attributing $^1O_2$ involvement based either on DPBF oxidation or DPBF conversion to DBB are still widely accepted as proof of $^1O_2$ production, when at best they must be considered inferential evidence.

These arguments are not limited to the benzofurans. Almost identical experiments have been carried out with substituted furans. These experiments go back to the observation of Martel (1957) that light catalyzed the oxidation of diphenylfuran (DPF) and tetraphenylfuran (TPF) to "photooxides" which on reduction yielded *cis*-dibenzoylethylene (DBE) and *cis*-dibenzoylstilbene (DBS) respectively as shown in Scheme 2.

Gollnick and Schenk (1967) in their comprehensive review on oxidation mechanisms quote an unpublished dissertation (Erbrich, 1952) dealing with the production of a variety of *cis*-diaroylethylenes formed by the photosensitized oxidation of aryl-substituted furans. However, this same report (Erbrich, 1952) contained evidence that thermal autoxidation of DPF generated *cis*-DBE, whereas thermal autoxidation of TPF formed *trans*-DBS. Lutz *et al.* (1962) observed that TPF could be converted to *cis*-DBS

$$H_5C_6\text{-(furan)-}C_6H_5 \ (\text{DPF}) \xrightarrow[O_2]{h\nu} H_5C_6\text{-(endoperoxide, O-O)-}C_6H_5 \xrightarrow{\text{reduction}} H_5C_6\text{-CO-CH=CH-CO-}C_6H_5 \ (cis\text{-DBE})$$

$$(H_5C_6)_4\text{-furan} \ (\text{TPF}) \xrightarrow[O_2]{h\nu} (H_5C_6)_4\text{-endoperoxide (O-O)} \xrightarrow{\text{reduction}} H_5C_6\text{-CO-C}(C_6H_5)\text{=C}(C_6H_5)\text{-CO-}C_6H_5 \ (cis\text{-DBS})$$

Scheme 2

by *both* photosensitized and chemical oxidations. Boyer *et al.* (1975) found that chloroperoxybenzoic acid converted DPF to *cis*-DBE. We have observed (Krinsky and Jong, 1977) a Ce^{4+} catalyzed conversion of DPF to *cis*-DBE in the *absence* of peroxy radicals. Takayama *et al.* (1977) observed the formation of *cis*-DBE from DPF using a modified Fenton's reagent (Brodie *et al.*, 1954) to produce hydroxyl (OH·) radicals, and Rosen and Klebanoff (1977) have observed the conversion of DPF to *cis*-DBE in the presence of hypochlorous acid.

Anthracenes and substituted anthracenes have also been studied extensively as $^1O_2$ acceptors and their oxidation products have been analyzed. In their comprehensive review, Gollnick and Schenck (1967) list over a 150 aromatic hydrocarbons and their derivatives with respect to their photosensitized oxidation products. In almost all cases which they cite, anthracenes initially form endoperoxides when exposed to $^1O_2$ generated in the presence of light and a photosensitizing dye. The formation of endoperoxides can also be carried out using $^1O_2$ generated by microwave discharge as first reported by Corey and Taylor (1964) who observed the exclusive formation of the 9,10-endoperoxide when anthracene was exposed to these conditions. Chemically generated $^1O_2$ has been used by McKeown and Waters (1966) to convert diphenylanthracene (DPA) to the corresponding 9,10-endoperoxide of DPA. Of particular interest has been the use of this latter reaction to analyze the mechanism of the decomposition of peroxy radicals. Russell (1957) first presented evidence that peroxy radicals decomposed to generate a ketone, an alcohol and oxygen, presumably through the formation of the transition state described in Scheme 3. Howard and Ingold (1968) argued that the decomposition of this transition state would not violate

$$2\ \rangle C(H)\text{-O-O}\cdot \rightleftharpoons [\text{cyclic tetroxide transition state: } \rangle C(H)\text{-O-O-O-O-CH}\langle] \longrightarrow \rangle C{=}O + HO\text{-CH}\langle + O_2$$

Scheme 3

the Wigner spin conservation rule if the oxygen were eliminated in a singlet state or if the ketone would be formed in its excited triplet state. In order to analyze these two possibilities, Howard and Ingold (1968) studied the Ce^{4+} catalyzed formation of peroxy radicals in the presence of DPA. They isolated and characterized the 9,10-endoperoxide of DPA as the principle product of this reaction and based on this analysis concluded that the "Russell" mechanism proceeded via the elimination of $^1O_2$. Shortly thereafter, Kellogg (1969) studied the mechanism of the chemiluminescence (CL) which arises from peroxy radicals and concluded that the reaction produces triplet states specifically, with a ratio of triplet states to singlet states greater than 10^4. As such, he proposed that the Russell mechanism should be drawn with the transition state shown in Scheme 4.

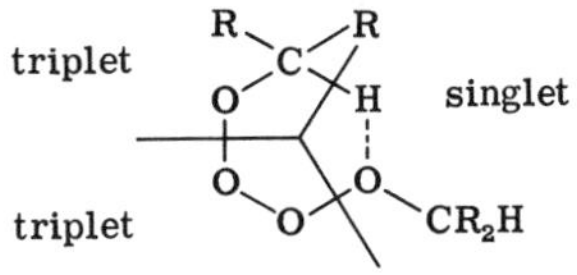

Scheme 4

Kellogg (1969) concluded, based on the rate constants for oxygen quenching of CL, that the excited triplet carbonyl is produced with an efficiency of the order of unity.

These rather contradictory conclusions may be explained by our observations (Krinsky and Jong, 1977) that products characteristic of $^1O_2$ reactions can be formed in the Howard and Ingold (1968) system in the *absence* of peroxy radicals. Krinsky and Jong (1977) demonstrated the facile conversion of DPF to *cis*-DBE in the presence of Ce^{4+} and linoleic acid hydroperoxide (LAHPO), as reported earlier by Howard and Ingold (1968). However, we observed that the conversion proceeded as well in the presence of Ce^{4+} in the complete *absence* of peroxy radicals. This reaction could not be induced to occur when Ce^{4+} was replaced with either Fe^{3+}, Fe^{2+}, or Ti^{4+}. The resolution of this controversy is important inasmuch as a number of individuals have utilized the Howard and Ingold (1968) interpretation of the Russell mechanism to explain the formation of $^1O_2$ in biological systems that may be generating lipid hydroperoxides. King *et al.* (1975) believe that the formation of *cis*-DBE from DPF in the presence of supplemental liver microsomes may actually be due to the breakdown of lipid hydroperoxides. In addition, Nakano and his associates (1975, 1976; Sugioka and Nakano, 1976) also rely on this interpretation as evidence that $^1O_2$ is produced during lipid peroxide decomposition. In fact, these workers (Nakano *et al.*, 1976) have treated various secondary hydroperoxides such as LAHPO or *sec*-butyl hydroperoxide with Ce^{4+} and have observed CL which they interpret as being due to $^1O_2$ emission. Although their emission spectrum shows certain similarities to what one might expect from the emission of $^1O_2$, it is not in total agreement. This CL may be related to the observation of Kochi (1962) who described the decomposition of alkyl hydroperoxides by various reducing metals which formed alkyl radicals, ketones, and alcohols potentially capable of generating CL. Dixon and Norman (1962) also report that Ti^{3+} can form radicals from peroxides and supports the concept of reducing metals generating radical products from peroxides.

Attempts to utilize anthracenes as $^1O_2$ acceptors in biological systems have not been as extensive as the furan work quoted above. Chan (1971b) observed that horseradish peroxidase in the presence of H_2O_2 converted anthracene to an anthraquinone but this system was unable to oxidize DPA. Sternson and Wiley (1972) were not able to observe any oxidation of anthracene in the presence of a microsomal system. Chen and Tu (1976) have used a microsomal cytochrome *P*-450 oxygenase to demonstrate the conversion of dimethylanthracene (DMA) to the 9,10-endoperoxide of DMA which could subsequently be reduced by an NADPH requiring microsomal system to the corresponding diol.

C. Chemiluminescence (CL)

Various aspects of the CL which accompanies the deactivation of either products of $^1O_2$ reactions or $^1O_2$ itself appear in contributions in this volume by Kasha, Ogryzlo, and Schaap in Chapters 1, 2, and 6 of this volume respectively. Unfortunately, the use or misuse of CL by biologists and biochemists has led to considerable confusion in attempting to define the relationship of $^1O_2$ to biological systems. One of the basic problems has been the ready availability of a device for measuring low level CL. This device is the liquid scintillation counter, a common laboratory instrument, which when placed in the out-of-coincidence mode, can be used as a sensitive light detection device. The spectral sensitivity of the machine, however, is limited by the sensitivity of the photomultiplier tube that is used. For many scintillation counters a tube comparable to the RCA type 4501, V3 series is used which has a sensitivity that drops off very sharply at 590 nm. The CL which accompanies the molecular pair interaction of two $^1O_2$ molecules emits maximally at 634 and 703 nm (Khan and Kasha, 1970). As such, most scintillation counters cannot be used to detect $^1O_2$ dimol emission. A much better tube is the red-sensitive photomultiplier tube RCA 4832 which has a sensitivity in the red region of the spectrum. The relative sensitivities of this tube and a typical scintillation counter tube are shown in Fig. 2. This red sensitive tube was used by Deneke and Krinsky (1976, 1977) in conjunction with specific interference filters to select out the wave lengths of light attributed to $^1O_2$ dimol emission. What is necessary, therefore, in

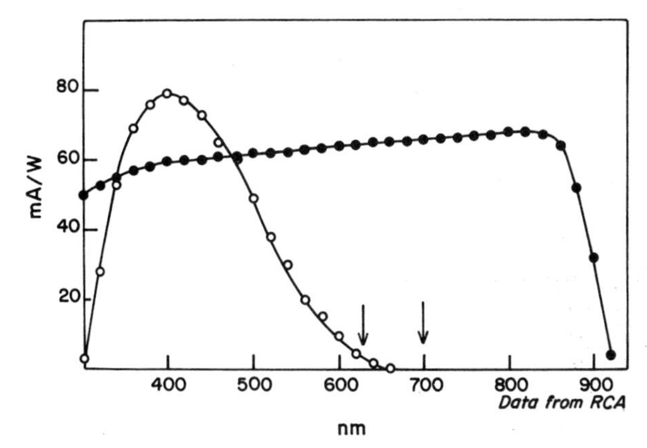

Fig. 2. Relative spectral sensitivity of a red-sensitive photomultiplier (●) (RCA4832) used by Deneke and Krinsky (1976, 1977) to detect $^1O_2$ paired emission, as compared to the relative sensitivity of a typical liquid scintillation counter photomultiplier (○). Arrows indicate the two peaks of $^1O_2$ paired emission at 634 and 730 nm.

order to characterize $^1O_2$ is a carefully constructed emission spectrum of the CL reactions. This has been attempted by Cheson *et al.* (1976) and by Nakano *et al.* (1975). In the work reported by Cheson *et al.* (1976), light emission from polymorphonuclear leukocytes phagocytosing opsonized zymosan particles was studied. These workers observed that the light produced during the phagocytic process could be inhibited by both superoxide dismutase (SOD) and catalase. Based on a comparison between the effects of these enzymes on the CL from phagocytosing polymorphonuclear leukocytes and the CL obtained from a O_2^- generating system which contained xanthine oxidase and purine, the authors concluded that the light reaction is due to the oxidation of portions of the zymosan by a variety of oxidizing agents which included H_2O_2 and possibly OH·. Similar conclusions had been reached by Rosen and Klebanoff (1976). Cheson *et al.* (1976) tried to obtain a corrected emission spectrum by using colored plastic filters in the scintillation counter as a means of identifying the light emission at broad wavelength regions. Their corrected emission spectrum showed a rather broad emission between approximately 475 and the long wavelength cutoff of their instrument, 624 nm. It was, of course, unfortunate that the photomultiplier that they used was not able to detect the intense $^1O_2$ bands which appear at 634 and 703 nm. They therefore could not totally rule out the possibility that some of the CL arises from the molecular pair emission of $^1O_2$ states.

In the case of Nakano *et al.* (1975), a single photon technique was used to detect CL and they compared the light emitted from a chemical reaction utilizing secondary hydroperoxides as sources of CL with that obtained from the classical H_2O_2/OCl^- system and found some similarities in the CL spectra. It is obvious from the published spectra of both of these publications that there is a large amount of background material which may arise from related reactions but cannot be attributed to $^1O_2$ alone.

D. Deuterium Enhancement

In D_2O, the lifetime at $^1O_2$ is enhanced up to a factor of approximately 10 (Merkel *et al.*, 1972). The use of this phenomenon in detecting $^1O_2$ reactions has been used by a number of investigators but the reader is referred to the contribution of Kearns in Chapter 4 in this volume for a more detailed review.

III. BIOLOGICAL SYSTEMS

A. Photodynamic Effects

Although photodynamic effects, mediated through light, O_2, and a photosensitizing dye, were the first example of the interaction of $^1O_2$ with biological systems (Blum, 1941), this area is quite beyond the scope of this chapter. Instead, the reader is referred to several excellent recent reviews dealing with photodynamic effects. These reviews include contributions by Douzou (1973), Foote (1976), Gollnick (1976), Grossweiner (1976), Knowles (1976), Krinsky (1976), and Spikes (1975).

B. Polymorphonuclear Leukocytes

One of the most intriguing problems that has confronted scientists is the mechanism whereby organisms protect themselves against foreign bodies and in particular against microbial invasion. There are numerous microbicidal systems which operate in higher organisms, consisting of both extracellular and intracellular phases. In this section, we will discuss the intracellular processes that lead to microbicidal action and how these processes are related to the oxidants produced by the host cells.

1. Microbicidal Activity

Intracellular microbicidal activity is carried out by a variety of phagocytic cells. Higher organisms possess two circulating phagocytic cells, the polymorphonuclear leukocyte and the mononuclear leukocyte. The former is characterized by possessing a multilobed nucleus and a relatively short life time which may be the order of magnitude of only a few days, whereas the mononuclear leukocyte, which has a single nucleus, has a considerably longer lifetime. Mononuclear leukocytes ultimately become tissue macrophages. Much work has been carried out with polymorphonuclear leukocytes and it is now becoming evident that many of the processes which occur in these cells are also applicable to other types of phagocytic cells and may serve as examples of a general type of intracellular microbicidal action.

The process whereby invading microorganisms are recognized and ingested by phagocytic cells has been described in several recent reviews (Karnovsky, 1973; Klebanoff, 1975a; Sbarra *et al.*, 1976). The entire process in a polymorphonuclear leukocyte is depicted schematically in Fig. 3. Figure 3 illustrates the various stages that are involved in the intracellular killing of microorganisms. Stage 1 consists of the recognition of the organism by the phagocytic cell and is frequently facilitated by the absorption of serum proteins called opsonins by the invading organism. This opsonization process permits the phagocytic cell to recognize the invading organism. In addition, factors can be released by invading bacteria that initiate a chemotactic response on the part of the phagocytic cell leading to its movement

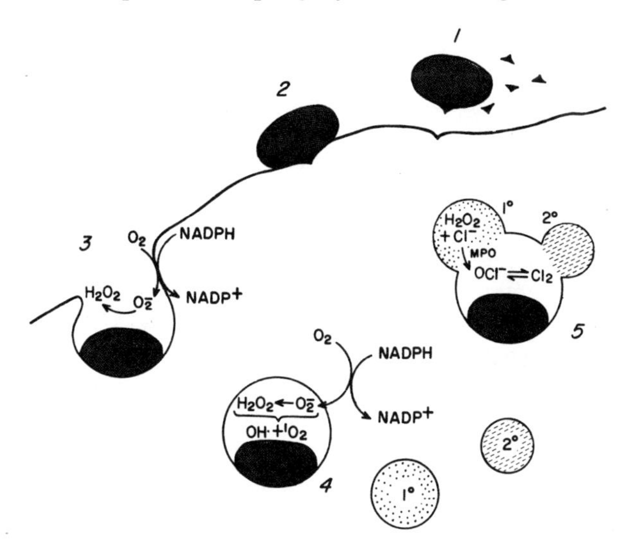

Fig. 3. Schematic representation of polymorphonuclear leukocyte function. (*1*) Opsonins bind to invading organism which serve as recognition factors by the host cell. (*2*) Binding of opsonized bacterium to surface of polymorphonuclear leukocyte. (*3*) Initiation of phagocytosis by pseudopodia emanating from the surface of the phagocytic cell. During this process, NADPH oxidases are activated, generating O_2^- and subsequently, H_2O_2. (*4*) Fusion of pseudopodia to form a phagosome, with continued generation of O_2^- and H_2O_2, and approach of both primary (1°) and secondary (2°) granules. (*5*) Fusion of granules with phagosome and discharge of granule enzymes and proteins within phagosomal space. The myeloperoxidase-catalyzed oxidation of Cl^- is depicted as one type of granule-induced reaction.

to the site of the invading organisms. Following the recognition of the opsonized microorganism, phagocytosis is initiated by the formation of pseudopodia from the surface of the phagocytic cell. These pseudopodia gradually surround the invading microorganism and ultimately will totally enclose it in a membrane-bound vesicle called a phagosome. It is during this process that the metabolic activity of the phagocytic cell alters. A respiratory burst is observed which consists of 10- to 20-fold increase in the rate of oxygen consumption. This enhanced oxygen consumption is cyanide insensitive and is mediated through the enhanced activity of an NADPH oxidase. The products of the reaction are initially the superoxide radical, O_2^-, and its breakdown product, H_2O_2. The production of H_2O_2 from O_2^- can occur either spontaneously [Eq. (1)] or when catalyzed by the enzyme, superoxide dismutase [Eq. (2))].

$$O_2^- + O_2^- \xrightarrow[+2H^+]{\text{spontaneously}} H_2O_2 + O_2 \tag{1}$$

$$O_2^- + O_2^- \xrightarrow[+2H^+]{\text{superoxide dismutase}} H_2O_2 + O_2 \tag{2}$$

Although much controversy has been generated over the nature of the enzymes responsible for the respiratory burst there is now little doubt as to the nature of the products. Iyer *et al.* (1961) were the first workers to demonstrate the enhanced production of H_2O_2 following phagocytosis and Babior and his associates (Babior *et al.*, 1973) have demonstrated that O_2^- is produced when resting polymorphonuclear leukocytes are stimulated to initiate the process of phagocytosis. That these products are important for the microbicidal activity of the host cell has been adequately demonstrated by the use of polymorphonuclear leukocytes from patients suffering from chronic granulomatous disease. These cells do not show a respiratory burst of H_2O_2 production when stimulated (Karnovsky, 1973) nor do they produce O_2^- (Curnutte *et al.*, 1974). In addition, these patients suffer from repetitive infections and frequently do not survive beyond childhood.

In the next stage of the process, the phagosome enclosed within the polymorphonuclear leukocyte fuses with specific granules which contain a variety of enzymes and proteins that have been reported microbicidal for varying organisms. For example, the primary granules contain the enzyme myeloperoxidase and the pioneering work of Klebanoff (1967, 1968) has demonstrated the microbicidal activity of this enzyme *in vitro* (Section III,B,2,a). The secondary granules contain other proteins such as lysozyme, alkaline phosphatase, and aminopeptidase (Cline, 1975). Many of these proteins have been proven to be microbicidal for one microorganism or another. It therefore appears that there are numerous microbicidal mechanisms available to the polymorphonuclear leukocytes for carrying out its principle function of destroying invading microorganisms.

Klebanoff (1975a) has reviewed the microbicidal mechanisms which occur in polymorphonuclear leukocytes and has classified them as either oxygen-dependent or oxygen-independent systems. From the point of view of workers interested in the biological role of $^1O_2$, the polymorphonuclear leukocyte was first involved with $^1O_2$ by Steele and his associates (Allen *et al.*, 1972). These workers were pursuing observations made earlier in their laboratories (Howes and Steele, 1971, 1972) which dealt with the appearance of CL during the metabolism of liver microsomes. This chemiluminescence was attributed to the generation of $^1O_2$. Although the evidence was far from definitive, Howes and Steele (1971, 1972) made an important contribution in bringing to the attention of biochemists the possibility that excited state molecules might function as intermediates or products in enzymatic reactions. It was, therefore, only natural for this same laboratory (Allen *et al.*, 1972) to attribute the CL they observed from phagocytosing polymorphonuclear leukocytes to the production of $^1O_2$.

One of the reasons Allen *et al.* (1972) suggested that $^1O_2$ was a product of the metabolism of phagocytosing polymorphonuclear leukocytes was that the exact basis for the microbicidal action of these cells was still not known. Since their report, a great deal of work has appeared on this and related areas but we still do not completely understand how polymorphonuclear leukocytes are able to kill invading organisms. There are probably several oxidant mechanisms which the cell utilizes to carry out this process.

2. Myeloperoxidase Activity

Agner (1972) first demonstrated the presence of a peroxidase in polymorphonuclear leukocytes which was ultimately called myeloperoxidase. Myeloperoxidase (MPO) is an example of that class of enzymes which, in the presence of H_2O_2, catalyze the oxidation of a wide variety of substrates.

a. Microbicidal Activity of Myeloperoxidase. MPO was first demonstrated to have microbicidal activity by Klebanoff (1967, 1968). The importance of this observation is based on the fact that MPO is one of the principal enzymes located in the primary granules of the polymorphonuclear leukocyte. When Klebanoff (1967, 1968) determined that MPO had microbicidal activity *in vitro*, he found that an oxidizable cofactor was necessary for the microbicidal activity in addition to H_2O_2. The cofactor could be any of a number of halides which had varying activities. The most potent was Br^-, followed by Cl^- and I^-. For maximum activity, however, concentrations of Br^- and I^- were required that were not physiological. The maximum activity for Cl^- was found to occur at approximately 100 m*M* which coincides with the physiological concentration of Cl^- both in serum and in the polymorphonuclear leukocyte.

b. Enzymatic Properties of Myeloperoxidase. Several properties of MPO, particularly those dealing with the nature of the oxidants produced in MPO-catalyzed reactions, will be described in Section III,B,3. The enzyme can also develop CL comparable to that seen in the intact polymorphonuclear leukocyte. Allen (1975a,b) has reported on the CL that can be observed from intact phagocytosing polymorphonuclear leukocytes in the presence of various halides. The effect of the various halides in the intact systems is in the order $Br^- > Cl^- > I^- > F^-$. This same order was obtained in the *in vitro* system using human MPO isolated from similar cells in the presence of H_2O_2. Based on the concentrations necessary for maximal CL, Allen (1975a) concluded that only the chloride ion was physiologically active in this system. In a continuation of his studies on the isolated MPO/H_2O_2/Cl^- system, Allen (1975b) found that at a physiological concentration of Cl^- (100 mEq/liter) the pH optimum was between pH 4.4–5.0, which coincides with the estimates for the pH of the phagosome (Jensen and Bainton, 1973).

3. Oxidants Produced by Myeloperoxidase

a. Active Chloride. Agner, who did much of the pioneering work in isolating and characterizing MPO isolated from polymorphonuclear leukocytes first suggested (1958) that one of the principal products of MPO activity would be hypochlorous acid (HOCl). The data for this hypothesis was presented in a posthumous publication (Agner, 1972) in which Agner argued that the products of the MPO/H_2O_2/Cl^- reaction were analogous to the products observed when various drugs were treated with HOCl. In addition, Agner reported (1972) that the mixture of MPO/H_2O_2/Cl^- was bactericidal and that equivalent results could be obtained if he replaced the MPO/H_2O_2/Cl^- with preformed HOCl. Based on these observations and the knowledge that MPO contains two iron atoms, one of which can bind Cl^- while the other binds H_2O_2, Agner (1972) formulated the reaction sequence shown in Scheme 5.

$$\text{MPO}\left[\begin{matrix}Fe^{3+}\\ \\ Fe^{3+}\end{matrix}\right. \xrightarrow[Cl^-]{H_2O_2} \text{MPO}\left[\begin{matrix}Fe^{3+}\cdots H_2O_2\\ \\ Fe^{3+}\cdots Cl^-\end{matrix}\right. \longrightarrow \text{MPO}\left[\begin{matrix}Fe^{3+}\\ \\ Fe^{3+}\end{matrix}\right. + HOCl + OH^-$$

Scheme 5

The nature of the active chloride produced by the MPO/H_2O_2/Cl^- system has been advanced by the studies coming from the laboratories of Zgliczynski and his associates (Zgliczynski *et al.*, 1968; Stelmaszynska and Zgliczynski, 1974) and those of Sbarra and his associates (Paul *et al.*, 1970; Strauss *et al.*, 1971; Selvaraj *et al.*, 1974). Recently, these two groups have collaborated (Zgliczynski *et al.*, 1977) and have presented data suggesting that the product of chloride oxidation catalyzed by MPO in the presence of H_2O_2 is the chlorinium ion (Cl^+). The concept that the $MPO/H_2O_2/Cl^-$ system produces Cl^+ was first proposed by Zgliczynski *et al.* (1971) and is based on the formation of a stable chloramine from the β-amino acid, taurine, as seen in Scheme 6. In the case of α-amino acids, the action of $MPO/H_2O_2/Cl^-$ would result in the formation of an unstable chloramine

$$HOSO_2CH_2CH_2NH_2 + H_2O_2 + Cl^- \rightarrow HOSO_2CH_2CH_2NHCl + H_2O + OH^-$$

Scheme 6

that would degrade with the production of NH_3, CO_2, Cl^-, and a free aldehyde (Scheme 7). Subsequent work by Selvaraj *et al.* (1974) indicated that this

$$RCH_2NH_2COOH + H_2O_2 + Cl^- \longrightarrow [RCH_2NHClCOOH] \longrightarrow RCHO + NH_3 + CO_2 + Cl^-$$

Scheme 7

enzyme system could not only decarboxylate α-amino acids but could also bring about the cleavage of peptide bonds. This was investigated by incorporating radioactive amino acids into the cell walls of bacteria such as *Escherichia coli.*

In their most recent paper Zgliczynski *et al.* (1977) used both diethanolamine and taurine as acceptors for the active chloride species generated by the $MPO/H_2O_2/Cl^-$ system. They were able to demonstrate that this chlorination can occur not only in acid pH but also in neutral pH.

Despite the strong arguments presented by both the groups of Zgliczynski and Sbarra that Cl^+ is formed, the conclusion is based solely on the appearance of either stable chloramines or the breakdown products of unstable chloramines. Other investigators do not agree that Cl^+ is the primary product of the action of $MPO/H_2O_2/Cl^-$. In fact, Harrison and Schultz (1976) present evidence that the reaction of $MPO/H_2O_2/Cl^-$ leads to the peroxidation of Cl^- generating HOCl. Inasmuch as HOCl is in equilibrium with Cl_2, the assay they use for HOCl is based on the color reaction between *o*-tolidine and Cl_2. Harrison and Schultz (1976) verify the appearance of Cl_2 qualitatively by detecting the odor of Cl_2 when H_2O_2 is added to a solution containing MPO and Cl^- at pH values between 3.5 and 6.0. For a quantitative assay of HOCl formation, they determine the amount of radioactive volatile product formed when they used $^{36}Cl^-$ as their starting material. Under these circumstances, they found the appearance of a volatile product was absolutely dependent on the presence of both H_2O_2 and MPO. They were also able to demonstrate that MPO did not utilize chlorite for chlorination reactions, in contrast to the actions of chloroperoxidase and horseradish peroxidase. In fact, chlorite addition inactivated MPO.

The concern with identifying the specific form of active chloride produced in the reaction of $MPO/H_2O_2/Cl^-$ will become more apparent in the following paragraphs.

b. Active Oxygen. The rediscovery by Seliger (1960) that the reaction of H_2O_2 and OCl^- resulted in the appearance of CL was subsequently attributed to the formation of an excited state of oxygen (Khan and Kasha, 1963, 1970; Arnold *et al.*, 1964; Browne and Ogryzlo, 1964). Much interest has evolved around the possibility that this chemical reaction [Eq. (3)] may in fact be duplicated in an *in vivo* system.

$$H_2O_2 + OCl^- \rightarrow {}^1O_2 + Cl^- + H_2O \qquad (3)$$

The first proposal that polymorphonuclear leukocytes could generate ${}^1O_2$ was made by Allen *et al.* (1972). These workers had reported earlier (Howes and Steele, 1972) that a microsomal system produced CL during the oxidation of NADPH and they suggested that the CL could be attributed to ${}^1O_2$ formed as shown in Scheme 8.

$$O_2^- \text{ and/or } H_2O_2 + Cl^- \xrightarrow{MPO} {}^1O_2 \longrightarrow CL \text{ (possibly through dioxetane intermediates)}$$

Scheme 8

By the time of their next publication (Allen *et al.*, 1974) this group had suggested another possible source of the CL, namely the formation of excited carbonyl products of ${}^1O_2$ mediated reactions which could relax to the ground state by emitting photons. Such reactions had been described earlier by McCapra and Hann (1969). It is interesting that even before evidence was available Allen *et al.* (1972) had suggested that the stimulation of polymorphonuclear leukocyte metabolism by the ingestion of opsonized bacteria would result, through the action of an NADPH oxidase, in the formation of O_2^- and/or H_2O_2. The actual formation of ${}^1O_2$ was then proposed to occur either from spontaneous O_2^- disproportion [Eq. (1)] or through the H_2O_2/OCl^- reaction [Eq. (3)]. The work of Steele and his associates (Allen *et al.*, 1972) was reviewed by Maugh (1973) and in this way brought to the attention of numerous research workers.

In an attempt to make the CL assay more sensitive, Allen and Loose (1976) added luminol (5-amino-2,3-dihydro-1,4-phthalazinedione) to alveolar and peritoneal macrophages from rabbits and polymorphonuclear leukocytes from both rabbits and humans. Luminol reacts with a large number of oxidizing agents and forms the electronically excited aminophthalate anion which upon relaxation to the ground state emits a photon, readily detectable in a liquid scintillation counter. In an attempt to identify the species responsible for the oxidation of luminol, Allen and Loose (1976) introduced either sodium benzoate, an effective quencher of OH· (Neta and Dorfman, 1968) or SOD to catalyze the disproportionation of O_2^- to $O_2 + H_2O_2$ [Eq. (1)]. In the peritoneal macrophages, both sodium benzoate and SOD inhibited CL. In contrast, only SOD inhibited CL in the rabbit alveolar macrophages. In the case of rabbit polymorphonuclear leukocytes, both SOD and sodium benzoate decreased the CL, whereas in human polymorphonuclear leukocytes, sodium benzoate was effective and SOD decreased CL only slightly. From this data, Allen and Loose (1976) concluded that human polymorphonuclear leukocytes generate OH· but the generation of O_2^- is not necessary for the oxidation of luminol to an electronically excited state.

One shortcoming with the use of CL measurements from either macrophages or polymorphonuclear leukocytes is that the phenomenon is qualitative in nature unless one knows the exact spectral distribution of the light that is emitted. Because of the low level of illumination this is a technically difficult task. Some of the approaches and shortcomings have been discussed in Section II,C.

An entirely different approach was taken by Krinsky (1974) who tried to take advantage of the well documented effect of carotenoid pigments in quenching $^1O_2$ *in vitro* (Mathews-Roth *et al.*, 1974; Foote, 1976) and by the assumption that the carotenoid pigments function the same way *in vivo* (Krinsky, 1971). In order to take advantage of the specificity of carotenoid pigments for quenching $^1O_2$, Krinsky (1974) used two strains of the coccus, *Sarcina lutea* which had been demonstrated to differ in their response to photosensitizing conditions (Mathews and Sistrom, 1960). The yellow carotenoid-containing wild-type strain was resistant to the harmful effects of light, dye, and O_2, whereas the colorless, carotenoidless mutant strain was readily killed under photosensitizing conditions (Mathews and Sistrom, 1960). A typical photodynamic effect, with carotenoid protection, is illustrated in Fig. 4A (Mathews-Roth and Krinsky, 1970). Krinsky (1974) exposed human polymorphonuclear leukocytes to both the wild-type strain and the carotenoidless mutant strain and measured survival of these organisms at various times after mixing the bacteria with the cells. The data he obtained is represented in Fig. 4B. Wild-type *S. lutea* showed no significant killing during a 90-min incubation at 37°C in the presence or absence of human serum whereas the colorless mutant strain in the presence of serum was rapidly killed under these circumstances. As can be seen, the results are very similar to data obtained when the lethal photodynamic effect of endogenous or exogenous photosensitizers was studied (Fig. 4A). The photosensitizer generates $^1O_2$ which is presumed to act as the lethal agent in this system (Wilson and Hastings, 1970). From the data presented in Fig. 4B, it can be inferred that the carotenoid pigments in the wild-type strain of *S. lutea* are acting as protective agents against the killing brought about by

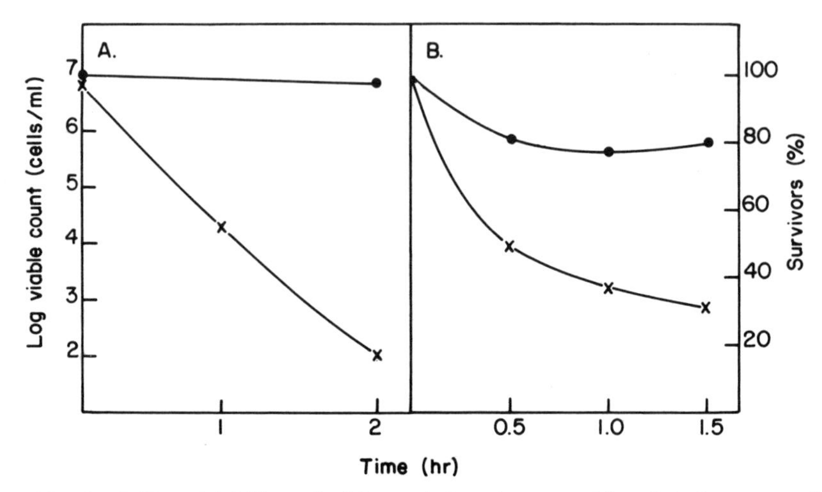

Fig. 4. Differential killing of wild-type (●) and carotenoidless (×) (mutant strain 93a) strain of *Sarcina lutea*. (A) Photosensitized killing induced by light (1000 foot candles), air and 2.5×10^{-6} *M* toluidine blue [reproduced by permission of Mathews-Roth and Krinsky (1970) and Pergamon Press]. (B) Killing of *S. lutea* by human polymorphonuclear leukocytes. The digests contained leukocytes and homologous serum, the latter being a necessary adjunct to the killing process. Reproduced by permission of Krinsky (1974) and *Science*.

the bactericidal action of human polymorphonuclear leukocytes. Although these data do not constitute absolute evidence for $^1O_2$ involvement in the microbicidal activity of polymorphonuclear leukocytes, they are suggestive that $^1O_2$ may be one of the microbicidal agents active in these cells.

Klebanoff (1975b) has also studied the effect of DABCO, a specific $^1O_2$ quencher, in altering MPO-catalyzed microbicidal action. He observed that at a concentration of 10^{-5} *M*, DABCO was able to prevent the microbicidal action of MPO/H_2O_2/Cl^- but that a much higher concentration, 10^{-2} *M* was necessary for overcoming the microbicidal action of MPO/H_2O_2/I^-. This concentration of DABCO was the same as that necessary for preventing the photodynamic killing of *E. coli* using eosin as the photosensitizing dye.

Another aspect of active oxygen production by the MPO/H_2O_2/halide system was investigated by Klebanoff (1974). In this investigation, Klebanoff (1974) replaced the H_2O_2 with a generating system that consisted of xanthine oxidase and xanthine and then studied the effects of various radical quenchers on the microbicidal activity. The system, MPO/xanthine oxidase/xanthine/Cl^- was microbicidal, and this activity could be abolished by the addition of either catalase, cytochrome *c*, ethanol or benzoate. The cytochrome *c* inhibition was due to the consumption of O_2^-, thus preventing H_2O_2 production. Catalase removed the H_2O_2, and the ethanol and benzoate presumably acted as OH· scavengers. Klebanoff (1974) concluded that both O_2^- and

OH· were involved in the formation of H_2O_2, and quenchers of these two radical species would therefore inhibit microbicidal activity.

More definitive evidence of $^1O_2$ involvement has appeared in a publication by Rosen and Klebanoff (1977). These workers used a combination of product analyses, $^1O_2$ quenchers, and deuterium oxide effects to study the action of MPO/H_2O_2/halide. In terms of product analyses, Rosen and Klebanoff (1977) studied the effects of the enzyme system on 2,5-diphenylfuran (DPF) which forms *cis*-dibenzoylethylene (*cis*-DBE) in the presence of $^1O_2$. Although this reaction has been considered to be specific for $^1O_2$ (King *et al.*, 1975) there are now numerous reports which indicate that other chemical oxidants can also yield *cis*-DBE from DPF. Rosen and Klebanoff (1977) demonstrated the conversion of DPF to *cis*-DBE with the MPO/H_2O_2/halide system and also observed that a variety of $^1O_2$ quenchers inhibited this reaction. Among the quenchers tested were β-carotene (Foote and Denny, 1968), bilirubin (Foote and Ching, 1975), 1,4-diazabicyclo-[2.2.2]octane (DABCO) (Ouannès and Wilson, 1968), and histidine (Hodgson and Fridovich, 1974). In all cases tested, these $^1O_2$ quenchers inhibited the formation of *cis*-DBE from DPF in the enzyme-catalyzed system. Finally, they were able to demonstrate, by substituting D_2O in their system, an enhanced conversion of DPF to *cis*-DBE. This action of D_2O on prolonging the life time of $^1O_2$ in aqueous solutions is referred to in Section II,D. The three properties demonstrated by the MPO/H_2O_2/halide system of yielding a $^1O_2$ product, being inhibited by $^1O_2$ quenchers, and being stimulated by the presence of D_2O are all strong supportive evidence that $^1O_2$ is indeed produced by the myeloperoxidase system.

In addition, Rosen and Klebanoff (1977) also studied the effect of hypochlorous acid (HOCl) with respect to converting DPF to *cis*-DBE in the presence of $^1O_2$ quenchers and D_2O. They have observed that HOCl could indeed convert DPF to *cis*-DBE and that this conversion was both inhibited by $^1O_2$ quenchers and stimulated by the presence of D_2O. They were able to demonstrate that the DPF conversion by HOCl was stimulated in the presence of Cl^- and actually inhibited by the addition of H_2O_2. They concluded that HOCl is decomposed in the presence of excess Cl^- to form a product which reacts similarly to $^1O_2$. The inhibition in the extent of reaction following the addition of H_2O_2 to the HOCl system would argue against the formation of $^1O_2$ by the classical H_2O_2/OCl^- reactions. Rosen and Klebanoff (1977) attribute the formation of $^1O_2$ to the spontaneous decomposition of HOCl which can occur at a mildly acid pH. It will of course be interesting to see if the suggestion of Rosen and Klebanoff that HOCl can induce the production of $^1O_2$ will be supported by the use of other criteria. The detailed mechanism of the production of $^1O_2$ in the H_2O_2/OCl^- reaction was studied by Cahill and Taube (1952) who carefully demonstrated that the O–O bond was not broken in the process of generating $^1O_2$.

4. Non-Myeloperoxidase-Produced Oxidants

It has become increasingly difficult to sort out all the possible interactions that can occur among the group of oxidants which have been reported to be produced by phagocytosing polymorphonuclear leukocytes. These oxidants include H_2O_2, O_2^-, $OH\cdot$, and $^1O_2$. They can all be demonstrated to play some role in microbicidal activity, although work is still required to differentiate between primary and secondary effects. It is not even possible to discuss them separately, as almost all experiments involve at least two of these oxidants, and frequently may involve three or more. Nevertheless, I will try to organize some of the available data in terms of the four oxidants mentioned above.

a. Superoxide (O_2^-). The first demonstration that O_2^- is a major product of phagocytosing polymorphonuclear leukocytes was made by Babior and his associates (Babior *et al.*, 1973). It appears to be generated through the action of an NADP oxidase and can be measured in the medium outside the polymorphonuclear leukocyte. In the case of patients with chronic granulomatous disease, little or no O_2^- is produced during phagocytosis by the polymorphonuclear leukocyte. This lack of O_2^- production is accompanied by a failure to kill ingested bacteria (Curnutte *et al.*, 1974) and this is presumably the basis for the chronic infections in these patients (Karnovsky, 1973).

Additional evidence as to the importance of O_2^- in the overall microbicidal effect was obtained by Johnston *et al.* (1975). These workers observed that the direct addition of SOD to phagocytosing polymorphonuclear leukocytes inhibited the killing of a variety of microorganisms and therefore concluded that O_2^- played some role in the process.

Yost and Fridovich (1974) have demonstrated that the death of various strains of *E. coli* exposed to human blood is a function of the SOD content of the different strains and this evidence also indicated that O_2 plays some role in microbicidal activity.

Despite this evidence, it is now generally accepted that O_2^- by itself is not particularly microbicidal. Evidence for this position was presented by Babior *et al.* (1975) who used a O_2^- generating system (xanthine oxidase/purine) to demonstrate microbicidal activity against both *Staphylococcus epidermidis* and *E. coli*. These authors found that *both* SOD and catalase protected *S. epidermidis*, and *only* catalase protected *E. coli*, whereas SOD was without effect in the latter organism. These authors concluded that O_2^- alone was not a microbicidal agent.

b. Hydroxyl Radical ($OH\cdot$). The evidence for the participation of $OH\cdot$ in microbicidal activity depends, to a large extent, on the validity of the

Haber–Weiss reaction. The reactions, formulated by Haber and Weiss (1934), involve the formation of OH· by electron transfer from a transition metal such as Fe^{2+} to H_2O_2 [Eq. (4)] followed by additional reactions [Eqs. (5–7)] which results in the interaction of O_2^- and H_2O_2 to yield OH· [Eq. (6)]. Recently, several workers (McClune and Fee, 1976; Halliwell,

$$Fe^{2+} + H_2O_2 \longrightarrow Fe^{3+} + OH^- + OH\cdot \qquad (4)$$

$$Fe^{2+} + 2H^+ + O_2^- \longrightarrow Fe^{3+} + H_2O_2 \qquad (5)$$

$$O_2^- + H_2O_2 \longrightarrow O_2 + OH^- + OH\cdot \qquad (6)$$

$$OH\cdot + H_2O_2 \longrightarrow H_2O + H^+ + O_2^- \qquad (7)$$

1976) have presented evidence arguing *against* the occurrence of the Haber–Weiss reaction in aqueous systems. Whether these same arguments are applicable to heterogenous systems such as exist in biological material has as yet not been investigated.

Nevertheless, evidence has accumulated that *both* O_2^- and H_2O_2 are essential for microbicidal activity. In the experiments of Johnston *et al.* (1975) referred to in the previous section, both catalase and SOD were effective in inhibiting microbicidal activity, indicating the involvement of both O_2^- and H_2O_2 in the process. In addition, Johnston *et al.* (1975) found that the OH· scavengers, benzoate and mannitol, were also effective in inhibiting bacterial killing, and they therefore concluded that the OH· radical was involved in the process. Even earlier, McCord (1974) had suggested, based on similar criteria, that OH· was produced by polymorphonuclear leukocytes in synovial fluid and this was the agent responsible for inflammatory effects.

Also, the results of Babior *et al.* (1975) reported in the previous section can be interpreted as demonstrating a microbicidal action of OH·, since both SOD and catalase protected *S. epidermidis* against killing by a xanthine oxidase/purine system. The technique of using both SOD and catalase to inhibit microbicidal action was also used by Klebanoff (1974), as reported in Section III,B,3,b. In this instance also, Klebanoff (1974) concluded that OH· was involved. An opposing point of view was presented by Drath and Karnovsky (1974) who were unable to observe an inhibition of microbicidal activity in an MPO/H_2O_2/halide system using either SOD or mannitol, although radical quenchers such as thiosulfate and metabisulfate were effective inhibitors.

The most compelling evidence for the production of OH· by polymorphonuclear leukocytes comes from the recent observations of Tauber and Babior (1977). These workers utilized the reaction described by Beauchamp and Fridovich (1970) in which OH· specifically converts methional to ethylene which can be readily analyzed by gas–liquid chromatography (Scheme 9). Using this system, Tauber and Babior (1977) have demon-

$$CH_3SCH_2CH_2CHO + OH\cdot \longrightarrow \tfrac{1}{2}\,(CH_3S)_2 + HCOOH + H_2C{=}CH_2$$

Scheme 9

strated that a product with the characteristics of $OH\cdot$ can be released from phagocytosing polymorphonuclear leukocytes. They also studied polymorphonuclear leukocytes from patients with chronic granulomatous disease and found no ethylene production under these circumstances. Testing various inhibitors in their system, they found that benzoic acid was able to inhibit ethylene production, although neither ethanol nor mannitol were effective. SOD was also effective in reducing ethylene production whereas catalase had no effect. They concluded that a highly oxidizing species, possibly $OH\cdot$, was produced from phagocytosing polymorphonuclear leukocytes and that O_2^- generated during the respiratory burst was involved in the production of this species. The failure of either ethanol or mannitol to affect ethylene production raises some question as to whether $OH\cdot$ is the actual oxidizing species but the results may reflect the experimental conditions used by these authors. Their other observation that catalase does not inhibit the production of ethylene may indicate that either the H_2O_2 necessary for the Haber–Weiss reaction is not available to the exogenous catalase or that the concentration of H_2O_2 is too low to be effectively removed by catalase but still high enough to interact with O_2^- to generate the $OH\cdot$.

c. Singlet Oxygen ($^1O_2$). There has been no *direct* demonstration of $^1O_2$ production in polymorphonuclear leukocytes. That is not to say that $^1O_2$ is not produced in these cells, for there has not been any direct demonstration of $^1O_2$ in any biological system. Nevertheless, there is evidence that $^1O_2$ may be produced from some of the intermediates which can be found in the cells.

Stauff *et al.* (1963; Sander and Stauff, 1971) have proposed that the spontaneous disproportion of O_2^- generates $^1O_2$, rather than ground state oxygen [Eq. (8)] and this concept has received support from the calculations

$$O_2^- + O_2^- \xrightarrow[+2H^+]{\text{spontaneous}} H_2O_2 + {}^1O_2 \tag{8}$$

of Knoppenol (1976). A. U. Khan (1970, 1976) has also proposed that O_2^- is a precursor of $^1O_2$, although he believes this occurs by a direct electron transfer reaction [Eq. (9)]. Nilsson and Kearns (1974), however, based on

$$O_2^- \rightarrow e^- + {}^1O_2 \tag{9}$$

experiments using D_2O, reject the suggestion that O_2^- serves as a direct precursor of $^1O_2$.

There are also suggestions that the Haber–Weiss reaction generates $^1O_2$. This idea was put forth by Kellogg and Fridovich (1975) while studying lipid peroxidation, and the reaction is depicted in Eq. (10).

$$O_2^- + H_2O_2 \rightarrow OH\cdot + OH^- + {}^1O_2 \tag{10}$$

An alternative approach was made by Arneson (1970) who proposed $OH\cdot$ involvement in $^1O_2$ production according to reaction (11).

$$O_2^- + OH\cdot \rightarrow {}^1O_2 + OH^- \tag{11}$$

Finally, Smith and Kulig (1976) present evidence that the spontaneous dismutation of H_2O_2 generates a small amount of $^1O_2$, as seen in Eq. (12).

$$H_2O_2 + H_2O_2 \rightarrow 2\,H_2O + {}^1O_2 \tag{12}$$

These latter experiments were based on the appearance of breakdown products of 3β-hydroxy-5α-cholest-6-ene 5-hydroperoxide, the unique $^1O_2$ product of cholesterol (Schenck, 1957; Kulig and Smith, 1973).

The remaining evidence with respect to $^1O_2$ production in polymorphonuclear leukocytes as related to the MPO/H_2O_2/halide system has been discussed in Section II,B,3,b.

C. Enzyme Systems

One of the earliest suggestions that $^1O_2$ might be involved in an enzymatic reaction was made by Krishnamurty and Simpson (1970). These investigators worked with the fungus, *Aspergillus flavus*, which produces an inducible oxygenase, quercetinase, that oxidizes the heterocyclic ring of the aglycone, quercetin. The product of this cleavage is a depside, 2-protocatechuoylphloroglucinolcarboxylic acid. Through the use of $^{18}O_2$, the authors were able to demonstrate that the enzyme quercetinase is a dioxygenase, for both atoms of $^{18}O_2$ were incorporated into the depside. Matsuura *et al.* (1967) obtained the same depside following the photosensitized oxidation of quercetin. Inasmuch as the photosensitized oxidation, which was mediated by $^1O_2$ gave rise to the same product as the enzyme-catalyzed reaction, Krishnamurty and Simpson (1970) concluded that the enzyme utilizes an "activated oxygen" to form an unstable cyclic peroxide intermediate which rapidly decomposes to give rise to the depside. They proposed that the reactions illustrated in Scheme 10 would account for their observations.

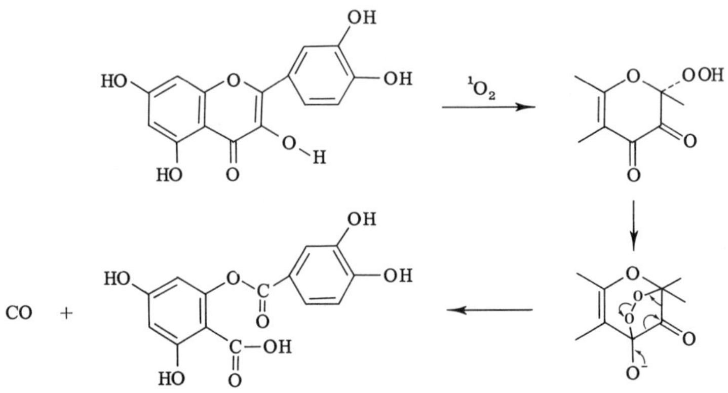

Scheme 10

Since the appearance of this publication, many other enzymes and enzyme systems have been implicated as either utilizing, interacting with, or producing $^1O_2$. Although the evidence is frequently indirect, it is worth describing these reactions at this time in the hope of being able to point out areas where additional investigations may enable us to reach more definitive conclusions.

1. Xanthine Oxidase

Stauff and his associates (Stauff *et al.*, 1963; Stauff and Wolf, 1964) reported that xanthine oxidase incubated with one of its substrates, xanthine, gives rise to CL. These authors suggested that the CL arose from $^1O_2$ formed by the spontaneous disproportion of O_2^- as demonstrated in Eq. (8). More recently, Stauff *et al.* (1973) have suggested that the CL observed in these systems may actually be due to the recombination of carbonate or bicarbonate radicals.

Arneson (1970) also studied the CL which arises from the xanthine oxidase system and found that H_2O_2, a compound normally generated during the dismutation of O_2^-, had a very stimulatory effect on CL, being active at a concentration as low as 10^{-5} *M*. Since H_2O_2 would be expected to react with O_2^- in a Haber–Weiss reaction, Arneson (1970) proposed that reaction (11) could occur generating $^1O_2$. Arneson suggested that the reaction (11) may have a higher yield of excited states than reaction (7) and may serve to explain why H_2O_2 stimulates CL. The failure to analyze the nature of the light emitted in the studies of Stauff and Arneson makes it extremely tenuous to attribute the CL to $^1O_2$.

Pederson and Aust (1973) utilized xanthine oxidase to study the effects of O_2^- on rat liver microsomal lipids. These authors found that in the presence of solubilized iron (0.2 m*M* $FeCl_3$/0.1 m*M* EDTA) there is a very marked increase in the extent of lipid peroxidation as assayed by the production of malondialdehyde. These authors found that the lipid peroxidation could be inhibited by the addition of either SOD or by adding DPBF. The addition of DPBF to this system did not affect the rate of O_2^- production but did yield DBB when it inhibited lipid peroxidation. Although DPBF can be converted to DBB by reactions which do not involve $^1O_2$ (Section II,B), Pederson and Aust (1973) used this evidence as support for their contention that O_2^-, generated by the action of xanthine oxidase, could decompose to yield $^1O_2$ which would then initiate lipid peroxidation.

A very similar series of experiments were reported by Kellogg and Fridovich (1975) using xanthine oxidase and measuring the peroxidation of linolenic acid. These authors reported that both SOD and catalase inhibited lipid peroxidation which indicated a requirement for both O_2^- and H_2O_2 in the overall process. Since these two compounds are involved in the Haber–Weiss reaction, Kellogg and Fridovich (1975) studied the effect of OH· scavengers. They reported that *tert*-butyl alcohol and mannitol were without effect on the overall reaction, whereas DABCO inhibited lipid peroxidation. Based on these observations, Kellogg and Fridovich (1975) proposed that the Haber–Weiss reaction gives rise not only to OH· but to $^1O_2$ as well [Eq. (10)] and it was this latter species that initiated linolenic acid peroxidation. To test their hypothesis, they studied the effects of xanthine oxidase on DMF. In their xanthine oxidase system, DMF was converted to a product which had similar chromatographic properties to the product formed in a photosensitized oxidation, presumably diacetylethylene [Eq. (13)]. As in the case of linolenic acid peroxidation, both SOD

H₃C–(furan ring, O)–CH₃ ⟶ H₃C–C(=O)–CH=CH–C(=O)–CH₃ (13)

DMF cis-diacetylethylene

and catalase inhibited the conversion of DMF to the new product whereas the reaction was not inhibited by the addition of OH· scavengers. Finally, Kellogg and Fridovich (1975) were able to show that DMF and β-carotene inhibited the formation of lipid peroxides in the presence of xanthine oxidase and offered that as final evidence that $^1O_2$ was generated from the products of xanthine oxidase.

Other workers suggest that OH· is the oxidant species generated from xanthine oxidase products. Fong *et al.* (1976) have presented evidence that OH· may be a product of enzymatic reactions. They had demonstrated earlier (Fong *et al.*, 1973) that Fe^{3+}, in a complex form, inhibited the O_2^- mediated reduction of cytochrome *c*, using xanthine oxidase as the source of O_2^-. They suggested at that time that the reduced Fe^{2+} could then interact with H_2O_2 generated by the xanthine to produce OH· as shown in Eqs. (14) and (4).

$$Fe^{3+} + O_2^{-} \rightarrow Fe^{2+} + O_2 \tag{14}$$

$$Fe^{2+} + H_2O_2 \rightarrow Fe^{3+} + OH^{-} + OH\cdot \tag{4}$$

In their more recent report, Fong *et al.* (1976) present evidence that OH· can be produced either by the H_2O_2-catalyzed oxidation of complex Fe^{2+}, or through the Haber–Weiss reaction when H_2O_2 is available in the presence of O_2^{-}. They contend that OH· is the species responsible for the initiation of lipid peroxidation and the damaging effects to cells and that O_2^{-}, although a radical itself, is not a direct promoter of lipid peroxidation.

Through the use of an ingenious new technique which involves trapping $^1O_2$ in the form of the transannular peroxide of rubrene (Scheme 11), a

H_5C_6 C_6H_5 $h\nu/O_2$ heat H_5C_6 C_6H_5

H_5C_6 C_6H_5 H_5C_6 C_6H_5

rubrene

Scheme 11

reaction attributed exclusively to $^1O_2$ (Wilson, 1966), Khan (1976) has obtained preliminary evidence in which he concludes that the xanthine/xanthine oxidase system generates $^1O_2$. In this same publication, Khan (1976) presents a theoretical analysis of both the generation and quenching of $^1O_2$ from O_2^{-} through the process of electron transfer reactions.

To summarize therefore the data available from several investigations using xanthine oxidase, it seems clear that a number of investigators have been able to obtain evidence which suggests that $^1O_2$ is a product of the overall reaction. It is safe to conclude, however, that most of the authors believe that the $^1O_2$ arises in a secondary fashion, either by means of an electron transfer reaction from O_2^{-}, or through the generation of $^1O_2$ by means of the Haber–Weiss reaction.

2. *Microsomal Lipid Oxidase Reactions*

Although a great deal of work has been carried out in trying to understand the relationship between microsomal lipid peroxidation and $^1O_2$, the situation still remains confused. Part of the confusion can be attributed to investigators using criteria for determining $^1O_2$ which unfortunately are not unique for $^1O_2$ reactions. In other cases, variations in experimental procedures have also lead to different results from different laboratories.

The first suggestion that liver microsomes evolved $^1O_2$ was made by

Howes and Steele (1971) who reported CL when NADPH was used as a substrate by this system. They also looked at lipid peroxide breakdown as measured by the thiobarbituric (TBA) assay but could find no correlation between MDA production and CL. Based on this they concluded that the reactions which generate CL precede the breakdown of lipid peroxides to TBA reactive material. Inasmuch as Rawls and van Santen (1970a,b) had demonstrated that $^1O_2$ could initiate fatty acid peroxidation, Howes and Steele (1971) suggested that their observations would be consistent with $^1O_2$ originating during the oxidation of NADPH by liver microsomes.

Shortly thereafter, Sternson and Wiley (1972) attempted to get more direct evidence for the possible role of $^1O_2$ in microsomal oxidation. These authors used a variety of aromatic substrates that presumably would give specific $^1O_2$ products when added to liver microsomes. In no case were they able to obtain the products characteristic of $^1O_2$ reactions although they could obtain hydroxylation products that were not necessarily attributable to $^1O_2$. In addition, they used a number of inhibitors of $^1O_2$, including carotene, DABCO, and ethyl sulfide. None of these quenchers affected the kinetics of the hydroxylation reactions of their aromatic substrates. Based on these observations, Sternson and Wiley (1972) concluded that the formation of $^1O_2$ during the aromatic hydroxylation by liver microsomes was very doubtful. Further evidence against the participation of $^1O_2$ in biological systems was presented by Smith and Teng (1974). Smith and Kulig (1976) had demonstrated, as had Schenck earlier (1957), that the attack of $^1O_2$ on cholesterol yielded the 3β-hydroxy-5α-cholest-6-ene 5-hydroperoxide as opposed to the epimeric 7-hydroperoxides of cholesterol. Smith and Teng (1974) incubated various rat liver subcellular fractions with [1,2-^{3}H]cholesterol in the presence of an NADPH-generating system and analyzed the oxidation products of cholesterol. In no case were they able to observe either the 5α-hydroperoxide derivative of cholesterol or its unique pyrolysis product, cholesta-4,6-diene-3-one. They therefore concluded that the sterol oxidations were similar to free radical oxidations and did not involve $^1O_2$. This work has now been extended (Teng and Smith, 1976) to indicate that the NADPH-dependent microsomal lipid peroxidation system acts as a dioxygenase, yielding initially the epimeric 7-hydroperoxides which can then be transformed into the corresponding 3β,7-diol derivatives of cholesterol, as well as the 7-keto derivative. These reactions are distinct from the microsomal 7α-hydroxylase which appears to be involved in bile acid biosynthesis (Johansson, 1971).

Using different techniques, Pederson and Aust (1975) also argued against the involvement of $^1O_2$ in NADPH-dependent lipid peroxidation reactions occurring in liver microsomal lipids. Their argument was based on the fact that DPBF had no effect on the NADPH-dependent lipid peroxidation reactions although it did inhibit a peroxidation initiated by xanthine/xanthine oxidase (Pederson and Aust, 1973). Since they also found that OH· scavengers and catalase had no effect on the NADPH-dependent lipid peroxidation and that SOD gave only a marginal inhibition, they concluded that the reaction did not involve either O_2^-, $^1O_2$, OH·, or H_2O_2. Instead, they suggested that alkoxyl radicals were formed by the decomposition of lipid hydroperoxides.

Wang and Kimura (1976) have also raised questions as to the nature of the oxidant that induces lipid peroxidation in biological systems. As had been reported earlier, the appearance of lipid peroxides in stored tissues was accompanied by a loss of microsomal cytochrome *P*-450. These authors studied this reaction in isolated bovine adrenal cortex mitochondria both in the absence and presence of Fe^{2+}. They observed that both the degradation of cytochrome *P*-450 and lipid peroxidation occurred simultaneously and were not inhibited by the addition of either catalase, SOD, DABCO, or ethanol. Based on these observations they concluded, in agreement with Peterson and Aust (1975), that neither H_2O_2, O_2^-, or OH· participated in the lipid peroxidation and cytochrome *P*-450 destruction. Wang and Kimura (1976) were however, able to obtain simultaneous lipid peroxidation and cytochrome *P*-450 destruction by adding cumene hydroperoxide to their incubation mixture. In order for them to observe the two reactions, however, oxygen had to be present in association with the cumene hydroperoxide. Based on their findings it would appear that a hydroperoxide, generated in the presence of Fe^{2+}, can initiate both lipid peroxidation and cytochrome *P*-450 destruction. For a more extended discussion of the relationship between formation and lipid peroxidation, the reader is referred to the excellent review of McCay and Poyer (1976).

However, as will be seen below, other workers have suggested that these radicals can in fact be responsible for the production of $^1O_2$ products. Nakano and his associates have been using spectroscopic evidence for the presence of $^1O_2$ in microsomal lipid peroxidation systems. These workers used a single photon counting method for determining very low levels of CL arising from various enzymatic reactions (Shimizu *et al.*, 1973). In their initial publication (Nakano *et al.*, 1974) they observed CL which they attributed to $^1O_2$ formation from either NADPH–cytochrome *c* reductase, L-amino acid oxidase or xanthine oxidase. They excluded the Haber–Weiss reaction by including catalase in their reaction mixture and concluded that $^1O_2$ was generated during the spontaneous dismutation of O_2^-. When they began investigating liver microsomes (Nakano *et al.*, 1975) they also observed an NADPH-dependent CL which was not inhibited by the addition of either SOD or catalase. They measured an emission spectrum of this low level CL using filters to produce the spectrum shown in Fig. 5. They also obtained

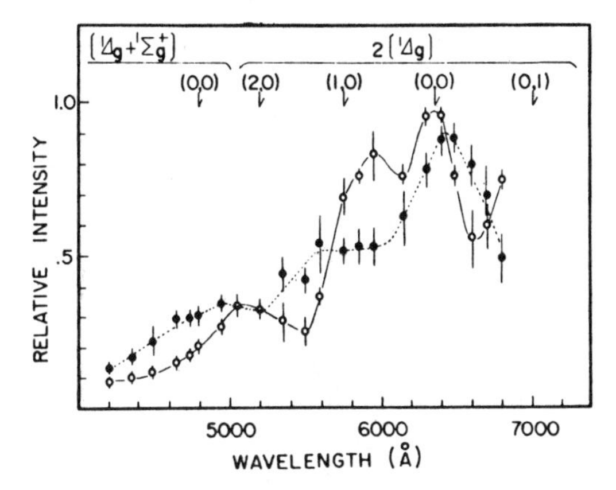

Fig. 5. Emission spectra from the microsomal lipid peroxidation system (——) and the $H_2O_2/OCl^-/NH_3$ system (----). The microsomal system contained 1.5 mg microsomes/ml. The chemical system contained 1% NaOCl, 0.1% H_2O_2, and 0.1 *N* NH_4OH, in a final volume of 20 ml. This latter reaction was initiated by the addition of H_2O_2. Twenty-seven colored filters were employed to cover the 275–670 nm region. The points represent averages of either two (○) or four (●) samples; the range is represented by the solid bars. Reproduced by permission of Nakano *et al.* (1975) and the *Journal of Biological Chemistry*.

an emission spectrum of a $H_2O_2/OCl^-/NH_3$ mixture, which generates $^1O_2$, and observed some similarities in these two emission spectra. However, it should be pointed out that the coincidence is not absolute and the resolution of each point on their graph is 30 nm, which allows relatively large errors. They found a 30% increase in light emission during the first 20 min when the reaction was carried out in D_2O, and observed that DMF produced a 35% inhibition of light emission, without inhibiting lipid peroxidation activity. Based on these observations Nakano *et al.* (1975) concluded that the generation of $^1O_2$ did not affect lipid peroxidation and suggested that $^1O_2$ production might be a result of lipid peroxidation. They concluded that the $^1O_2$ might arise through the self-reactions of lipid peroxy radicals, based on the observations of Howard and Ingold (1968). They have presented additional evidence (Sugioka and Nakano, 1976) that the emission is due to $^1O_2$ for they found that low concentrations of β-carotene were able to quench the CL without affecting malondialdehyde production. They also reported that during the early stages of lipid peroxidation, the intensity of the CL was proportional to the square of the concentration of lipid peroxide and used this observation as evidence that $^1O_2$ was indeed generated through the self-reaction of lipid peroxy radicals, the so called Russell (1957) mechanism (Scheme 3).

McCay and his associates (King *et al.*, 1975; McCay *et al.*, 1976) have also been studying lipid peroxidation in liver microsomes. Based on the formation of DBE from DPF, they concluded that liver microsomes in the presence of NADPH generate $^1O_2$. They make a strong argument that both the lipid peroxidation and the formation of $^1O_2$ are related to the amount of Fe^{3+} present in the system and also feel that the $^1O_2$ arises from the breakdown of lipid peroxides. Since they had demonstrated earlier (Fong *et al.*, 1973) that lipid peroxidation in these systems arose as a result of free radical formation, they suggested that the $^1O_2$ was formed from the breakdown of lipid peroxides. Again, Fong *et al.* (1973) had demonstrated that Fe^{3+} actually promoted NADPH-dependent lipid peroxidation in microsomes, presumably through the increased formation of OH . They suggest that the differences in their data with those of Smith and Teng (1974) and Sternson and Wiley (1972) may be explained on the basis of the amount of Fe^{3+} present in the different systems. King *et al.* (1975) also investigated the question of whether xanthine oxidase generated $^1O_2$ and concluded, based on their failure to observe DBE formation in the presence of DPF that $^1O_2$ was not produced in this system. In fact they point out quite correctly that other $^1O_2$ acceptors such as DPBF, which had been used by Pederson and Aust (1973) as evidence for $^1O_2$ involvement in xanthine oxidase reactions, can react with ground state oxygen in the dark to yield the presumed $^1O_2$ product, DBB (Mayeda and Bard, 1973). The requirement for Fe^{3+} in their system for observing both $OH\cdot$ and $^1O_2$ by a system that generates O_2^- is intriguing. One possibility is that Fe^{3+} interacts directly with O_2^- and generates $^1O_2$, as seen below:

$$Fe^{3+} + O_2^- \rightarrow Fe^{2+} + {}^1O_2 \tag{15}$$

This reaction could certainly help explain some of the discrepancies reported by different laboratories.

A somewhat different interpretation of the role of liver microsomes and its relationship to $^1O_2$ can be derived from the data of Chen and Tu (1976). These authors investigated the oxidation of polycyclic hydrocarbons by rat liver microsomal cytochrome *P*-450 oxygenase systems and observed that, in the presence of NADPH, 9,10-dimethyl-1,2-benzanthracene (DMBA, **1**) is converted to a transannular 1,4-peroxide. During continued incubation with this system, the transannular product is reduced to form *cis*-9,10-dihydroxy-9,10-dimethyl-1,2-benzanthracene (**2**). The transannular peroxide is a known photoproduct, presumably arising by the interaction of $^1O_2$ with DMBA, and has been observed earlier in enzyme systems but attributed to the photooxidation reaction rather than to the enzymatic reaction (Boyland and Sims, 1965). Chen and Tu (1976) took great care to insure that photooxidation did not occur during their enzymatic reactions and found that various inhibitors of the microsomal cytochrome *P*-450 system effectively reduced the conversion of compound (**1**) to compound (**2**). The structures of these compounds are shown below in Scheme 12.

Scheme 12

The conclusion that can be drawn from their observations is that the cytochrome *P*-450 acts as a dioxygenase rather than as a monooxygenase (Hayaishi, 1974) or a mixed-function oxidase (Mason, 1957), and adds oxygen in a form equivalent to the addition of $^1O_2$. Baldwin *et al.* (1968) had reached similar conclusions earlier.

Another possible relationship between cytochrome *P*-450 and $^1O_2$ can be derived from the observations of Sligar *et al.* (1974) who worked with the enzyme from *Pseudomonas putida* that is responsible for the hydroxylation of camphor. Using the purified enzymes, Sligar *et al.* (1974) have observed CL during the autoxidation of the oxygenated from of cytochrome *P*-450 (Fe^{2+}). The CL was abolished by free radical scavengers, $^1O_2$ quenchers, and by SOD. The authors concluded that the autoxidation proceeds with the formation of O_2^- which can then undergo a spontaneous dismutation to generate $^1O_2$. They assumed that the CL they observed was directly related to $^1O_2$ formation. CL has also been used by Goda *et al.* (1973) and Schaap *et al.* (1976) as an indicator of the formation of $^1O_2$ in bovine adrenodoxin reductase–adrenodoxin system. The process involves the oxidation of NADPH by O_2. CL was inhibited by the addition of either SOD or DABCO. Goda *et al.* (1973) concluded that $^1O_2$ was formed during the spontaneous dismutation of O_2^-.

3. Prostaglandin Biosynthesis

During the biosynthesis of prostaglandins from fatty acids such as eicosa-8,11,14-trienoic acid, a cyclooxygenase adds a molecule of oxygen to form a cyclic endoperoxide intermediate. Samuelsson (1965) was the first worker to point out the similarity of this reaction to photochemical reactions which utilize $^1O_2$ to form endoperoxides. There have now been several groups that have looked at the mechanism of prostaglandin biosynthesis to see whether $^1O_2$ is involved in this process. Panganamala *et al.* (1974a) initiated the investigation of the possible role of $^1O_2$ in prostaglandin biosynthesis by studying the effects of various $^1O_2$ quenchers on this process. They used bovine vesicular glands and measured the extent of formation of prostaglandin E from eicosa-8,11,14-trienoic acid and found that both DPBF and tetraphenylcyclopentadienone effectively inhibited this process. However, they were unable to show a stimulation when the reaction was run in D_2O and in fact found a partial inhibition of prostaglandin E biosynthesis in this solvent. They also found that SOD did not affect this biosynthesis and concluded that either $^1O_2$ or OH· was involved in prostaglandin biosynthesis. In a subsequent paper (Panganamala *et al.*, 1974b) this same group studied the action of other inhibitors on this process and found that the biosynthesis of prostaglandin E was inhibited by catalase, 3-amino-1,2,4-triazole, NaN_3, and chlorpromazine, a OH· scavenger. DABCO did not inhibit prostaglandin E biosynthesis nor did it inhibit platelet aggregation, which had been used as a measure of the conversion of eicosa-8,11,14-trienoic acid into prostaglandin E. At that time, this group favored the OH· as the principal oxidant in prostaglandin biosynthesis as opposed to $^1O_2$. In their most recent publication (Panganamala *et al.*, 1976) the authors worked with additional OH· scavengers and found a differential effect at either high or low concentrations of scavengers. Thus, at high concentrations of OH· scavengers they observed an inhibition of prostaglandin biosynthesis whereas low concentrations stimulated the biosynthesis of these compounds. They concluded that these differential effects can be explained by a mechanism involving OH· utilization in both the consumption of H_2O_2 and in the generation of $^1O_2$. They suggest that the function of the OH· is to generate O_2^- which can then be utilized in the Haber–Weiss reaction to generate $^1O_2$, as first proposed by Kellogg and Fridovich (1975) [Eq. (10)].

Samuelsson and his associates (Marnett *et al.*, 1974, 1975) have also been concerned with the possible involvement of $^1O_2$ in the biosynthesis of prostaglandins in the microsomal fraction of sheep vesicular glands. In their initial publication dealing with this subject (Marnett *et al.*, 1974), they observed a CL associated with the reactions responsible for prostaglandin biosynthesis. They were able to inhibit the CL using either indomethacin, aspirin, or dimercaptopropanol, all inhibitors of prostaglandin synthetase, as well as by the addition of SOD and β-carotene. They were also able to observe the oxidation of DPBF during the conversion of arachidonic acid to prostaglandin. They were very cautious in attempting to attribute the CL to $^1O_2$ in as much as they observed strong CL when they used 5,8,11,14-eicosatetraynoic acid in their system. This compound is a potent inhibitor of prostaglandin synthetase yet was able to induce CL in their system comparable to that seen with arachidonic acid. They therefore concluded that the CL could arise even in reactions which do not involve $^1O_2$. Marnett *et al.* (1975) have presented additional evidence with respect to the process involved in oxidizing substrates other than unsaturated fatty acids by the microsomal fraction of sheep vesicular glands. In addition to DPBF, they studied the cooxygenation of compounds such as luminol, benzo[*a*]pyrene, and cholesterol. In the case of DPBF, luminol, and benzo[*a*]pyrene, the co-oxygenation could be inhibited completely by the addition of indomethacin and dimercaptopropanol when the system was supplemented with arachidonic acid. If they substituted the arachidonic acid with either the hydroperoxy endoperoxide, prostaglandin G_2, or 15-hydroperoxy-5,8,11,13-eicosatetra-enoic acid, they could still observe cooxygenation which was not inhibited by indomethacin or dimercaptopropanol. They therefore concluded that the oxygenation of organic substrates by prostaglandin synthetase was due to an interaction between various hydroperoxide intermediates which arise normally during prostaglandin biosynthesis and the microsomal enzyme system responsible for carrying out the biosynthesis. The oxidation product of DPBF was characterized as DBB. An attempt at characterizing the oxida-tion product when benzo[*a*]pyrene was utilized as a cosubstrate indicated that the product formed was less polar than 3-OH-benzo[*a*]pyrene, based on migration on a thin-layer chromatographic plate, and could conceivably be either an epoxide or an endoperoxide of benzo[*a*]pyrene. They attempted to resolve the question as to whether $^1O_2$ was involved in the process by studying the oxidation products of cholesterol in their system, using the analysis described by Smith and Teng (1974). Unfortunately, they were never able to isolate any oxidation product of cholesterol and therefore, could not conclude whether $^1O_2$ or radical processes were occurring in their system. Aware of the fact that neither DPBF oxidation nor CL can be considered unique to $^1O_2$ reactions, they state that they have no conclusive evidence for the function of $^1O_2$ reactions in the cooxygenation of various substrates by the prostaglandin synthetase system.

A somewhat different mechanism for the cyclooxygenase action of prostaglandin synthetase has been proposed by Rahimtula and O'Brien (1976). They studied both the oxidation of $^1O_2$ acceptors such as bilirubin, DBF, and DPBF and the effects of D_2O on the extent of the prostaglandin synthetase activity. They found that bilirubin, an effective $^1O_2$ substrate, was readily oxidized by their enzyme system in the presence of eicosa-8,11,14-trienoic acid and that this oxidation was not affected by the addition of either indomethacin or by the radical quencher, butylated hydroxyanisole. D_2O, which extends the lifetime of $^1O_2$ in aqueous systems, inhibited the action of the prostaglandin synthetase. Their conclusion is that the prosta-glandin synthetase consists of two activities, a peroxidase activity which is involved in the generation of $^1O_2$ followed by the subsequent attack of this $^1O_2$ on unsaturated fatty acid substrates to form an unconjugated hydro-peroxide. This hydroperoxide then forms a peroxy radical which can undergo a cyclization reaction yielding the cyclic endoperoxide intermediate in prostaglandin biosynthesis, as described by Samuelsson *et al.* (1971). The reaction sequence proposed by Rahimtula and O'Brien (1976) can be described in Scheme 13.

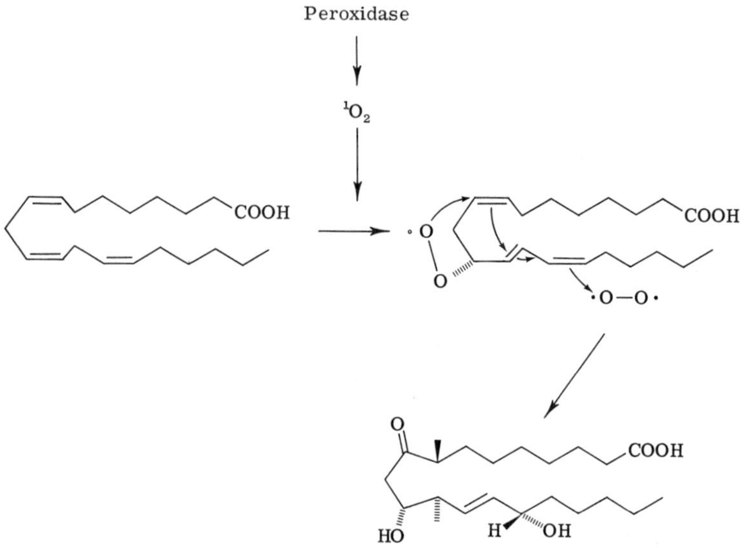

Peroxidase

$^1O_2$

COOH

COOH

$\cdot O - O \cdot$

COOH

HO H OH

Scheme 13

It would be interesting to know if the approach of Marnett *et al.* (1975) of utilizing either the endoperoxide or hydroperoxide intermediates would be able to bring about the cooxygenation of bilirubin. Rahimtula and O'Brien (1976) attributed the failure of D_2O to affect their oxidation of bilirubin to the formation of a $^1O_2$-metal complex within the lipid phase of the membrane and, although inaccessible to D_2O, would still be in a position to oxygenate reactive quenchers.

4. Lipoxidase

Although there are certain similarities in the products that are formed when unsaturated fatty acids are treated either with the enzyme lipoxidase or with $^1O_2$, a careful analysis of the products excludes $^1O_2$ as being involved in the action of lipoxidase. Nevertheless, this enzyme has been frequently invoked as functioning by means of a $^1O_2$ intermediate. Superficially, lipoxidase, $^1O_2$, and a free radical attack on unsaturated fatty acids can give rise to a variety of hydroperoxides. Very careful analyses of the products of $^1O_2$ attack on unsaturated fatty acids and radical attack on unsaturated fatty acids indicated that the products differ. $^1O_2$ attack forms hydroperoxides that have both conjugated and unconjugated polyene systems, whereas radical attack or autoxidation yields only hydroperoxides with conjugated polyene systems (Hall and Roberts, 1966; Cobern *et al.*, 1966; Rawls and van Santen, 1970a,b; N. A. Khan, 1970; Clements *et al.*, 1973). In contrast, soybean lipoxidase when acting on polyunsaturated fatty acids yields only conjugated hydroperoxides, comparable to radical attack, and produces *no* unconjugated hydroperoxides (Hamberg and Samuelsson, 1967). Despite these differences Chan (1971a) proposed, on the basis of the similarity between lipoxidase attack on dienoid compounds and the lipoxidase–ethyl linoleate catalyzed oxidation of tetraphenylcyclopentadienone to its endoperoxide and DPBF to DBB that an intermediate is generated during the lipoxidase-catalyzed oxidation of ethyl linoleate that is a close analog of $^1O_2$. The basis for the argument was that the tetracyclone endoperoxide formed was a specific $^1O_2$ product (Chan, 1970) but this argument was refuted by Baldwin *et al.* (1971) who found that the product was a known keto lactone and not the endoperoxide. Based on these observations, Baldwin *et al.* (1971) suggested that the proposal of Chan (1971a) that $^1O_2$ was involved in the mechanism of lipoxidase action should be reconsidered.

Further evidence arguing against the involvement of $^1O_2$ in the action of lipoxidase was presented by Teng and Smith (1973, 1976) who studied the products of cholesterol oxidation during the action of soybean lipoxidase on ethyl linoleate. They concluded, based on the appearance of the epimeric 7-hydroperoxides of cholesterol without any detectable observation of the 5α-hydroperoxide derivative of cholesterol that $^1O_2$ could not be involved as an intermediate in this reaction. Nilsson and Kearns (1974) took advantage of the deuterium isotope effect on $^1O_2$ lifetime which they had reported earlier (Nilsson *et al.*, 1972) and monitored the oxidation of linoleic acid using soybean lipoxidase. These authors found no effect on the rate of oxidation when D_2O was substituted for H_2O and therefore concluded that $^1O_2$ played no role in this reaction. Despite these observations, Finazzi Agrò *et al.* (1973) have concluded, based on the oxidation of DPBF in the presence of lipoxidase and linoleic acid, that $^1O_2$ is involved in this enzymatic system. These results of Finazzi Agro *et al.* (1973) might be explained by the observation of Faria de Oliveira *et al.* (1974) who also studied lipoxidase–linoleate systems effect on the bleaching of cytochrome *c*. They utilized the D_2O technique and made the following two observations. First, they found that D_2O was without effect on the rate of formation of conjugated hydroperoxides in their linoleate–lipoxidase system, therefore concluding that $^1O_2$ was not involved in the initial process of the attack of lipoxidase on the unsaturated fatty acid. On the other hand, they found a very pronounced increase in the rate of cytochrome *c* bleaching in the presence of D_2O in this same system and concluded that they were observing the production of $^1O_2$ by the lipoxidase–linoleate system. They concluded that $^1O_2$ could not originate from the disproportionation of peroxy radicals but rather might arise from either a planar-oriented tetroxide or perepoxide intermediate which could be equivalent to $^1O_2$ (Smith and Lands, 1972).

5. Other Oxidases

Horseradish peroxidase was used by Chan (1971b) to catalyze the oxygenation of 1,3-dienes such as DPBF, TPF, and anthracene. He found products such as DBB, DBS, and anthraquinone, all presumed $^1O_2$ products of the 1,3-dienes mentioned above. Chan (1971b) attributed these results either to $^1O_2$ or to the participation of an enzyme-bound metal peroxy complex that could mimic $^1O_2$ reactions. Teng and Smith (1973, 1976) tested horseradish peroxidase in their cholesterol system and found only the epimeric 7-hydroperoxycholesterols as products and no trace of the $^1O_2$ product, 5α-hydroperoxycholesterol. These products are analogous to those obtained in a free radical oxidation of cholesterol (Smith *et al.*, 1973).

Piatt *et al.* (1977) have attempted to obtain direct evidence as to whether a peroxidase can generate $^1O_2$ during the oxidation of a halide in the presence of H_2O_2. To accomplish this, these workers utilized commercial lactoperoxidase (LPO) as the enzyme and Br^- as the halide. They found that the $LPO/H_2O_2/Br^-$ system could effectively carry out the oxidation of DPBF and DPF. On thin-layer chromatographic analysis these compounds yielded only DBB or DBE, respectively. The same products were obtained if the reaction mixture contained OBr^- and H_2O_2 instead of the LPO. They also observed CL with the $LPO/H_2O_2/Br^-$ system which was inhibited in the presence of DPF or methionine and stimulated in a D_2O buffer. Finally, by analyzing the rate of oxidation of DPBF at 420 nm, they concluded that the $^1O_2$ quenchers, DABCO, methionine, tryptophan, and histidine, were effective inhibitors of the reaction, whereas the radical scavengers, 2,6-di-*tert*-butyl phenol, butylated hydroxytoluene, mannitol, benzoate, and *tert*-butanol were ineffective. They concluded that the oxidation of DPBF and DPF and the CL were due to the oxidation of Br^- by the LPO generating OBr^- and this could subsequently react with H_2O_2 to generate $^1O_2$ as shown in the Eqs. (16) and (17).

$$Br^- + H_2O_2 \xrightarrow{LPO} OBr^- + H_2O \tag{16}$$

$$OBr^- + H_2O_2 \longrightarrow Br^- + H_2O + {}^1O_2 \tag{17}$$

Further support for the concept that both the oxidation of DPBF and the CL could be attributed to $^1O_2$ was their observation that the rate of oxidation was doubled in a D_2O medium and that the CL was stimulated in this medium.

As opposed to the effect of peroxidases, catalase, which catalyzes the disproportion of H_2O_2 [Eq. (18)], has been studied by Porter and Ingraham

$$2\,H_2O_2 \xrightarrow{\text{catalase}} 2\,H_2O + O_2 \tag{18}$$

(1974) with respect to the possibility that $^1O_2$ is evolved in the process. They used a spectrophotometric assay for measuring the oxidation of DPF in the presence of D_2O. Under these circumstances they found that less than 0.5% of the total oxygen formed in the decomposition of H_2O_2 by catalase could be released as $^1O_2$ which was the limit of the sensitivity of their assay. They therefore concluded that only triplet state oxygen is released during the disproportionation of H_2O_2 by catalase.

6. Superoxide Dismutase (SOD, Erythrocuprein)

Several years ago, a group of papers appeared presenting evidence that SOD acted directly on $^1O_2$ to generate ground state triplet oxygen and the authors of these papers suggested that this decontaminating activity of SOD was its principle function in nature (Finazzi Agrò *et al.*, 1972; Weser and Paschen, 1972; Paschen and Weser, 1973). The conclusions were based on the apparent inhibition of CL by SOD, or by the inhibition of DPBF oxidation by SOD in a system generating O_2^- (Mayeda and Bard, 1974). In order to obtain more direct evidence with respect to the question of SOD affecting $^1O_2$, Schaap *et al.* (1974) utilized sensitized reactions to generate $^1O_2$. When these workers studied the effect of SOD on the rate of the photosensitized oxidations, they were unable to detect any effect whatsoever of SOD at concentrations up to 1 mg/ml (10^{-5} *M*) and concluded that SOD did not quench $^1O_2$. Similar evidence against the role of SOD in quenching singlet oxygen was presented by Michelson (1974) and by Goda *et al.* (1974).

Although this seemed to settle the controversy in favor of SOD not affecting $^1O_2$, Weser *et al.* (1975) have presented additional evidence that the oxidation of DPBF in a photosensitized system to which D_2O had been added could be sepcifically inhibited by SOD. They observed no inhibition using the apoprotein of SOD or various chelates of Cu^{2+} which had been shown earlier by these authors to be as effective as SOD in carrying out the dismutation of O_2^- (Brigelius *et al.*, 1974, 1975). In an attempt to obviate the argument that the histidine residues in SOD were acting as $^1O_2$ quenchers, Weser *et al.* (1975) treated the SOD in the photosensitized system for one hour prior to the addition of DPBF. Under these circumstances they claimed that all of the histidine residues were destroyed but they found no difference whatsoever in the activity of the SOD or in any physical properties such as electronic absorption spectra, X-ray photoelectron spectroscopy, electron paramagnetic resonance data, or circular dichroism data. They conclude that SOD is remarkably resistant to the actions of $^1O_2$ and that this resistance plays an important role in the ability of the enzyme to convert $^1O_2$ to the ground-state triplet oxygen.

D. Miscellaneous Oxidant Effects

There are a number of other systems in which there is indirect evidence that $^1O_2$ may be involved. Some of these systems may offer new areas for exploring the interrelationship between $^1O_2$ and biological systems.

1. Paraquat Toxicity

Paraquat (1,1′-dimethyl-4,4′-bipyridilium dichloride, methyl viologen) is a broad spectrum herbicide which is effective not only against plants but is also highly toxic to mammalian species (Murray and Gibson, 1973). The toxicity of paraquat was studied both *in vitro* and *in vivo* by Bus *et al.* (1974, 1975, 1976) who concluded that the toxicity was due to $^1O_2$ damage. They observed that paraquat could be reduced by a NADPH–cytochrome *c* reductase system and that the reduced paraquat would then be reoxidized by molecular oxygen, generating O_2^-. Subsequent lipid peroxidation could then be prevented by the addition of SOD in the *in vitro* system (Bus *et al.*, 1974). They were also able to demonstrate that DPBF inhibited the peroxidation and concluded that $^1O_2$ was formed through the spontaneous dismutation of O_2^-. This $^1O_2$ could then initiate the formation of lipid hydroperoxides from polyunsaturated lipids, ultimately leading to the formation of lipid free radicals and membrane damage. Bus *et al.* (1975) were also able to demonstrate that mice deficient in either selenium or vitamin E were more susceptible to the toxic effect of paraquat than control animals. This increased susceptibility was due presumably to a deficiency in GSH peroxidase accompanying the selenium deficiency and a deficiency of vitamin E in quenching lipid free radicals.

2. Cytotoxic Agents

During an investigation dealing with the cytotoxic action of 6-hydroxydopamine, 6-aminodopamine, 6,7-dihydroxytryptamine and dialuric acid, Cohen and Heikkila (1974) concluded that both O_2^- and H_2O_2 were being formed during the autoxidation of these compounds. In addition, Cohen and Heikkila (1974) were able to demonstrate, through the formation of ethylene from methional, that OH· was also produced during the autoxidation of these agents. Since either catalase or SOD inhibited ethylene production, they concluded that the Haber–Weiss reaction was responsible for the production of the OH· radicals. In a closely related study, Heikkila *et al.* (1974) demonstrated that alloxan-induced diabetes could be prevented by the simultaneous administration of ethanol, a known OH· quencher. Alloxan is the oxidized form (quinone) of dialuric acid. However, as with any system in which the Haber–Weiss reaction has been invoked, the possibility exists that $^1O_2$ is either formed directly in the Haber–Weiss reaction or may result from the associated reaction which involves quenching of the O_2^- by OH· [Eq. (11)].

A related example of what can be considered cytotoxicity was reported by Goldstein and Weissmann (1977) who reported damage to liposome membranes following the generation of O_2^-. In their system, xanthine oxidase and hypoxanthine were used to generate O_2^- and Goldstein and Weissmann (1977) followed the liposomal damage by the release of sequestered chromate ions. These authors observed that both SOD and catalase were effective in protecting the liposomal membrane against the effect of O_2^- and, by inference, H_2O_2. Although the authors did not state this in their conclusion, it is apparent that they have observed a phenomenon which can not be attributed to either O_2^- or H_2O_2 alone but must result from an interaction of these two species. Again we can invoke the Haber–Weiss reaction as one possibility for producing oxidants that can damage the liposomal membrane. Whether these oxidants are OH· or $^1O_2$ requires additional studies. It should be pointed out however, that Anderson and Krinsky (1973) were able to protect liposomal membranes against photosensitized damage, which involved $^1O_2$ production, by incorporating $^1O_2$ quenchers such as β-carotene directly into the liposomal membrane.

3. *Red Blood Cell Damage*

Although there are many examples whereby red blood cells can be damaged by $^1O_2$, the bulk of these are the result of the photosensitized production of $^1O_2$ using either endogenous or exogenous photosensitizers. The possibility that $^1O_2$ may damage red blood cells directly in a nonphotosensitized system has been proposed by Carrell *et al.* (1975) in attempting to explain the various forms of damage to red blood cells associated with oxidative hemolytic diseases. Carrell *et al.* (1975) also invoked the Haber–Weiss reaction as generating either $^1O_2$ or OH· which can subsequently lead to both the denaturation of hemoglobin and damage to the red blood cell membrane, both effects culminating in a hemolytic condition. Some experimental evidence for the existence of the Haber–Weiss reaction in red blood cells treated with a source of O_2^- has appeared recently (Michelson and Durosay, 1977). Further details of some of the studies in this area appear in Chapter 11 by Kaplan and Trozzolo in this volume.

IV. SUMMARY

In this review I have attempted to bring together much of the information that deals with the possible involvement of $^1O_2$ in biological systems.

Unfortunately, while there is much data available the conclusions that can be drawn are few in number. One of the problems arises from the fact that $^1O_2$, although long lived with respect to electronically excited species, has an extremely short lifetime from the point of view of biological systems and therefore is difficult to analyze directly. It is possible that some of the newer techniques, such as that described by Lion *et al.* (1976), in which a stable nitroxide radical was produced from a $^1O_2$-catalyzed reaction may prove useful in quantifying the amount of $^1O_2$ present in biological systems. An attempt has already been made to carry out this by Cannistraro and van de Vorst (1977).

The use of specific reaction products as means of identifying $^1O_2$ in biological systems becomes more and more limited as other reactions give rise to products that are identical to those originally assumed to be unique for $^1O_2$ reactions. For example, the substituted furans can no longer be considered to be absolute proof of the presence of $^1O_2$ based on their reaction products. From the point of view of biological systems, the best substrate for identifying $^1O_2$ appears to be cholesterol, based on the work of Smith and his associates (Kulig and Smith, 1973; Smith and Kulig, 1976) who used the oxidation products of cholesterol to differentiate between $^1O_2$ reactions and radical reactions. Lamola *et al.* (1973) used this technique to demonstrate $^1O_2$ involvement in the photosensitized oxidation of red blood cell ghosts and Suwa *et al.* (1977) have used the same technique to demonstrate the formation of $^1O_2$ in the bimolecular lipid layers of liposomes.

As the techniques for determining $^1O_2$ become more specific and more quantitative, we can look forward to a better understanding of the role of $^1O_2$ in biological systems. Once this role is defined, we should be able to use both $^1O_2$ generators and $^1O_2$ quenchers to alter the rates and directions of biochemical reactions.

ACKNOWLEDGMENTS

The work in the author's laboratory has been supported by research grants from both the National Science Foundation and the National Institutes of Hèalth.

REFERENCES

Agner, K. (1958). *Proc. IV Int. Cong. Biochem.* (Vienna) **15,** 64.

Agner, K. (1972). *In* "Structure and Function of Oxidation-Reduction Enzymes" (Å. Åkeson and A. Ehrenberg, eds.), pp. 329–335. Pergamon, Oxford.

Allen, R. C. (1975a). *Biochem. Biophys. Res. Commun.* **63,** 675–683.

Allen, R. C. (1975b). *Biochem. Biophys. Res. Commun.* **63,** 684–691.

Allen, R. C., and Loose, L. D. (1976). *Biochem. Biophys. Res. Commun.* **69,** 245–252.

Allen, R. C., Stjernholm, R. L., and Steele, R. H. (1972). *Biochem. Biophys. Res. Commun.* **47,** 679–684.

Allen, R. C., Yerich, S. J., Orth, R. W., and Steele, R. H. (1974). *Biochem. Biophys. Res. Commun.* **60,** 909–917.

Anderson, S. M., and Krinsky, N. I. (1973). *Photochem. Photobiol.* **18,** 403–408.

Anderson, S. M., Krinsky, N. I., Stone, M. J., and Clagett, D. C. (1974). *Photochem. Photobiol.* **20,** 65–69.

Arneson, R. M. (1970). *Arch. Biochem. Biophys.* **136,** 352–360.

Arnold, S. J., Ogryzlo, E. A., and Itzke, H. (1964). *J. Chem. Phys.* **40,** 1769–1770.

Babior, B. M., Kipnes, R. S., and Curnutte, J. T. (1973). *J. Clin. Invest.* **52,** 741–744.

Babior, B. M., Curnutte, J. T., and Kipnes, R. S. (1975). *J. Lab. Clin. Med.* **85,** 234–244.

Baldwin, J. E., Basson, H. H., and Krauss, H., Jr. (1968). *Chem. Commun.* pp. 984–985.

Baldwin, J. E., Swallow, J. C., and Chan, H. W.-S. (1971). *Chem. Commun.* pp. 1407–1408.

Beauchamp, C., and Fridovich, I. (1970). *J. Biol. Chem.* **245,** 4641–4646.

Bland, J. (1976). *J. Chem. Educ.* **53,** 274–279.

Blum, H. F. (1941). "Photodynamic Action and Diseases Caused by Light." Van Nostrand-Reinhold, New York.

Boyer, R. F., Lindstrom, C. G., Darby, B., and Hylarides, M. (1975). *Tetrahedron Lett.* pp. 4111–4115.

Boyland, E., and Sims, P. (1965). *Biochem. J.* **95,** 780–787.

Brigelius, R., Spöttl, R., Bors, W., Lengfelder, E., Saran, M., and Weser, U. (1974). *FEBS Lett.* **47,** 72–75.

Brigelius, R., Hartmann, H. J., Bors, W., Lengfelder, E., Saran, M., and Weser, U. (1975). *Hoppe-Seyler's Z. Physiol. Chem.* **365,** 739–745.

Brodie, B. B., Axelrod, J., Shore, P. A., and Udenfriend, S. (1954). *J. Biol. Chem.* **208,** 741–750.

Browne, R. J., and Ogryzlo, E. A. (1964). *Proc. Chem. Soc., London* p. 117.

Bus, J. S., Aust, S. D., and Gibson, J. E. (1974). *Biochem. Biophys. Res. Commun.* **58,** 749–755.

Bus, J. S., Aust, S. D., and Gibson, J. E. (1975). *Res. Commun. Chem. Pathol. Pharmacol.* **11,** 31–38.

Bus, J. S., Aust, S. D., and Gibson, J. E. (1976). *Environ. Health Perspect.* **16,** 139–146.

Cahill, A. A., and Taube, H. (1952). *J. Am. Chem. Soc.* **74,** 2312–2318.

Cannistraro, S., and van de Vorst, A. (1977). *Biochem. Biophys. Res. Commun.* **74,** 1177–1185.

Carrell, R. W., Winterbourn, C. C., and Rachmilewitz, E. A. (1975). *Br. J. Haematol.* **30,** 259–264.

Chan, H. W.-S. (1970). *Chem. Commun.* pp. 1550–1551.

Chan, H. W.-S. (1971a). *J. Am. Chem. Soc.* **93,** 2357–2358.

Chan, H. W.-S. (1971b). *J. Am. Chem. Soc.* **93,** 4632–4633.

Chen, C., and Tu, M.-H. (1976). *Biochem. J.* **160,** 805–808.

Cheson, B. D., Christensen, R. L., Sperling, R., Kohler, B. E., and Babior, B. M. (1976). *J. Clin. Invest.* **58,** 789–796.

Clements, A. H., Van Den Engh, R. H., Frost, D. J., Hoogenhout, K., and Nooi, J. R. (1973). *J. Am. Oil Chem. Soc.* **50,** 325–330.

Cline, M. J. (1975). "The White Cell." Harvard Univ. Press, Cambridge, Massachusetts.

Cobern, D., Hobbs, J. S., Lucas, R. A., and Mackenzie, D. J. (1966). *J. Chem. Soc. C* pp. 1897–1902.

Cohen, G., and Heikkila, R. E. (1974). *J. Biol. Chem.* **249,** 2447–2452.

Corey, E. J., and Taylor, W. C. (1964). *J. Am. Chem. Soc.* **86,** 3881–3882.

Curnutte, J. T., Whitten, D. M., and Babior, B. M. (1974). *N. Engl. J. Med.* **290,** 593–597.

Davidson, R. S., and Trethewey, K. R. (1976). *J. Am. Chem. Soc.* **98,** 4008–4009.

Deneke, C. F., and Krinsky, N. I. (1976). *J. Am. Chem. Soc.* **98,** 3041–3042.

Deneke, C. F., and Krinsky, N. I. (1977). *Photochem. Photobiol.* **25,** 299–304.
Dixon, W. T., and Norman, R. O. C. (1962). *Nature (London)* **196,** 891–892.
Douzou, P. (1973). *Adv. Radiat. Res., Phys. Chem.* **1,** 217–228.
Drath, D. B., and Karnovsky, M. L. (1974). *Infect. Immun.* **10,** 1077–1083.
Dufraissе, C., and Ecary, S. (1946). *C. R. Hebd. Seances Acad. Sci.* **223,** 735–737.
Erbrich, B. (1952). Dissertation, University of Göttingen.
Faria, de Oliveira, O. M. M., Sanioto, D. L., and Cilento, G. (1974). *Biochem. Biophys. Res. Commun.* **58,** 391–396.
Finazzi Agrò, A., Giovagnoli, C., DeSole, P., Calabrese, L., Rotilio, G., and Mondoví, B. (1972). *FEBS Lett.* **21,** 183–185.
Finazzi Agrò, A., DeSole, P., Rotilio, G., and Mondoví, B. (1973). *Ital. J. Biochem.* **22,** 217–230.
Fong, K.-L., McCay, P. B., Poyer, J. L., Keele, B. B., and Misra, H. (1973). *J. Biol. Chem.* **248,** 7792–7797.
Fong, K.-L., McCay, P. B., Poyer, J. L., Misra, H. P., and Keele, B. B. (1976). *Chem.-Biol. Interact.* **15,** 77–89.
Foote, C. S. (1968a). *Science* **162,** 963–970.
Foote, C. S. (1968b). *Acc. Chem. Res.* **1,** 104–110.
Foote, C. S. (1976). *In* "Free Radicals in Biology" (W. A. Pryor, ed.) Vol. 2, pp. 85–133. Academic Press, New York.
Foote, C. S., and Ching, T.-Y. (1975). *J. Am. Chem. Soc.* **97,** 6209–6214.
Foote, C. S., and Denny, R. W. (1968). *J. Am. Chem. Soc.* **90,** 6233–6235.
Foote, C. S., and Wexler, S. (1964a). *J. Am. Chem. Soc.* **86,** 3879–3880.
Foote, C. S., and Wexler, S. (1964b). *J. Am. Chem. Soc.* **86,** 3880–3881.
Fujimori, E., and Livingston, R. (1968). *Nature (London)* **180,** 1036–1038.
Goda, K., Chu, J.-W., Kimura, T., and Schaap, A. P. (1973). *Biochem. Biophys. Res. Commun.* **52,** 1300–1306.
Goda, K., Kimura, T., Thayer, A. L., Kees, K., and Schaap, A. P. (1974). *Biochem. Biophys. Res. Commun.* **58,** 660–666.
Goldstein, I. M., and Weissmann, G. (1977). *Biochem. Biophys. Res. Commun.* **75,** 604–609.
Gollnick, K. (1976). *In* "Radiation Research: Biomedical, Chemical and Physical Perspectives" (O. F. Nygaard, H. I. Adler, and W. K. Sinclair, eds.), pp. 590–611. Academic Press, New York.
Gollnick, K., and Schenck, G. O. (1967). *In* "1,4-Cycloaddition Reactions" (J. Hamer, ed.), pp. 255–344. Academic Press, New York.
Gorman, A. A., Lovering, G., and Rodgers, M. A. J. (1976). *Photochem. Photobiol.* **23,** 399–403.
Grossweiner, L. (1976). *Curr. Top. Radiat. Res. Q.* **11,** 141–199.
Guyot, A., and Catel, J. (1906). *Bull. Soc. Chim. Fr.* [3] **35,** 1124–1134.
Haber, F., and Weiss, J. (1934). *Proc. R. Soc. London, Ser. A* **147,** 332–351.
Hall, G. E., and Roberts, D. G. (1966). *J. Chem. Soc. B* pp. 1109–1112.
Halliwell, B. (1976). *FEBS Lett.* **72,** 8–10.
Hamberg, M., and Samuelsson, B. (1967). *J. Biol. Chem.* **242,** 5329–5335.
Harrison, J. E., and Schultz, J. (1976). *J. Biol. Chem.* **251,** 1371–1374.
Hayaishi, O. (1974). *In* "Molecular Mechanisms of Oxygen Activation" (O. Hayaishi, ed.), pp. 6–10. Academic Press, New York.
Heikkila, R. E., Barden, H., and Cohen, G. (1974). *J. Pharmacol. Exp. Ther.* **190,** 501–506.
Hodgson, E. K., and Fridovich, I. (1974). *Biochemistry* **13,** 3811–3815.
Howard, J. A., and Ingold, K. U. (1968). *J. Am. Chem. Soc.* **90,** 1056–1058.
Howes, R. M., and Steele, R. H. (1971). *Res. Commun. Chem. Pathol. Pharmacol.* **2,** 619–626.
Howes, R. M., and Steele, R. H. (1972). *Res. Commun. Chem. Pathol. Pharmacol.* **3,** 349–357.
Iyer, G. Y. N., Islam, D. M. F., and Quastel, J. H. (1961). *Nature (London)* **192,** 535–541.

Jensen, M. S., and Bainton, D. F. (1973). *J. Cell. Biol.* **56,** 379–388.
Johansson, G. (1971). *Eur. J. Biochem.* **21,** 68–79.
Johnston, R. B., Jr., Keele, B. B., Jr., Misra, H. P., Lehmeyer, J. E., Webb, L. S., Baehner, R. L., and Rajagopalan, K. V. (1975). *J. Clin. Invest.* **55,** 1357–1372.
Karnovsky, M. L. (1973). *Fed. Proc., Fed. Am. Soc. Exp. Biol.* **32,** 1527–1533.
Kearns, D. R. (1971). *Chem. Rev.* **71,** 395–427.
Kellogg, E. W., III, and Fridovich, I. (1975). *J. Biol. Chem.* **250,** 8812–8817.
Kellogg, R. E. (1969). *J. Am. Chem. Soc.* **91,** 5433–5436.
Khan, A. U. (1970). *Science* **168,** 476–477.
Khan, A. U. (1976). *J. Phys. Chem.* **80,** 2219–2228.
Khan, A. U., and Kasha, M. (1963). *J. Chem. Phys.* **39,** 2105–2106.
Khan, A. U., and Kasha, M. (1970). *J. Am. Chem. Soc.* **92,** 3293–3300.
Khan, N. A. (1970). *Oleagineux* **25,** 281–288.
King, M. M., Lai, E. F., and McCay, P. B. (1975). *J. Biol. Chem.* **250,** 6496–6502.
Klebanoff, S. J. (1967). *J. Exp. Med.* **126,** 1063–1078.
Klebanoff, S. J. (1968). *J. Bacteriol.* **95,** 2131–2138.
Klebanoff, S. J. (1974). *J. Biol. Chem.* **249,** 3724–3728.
Klebanoff, S. J. (1975a). *Semin. Hematol.* **12,** 117–142.
Klebanoff, S. J. (1975b). *In* "The Phagocytic Cell in Host Resistance" (J. A. Bellanti and D. H. Dayton, eds.), pp. 45–59. Raven, New York.
Knowles, A. (1976). *In* "Radiation Research: Biomedical, Chemical and Physical Perspectives" (O. F. Nygaard, H. I. Adler, and W. K. Sinclair, eds.), pp. 612–622. Academic Press, New York.
Kochi, J. K. (1962). *J. Am. Chem. Soc.* **84,** 1193–1197.
Koppenol, W. H. (1976). *Nature (London)* **262,** 420–421.
Krinsky, N. I. (1971). *In* "Carotenoids" (O. Isler, ed.), pp. 669–716. Birkhaüser Verlag, Basel.
Krinsky, N. I. (1974). *Science* **186,** 363–365.
Krinsky, N. I. (1976). *In* "The Survival of Vegetative Organisms" (T. G. R. Gray and J. R. Postgate, eds.), pp. 209–239. Cambridge Univ. Press, London and New York.
Krinsky, N. I. (1977). *Trends Biochem. Sci.* **2,** 35–38.
Krinsky, N. I., and Jong, P. (1977. *Abstr., Int. Conf. Singlet Oxygen, August 21–26, 1977* p. K-15.
Krishnamurty, H. G., and Simpson, F. J. (1970). *J. Biol. Chem.* **245,** 1467–1471.
Kulig, M. J., and Smith, L. L. (1973). *J. Org. Chem.* **38,** 3639–3642.
Lamola, A. A., Yamane, T., and Trozzolo, A. M. (1973). *Science* **179,** 1131–1133.
Le Berre, A., and Ratsimbazafy, R. (1963). *Bull. Soc. Chim. Fr.* pp. 229–230.
Lion, Y., Delmelle, M., and van de Vorst, A. (1976). *Nature (London)* **263,** 442–443.
Lutz, R. E., Welstead, W. J., Jr., Bass, R. G., and Dale, J. I. (1962). *J. Org. Chem.* **27,** 1111–1112.
McCapra, F., and Hann, R. A. (1969). *Chem. Commun.* pp. 442–443.
McCay, P. B., and Poyer, J. L. (1976). *In* "The Enzymes of Biological Membranes" (A. Martonosi, ed.), Vol. 4, pp. 239–256. Plenum, New York.
McCay, P. B., Fong, K.-L., King, M., Lai, E., Weddle, C., Poyer, L., and Hornbrook, K. R. (1976). *In* "Lipids" (R. Paoletti, G. Porcellati, and G. Jacini, eds.), Vol. 1, pp. 157–168. Raven, New York.
McClune, G. S., and Fee, J. A. (1976). *FEBS Lett.* **67,** 294–298.
McCord, J. M. (1974). *Science* **185,** 529–531.
McKeown, E., and Waters, W. A. (1966). *J. Chem. Soc. B* pp. 1040–1046.
Marnett, L. S., Wlodawer, P., and Samuelsson, B. (1974). *Biochem. Biophys. Res. Commun.* **60,** 1286–1294.
Marnett, L. S., Wlodawer, P., and Samuelsson, B. (1975). *J. Biol. Chem.* **250,** 8510–8517.

Martel, J. (1957). *C. R. Hebd. Seances Acad. Sci.* **244,** 626–629.
Mason, H. S. (1957). *Science* **125,** 1185–1188.
Mathews, M. M., and Sistrom, W. R. (1960). *Arch. Mikrobiol.* **35,** 139–146.
Mathews-Roth, M. M., and Krinsky, N. I. (1970). *Photochem. Photobiol.* **11,** 419–428.
Mathews-Roth, M. M., Wilson, T., Fujimori, E., and Krinsky, N. I. (1974). *Photochem. Photobiol.* **19,** 217–222.
Matsuura, T., Matsushima, H., and Sakamoto, H. (1967). *J. Am. Chem. Soc.* **89,** 6370–6371.
Maugh, T. H., II. (1973). *Science* **182,** 44–45.
Mayeda, E. A., and Bard, A. J. (1973). *J. Am. Chem. Soc.* **95,** 6223–6226.
Mayeda, E. A., and Bard, A. J. (1974). *J. Am. Chem. Soc.* **96,** 4023–4024.
Merkel, P. B., Nilsson, R., and Kearns, D. R. (1972). *J. Am. Chem. Soc.* **94,** 1030–1031.
Michelson, A. M. (1974). *FEBS Lett.* **44,** 97–100.
Michelson, A. M., and Durosay, P. (1977). *Photochem. Photobiol.* **25,** 55–63.
Murray, R. E., and Gibson, J. E. (1972). *Exp. Mol. Pathol.* **17,** 317–325.
Nakano, M., Noguchi, T., Tsutsumi, Y., Sugioka, K., Shimizu, Y., Tsuji, Y., and Inaba, H. (1974). *Proc. Soc. Exp. Biol. Med.* **147,** 140–143.
Nakano, M., Noguchi, T., Sugioka, K., Fukuyama, H., and Sato, M. (1975). *J. Biol. Chem.* **250,** 2404–2406.
Nakano, M., Takayama, K., Shimizu, Y., Tsuji, Y., Inaba, H., and Migita, T. (1976). *J. Am. Chem. Soc.* **98,** 1974–1975.
Neta, P., and Dorfman, L. M. (1968). *Adv. Chem. Ser.* **81,** 222–230.
Nilsson, R., and Kearns, D. R. (1974). *J. Phys. Chem.* **78,** 1681–1683.
Nilsson, R., Merkel, P. B., and Kearns, D. R. (1972). *Photochem. Photobiol.* **16,** 117–124.
Ouannés, C., and Wilson, T. (1968). *J. Am. Chem. Soc.* **90,** 6527–6528.
Panganamala, R. V., Brownlee, N. R., Sprecher, H., and Cornwell, D. G. (1974a). *Prostaglandins* **7,** 21–28.
Panganamala, R. V., Sharma, H. M., Sprecher, H., Geer, J. C., and Cornwell, D. G. (1974b). *Prostaglandins* **8,** 3–11.
Panganamala, R. V., Sharma, H. M., Heikkilla, R. E., Geer, J. C., and Cornwell, D. G. (1976). *Prostaglandins* **11,** 599–607.
Paschen, W., and Weser, U. (1973). *Biochim. Biophys. Acta* **327,** 217–222.
Paul, B. B., Jacobs, A. A., Strauss, R. R., and Sbarra, A. J. (1970). *Infect. Immun.* **2,** 414–418.
Pederson, T. C., and Aust, S. D. (1973). *Biochem. Biophys. Res. Commun.* **52,** 1071–1078.
Pederson, T. C., and Aust, S. D. (1975). *Biochim. Biophys. Acta* **385,** 232–241.
Piatt, J. F., Cheema, A. S., and O'Brien, P. J. (1977). *FEBS Lett.* **74,** 251–254.
Politzer, I. R., Griffin, G. W., and Laseter, J. (1971). *Chem.-Biol. Interact.* **3,** 73–93.
Porter, D. J. T., and Ingraham, L. L. (1974). *Biochim. Biophys. Acta* **334,** 97–102.
Rahimtula, A., and O'Brien, P. J. (1976). *Biochem. Biophys. Res. Commun.* **70,** 893–899.
Rawls, H. R., and van Santen, P. J. (1970a). *J. Am. Oil Chem. Soc.* **47,** 121–125.
Rawls, H. R., and van Santen, P. J. (1970b). *Ann. N. Y. Acad. Sci.* **171,** 135–137.
Rosen, H., and Klebanoff, S. J. (1976). *J. Clin. Invest.* **58,** 50–60.
Rosen, H., and Klebanoff, S. J. (1977). *J. Biol. Chem.* **252,** 4803–4810.
Russell, G. A. (1957). *J. Am. Chem. Soc.* **79,** 3871–3877.
Samuelsson, B. (1965). *J. Am. Chem. Soc.* **87,** 3011–3013.
Samuelsson, B., Granström, E., Green, K., and Hamberg, M. (1971). *Ann. N. Y. Acad. Sci.* **180,** 138–161.
Sander, U., and Stauff, J. (1971). *An. Asoc. Quim. Argent.* **59,** 149–155.
Sbarra, A. J., Selvaraj, R. J., Paul, B. B., Poskitt, P. F. K., Zgliczynski, J. M., Mitchell, G. W., Jr., and Louis, F. (1976). *Int. Rev. Exp. Pathol.* **16,** 249–271.

Schaap, A. P., Thayer, A. L., Faler, G. R., Goda, K., and Kimura, T. (1974). *J. Am. Chem. Soc.* **96,** 4025–4026.

Schaap, A. P., Goda, K., and Kimura, T. (1976). *In* "Excited States of Biological Molecules" (J. B. Birks, ed.), pp. 78–93. Wiley, New York.

Schenck, G. O. (1957). *Angew. Chem.* **69,** 579–599.

Seliger, H. H. (1960). *Anal. Biochem.* **1,** 60–65.

Selvaraj, R. J., Paul, B. B., Strauss, R. R., Jacobs, A. A., and Sbarra, A. J. (1974). *Infect. Immun.* **9,** 255–260.

Shimizu, Y., Inaba, H., Kumaki, K., Mizuno, K., Hata, S., and Tomioka, S. (1973). *IEEE Trans. Instrum. Meas.* **im-22,** 153.

Sligar, S. G., Lipscomb, J. D., Debrunner, P. G., and Gunsalus, I. C. (1974). *Biochem. Biophys. Res. Commun.* **61,** 290–295.

Smith, L. L., and Kulig, M. J. (1976). *J. Am. Chem. Soc.* **98,** 1027–1029.

Smith, L. L., and Teng, J. I. (1974). *J. Am. Chem. Soc.* **96,** 2640–2641.

Smith, L. L., Teng, J. I., Kulig, M. J., and Hill, F. L. (1973). *J. Org. Chem.* **38,** 1763–1765.

Smith, W. L., and Lands, W. E. M. (1972). *J. Biol. Chem.* **247,** 1038–1047.

Spikes, J. (1975). *Ann. N. Y. Acad. Sci.* **244,** 496–508.

Stauff, J., and Wolf, H. (1964). *Z. Naturforsch. Teil B* **19,** 87–96.

Stauff, J., Schmidkunz, H., and Hartmann, G. (1963). *Nature (London)* **198,** 281–282.

Stauff, J., Sander, U., and Jaeschke, W. (1973). *In* "Chemiluminescence and Bioluminescence" (M. J. Cormier, D. M. Hercules, and J. Lee, eds.), pp. 131–141. Plenum, New York.

Stelmaszynska, T., and Zgliczynski, J. M. (1974). *Eur. J. Biochem.* **45,** 305–312.

Sternson, L. A., and Wiley, R. A. (1972). *Chem.-Biol. Interact.* **5,** 317–325.

Strauss, R. R., Paul, B. B., Jacobs, A. A., and Sbarra, A. J. (1971). *Infect. Immun.* **3,** 595–602.

Sugioka, K., and Nakano, M. (1976). *Biochim. Biophys. Acta* **423,** 203–216.

Suwa, K., Kimura, T., and Schaap, A. P. (1977). *Biochem. Biophys. Res. Commun.* **75,** 785–792.

Takayama, K., Noguchi, T., and Nakano, M. (1977). *Biochem. Biophys. Res. Commun.* **75,** 1052–1058.

Tauber, A. I., and Babior, B. M. (1977). *J. Clin. Invest.* **60,** 374–379.

Teng, J. I., and Smith, L. L. (1973). *J. Am. Chem. Soc.* **95,** 4060–4061.

Teng, J. I., and Smith, L. L. (1976). *Bioorg. Chem.* **5,** 99–119.

Wang, H.-P., and Kimura, T. (1976). *Biochim. Biophys. Acta* **423,** 374–381.

Wasserman, H. H., Scheffer, J. R., and Cooper, J. L. (1972). *J. Am. Chem. Soc.* **94,** 4991–4996.

Weishaupt, K. R., Gomer, C. J., and Dougherty, T. J. (1976). *Cancer Res.* **36,** 2326–2329.

Weser, U., and Paschen, W. (1972). *FEBS Lett.* **27,** 248–250.

Weser, U., Paschen, W., and Younes, M. (1975). *Biochem. Biophys. Res. Commun.* **66,** 769–777.

Wilson, T. (1966). *J. Am. Chem. Soc.* **88,** 2898–2902.

Wilson, T., and Hastings, J. W. (1970). *Photophysiology* **5,** 49–95.

Yost, F. Y., Jr., and Fridovich, I. (1974). *Arch. Biochem. Biophys.* **161,** 395–401.

Zgliczynski, J. M., Stelmaszynska, T., Ostrowski, W., Naskalski, J., and Sznajd, J. (1968). *Eur. J. Biochem.* **4,** 540–547.

Zgliczynski, J. M., Stelmaszynska, T., Domanski, J., and Ostrowski, W. (1971). *Biochim. Biophys. Acta* **235,** 419–424.

Zgliczynski, J. M., Selvaraj, R. J., Paul, B. B., Stelmaszynska, T., Poskitt, P. K. F., and Sbarra, A. J. (1977). *Proc. Soc. Exp. Biol. Med.* **154,** 418–422.

Author Index

Numbers in parentheses are reference numbers and indicate that an author's work is referred to although the name is not cited in the text. Numbers in italics show the page on which the complete reference is listed.

A

Abbott, S. R., 93, *112*, 525, *569*
Abe, S., 532, 536, *573*
Abrahamson, E. W., 36, 44, 45, *57*
Acharya, S. P., 397(1), *419*
Acher, A. J., 345(217), *424*
Achhammer, B. G., 577, 584, *592*
Adam, W., 103, 104, *112*, *114*, 198, 199, 200, 219, 220, *238*, *241*, *242*, 245, 246, 248, 251, 252, 254, 263, 265, 268, 269, *283*, *285*, *286*, 413(2, 3, 3a, 3b), *419*
Adams, D. R., 121, 123, *136*, 143, 150, *167*, 290, *419*
Adams, J. H., 580, *592*
Adams, W. R., 287(6), 360(5), 388(5), *419*, 542, *569*
Adelman, A. H., 221, 222, *240*, 538, *573*
Agner, K., 609, 610, *636*
Akiba, K., 79, *112*
Akutagawa, M., 184, *238*
Alford, J. A., 254, *285*
Algar, B. E., 513, 514, *569*, *573*
Allen, R. C., 609, 610, 612, 613, *636*, *637*
Alscher, A., 549, 550, 551, *573*
Altenbach, H.-J., 548, *573*
Alzérreca, A., 413(3a), *419*
Anderson, S. M., 599, 635, *637*
Ando, W., 65, 67, 68, 69, 70, 87, 98, 108, *112*, 183, 199, *238*, *242*, 267, 268, *283*, 297(7), 325(54), 326(54), 342(54), 344(57), 346(7), 347(7), 371(7), 375(54), 390(54), 392(57), 418(54), *419*, *420*, 443, *506*, 548, *570*
Aoyama, H., 184, *238*
Appel, R., 433, *509*
Arai, T., 199, *238*, 268, *283*, 532, 534, 553, 556, 557, 562, *573*
Arbuzov, Y. A., 576, *592*
Ardenne, R., 199, *241*
Arneson, R. M., 86, *112*, 619, 620, *637*
Arnold, J. S., 61, *112*
Arnold, S. J., 24, 26, *32*, 36, 39, 44, *56*, *57*, 60, 612, *637*
Arudi, R. L., 372(38a), *420*
Asaka, T., 475, *508*
Asao, T., 382(7a), *419*, 522, *571*
Ashford, R. D., 49, 50, 51, *57*, 218, 235, *238*, 298(8), 301, *419*
Asker, W., 537, *571*
Atkinson, R. S., 158, *167*, 189, *238*
Audubert, R., 525, *569*
Auerbach, J., 451, *506*
Aust, S. D., 601, 620, 621, 623, 624, 626, 634, *637*, *640*
Avram, M., 584, *592*
Axelrod, J., 602, *637*
Ayer, D. E., 367(252a), 368(252a), *426*

B

Babior, B. M., 605, 608, 616, 617, *637*, *641*
Bacon, R. C., 577, 580, *595*

Bader, L. W., 36, *57*
Baehner, R. L., 616, 617, *639*
Baer, R. L., 591, *593*
Bagli, J. F., 71, *113*, 316(182, 183), 318, 402 (182, 183), 408(182, 183), *424*
Bailey, P. S., 94, 106, *112*, 254, *283*
Bainton, D. F., 610, *639*
Baldwin, J. E., 90, *112*, 221, 232, *238*, 436, *506*, 530, 549, *569*, 601, 627, 631, *637*
Ballhausen, C. J., 24, *32*
Ballotta, C., 441, *506*
Balny, C., 62, *112*
Balogh, V., 445, 449, *507*
Banerjee, A. K., 400(9), 401(9), *419*
Banthorpe, D. V., 348(10), 366(10), *419*
Baranne-Lafont, J., 519, 526, *572*
Bard, A. J., 84, 111, *113*, 600, 626, 633, *640*
Barden, H., 634, *638*
Barnes, M. F., 365(11), *419*
Barnett, G., 266, *283*
Baron, W. J., 270, *283*
Barrett, H. C., 394(12), *419*
Bartlett, P. D., 95, 96, 97, 99, *112*, *114*, 178, 188, 191, 192, 195, 202, 203, 215, 224, 232, 233, *238*, *239*, *241*, *242*, 245, 246, 247, 252, 253, 254, 262, 270, 271, 274, 275, 276, 277, 280, 283, *283*, *284*, *285*, *286*, 332(14, 15, 226), 345(13), 370(67c), 371(67c), *419*, *420*, *425*
Barton, D. H. R., 254, *283*
Basolo, F., 90, *114*
Bass, R. G., 434, 437, 439, *507*, 601, *639*
Basselier, J.-J., 193, 194, *238*, 249, *283*, 439, 441, *506*, 517, 519, 520, 526, 528, 529, 541, 543, *569*, *570*, *571*, *572*
Basson, H. H., 221, *238*, 530, *569*, 627, *637*
Bateman, L., 586, *592*
Battersby, A. R., 552, *573*
Baumstark, A. L., 193, *242*, 248, 253, 260, 261, 262, 270, 271, 275, 277, 280, 283, *283*, *286*
Bayer, P., 551, *572*
Bayes, K. D., 344(284), *427*
Beauchamp, C., 617, *637*
Beauvois, C., 528, *570*
Becherer, J., *242*
Becker, J., 395(197), *424*
Becker, K. H., 44, *57*
Begeman, C. R., 582, *592*
Beheshti, I., 187, 203, 206, 233, *240*, 254, *285*, 365(148a), *423*
Bekowies, P. J., 87, 88, 89, 90, *114*, 342(205), 345(205), 369(205), 372(205), *424*
Bell, R. A., 366(16), 400(16), 401(16), *419*
Bellarmine Grdina, S. M., 342(200a), 351 (200a), 352(200a), *424*
Bellesia, F., 370(17), 385(17), *419*
Bellin, J. S., 73, *112*, 416(18), *419*
Bellus, D., 140, 141, 157, 164, 165, *167*
Belyakov, V. A., 255, 256, *283*
Bender, M. L., 272, *283*
Bensasson, R., 151, 154, *167*
Benson, R., 135, *136*
Benson, S. W., 100, *112*, 266, *283*
Berchtold, G. A., 76, 105, *112*, 522, 548, 549, *570*, *572*
Berends, W., 490, 496, 497, *509*
Bergman, W., 91, *112*
Bergmann, W., 524, *569*
Bergsma, D., 162, *167*, 591, *592*
Bernardi, F., 225, *239*
Bernauer, K., 180, *240*
Bernheim, F., 448, *506*
Berson, J. A., 72, *112*
Berthelot, J., 177, 188, 190, 192, 194, 204, 237, *241*, 249, *285*
Beugelmans, R., 403(208), 404(208), *424*
Bevilacqua, E. M., 587, *592*
Bezman, S. Z., 274, *284*
Bhatnagar, A. K., 232, *238*
Bick, I. R. C., 538, *569*
Bilheimer, D., 591, *594*
Binsch, G., 225, *241*
Birch, A. J., 475, *506*
Bird, P. H., 274, *284*
Bisacchi, G. S., 246, *285*, 450, *507*
Bito, T., 349(28, 29), *419*
Bixon, M., 126, *136*
Black, R. M., 577, *594*
Bland, J., 75, *112*, 598, *637*
Block, E., 102, *113*
Blossey, E. C., 186, *241*, 344(19, 227), 370(19), 372(19), *419*, *425*
Blum, H. F., 598, 606, *637*
Blyumberg, E. A., 215, *239*
Boche, J., 480, *508*
Boden, R. M., 344(20), *419*
Böhm, M., *242*
Böse, D., 553, 554, *572*
Bogan, D. J., 51, 53, *57*, 217, *238*
Bolland, J. L., 587, *592*

Bollyky, L. J., 103, *112*, 189, 219, *238*
Bolsman, T. A. B. M., 443, *507*
Bolton, P. H., 135, *136*
Bonnett, R., 162, *168*, 465, 466, *506*, 591, *592*
Bonstead, I., 582, *592*
Bors, W., 633, *637*
Boschung, A. F., 195, 202, 214, 215, *239*, *240*, 247, 254, *284*, 293(112, 114), 295(112, 114), 310(112, 114), 312(112, 114), 313 (112, 114), 330(114), 341(116), 360(112, 114), 361(113), 370(67c), 371(67c), 388 (112, 114), 390(113), *420*, *422*, 443, *507*
Bose, A. K., 558, *572*
Boss, C. R., 583, *593*
Bourdon, J., 514, *573*, 577, 578, 588, *592*, *595*
Bowen, E. J., 110, 111, *112*, 512, 517, *569*
Boyer, R. F., 79, *112*, 583, *594*, 601, 602, *637*
Boyland, E., 626, *637*
Brabham, D. E., 30, *32*, 160, *168*
Brady, W. T., 232, *238*
Braithwaite, M., 44, 45, 46, *57*
Branham, R. V., 466, *508*
Breck, A. K., 42, *57*, 161, *168*, 580, *592*
Breitenback, J. W., 526, *569*
Breliére, C., 528, *572*
Bremner, J. B., 538, *569*
Brennan, M. E., 101, *112*
Brenner, M., 72, *112*, 307(58), 377(21), 378 (21), 381(21, 58), *419*, *420*
Brewer, D., 143, 152, 153, 157, 166, *170*, *171*, 236, *238*, 290(285), 317(286), *427*, 535, *574*, 588, *595*
Bricout, D., 206, 212, *241*, 307(215), 370(214), *424*, 540, *572*
Brigelius, R., 633, *637*
Briggs, P. J., 582, *592*
Brill, W. F., 328(22), *419*
Brimage, D. R. G., 157, *167*
Brkic', D., 344(23, 23a), *419*
Broadbent, A. D., 344(24, 72), *419*, *421*
Brocker, U., 550, *573*
Brockman, J. A., Jr., 245, *284*
Brockmann, H., 516, *569*
Brodie, B. B., 602, *637*
Broida, H. P., 38, *57*
Brossi, A., 537, *570*
Brown, R. S., 463, *508*
Browne, R. J., 26, *32*, 61, 62, *112*, 612, *637*
Brownlee, N. R., 627, *640*
Brunken, J., 174, *242*
Buchwald, G., 302(249), 325(249), 374(249, 250), 375(249), 390(250), *425*
Büchi, G., 394(12), *419*
Bullock, M. W., 245, *284*
Burford, A., 183, 184, *240*
Burgess, A. R., 577, 580, *592*
Burnell, R. H., 475, *506*
Burns, P. A., 193, 194, 196, 197, 210, 211, 223, 224, *238*, *239*, *241*, *242*, 249, 254, *284*, 307 (27), 308, 309, 389(25), 391(27), *419*, 539, 540, *569*, *570*
Burris, W. A., 528, *570*
Burstain, I. G., 431, 432, 438, 449, *506*
Bus, J. S., 634, *637*
Buso, O., 590, *593*, *594*
Butsugan, Y., 77, *113*, 349(28, 29), *419*, *423*
Byers, G. W., 156, *168*
Bystritskaya, E. V., 582, *592*

C

Cahill, A. E., 63, *112*, 615, *637*
Cahnmann, H. J., 553, 558, 559, 566, *573*
Caille, J., 194, *238*, 249, *283*, 439, 441, *506*
Calabrese, L., 164, *168*, 633, *638*
Calvert, J., 577, *592*
Cambie, R. C., 398(30), 399(30), *419*
Campbell, B. S., 278, *284*
Camurati, F., 350(50), *420*
Cannistraro, S., 636, *637*
Capdevielle, P., 212, *241*
Caporale, G., 441, *506*
Carlberg, D., 225, *239*
Carlson, G. L., 588, *593*
Carlsson, D. J., 42, *57*, 148, 152, 154, 157, 161, *168*, 578, 579, 580, 582, *592*, *593*
Carnahan, E., 245, 270, *286*
Carr, R. V. C., 254, *286*, 362(49), *420*
Carrell, R. W., 635, *637*
Carter, T. P., Jr., 254, *283*
Cassar, L., 274, *284*
Catel, J., 600, *638*
Cauzzo, G., 470, *506*
Cavill, G. W. K., 350(31), *419*
Chabaud, J.-P., 409(172), *423*
Chaineaux, J., *426*
Chalvet, O., 517, 518, *569*
Chambers, R. W., 405(118), *422*
Chan, H. W.-S., 90, *112*, 350(32), *419*, 436, *506*, 601, 604, 631, 632, *637*

Chan, Y.-M., 384(48a), *420*
Chang, B. C., 276, *284*
Chang, W.-H., 101, *112*
Chang, Y. C., 72, 73, *112*, 140, 141, 143, 144, 145, 149, 154, 155, 157, 163, 166, 167, *168*, 183, 184, 210, 235, *239*, *240*, 337(63), *420*, 500, *509*, 587, *593*
Chapelle, A., 475, *506*
Charifi, M., 206, *241*, 370(213), *424*
Charlesby, A., 582, *592*
Chatzidakis, A., 537, *569*
Cheema, A. S., 601, 632, *640*
Chen, C., 604, 626, *637*
Chen, F. Y.-H., 476, *507*
Cheng, H., 68, 69, 71, 88, 89, *113*, 291(99), 292(99), 294(99), 300(99), 333(99), 342(99), *421*
Cherton, J.-C., 194, *238*, 249, *283*, 439, 441, *506*
Cheson, B. D., 605, *637*
Chiba, J., 417(181d), *424*
Chien, J. C. W., 583, *593*
Ching, T.-Y., 148, 149, 150, 153, 159, 160, 162, 163, 164, *168*, 234, *239*, 415(33), *420*, 463, *506*, 560, *570*, 615, *638*
Chinh, N., 236, *242*, 436, *509*
Cho, S.-C., 307(271), 382(271), *426*
Choi, S. C., 232, *238*
Chopard-Dit-Jean, L. H., 537, *570*
Chow, M.-F., 207, 219, *239*, *241*, *242*
Christensen, R. L., 605, *637*
Chrysochoos, J., 160, *168*, 553, 560, *569*
Chu, J.-W., 85, *113*, 627, *638*
Chu, J. Y.-C., 382(124), *422*, 551, *571*
Chuang, V. T., 330(189), 403(189), *424*
Chujo, Y., 553, 558, 559, 566, *573*
Chung, S.-K., 302(34), 384(34), *420*
Churakova, N. I., 590, *593*
Ciamician, G., 447, *506*
Cicchetti, O., 577, *593*
Cilento, G., 631, *638*
Clagett, D. C., 42, *57*, 491, *506*, 599, *637*
Clardy, J., 153, 158, *170*, 317(204), 318(204), *424*
Clements, A. H., 350(35), *420*, 631, *637*
Cline, M. J., 608, *637*
Cobern, D., 350(36), *420*, 631, *637*
Coggiola, I. M., 350(31), *419*
Cohen, G., 634, *637*, *638*
Cohen, N., 396(151, 152), *423*
Cohen, N. C., 517, *572*
Cole, T. M., Jr., 538, *573*
Collington, E. W., 434, *508*
Collins, D. J., 356(37), *420*
Combrisson, S., 212, *241*
Conia, J. M., *242*, 334(218), 343(218b, 218c), 351(218a, 218c), 352(218, 218a, 218b, 218c), 353(218c), 355(218c), 361(218a, 218c), 364(218c), *425*
Connor, J. A., 90, *112*
Conrad, W. E., 276, *284*
Cook, C. F., 152, 161, *169*
Cook, T. J., 580, *593*
Coombs, R. V., 392(281), *426*
Cooper, J. L., 88, 91, 92, *114*, 345(279), *426*, 525, 526, *573*, 600, *641*
Corey, E. J., 42, *57*, 63, 94, *112*, 600, 602, *637*
Cornwell, D. G., 627, 628, *640*
Cotton, F. A., 273, *284*
Cowell, G. W., 514, *569*
Crandall, D. C., 449, 452, 465, 466, 467, *507*
Craney, B., 75, *112*
Cremer, R. J., 463, *506*, 591, *593*
Criegee, R., 177, *238*, 245, 282, *284*
Crumrine, D. S., *286*
Cueto, O., 413(3), *419*
Cunneen, J. I., 585, *593*
Cuong, N. K., 517, *570* (*see also* Kim Cuong, N.)
Curnutte, J. T., 608, 616, 617, *637*
Curry, N. U., 158, *169*
Curvall, M., 357(275a), *426*

D

Dässler, H. G., 361(39), *420*
Dak, J. I., 434, 437, 439, *507*
Dale, J. I., 601, *639*
Dalle, J. P., 157, *168*, 307(170, 171, 173), 378(170, 171), 379(170), 381(38, 170, 171, 173), *420*, *423*, 587, *593*
Dalton, J. C., 260, *284*
Dalvas, J., 591, *595*
Daly, J. W., 531, *569*
Danen, W. C., 372(38a), *420*
Daniels, P. J. L., 330(189), 403(189, 190), *424*
Darby, B., 79, *112*, 601, 602, *637*
Darling, T. R., 249, *284*
Darnall, K. R., 82, *114*

Daudel, R., 517, 518, *569*
Davidson, J. A., 36, 43, 44, 45, 46, 47, *57*
Davidson, R. S., 141, 142, 157, 158, *167*, *168*, 215, *238*, 599, *637*
Davis, J. C., Jr., 106, *112*
Dean, P. M., 91, *113*, 221, *240*, 512, *571*
de Bruijn, H., 35, 42, *57*
Debrunner, P. G., 627, *641*
Defoin, A., 517, 526, 530, *572*
Dehm, D., 566, *572*
Dekkers, H. P. J. M., 201, *242*
De La Faille, H. B., 591, *593*
DeLeo, V. A., 165, *168*
Deletang, C., 519, 520, 526, 529, 530, *572*
Delevan, W. M., 582, *595*
del Fiero, J., 413(3b), *419*
Delmelle, M., 636, *639*
Delvan, W. M., 36, *58*
de Mayo, P., 393(97), *421*, 448, *506*
Demole, E., 377(40), *420*
Dempster, C. J., 232, *238*
Deneke, C. F., 158, *168*, 604, 605, *637*, *638*
Denk, W., 289(333), 310(233), 313(233), 361 (233), 362(233), 369(233), 371(233), 374 (233), 375(233), 390(233), *425*
Denney, D. B., 275, 276, 277, 278, *284*
Denney, D. Z., 275, 276, 278, *284*
Denny, R. W., 70, 71, 72, 73, 86, *112*, 116, *136*, 140, 141, 145, 147, 149, 151, 153, 154, 155, 157, 166, 167, *168*, 204, 236, 237, *238*, *239*, 287(42), 290(62), 291(56, 61), 292 (62), 297(61), 308(61), 319(42), 332(42), 342(41, 61, 62), 355(41, 61), 380(189), 403 (189), *420*, *424*, 431, 432, 438, 449, *506*, 512, 515, 563, *569*, 587, *593*, 615, *638*
De Sole, P., 164, *168*, 601, 631, 633, *638*
Devaquet, A., 225, *238*
Dewar, M. J. S., 51, 52, *57*, 184, 202, 227, 228, 231, 232, 233, *238*, 247, 264, *284*, 341(43), *420*, 487, *506*
Dexter, D. L., 126, *136*
Dicke, F., 516, *569*
Di Giorgio, J. B., 330(189), 402(185), 403(189, 190), *424*
Dimroth, V. K., 537, *569*
Ding, J.-Y., 178, *240*, 249, 252, 265, 274, *284*, *285*
Dixon, W. T., 604, *638*
Dlin, E. P., 280, *285*
Dmitrienko, G. I., 476, *507*
Dobashi, S., 214, *240*, 307(157), 351(157), 356(157a), *423*, 443, *507*, 540, 541, *571*
Döpp, D., *286*
Dörhöfer, G., 302(85), 325(85), 326(85), 327 (85), 332(85), 342(85), 344(85), 373(85), 375(85), 395(85), *421*, 553, 560, *570*
Doering, W. von E., 177, *239*
Doleiden, F. H., 149, 159, 166, 167, *168*, 294 (44), *420*, 560, *570*, 592, *593*
Domanski, J., 611, *641*
Dorfman, L. M., 613, *640*
Dorhofer, G., 578, *593*
Doshan, H., 248, 256, 257, *286*
Doskotch, R. W., 384(44a, 48a), *420*
Dougherty, T. J., 601, *641*
Doumaux, A. R., 443, *509*
Douzou, P., 62, *112*, 606, *638*
Draft, S., 441, *507*
Drath, D. B., 617, *638*
Druckrey, E., 500, 501, *509*
Dubini, R., 356(270), *426*
Dufraisse, C., 91, *113*, 221, *239*, *240*, 415(45), *420*, 434, 455, 456, 481, *506*, 512, 517, 524, 528, 541, 563, *569*, *570*, *571*, 600, *637*
Dumas, J. L., 344(46), *420*
Duncan, C. K., 36, *57*, 117, 134, *136*, 152, *168*, 582, *594*
Dunn, J. R., 587, *593*
Dunphy, P. J., 349(47), *420*
Dupont, R., 517, 526, *572*
Duran, N., 269, *283*
Durley, R. C., 365(11), *419*
Durosay, P., 165, *170*, 635, *640*
Duynstee, E. F. J., 588, *593*
Dzakpasu, A. A., 182, 225, 233, 234, 237, *239*, 246, 250, 267, *284*, 487, *506*

E

Eastman, R. H., 257, *285*
Ebsworth, E. A. V., 90, *112*
Ecary, S., 434, *506*, 600, *637*
Eddy, K. L., 135, *137*, 152, 156, *170*
Edelman, R., 276, *284*
Egerton, G. S., 579, 588, *593*
Eggert, H., 289(233), 310(233), 313(233), 361 (233), 362(233), 369(233), 371(233), 374 (233), 375(233), 390(233), *425*
Eguchi, S., 354(223), *425*
Ehrig, V., 413(3), *419*

Eichenauer, H., *242*
Eickman, N., 153, 158, *170*, 317(204), 318 (204), *424*
Eisfeld, W., 304, 305, 314(242), 328(239), 392 (242), 407(239), 409(48, 252), 411(48), *420*, *425*, *426*
El-Feraly, F. S., 384(44a, 48a), *420*
El-Shafei, Z. M., 205, *240*
Elias, L., 36, *57*
Ellis, J. W., 25, *32*
Emge, D. E., 108, 109, *114*
Engel, P. S., 257, *284*
Enggist, P., 377(40), *420*
Ensley, H. E., 362(49), *420*
Enzell, C. R., 357(275a), *426*
Epiotis, N. D., 225, *239*
Epstein, S. S., 133, *136*
Erbrich, B., 601, *638*
Ericksen, J., 179, 182, 225, *239*
Erickson, R. E., 106, *112*
Eskins, K., 160, *169*, 471, *506*, 559, 560, 562, *570*
Etienne, A., 481, *506*, 528, 541, *570*
Evans, D. F., 130, *136*
Evans, N. A., 73, *112*, 468, *506*
Evenson, K. M., 38, *57*
Evleth, E. M., 266, *284*
Eyring, H., 23, *32*

F

Fahrenholtz, S. R., 93, *114*, 149, 159, 166, 167, *168*, 294(44), *420*, 434, 435, 457, *509*, 560, *570*, 578, 580, 592, *593*, *595*
Fairchild, E. H., 384(44a, 48a), *420*
Faler, G. R., 51, *58*, 84, *114*, 196, 200, 201, 232, 234, 237, *239*, *241*, 245, 247, 252, 263, *285*, 633, *641*
Fanta, W. I., 392(149), *423*
Faria de Oliveira, O. M. M., 631, *638*
Farid, S., 441, *507*
Farmilo, A., 143, 150, 151, 152, 153, 161, *168*
Fedeli, E., 350(50), *420*
Fee, J. A., 617, *639*
Fehrenfeld, F. C., 38, *57*
Felder, B., 161, 163, *168*
Feler, G., 266, *284*
Fenical, W., 188, 209, *239*, *240*, 332(51, 52, 120), 336(52), 341(120), 345(52), 371(52), 372(52), 388(51, 52), 389(51), *420*, *422*, 540, *570*
Feriozi, D., 153, 157, 166, 167, *171*, 317(286), *427*, 535, *574*
Fieser, L. F., 174, *239*
Filby, J. E., 178, *240*, 252, 254, 261. 265, 274, *284*, *285*
Fillippova, T. V., 215, *239*
Filseth, S. V., 44, *57*, 125, *136*
Finazzi Agró, A., 164, *168*, 601, 631, 633, *638*
Findlay, F. D., 36, *57*, 125, *136*
Finlayson, N., 39, *57*
Fisch, M. H., 157, *168*, 535, *570*
Fischer, C. M., 254, *283*
Fischer, H., 447, *508*
Fischer, S. F., 126, *137*
Fisher, G. S., 348(125), 374(126), *422*
Fitzpatrick, T. B., 164, *169*, 591, *594*
Fleischer, A. S., 591, *593*
Flippen, J. L., 416(181c), 417(181c), *424*
Flood, J., 152, *168*, 582, *593*
Floyd, M. B., 416(277), *426*, 476, 482, 484, 485, 489, 498, 499, *509*
Förster, T., 126, *136*
Folin, M., 590, *593*, *594*
Foner, S. N., 36, *57*
Fong, K.-L., 621, 622, 625, 626, *638*, *639*
Font, J., 189, *239*
Fontana, A., 469, *506*
Foote, C. S., 42, *57*, 62, 63, 64, 65, 67, 68, 69, 70, 71, 72, 73, 86, 87, 88, 89, 98, 108, 110, *112*, *113*, 116, 119, 123, *136*, 140, 141, 143, 144, 145, 147, 148, 149, 150, 151, 153, 154, 155, 156, 157, 159, 160, 161, 162, 163, 164, 166, 167, *168*, *169*, 176, 178, 179, 180, 181, 182, 193, 194, 196, 210, 211, 215, 218, 225, 233, 234, 235, 236, 237, *238*, *239*, *240*, *242*, 246, 248, 249, 250, 254, 257, 267, *284*, *285*, 287(55), 290(62), 291(56, 61, 99), 292(59, 60, 62, 99), 293(59), 294(99), 297(61), 300 (59, 99), 301(59), 307(27, 58), 308(61), 309, 324, 325(54), 326, 332(55, 166), 333 (55, 99), 334(59), 337(63), 342(54, 55, 59, 61, 62, 99), 344(53, 57, 60), 352(63a), 353 (63a), 354(144a), 355(61), 369(59), 371 55, 59), 372(55), 375(54, 55), 381(58), 383 (59), 389(25), 390(54, 55), 391(27), 392 (57), 415(33), 418(54), *419*, *420*, *421*, *423*, 431, 432, 438, 449, 463, 487, *506*, 514, 539,

540, 548, 552, 553, 555, 560, 565, *569*, *570*, *573*, 576, 586, 587, 589, 590, *593*, 598, 599, 600, 613, 615, *638*,
Forbes, E. J., 344(65), *420*, 545, 546, 547, *570*
Ford, R. A., 289, *421*
Fordham, W. D., 348(10), 366(10), *419*
Forster, C. H., 531, 548, 549, *570*
Forster, J. W., *570*
Fortin, C. J., 125, *136*
Forzatti, P., 344(23, 23a), *419*
Foster, C. H., 76, 105, *112*
Fourrey, J. L., 357(65), 358(65), 359(65), 398 (65), *420*
Fox, J. E., 411(66, 67), *420*
Fox, R. B., 578, *593*
Fracheboud, M., 365(165), 394(255), *423*, *426*
Franck, R. W., 451, *506*
Franken, T., 302(85), 325(85), 326(85), 327 (85), 332(85), 342(85), 344(85), 373(85), 375(85), 395(85), *421*, 512, *570*, 578, *593*
Frankevich, Ye. L., 295(273), *426*
Fraser, A. R., 274, *284*
Fridovich, I., 85, *113*, 164, *169*, 615, 616, 617, 619, 621, 628, *637*, *638*, *639*, *641*
Friedrich, E., *242*, 414(67a), *420*
Frimer, A., 163, 167, *170*, 195, *239*, 254, *284*, 352(67b), 353(67b), 370(67c), 371(67c, 67d), *420*, *421*
Fristad, W. E., 317(203), 386(203), 387(203), *424*
Frosch, R. P., 127, *137*
Frost, D. J., 350(35), *420*, 631, *637*
Fueno, T., 226, *242*
Fuhr, H., 87, 88, 89, 90, *113*
Fujimori, E., 154, 166, 167, *169*, 599, 613, *638*, *640*
Fujimoto, H., 225, 227, *239*, *242*, 341(104), 351(104), *422*
Fujimoto, T. T., 143, 144, 163, 167, *168*, 210, 235, *239*, 293(68), 337(63), 342(68), 355 (68), 389(68), *420*, *421*
Fujita, E., 400(70), 401(69, 70), *421*
Fujita, T., 400(70), 401(69, 70), *421*
Fukui, K., 51, *57*, 225, 227, 232, *239*, *242*, 341 (103, 104), 351(104), *422*, 519, *570*
Fukumoto, K., 452, *507*
Fukutome, H., 226, *242*
Fukuyama, H., 81, *113*, 603, 605, 624, 625, *640*
Furakawa, K., 580, *593*
Furue, H., 161, 166, 167, *168*
Furukawa, K., 40, 56, *57*, 151, 152, 153, 154, 158, *169*, 580, *593*
Furutachi, N., 334(71), 345(71), 407(71), 410 (71), 412(71), *421*
Fuson, R. C., 177, *239*

G

Gadola, M., 362(257), 363(257), *426*
Gak, Y. V., 582, *594*
Galen, T. J., 42, *57*, 491, *506*
Galiazzo, G., 590, *593*, *594*
Gallo, U., 589, *593*
Gann, R. G., 51, 53, *57*, 217, *238*
Gardini, G. E., 531, *570*
Garner, A., 150, *169*
Garvin, D., 54, *57*
Gauthier, M. J. R., 36, 44, 46, *57*
Gay, F. P., 577, *593*
Geacintov, N. E., 135, *136*
Gee, G., 586, *592*
Geer, J. C., 628, *640*
Geller, G. G., 149, 153, 159, *168*
Gennari, G., 590, *594*
Geriozi, D., 588, *595*
Gerstenberger, A., 490, *509*
Gertler, S., 465, 466, 467, *507*
Ghiron, C. S., 590, *593*
Giachino, G. G., 132, *136*
Giacomelli, C., 441, *506*
Gibson, D. T., 531, *570*, *574*
Gibson, J. E., 634, *637*, *640*
Gilpen, R., 53, *57*
Giovagnoli, C., 164, *168*, 633, *638*
Glass, C., 563, *570*
Gleason, W. S., 342(73), 344(24, 72, 73), 371 (73), 372(73), 417(73), *419*, *421*
Gleiter, R., 153, 158, *170*, *242*, 317(204), 318 (204), *424*
Goasdoue, C., 222, *240*, 543, *571*
Goda, K., 84, 85, *113*, *114*, 164, *169*, 627, 633, *638*, *641*
Goddard, W. A., III, 224, 231, 232, *239*, *242*, 341, 351(93b), *421*
Goddard, W. R., 52, *57*
Golan, D. E., 193, *242*, 248, 260, 261, *286*
Goldschmidt, C. R., 514, *572*
Goldstein, B. D., 275, *284*, 591, *593*

Goldstein, I. M., 635, *638*
Gollnick, K., 67, 71, 72, 98, *113*, 140, 141, 143, 144, 145, 149, 151, 157, 163, *169*, *170*, 174, 175, 208, 210, *239*, 287(83, 90), 288(76, 83), 289(76, 78, 83, 90, 248), 290(77, 248), 291(77, 78, 82, 83), 292(82, 248), 293(82, 236), 294(82), 295(83, 248), 297(77), 300 (82), 301(82), 302(83, 84, 85, 249), 304 (236), 307(81, 83), 308, 309(83), 310(76, 79, 83), 313(80, 83, 248), 314(83), 316 (236), 317(90, 241), 318(236), 319(76, 83), 325(83, 85, 87, 249), 326, 327(85), 330 (275), 332(76, 82, 83, 85), 337(89, 90), 339 (75), 340(75, 86), 341(275), 342(85, 248), 344(85, 87), 345(86), 361(80), 363(74, 83, 85), 374(249, 250), 375(85, 249), 388(83), 389(80), 390(76, 79, 80, 250), 395(84, 85), 398(78), 406(236), 407(236), 408(236), 418 (76), *421*, *425*, *426*, 512, 517, 525, 526, 535, 552, *570*, *571*, 578, *593*, 601, 602, 606, *638*
Gomer, C. J., 601, *641*
Goodrich, J. E., 580, *592*
Goodyear, W. F., 276, *284*
Gordon, A. J., 289, *421*
Gordon, M. P., 496, *508*
Gorenstein, D., 276, *284*
Gorman, A. A., 290, *421*, 600, *638*
Goto, G., 359(169), *423*
Goto, T., 184, *239*
Gramain, J. C., 157, *168*, 535, *570*
Grams, G. W., 160, *169*, 553, 559, 560, *570*
Granström, E., 630, *640*
Grassie, N., 583, 584, *593*
Gray, E., 157, *167*
Gray, E. W., 26, *32*, 40, 56, *57*, 151, 154, *169*
Gray, H. B., 24, *32*
Gray, W. W., 580, *593*
Greibrokk, T., 219, *239*
Grevels, F.-W., 347(136), *422*
Greider, A., 365(165), *423*
Grien, K., 630, *640*
Griffin, A. C., 202, 231, 233, *238*, 247, *284*
Griffin, G. W., 206, 207, *239*, *241*, 598, *640*
Griffith, J., 566, *570*
Griffith, J. S., 8, 24, *32*
Griffiths, J., 344(64), *420*, 545, 546, 547, *570*
Groh, P., 26, *32*, 61, *113*
Gross, H., 537, *572*
Gross, S., 156, *168*
Grossweiner, L. I., 133, *136*, 160, *168*, *171*, 553, 560, *569*, *570*, *571*, *574*, 589, 590, *593*, *594*, 606, *638*
Groth, W., 44, *57*
Grunwald, E., 235, *240*
Guénard, D., 403(208), 404(208), *424*
Guillaume, J., 517, *572*
Guillet, J. E., 578, *593*
Guillory, J. P., 152, 161, *169*
Guiraud, H. J., 86, 87, *113*, 163, 167, *169*
Gulati, A. S., 277, *285*
Gunsalus, I. C., 627, *641*
Gupta, V. D., 280, *285*
Gurvich, L. V., 295(273), *426*
Guyot, A., 600, *638*

H

Haber, F., 617, *638*
Haddon, R. C., 231, *238*
Häflinger, O., 537, *570*
Haines, R. M., 177, *239*
Haisch, D., 163, *169*, 337(86), 339(75, 92), 340 (75, 86), 345(86, 92), *421*
Hall, C. D., 275, 276, *284*
Hall, G. E., 350(93), *421*, 631, *638*
Hallett, B. P., 75, *113*, 491, *506*
Hallett, F. R., 75, *113*, 491, *506*
Halliwell, B., 617, *638*
Halpern, J., 274, *284*
Hamberg, M., 501, *506*, 630, 631, *638*, *640*
Hammond, G. S., 231, *239*, 259, *286*
Hampson, R. F., Jr., 54, *57*
Hann, R. A., 177, 187, 206, *240*, 522, *571*, 612, *639*
Happ, J. W., 522, *571*
Hara, H., 560, *570*
Harada, N., 541, 542, 543, *570*
Harber, L. C., 165, *168*, *169*, 591, *593*, *594*
Harding, L. B., 52, *57*, 224, 231, 232, *239*, *242*, 341, 351(93b), *421*
Harding, M. J. C., 416(282), *426*, 481, 485, *509*, 563, *574*
Haq, M. Z., 177, *241*
Harris, D. L., 396(221), *425*
Harris, M. S., 193, *242*, 248, 260, 261, *286*
Harrison, J. E., 611, *638*
Hart, E. J., 586, *593*
Hart, H., 92, *114*, 519, 520, 564, *570*, *574*
Hartmann, G., 61, 85, 86, *114*, 618, 620, *641*

Hartmann, H., 289(78), 291(78), 398(78), *421*
Hasselmann, J., 541, *570*
Hastings, J. W., 576, *595*, 598, 613, *641*
Hasty, N. M., 51, *57*, 163, 167, *169*, 209, 210, 215, *239*, 248, *284*, 290(95), 296, 337(94), 341(95), 345(94), 346(94, 95), *421*
Hata, S., 624, *641*
Hatsui, T., *242*, 397(269d), *426*, 544, *573*
Hautala, R. R., 245, 270, *286*
Hawkins, C., 566, *570*
Hayashi, O., 531, 559, *571*, *572*, 627, *638*
Hayashi, T., 469, *507*
Haynes, R. K., 254, *283*
Hayward, R. C., 398(30), 399(30), *419*
Heacock, J. F., 577, *593*
Heather, J. B., 434, *506*
Hehre, W. J., 231, *239*, 341, *421*
Heikkilla, R. E., 628, 634, *637*, *638*, *640*
Hellwinkel, D., 275, *284*
Helmlinger, D., 393(97), *421*
Helms, G., 307(246), 310(246), 363(246), 392 (246), 393(246), *425*
Henderson, W. A., Jr., 161, 166, 167, *171*, 567, 568, *571*, *573*, 587, 588, *595*
Henne, A., *242*
Henrichs, P. M., 156, *168*
Henry, B. R., 126, *136*
Hercules, D. M., 93, *112*, 525, *569*
Herkstroeter, W. G., 135, *137*, 151, 152, 154, 156, *169*, *170*
Herring, J. W., 221, *241*
Herron, J. T., 301, *421*
Hertel, L. W., *242*
Herz, W., 179, 214, *241*, 391(272a), *426*
Herzberg, G., 2, 24, 25, *32*, 129, *136*
Herzberg, L., 25, *32*
Heskins, M., 578, *593*
Hess, J., 201, *240*
Hiatt, R., 215, *239*, 455, *506*
Hiatt, R. R., 236, *239*
Hiegel, G. A., 392(426), *426*
Higgins, G. M. C., 585, *593*
Higgins, R., 67, 68, 69, 70, 71, 88, 89, *112*, *113*, 291(99), 292(99), 294(99), 300(99), 333 (99), 342(99), *421*, 548, *570*
Higley, D. P., 435, *508*
Hill, F. L., 316(263), *426*, 632, *641*
Hinder, M., 397(198), *424*
Hino, T., *242*, 416(178, 181, 181b, 181c, 220a), 417(179, 180, 181a, 181b, 181c, 181d, 220a), *423*, *424*, *425*, 469, 472, 473, 474, 475, 477, 479, *508*, 563, *572*
Hirama, M., 216, *240*, 397(110), 398(110, 298), *422*, *426*
Ho, M. S., 202, 233, *238*, 245, 247, 252, *283*, *284*
Hobbs, J. J., 356(37), *420*
Hobbs, J. S., 350(36), *420*, 631, *637*
Hochstetler, A. R., 76, *113*, 393(150), 396(152, 153), *423*
Hodge, V. F., 75, *114*, 205, *241*, 246, *285*, 479, *508*, 522, 563, *572*
Hodgson, E. K., 164, *169*, 615, *638*
Hodson, W. G., 534, *574*
Höfle, G., 232, *238*
Hoff, E. F., Jr., 232, *238*
Hoffman, B. M., 90, *114*
Hoffman, J. M., Jr., 505, *506*
Hofmann, A. W., 575, *593*
Hoffmann, R., 232, *242*, 246, 273, *286*
Hoft, E., 447, *506*
Hoggenhout, K., 350(35), *420*
Hollahan, J. R., 588, *593*
Hollinden, G. A., 301, *422*
Hollins, R. A., 405(117, 118), 418(117), *422*
Honnaka, A., 215, *240*
Hoogenhout, K., 631, *637*
Hora, J., 307(274), 380(274), 381(274), *426*
Horinaka, A., 77, *113*, 294(161), 301(161), 383 (101, 161), 384(101), 385(101), 396(101), *422*, *423*
Hornbrook, K. R., 625, *639*
Howard, J. A., 80, 81, *113*, 436, *507*, 580, *593*, 602, 603, *638*
Howe, G. R., 215, *239*
Howes, R. M., 609, 612, 623, *638*
Hsia, D. Y. Y., 591, *592*
Hsia, Y.-Y., 162, *167*
Hsu, J., 591, *593*
Huang, C.-T., 384(44a), *420*
Huang, P.-K. C., 491, *506*
Huber, J. E., 180, *239*, 246, *284*
Hudson, R. F., 225, *240*
Hudson, R. L., 36, *57*
Hückel, E., 12, 18, 20, 21, *32*
Huet, F., *242*
Huffman, J. W., 397(102), 402(102), *422*
Hughes, H., 587, *592*
Huie, R. E., 301, *421*
Huisgen, R., 232, 235, 237, *239*

Hursthouse, M. B., 307(274), 380(274), 381 (274), *426*
Huyffer, P. S., *286*
Hycon, S. B., 307(105, 106, 107), 380(105, 107, 108), 382(106, 108), *422*
Hylarides, M., 79, *112*, 601, 602, *637*

I

Ichikawa, H., 307(105), 380(105), *422*
Ihara, M., 452, *507*
Ikari, M., 482, *507*
Ikeuchi, K., 155, *170*
Imada, J., 359(169), *423*
Inakagi, S., 341(103, 104), 351(104), *422*
Imuta, M., 159, *170*, 430, 467, 472, 479, 487, *508*, 519, 532, 533, 534, 535, 555, 563, *571*, *573*
Inaba, H., 81, *113*, 603, 624, *640*, *641*
Inagaki, S., 51, *57*, 227, 232, *239*, *242*
Inamoto, Y., 236, *239*
Inglett, G. E., 559, 560, *570*
Ingold, K. U., 80, 81, *113*, 254, *285*, 580, *593*, 602, 603, *638*
Ingraham, L. L., 632, *640*
Inhoffen, H. H., 458, *506*
Inoue, K., 497, *507*
Ireland, R. E., 366(16), 400(16), 401(16), *419*
Irie, M., 488, *507*
Irina, J., 538, *571*
Irwin, K. C., 586, *594*
Ishibe, N., 199, 220, *239*
Ishii, Y., 236, *239*
Ishikawa, K., 207, *239*
Islam, D. M. F., 608, *638*
Ismail, A. F. A., 205, *240*
Isoe, S., 307(105, 106, 107), 380(105, 107, 108), 382(106, 108), *422*
Itô, S., 216, *240*, 397(109, 110, 268), 398(111, 269), *422*, *426*
Ito, T., 133, *136*, 165, *169*
Ito, Y., 219, *241*, *242*, 547, *571*
Itzke, H., 612, *637*
Ivanov, V. B., 140, *170*
Iverson, L., 591, *594*
Ives, J. L., 185, *241*
Iyer, G. Y. N., 608, *638*

J

Jacini, G., 350(50), *420*
Jackson, C., 162, *167*, 591, *592*
Jackson, H. L., 177, *239*
Jaeschke, W., 620, *641*
Jacobs, A. A., 610, 611, *640*, *641*
Janzen, G., 522, *571*
Jefford, C. W., 202, 207, 214, 215, *240*, *242*, 247, *284*, 293(112, 114), 295(112, 114), 310(112, 114), 312, 313(112, 114), 330 (114), 341(116), 360(112, 114, 116b), 361 (113), 368(115), 369(115), 372(115), 387 (116b), 388(112, 114, 116a), 390(113), 414 (116a), *422*, 430, 443, *507*
Jensen, M. S., 610, *639*
Jerina, D. M., 531, *569*, *574*
Jewett, J. G., 195, *239*, 254, *284*, 370(67c), 371 (67c), *420*
Jindal, S. L., 162, *170*
Johansson, G., 623, *639*
Johnson, A. W., 552, *571*
Johnston, R. B., Jr., 616, 617, *639*
Jones, D. H., 275, 276, 277, *284*
Jones, L. T. N., 579, *593*
Jong, P., 602, *639*
Jori, G., 470, *506*, 590, *593*, *594*
Jortner, J., 126, *136*
Joschek, H.-I., 552, *571*
Jura, W. H., 534, *574*

K

Kacher, M. L., 162, 166, 167, *169*
Kaguchi, H., *508*
Kaiser, J. K., 217, *240*, 333(123), 364(122, 123), *422*
Kajiwara, T., 62, 81, 83, *113*, 122, 123, *136*
Kaki, A. H., 582, *594*
Kallia, R. E., 531, *570*
Kametani, T., 452, *507*
Kanamori, H., 544, *573*
Kanazawa, R., 434, *507*
Kaneko, T., 416(178), *423*, 473, *508*
Kanno, S., 394(128), *422*
Kaplan, M. L., 42, *57*, 94, 95, 97, 98, 99, 100, 101, 108, *113*, *114*, 345(174, 175), 372 (175), *423*, 548, *571*, 576, 578, 586, 587, 588, *594*

Karlsson, K., 357(275a), *426*
Karmitz, P., 587, *594*
Karnovsky, M. L., 607, 608, 616, 617, *638*, *639*
Karpukhin, O. N., 582, *592*, *594*
Kasche, V., 560, *571*
Kasha, M., 24, 25, 26, 27, 28, 30, 31, *32*, 60, 61, 62, 64, *113*, 122, 126, *136*, 514, *571*, 604, 612, *639*
Kasper, B., 361(141a), 390(141a), *423*
Kass, E. H., 165, *169*, 591, *594*
Kastell, A., 526, *569*
Kas'yan, L. F., 215, *239*,
Katayama, H., 400(69), 401(69, 70), *421*
Kato, S., 76, *114*, 491, *507*, 519, 532, 553, 556, 558, 560, 561, *571*, *572*
Kato, T., 394(128), *422*
Katritzky, A. R., 447, *506*
Katsui, G., 359(200), *424*
Katsumura, S., 307(105, 107), 380(105, 107, 108), 382(108), *422*
Katz, T. J., 245, 270, *286*, 445, 449, *507*
Kaufman, F., 39, 40, *57*
Kautsky, H., 35, 42, *57*, 60, 63, *113*
Kauzmann, W., 23, *32*, 128, *136*
Kaversneva, E. D., 590, *593*
Kawamoto, A., 543, *571*
Kawaoka, K., 27, *32*
Kayama, K., 24, *32*
Kayser, R. H., 153, 157, 166, 167, *171*, 236, *242*, 317(286), *427*, 436, *509*, 535, *574*, 588, *595*
Kear, K. E., 36, 44, 45, *57*
Kearns, D. R., 8, 24, 25, 26, 27, *32*, 36, 45, 51, 52, *57*, 60, 62, 81, 82, 83, *113*, 140, 143, 145, 146, 148, 149, 150, 151, 152, 153, 156, 163, 164, 167, *168*, *169*, *170*, 188, 201, 209, 210, 215, 224, 225, 226, 227, *239*, *240*, 248, *284*, 287(121), 290(95, 148, 167, 168), 291 (168), 296, 298, 318(121), 332(51, 52, 120, 121), 337, 341(95, 120), 344(191), 345(51, 52, 94), 346(94, 95), 371(52), 372(52), 388 (51, 52), 389(51), 405(117, 118), 418, *420*, *421*, *422*, *423*, *424*, 497, *507*, 511, 514, 515, 540, *570*, *571*, 576, 582, 590, *594*, 598, 606, 618, 631, *639*, *640*
Keck, G. E., 205, *242*, 250, 259, *286*
Keehn, P. M., 522, 538, *573*
Keele, B. B., Jr., 616, 617, 621, 622, 626, *638*, *639*
Kees, K., 101, 102, *114*, 164, *169*, 345(228), *425*, 633, *638*
Kelleher, P. G., 42, *57*, 578, 586, 587, 588, *594*
Keller, R. A., 143, 153, *171*, 290(285), *427*
Kellogg, E. W., III, 619, 621, 628, *639*
Kellogg, R. E., 81, *113*, 603, *639*
Kellogg, R. M., 217, *240*, 333(123), 364(122, 123), *422*
Kemper, R., 111, *114*, 152, *170*, 219, *241*
Kende, A. S., 382(124), *422*, 551, *571*
Kenner, R. D., 135, *136*, 152, *169*
Kenney, R. L., 348(125), 374(126), *422*
Kepka, A. G., 133, *136*, 560, *571*, 589, 590, *593*, *594*
Kerckow, A., 282, *284*
Khalil, M. F., 475, *506*
Khan, A. U., 24, 25, 26, 27, 28, 31, *32*, 36, *57*, 60, 61, 62, 64, 82, 83, 85, 111, *113*, 122, 135, *136*, 152, *169*, 405(117, 118), 418 (117), *422*, 511, 514, *571*, *576*, 577, 582, *594*, 612, 618, 622, *639*
Khan, N. A., 350(127), *422*, 631, *639*
Kimball, G. E., 23, *32*
Kim Cuong, N. (*also* Cuong, N. K.), 517, 519, 520, 526, 527, 529, 530, *572*
Kimura, T., 84, 85, *113*, *114*, 164, *169*, 624, 627, 633, 636, *638*, *640*, *641*
King, A. D. 290(154a), *423*
King, M. M., 601, 603, 615, 625, 626, *639*
Kinkel, K. G., *425*
Kipnes, R. S., 608, 616, 617, *637*
Kirby, A. J., 277, *284*
Kirk, D. I., 461, *507*
Kirrmann, K. A., 26, *32*, 61, *113*
Kirschner, S., 228, *238*, 264, *284*
Kiselev, V. D., 235, *240*
Kitahara, Y., 382(7a), 394(128), *419*, *422*, 544, 547, *571*, *572*
Kitzing, R., 445, 447, *509*
Klebanoff, S. J., 602, 605, 607, 608, 609, 614, 615, 617, *639*, *640*
Kleiman, R., 562, *570*
Klein, E., 217, *240*, 302(132), 347(194, 244), 348(130), 354(129, 132), 355(129, 132), 356(133), 360(132), 362(132), 390(131), 418, *422*, *424*, *425*
Kleo. J., 151, 154, *169*
Kline, G. M., 577, *592*
Klopman, G., 225, *240*

Klug, H.-H., 550, 551, *573*
Kneser, H. O., 25, *32*
Knowles, A., 497, *507*, 606, *639*
Knowles, W. S., 94, 106, *113*
Kobal, V. M., 531, *574*
Kobayashi, K., 133, *136*, 165, *169*, 475, *508*
Koch, E., 48, *57*, 99, 100, *113*, 123, *136*, 233, *240*, 289(134), 294(134), 301, *422*, *425*, 431, *509*, 515, 541, *571*
Kochi, J. K., 604, *639*
Kocór, M., 358(135), *422*
Kodato, S., 416(181c), 417(181c), *424*
Kodrat'yev, V. N., 295, *426*
Köller, H., 307(240), 308, 344(247), 392(240), *425*
Koerner von Gustorf, E., 344(247), 347(136, 235), *422*, *425*
Kohler, B. E., 605, *637*
Kohler, E. P., 177, *240*
Kohmoto, S., *242*
Koizumi, M., 567, *572*
Kondo, K., 213, 214, 216, *240*, *242*, 302(155), 307(156, 157), 348(137, 155), 351(157), 355(156), 356(157a), *422*, *423*, 443, 480, *507*, 540, 541, *571*
Kopecky, K. R., 52, *57*, 105, *113*, 178, 226, *240*, 244, 245, 246, 252, 254, 264, 265, 274, *284*, *285*, 291(141), 294(138), 295 (138), 330(141), 332(138, 139, 140), 340 (141), 371(139), *422*, *423*, 578, *594*
Kopelman, A. E., 463, *508*
Koppenol, W. H., 618, *639*
Kornhauser, A., 491, *506*
Koshland, D. E., Jr., 482, *508*, 590, *595*
Kosugi, H., 543, *571*
Kotani, M., 24, *32*
Kotsonis, F., 290(77), 291(77), 297(77), 308, *421*
Kotsuki, H., 434, *507*
Kozhevnikov, I. V., 273, *285*
Krauch, C. H., 441, 503, *507*, *509*
Krauss, H., Jr., 221, *238*, 530, *569*, 627, *637*
Kravchenko, N. A., 590, *593*
Krebs, A., 111, *114*, 152, *170*, 219, *241*
Krinsky, N., 154, 164, 166, 167, *169*
Krinsky, N. I., 158, *168*, 491, *506*, 598, 599, 602, 603, 604, 605, 606, 613, 614, 635, *637*, *638*, *639*, *640*
Krishnamurty, H. G., 619, *639*
Kropf, H., 361(141a), 390(141a), *423*
Krubiner, A. M., 367(142), *423*
Krupenie, P. H., 26, *32*
Kubo, M., 24, *32*, 44, *57*
Kuck, V. J., 36, *58*, 582, *595*
Kuhn, H. J., 157, *169*, 324, *424*, 535, *571*
Kulig, M. J., 110, *114*, 316(263), 407(143), *423*, *426*, 619, 623, 626, 632, 636, *639*, *641*
Kumaki, K., 624, *641*
Kurien, K. C., 538, *571*
Kutsch, W. A., 87, *114*
Kwei, K. P. S., 588, *594*

L

Labovitz, J., 316(224), 317(224), *425*
Lacombe, L., 212, *241*, 307(215), *424*, 540, *572*
Laffer, M. H., 293(112), 295(112), 310(112), 312(112), 313(112), 360(112), 361(113), 388(112), 390(113), *422*
Lai, E. F., 601, 603, 615, 625, 626, *639*
Lamola, A. A., 149, 159, 166, 167, *168*, 294 (44), *420*, 560, *570*, 591, 592, *593*, *594*, 636, *639*
Land, E. J., 151, 154, *167*
Landis, M. E., 215, *238*, 253, 262, 270, 271, 275, 277, 280, 283, *283*, *285*, *286*
Lands, W. E. M., 632, *641*
Lane, A. G., 94, *112*
Larsen, D. L., 519, 520, 521, 522, 525, 526, *571*, *573*
Laseter, J., 598, *640*
Lavrov, I. A., 590, *593*
Lawrence, R. V., 307(254), *426*
Learer, I. H., *571*
Le Berre, A., 600, *639*
LeBlanc, J. R., 342(262), *426*
LeBras, J., 524, *569*
Lecas-Nawrocka, A., 462, *508*
Lechevallier, A., *242*
Lechtken, P., 245, 248, 261, 263, 264, 269, 270, *285*, *286*
Leclerc, G., 254, *283*
Lee, D. C.-S., 224, *240*, 270, *285*
Lee, G. A., 566, *572*
Lee, J., 117, 119, *137*, 148, 150, 155, 158, 160, *168*, *169*, 290(154a), *423*
Lee, K.-W., 123, *136*, 156, 166, 167, *168*, *169*, 187, *240*
Leete, E., 476, *507*
Leffler, J. E., 231, *240*
Legg, K. D., 187, *240*
Lehmeyer, J. E., 616, 617, *639*

Leitich, J., 551, *572*
Lengfelder, E., 633, *637*
Lennard-Jones, J. E., 12, *32*
Lenz, G. R., 500, 502, *509*
Le Perchec, P., 334(218), 343(218b, 218c), 351 (218a, 218c), 352(218, 218a, 218b, 218c), 353(218c), 355(218c), 361(218a, 218c), 364(218c), *425*
Lerdal, D., 196, 210, *239*, *242*, 254, *284*, 354 (144a), *423*, 539, *570*
Lerman, C. L., 275, *283*
LeRoux, J. P., 193, 222, *238*, *240*, 439, *506*, 543, *571*
Lever, O. W., Jr., 549, *569*
Lewis, G. N., 3, 5, 25, *32*
Lewis, P., 36, *57*
Li, K., 249, *284*
Li, W.-K., 231, *238*
Liao, C. C., 153, 158, *170*, 317(203, 204), 318 (204), 386(203), 387(203), *424*
Liberles, A., 436, 439, 448, 455, 461, *509*
Libman, J., 254, *286*
Lier, J. E. Van, *423*
Lightner, D. A., 246, *285*, 443, 448, 449, 450, 452, 453, 454, 455, 461, 463, 464, 465, 466, 467, *507*, *508*, 592, *594*
Lin, J.-C., 104, *112*
Lin, J. W.-P., 97, 98, 99, 107, 108, *113*, 176, 180, 181, 182, 218, 225, 233, 234, 237, *239*, 345(175, 177), 372(175), *423*, 487, *506*, 580, *594*
Lind, H., 157, 164, *167*
Lindberg, J. C., 177, *241*
Lindblad, B., 559, *571*
Lindner, J. H. E., 149, 157, *169*, 535, *570*, *571*
Lindqüist, L., 560, *571*
Lindstedt, G., 559, *571*
Lindstedt, S., 559, *571*
Lindstrom, C. G., 79, *112*, 601, 602, *637*
Lion, Y., 636, *639*
Liotta, D. C., 153, 158, *170*, 302(202), 303 (202), 317(203, 204), 318(204), 386(203), 387(203), *424*
Lipscomb, J. D., 627, *641*
Litt, F. A., 214, *240*, 330(147), 332(147), 385 (147), 407(146), *423*
Liu, J.-C., 198, 199, 200, 219, 220, *238*, *241*, 245, 251, 252, 254, 263, 269, *283*, *285*, 413 (2, 3a), *419*
Liu, J. W.-P., 246, 250, 267, *284*
Liu, K.-C., 111, *114*, 152, *170*
Livingston, R., 89, *114*, 589, *595*, 599, *638*
Lloyd, R. A., 111, *112*
Lockwood, P. A., 178, *240*, 252, 254, 265, 274, *284*, *285*
Long, C. A., 25, *32*, 123, 124, 133, *137*, 290 (148), *423*
Loose, L. D., 612, 613, *636*
Lopez Nieves, M. I., 413, *425*
Lopp, I. G., 522, *571*
Louis, F., 607, *640*
Lovering, G., 600, *638*
Low, L. K., 449, 453, *507*
Lucas, R. A., 350(36), *420*, 631, *637*
Ludwig, G. D., 591, *594*
Lukovnikov, A. F., 582, *594*
Lumma, W. C., Jr., 107, 108, *113*, 345(177), *423*, 580, *594*
Lundeen, G. W., 221, 222, *240*
Lutz, R. E., 434, 437, 439, *507*, 601, *639*
Lutz, W., 242, 414(67a), *420*
Lyons, A., 245, 270, *286*

M

MacCallam, J. R., 583, 588, *594*
McCapra, F., 177, 178, 183, 184, 187, 202, 203, 206, *240*, 254, 264, *285*, 365(148a), *423*, 522, *571*, 612, *639*
McCay, P. B., 601, 603, 615, 621, 622, 624, 625, 626, *638*, *639*
McClune, G. S., 617, *639*
McClure, D. E., 99, *114*, 321(267), 323(267), 342(267), *426*
McCord, J. M., 617, *639*
McDonagh, A. F., 162, *169*, 464, 465, *507*, 591, *594*
Machin, P. J., 186, *240*
McKellar, J. F., 582, *592*
McKennis, J. S., 274, *283*
Mackenzie, D. J., 350(36), *420*, 631, *637*
McKeown, E., 64, 78, 79, 110, *113*, 602, *639*
MacKnight, M. L., 589, *595*
McLean, M. J., 91, *112*, 524, *569*
McLean, S., 476, *507*
MacMillan, J., 365(11), *419*
McNeill, D., 479, *508*
Maeda, K., 469, *507*
Magnus, I. A., 591, *595*
Magnus, P. D., 254, *283*
Magous, R., 157, *168*, 587, *593*
Mahoney, R. D., 73, *112*, 416(18), *419*

Maier, G., 217, *240*, 244, 251, *285*
Maki, A. H., 36, *57*
Mallet, L., 60, *113*
Mallon, C., 236, *242*, 436, *509*
Mallory, F. B., 577, *593*
Manchas, M. S., 558, *572*
Mander, L. N., 366(16), 400(16), 401(16), *419*
Mani, J.-C., 307(170, 171, 173), 378(170, 171), 379(170), 381(38, 170, 171, 173), *420*, *423*, 538, *573*
Manske, R. H., 184, *240*, 268, *285*
Mariano, P. S., 538, *573*
Mark, H., 558, *572*
Marnett, L. S., 601, 628, 630, *639*
Marshall, J. A., 76, *113*, 392(149), 393(150), 396(151, 152, 153), *423*
Marsi, K. L., 275, 276, *284*
Martel, J., 435, 448, 481, 503, *506*, *507*, 601, *640*
Martin, A., 400(9), 401(9), *419*
Martin, R. L., 147, 152, 153, 157, 166, 167, *171*, 236, *242*, 317(286), *427*, 436, *509*, 535, *574*, 588, *595*
Masamume, S., 401(154), *423*
Maseles, F. C., 531, *570*
Mason, H. S., 627, *640*
Massey, A. G., 272, *285*
Masure, M., 450, 461, *508*
Mateescu, G., 584, *592*
Matheson, I. B. C., 117, 119, *137*, 148, 150, 155, 158, 162, *169*, 290, *423*
Matheson, L. A., 583, *594*
Matheson, M. S., 586, *593*
Mathews, M. M., 613, *640*
Mathews-Roth, M. M., 154, 165, 166, 167, *168*, *169*, 591, *594*, 599, 613, 614, *640*
Mathis, P., 151, 154, *169*
Matreyek, W., 579, *595*
Matsugo, S., 430, 472, 473, 487, *508*, 563, *573*
Matsumoto, M., 213, 214, 216, *240*, *242*, 302 (155), 307(156, 157), 348(137, 155), 351 (157), 355(156), 356(157a), *422*, *423*, 443, 480, *507*, 540, 541, *571*
Matsuo, K., 558, *571*
Matsushima, H., 187, *240*, 417(158, 159), *423*, 553, 558, 560, 561, 565, *571*, 619, *640*
Matsushita, S., 350(269a, 269b, 284a), *426*, *427*
Matsuura, H., 619, *640*
Matsuura, T., 76, 77, 92, 105, *113*, *114*, 159, *170*, 187, 204, 215, 237, *240*, *241*, *242*, 294 (161), 299(220), 301(161), 356(220), 357 (220), 383(101, 161), 384(101), 385(101), 396(101), 416(220a), 417(158, 159, 220a), *422*, *423*, *425*, 430, 467, 469, 472, 473, 479, 482, 487, 491, 492, 493, 495, 496, 497, 502, *507*, *508*, 519, 532, 533, 534, 535, 536, 552, 553, 554, 555, 556, 557, 558, 559, 560, 561, 562, 563, 565, 566, *571*, *572*, *573*
Maudinas, B., 151, 154, *167*
Maugh, T. H., II, 612, *640*
Maumy, M., 212, *241*, 410(163, 164), 413(162), *423*, 519, *572*
Maurer, B., 365(165), *423*
Mautner, G. N., 497, *507*
Mayeda, E. A., 84, 111, *113*, 600, 626, 633, *640*
Mayo, F. R., 215, *240*, 244, *285*, 582, *594*
Mazur, S., 178, 181, 182, 193, 194, 196, 209, 210, 225, 237, *238*, *239*, *240*, 248, 254, *284*, *285*, 332(166), *419*, *423*, 539, *569*, *570*
Meadus, F. W., 254, *285*
Meckler, A., 24, *32*
Medvedev, V. A., 295(273), *426*
Meguerian, G., 276, *283*
Mehrotra, R. C., 280, *285*
Mellier, M.-T., 517, *571*
Melnick, B., 443, *507*
Mendelson, W. L., 328(184, 187), 404(188), 405(184, 187), 406(186, 187), 408(184, 187), 418(187), *424*, 438, *508*
Mendenhall, G. D., 42, *57*, 95, 97, 99, *112*, 148, 152, 154, 161, *168*, 178, 188, 195, *238*, 332 (15), 345(13), *419*, 436, *507*, 582, *592*
Menzies, I. D., 254, *283*
Merkel, P. B., 25, *32*, 45, *57*, 116, 117, 118, 119, 120, 121, 128, 130, 131, 133, 135, *137*, 143, 145, 146, 148, 149, 150, 151, 153, 156, 163, 164, 167, *169*, *170*, 209, 210, *239*, 290(167, 168), 291(168), 337(94), 345(94), 346(94), *421*, *423*, 497, *507*, 514, 515, *571*, 606, 631, *640*
Metzger, W., 447, *508*
Mevis, M. E. A. H., 588, *593*
Meyer, K.-H., 347(235), *425*
Meyer, T. H., 382(260), *426*
Meyers, A. I., 434, *508*
Michelson, A. M., 164, 165, *169*, *170*, 633, 635, *640*

Migita, T., 81, *113*, 183, 199, *238*, 267, 268, *283*, 297(7), 346(7), 347(7), 371(7), *419*, 443, *506*, 603, *640*
Mill, T., 582, 586, *594*
Miller, A. H., 458, *509*
Miller, J. A., 580, *593*
Miller, J. G., 235, *240*
Miller, S. I., 552, *571*
Miller, W. A., 576, *594*
Mirbach, M. J., *242*
Mishima, T., 469, *507*
Misra, H. P., 616, 617, 621, 622, 626, *638*, *639*
Mitchell, G. W., 607, *640*
Mittal, R. S. D., 434, *506*
Mizuno, K., 624, *641*
Mizuno, Y., 24, *32*
Moffitt, W., 24, *32*
Mondovi, B., 164, *168*, 601, 631, 633, *638*
Monroe, B. M., 148, 157, 158, 166, 167, *170*, 295(168a), *423*
Montgomery, F. C., 209, *241*, 251, 257, 258, 264, 265, 266, 270, *285*, 563, *572*
Montgomery, L. K., 232, *238*
Moore, T. A., 519, 520, *573*
Moosmayer, A., 552, *571*
Morand, J., 579, 585, 586, *594*
Moreau, C., 91, *113*
Moreau, M., 519, 526, 527, *572*
Morgan, J. E., 448, *506*
Mori, A., 382(168b), *423*, 565, *571*
Moriarty, R. M., 361(113), *422*, 443, *507*
Morimoto, H., 359(169), *423*
Morita, N., 522, *571*
Moryzanov, B. N., 280, *285*
Moureu, C., 221, *240*, 512, *571*
Mousseron-Canet, M., 157, *168*, 307(170, 171, 173), 378(170, 171), 379(170), 381(38, 170, 171, 173), 409(172), *420*, *423*, 587, *593*
Müller, B. L., 362(257), 363(257), 378(258), *426*
Müller, E., 552, *571*
Muetterties, E. L., 276, *285*
Mukai, T., *242*
Müller, B. L., 342(258a), *426*
Mulliken, R. S., 12, *33*
Mumford, C., 178, 226, *240*, 244, 245, 246, 252, 264, 265, 274, *284*, *285*, 332(139, 140), 371(139), *422*, *423*
Muroi, T., 397(109, 268), *422*, *426*
Murray, P. E., 634, *640*
Murray, R. W., 94, 95, 97, 99, 100, 101, 102, 106, 107, 108, 109, *113*, *114*, 162, *170*, 319, 345(174, 175, 176, 177), 372(175), *423*, 435, *508*, 548, *571*, 580, 586, *594*
Musajo, L., 441, *508*
Musso, H., 552, *571*
Mustafa, A., 537, *571*
Muszkat, K. A., 160, *170*
Muto, M., 349(28, 29), *419*
Myers, G. H., 44, *57*

N

Nahavandi, F., 436, *509*
Nakadaira, Y., 334(71), 345(71), 407(71), 410 (71), 412(71), *421*
Nakagawa, M., *242*, 416(178, 181, 181b, 181c, 220a), 417(179, 180, 181a, 181b, 181c, 181d, 220a), *423*, *424*, *425*, 469, 472, 473, 474, 475, 476, 477, 479, *508*, 563, *572*
Nakai, C., 559, *572*
Nakajima, M., 584, *594*
Nakanishi, K., 334(71), 345(71), 407(71), 410 (71), 412(71), *421*
Nakashima, R., 187, 215, *240*
Nakano, M., 81, *113*, 602, 603, 605, 624, 625, *640*, *641*
Nakano, T., 400(9), 401(9), *419*
Nakashima, R., 294(161), 301(161), 383(101, 161), 384(101), 385(101), 396(101), 417 (161), *422*, *423*, 553, 557, 558, 565, *571*
Nakatani, M., 482, *508*
Narwid, T. A., 434, *508*
Naskalski, J., 610, *641*
Neckers, D. C., 186, *241*, 344(19, 227), 370 (19), 372(19), *419*, *425*
Nemoto, M., 567, *572*
Nesbit, M. R., 447, *506*
Ness, S., 93, *112*, 525, *569*
Neta, P., 612, *640*
Neumüller, O.-A., 293(236), 304(236), 305 (252), 316(236), 318(236), 325(251), 328 (239), 374(251), 375(251), 406(236), 407 (236, 238), 408(236, 238), 409(252), *425*, *426*
Nguyen, K. C., 517, 526, *572*
Nichols, P. C., 220, *242*

Nickon, A., 71, *113*, 204, 214, *238*, *240*, 287 (42), 316(182, 183), 318, 319(42), 328(184, 187), 330(147, 189), 332(42, 147), 385 (147), 402(182, 183, 185), 403(189, 190), 404(188), 405(184, 187), 406(185, 187), 408(182, 183, 184, 187), 418, *420*, *423*, *424*, 438, *508*, 512, 515, 563, *569*
Niki, H., 44, *58*, 125, *137*
Nilsson, R., 36, *57*, 83, *113*, 121, 131, 133, 134, *137*, 143, 145, 146, 164, 165, *169*, *170*, 290 (167), 344(191), *423*, *424*, 497, *507*, 590, *594*, 606, 618, 631, *640*
Nishida, T., 357(275a), *426*
Nishinaga, A., 92, 105, *113*, 456, *508*, 532, 534, 553, 554, 556, 557, 558, 562, *571*, *573*
Noguchi, T., 81, *113*, 602, 603, 605, 624, 625, *640*, *641*
Nolen, R. L., 434, *508*
Nooi, J. R., 350(35), *420*, 631, *637*
Norman, R. O. C., 604, *638*
Norris, R. D., 246, *285*, 450, 461, 463, *507*, 592, *594*
Noxon, J. F., 44, *57*
Nozaki, O., 559, *572*
Numan, H., 201, *240*, *242*
Nye, M., 393(97), *421*

O

O'Brien, P. J., 44, *57*, 601, 629, 630, 632, *640*
Oda, M., 544, 547, *571*, *572*
Odani, M., 199, 220, *239*
Odell, G. B., 162, *170*, 463, *508*
Ogata, Y., 563, *573*
Ogryzlo, E. A., 24, 26, 27, *32*, *33*, 36, 39, 40, 43, 44, 45, 46, 47, 48, 49, 50, 51, 55, 56, *56*, *57*, *58*, 60, 61, 62, *112*, 151, 152, 153, 154, 158, *169*, *170*, 218, 235, *238*, 298 (8), 301, 317(192), *419*, *424*, 535, *572*, 580, 587, *593*, 612, *637*
Ohloff, G., 302(249, 256), 310(79), 325(249, 251, 347(194, 195, 199, 244, 249, 250), 364 (193, 199), 365(165, 199), 373(196), 374 (251), 375(249, 251), 378(258), 390(79, 250), 394(199, 255), 395(197, 256), 397 (198), *421*, *423*, *424*, *425*, *426*, 504, *509*, 541, 542, *573*
Ohmae, M., 359(200), *424*
Ohno, M., 354(223), *425*
Ohsawa, T., 452, *507*
Ohta, S., 565, *571*
Ohyoshi, N., 416(181), *424*, 479, *508*
Oine, T., 566, *572*
Oishi, Y., 558, *571*
Okada, K. O., *242*
Okajima, H., 416(181c), 417(179, 181a, 181c), *423*, *424*, 472, 477, *508*
Okajima, Y., 475, *508*
Oku, A., 519, 520, *570*
Oldekop, Y. A., 280, *285*
Olesen, J. A., 157, *168*, 535, *570*
Oliveto, E. P., 367(142), *423*
Omote, Y., 184, *238*
Omura, K., 553, 557, 558, *571*
O'Neal, H. E., 245, 251, 264, 265, 266, *285*
Orell, J., 248, 263, *286*
Orfanopoulos, M., 342(200a), 351(200a), 352 (200a), *424*
Orito, K., 184, *240*, 268, *285*
Orland, C. M., Jr., 558, *572*
Ors, J. A., 233, *241*, 514, 515, 518, 519, *573*
Orth, R. W., 612, *637*
Orton, G., 342(283), 417(283), *427*
Osborn, J. A., 274, *284*
Ostrow, J. D., 466, *508*
Ostrowski, W., 610, 611, *641*
Ottolenghi, M., 514, *572*
Ouannés, C., 89, *113*, 140, 149, 152, 153, 157, *170*, 317(201), *424*, 535, *572*, 615, *640*
Ouchi, M. D., 42, *58*, 344(229), 372(229), *425*
Ozaki, A., 273, *286*

P

Padwa, A., 566, *571*
Pagnoni, U. M., 370(17), 385(17), *419*
Pak, C.-S., 449, *507*
Panganamala, R. V., 628, *640*
Paquette, L. A., 153, 158, *170*, *242*, 302(202), 303(202), 317(203, 204), 318, 386(203), 387(203), *424*
Parker, T. L., 179, 182, 225, *239*
Partridge, R. H., 582, *594*
Paschen, W., 164, *170*, 600, 633, *640*, *641*
Pasquon, I., 344(23, 23a), *419*
Pathak, M. A., 164, *169*, 591, *594*
Patterson, E. L., 245, *284*
Patterson, L. K., 135, *137*, 143, 145, *170*

Paul, B. B., 607, 610, 611, *640*, *641*
Paulig, G., 245, *284*
Pauling, L., 14, 23, *33*, 245, *285*
Pavia, D. C., 568, *573*
Pearson, E. A., 27, *33*
Pederson, T. C., 601, 620, 621, 623, 624, 626, *640*
Perez, S. R., 149, 159, 162, 166, 167, *170*, 233, 234, *241*, 514, 515, 518, 519, 560, *573*
Perlmutter, H. D., 72, *112*
Perryman, P. W., 463, *506*, 591, *593*
Peters, J. W., 72, 87, 88, 89, 90, *112*, *113*, *114*, 140, 141, 145, 149, 157, 159, 161, 166, 167, *168*, 342(205), 345(205), 369(205), 372(205), *424*
Peterson, E. R., 123, *136*, 166, 167, *168*
Pflederer, J. L., 205, *242*
Pfoertner, K., 180, *240*, 553, 554, *572*
Piatt, J. F., 601, 632, *640*
Pierce, J. V., 245, *284*
Pinkus, A. G., 177, *241*
Pitts, J. N., Jr., 54, *58*, 82, 87, 88, 89, 90, *113*, *114*, 342(73, 205), 344(24, 72, 73, 266), 345(205), 369(205), 371(73), 372(73, 205), 417(73), *419*, *421*, *424*, *426*, 491, *508*, 514, 538, *569*, *573*, 576, 577, *592*, *594*
Pitzer, K. S., 128, 129, *137*
Pivovarov, A. P., 582, *594*
Politzer, I. R., 206, 207, *239*, *241*, 598, *640*
Polonsky, J., 357(65), 358(65), 359(65), 398 (65), *420*
Pomeranz, S. B., 135, *136*
Pon-Fitzpatrick, M., 165, *168*
Ponce, C., 518, *569*
Pople, J. A., 231, *241*, 341, *424*
Porter, D. J. T., 632, *640*
Porter, G., 135, *137*, 143, 145, *170*
Poskitt, P. F. K., 607, 610, 611, *640*, *641*
Pospisil, J., 156, *170*
Potashnik, R., 563, *572*
Potzinger, P., 324, *424*
Pouchot, O., 415(45, 212), 416(211), *420*, *424*, 455, 456, *506*, *508*, 563, *570*
Poupko, R., 83, *114*
Powell, R. L., 276, *284*
Poyer, J. L., 621, 622, 624, 625, 626, *638*, *639*
Price, C. C., 568, *573*
Pross, A. W., 577, *594*
Prue, J. E., 272, 273, *285*
Pusset, J., 403(208), 404(208), *424*

Q

Quastel, J. H., 608, *638*
Quistad, G. B., 443, 448, 449, 450, 452, 453, 454, 455, 464, 465, 466, 467, *507*, *508*

R

Rabek, J. F., 140, *170*, 577, 583, 584, *594*
Rachmilewitz, E. A., 635, *637*
Radlick, P., 163, 167, *169*, 188, 209, 210, *239*, *240*, 332(51, 52, 120), 336(52), 337(94), 341(120), 345(52, 94), 346(94), 371(52), 372(52), 388(51, 52), 389(51), 405(117, 118), 418(117), *420*, *421*, *422*, 540, *570*
Radziszewski, B., 481, *508*
Rahimtula, A., 629, 630, *640*
Rajagopalan, K. V., 616, 617, *639*
Rajee, R., *242*
Ramamurthy, V., 111, *114*, 152, *170*, 219, *241*, *242*
Ramasseul, R., 415(209), *424*, 450, 455, 457, *508*, 563, *572*
Ramirez, F., 277, *285*
Rånby, B., 140, *170*, 577, 583, 584, *594*
Ranby, B. J., 577, 583, 584, *594*
Ranico, R., 517, *572*
Ranjon, A., 415(45, 212), 416(211), *420*, *424*, 434, 455, 456, 462, 482, 485, *506*, *508*, 563, *570*
Rankin, C. T., 583, 588, *594*
Ranney, G., 258, *285*
Rapoport, H., 359(264), *426*
Rassat, A., 415(209), *424*, 450, 455, 457, *508*, 563, *572*
Ratsimbazafy, R., 600, *639*
Rautenstrauch, V., 342(258a), *426*
Rawls, H. R., 42, *58*, 350(210), *424*, 623, 631, *640*
Ray, W. J., Jr., 482, *508*, 590, *595*
Razmara, F., 436, *509*
Razuvaev, G. A., 280, *285*
Rehm, D., 152, *170*
Reich, H. J., 294(138), 295(138), 332, *422*, 578, *594*
Reiche, A., 537, *572*
Reid, R. W., 252, 254, *284*
Reid, S. T., 448, *506*
Reiney, M. J., 584, *592*
Reinhart, F. W., 584, *592*

Richards, D. H., 463, *506*, 591, *593*
Richardson, A. H., 245, *285*
Richardson, H., 582, *594*
Richardson, W. H., 75, *114*, 205, 209, *241*, 246, 251, 257, 258, 264, 265, 266, 270, *285*, 479, *508*, 522, 563, *572*
Rieker, A., 456, *508*, 552, *571*
Riesenfeld, E. H., 87, *114*
Rigaudy, J., 177, 212, 214, *241*, 410(163, 164), 413(162), *423*, 512, 517, 518, 519, 520, 526, 527, 528, 529, 530, *569*, *570*, *572*
Rimbault, C. G., 207, *240*, *242*, 341(116), 360 (116b), 361(113), 368(115), 369(115), 372 (115), 387(116b), 388(116a), 390(113), 414(116a), *422*, 430, *507*
Rio, G., 177, 188, 190, 192, 194, 204, 206, 212, 237, *241*, 249, *285*, 307(215), 370(213, 214), 415(45, 212), 416(211), *420*, *424*, 434, 450, 455, 456, 461, 462, *506*, *508*, 522, 528, 540, 563, *570*, *572*
Ritter, A., 551, *572*
Robbins, W. K., 257, *285*
Roberts, D. G., 350(93), *421*, 631, *638*
Roberts, D. R., 266, *285*
Robin, M. B., 295, *424*
Robinson, G. W., 127, *136*
Rodgers, M. A. J., 290(91a), *421*, 600, *638*
Rodrighiero, G., 441, *506*, *508*
Rodrigo, R., 184, *240*, 268, *285*
Rodriguez, O., 219, *241*
Rohwedder, W., 562, *570*
Rojahn, W., 217, *240*, 302(132), 348(130), 354 (129, 132), 355(129, 132), 356(133), 360 (132), 362(132), 390(131), 418, *422*
Rokach, J., 479, *508*
Rondest, J., 357(65), 358(65), 359(65), 398 (65), *420*
Rooney, C. S., 479, *508*
Rosen, H., 602, 605, 615, *640*
Rosenkranz, P., 133, *137*, 590, *595*
Rosenthal, I., 83, 86, 87, 88, 89, 90, *113*, *114*, 163, 167, *170*, 342(73), 344(73), 345(217), 371(73), 372(73), 417(73), *421*, *424*, 491, *508*
Roswell, D. F., 256, *286*
Rot, D., 352(67b), 353(67c), *420*
Rotilio, G., 164, *168*, 601, 631, 633, *638*
Rousseau, G., *242*, 334(218), 343(218b, 218c), 351(218a, 218c), 352(218, 218a, 218b, 218c), 353(218c), 355(218c), 361(218a, 218c), 364(218c), *425*
Rubottom, G. M., 413(219), *425*
Rudakov, E. S., 273, *285*
Rugg, F. H., 577, 580, *595*
Runquist, O., 480, *508*
Russel, K. E., 42, *57*
Russell, G. A., 80, *114*, 602, 625, *640*
Russell, K. E., 161, 166, 167, *168*, 580, 582, *592*, *593*
Russell, K. F., 152, *168*

S

Safe, S., 475, *508*
Saiki, T., 183, *238*, 267, *283*, 443, *506*
Saito, I., 76, 77, 92, 105, *113*, *114*, 159, *170*, 204, 237, *241*, *242*, 278, 279, *286*, 299 (220), 356(220), 357(220), 416(220a), 417 (220a), *425*, 430, 439, 462, 467, 469, 472, 473, 479, 487, 491, 492, 493, 495, 496, 497, 502, *507*, *508*, *509*, 519, 532, 533, 534, 535, 536, 553, 554, 556, 557, 558, 559, 560, 561, 562, 563, 566, *571*, *572*, *573*
Saito, J., 204, 237, *241*
Sakai, M., 396(221), *425*
Sakamoto, H., 417(158), *423*, 565, *571*, 619, *640*
Sakan, T., 307(105, 106, 107), 380(105, 107, 108), 382(106, 108), *422*
Sam, T. W., 104, *114*, 302(222), 364(222), 365 (222), 385(222), *425*
Sammes, P. G., 186, *240*
Samuelsson, B., 501, *506*, 601, 627, 628, 630, 631, *638*, *639*, *640*
Sander, U., 618, 620, *640*, *641*
Sanders, F., 245, *284*
Sanioto, D. L., 631, *638*
Santamaria, L., 589, *593*
Santus, R., 590, *595*
Saran, M., 633, *637*
Sasaki, T., 354(223), *425*
Sasson, I., 316(224), *425*
Sastry, K. S., 496, *508*
Sato, M., 81, *113*, 603, 605, 624, 625, *640*
Sato, T., 371(225), 375(225), 396(225), 397 (268), *425*, *426*
Saucy, G., 367(142), *423*
Saunder, K. J., 585, *595*
Savides, C., 583, *595*
Savige, E. W., 469, *508*
Sawaki, Y., 563, *573*
Sbarra, A. J., 607, 610, 611, *640*, *641*

Schaap, A. P., 25, *33*, 51, *58*, 84, 85, 96, 97, 101, 102, *112*, *113*, *114*, 164, *169*, 178, 186, 188, 191, 192, 193, 195, 197, 200, 201, 223, 224, 232, 233, *238*, *241*, *242*, 245, 246, 247, 248, 249, 252, 253, 255, 263, 269, *283*, *285*, *286*, 332(14, 15, 226), 344(19, 227), 345(228), 370(19), 372(19, 228), *419*, *425*, 627, 633, 636, *638*, *641*
Schade, G., 155, 163, *169*, *170*, 210, *239*, 289 (78), 291(78), 302(84, 85), 310(79), 313 (80), 325(85, 87), 326, 327(85), 332(85), 337(86), 339(75), 340(75, 86), 342(85), 344(85, 87), 345(86), 361(80), 373(85), 374(250), 389(80), 390(79, 80, 250), 395 (84, 85), 398(78), *421*, *425*, 512, *570*, 578, *593*
Schäfer-Ridder, M., 550, 551, *573*
Schänzer, W., 347(235), *425*
Schaffer, O., 537, *569*
Schaffer, R., 162, *170*
Schaffner, K., *242*
Scheffer, J. R., 42, *58*, 88, 91, 92, *114*, 344(229), 345(276, 279), 372(229), *425*, *426*, 525, 526, *573*, 600, *641*
Scheffler, S., 552, *571*
Schenck, G. O., 67, 71, 72, 98, *113*, 140, 155, *170*, 174, 176, 195, *239*, *241*, 288(76, 230, 231), 289(76, 233, 248), 290(248), 292 (245, 248), 293(236), 295(248), 301(243), 302(249), 304, 305(252), 307(81, 240, 246), 308(240, 248), 310(76, 79, 233, 246), 313(248), 314(242), 316(236), 317(241), 318(236), 319(76, 245), 325(249, 251), 328 (237, 239), 332(76), 342(245, 248), 344 (237, 247), 347(136, 194, 235, 244), 361 (233, 234), 362(233), 363(246), 364(237), 369(233, 234), 370(234), 371(233, 234, 245), 374(233, 234, 249, 250, 251), 375 (233, 234, 249, 251), 377(245), 383(234), 390(76, 79, 233, 234, 250), 392(237, 240, 242, 246), 393(246), 406(236), 407(236, 238, 239), 408(236, 238), 409(252), 418 (76), *421*, *422*, *424*, *425*, *426*, 431, 432, 433, 435, 438, 449, 503, *506*, *508*, *509*, 512, 517, 526, 551, *570*, *573*, 601, 602, 619, 623, *638*, *641*
Schiff, H. I., 36, 44, 45, 46, 53, *57*
Schlag, E. W., 126, *137*
Schlessinger, R. H., *241*, 369(259), *426*, 503, 505, *506*, *509*
Schmid, G. M., 517, 518, *569*
Schmid, R., 463, *509*
Schmidkunz, H., 26, *33*, 61, 85, 86, *114*, 618, 620, *641*
Schmidt, H., 133, *137*, 590, *595*
Schmidt, W., 505, *509*
Schneider, O., 537, *570*
Schneider, S., 126, *137*
Schneider, W. P., 367(252a), 368(252a), *426*
Schnuriger, B., 514, *573*, 577, 578, 588, *592*, *595*
Schönberg, A., 199, *241*, 288, *426*, 517, *573*
Scholl, M. J., 415(212), *424*, 456, *508*
Schomburg, G., 551, *572*
Schore, N. E., *286*
Schothurst, A. A., 591, *593*, *595*
Schroeter, S., 302(249), 310(79), 325(249, 251), 374(249, 250, 251), 375(249, 251), 390 (79, 250), *421*, *425*, *426*
Schueller, K. E., 232, *238*
Schuller, W. H., 307(254), 366(254), *426*
Schulman, J., 445, 449, *507*
Schulte-Elte, K. H., 176, *241*, 292(245), 302 (256), 319, 342(245, 258a), 347(195), 362 (257), 363(257), 371(245), 377(245), 378 (258), 394(255), 395(197, 256), *424*, *425*, *426*, 431, 432, 438, 449, *506*, 541, 542, *573*
Schultz, A. G., *241*, 369(259), *426*
Schultz, J., 611, *638*
Schumaker, R., 161, 163, *168*
Schunn, R. A., 276, *285*
Schurath, U., 44, *57*
Schuster, G. B., 200, *241*, 245, 248, 252, 263, *285*, 286
Schwartz, N., 402(185), *424*
Scott, A. I., 302(34), 384(34), 411(66, 67), *420*, 552, *573*
Seeley, G. R., 382(260), *426*
Seibles, T. S., 482, *509*
Seiki, T., 583, *595*
Seliger, H. H., 60, 62, *114*, 612, *641*
Selvaraj, R. J., 607, 610, 611, *640*, *641*
Serebryakov, F. P., 590, *593*
Serratosa, F., 189, *239*
Shapiro, M. J., 246, 247, 253, 274, 277, *283*, *285*
Shaply, J. R., 274, *284*
Sharma, H. M., 628, *640*
Sharp, D. B., 201, 224, *241*, 247, *285*, 332, 342 (262), *426*
Shaw, R., 100, *112*
Sheeto, J., 181, *241*

Sheinson, R. S., 51, 53, *57*, 217, *238*
Shen, M., 220, *242*
Shih, L. S., 278, *284*
Shimazu, H., 553, 558, 559, 566, *573*
Shimizu, N., 203, 215, *241*, 262, *283*
Shimizu, Y., 81, *113*, 603, 624, *640*, *641*
Shimooda, I., *242*, 397(269d), *426*
Shlyapintokh, V. Y., 140, *170*
Shore, P. A., 602, *637*
Shortridge, T. J., 232, *238*
Sih, C. J., 434, *506*
Silber, P., 447, *506*
Simamura, O., 79, *112*
Simon, M. I., 490, 495, *509*
Simopoulos, A. P., 162, *170*
Simpson, F. J., 619, *639*
Simpson, G. A., 251, 269, *283*
Sims, P., 626, *637*
Singer, L. A., 187, *240*
Singh, P., 163, 167, *170*
Skold, C. N., 503, *509*
Sligar, S. G., 627, *641*
Slobodetskaya, E. M., 582, *594*
Sloneker, J., 562, *570*
Slusser, P., 209, *241*, 270, *285*
Small, R. D., 149, 159, 162, 166, 167, *170*
Small, R. D., Jr., 162, *170*, 560, *573*
Smetana, R. D., 102, 107, 108, *113*
Smith, C. P., 277, *285*
Smith, E. B., 577, *594*
Smith, J. J., 577, 580, *595*
Smith, L. L., 110, *114*, 316(263), 407(143), *423*, *426*, 619, 623, 626, 629, 631, 632, 636, *639*, *641*
Smith, W. F., Jr., 135, *137*, 149, 150, 152, 153, 156, 157, *170*, 535, *573*
Smith, W. L., 632, *641*
Snelling, D. R., 36, 44, 46, 53, *57*, *58*, 125, *136*
Snipes, W., 75, *113*, 491, *506*
Snyder, C. D., 359(264), *426*
Sobby, M. E. D., 537, *571*
Sommerfeld, C.-D., 548, *573*
Song, P.-S., 519, 520, *573*
Sonnenberg, J., 416(265), *426*, 481, 485, *509*
Sosnovsky, G., 552, *573*
Sparfel, D., 520, 526, 529, 530, *572*
Sperling, R., 605, *637*
Spikes, J. D., 89, *114*, 589, 590, *593*, *595*, 604, *641*
Spöttl, R., 633, *637*
Sprecher, H., 627, 628, *640*
Sprecher, M., 352(67b), 353(67b), *420*
Sprung, J. L., 344(266), *426*
Stary, F. E., 108, 109, 110, *114*
Stauff, J., 26, *33*, 61, 85, 86, 111, *114*, 618, 620, *640*, *641*
Steele, R. H., 609, 612, 623, *637*, *638*
Steer, R. P., 82, *114*, 344(266), *426*
Steinberger, R., 273, *285*
Steiner, G., 237, *239*
Steinmetzer, H.-C., 200, *241*, 245, 248, 251, 252, 254, 263, 265, *283*, *285*, *286*
Stelmaszynska, T., 610, 611, *641*
Stephens, E. R., 82, *114*
Stephens, H. N., 177, *241*
Stephenson, L. M., 99, *114*, 321(267), 323 (267), 342(200a, 267), 351(200a), 352 (200a), *426*
Sternson, L. A., 604, 623, 626, *641*
Stevens, B., 149, 159, 162, 166, 167, *170*, 233, 234, *241*, 512, 513, 514, 518, 519, 560, *569*, *573*
Stevens, M. P., 436, *509*
Stewart, J. C. M., 162, *168*, 465, 466, *506*, 591, *592*
Stigall, D. L., 563, *572*
Stiller, K., 416(277), *426*, 482, 484, 485, 489, *509*
Stjernholm, R. L., 609, 612, *637*
Stokstad, E. L. R., 245, *284*
Stone, M. J., 599, *637*
Storianetz, W., 541, *573*
Story, P. R., 254, *285*
Strating, J., 103, *114*, 200, 220, 233, *242*, 245, 246, 252, *286*
Strauss, R. R., 610, 611, *640*, *641*
Strehlow, W., 503, *509*
Strickland, R. C., 434, *508*
Strickler, H., 397(198), *424*
Strutt, R. J., 36, *58*
Stuhl, F., 44, *58*, 125, *137*
Sugioka, K., 81, *113*, 603, 605, 624, 625, *640*, *641*
Suld, G., 568, *573*
Sunami, M., 199, 220, *239*
Sundberg, R. J., 467, *509*
Suprunchuk, T., 42, *57*, 148, 152, 154, 157, 161, *168*, 582, *592*, *593*
Sussenbach, J. S., 490, 496, 497, *509*
Sustmann, R., 225, *241*

Sutherland, J. K., 104, *114*, 302(222), 364 (222), 365(222), 385(222), *425*
Suurmond, D., 591, *593*, *595*
Suwa, K., 636, *641*
Suzuki, J., 199, *238*, 268, *283*, 297(7), 346(7), 347(7), 371(7), *419*
Suzuki, S., 541, 542, 543, *570*
Suzuki, T., 394(128), *422*
Swallow, J. C., 90, *112*, 436, *506*, 601, 631, *637*
Swanbeck, G., 165, *170*
Swenton, J. S., 259, *286*
Sysak, P. K., 321(267), 323(267), 342(267), *426*
Sznajd, J., 610, *641*

T

Taimr, L., 156, *170*
Takahashi, Y., 430, 472, 473, 487, *508*, 532, 536, *573*
Takayama, K., 81, *113*, 602, 603, *640*, *641*
Takei, S., 583, *594*
Takeshita, H., 216, *240*, *242*, 382(168b), 397 (109, 110, 268, 268d), 398(111, 269), *422*, *423*, *426*, 544, 565, *571*, *573*
Tamm, S. W., 552, *571*
Tanaka, K., 273, *286*
Tanemura, M., 394(128), *422*
Tang, C. W., 48, *58*, 317(192), *424*, 535, *572*, 587, *594*
Tanielian, C., *426*
Taube, H., 63, *112*, 615, *637*
Tauber, A. I., 617, *641*
Taylor, A., 475, *508*
Taylor, C. L., 42, *57*, 161, *168*, 580, *592*
Taylor, F. B., 133, *136*
Taylor, W. C., 42, *57*, 63, 94, *112*, 600, 602, *637*
Taylor, W. I., 476, *509*, 552, *573*
Teng, J. I., 316(263), *426*, 623, 626, 629, 631, 632, *641*
Terao, J., 350(269a, 269b), *426*
Terao, S., 182, *242*, 245, 250, 266, *286*, 487, *509*
Thayer, A. L., 84, 101, 102, *114*, 164, *169*, 186, 193, *241*, *242*, 344(19, 227), 370(19), 372 (19), *419*, *425*, 633, *638*, *641*
Theilacker, W., 505, *509*
Thiel, W., 51, 52, *57*, 184, 202, 227, 231, 232, 233, *238*, 247, *284*, 341(43), *420*, 487, *506*
Thomas, A. F., 356(270), 364(193), *424*, *426*
Thomas, M., 153, 159, 160, 166, 167, *168*, *170*, 555, *573*
Thomas, R. G. O., 44, 46, 56, *58*
Thompson, J. A., 254, *283*
Thompson, Q. E., 94, 99, 101, 105, *113*, *114*
Thompson, R. B., 177, *240*
Thrush, B. A., 44, 46, 56, *58*
Timmons, R. B., 301, *422*
Tokoroyama, T., 434, *507*
Toledo, M. M., 158, 162, *169*
Tomida, I., 584, *594*
Tomioka, S., 624, *641*
Tomita, M., 488, *507*
Toniolo, C., 469, *506*
Tontapanish, N., 191, *241*, 249, *285*
Topp, M. R., 135, *137*, 143, 145, *170*
Toube, T. B., 307(274), 380(274), 381(274), *426*
Trave, R., 370(17), 385(17), *419*
Trecker, D. J., 360(5), 388(5), *419*, 542, *569*
Trethewey, K. R., 141, 142, 157, *168*, 215, *238*, 599, *637*
Trifiro, F., 344(23, 23a), *419*
Trozzolo, A. M., 93, *114*, 149, 159, 166, 167, *168*, 294(44), *420*, 434, 435, 457, *509*, 560, *570*, 576, 577, 578, 579, 581, 586, 591, 592, *593*, *594*, *595*, 636, *639*
Tryon, M., 577, 583, 584, *592*, *595*
Tschurdy, D. P., 591, *595*
Tsuji, K., 583, *595*
Tsuji, Y., 81, *113*, 603, 624, *640*
Tsukida, K., 155, *170*, 307(271), 382(271), *426*
Tsunetsugu, J., 330(189), 403(189), *424*
Tsutsumi, Y., 624, *640*
Tu, M.-H., 604, 626, *637*
Turchi, I. J., 202, 231, 233, *238*, 247, *284*
Turck, G., 490, *509*
Turner, A. H., 81, *114*
Turner, D. W., 295, *426*
Turner, J. A., 179, 214, *241*, 391(272a), *426*
Turro, N. J., 111, *114*, 152, *170*, 207, 219, *239*, *241*, *242*, 245, 248, 252, 260, 261, 263, 264, 269, 270, *283*, *284*, *285*, *286*, 577, *595*

U

Uda, H., 541, 542, 543, *570*, *571*
Udenfriend, S., 602, *637*
Uemo, H., 541, 542, 543, *570*
Uhde, G., 292(60), 344(60), 373(196), *420*, *424*
Ukita, T., 488, *507*
Ullman, E. F., 163, 167, *170*, 567, 568, *571*, *573*

Unger, L. R., 342(283), 417(283), *427*
Urry, W. H., 181, *241*
Usubillaga, A., 400(9), 401(9), *419*
Usui, Y., 567, *572*
Utsumi, S., 542, *570*

V

Valentine, D., Jr., 259, *286*
Van Den Engh, R. H., 350(35), *420*, 631, *637*
van de Sande, J. H., 52, *57*, 105, *113*, 178, *240* 245, 252, *284*, 291(141), 330(141), 332 (139), 340, 371(139), *422*, *423*
van de Vorst, A., 636, *637*, *639*
Van Santen, P. J., 42, *58*, 350(210), *424*, 623, 631, *640*
van Stevenick, J., 591, *593*, *595*
Van Tamelen, E. E., 538, *573*
Van Verth, J. E., 497, *509*
Van Vunakis, H., 490, 495, *509*
Vassil'ev, R. V., 255, 256, *283*
Vedeneyev, V. I., 295, *426*
Velluz, L., 221, *239*, 517, 524, *569*
Vermeglio, A., 151, *169*
Vial, C., 364(193), *424*
Vickers, R. S., 565, *573*
Vidaud, P. H., 47, 54, *58*
Vilhuber, H. G., 330(189), *424*
Villarasa, L., *239*
Ville, T. E. de, 307(274), 380(274), 381(274), *426*
Vinick, F. J., 436, 438, 500, *509*
Vogel, E., 548, 549, 550, 551, *573*
von Engel, A., 47, 54, *58*
von Saltza, M. H., 245, *284*
Vos, A., 201, *240*

W

Wacker, A., 441, 490, *507*, *509*
Waddell, W. H., 270, *286*
Wagenaar, A., 220, *242*
Wagner, H. U., 330(275), *426*
Wagner, P. J., 262, *286*
Wahlberg, I., 357(275a), *426*
Wall, L. A., 583, 584, *592*, *595*
Wall, R. G., 72, *112*
Wallbillich, G. E. H., 232, *238*
Wallis, T. G., 153, 158, *170*, 317(204), 318 (204), *424*
Walrant, P., 590, *595*
Walter, J., 23, *32*
Walter, M., 537, *570*
Wamser, C. C., 221, *241*
Wan, J. K. S., 42, *57*, 152, 161, *168*, 580, 582, *592*, *593*
Wang, H.-P., 624, *641*
Warburg, E., 36, *58*
Warren, S. G., 277, *284*
Wasserman, A., 49, *58*
Wasserman, E., 36, *58*, 95, *114*, 582, *595*
Wasserman, H. H., 88, 91, 92, *114*, 182, 185, *241*, *242*, 245, 250, 266, 278, 279, *286*, 345 (276, 279), 360(278), 416(277), *426*, 436, 438, 439, 443, 445, 447, 448, 455, 458, 459, 461, 462, 476, 482, 483, 484, 485, 486, 487, 489, 495, 497, 498, 499, 500, 501, 502, 503, 505, *509*, 519, 520, 521, 522, 525, 526, 538, 568, *573*, 600, *641*
Watanabe, H., 416(181c), *424*
Watanabe, K., 199, *238*, 268, *283*, 297(7), 346 (7), 347(7), 371(7), *419*
Waters, W. A., 64, 78, 79, 81, 110, *113*, *114*, 602, *639*
Wayne, R. P., 54, *58*, 82, *114*, 576, 577, 579, *593*, *594*, *595*
Weaver, L., 72, *112*, 140, 141, 145, 149, 157, *168*, *170*
Webb, L. S., 616, 617, *639*
Weddle, C., 625, *639*
Weedon, B. C. L., 307(274), 380(274), 381 (274), *426*
Wehrly, K., 147, *171*
Wei, C. C., 248, 256, 257, *286*
Wei, K. S., 538, *573*
Weil, L., 482, *509*, 560, *574*
Weiner, P. K., 231, *238*
Weinstein, M., 160, *170*
Weir, N. A., 583, 584, *593*
Weishaupt, K. R., 601, *641*
Weiss, J., 617, *638*
Weissmann, G., 635, *638*
Welge, K. H., 44, 53, *57*, *58*, 125, *136*
Weller, A., 152, *170*
Welstead, W. J., Jr., 434, 437, 439, *507*, 601, *639*
Wennerstein, G., 165, *170*
Went, L. D., 591, *595*

Went, L. N., 591, *593*
Werstiuk, E., 330(189), 403(189), *424*
Weschler, C. J., 90, *114*
Weser, U., 164, *170*, 600, 633, *637*, *640*, *641*
Westfelt, L., 393(97), *421*
Westheimer, F. H., 273, 276, *284*, *285*
Wexler, S., 42, *57*, 62, 63, 64, 65, 67, 68, 69, 70, 87, 98, 108, *112*, 325(54), 326(54), 342(54), 344(53, 57, 280), 371(280), 372(280), 375(54, 280), 390(54), 392(57), 418(54), *420*, *426*, 431, 432, 438, 449, *506*, 548, *570*, 600, *638*
Wharton, P. S., 392(281), *426*
White, D. M., 416(265), *426*, 481, 485, *509*
White, D. W., 276, *284*
White, E. H., 248, 256, 257, *286*, 416(282), *426*, 450, 481, 485, *509*, 563, *574*
White, R., 274, *284*
Whited, E. A., 254, *285*
Whitten, D. M., 608, 616, *637*
Whittle, E., 344(24, 72), *419*, *421*
Widdowson, D. A., 402(185), *424*
Wiecko, J., 248, 256, 257, *286*
Wieringa, J. H., 103, *114*, 200, 201, 233, *240*, *242*, 245, 246, 252, *286*
Wildes, P. D., 248, 256, 257, *286*
Wiles, D. M., 42, *57*, 148, 152, 154, 157, 161, *168*, *170*, 578, 579, 580, 582, *592*, *593*
Wiley, R. A., 604, 623, 626, *641*
Wilkinson, F., 121, 123, *136*, 143, 150, 151, 152, 153, 154, 161, 166, 167, *167*, *168*, *169*, *170*, 290, *419*
Wilkinson, G., 273, *284*
Willhalm, B., 347(195), 397(198), *424*, 504, *509*, 542, *573*
Williams, F. W., 51, 53, *57*, 217, *238*
Williams, J. R., 342(283), 417(283), *427*
Wilms, K., 549, 550, 551, *573*
Wilson, E. B., Jr., 23, *33*
Wilson, L. A., 276, *284*
Wilson, T., 27, *33*, 89, *113*, 149, 152, 153, 154, 157, 166, 167, *169*, *170*, 178, 193, 221, 224, *240*, *242*, 248, 253, 255, 260, 261, 262, 269, 270, *283*, *285*, *286*, 317(201), *424*, 514, 525, 530, 535, *572*, *574*, 576, *595*, 598, 599, 613, 615, 622, *640*, *641*
Windaus, A., 174, *242*
Winer, A. M., 87, 88, 89, 90, *114*, 342(205), 344(284), 369(205), 372(205), *424*, *427*
Winslow, F. H., 577, 579, 581, *595*
Winstein, S., 396, *425*
Winterbourn, C. C., 635, *637*
Wiriyachirta, P., 538, *569*
Witkop, B., 416(181c), 417(181c), *424*, 531, *569*
Witzke, H., 26, *32*, 36, *56*, 60, 61, *112*
Wlodawer, P., 601, 628, 630, *639*
Wohlers, H. E., 87, *114*
Wojciechowska, W., 358(135), *422*
Wolbarsht, M. L., 150, *169*
Wolf, H., 620, *641*
Wong, S.-Y., 181, 218, 225, *239*
Wood, R. W., 36, 37, *58*
Woodward, R. B., 232, *242*, 246, 273, *286*
Worman, J. J., 220, *242*
Wriede, P. A., 260, *284*
Wright, J. J., 475, *506*
Wuesthoff, M. T., 431, 432, 438, 449, *506*
Wyatt, J. F., 157, 164, *167*
Wynberg, H., 103, *114*, 200, 201, 233, *240*, *242*, 245, 246, 252, *286*

Y

Yager, W. A., 36, *58*, 95, *114*, 582, *595*
Yagihara, M., 382(7a), *419*
Yamabe, S., 227, *239*
Yamaguchi, K., 226, *242*
Yamamoto, H., 184, *238*
Yamanashi, B. S., 150, *169*
Yamane, M., 553, 558, 559, 566, *573*
Yamane, T., 591, *594*, 636, *639*
Yamauchi, R., 350(284a), *427*
Yang, N. C., 254, *286*
Yany, F., 219, *241*, 251, *283*, 413(3a), *419*
Yaron, M., 47, 54, *58*
Yates, R. L., 225, *239*
Yeats, R. B., 393(97), *421*
Yekta, A., 269, 270, *285*, *286*
Yelvington, M. B., 209, *241*, 251, 257, 258, 264, 265, 266, 270, *285*, 563, *572*
Yerich, S. J., 612, *637*
Yokota, M., 155, *170*, 307(271), 382(271), *426*
Yoshida, H., 77, *113*, *423*
Yoshida, S., 349(28, 29), *419*
Yoshikawa, K., 416(178), 417(180), *423*, 473, 474, 475, *508*, 563, *572*
Yoshikawa, M., 383(101), 384(101), 385(101), 396(101), *422*

Yoshimura, N., 92, 105, *113*, 532, 534, 553, 554, 556, 557, 562, *571*, *573*
Yost, F. Y., Jr., 616, *641*
Younes, M., 164, *170*, 600, 633, *641*
Young, D. W., 411(66, 67), *420*
Young, M. R., 348(10), 366(10), *419*
Young, R. H., 92, *114*, 143, 147, 152, 153, 157, 166, 167, *170*, *171*, 236, *242*, 290, 317 (286), *427*, 436, *509*, 535, 564, *574*, 588, *595*

Z

Zaklika, K. A., 187, 193, 197, 206, 223, 224, *240*, *241*, *242*
Zaret, E. H., 552, *573*
Zgliczynski, J. M., 607, 610, 611, *640*, *641*
Zia, A., 44, *57*, 125, *136*
Ziegler, K., 174, *241*
Ziffer, H., 531, *574*
Zimmerman, H. E., 205, *242*, 250, 259, *286*
Zincke, H., 282, *284*
Zwanenburg, B., 220, *242*
Zweig, A., 161, 166, 167, *171*, 534, *574*, 587, 588, *595*
Zwicker, E. F., 160, *171*, 553, *570*, 589, *593*, 553, *574*

Subject Index

A

Acceptor half-value concentration, 289, 290, *see also* β-value
Acenaphthene, singlet oxygen reaction, 521
Acenes
  aza analogs, 515
  endoperoxides, 517–531
  peroxides, thermal decomposition of, as source of singlet oxygen, 344, 345
  photocarcinogenesis with, 514
  reactivity index (β-value), 514
  singlet oxygen reactions, 514–528
Acetylenes, singlet oxygen reactions, 219
N-Acetyltyramine, singlet oxygen reaction, 558
Activation energy, for gas phase and liquid phase ene reactions, 298, 300, 301
Adamantylideneadamantane
  dioxetane production in, 103
  singlet oxygen reaction, 200
Adrenodoxin, 627
Adrenodoxin reductase, 627
Alkaline phosphatase, 608
Alkenes, singlet oxygen reactions, 246–254
Alkylbenzenes, singlet oxygen reactions, 531
Allenes, singlet oxygen reactions, 219
Allenic alcohols, from ene reactions, 377, 380–382
Alloxan, 634
Allylic alcohols
  epimeric, 354, 358, 365, 373–380, 388, 390, 391, 393–396, 400–403, 406, 410
  reactions with singlet oxygen, 328, 405, 406, 408
  survey, 342–417
  axial, in ene reactions, 318, 319, 325, 326, 329
  quasi-axial, in ene reactions, 324, 333
Allylic hydroperoxides
  Δ-isomeric from fatty acid esters, 350, 351
  epimeric, 164, 333–334. 396, 404, 407, 410, 413
  equilibrium conformation, 321
  with exocyclic double bonds, 309, 324, 326
  mechanism of formation, 287–288
  methods of reduction, 342–417
  reactions with singlet oxygen, 303–307, 348, 366, 392, 409, 410
  survey, 342–417
Allylic isomerization, 328, 390, 401, 404–407, 411
Amines
  catalysts for dioxetane decomposition, 270
  quenching of singlet oxygen, 157–159, 166, 535, 587
Amino acids, 589, 590
  photodynamic effect, 589, 590
L-Amino acid oxidase, 624
5-Amino-2,3-dihydro-1,4-phthalazinedione, 612, 613, 628, 629
6-Aminodopamine, 634
Aminopeptidase, 608
3-Amino-1,2,4-triazole, 628
Anisole, singlet oxygen reaction, 533, 534
6-(*p*-Anisyl)-3,4-dihydro-2*H*-pyran, singlet oxygen reaction, 189
[14]Annulene, singlet oxygen reaction, 550
[$4n + 2$]Annulene, singlet oxygen reaction, 549

Anthracenes, 602, 604, 632
aza analogs, 515
β-elimination of 9,10-endoperoxide, 528
derivatives, 512, 514, 515
singlet oxygen reactions, 514–516
Anthracene bisepoxide, 526
Anthracene compounds
oxygenation of
with hydrogen peroxide-bromine, 78
with triphenylphosphite ozonide, 94
Anthracene 9,10-endoperoxide, 515, 526
reaction with base, 528
Anthraquinone, 604, 632
2-(2-Anthryl)-1,4-dioxene, singlet oxygen reaction, 196–197
Arachidonic acid, 628, 629
Arene oxides, singlet oxygen reactions, 548
4-Arylazo-1-naphthols, singlet oxygen reactions, 566, 567
Arylfurans, singlet oxygen reactions, 434
Arylpyrroles, singlet oxygen reactions, 455
Ascaridole, 307, 337, 339
from α-terpinene, 174
Aspirin, 628
Autoxidation
of cyclohexene, 177
of enols, 177
formation of α-hydroperoxy ketones from enols, 177
of olefins, *see* Radical reactions
of tetrakis(dimethylamino)ethylene, 181
1-Azaanthracenes, singlet oxygen reactions, 516
2-Azaanthracenes, singlet oxygen reactions, 516
5-Azanaphthacene, singlet oxygen reaction, 516
9-Azanaphthacene, singlet oxygen reaction, 516
Azide, quenching of singlet oxygen, 144, 163, 167
Azide ion, trapping of intermediates with, 209, 210
Azido alcohols, 337, 338, 344, 345, 372, 388
Azido hydroperoxides, 345, 346
mechanism of formation, 336–340
Azodioxide, quenching of singlet oxygen, 163
Azomethine dye, quenching of singlet oxygen, 156, 157
Azulene, singlet oxygen reaction, 514

B

1,2-Benzanthracene, singlet oxygen reaction, 515
Benzene, singlet oxygen reaction, 584
Benzene oxide, oxepine tautomeric system, single oxygen reaction, 548
trans-Benzene trioxide, formation from benzene oxide, 548–549
Benzhydrylidenecyclobutane, singlet oxygen reaction, 540
1,2-Benznaphthacene, singlet oxygen reaction, 516
Benzofurans, singlet oxygen reactions, 194, 441–443
Benzo[*a*]pyrene, 628, 629
Benzopyrones, singlet oxygen reactions, 567, 568
Benzoxazines, singlet oxygen reactions, 471, 472
1-Benzoxepin, singlet oxygen reaction, 549
Benztropolone, methyl ether, singlet oxygen reaction, 546
4-Benzyl-2,6-di-*tert*-butylphenol, singlet oxygen reaction, 556
Benzylideneacridane, singlet oxygen reaction, 188
β-value, 68–69
definition, 289
effect of solvent on, in photoxygenation, 70, 71
measure of substrate reactivity, 68
Biadamantylidenes, singlet oxygen reactions, 246, 247
Bicarbonate radicals, 620
9,9′-Bifluorenylidene
oxygenation of, with hydrogen peroxide–hypochlorite, 75
photooxidative cleavage of, 205
singlet oxygen reaction, 522, 524
Bile acid, 623
Bilirubin, 591, 592
quenching of singlet oxygen, 162, 615, 629, 630
reaction with singlet oxygen, 463, 592
sensitizer in neonatal jaundice, 591
Biliverdin, quenching of singlet oxygen, 162

7,7′-Binorbornylidene, singlet oxygen reaction, 202
Bipolymers, 589, *see also* specific substances
photodegradation, 589–592
4,5-Bis(*N*,*N*-dimethylamino)-*o*-xylene, singlet oxygen reaction, 536
Bisepoxides, 338
formation in singlet oxygen reactions, 434, 437, 447, 480
Bishydroperoxide, 409
Bis-1,4-peroxides, 372
4-Bromo-2,6-di-*tert*-butylphenol, singlet oxygen reaction, 556
Bromo hydroperoxides, 178, 245–246, 340–341
sec-Butyl hydroperoxides, 603
sec-Butylperoxy radical, singlet oxygen from self reaction of, 80

C

Camphenylideneadamantane, singlet oxygen reaction, 203
Camphor, 627
Caoutchouc decay of, 576
Carbonate radicals, 620
Carotene, quenching of singlet oxygen, 140, 154, 155, 166, 613, 615, 621, 623, 625, 628, 635
β-Carotene, 291, 307, 382
quencher of singlet oxygen, 72–73
Δ^3-Carene, singlet oxygen reaction, 310, 390
$\Delta^{4(10)}$-Carene, singlet oxygen reaction, 313, 361
Carotenoids, 307, 377–382
Caryophyllene, singlet oxygen reaction, 300, 395
Catalase, 605, 614, 616–618, 621, 624, 627, 632–635
Cellulose, 588
singlet oxygen reaction, 588
Ceric ion, 602, 603
Chemiluminescence, 603–606, 609, 610, 612, 613, 620, 623, 624, 627, 628, 632, 633
from acene endoperoxides, 525, 530
circularly polarized, from 1,2-dioxetane, 201
from dibenzal diperoxide, 525
from 1,2-dioxetanes, 178, 187, 188, 255–263
by singlet oxygen, 25–27
8-Chloro-1-methoxynaphthalene, singlet oxygen reaction, 521
Chloroperoxidase, 611
m-Chloroperoxybenzoic acid, 601, 602
Chlorpromazine, 628
2,4-Cholestadiene, oxygenation of, with hydrogen peroxide–hypochlorite, 64
Cholesta-4,6-dien-3-one, 623
Δ^5-cholestenes
electronic effects in ene reaction, 304
singlet oxygen reactions, 293, 304, 316, 318, 406–408
Cholesterol, 619, 623, 629, 631, 632, 636
singlet oxygen reaction, 592
Cholesterol 3β,7-diol, 623
Cholesterol 7-hydroperoxides, 623, 631, 632
Chromium pentoxide etherate, decomposition of, as possible source of singlet oxygen, 90
Chronic granulomatous disease, 608, 616, 618
Cumene hydroperoxide, 629
Curtin–Hammett principle, 303, 321
Cyanine dye, quenching of singlet oxygen, 156
8-Cyanoheptafulvene, singlet oxygen reaction, 545
Cyclic acetylenes, oxidation of, as possible singlet oxygen sources, 111
1,2-Cycloaddition of singlet oxygen, 296, 299, 309, 328, 341, 397
competition with ene reaction, 290, 308, 346, 347, 357, 361, 368, 369, 370, 371, 388, 389, 396, 404
electronic effects on, 196, 235–237
linear free energy relationships of, 196, 235, 236
mechanisms of, 224–237
MINDO/3 calculations for, 227–231
molecular orbital calculations for, 224–237
orbital correlation diagrams of, 225
solvent effects on, 296, 297, 346, 347, 370, 371
substituent effects on, 196, 204, 235–237
to thioethylenes, 192–199
to vinyl ethers, stereochemistry of, 190–192, 232
1,4-Cycloaddition of singlet oxygen, 300–301, 307, 348, 366, 377, 391, 392, 394, 410
activation energy for, 307

activation entropy for, 307
competition with ene reaction, 307, 308, 345, 348, 351, 352, 358, 372, 381, 392, 394, 404
to 1,1-diphenyl-2-methoxyethylene, 196
stereospecific, 317–318
to styrenes, 212
switch-over to ene reaction by introduction of methyl group, 316
to vinylanthracenes, 197
to vinylnaphthalenes, 213
to vinylthiophenes, 214
Cyclobutenes, singlet oxygen reactions, 208
Cycloheptatriene, singlet oxygen reaction, 551
Cyclohepta-2,4,6-triene-1-carboxylate, singlet oxygen reaction, 551
1,3-Cyclohexadiene
oxygenation of
with hydrogen peroxide–hypochlorite, 63
with triphenylphosphite ozonide, 94
Cyclohexene, autoxidation, 177
Cyclohexylidenecyclohexanes
ene reactions of, 333–334, 364
singlet oxygen reactions, 176, 217
1,5-Cyclooctadiene, oxygenation of, with hydrogen peroxide–hypochlorite, 77
Cyclooxygenase, 627
Cyclopentadiene, singlet oxygen reaction, 541
Cyclophanes, naphthalene analogs, singlet oxygen reactions, 519
Cyclopropanone, intermediate in fulvene-singlet oxygen reaction, 542
Cyclopropenes, singlet oxygen reactions, 207
Cytochrome *c*, 614, 621, 631
Cytochrome *P*-450, 604, 624, 626, 627

D

DBB, *see o*-Dibenzoylbenzene
DBE, *see cis*-Dibenzoylethylene
DBS, *see cis*-Dibenzoylstilbene
Delayed fluorescence, *see* Fluorescence, delayed
Depside, 619
Determinantal wave functions, oxygen, 16–18, 22
7-Deuteriocholesterol, singlet oxygen reaction, 318, 319
(*R*)(−)-*cis*-2-Deuterio-5-methyl-3-hexene, singlet oxygen reaction, 321, 335
Deuterium isotope effect
allylic H/D-abstraction, 323, 330
H/D-abstraction by CH_3OH, 323
Deuterium oxide, 606–615, 618, 625, 628–633
Diabetes, 634
Diacetylfilicinic acid, singlet oxygen reaction, 564
α,α′-Dialkylstilbenes, substituted, singlet oxygen reactions, 291, 297, 356–357
Dialkyl sulfides, oxidation of, with triphenylphosphite ozonide, 102
Dialuric acid, 634
1,1-Dianisyl-2-methoxyethylene, singlet oxygen reaction, 196
1,1′-Dianthracene, singlet oxygen reaction, 516
9,9′-Dianthracene, singlet oxygen reaction, 516
2,3-Di(2-anthryl)-1,4-dioxene, 197
1,4-Diazabicyclo[2.2.2]octane
quenching of singlet oxygen, 140, 153, 157, 158, 166, 201, 202, 215, 614, 615, 621, 623, 624, 627, 628, 632
from potassium perchromate, 89
from superoxide ion, 85
1,3-Diazaanthracene, singlet oxygen reaction, 516
1,2,5,6-Dibenzanthracene, singlet oxygen reaction, 515, 516
1,2,5,6-Dibenznaphthacene, singlet oxygen reaction, 516
1,4-Dibenzyloxyanthracene, singlet oxygen reaction, 517
o-Dibenzoylbenzene, 600, 601, 621, 626, 629, 631, 632
cis-Dibenzoylethylene, 601–603, 615, 626, 632
cis-Dibenzoylstilbene, 601, 632
1,4-Dibenzylanthracene 1,4-peroxide, 526
3,5-Di-*tert*-butylcatechol, singlet oxygen reaction, 560
3,5-Di-*tert*-butylcatechol, dimethyl ether, singlet oxygen reaction, 534, 560
3,6-Di-*tert*-butylguaiacol, singlet oxygen reaction, 560
4,6-Di-*tert*-butylguaiacol, singlet oxygen reaction, 560
2,5-Di-*tert*-butylhydroquinone, singlet oxygen reaction, 557
2,5-Di-*tert*-butylhydroquinone, dimethyl ether, singlet oxygen reaction, 534, 557

2,6-Di-*tert*-butyl-4-hydroxymethylphenol, singlet oxygen reaction, 556
2,5-Di-*tert*-butyl-4-methoxyphenol, singlet oxygen reaction, 557, 558
4,6-Di-*tert*-butyl-3-methoxyphenol, singlet oxygen reaction, 556
2,6-Di-*tert*-butyl-4-methylphenol, singlet oxygen reaction, 553, 554, 556
2,6-Di-*tert*-butylphenol
 oxygenation of, with triphenylphosphite ozonide, 105
 singlet oxygen reaction, 553, 554
  4-substituted, 553, 554, 556
4,6-Di-*tert*-butylresorcinol, singlet oxygen reaction, 534, 562
4,6-Di-*tert*-butylresorcinol monomethyl ether, oxygenation of, with hydrogen peroxide–hypochlorite, 76, 77
9,10-Dichloroanthracene, singlet oxygen reaction, 517
1,4-Dichloro-9,10-diphenylanthracene, singlet oxygen reaction, 517
1,1-Dicyclopropylethylenes, singlet oxygen reactions, 335, 352
Dienes
 acyclic, in ene reactions, 345–351
 cyclic, in ene reactions, 263, 377, 383–385, 391, 392
1,3-Dienes, singlet oxygen reactions, 174
1,4-Diethoxyanthracene, singlet oxygen, 517
1,4-Diethoxy-9,10-diphenylanthracene, singlet oxygen reaction, 517
cis-Diethoxyethylene, singlet oxygen reaction, 178, 191
cis- and *trans*-Diethoxyethylene, oxygenation with triphenylphosphite ozonide, 96
1,4-Diethyl-5,8-dimethylnaphthalene, singlet oxygen reaction, 521
Dihydrohexamethyl(Dewar)benzene, singlet oxygen reaction, 179
1,2-Dihydronaphthalenes, singlet oxygen reactions, 211, 308, 539
Dihydropyran, singlet oxygen reaction, 195
cis-9,10-Dihydroxy-9,10-dimethyl-1,2-benzanthracene, 626
1,4-Dihydroxynaphthalene
 dimethyl ether, singlet oxygen reaction, 520
 diphosphate, singlet oxygen reaction, 520
3-(3,4-Dihydroxyphenyl)alanine (DOPA), singlet oxygen reaction, 560
6,7-Dihydroxytryptamine, 634
1,4-Diisopropyl-5-methylnaphthalene, singlet oxygen reaction, 521
Dimercaptopropanol, 628, 629
Di-π-methane rearrangement, from 1,2-dioxetane thermolysis, 205, 206
1,4-Dimethoxyanthracene, singlet oxygen reaction, 517
9,10-Dimethoxyanthracene, singlet oxygen reaction, 515
1,2-Dimethoxybenzenes, singlet oxygen reactions, 534
1,3-Dimethoxybenzenes, singlet oxygen reactions, 534
1,4-Dimethoxybenzenes, singlet oxygen reactions, 534
2,4-Dimethoxy-*N*,*N*-dimethylaniline, singlet oxygen reaction, 536
1,4-Dimethoxy-9,10-diphenylanthracene, singlet oxygen reaction, 517, 524
2,3-Dimethoxy-9,10-diphenylanthracene, singlet oxygen reaction, 517
1,4-Dimethoxy-9,10-diphenylanthracene 1,4-peroxide, reaction with acid, 530
1,4-Dimethoxy-5,8-diphenylnaphthalene, singlet oxygen reaction, 520
9,10-Dimethoxyphenanthrene, singlet oxygen reaction, 190
α,α'-Dimethoxystilbene, singlet oxygen reaction, 189
1,1-Dimethoxy-2,4,6-tri-*tert*-butylphosphorine, singlet oxygen reaction, 537
1,1-Dimethyl-2-alkylethylene, singlet oxygen reaction, 319
1-Dimethylaminoanthracene, singlet oxygen reaction, 517
1-(*N*,*N*-Dimethylamino)-9,10-diphenylanthracene 1,4-peroxide, thermal reaction, 530–531
1-Dimethylamino-4-methyl-9,10-diphenylanthracene, singlet oxygen reaction, 517
N,*N*-Dimethylanilines, substituted, singlet oxygen reactions, 535
Dimethylanthracene, 604
9,10-Dimethylanthracene, singlet oxygen reaction, 514, 515
Dimethylanthracene 9,10-endoperoxide, 604
 thermolysis, 528
9,10-Dimethylanthracene trioxide, as possible source of singlet oxygen, 106

9,10-Dimethyl-1,2-benzanthracene, 626
singlet oxygen reaction, 513
10,10′-Dimethyl-9,9′-biacridylidene, singlet oxygen reaction, 177, 187, 522, 524
2,3-Dimethyl-2-butene
oxygenation of
with hydrogen peroxide–hypochlorite, 63, 67
with potassium perchromate, 87
with triphenylphosphite ozonide, 94
1,2-Dimethyl-1-cyclohexene
oxygenation of, with triphenylphosphite ozonide, 97
singlet oxygen reaction, 291, 336, 337, 371, 372
1,4-Dimethyl-9,10-diphenylanthracene, singlet oxygen reaction, 517
2,3-Dimethyl-9,10-diphenylanthracene, singlet oxygen reaction, 517
6,6-Dimethylfulvene, singlet oxygen reaction, 541
Dimethylfuran, 621, 625
oxygenation of, with hydrogen peroxide–hypochlorite, 63, 64
singlet oxygen reaction, 621, 625
2,5-Dimethyl-2,4-hexadiene, 215
reaction rate constants, 296
1,4a-Dimethyl-5,6,7,8,8a-*trans*-hexahydronaphthalene, singlet oxygen reaction, 316
1,2-Dimethylnaphthalenes, singlet oxygen reactions, 519
2,4-Dimethyl-2-pentene, singlet oxygen reaction, 319
α,α′-Dimethylstilbenes, singlet oxygen reactions, 297, 356–357
Dimol emission, solvent isotope effect on, 122
1,4-Dioxene, singlet oxygen reaction, 188, 193
1,2-Dioxetanes, *see also* 1,2-Cycloaddition of singlet oxygen
from adamantylideneadamantane, 198
amine-catalyzed decomposition, 270
from 2-(3-anthryl)-1,4-dioxene, 197
from arylcyclopropenes, 207
from benzofurans, 194
from benzylideneacridane, 188
β-elimination in, 182, 183
from biadamantylidene, 246, 247, 252, 263
from camphenylideneadamantane, 203
catalysts for decomposition of, 269–275
chain decomposition, 269
chemiluminescence from decomposition of, 255–263
from *cis*- and *trans*-diethoxyethylene, 178, 190–192
from *cis*- and *trans*-α,α′-dimethoxystilbene, 192
from 10,10′-dimethyl-9,9′-biacridylidene, 187
from 2,5-dimethyl-2,4-hexadiene, 215
from 1,4-dioxene, 193
from 1,3-dioxole, 193
eclipsing strain in, 245
from enamines, 180–188
from *cis*- and *trans*-ethoxyphenoxyethylene, 191
excited carbonyl products from, 255–263
formation
in gas phase, 217–218
in singlet oxygen reactions, 246–253, 290, 297, 333, 438, 440, 441, 464, 465, 480, 487, 500, 522, 524, 530, 536, 540, 544, 560, 563, 565
gas phase reactions of, 52
from indenes, 210, 211
mechanisms for chemiluminescent decomposition of, 263–266
metal ion catalyzed decomposition, 270–275
from 1-methylene-4,4-diphenyl-2,5-cyclohexadienes, 205
optically active, 201, 203
physical properties of, 244–246
preparation of, 245–254
reactions with boron trifluoride, 280–283
with divalent sulfur compounds, 278–279
with stannous compounds, 280
with trivalent phosphorus compounds, 274–277
rearrangements accompanying decomposition of, 266–269
from rearrangement of endoperoxides, 221–224
table, 248–253
from tetramethoxyethylene, 178, 193
thermal rearrangements of, 266–269
from tri-*tert*-butylcyclobutadiene, 217
from trimethylcyclobutadiene with triplet oxygen, 243, 244
X-ray structure of, 201
Dioxetanones, from ketenes, 220

1,2-Dioxetenes, from strained acetylenes, 218
1,3-Dioxole, singlet oxygen reaction, 193
Dioxygen, electronic structures of, 16, 22
Diperoxychromium(VI) oxide etherate, 601
Diphenylanthracene, 602–604
1,4-Diphenylanthracene, singlet oxygen reaction, 517
9,10-Diphenylanthracene, singlet oxygen reaction, 514, 515, 517, 524
9,10-Diphenylanthracene 9,10-endoperoxide, 602–603
  reaction with acid, 530
  as singlet oxygen source, 525
Diphenylbenzopyrylium oxide, singlet oxygen reaction, 568, 569
1,2-Diphenylcyclopentene, singlet oxygen reaction, 206
2,3-Diphenyl-1,4-dioxene, singlet oxygen reaction, 188
2,6-Diphenyl-4-diphenylmethylene-4*H*-thiopyran, singlet oxygen reaction, 199
Diphenylfuran, 232, 233, 601–603, 615, 626
Diphenylisobenzofuran, quenching of singlet oxygen, 164
1,3-Diphenylisobenzofuran, 435, 436, 600, 623, 626–629, 631–634
  as detector of singlet oxygen, 84
  singlet oxygen reaction, 117, 118, 435, 436, 515
Diphenylketene, singlet oxygen reaction, 219
1,1-Diphenyl-2-methoxyethylene
  1,4-cycloaddition, 196
  singlet oxygen reaction, 539
1,2-Diphenyl-1-morpholinoethylene, singlet oxygen reaction, 182
5,6-Diphenyluracil, singlet oxygen reaction, 565
Diradicals, as intermediates in ene reactions, 332
Diterpenes, ene reactions of, 366, 396
DMA, *see* Dimethylanthracene
DMF, *see* Dimethylfuran
Double bond shift, in ene reaction, 287, 288, 303–307
Double-molecule sensitization, singlet oxygen, 25, 28
DPA, *see* Diphenylanthracene
DPBF, *see* 1,3-Diphenylisobenzofuran
DPF, *see* Diphenylfuran
Dyes, sensitization of, 24–31

E

8,11,14-Eicosatrienoic acid, 627–629
5,8,11,14-Eicosatetraynoic acid, 628
Elastase, 590
  photodynamic effect, 590
Electric discharge, generation of singlet oxygen, 218
Electrolysis of azide ion in presence of singlet oxygen acceptors and triplet oxygen, 337, 338
Electron repulsion, in oxygen states, 18–20
Electron transfer mechanisms, in singlet oxygen reactions, 182, 225
Enamines, singlet oxygen reactions, 180–188, 563
Enamino ketones, singlet oxygen reactions, 184
Endoperoxides, 345, 348, 351, 352, 358, 366, 372, 377, 381, 391, 392, 394, 410, *see also* specific compounds
  of acenes, 517–531
  acid-catalyzed reactions of, 528, 530
  activation energy of thermal decomposition, 525–526
  from 2-(2-anthryl)-1,4-dioxene, 196, 197
  base-catalyzed reactions of, 528
  chemiluminescence, 525
  from 1,3-dienes, 212
  from ergosterol, 174
  formation in singlet oxygen reactions, 431, 432, 434, 443–445, 449, 452, 457, 464, 479, 480, 495, 500, 501
  from polyarylfulvenes, 222
  rearrangement to 1,2-dioxetanes, 221–224
    with silica gel, 223
  of rubrene, reaction with acid, 529
  singlet oxygen formation from, 524, 549
  of styrene-type compounds, 539
  of substituted benzenes, 531–538
  from vinylanthracenes, 197
  from vinylnaphthalenes, 213
  from vinylthiophenes, 214
Ene reactions
  competition with 1,2-cycloaddition of singlet oxygen, 307, 308, 346, 347, 361, 368–371, 388, 389, 396, 404
  with 1,4-cycloaddition of singlet oxygen, 307, 308, 345, 351, 352, 358, 372, 381, 392, 394, 404

conformational effects in, 318–327, 329, 333, 334
electronic effects in, 297–309, 329
gas phase, 49
general aspects of, 287, 288
mechanisms of, 287, 288, 296–298, 318, 329–341
potential energy changes in, 331, 332
rate constants
for Δ^3-carene, 310
for $\Delta^{4(10)}$-carene, 314
for Δ^5-cholestenes, 304
correlation with CT absorption maxima, 299
determination of, 288–290
for 1,1-dimethyl-2-alkylethylenes, 319
for $\Delta^{8(14)}$-ergostenes, 305
for 1-methylcycloalkenes, 324
for methylenecyclopentane, 313
for 2-methyl-2-pentene, 300, 301
for monoolefins, 291–295, 299
for norbornenes, 312, 313
for α-pinene, 311, 313
for β-pinene, 313
secondary products in, 328, 343, 348, 366, 392, 404, 406, 407, 410
solvent effects in 290–297
steric effects in, 305, 309–319, 320, 329
substitution effects in, 319, 320
survey (tabular), 342–417
transition state in, 298, 330–333
Energy pooling reactions, gas phase, 54–56
Enols
autoxidation, 177
singlet oxygen reactions, 563–569
Enol ethers, ene reactions of, 342, 346, 347, 351, 352, 371, 403, 413, 416
Eosin, 614
1,2-Epoxides, 329, 338, 363, 377, 378, 380, 382, 391, 400, 405, 408
formation in singlet oxygen reactions, 201–203, 232, 247, 254, 434, 437–439, 444, 456
$\Delta^{8(14)}$-Ergostenes
electronic effects in ene reactions, 305
singlet oxygen reactions, 304, 305
Ergosterol, singlet oxygen reaction, 174
Erythropoietic protoporphyria (EPP), 590, 591
β-carotene in treatment of, 591
photochemolysis, 591
photosensitivity, 591
4-Ethoxycarbonyl-2,6-di-*tert*-butylphenol, singlet oxygen reaction, 555
β-Ethoxystyrene, singlet oxygen reaction, 188
Ethylene–carbon monoxide copolymer, 579
Ethylenes, substituted, in ene reactions, 342–359
Ethyl sulfide
quenching of singlet oxygen, 623
1-Ethylthio-2-ethyl-1-hexene, singlet oxygen reaction, 297, 347
4-Ethyl-2,6,7-trioxa-1-phosphabicyclo[2.2.2]-octane ozonide, singlet oxygen from, 101

F

Fatty acids
ester, ene reaction of, 350, 351
unsaturated, 634
Fenton's reagent, 602
9,9′-Fluorenylideneanthrone
photooxidative cleavage of, 205
Fluorescein, oxygen-sensitized chemiluminescence of, 29
Fluorescence, delayed, 29
sensitization by, 29
triplet-triplet induced, 30, 31
Free radicals, 626, 630–632, 634, 636
Frontier orbitals, interaction in ene reactions, 331
Fulvenes, singlet oxygen reactions, 541
[2.2](2.5)Furanonaphthane, singlet oxygen reaction, 522, 523
Furanophanes, singlet oxygen reactions, 443, 444
Furans, singlet oxygen reactions, 431–447
Furocoumarins, singlet oxygen reactions, 441

G

Gas phase quenching rate constants
in singlet oxygen ($^1\Sigma_g^+$) reactions, 43–46
with alcohols, 46
with alkanes, 46
with amines, 46
with halides, 46
with H_2, 44–46
with HBr, 44–46
with H_2O, 45, 46

Gas phase reaction rate constants
  in singlet oxygen ($^1\Delta_g$) reactions, 46–54
  with amines, 47
  with H_2, 46, 47
  with H_2O, 46, 47
  with sulfides, 46
Germacrene
  oxygenation of, with triphenylphosphite ozonide, 104
  singlet oxygen reaction, 300, 385
Glutathione peroxidase, 634
Gutta-Percha, 576, 585

H

Haber–Weiss reaction, 617–622, 624, 628, 634, 635
Half-wave oxidation potential
  of methoxybenzenes, 534
  of substituted phenols, 555
Hammett correlation for ene reactions, 297, 298, 355
Helianthrene, singlet oxygen reaction, 516
Hemoglobin, 635
Hemolysis, 635
Hexacenes, singlet oxygen reactions, 512
Hexamethoxybenzene, singlet oxygen reaction, 532, 534
Histidine, 614, 632, 633
  quenching of singlet oxygen, 164
Hock reaction, of allylic hydroperoxides, 175–176, 179, 181
2,3-Homotropone, singlet oxygen reaction, 547
Horseradish peroxidase, 601, 604, 611, 632
Hydrogen peroxide, 604–606, 608–622, 624, 625, 628, 632–635
  dismutation, as singlet oxygen source, 110
Hydrogen peroxide–bromine system, 78, 79
  chemiluminescence, 78
Hydrogen peroxide–hypochlorite system, 60–78
  chemiluminescence in, 60–62
  mechanism of production of singlet oxygen, in, 63, 64
  product selectivities in oxygenation, with, 65
  quenching of singlet oxygen in, 72
  source of singlet oxygen, 62–78
  spectroscopic investigations in, 60–62
Hydroperoxides, 580, 584, 588, *see also* Ene reactions, Allylic hydroperoxides
  decomposition of, 584
  from olefins, 174
  products in reactions of singlet oxygen with polydienes, 588
15-Hydroperoxy-5,8,11,13-eicosatetraenoic acid, 629
Hydroperoxy nitrones, 415, 416
Hydrotrioxides, singlet oxygen from, 107–110
9-Hydroxy-10-alkylanthracenes, singlet oxygen reactions, 563
3β-Hydroxy-5α-cholest-6-ene 5-hydroperoxide, 619, 623, 631, 632
6-Hydroxydopamine, 634
3-Hydroxyflavones, singlet oxygen reactions, 565, 566
3β-Hydroxy-5α-hydroperoxy-Δ^6-cholestene, 591
  as product of singlet oxygen on cholesterol, 591
p-Hydroxyphenethyl alcohol, singlet oxygen reaction, 558
p-Hydroxyphenylacetic acid, singlet oxygen reaction, 558
p-Hydroxyphenylpyruvic acid, singlet oxygen reaction, 559, 566
Hydroxyl radical, 605, 612–619, 621, 622, 624, 626, 628, 634, 638
Hypoxanthine, 635

I

Imidazoles, singlet oxygen reactions, 481–490
1,6-Imino[10]annulene, singlet oxygen reaction, 550
Indenes, singlet oxygen reactions, 208–214, 539, 540
Indoles, singlet oxygen reactions, 467–481
Indomethacin, 628, 629
Inductive effect, negative, of allylic OR groups, 304
Inflammatory effects, 617
Iodide
  quenching of singlet oxygen, 163, 167
Ionization potentials, 318, 325
  correlation with rate of ene reaction for monoolefins, 299, 300
  for various monoolefins, 291–295

π-Ionization potentials of methoxybenzenes, 534
β-Ionone, 307, 379, 380
Iron, 603, 617, 620–622, 624, 626
Isocaryophyllene, singlet oxygen reaction, 300, 395
Isoeugenol, singlet oxygen reaction, 562
6-Isopropylfulvene, singlet oxygen reaction, 543
Isosylvestrene, singlet oxygen reaction, 300, 373, 374
Isotetraline, 307, 392

J

Jaundice, neonatal, 591, 592
  bilirubin in, 591
  phototherapy, 591

K

Ketones, singlet oxygen reactions, 103
α-Keto acids, singlet oxygen reactions, 443
7-Ketocholesterol, 623
Ketones
  excited from dioxetane cleavage, 255–263
  photochemistry, 578

L

Lactoperoxidase, 601, 632
Leukocytes, 616, 617
  polymorphonuclear, 605–610, 612–614, 616–619
Lewis acids, reactions with 1,2-dioxetanes, 270–275, 280–283
Lewis structures for molecular oxygen, 2–4
Limonene, singlet oxygen reaction, 326, 327, 375–377
(+)-Limonene, oxygenation of, with hydrogen peroxide–hypochlorite, 65
Linoleic acid, 603, 621, 631
Linoleic acid hydroperoxide, 603
Lipid hydroperoxides, 603, 624, 626, 629, 634
Lipid peroxidation, 619, 621–626, 634
Lipid peroxy radicals, 625
Liposomes, 635, 636
Lipoxidase, 630, 631
Lipoxygenase, 601
Luminescence, from dioxetane decomposition, 255, 256
Luminol, *see* 5-Amino-2,3-dihydro-1,4-phthalazinedione
Lumisteryl acetate, oxygenation of, with hydrogen peroxide–hypochlorite, 64
Lysozyme, 590, 608
  eosin-sensitized photoinactivation, 590
  photodynamic effect, 590

M

Macrophages, 612, 613
Malondialdehyde, 621, 623, 625
Markovnikov effect, absence of, in ene reactions, 332, 334
Menthofuran, singlet oxygen reaction, 432
Mercuric oxide, for removal of oxygen atoms, 39
Mesonaphthodianthrene, singlet oxygen reaction, 516
Metal chelates, as stabilizers for rubber, 587
Metal complex
  quenching ef singlet oxygen, 160, 161, 166
Metal ions, catalysts for dioxetane decomposition, 270–275
1,6-Methano[10]annulene, singlet oxygen reaction, 550
Methional, 617, 618, 634
Methionine, 632
  quenching of singlet oxygen, 164
  singlet oxygen reaction, 590
2-Methoxyanthracene, singlet oxygen reaction, 517
9-Methoxyanthracene singlet oxygen reaction, 515
Methoxybenzenes, singlet oxygen reactions, 531–535
1-Methoxy-9,10-diphenylanthracene, singlet oxygen reaction, 517
1-Methoxynaphthalene, singlet oxygen reaction, 521
β-Methoxystyrene, singlet oxygen reaction, 181
9-Methylanthracene, singlet oxygen reaction, 515, 517
2-Methyl-4-aryl-2-butenes, singlet oxygen reactions, 334, 355
2-Methyl-2-butene, oxygenation with triphenylphosphite oxonide, 97

1-Methylcycloalkenes, singlet oxygen reactions, 324
trans,trans-1-Methylcyclodeca-1,6-triene, singlet oxygen reaction, 300
Methylene blue, oxygen-sensitized, chemiluminescence of, 30
1-Methylene-4,4-diphenyl-2,5-cyclohexadienes, singlet oxygen reactions, 205
6-Methyl-6-ethylfulvene, singlet oxygen reaction, 543
1-Methyl-4a,5,6,7,8,8a-*trans*-hexahydronaphthalene, singlet oxygen reaction, 316
3-Methylindole, oxygenation of, with hydrogen peroxide–hypochlorite, 74–75
1-Methylnaphthalene, singlet oxygen reaction, 521
10-Methyl-$\Delta^{1(9)}$-octalin, singlet oxygen reaction, 314, 392
2-Methyl-2-pentene, singlet oxygen reaction, 70–71, 317, 319
9-Methyl-10-phenylanthracene, oxygenation of, with hydrogen peroxide–hypochlorite, 64
N-Methylpyridinium 3-oxide, singlet oxygen reaction, 565
6-Methyl-6-phenylfulvene, singlet oxygen reaction, 543
β-Methylstyrenes, singlet oxygen reactions, 308
6-Methyl-1,2,3,4-tetraphenylfulvene, singlet oxygen reaction, 543
2-Methylthio-9,10-diphenylanthracene, singlet oxygen reaction, 517
Methyl Violagen, 634
Microbicidal activity, 606–619
Microcrystalline cellulose, for substrate support, 40
Microsomal enzymes, 601, 603, 604, 609, 612, 622–626, 628, 629
Microsomal 7α-hydroxylase, 623
Microsomal lipids, 620
Microwave discharge
  in carbon dioxide, as source of singlet oxygen, 326, 344
  in oxygen, as source of singlet oxygen, 326, 337, 342, 344, 371, 372, 376
  production of singlet oxygen, 587
Microwave generators, 37
Molecular orbitals, oxygen, 2, 6, 12, 13, 15, 16, 22
Monoolefins, ionization potentials and rate constants of ene reactions, 291
Mucondialdehyde, 584
Myeloperoxidase, 608–612, 614–617, 619
α-Myrcene, singlet oxygen reaction, 300–303
β-Myrcene, 300–303, 348

N

NADPH, 604, 612, 623, 624, 626, 627
NADPH-cytochrome *c* reductase, 624, 634
NADPH oxidase, 608, 612, 616
Naphthacene, singlet oxygen reaction, 513, 515, 521
Naphthalene
  singlet oxygen reaction, 519
Naphthalene *cis*-diepoxide, formation of, 550, 551
Naphthalene 1,4-endoperoxides, 526, 527, 551
Naphthalene pentaoxide, formation of, 551
Neoabietic acid, 307, *see also* Diterpenes
Nickel chelates, 582, 583
  as peroxide decomposers, 583
  quenching of singlet oxygen, 161, 166, 582
Nicotine, quenching of singlet oxygen, 140
Nitrogen compounds, in ene reactions, 401, 415, 416
Nitrogen dioxide, for removal of oxygen atoms, 39
Nitrogen oxides, 415, 416
Nitrone, quenching of singlet oxygen, 163
Nitroso compounds, quenching of singlet oxygen, 163, 167
Nitroxide radicals, 636
Nopadiene, singlet oxygen reactions, 307, 392
Norbornene, photooxidative cleavage of, 215
Norbornane, derivatives, singlet oxygen reactions, 293, 312, 313, 360, 388
Norcarans, 317, 386, 387
Norrish Type I process, 581
Norrish Type II process, 577, 578, 581
Nucleic acids, singlet oxygen reactions, 490
Nucleic acid constituents
  oxygenation of, with hydrogen peroxide–hypochlorite, 75
    with potassium perchromate, 90

O

$\Delta^{9,10}$-Octalin, oxygenation of, with hydrogen peroxide–hypochlorite, 67

Octamethylnaphthalene, singlet oxygen reaction, 520
O-heterocyclic compounds, in ene reactions, 356, 357, 370, 382, 384, 396, 397, 400, 401
Olefins
  aryl-substituted, singlet oxygen reactions, 204–208
  autoxidation, *see* Radical reactions
  ene reactions of, 291–294, 324, 342, 351, 360, 367, 368, 386, 396, 402
  hindered, singlet oxygen reactions, 200–204
  "overoxidation" of, substitution effect of, 303–305
  singlet oxygen reactions, 586
Opsonins, 607, 608, 612
Orbital interaction, 519
Oxazoles, singlet oxygen reactions, 498–502
Oxepin
  oxygenation of, with hydrogen peroxide–hypochlorite, 76
    with triphenylphosphite ozonide, 105
2(3*H*)-Oxepinones, from fulvene-singlet oxygen reactions, 543
1,6-Oxido[10]annulene, singlet oxygen reaction, 551
Oxygen
  complex π-orbitals for, 6, 8, 9
  complex state functions for, 16–18
  degeneracies of electronic states in, 21, 22
  determinantal wave functions for, 16–18, 22
  electron repulsion in molecular states of, 18–20
  molecular, 23, 576
    three-electron bond structure of, 14
  molecular orbital configuration of, 12, 13, 15
  potential energy diagram for molecular states, 2
  primitive electron representations of molecular, 3, 4
  real π-orbitals for, 5, 6
  real state functions for, 22–24
  spin and orbital angular momenta in, 14, 15
  transfer in singlet oxygen reactions, 445
Oxygenases, singlet oxygen reactions as models for, 531, 559, 565
Oxygenation
  pH effect on product yield, 69
  solvent effect on product yield, 69
  substituent effect on rate and product distribution, 70
Ozone
  concentration in discharge products, 54
  reaction with $O_2(^1\Delta_g)$, 53, 54
    with $O_2(^1\Sigma_g^+)$, 53, 54
Ozonides, thermal decomposition of, as sources of singlet oxygen, 345, 371, 372, 385

P

Papain
  photodynamic effect, 590
[2.2]Paracyclophane, singlet oxygen reaction, 522
Paralocalization energy, of anthracene derivatives, 518
Paraquat, *see* Methyl Violagen
Pentacenes, singlet oxygen reactions, 512, 515, 516
Pentamethoxybenzene, singlet oxygen reaction, 533, 534
6,6-Pentamethylenefulvene, singlet oxygen reaction, 543
1,2,4,5,8-Pentamethylnaphthalene, singlet oxygen reaction, 521
Peptides, 589
  photodynamic effect, 589, 590
Peracids, singlet oxygen from decomposition of, in alkaline solution, 79
Perepoxides, 297, 323, 332–341
  from 2,5-dimethyl-2,4-hexadiene, 215
  formation in singlet oxygen reactions, 247, 254, 430
  from indene, 209
  intermediacy in 1,2-cycloaddition of singlet oxygen to olefins, 201–203, 224–237
Peroxidase, 629, 630, 632
Peroxides, *see* Allylic hydroperoxides, Bromo hydroperoxides, Azido hydroperoxides, Ene reactions, 1,2-Cycloaddition of singlet oxygen, 1,4-Cycloaddition of singlet oxygen, Endoperoxides, Hydroperoxides
1,4-Peroxido-hydroperoxides, 348, 366, 392, 410
Peroxyacylnitrates, singlet oxygen from base-induced decomposition of, 82
α-Peroxylactones, 254
  formation from ketenes, 219–220
Phagocytosis, 608
Phenafulvenes, 10,10′-disubstituted, singlet oxygen reactions, 522

Phenanthrene, singlet oxygen reaction, 519
derivatives, 512, 522–524
Phenols
quenching of singlet oxygen, 159, 160, 166, 552
singlet oxygen reactions, 552–563
Phenoxy radicals, formation from phenols 552, 553, 560, 561
9-Phenylanthracene, singlet oxygen reaction, 515, 517
9-Phenylfulvene, singlet oxygen reaction, 543
3-Phenylisocoumaranone, singlet oxygen reaction, 566, 567
Phloretic acid, singlet oxygen reaction, 558
1-Phospha-2,8,9-trioxaadamantane ozonide, singlet oxygen from 84, 85, 101
Phosphines, reactions with 1,2-dioxetanes, 274–277
Phosphoglucomutase, 590
photodynamic effect, 590
Phosphoranes
from dioxetanes and phosphorus compounds, 275–277
thermal decomposition of, 277
Photodynamic effects, 589, 613, 614, *see also* Photosensitizing conditions
imidazole photooxidation in, 488–490
singlet oxygen in, 72, 75, 89, 101–102
Photon counting, 624
Photoperoxides, decomposition of, 91–93
Photosensitizing conditions, 613, 619, 621, 633, 635, 636
α-Pinene
oxygenation of, with hydrogen peroxide–hypochlorite, 67
singlet oxygen reaction, 174, 310, 390, 391
β-Pinene, singlet oxygen reaction, 313, 361
Piperidine *N*-oxyl, quenching of singlet oxygen, 163, 164
Platelet aggregation, 628
cis-Polybutadiene, singlet oxygen reaction, 586, 587
trans-Polybutadiene, singlet oxygen reaction, 586, 587
Polycyclic aromatic hydrocarbons, *see* Acenes
Polydienes, 585–588
oxidative photodegradation, 585–588
singlet oxygen reactions, 586
Polyenes, *see also* Carotene, Carotenoid
singlet oxygen reactions, 350, 351, 359
Polyethylene, 577–580
chain scission, 578
effect of thickness, 579
mechanism of photodegradation, 577–580
olefinic impurities, 577
oxidative photodegradation, 577–580
surface effect, 579
1,4-Polyisoprenes, 585
Polymers, 575–595
degradation of, 575–595
singlet oxygen reactions, 42
Polymer-bound sensitizers, 186, 219
Polynuclear aromatic compounds (PNA), 582
Polypropylene
oxidative photodegradation, 580–583
polymeric carbonyl products, 580, 581
preoxidized, 580, 581
Polystyrene, 583–585
chain scission, 583
discoloration, 583, 584
light stability, 583
oxidative photodegradation, 583–585
Poly(vinyl chloride), 588
photosensitized oxidation, 588
reaction with singlet oxygen, 588
Potassium perchromate
as singlet oxygen source, 87–90, 342, 345, 369, 371, 372
Progesterone, singlet oxygen reaction, 180
Prostaglandins, 627–630
Prostaglandin synthetase, 628, 629
Psoralens, singlet oxygen reactions, 441
Purines, singlet oxygen reactions, 490–498
Pyrroles, singlet oxygen reactions, 447
Pyrrolophanes, singlet oxygen reactions, 459

Q

Quenching of singlet oxygen
by acceptor, 149
by amines, 157–159, 166, 535, 587
by azide, 144, 163, 167
by azodioxides, 163
by azomethine dyes, 156, 157
by bilirubin, 162, 615, 629, 630
by biliverdin, 162
biological applications, 164, 165
in biological systems, 599–615
by carotenes, 140, 154, 155, 166, 613, 615, 621, 623, 625, 628, 635
chemical, 139
by cyanine dyes, 156

by DABCO, *see* Quenching of singlet oxygen, by 1,4-diazabicyclo[2.2.2]octane
definitions, 139, 140
dependence on k_A[A], 146
on source of singlet oxygen, 146, 147
in deuterated *vs.* protiated solvent, 146
by 1,4-diazabicyclo[2.2.2]octane, 140, 153, 157, 158, 166, 201, 202, 215, 614, 615, 621, 623, 624, 627, 628, 632
by diphenylisobenzofuran, 164
directly generated by laser photolysis, 148
"energy pooling" process, 154
in erythropoetic protoporphyria, 165
by ethyl sulfide, 623
fluorescent acceptor technique, 147, 148
gas phase methods, 151
by heavy atoms, 153, 154
by histidine, 164
by hydrazide functionality, 317
intercepts of plots, 144, 145
by iodide, 163, 167
kinetics, 141–151
by laser Q-switch dyes, 135
mechanisms of, 151–154
charge-transfer, 140, 152, 157, 159, 161–163
energy-transfer, 140, 151–153, 155, 161
by metal complexes, 160, 161, 166
by methionine, 164
by miscellaneous compounds, 162–164
by Ni(II) chelates, 161, 166, 582
by nicotine, 140
by nitrones, 163
by nitroso compounds, 163, 167
nondifferential technique, 148
oxygen dependence, 145
by phenols, 159, 160, 166, 552
in photosensitized hemolysis, 165
by photostabilizers, 161, 166
physical, 139, 166
by piperidine *N*-oxyl, 163, 164
rate constants, 165–167
separation of k_Q from k_R, 149, 150
singlet oxygen decay rate dependence, 146
steady state kinetic techniques, 144–150
by stilbene quinones, 156
by sulfides, 161, 162, 166
by superoxide, 162, 167
by superoxide dismutase, 164
by systems with extensive conjugation, 155–157, 166
time-resolved techniques, 150
by tocopherols, 159, 160, 166, 559, 560, 591
total, 140, 166
by triethylamine, 157–159, 166
by tryptophan, 164, 632
types of quenchers, 154–164
Quercetase, 619
Quercetin, 619
singlet oxygen reaction, 565
p-Quinols, from phenol-singlet oxygen reactions, 558

R

Radicals, quenching by
benzoate, sodium, 612–614, 617, 618, 632
butylated hydroxyanisole, 629
butylated hydroxytoluene, 632
2,6-di-*tert*-butylphenol, 632
ethanol, 614, 618, 624, 634
mannitol, 617, 618, 621, 632
metabisulfate, 617
tert-butyl alcohol, 621, 632
thiosulfate, 617
Radical reactions of olefins with triplet oxygen, 288, 311, 323, 326, 328
Radiofrequency generators, 37
Rate constants for reactions of singlet oxygen
determination of, 288–290
in ene reactions, *see* Ene reactions
with *N,N*-dimethylisobutenylamine, 182
with olefins, 234, 300, 301
with tetraethoxyethylene, 196
Reactivity index (β-value)
for acenes, 515, 519
for phenols, 555
Red blood cells, 36, 635
Red blood cell ghosts, 591
Reduction of ene reaction primary products, 342–417
Regiospecificity, of azido hydroperoxide formation, 337
Rose bengal
oxygen-sensitized chemiluminescence of, 29
polymer-bound, 186, 219
Rubber, 575
degradation of, 575
natural, vulcanization, 585
Rubrene, 622
oxygen-sensitized chemiluminescence of, 27–29

singlet oxygen reaction, 512–514, 525
Russell mechanism, 603, 625

S

Schenck reaction, 288, *see also* Ene reactions
Sensitization mechanisms for singlet oxygen, 24–31
Sensitizers
effect of, on product ratio, 326, 344, 368, 369, 405, 406
polymer-bound, 342, 344, 345, 362, 363, 370–372
ST splitting of, 512, 573
triplet energy, 512, 513
use in sensitized photooxygenation of olefins, 343
Silicon compounds, 413
Silicones, 588
polymers containing, 588
singlet oxygen reactions, 588
Simultaneous transitions, for singlet oxygen, 25–27
Singlet oxygen
annihilation of, 134, 135
chemical probes for detection of, 316, 325
chemically generated, 525, 547, 550, 552, 554, 564
chemiluminescence of, 25–27
efficiency (α) of production, 288, 289
electronic states of, 2, 25
electronic structure of, 15, 16, 22
as electrophile, 297, 330, 331
energy pooling reactions in $^1\Delta_g$ state, 55, 56
identification, 599–606
International Conference, 576
lifetimes
of $^1\Delta_g$ state, 577
gas phase quenching, relation to solution quenching, 124, 125
laser flash photolysis, 121
radiationless decay, 116, 117
role of Franck-Condon factors, 129, 130
of spin-orbit coupling, 128
in solution, 290
solvent deuterium isotope effect on, 116, 122, 131
solvent effect on, 120, 121, 290
temperature dependence of, 123
theoretical analysis of radiationless decay, 125–129
methods of generation, 343
microwave discharge-generated, 553
reactions, *see also* specific compounds
electronic effects of aryl groups on, 309
gas phase, with olefins, 217, 218
kinetics of, 512
at low temperature, 522, 540, 542, 543, 546, 565
as models of oxygenase reactions, 531, 558, 566
regiospecificity in, 514
relative rates for methoxybenzene, 534
solvent effects in, 545
solvent isotope effects on, 133, 553
substituent effect in, 515, 546
reactivity index (β-value) for acenes, 514
regeneration from acene endoperoxides, 524
relation to near-infrared absorption of solvent, 128–130
role in enzymatic reactions, 133, 134
in photodynamic inactivation, 133
sensitization mechanisms for, 24–31
$^1\Sigma_g^+$
reaction with ozone, 53–55
removal from singlet oxygen streams, 39, 41, 42
simultaneous transitions for, 25–27
solvent isotope test, 122
sources, 598, 599
stereospecific or stereoselective attack of, 309–318
Singlet sensitizers, quenching of, 142, 157, 158
Sodium lamps, application in sensitized photooxygenation of olefins, 351, 352, 355, 378, 381, 409
Solvent effects
in ene reactions, 290–297, 346, 347, 368–371
in singlet oxygen reactions with arylcyclopropenes, 206–207
with camphenylideneadamantane, 203
with dihydropyran, 195
with 2,5-dimethyl-2,4-hexadiene, 215
with *N,N*-dimethylisobutenylamine, 182
with 1,2-diphenylcyclobutene, 206
with 1-ethylthio-2-ethyl-1-hexene, 297, 347
with olefins, 233
with tetraethoxyethylene, 195–196
Spiro[2.4]hepta-4,6-diene, singlet oxygen reaction, 544

Stereoselectivity of singlet oxygen attack, 309–318, 326, 329
Stereospecificity
in formation of azido hydroperoxides, 336, 337
of singlet oxygen attack, 310, 329, 330
Steric effects
of axial hydrogen in ene reactions, 325
in ene reactions, 309, 318, 329, 352, 386, 387, 413
Steroids, in ene reactions, 304, 305, 307, 318 358, 367, 402–413
Stilbene quinone, quenching of singlet oxygen, 156
Sulfides
quenching of singlet oxygen, 161, 162, 166
reactions with 1,2-dioxetanes, 278, 279
Sulfuranes
from dioxetanes and divalent sulfur compounds, 278
thermal decomposition of, 278
Sulfur compounds, in ene reactions, 347, 351, 352, 371
Superoxide, 338, 608, 612–614, 616–622, 624, 626–628, 633–635
anion radical
intermediacy in singlet oxygen reactions, 182, 225
photosensitized formation of, 174
ion, 81, 87
electrochemical production of, 84
intermediate in adrenodoxin reductase-adrenodoxin system with NADPH, 85
quencher of singlet oxygen, 83, 86
singlet oxygen from dismutation of, 81, 84
quenching of singlet oxygen, 162, 167
Superoxide dismutase, 605, 608, 613, 616–618, 621, 624, 627–628, 633–635
inhibitor of singlet oxygen production, 81, 84
quenching of singlet oxygen, 164
Synovial fluid, 617

T

α-Terpinene
oxygenation of, with triphenylphosphite ozonide, 95
singlet oxygen reaction, 307, 337, 338, 339
α-Terpinolene, singlet oxygen reaction, 300, 362
1,2,3,4-Tetrachloronaphthalene, singlet oxygen reaction, 521
Tetrakis(dimethylamino)ethylene autoxidation, 181
Tetramethoxybenzenes, singlet oxygen reactions, 532–534
Tetramethyl-1,2-dioxetane, reaction with boron trifluoride, 280–283
Tetramethylethylene ene reaction, 291, 336, 340, 344
Tetramethylnaphthalenes, singlet oxygen reations, 520
Tetraphenylcyclopentadienone, 627
oxygenation of, with hydrogen peroxide–hypochlorite, 63
with triphenylphosphite ozonide, 94
Tetraphenylethylene, singlet oxygen reaction, 176, 177
1,2,3,4-Tetraphenylfulvene, singlet oxygen reaction, 541
Tetraphenylfuran, 601, 632
1,2,3,4-Tetraphenylnaphthalene, singlet oxygen reaction, 521
Tetrathioethylene, singlet oxygen reaction, 104
Thiazine dyes, oxygen-sensitized chemiluminescence of, 30, 31
Thiobarbituric acid, 623
Thioethylene, 1,2-cycloaddition, 192–199
Thiones, singlet oxygen reactions, 220
Thiophenes, singlet oxygen reactions, 503–505
Thujopsene, singlet oxygen reaction, 216
Titanium, 603, 604
Tocopherol, quenching of singlet oxygen, 159, 160, 166, 559, 560, 591
α-Tocopherol, singlet oxygen reaction, 559
TPF, *see* Tetraphenylfuran
Trautz reaction, as possible source of singlet oxygen, 110
1,2,4-Triacetoxynaphthalene, singlet oxygen reaction, 521
Trienes, in ene reactions, 347, 348, 350, 382
2,4,6-Tri-*tert*-butylphenol, singlet oxygen reaction, 555, 556

2,4,6-Tri-*tert*-butylphosphorine, singlet oxygen reaction, 537
Triethylamine, quenching of singlet oxygen, 157–159, 166
Triethylphosphite ozonide, singlet oxygen from, 101
1,2,3-Trimethoxybenzene, singlet oxygen reaction, 534
1,2,4-Trimethoxybenzene, singlet oxygen reaction, 533, 534
1,3,5-Trimethoxybenzene, singlet oxygen reaction, 534
Trimethyl-1,2-dioxetane, 178
1,2,3-Trimethylnaphthalene, singlet oxygen reaction, 520
1,2,4-Trimethylnaphthalene, singlet oxygen reaction, 520
1,4,5-Trimethylnaphthalene 9,10-endoperoxide, activation energy for thermal decomposition, 525–526
2,3,6-Trimethylphenol, singlet oxygen reaction, 554, 555
α,β,β-Trimethylstyrenes, singlet oxygen reactions, 291, 297, 309, 355
Triphenylcarbonium ion, singlet oxygen reaction, 538
1,3,5-Triphenylformazan, oxygenation of, with hydrogen peroxide-hypochlorite, 73
2,3,6-Triphenylphenol, singlet oxygen reaction, 554
Triphenylphosphite ozonide, 93–99, 177, 187, 220
  bimolecular oxygenations with, 95–99
  deuterium isotope effects in reactions of, 105, 106
  ESR detection of singlet(delta) oxygen from, 95
  gas phase reactions of singlet oxygen from, 95
  mechanism of decomposition of, 99, 100
  singlet oxygen from, 93–99
2,4,6-Triphenylpyrylium 3-oxide, singlet oxygen reaction, 568
Triplet energy pooling, 27, 28
Triplet oxygen
  electron repulsion in, 18–20
  molecular orbital configuration of, 6, 15, 16, 22
Triplet sensitizers, quenching of, 142–148, 158
  by acceptor, 147–148
  dependence on k_A[A], 146
  distinction of, from singlet oxygen quenching, 143
  by ground state oxygen, 118
  intercepts of plots, 144, 145
  other steady-state kinetic techniques, 147, 148
  oxygen dependence, 145
  by singlet oxygen, 134, 135
    decay rate dependence, 146
Triplet state, sensitization of, 27–31
Triplet-triplet annihilation, sensitization of, 29, 30
Tropolone, methyl ether, singlet oxygen reaction, 545
Tropone, singlet oxygen reaction, 547
Tryptophan
  quenching of singlet oxygen, 164, 632
  singlet oxygen reaction, 469, 470, 477, 589, 590
Type I photosensitized oxygenation
  of aromatic amines, 535
  of methylbenzenes, 538
  of phenols, 552, 557
Tyrosine, singlet oxygen reaction, 560

V

Vinylanthracenes
  1,4-cycloaddition, 197
Vinyl ethers
  1,2-cycloaddition, 190–192, 232
  singlet oxygen reactions, 178, 188–198
Vinylnaphthalenes
  1,4-cycloaddition, 213
  singlet oxygen reaction, 540
9-Vinylphenanthrene, singlet oxygen reaction, 540
Vinylthiophenes
  1,4-cycloaddition, 214
Violanthrone, oxygen-sensitized chemiluminescence of, 27–29
Vitamin D_3, oxygenation of, with hydrogen peroxide-hypochlorite, 75

X

Xanthine, 614, 620, 622, 624
Xanthine oxidase, 85, 86, 601, 605, 614, 616–617, 620–622, 624, 626, 634
Xanthine-xanthine oxidase system, as source of singlet oxygen, 85, 86

Y

Yeast alcohol dehydrogenase, 590
  photosensitized oxidation, 590
Ylides, singlet oxygen reactions, 220
Ynamines, singlet oxygen reactions, 218

Z

Zinc oxide, as sensitizer, 586
Zwitterions, 181, 184, 185, 224–237, 430, 438, 439, 473, 479, 481
  as intermediates in ene reactions, 297, 332
Zymosan particles, 605

ORGANIC CHEMISTRY

A SERIES OF MONOGRAPHS

1. Wolfgang Kirmse. CARBENE CHEMISTRY, 1964; 2nd Edition, 1971

2. Brandes H. Smith. BRIDGED AROMATIC COMPOUNDS, 1964

3. Michael Hanack. CONFORMATION THEORY, 1965

4. Donald J. Cram. FUNDAMENTALS OF CARBANION CHEMISTRY, 1965

5. Kenneth B. Wiberg (Editor). OXIDATION IN ORGANIC CHEMISTRY, PART A, 1965; Walter S. Trahanovsky (Editor). OXIDATION IN ORGANIC CHEMISTRY, PART B, 1973; PART C, 1978

6. R. F. Hudson. STRUCTURE AND MECHANISM IN ORGANO-PHOSPHORUS CHEMISTRY, 1965

7. A. William Johnson. YLID CHEMISTRY, 1966

8. Jan Hamer (Editor). 1,4-CYCLOADDITION REACTIONS, 1967

9. Henri Ulrich. CYCLOADDITION REACTIONS OF HETEROCUMULENES, 1967

10. M. P. Cava and M. J. Mitchell. CYCLOBUTADIENE AND RELATED COMPOUNDS, 1967

11. Reinhard W. Hoffmann. DEHYDROBENZENE AND CYCLOALKYNES, 1967

12. Stanley R. Sandler and Wolf Karo. ORGANIC FUNCTIONAL GROUP PREPARATIONS, VOLUME I, 1968; VOLUME II, 1971; VOLUME III, 1972

13. Robert J. Cotter and Markus Matzner. RING-FORMING POLYMERIZATIONS, PART A, 1969; PART B, 1; B, 2, 1972

14. R. H. DeWolfe, CARBOXYLIC ORTHO ACID DERIVATIVES, 1970

15. R. Foster. ORGANIC CHARGE-TRANSFER COMPLEXES, 1969

16. James P. Snyder (Editor). NONBENZENOID AROMATICS, VOLUME I, 1969; VOLUME II, 1971

17. C. H. Rochester. ACIDITY FUNCTIONS, 1970

18. Richard J. Sundberg. THE CHEMISTRY OF INDOLES, 1970

19. A. R. Katritzky and J. M. Lagowski. CHEMISTRY OF THE HETEROCYCLIC *N*-OXIDES, 1970

20. Ivar Ugi (Editor). ISONITRILE CHEMISTRY, 1971

21. G. Chiurdoglu (Editor). CONFORMATIONAL ANALYSIS, 1971

22. Gottfried Schill. CATENANES, ROTAXANES, AND KNOTS, 1971

23. M. Liler. REACTION MECHANISMS IN SULPHURIC ACID AND OTHER STRONG ACID SOLUTIONS, 1971

24. J. B. Stothers. CARBON-13 NMR SPECTROSCOPY, 1972

25. Maurice Shamma. THE ISOQUINOLINE ALKALOIDS: CHEMISTRY AND PHARMACOLOGY, 1972

26. Samuel P. McManus (Editor). ORGANIC REACTIVE INTERMEDIATES, 1973

27. H. C. Van der Plas. RING TRANSFORMATIONS OF HETEROCYCLES, VOLUMES 1 AND 2, 1973

28. Paul N. Rylander. ORGANIC SYNTHESES WITH NOBLE CATALYSTS, 1973

29. Stanley R. Sandler and Wolf Karo. POLYMER SYNTHESES, VOLUME I, 1974; VOLUME II, 1977

30. Robert T. Blickenstaff, Anil C. Ghosh, and Gordon C. Wolf. TOTAL SYNTHESIS OF STEROIDS, 1974

31. Barry M. Trost and Lawrence S. Melvin, Jr. SULFUR YLIDES: EMERGING SYNTHETIC INTERMEDIATES, 1975

32. Sidney D. Ross, Manuel Finkelstein, and Eric J. Rudd. ANODIC OXIDATION, 1975

33. Howard Alper (Editor). TRANSITION METAL ORGANOMETALLICS IN ORGANIC SYNTHESIS, VOLUME I, 1976; VOLUME II, 1978

34. R. A. Jones and G. P. Bean. THE CHEMISTRY OF PYRROLES, 1976

35. Alan P. Marchand and Roland E. Lehr (Editors). PERICYCLIC REACTIONS, VOLUME I, 1977; VOLUME II, 1977

36. Pierre Crabbé (Editor). PROSTAGLANDIN RESEARCH, 1977

37. Eric Block. REACTIONS OF ORGANOSULFUR COMPOUNDS, 1978

38. Arthur Greenberg and Joel F. Liebman, STRAINED ORGANIC MOLECULES, 1978

39. Philip S. Bailey. OZONATION IN ORGANIC CHEMISTRY, VOL. I, 1978

40. Harry H. Wasserman and Robert W. Murray (Editors). SINGLET OXYGEN, 1978